TOOLEY'S DICTIONARY OF MAPMAKERS

6 *Terrarum tractus, et latas metior oras,*
Ingenio gaudens subtili, et acumine mentis.

TOOLEY'S DICTIONARY OF MAPMAKERS

Revised Edition

A - D

Editor
Josephine French

Consulting Editors
Valerie Scott, Mary Alice Lowenthal

Picture research
Valerie Scott

MAP COLLECTOR PUBLICATIONS
in association with Richard Arkway Inc.
1999

This revised edition is dedicated to all the unbelievers who thought we would never get past Aa.

48 High Street, Tring, Herts HP23 5BH, England

Designed by Roy Walker MCSD

Printed in England by Redwood Books Limited, Trowbridge, Wiltshire

ISBN 0-906430-14-3

INTRODUCTION

To the reader of this volume: May you bless its usefulness and not curse its shortcomings!

When I signed a contract in 1992 to buy the copyright of *Tooley's Dictionary of Mapmakers* from C. Broekema I was intending to bring out a new edition quite soon. What began as a few basic corrections turned into more than six years work involving thousands of hours of research. At last the task is nearing completion.

To quote Robert Plot in his book *The Natural History of Oxford-shire* (1705) "...this ESSAY has swell'd to so much greater a bulk than ever I expected it could possibly have done...". R.V. Tooley claimed 20,000 entries in his original work and this has expanded almost daily over the past seven years.

Through using the original *Tooley's* in my role as Editor of *The Map Collector* (the magazine was closed in 1996 after being in publication for nearly 20 years), I knew the revision was going to be a hard task but it turned out to be a massive undertaking involving experts around the world and a number of editors. However, my main thanks go to Jo French who has stuck by me when many others have fallen by the wayside. Without her I am not sure that this new edition would have seen the light of day. Mary Alice Lowenthal has been another stalwart in recent years. Without her the later volumes would not be as well advanced as they are.

Thanks are due also to Derek Allen who helped us with the checking, John Goss who gave us the benefit of his expertise early in the project, and Francis Herbert for assistance with final proof checking. Then there are all the researchers who answered our cries for help and responded to specific requests for information. These names are listed on p.xii.

A word to the sceptics who told me a revision of the *Dictionary* was an "impossible" task. As I explained to them, the first complete edition, published in 1978 (subsequently reprinted in paperback) and its *Supplement* of 1985, were never intended to be definitive, but a first point of reference. Although in her Foreword to the first edition the late Dr Helen Wallis, Map Librarian at The British Library from 1973-1986, praised the *Dictionary* as "the first comprehensive work of this kind in English", Tooley himself stated more modestly that his aim was to "give information in the most compact form possible on persons associated with the production of maps from the earliest times to the year 1900". As such it has been an amazing success; perhaps the most successful, or at least most used, work of reference in the history of cartography.

Working on this edition we, like Tooley, had to accept that we could not achieve total comprehensiveness. An attempt to list every person involved with producing maps in every country of the world since records began would have been folly and unattainable. Our aim was to revise a work deemed 'useful' by some and 'indispensable' by others, improve it, and correct it using the additional research available today. With amendments, many totally new entries, and the addition of references, our new edition has almost doubled in size.

Accessibility and usefulness have been our aims throughout. Where possible the 21,450 original entries have been checked and augmented. However, all along we felt it was important to respect the spirit of Tooley's work and this has meant leaving in entries which we were unable to verify. Tooley told me he had compiled the *Dictionary* from his own card index and, if this was the case, he may have handled maps which no longer exist or are so rare they are hardly ever seen. This may seem idiosyncratic to some researchers today but was our personal decision. We have extended the time frame only to include a few unusual or collectable 20th-century works and their makers. One change we did make was to eliminate Tooley's abbreviations which we felt were confusing, especially to readers not fluent in the English language.

As Dr Wallis pointed out in the first edition, the majority of people listed in this dictionary would not have called themselves a 'cartographer'. Only the later ones could call themselves that as the word 'cartography' ('cartografia') was not coined until 1839. We have done our best to credit all those associated with the production of maps by including draughtsmen, editors, engravers, printers and publishers and some booksellers, as did Tooley. We have not attempted to include all the English surveyors as these are available in Sarah Bendall's 1997 revision of Peter Eden's *Dictionary of Land Surveyors and Local Mapmakers of Great Britain and Ireland 1530-1850.*

One advantage of the time it has taken to compile this revision is that I have put together a superb selection of portraits. There is no other reference I know of which brings together so many of the likenesses of these men of the past. Some portraits did appear in the first edition and also *Map Collectors' Circle* but only a handful in comparison. Here I must thank those who have brought portraits to my attention and in particular Rodney Shirley, who has a collection of originals to which he has allowed me access.

We have, of course, had problems preparing this volume. One of the greatest has been a lack of funding to help us travel to some of the countries we would like to have visited to carry out research. This has meant that we have had to rely on the kindness of researchers in these countries and on published books about the history of cartography. Latterly, we have been assisted by lists on the internet and this will no doubt change our perception in any future editions. We have also added to the usefulness of the book by including references to monographs, or articles in books or magazines. Although in most cases these references indicate where we got some of our information, this is not the reason they are there. Our purpose for references is to direct the reader to books or articles that give substantial further information on the mapmaker or a significant work, not to give authority to the information we have included. That would have been impossible. The many major reference works on which we have relied will be credited in the bibliography to be included in a later volume.

The original *Tooley's* did not attempt to put in correct titles of maps but where possible this has been done, in addition to adding, where known, the book or atlas in which the maps first appeared. However, it should be remembered that quite often different editions of the same map carried titles that differed, not only in language, but often in phraseology, wording, or emphasis. This may have led to a few discrepancies. *Readers should note that if a title is not in italics this is because we have not been able to verify it.* We have not attempted to include all works by mapmakers, especially the most prolific; neither is there any particular rule governing what is or is not listed. Where possible we have cited atlases rather than single maps in atlases. However, there are bound to be instances where we have missed the fact that a map is actually part of a larger work. We have found that reference works differ greatly in their use of square brackets with dates. For this reason we have been unable to adopt the convention of using square brackets for supplied dates. The dates given are for the earliest issue to which we have been able to find reference, and that date does not necessarily appear on the map itself.

We have tried to group family members together and clarify the complicated relation-

19. SCVLPTVRA IN ÆS.
Sculptor noua arte, bracteata in lamina Scalpit figuras, atque prælis imprimit.

ships between companies and individuals. This has proved one of the hardest tasks but should be immensely useful.

The spelling and use of names has also proved a problem. This reflects past inconsistencies, language differences, and the way names are presented in the reference works we have used. Names including prefixes such as *de la*, proved particularly difficult as standardisation was simply not appropriate. We have opted to accept the idiosyncracies of usage, bowing to the conventions of the country of origin of the mapmaker for preference. Wherever possible we have provided cross references for guidance.

Literally thousands of reference books and articles have been checked to bring you this new edition and to those authors and compilers who think indexes are unnecessary we could issue a few choice words! We realise that there may be works of reference we have missed and even a few we have misinterpreted.

Finally, a plea to users of this volume. Do send us any additions and corrections so that they can be incorporated into any future editions.

Valerie G. Scott, Tring, England. 1999

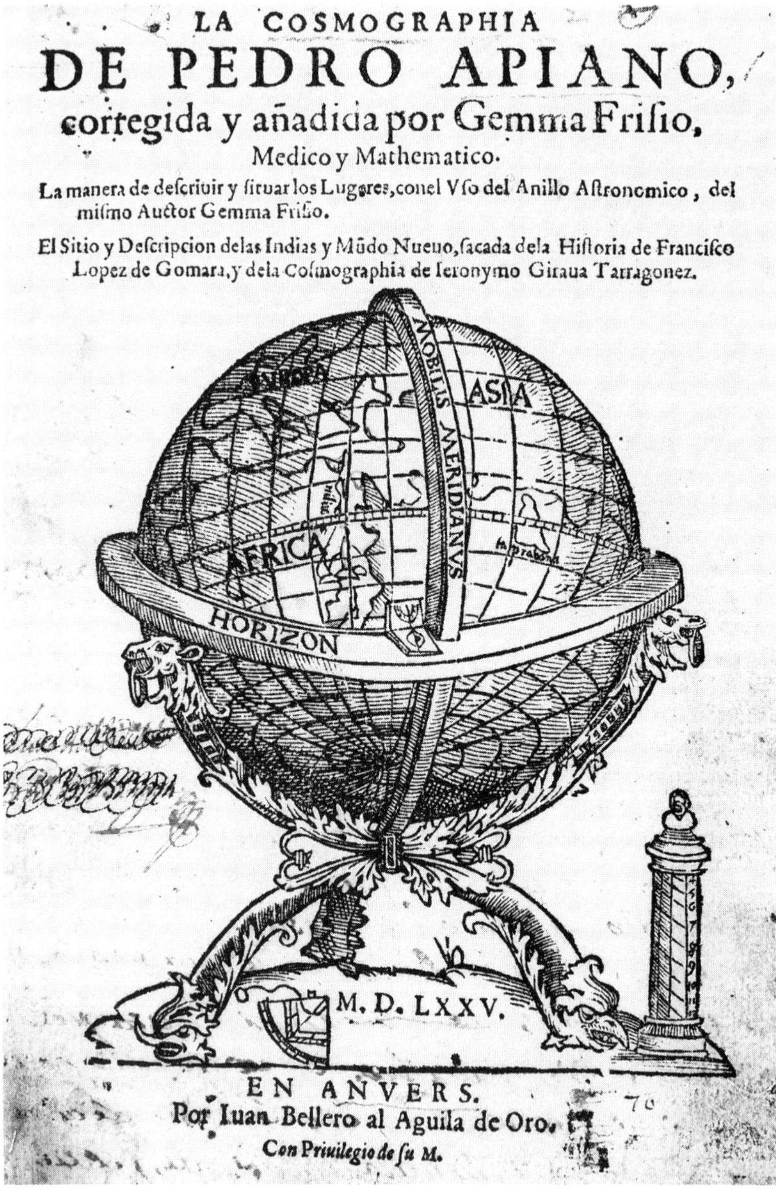

LA COSMOGRAPHIA
DE PEDRO APIANO,
corregida y añadida por Gemma Friſio,
Medico y Mathematico.
La manera de deſcriuir y ſituar los Lugares, con el Vſo del Anillo Aſtronomico, del miſmo Auctor Gemma Friſio.
El Sitio y Deſcripcion delas Indias y Mũdo Nueuo, ſacada dela Hiſtoria de Franciſco Lopez de Gomara, y dela Coſmographia de Ieronymo Giraua Tarragonez.
M.D.LXXV.
EN ANVERS.
Por Iuan Bellero al Aguila de Oro.
Con Priuilegio de ſu M.

PREFACE

Ronald Vere Tooley 1898—1986
An appreciation

It would be difficult to find another person who has left such an indelible impression on the world of the antiquarian book and map trade as did R.V. 'Mick' Tooley. During his sixty-six years of activity as an antiquarian book and map seller (many of them in the firm of Francis Edwards Ltd. of Marylebone High Street, London) he became a well known figure in the salerooms of the auctioneers, Sotheby's, the Map Library of the British Library, the Map Room and Library of the Royal Geographical Society, and, above all, in the circle of collectors, antiquarian book and map sellers in which I have moved for many years.

Tooley was born in London where he went to school until the outbreak of the First World War. He volunteered, and was accepted, into the army by giving his age as eighteen when he was, in fact, only sixteen. He joined the Queen's Westminster Rifles, saw service in France and took part in many grim battles where the Regiment suffered heavy casualties. In 1919, after the Armistice, he was demobilised and sought a post in the antiquarian book trade. He was apprenticed to the firm of Francis Edwards Ltd. in London where the founder of the firm still managed the business. At first, Tooley's interest centred on travel books and illustrated topography which resulted in publication of his first major work, *Some English books with coloured plates 1790--1860* in 1935.

In 1929 the Wall Street crash had a disastrous effect on the antiquarian book trade. Tooley had already begun to specialise in the field of old atlases and maps and worked through the Depression years when book prices dropped to almost nothing. For example, an atlas by John Speed would fetch a couple of pounds. Tooley had to leave the firm of Francis Edwards and opened up a small antiquarian business called 'The Atlas Bookshop' in Cranbourne Street where he sold old maps. A year later he moved the business to Cecil Court where his name is still carried in the business named Tooley Adams and run by his stepson, Douglas Adams.

Tooley was not a scholar in the academic sense of the word but he gathered an immense store of practical knowledge on maps and atlases which earned him the friendship of such renowned specialists as Dr Edward Lynam, R.A. Skelton, G.R. Crone, Mademoiselle de la Roncière, A. de Smet, and Alexander Vietor. He had boundless energy, a retentive mind, and, above all, a wealth of ideas which inspired collectors and mapsellers alike. In 1939 he published 'Maps in Italian atlases of the sixteenth century' in *Imago Mundi* Vol. III, a work which is still widely quoted.

During the Second World War Tooley worked as a telephone operator and sold maps in his spare time. In 1949 he rejoined the firm of Francis Edwards Ltd., where John P. Love was then Director. Tooley was put in charge of the Map Department housed on the top floor of the building in Marylebone High Street. In 1949 he achieved a lasting reputation through the publication of *Maps and Map Makers* (published by B.T. Batsford Ltd.). This is a basic reference book for collectors and dealers and has gone into many editions. It was instrumental in bringing old maps to the attention of a large public both in the British Isles and elsewhere. In later years Tooley sometimes styled himself 'Father of Mapselling', and the title is well deserved. It must have been about this time when I first made his acquaintance, an acquaintance which soon ripened into friendship. Francis Edwards was the nucleus of a small circle of map sellers such as Robert Cooper, Roger Baynton-Williams, and a few others. During the war, maps of the occupied countries of Europe had been unsaleable and stocks of these maps were correspondingly high and prices low. Maps of Germany were priced at a few shillings, and those of France and Italy even less. It was to London that map sellers from the Continent gravitated to replenish their stocks, and many collectors found treasure troves at Francis Edwards where Tooley helped to build collections which were to become famous. These included Eran Laor's collection of maps of Palestine, now in the Jewish National Library in Jerusalem.

Tooley himself collected maps of certain areas such as Africa and Australia. His African collection (later acquired by Dr D. Schrire) gave birth to *Collectors' guide to maps of the African continent and Southern Africa* (1967), and his Australia collection led to *The Mapping of Australia* (1970). His collection of maps of Australia went to the National Library of Australia at Canberra.

In 1963 he founded the *Map Collectors' Circle*, a series of monthly monographs on maps of specific periods or areas. Between 1963 and 1974 110 such monographs were published under Tooley's editorship. During these prolific years the top floor of Francis Edwards was frequently thronged by colleagues and collectors, and frequently business was conducted over lunch at The Rising Sun, a convenient local pub. Among Tooley's assistants was Tony Campbell, now Map Librarian at the British Library, who also contributed monographs on maps of Japan and the Visscher map of New England to the series. Tooley himself wrote many of the monographs including 'California as an Island' (1963) and 'Leo Belgicus' (1964). Other contributors included R.A. Skelton, A. Fordham, Coolie Verner, H. Malinowski and R.P. Hargreaves.

So far, I have not mentioned the *Dictionary of Mapmakers* of which you are now holding a copy of the second edition. Various lists of mapmakers had already been published and Professor W. Bonacker in Germany was working on a comprehensive catalogue of mapmakers. When I suggested to Tooley that he would be the obvious person to compile and publish such a dictionary for the English-speaking world, he looked at me in dismay and said, "but that is what I have been working on for the past thirty years." In subsequent discussions it turned out that he had built up a formidable card index of cartographers, mostly from maps and atlases which had passed through his hands, and supplemented by the study of existing handbooks and booksellers' catalogues. He started by publishing the names in parts in the *Map Collectors' Circle* but had not finished the alphabet when publication of that series was discontinued.

I then offered to publish an enlarged edition in book form, supplemented by new knowledge and additions suggested by readers of the series. He agreed to my proposal and, since I did not have the necessary channels of distribution in the United States, we agreed to bring in an American firm as co-publisher. For this, I invited my friend Alan R. Liss to participate. As a result, the first complete edition appeared in 1979 followed by a *Supplement* in 1985, bringing the total number of entries to 24,000.

By this time Tooley had retired from the post as Director of his department at Francis Edwards and had moved to Norfolk where he bought a house in Mattishall, a village not far from Norwich. His 75th birthday had been commemorated by publication of a collection of essays entitled *My head is a map*. Throughout his retirement he remained active, corresponding with his many friends and developing new ideas for future publications. One letter to me read, 'I have a few books accepted by publishers but slow in making their appearance.' Until two days before his death in 1986 he remained a regular visitor to the Map Library of the British Library. Tooley was a man with many friends and those that are still alive will remember him with affection. He valued friendship and was always ready with counsel and advice. His last letter to me, written a few weeks before his death, ends with the words, 'how nice to have old friends.'

C. Broekema, Amsterdam.

List of contributors

Thanks are due to the following for their assistance:-

Jeffrey Abbott, London, UK
Cyrus Ala'i, London, UK
John Bartholomew, Edinburgh, Scotland
Roger Baskes, Chicago, IL, USA
Alexander Bely, Minsk, Republic of Belarus
Sarah Bendall, Oxford, UK
Eugene Burden, Ascot, Berks, UK
Philip Burden, Herts, UK
Zachary Calhoun, CA, USA
Jacqueline Carpine-Lancre, Monaco
Raymond Carroll, Spalding, Lincs, UK
M.F. Cartwright, Cape Town, South Africa
Michael Conzen, Chicago, IL, USA
Andrew Cook, London, UK
Jeremy Crampton, Fairfax, VA, USA
Bjorn Dahl, Copenhagen, Denmark
Oswald Dilke, Leeds, UK (deceased)
Johannes Dörflinger, Vienna, Austria
Paul Erling, Chicago, IL, USA
Susannah Fisher, Southampton, UK
Josep Gasset-Argemi, Barcelona, Spain
John Goss, Bletchley, Bucks, UK
László Gróf, Oxford, UK
Warren Heckrotte, Oakland, CA, USA
Francis Herbert, London, UK
Catherine Hofmann, Paris, France
Donald and Yolande Hodson, Tewin, Herts, UK
Ralph Hyde, London, UK
Ieuan E. Jones, Birmingham, UK
Suzuki Junko, Tokyo, Japan
Kit S Kapp, Osprey, FL, USA
James Kelly, Oxford, UK
Geoffrey King, Eccleshall, Staffs, UK
Peter van der Krogt, Utrecht, The Netherlands
Christopher W. Lane, PA, USA
Michael Layland, Vancouver, BC, Canada
Ira S. Lourie, Rockville, MA, USA
Montserrat Galera i Monegal, Barcelona, Spain
Alberto Montillo, The Phillippines
Maura O'Connor, Canberra, Australia
Marcus Oehrli, Wabern, Switzerland
John Roger Paas, Northfield, MN, USA
Mary Pedley, Ann Arbor, MI, USA
A.J. Peile, Bracknell, Berks, UK
Mireille Pastoureau, Paris, France
Monique Pelletier, Paris, France
G.S. Ritchie, Aberdeen, UK
Walter Ristow, Mitchellville, MA, USA
Iolo and Menai Roberts, Newcastle, Staffs, UK
Robert Ross, Calabasas, CA, USA
Keith Rossetti, Menorca, Spain
Wolfgang Scharfe, Berlin, Germany
Judith Scurfield, Melbourne, Australia
Kirsten Seaver, Stanford, CA, USA
Rodney Shirley, Buckingham, UK
Vladimir Shkurkin, San Pablo, CA, USA
David Smith, Bexley, Kent, UK
Richard Stephenson, Winchester, VA, USA
Harry Steward, Worcester, MA, USA
Jeffrey Stone, Aberdeen, Scotland
Tom Suarez, Hawthorne, NY, USA
E. Tallis, Farnham, Surrey, UK
Iain Taylor, Halifax, Nova Scotia, Canada
Vladimiro Valerio, Naples, Italy
David Webb, Atworth, Wiltshire, UK
Diana Webster, Edinburgh, Scotland
Margaret Wilkes, Edinburgh, Scotland
Joan Winearls, Toronto, Ontario, Canada
Eric Wolf, Falls Church, VA, USA
David Woodward, Madison, WI, USA
Laurence Worms, London, UK
Cordell Yee, Annapolis, USA

FOREWORD

Maps are made by humanity for humanity. They require ingenuity, science, and art in differing proportions in order to conceive spatial and chronological ideas or to perceive reality, to symbolise them, and to aid the release of their encoded data by other humans. Whether we are novices in their creation, discovery, or interpretation, or whether we are experienced professionals, both the remarkable diversity and the seemingly endless quantity of cartographic artefacts that have, or will, come to our attention is nothing less than wondrous. This is whether we consider indigenous (or 'native') map-making, 'traditional cartography' (with its manual and mechanical production methods), or the 'New Age' of electronic data compilation, storage, and manipulation into seamless (some would write 'soul-less') end products. Ironically, for an intensely humanity-based activity, the latter 'New Age' computerised cartography can mean that an individual's or a corporation's artistic style is often subjugated to digitised inflexibility and commercial expediency.

In a similar way countless other makers of maps, charts, and globes from prehistoric and historic times will remain anonymous. The 'History of Cartography Project' (co-founded by the late J. Brian Harley and continued by David Woodward) aims for both universal and historical coverage of our subject and Tooley attempted the same. There is still both a need and a space in the literature of cartography for general overviews and introductions. Whatever the inherent shortcomings and 'lacunae' in this 'New Tooley' (as it is known to librarians, collectors, dealers, and researchers worldwide) it is intended to be just that: an introduction. It will replace the earlier edition and its supplement and will continue, doubtless, to be a welcome *vade mecum* for those both in and out of the map world.

Francis Herbert
Curator of Maps
Royal Geographical Society (with the Institute of British Geographers), London

How to use the Dictionary

Order of entries

Entries appear in alphabetical order except in the case of entries under the same name. These are headed by families of that name, irrespective of the names of the individuals. Generations are indicated by bullet points [•].

After family entries, individuals are then listed in alphabetical order by their given names. Where an initial is known only, that individual precedes those whose given names begin with that initial. Where two or more individuals have the same given names, they are listed chronologically.

Companies follow individuals with the same name. Successions are indicated by bullet points [•].

NB. The examples on this page are fictitious.

Williams, family. Engravers of London.
• **Williams**, John (1720-1772).
•• **Williams**, Michael George (1743-1791).
•• **Williams**, John & Son (*fl.*1761-1769). Partnership of John Williams and his son Michael.
••• **Williams**, George (1764-1835). Son of Michael.
•• **Williams**, Adam James (1747-1810). Son of John and brother of Michael.

Williams, Adam (1826-1911).

Williams, G. (1803-1856).

Williams, George (1780-1850).

Williams, George (1820-1880).

Williams & Co. (*fl.*1760-1776).
• **Williams, Smith & Co.** (*fl.*1777-1781).

Content of entries

The mapmaker's name is followed by birth and death dates where known, and a small amount of biographical information if available.

This is followed by a few examples of work in chronological order, together with any available information about others involved.

Where possible references have been included. These give details of where further information on this mapmaker or his works can be found.

Williamson, George (1780-1850)[1]. Engraver 'at the sign of the burin, Cornhill'[2] London (1800-1820); 22 Fleet Street, London[3] (1821-1845[4]). John Rose's *The Road from London to Birmingham*[5] (3 sheets[6]), 1810[7], 1812[8]; *Atlas of England* (28 maps[9]), London[10], J. Smith[11] 1815, Oxford[12], 1817[12]; *Map of London, Westminster and Southwark*, London, J. Smith 1822; map of Ireland[13], 1823; *London*, 1840 (with D. Beech[14]). **Ref.** SMITH, J. *George Williamson: Engraver* (London 1996); WILLS, C. *London engravers* (1987) pp.10-16.[15]

Key to example

1. Birth and death dates unless otherwise indicated.
2. Address as shown on works at given date(s).
3. Supplied address, not necessarily on works.
4. Dates at this address.
5. Title of work.
6. Number of sheets for a single map.
7. Earliest known date for the work.
8. Later issue or edition (not always the same title).
9. Number of maps in a publication.
10. Place of publication.
11. Publisher.
12. Details of later issue or other edition.
13. Unverified title or description of an untitled map.
14. Collaborator.
15. Useful references in reverse chronological order.

The following abbreviations have been used

b.	born
d.	died
c.	*circa*
ed.	editor
fl.	flourished
MS	manuscript
q.v.	which see
Ref.	reference
SJ	Society of Jesus
RE	Royal Engineers
RN	Royal Navy
USN	US Navy

A., See **Aelst**, Nicolas van.

A., A. *Mapa de la República de Argentina*, 1868.

A., G.L. Georgius Lilius Anglus. See **Lily**, G.

A., J. *Plan de Montréal*, 1756.

A., J. See **Adriaansz**, Jasper.

A., J.L. de. *Atlas geográfica de España*, 1848.

A., K. See **Krishna**, Pundit.

A., M.J. See **Armstrong**, Mostyn John.

A.B., G. See **Apian**, Georg Bienewitz.

A.V. See **Musis**, Agostino de.

A.V.T. See **Avity**, Pierre de Montmartin.

Aa, Van der family. (*fl.* 17th-18th centuries). **Ref.** KROGT, P. van der *Stock catalogues of maps and atlases by Covens & Mortier* (facsimile with introduction), (Utrecht 1992); KROGT, P. van der *Advertenties voor kaarten, atlassen, globes e.d. in Amsterdamse kranten 1621-1811* (1985); KOEMAN, C. *Atlantes Neerlandici* (1967-1971) Vol.I pp.1-29.

• **Aa,** Pieter [Pierre, Petrus] van der (1659-1733). Publisher, editor and bookseller of Leiden: 'In de St. Pieterskerkkoorsteeg te Leiden'. Advertised numerous collections and libraries for sale from 1700 onwards. *Gedenkwaerdige Oost en Westindise Voyagien* (28 volumes in 12), first announced 1706; *Cartes des itineraires & voyages modernes*, 1707; *Les Indes Orientales et Occidentales*, 1710; *Atlas nouveau et Curieux...*, 1710; *Les dix-sept Provinces des Pais-Bas...* (9 sheets), *c.*.1710; *Le nouveau théatre du monde*, 1713 (by Gueudeville and Ferrarius); *Nouvel Atlas*, 1714; *Cartagene avec ses Ports et Fortresses*, 1715; *Forces de l'Europe &c.* (20 parts in 2 volumes), 1726; *Galérie agréable du monde* (66 parts), 1729 (said to have been issued in an edition of 100 copies only).

• **Aa**, Boudewijn [Boudewyn] van der (*fl.c.*1701). Printer and bookseller of Leiden: 'op 't Rapenburg, over d'Academie'. Worked for his brother Pieter, sold stocks of various Blaeu atlases.

• **Aa**, Hildebrand van der. Engraver. Brother of Pieter.

•• **Aa**, Janssoons van der (*fl.*1726-1735). Bookseller of Leiden, nephew of Pieter. Sold remainder stocks of his uncle's publications.

Aa, Cornelis van der (1749-1816). Bookseller at Haarlem and Utrecht. De Jong and M. Sallieth's *Atlas van de Zeehavens der Bataafsche Republiek*, F. Maaskamp 1805.

Aagaard, Ulrik Frederick (1694-1727). Danish cartographer and painter of Stavanger.

Aalst, Nicolas van. See **Aelst**, Nicolas van.

Aanum, Ole M. Reykjavik, 1801.

Aarnio, B. *Agrogeologisia karttoja*, Helsinki 1916.

Aaron ben Chaim, Rabbi of Grodno. Map *Eretz Israel* (in Hebrew), in *Moreh Derech* [A Guide Book], Vilna [Vilnius] 1836, and numerous reprints to 1839. **Ref.** WAJNTRAUB, E. & G. *Hebrew maps of the Holy Land* (1992) N°W.57 pp.141-143.

Abadie, Philibert. *Atlas historique...*, 1846.

Abaffy, *Lieutenant* —. (*fl.*1782-1784). Participated in first military survey of Hungary, mapped 11 sections in counties around Pest.

Abancourt, Charles-François Frérot d' (1756-1801). 'Ingenieur géographe' of Paris, died Munich. Map of Switzerland.

Abbadie, Antoine-Thomson d' (1810-1897). French geographer, worked in Ethiopia. *Géodésie d'une partie de la Haute-Ethiopie* (10 sheets), 1860-1863; *Géographie de l'Ethiopie*, 1890. **Ref.** BROC, N. *Dictionnaire illustré des explorateurs Français du XIXe siècle: I Afrique* (1988) p.1 [in French]; PASTOUREAU, M. 'Antoine d'Abbadie, 1810-1897' in: *Geographers*, 3 (1979), pp.29-33.

Abbatt, Richard. Orthographic projection of world, 1857; *World of the Antipodes*, 1857.

Abbe, Cleveland (1838-1916). Mathematician, astronomer, globemaker of New York.

Abbela y Casanego, —. See **Abella** y Casariego, Enrique.

Abbestee, Dirck [Pietersz.]. (*c.*1615 or 1617-1688). Dutch land surveyor of Callantsoog. *Karte van des Graeffelijckhts end d'Oochduijne... van Huijsduijne*, 1662 (with Jan van Heymenbergh); *Kaerte ende meetingen vande binnenlanden van Huijsduijne...*, 1669, both MS; *Karte van t'Koegras* (polder map), 1676.

Abbot, Charles, Lord Colchester. *Route British Embassy along Yang-tse-kiang*, Admiralty chart 1841.

Abbot, Gorham Dummer (1807-1874). Teacher in a New York school for 'Young Ladies'. Patented an inflatable globe, 1868.

Abbot, *Lieutenant* (later *Captain*) Henry L. American cartographer. *Routes proposed for Pacific Rail Road*, 1855; *Basins of the Mississippi & Tributaries*, 1861; *Army of the Potomac*, 1864.

Abbot, K.E. (*d.*1873). *Account of Persia*, 1849-1850.

Abbott, E. *Dallas County, Iowa*, New York 1874.

Abbott, *Captain* F. *Reconnaissance Sketch of Kabul and Its Environs*, 1842, London, War Office 1879.

Abbott, *Lieutenant* (later *Captain*) J. Revenue surveyor in India, worked with G.J. Fraser. *District of Bareilly*, 1833-1837, published Calcutta 1858; *District of Shahjuhanpoor...*, 1838-1839, published Calcutta, 1846.

Abbott, Moses. Plan of Tewkesbury, 1794 MS.

Abbott, *Lieutenant* S.A. Revenue surveyor. *...District of Bandah...*, 1840-1841, published 1865; *... District of Cawnpoor*, 1845; *...District of Benares...*, 1846; *...District of Jounpoor...*, 1846.

Abbott, William. English cartographer. *Map of the World shewing the submarine Cables*, 1872, 1873, 1875; *Submarine Telegraph Map of the World*, 1876; *...the Dominion of Canada...*, 1884.

Abbs, *Lieutenant* E.C., *RN*. *A Map of British Honduras*, London, E. Weller 1867 (after J.H. Faber and E.L. Rhys).

Abcouw [Abkoude], J. van (*fl.*1730-early 1740s). Bookseller of Leiden. Sold Jan de Lat's *Nieuw en Beknopt Kaert-boekje, vertonende de 17 Nederlandse Provintien, etc.*, 1736; J.D. Knapp's *Kaert van de Berbice*, 1743. **Ref.** KROGT, P. van der *Advertenties voor kaarten, atlassen, globes e.d. in Amsterdamse kranten 1621-1811* (1985) N°706 p.136, N°867 p.167 [in Dutch].

Abd al-Aziz [Abdulaziz] Abdulgani el-Erzincani (*fl.*19th century). Turkish cartographer who produced a large cloth map of Europe, Asia and North Africa, 1813.

Abd Allah [of Kailaspur]. *Atlas of India...*, 1917.

Abd al-Rahman, al-Khazini (*fl.*12th century). Persian author. Treatise on celestial globe: *Sphere that rotates by itself.*

Abd al-Rahman Efendi (*fl.* early 19th century). Turkish atlas of the World, Istanbul 1802-1803.

Abd al-Rahman Sufi-al. See **Sufi-al.**

Abdulaziz. See **Abd** al-Aziz.

Abe Tameto. *Kaisei Dai Nihon...* [A map of Japan without names of places], 1875; a map of Tokyo, 1876.

Abe Yasuyuki. Japanese mapmaker. *Bankoku Yochi Zenzu* [map of the World], Suzutei Morikuwa 1853.

Abe Yoshito (1805-1870). Japanese scholar. Preface for *Bankoku chikyū zenzu* [Japanese world map], *c.*1850.

Abeken, Heinrich A. *Karte der Sinai-Halbinsel...*, 1859.

Abel, Caspar (1676-1763). Theologian and historian of Hindenburg. *Preussische Staats-Geographie*, 1735.

Abel, Gottlieb Friedrich (*b.*1763). Royal court engraver at Stuttgart. One of the engravers for Johann Wilhelm Jaeger's *Grand Atlas d'Allemagne*, published 1789; Swabia for Johann Bohnenberger, 1798; *Grundriss von Stuttgart*, 1811.

Abela, Giovanni Francesco. *Descrittione di Malta*, 1647, believed to be the first printed description of Malta with maps.

Abelin, Hans. Draughtsman. Kempten, 1569.

Abelin, Johann Philipp. Associate of Theodore de Bry. Often quoted as working under the pseudonym of Johann Ludwig Gottfried, and credited with Gottfried's *Newe Welt und Americanische Historien*, and other works. **N.B.** Some sources claim that Gottfried and Abelin are one and the same. Others, including Philip Burden in *The mapping of North America* (1996) pp.273, 293, assert that they were not.

Abella y Casariego, Enrique. *Isla de Panay...* [Philippines] (2 sheets), Manila 1890.

Abendanon, Eduard Cornelius. Mining engineer. *Midden-Celebes-Expeditie. Geologische en geographische doorkruisingen van Midden-Celebes 1909-1910* (with 16 maps and profiles), Leiden, E.J. Brill 1916. **Ref.** KOEMAN C. *Atlantes Neerlandici* Vol. VI (1985) p.67.

Abendroth, E. *Charte der Elb-Mündungen...*, 1846.

Abensour, Léon. Compiled *Nouvel Atlas Larousse*, 1924.

Abent, Leonard. Plan of Passau for Braun and Hogenberg *q.v.*, 1581.

Abercrombie [Abercromby], *Major General* James, (1706-1781). In 1758 became supreme commander of all British forces in America and led an assault on Fort Ticonderoga, Essex county, New York. Not to be confused with Captain [Thomas] Abercrombie (below), who drew a plan of the scene of battle which he dedicated to General James.

Abercrombie, *Sir* Leslie Patrick. *Greater London Plan* (11 folding maps), 1945; *A Civic Survey & Plan for the City and Royal Burgh of Edinburgh*, 1949.

Abercrombie, *Captain* Thomas. Sketch of Lake St George, 1756; manuscript plans of forts and towns in New York Province, North America, 1756-1759; *Plan of Fort Ticonderoga on Lake Champlain*, 1758.

Aberdeen, A. *Orkney*, 1770 (with William Aberdeen).

Aberdeen, William. *Orkney*, 1770 (with A. Aberdeen, above); *Lerwick*, 1776.

Åbergh, Oscar. *Sverige och Norge*, 1871.

Aberl, L. *Island of St. John on the Lake of Bienne* [Switzerland], 1806.

Abernathy, G.E. *Oil and Gas Exploration Map, Labette County, Kansas*, 1839.

Abernethie, Thomas. American engraver. *A Sketch of the operation before Charleston...*, 1785; *A Sketch of Charleston Harbour*, 1785; *A Sketch of the Situation...of British Vessels under Sir Peter Parker*, 1785; maps for David Ramsay's *History of the*

Revolution of South Carolina, 1785; *Travelling Map Roads of South Carolina*, 1787 (with Walker); *Plan of the City of Charleston, South Carolina*, 1790. **Ref.** WHEAT & BRUN *Maps and charts published in America before 1800* (1978) Nos.593, 594, 596, 599 pp.128-129.

Abernethy, George. *Plan of the Outer Bar, Port Adelaide*, 1858.

Abernethy, James (1814-1896). Harbour engineer, produced maps of canals, docks, harbours, railways and water supplies.

Abert, *Lieutenant* J.W. *Territory of New Mexico*, Washington DC 1845-1847 (with Lieutenant W.G. Peck).

Abert, *Colonel* John James (*fl.*1824-1851). American cartographer; Chief, Topographical Bureau 1851. *Map of the Patuxent and St. Mary's Rivers, Maryland*, 1824; *Map of the Seat of War in East Florida*, 1838; *...Routes for the proposed Maryland Canal*, *c.*1839; *...Mississippi River*, 1843; *Military Reconnaissance of the Arkansas, Rio de Norte and Rio Gila*, 1847.

Abich [Abikh], Otto Wilhelm Hermann (1806-1886). Mineralogist and geologist of Berlin. *Der Krater des Aetna*, *c.*1834; *Der Krater des Vesuv...*, *c.*1834; atlases of volcanoes from 1839; *Karte des Ponza Inseln*, 1842; *Caucasian Atlas*, 1882; *Geologische Fragmente*, 1887.

Abilgaard, Søren (1718-1791). Drew maps for Erik Pontoppidan's *Danske Atlas*, 1763-1769.

Abington, William. Bookseller at the Three Silk Worms in Ludgate Street, London. Richard Blome's editions of Varenius' *Cosmography*, 1683.

Abkoude. See **Abcouw**, J. van.

Ablancourt, Nicolas Frémont d'. See **Frémont** d'Ablancourt, Nicolas.

Ablincourt, C.F. d'. Military engineer and surveyor in Netherlands. *Plan de Gorinchem*, *c.*1750? MS.

Abollado, *Don* Yldefonso de Aragon y. See **Aragon** y Abollado, Don Yldefonso.

Abraham bin Chija (*fl.*1100-1130). Jewish geographer and astronomer, of Barcelona.

Abraham, Helevi ben David (*fl.*12th century). Jewish geographer.

Abrahams, —. *Eiland Walcheren*, 1865.

Abreu Corgan, João. Brazil, 1747.

Abrial, —. *Département du Tarn*, 1873.

Abrille, F.V. See **Villa** Abrille, Faustino.

Abu al-Fida, Ismail, Imad al-Din of Damascus (1273-1331). Knowledgeable emir [prince], historian and geographer. *Taqwim al-Buldan* [Location of the Countries], 1321, translated and published at Göttingen 1802, also Paris 1848; *History by Abu al-Fida* (4 volumes), 1329.

Abu al-Hasan, al-Marakashi (*fl.*13th century). Moroccan astronomer in Cairo.

Abu Bakr Ibn-Bahram, al-Dimashqi (*fl.*17th century). Syrian geographer of Damascus.

Abu Mashar Jafar [Albumasar], Ibn-Muhammad Balkhi (805-885). A scholar of Kindi, born in Balkh of Greater Khurasan, mathematician and astronomer. *Qaranat al-Kawakib* [Inter-relation of the stars]; *Al-Madkhal fi Ilm al-Nujum* [Introduction to Astronomy].

Abu Rayhan. See **Biruni**.

Abu Ubayd Abdallah, Ibn Abd al-Aziz al-Bakri (11th-century, *d.*1094). From Andalusia. Known as the second greatest geographer of the Muslim west (the first being al-Idrisi). *Kitab al-Masatik Wa al-Mamalik* [The Book of Routes and Countries], 1068.

Aburrow, Charles. *...Diamond Fields, Griqualand West* [South Africa], 1886.

Ac ko mok ki. Blackfoot Indian chief, author of a map showing "...the rivers and other remarkable places to the Mis sis su ry river", 1801. **Ref.** HECKROTTE, Warren *The Map Collector* 39 (1987) p.20; MOODY, D.W. and KAYE, B. 'The Ac ko mok ki Map', in: *The Beaver*, Spring 1977, pp.4-15.

Acacius. See **Enenckel** von Hoheneck, *Baron* Georg Acacius.

Acad. Reg. Scient. et Eleg. Litt. Boruss [Prussian Royal Academy of Science and Letters]. Granted royal privilege by Frederick the Great, 1748, for compilation and production of maps. The organisation produced 70 titles up to c.1790 including: *Atlas geographicus*, Berlin 1753-1760.

Accademia cosmografica degli Argonauti, Venice. Considered to be the world's first organised geographical society. Founded by Vincenzo Maria Coronelli, *c.*1684.

Accarias, L. *Circuit d'Auvergne, coupe Gordon-Bennett*, *c.*1905.

Accerlebout, J. See **Aerlebout**, Josias.

Accinelli, Francesco Maria. Italian cartographer. ...*Contorni di Genova*..., 1746.

Acerlebout, J. See **Aerlebout**, Josias.

Achephol, —. ...*Dortmund*, 1872.

Achin, —. *Plan de la Ville de Paris*, *c.*1830; *Nouvelle Carte Routière du Royaume du France*, 1831; *Nouvelle Carte Routière des Environs de Paris*, 1840; ...*Départements de la France et de ses Colonies*, *c.*1845; *Plan de Paris*, *c.*1850; *Plan Itinéraire de Paris*, 1853; *Carte de France*, 1873.

Ackermann, E. *Konstanz und Umgebung*, 1891.

Ackermann [Ackerman], George. Lithographer of New York City, 379 Broadway and 120 Fulton Street. J.H. Colton's *United States*, 1849; *Mexico, Bird's eye view looking towards the west*, *c.*1850; *General Map of the Orange & Alexandria Rail Road* [USA], 1851; surveys western Utah, 1852; August Faul's *Map and profile of the Orange and Alexandria Rail Road*, New York, *c.*1854; *Sketch of General Riley's Route through the Mining Districts...1849*, [n.d.].

Ackermann, Rudolph [& Co]. (*fl.*1797-1843). London printsellers and publishers of 96 Strand, 101 Strand (from 1797), 96 Strand (1827). Publishers of numerous sporting and topographical works illustrated with

hand-coloured aquatints. *A Birds Eye View of Dartmoor Prison...*, 1810; *Microcosm of London or London in Miniature*, [n.d.]; Johann Isaac Gerning's *A Picturesque Tour of the Rhine*, 1820; *Topographical Plan of the ... City of Buenos Aires*, *c.*1820; Jean-Baptiste Poirson's *Mexico*, 1826; *Panoramic View round the Regent's Park*, 1831; *A New General Atlas*, 1838; *Bauerkeller's New Embossed Plan of London*, 1841. **N.B.**The work of father and son have not been distinguished above. **Ref.** FORD, John *Ackermann 1783-1983* (Arthur Ackermann, London 1983); TOOLEY, R.V. *Some English Books with Coloured Plates* (1954).

• **Ackermann**, Rudolph (1764-1834). Publisher and artist from Germany, founder of the Ackermann publishing company of London.

•• **Ackermann**, Rudolph *junior* (1796-1863). Son of Rudolph (above). Founded his own company at 191 Regent Street, London, in 1820. *Grand Panorama of Canterbury*, 1847; plan of the city of Exeter, *c.*1850.

Ackers, —. ...*Charters Towers Gold-Field, Queensland* (3 sheets), 1887.

Acland, H.W. Map of Oxford showing the localities in which cholera occurred, 1854.

Acland, J.B.A. Additions to Edward Stanford's *Map of the Province of Canterbury, New Zealand*, 1856.

Acosta, F. de P. *Mapa Político de los Estados Unidos de Venezuela*, 1870.

Acosta, *Colonel* Joaquín (1799-1852). American historian and geographer. *Mapa de la República de la Nueva Granada*, Paris 1847; *Mapa de Territorio de la Nueva Granada...*, Paris 1848.

Rudolf Ackermann Snr (1764-1834). Portrait by François Nicolas Mouchet. (Picture Library, National Portrait Gallery, London)

Acosta, José de, *SJ* (1539-1600). Spanish missionary, historian and cosmographer. *Historia natural de las Indias* [America] (30 maps, including map of New World), 1588 and other editions.

Acosta, Juan Luis de. 'Capitan y piloto mayor'. *Rio de Cagaya*, 1720 MS.

Acqua, Giovanni Domenico. See **Dall'**Acqua, Giovanni Domenico.

Acrelius. See **Akrel**, Frederik.

Acton, Oliver. Plan Bridewell Estate Thames, 1737.

Acugnal [Acuna], Christobal d' (1597-1655). *b*. Burgos. Returned with Pedro Teixeira from South America. *Nuevo descubrimiento del gran rio de las Amazonas*, 1641.

Acuna, P.P. d'. *Carte de la Terre Firme* [South America] (2 sheets), Paris 1703 (information used by Delisle, 1782).

Adair, James. *American Indian Nations*, 1775.

Adair, John (1660-1718). Scottish mathematician, surveyor and chartmaker, Fellow of the Royal Society 1688. Manuscript maps of Scottish shires, 1682-1688; *Turnings of the River Forth*, 1688; *Hydrographical Description of the Sea Coast of Scotland*, 1688-1703; *Strathern*, 1690; *...the Town and Water of Montrose*, 1693; chart of the Firth of Clyde, 1700 MS (Admiralty Library); *The Description of the Sea-Coasts and islands of Scotland...* (6 engraved charts), 1703; *...Pentland Firth*, *c*.1715 MS (Admiralty Library); posthumously, *...the River Clyde*, 1731; *...Lothians* (3 sheets), 1745. **Ref.** SIMPSON, Allen *The Map Collector* 62 (1993), pp.32-36; MOORE, John N. *The Map Collector* 30 (1985), pp.30-34.

Adair, Robert Alexander Shafto. *Military Sketch-Map of England and Wales*, 1855 (printed by Cowell's Anastatic Press).

Adair, Thomas. Plans of parishes and allotments in Victoria, New South Wales, [n.d.].

Adam of Bremen (*fl*.1069-1076). Historian and geographer. First writer to mention Norse discoveries in America (Vinland). In his *History* he devoted a section to geography, particularly of northern latitudes.

Adam, Alexandre. *Plan de la ville et de la Commune de Boulogne sur Mer*, 1863.

Adam, H. Würzburg, 1860.

Adam, Jakob (1748-1810). Engraver in Vienna, for Franz Anton Schraembl and others. Ungarn, 1780; *Mappa von der osterreich...*, 1781; world atlas, 1800.

Adam, V. Worked with Giraldon on maps for Pierre Lapie's *Atlas classique*, 1812; globe, 1815 (with Giraldon); *Océanie*, 1816.

Adami, Karl Christian Ludwig (1802-1874). Berlin cartographer and globe maker. Globe, 1852; Atlas, 1855; *Schul-Atlas*, 1868.

Adams, —. See **Asher** and Adams.

Adams, family of globe and instrument makers, active in the 18th century.
• **Adams**, George, *the elder* (1709-1772). Instrument maker. Terrestial globe, 1760, 1769; pair of 18-inch globes, 1766; *A Treatise Describing and Explaining the Construction and Use of new Celestial and Terrestrial Globes*, 1766; Celestial globe, 1769.
•• **Adams**, George, *the younger* (1750-1795). Son of George above. Instrument maker to George III. Fleet Street, London. *Wright's New Improved Terrestial Globe...*, 1782, 1790.
•• **Adams**, Dudley (1762-1825). 'Globe and Mathematical Instrument Maker to the King' [George III]. 60 Fleet Street, London. New edition of the 1766 18-inch globes of George Adams *the elder*, 1789; pair of terrestrial and celestial globes, 1797; *...Globe of the Earth correctly laid down...*, [n.d.].

Adams, Clement (1519-1587). Schoolmaster, tutor to the Royal Pages at Greenwich, and author. Cited by Richard Hakluyt as engraver of the English edition of Sebastian Cabot's map of the West Indies, 1549; edited Anthony Jenkinson's *Nova absolutaque Russiae, moscoviae, & Tartariae*, 1562. **Ref.** KARROW, R.W. *Mapmakers of the sixteenth century and their maps* (1993) pp.110-111, 318.

Adams, D.P. Charts of Hampton Roads and Chesapeake Bay for the United States Coast Survey, 1830.

Adams, Daniel (1773-1864). American geographer. *Atlas to Adams' Geography for the use of schools and academies*, Boston, West & Blake 1814 (engraved by Wightman); *School atlas to Adams' Geography*, Boston, Lincoln and Edmands, Boston 1817 and later (maps engraved by Annin and Smith).

Adams, F.D. *Geological Map of portions of Hastings...and Peterborough Counties* [Ontario, Canada], 1909.

Adams, George I. *Geological Map of part of the Eastern Choctaw Coalfield* [Indian Nation], 1900.

Adams, H.J. (*fl.*19th century). Lithographer, of London. Adams's [W.J.] *Pictorial Map of London for 1851.*

Adams, Henry Martin. Surveys of the Island of Antigua, 1882-1891 MSS; *Barbuda in the West Indies*, 1891 (copied from an earlier map of 1813).

Adams, J. *Present Seat of War in Turkey*, 1774.

Adams, J.Q. & Co. *The new international atlas of the World*, Boston 1914.

Adams, John. **N.B.** Please note that uncertainties arising from the number of entries in the name of John Adams for the late 18th and early 19th centuries, especially the two associated with Edmonton [Middlesex], remain unresolved.

Adams, John (1670-1738). Topographer and barrister of the Inner Temple from Shropshire. Attempted trigonometrical survey of England and Wales (1681-1684). *Angliae Totius Tabula* (distance map of England and Wales in 12 sheets), 1677; reduced version in 2 sheets, 1679; Carington Bowles edition, *c.*1770; *Index Villaris*, 1680, 1690, 1710 [a gazetteer of 24,000 place names in England and Wales]. **Ref.** SHIRLEY, R. *Printed maps of the British Isles 1650-1750*, pp.18-22 (1988); *Geographical Journal* 151 (1985), pp.21-39; RAVENHILL, William *Imago Mundi* 33 (1981) p.26; *Geographical Journal* 144 (1978), pp.424-437; *Geographical Journal* 97 (1941), pp.182-184.

Adams, John. Assistant master mason. *Plan of the Seige of St Philips Castle...Minorca*, 1782.

Adams, John. *Map of the Roads and Inland Navigation of Pennsylvania*, 1791 (with John Wallis).

Adams, John [of Edmonton]. *Directions for the Mediterranean Pilot...*, 1795.

Adams, John. Teacher of mathematics in Edmonton [Middlesex]. *The Young Lady's and Gentleman's Atlas...*, London, W. Darton and J. Harvey 1805; *A General Atlas for the use of Schools*, 1843.

Adams, John, father and son.
• **Adams**, John I (*fl.*1805-1845). Kent estate surveyor. Plan of an estate at Tenterden [Kent], 1806; ...*Hocklebury Estate...* [Kent], 1809; *Dover and its Environs*, 1828.
•• **Adams**, John II (*fl.c.*1815-1834). Son of John above. Kent estate surveyor.

Adams, John. *Bay of Cartagena and Plan of Cartagena*, London 1807.

Adams, John. Military surveyor and artist. Copied Charles Shipley's harbour and battery plans of Jamaica, 1808; *Plan of the Harbour, Town and Fort of St Johns in Newfoundland*, 1810 (copied from Hugh Debbieg's survey of 1762); ...*Map of Quebec and its environs...*, 1822, published James Wyld 1826 (engraved by E. Bennet); map of Mille Vaches Bay [Canada], 1831; *Map of Upper and Lower Canada...*, 1832.

Adams, M. Plan of Hounslow, Middlesex, 1836.

Adams, Robert (*c.*1540-1595). Surveyor of the Queen's Works 1594-1595, designer of houses and an expert on fortifications. Several manuscript maps extant: Flushing, 1585; Portsmouth, 1585; compiled charts for Petruccio Ubaldini's *Expeditionis Hispaniorum in Angliam vera descriptio*, 1588, 1590; Middleburgh, 1588; *Thamesis descriptio*, 1588; charts of the Spanish Armada, engraved by Augustine Ryther for publication in *Expeditionis Hispanorum in Angliam...*, 1590; *Poole Harbour*, 1590; *Sa Nicholas Iland by Plimouthe*, 1592; Plymouth, 1593.

Adams, Robert. Surveyor. *A Chart of Southampton Water*, 1810.

Adams, Samuel W. (*fl.*1839-1847). Estate and tithe surveyor, of Newton Ferrers. *Hayle River* [Cornwall], 1847 MS.

Adams, Sebastian C. *Chronological Chart of Biblical History*, 1883 (maps by J.S. Paine).

Adams, Thomas (*fl.*early 17th century). Succeeded his cousin George Bishop as printer and bookseller in St Paul's Churchyard, London. Business passed to Andrew Heb.

Adams, W. Hamilton. Publisher, of 52 Paternoster Row, London. *R.H. Stone's Penny Waistcoat Pocket Map of London*, 1868; *Dolling's Penny Waistcoat Pocket Map of London*, 1872.

Adams, Walter Edward (1861-1936). Engineer, of Sydney, New South Wales, Public Works Dept 1889-1901. *Birds eye view of general wharfage scheme west of Dawes Point*, *c.*1913.

Adams, William. *...deluge of the Fens,...on the 4th May 1862*, 1862.

Adams, William James [& Sons]. Publisher of 59 Fleet Street, London. *Adams's Guide Map to the Environs of London 25 Miles Round*, 1851 (engraved by B.R. Davies); *Adams's Pictorial Map of London for 1851*, 1851; *The Improved District Railway Map of London*, 1879, 1880; *The District Railway Miniature Map of London*, *c.*1881; *The District Railway Country Diagram*, *c.*1883.

Adams-Reilly, A. Surveys in the Alps, 1863-1868, including: *The chain of Mont Blanc*, 1865; *The Valpelline, Valtournanche and the Southern Valleys of the Chain of Monte Rosa*, 1868.

Adamy, Heinrich. *Wand-Karte von Schlesien*, 1850 and later editions; *Umgegend von Breslau*, *c.*1865; *...Grafschaft Glatz*, 1869; revised and completed *Special-Karte von Schlesien und der Grafschaft Glatz*, 1871 (with Prof. Dr. Sudebeck); *Schul-Karte von Schlesien*, 1874 and later editions.

Adanson, Jean Baptiste (1732-1804). Manuscript maps of Egypt, 1763-1784.

Adaro y Magro, Luis. *Atlas...de la Cuenca Hullera Asturiana*, 1926.

Adcock, G.H. *Viticultural Map of Victoria* [Australia], 1901.

Addison, J.W. *...South African Campaign*, 1900.

Addison, John (*fl.*1825-1845). Globe-maker. Drew and engraved Greenwood's *Kent*, 1839; *Malby's Celestial Globe Atlas* (21 sheets), 1845.

Addison, M. *Colombia*, 1913.

Addison, Thomas (*fl.*1771-1781). Surveyor of Lancashire. Speke Estate, 1781.

Adelbauer, Johann Ernst. Printer of Nuremberg. Kohler's *Schul Atlas*, 1719; Homann's *Grosser Atlas*, 1731.

Adelhauser, Hans. Plans of Gotha and Grodno, 1568. Used by Braun and Hogenberg, *q.v.*

Adeline, Jules. *Rouen disparu*, 1876; *Les Quais de Rouen autrefois & aujourd'hui*, 1878.

Adelung, Friedrich von. *Reisenden in Russland*, 1846.

Adenis, —. *Département de la Creuse*, 1867.

Aderholz, Georg Philip. *...Russische Provincien Lithauen, Wolhynien, Podolien &c.*, 1831.

Adgerus, C. See **Adgiers**, Cornelis.

Adgiers [Adgerus; Aedgerus; Aldger], Cornelis [Cornelius] (*c.*1520-*c.*1595). Dutch mathematician and geographer, from Leeuwarden, admitted as surveyor 1548 in Friesland. *Coloniensis dioecesis typus*, 1583 (used by Abraham Ortelius). **Ref.** MEURER, P. *Fontes cartographici Orteliani...* (1991) p.98 [in German].

Adie, Alexander John. Contributed to Johnston's plan of Edinburgh, 1850.

Adies, —. Borneo coast, 1789.

Adkins, I.F. *Jefferson County, Georgia*, 1879.

Adlard, Alfred. Engraver. *Adlard's Vignette Modern Atlas*, 1849-1850; *Ten miles round Ludlow*, 1850.

Adlard, H. Engraver of Thomas Bullock's map of Great Salt Lake City, London, Longman & Co. 1860.

Adlard, James. Printer, 39 Duke St., West Smithfield, London. Wilkes' *Encyclopaedia Londinensis*, 1810.

Adlard, J. & C. Printing company of Bartholomew Close [London], formed by James and Charles Adlard. *The Arbitrator or Metropolitan Distance-Map* [F.G. Harding's map of London], 1834.

Adlard, William. Printer of London, founding partner in Adlard & Brown[e].
• **Adlard & Browne**. Printers of Fleet Street, London. P. Russell & O. Price's *England Displayed*, 1769 (many of the maps had been used earlier in Thomas Read's *Small British Atlas*, 1753 and later).

Adlard & Jones [Adlard & Co.]. Publishers. Ave Maria Lane, London. Maps in Wilkes' *Encyclopaedia Londinensis*, 1810.

Adler, C. Publisher in Hamburg. *Handatlas für Volks- und Bürgerschulen*, 1866; *Handatlas*, 1870; *Schulatlas mit Texbeigabe*, 1888; *Schulatlas*, 1889; lithographer for *Neuester geographischer Atlas für Militär...*, Berlin 1890.

Adlum, J. Map of the roads and inland navigation in Pennsylvania, *c.*1795 (with J. Wallis).

Admiralty. See **Hydrographic** Office.

Adnitt & Naunton. Lithographers. T.W.M. Tisdale's plan of Shrewsbury, 1875.

Adriaansz. [Adriaanszoon; Adriaensz; Adriaenszoon], Jasper (*fl.*1539-1568). Surveyor of The Hague, worked in Zuid-Holland (*c.*1540-1557). Heukelum and Asperen, 1549 MS; Zuider-Zee, 1556 MS.

Adriazola, V.R. *Departemento de La Paz*, 1938.

Adrichom [Adrichem; Adrichomius], Christiaan van (1533-1585). Theologian and surveyor from Delft. He died in Cologne, and his work was published posthumously. *Briefe Description of Hierusalem...*, Cologne 1588 (Latin edition), reprinted in Dutch, English (1595), French, Spanish, Italian and German; *Theatrum Terrae Sanctae...* (1 general map on two joined sheets, and 11 regional maps), Cologne 1590, numerous reprints to 1722. **Ref.** MEURER, P. *Fontes cartographici Orteliani...* (1991) pp.98-99 [in German]; MEURER, P. *Atlantes colonienses* (1988) pp.64-65 [in German].

Adshead, Joseph. *Adshead's twenty-four Illustrated maps of the township of Manchester...* [UK] (23 sheets plus index), 1851.

Aedgerus, G. See **Adgiers**, Cornelis.

Aedone, A. d'. Sicily, 1757.

Aefferden [Afferden; Alferden], *Don* Francisco de (1653-1709). Belgian cartographer, of Antwerp. *El atlas Abbreviado*, Amberes [Antwerp], Juan Duren 1696, 1697, 1709, 1711 and 1721. **Ref.** KOEMAN, C. *Atlantes Neerlandici* (1967-1971) Vol.I pp.29-30.

Aegidius Bulionius Belga. See **Boileau** de Bouillon, Gilles.

Aegido, M. *Limburg*, 1603.

Aelst, Guilliam van. *Leo Belgicus*, 1645 (engraved title-page of Famiano Strada's *De Bello Belgico*, Dutch edition, Antwerp, 1645). **Ref.** HEIJDEN, H.A.M. van der, *Leo Belgicus* (1990) p.49.

Aelst [Aalst], Nicolas van (*c.*1527-1613). Flemish printseller and engraver resident at '5 Maire della Pace', Rome. Acquired some of Antonio Lafreri's and Antonio Salamanca's plates. Pozzuoli, [from 1550]; *Novissima Urbis Romae descriptio*, 1583; *Provincia Ulteriore d'Abruzzo,*, 1587; *La clarissima città di Catania*, 1590; *Napoli la Gentile*, *c.*1590; *Burgundiae Comitatus...*, 1596.

Aeneas Sylvius Piccolomini [later Pope Pius II] (1405-1464). Author of a geographical treatise entitled, *Historia rerum ubique gestarum locorumque descriptio* [*Cosmographia*], which is thought to have influenced Christopher Columbus.

Aerlebout [Accerlebout; Acerlebout; Arebault; Arelebout], Josias (*fl.*1639-1640). Surveyor. ...*Hadfield Chase in the counties of Yorke, Lincolne and Notinham...*, 1639, used in W. Dugdale's *History of Imbanking and Drayning...*, 1662.

Aernout [Aernoutszoon]. See **Arnoldi**, Arnoldo di.

Aertsen. See **Arsenius** family.

Aertssens. See **Arsenius** family.

Aescher, Johann. German cartographer and engraver. *Monumental-Plan von Berlin* (4 sheets), 1902.

Aeschler [Eszler], Jacobus. Co-editor Strasbourg Ptolemy, 1513, with Georg Ubelin.

Aethicus of Istria. See **Ethicus**.

Aeuer, H. (*fl.*mid 19th century). German globemaker at Eberfeld.

Afan de Rivera, Carlo (1779-1852). Involved in the development of the Officio Topografico di Palermo (founded in 1807); became director of the Deposito della guerra di Napoli, 1820. Wrote a topographical description of the Island of Sicily, for which he re-used and amended Samuel von Schmettau's map of Sicily, *c.*1813 (first published 1719-1721).

Affeita, Isidor. Surveyor and draughtsman. Siege of Kraków, 1657.

Afferden. See **Aefferden**, Don Francisco de.

Affleck, *Rear-Admiral* Philip, *RN*. Maps of Jamaica, *c.*1793.

Africanus, Giovanni Leo [al-Hasan ibn-Muhammad al-Wazzan]. See **Leo** Africanus.

Agas, John. Essex estate surveyor. *Great Coggeshall*, 1619 MS.

Agas [Aggas], Ranulph [Ralph] (*c.*1540-1621). English land surveyor of Stoke by Nayland in Suffolk. Usually credited with a London perspective map, *Civitas Londinium*, *c.*1563, *c.*1633; various estate plans, 1575-1593, including *Manerii de Layham in Comitatu Suffolk* (4 sheets); West Lexham [Norfolk], 1575; Bird's-eye plan Oxford, 1578, engraved by Ryther *c.*1588; *A Plan exhibiting the Remains of the Ancient City of Dunwich*, 1587, published 1754; treatise *A Preparative to Platting*, 1596. **Ref.** RAVENHILL, W. *The Map Collector* 35 (1986), pp.36-37.

Agassiz, A.R. Map of Tong-Kin &c. (5 sheets), 1890.

Agassiz, *Professor* Jean Louis Rodolphe (1807-1873). Pioneer geologist, zoologist and paleontologist in Neuchâtel and Cambridge [Massachusetts]. Studied hydrography of South Atlantic and Pacific. *De la succession et du développement des êtres organisés à la surface du globe terrestre*, Neuchâtel 1841; *Chart of Zoological Regions*, 1848. **Ref.** MARCON, Jules *Life, letters and works of Louis Agassiz* (Farnborough 1972); LURIE, Edward *Louis Agassiz. A life in science* (Chicago 1960).

Agathadaimon. See **Agathodaemon**.

Agatharchides of Cnidos (*fl.*148-105 BC). Greek historian and geographer, active in Alexandria, tutor to the young Ptolemy Soter (*c.*116 BC). Periplus of the Erythraean Sea in 5 books (fragments known).

Agathemerus [Agathemere] (*fl.c.*AD 185). Greek geographer. Wrote a brief account of Greek geographers. **Ref.** MULLER, C. *Geographi Graeci Minores* II pp.471 et seq.

Agathodaemon [Agathadaimon; Agathos Daimon] (*fl.c.*6th-century AD). Greek technician at Alexandria. According to notes bound into Book VIII of Ptolemy, he drew a map based on the *Geographia*.

Agde, Jean Cavalier d'. See **Cavalier**, Jean.

Agé, J.F. Jacobsz. de (*fl.c.*1789). Bookseller of The Hague (with J.W. Smit of Amsterdam). Published C.G. de Schaichowsky's *Carte des environs d'Amsterdam... vers la fin de septembre 1787* (engraved by G. Tischbein).

Ageminius, P. See **Paulus** Ageminius.

Aggas, Ralph. See **Agas**, Ranulph.

Aggere, P. See **Petrus** ab Aggere.

Agiéras, Marcel (1882-1958). French explorer. Map of the western Sahara, 1919; enlarged edition, 1923.

Aginelli, Nicolo. *Plan of Raab* [Györ, Hungary], 1566 (used by Braun & Hogenberg *q.v.*).

Aglio, Pietro Martire. River Po, 1758.

Agnelli, Federico. Engraver for Giovanni Battista Sesti's *Pianta della Città di Milano*, 1707.

Agnese, Battista (1514-1564). Genoese cartographer, worked in Venice. Portolan chartmaker credited with more than 70 known manuscript atlases, some with 10 or more maps. Portolan chart of the British Isles and the European coast from Hamburg along the Dutch and French coasts to Oléron, *c.*1540 (at University of Texas at Arlington). **Ref.** WAGNER, H.R. 'The manuscript atlases of Battista Agnese', in: *Papers of the Bibliographical Society of America* 25 (1931), pp.1-110.

Agnew, Daniel. *Fort McIntosh, Pennsylvania*, 1896.

Agnew, J.C. *...Estates in the vicinity of the City of Vernon*, 1907.

Agnew, John Robinson (*fl.*1860-1864). Aspiring globemaker of Philadelphia; registered several patents for terrestrial and celestial globes, 1860-1864. **Ref.** WARNER, D. *Rittenhouse* Vol.2 N°·1 (1987) p.22.

Agnew, *Captain* John Vans. *Corea*, 1894.

Agniet, J. See **Lagniet**.

Agnis, father & son. English estate surveyors of Essex.
• **Agnis**, Benjamin (*fl.c.*1720-*d.c.*1765). Althorne, 1728 MS; W. Mersea, 1756 MS.
•• **Agnis**, John (*d.c.*1808). Ardleigh, 1773 MS.

Agnolo, Jacopo d'. See **Angelus** Jacopo d'.

Agostini, Giovanni de. *L'Abissinia...*, 1935.

Agostini, Tommaso. Engineer and draughtsman. *Pianta di Firenze*, 1867, *c.*1877.

Agostino, Giovanni Battista. See **Codazzi**, Agostino.

Agote, Manuel de. *City of Macao*, Admiralty Chart 1805.

Agrigonio, *Fra* Bono (*fl.*16th century). Venetian cartographer.

Agrimensores. Roman land surveyors, in the later Roman Empire also known as *gromatici* from use of the *groma* or surveyor's cross. *Corpus Agrimensorum*, includes plans from 1st century AD; later miniature paintings and geometrical drawings up to 9th century are now in the Vatican. **Ref.** DILKE, O.A.W. *Greek and Roman maps* (1985) pp.81-101.

Agrippa, Marcus Vipsanius (*c.*63 BC-AD 12). Admiral and minister of the emperor Augustus. Prepared map of the World under Augustus, completed by the latter, with commentary. Map erected in Agrippa's colonnade, Rome, and may have influenced the so-called Peutinger Table.

Agueros, *Fray* Pedro Gonzalez. Chile, 1787 MS.

Aguesseau, chevalier — d'. Battle map of Flanders, 1774 (with de Beaurain).

Aguiar, Jorge de. Manuscript portolan chart of the Mediterranean and western Atlantic, 1492 (believed to be the earliest signed and dated Portuguese chart). **Ref.** PINHEIRO MARQUES, Alfredo *Imago Mundi* 41 (1989), pp.90-95.

Aguilar Paz, Jesus. *Mapa General de la República de Honduras*, 1933.

Aguirre, Domingo de. *Topografia del Real Sitio de Aranjuez* (16 sheets and 7 views), 1775.

Agutter, Benjamin. *Borrough... of St Albans*, 1634.

Aher, David (1780-1842). Irish surveyor and mining engineer. Engineer to the Grand Canal Company 1803-1805. *Map of the Bogs of District No. 12 in the Queen's County*, 1812, 1814; *The Bogs... situated in the Counties of Tipperary, Kilkenny & Queen's County* (2 sheets), 1814.

Ahmer, V. Engraver. *Eisenbahn Karte — Nord Amerikas*, *c.*1865.

Ahnert [Anert], Edouard Edouardovich (*b.*1865). Mining engineer and geologist. *Bogatstva nedr: Dal'nego vostoka*, Vladivostok 1928; *Poleznye iskopaemye Severnoi Man'chzhurii*, Kharbin 1928; Mineral resources map of North Manchuria, Peiping 1929.

Ahrenfeld, Alfred. *Plan of the London County Council, showing the New Streets from Southampton Row to the Strand proposed to be made...*, 1893.

Ahrens, H. *Monarchie Oesterreich-Ungarn*, *c.*1874.

Aigenler, Adam, *SJ* (1635-1673). Mathematician and cosmographer of Ingolstadt. *Tabula geographico-horologa universalis*, 1668.

Aigner, Hans von. Plans of Breslau [Wroclaw], 1857, 1866.

Aikin, James. Engraver for *A Plan of Charlestown from a survey of Edward Crisp in 1704* published in David Ramsey's *History of Carolina*, Charleston 1809. **N.B.** Possibly the same as James Akin, below.

Aikin, John (1747-1822). Man of Letters. *England delineated* (with a series of 43 county maps), 1790 and later editions (first published without maps in 1788); *A description of the country... round Manchester* (2 sheets), John Stockdale 1794; map of the environs of London, John Stockdale 1811.

Aikman, George. Engraver. Plan of Newcastle upon Tyne, 1833; *Canals & railways in USA*, 1838; worked also for A. & C. Black.

Ailly, Pierre d' [Alliaco, Petrus d'; Auliacus] (1350-1420). Cardinal of Cambrai, philosopher and teacher. Wrote *Ymago mundi...* (with world map), *c.*1409 or 1410, first printed in *De imagine mundi et alii tractatus* (woodcut), Leuven, Johannes de Westfalia *c.*1483 (this map conveyed Roger Bacon's belief that the sea distance from Portugal westwards to India was only half the land distance eastwards, and possibly influenced Columbus). **Ref.** CAMPBELL, Tony *Earliest printed maps 1472-1500* (1987) N° 15 pp.87-88; SHIRLEY, R. *The mapping of the World* (1983) N° 12 p.11.

Ainslie, *Lieutenant* C. *Chin Hills Rangoon*, 1802.

Ainslie, J. Engraver, Ordnance Survey Office Dublin. Waterford, 1842.

Ainslie, John and James (brothers). 'Booksellers and Stationers, St Andrews Street', Edinburgh. John Ainslie's *Scotland...*, 1789 (with William Faden). **Ref.** MOIR, D.G. *The early maps of Scotland* (1973) Vol.I pp.120-121, 125-126, 135.

• **Ainslie**, John (1745-1828). Surveyor, cartographer, engraver and publisher. Produced numerous regional and county maps in Scotland and England, as well as many coastal and river charts, canal plans, and large scale estate surveys. With Thomas Donald, worked for Thomas Jeffreys on a survey of Bedfordshire, 1765; they later surveyed Buckinghamshire, 1766-1768; Yorkshire, 1767-1770; Westmoreland, 1768-1770. Ainslie alone - *Jedburgh and its Environs* (4 sheets), 1770; *A map of Selkirk Shire or Ettrick Forest...*, 1773, 1801; *The Counties of Fife and Kinross...*, 1775; *Map of the environs of Edinburgh*, 1778 (retitled *Map of the Country round Edinburgh*, 1779, 1801, 1812); *County of Wigton of the Shire of Galloway* (4 sheets), 1782, 1801; *Ainslie's Travelling Map...*, 1783 (copied from James Dorret); surveyed the coasts of Scotland, charts published 1785 and 1786; ...*Water of Leith shewing the course of an intended canal*, 1787; *Scotland Drawn and Engraved...* (9 sheets), Ainslie 1789, Thomas Brown 1800; plans of the estates of the Earl of Eglinton, 1789-1790; *Firth of Clyde to the Crinan Canal*, 1792; *Map of the County of Forfar or Shire of Angus* (4 sheets), 1794, Faden 1801; *A plan of the town and harbour* [Glasgow], 1795; *Map of the County of Renfrew* (4 sheets), 1796 published 1800; *Roads of Great Britain*, 1797; *Old and New Town of Edinburgh and Leith*, 1804; *Town and harbour of Peterhead*, 1805; *Agricultural Map of Berwickshire*, 1813; *West and north coast of Scotland...*, 1822; and many others. **Ref.** ADAMS, I., in: *Dictionary of national biography: Missing persons* (1993) p. 6; MOIR, D.G. *The early maps of Scotland to 1850* Volume 2 (Edinburgh 1983) contains details of maps; *ibid.* Volume 1 (Edinburgh 1973) pp.120-121, 125-126.

• **Ainslie**, James. Co-publisher, with John Ainslie, of the latter's map of Scotland, 1789.

Ainsworth, A.B. *Allotments. Parish of Caramut*, 1858; *Parishes of Terang and Kailambete*, 1858.

Ainsworth, John. *Country between Mumbassa & Machakos*, 1892; ...*Ukamba Province* [East Africa], 1901.

Aird, *Lieutenant* D., *RN*. Hydrographer. Eastern Australia, 1845-1847.

Airey, John (*fl.*1867-1900). Worked for the Distances Section of the Mileage Department of the Railway Clearing House, and used his records to produce numerous railway maps of English regions, Wales, Scotland. *Railway Junction Diagrams*, 1867; district maps of London, 1869; Manchester & District, 1869; Scotland, 1875; *Airey's Railway Map of London and Its Suburbs*, 1875; *Airey's Map of County Durham and District*, 1876; *Airey's Railway Map of the Derbyshire and Nottinghamshire Districts*, 1877; *Railway Map of England and Wales* (2 sheets), 1877; ...*Railway Systems in England & Scotland* (2 sheets), 1880; South Wales, 1880; *Lancashire & District*, 1884; *John Airey's Railway Map of the Gloucestershire & Oxfordshire Districts...*, 1893 etc. **Ref.** GARNETT, David *The Map Collector* 26 (1984), pp.28-31.

Aitken, Alexander (*d.*1800). Canadian mathematician and land surveyor. York Harbour, 1793.

Aitken, Robert (1734-1802). Bookseller, printer and engraver of Philadelphia. Born Dalkeith, Scotland, moved to Philadelphia (1769) 'opposite the London Coffee House, Front Street Phila.' Maps for *Pennsylvania Magazine*, 1775; *A New and Correct Plan of the Town of Boston...*, 1775; *A Plan of Quebec...*, 1775; *Present Seat of War on the Borders of Canada*, 1775; *Montreal*, 1776; *Nova Scotia*, 1776; Pierre Eugène Du Simitière's *Map of the Maritime Parts of Virginia*, 1776. **Ref.** GUTHORN, P.J. *American maps and map makers of the American Revolution* (1966) p.7.

Aitken, William. Lothians, from John Adair's observations by J. Elphinstone, 1744.

Aitoff, D. *Plan de Paris*, 1886; lettering for Louis Vivien de Saint-Martin's *Australie*, 1889.

Aiton, William. Ayrshire, 1811; Bute, 1816.

Aitzing [Aitsinger; Aitzinger; Eitzing; Eitzinger; Eytzinger], Michael, *Freiherr* von (*c.*1530- 1598). Historian and cartographer from Obereitzing, Upper Austria. *Itinerarium Belgicum*, Cologne 1577; possibly responsible for *Itinerarium Orbis Christiani*, 1579-1580 (route map for pilgrims, said to be the first printed road atlas); best known for inventing the *Leo Belgicus*, a map of the Seventeen Provinces in the shape of a lion: *De Leone Belgico*, Cologne 1583. **Ref.** HEIJDEN, H.A.M. *Leo Belgicus* (1990) pp. 16-17, 21-22; HOPPEN, Stephanie *The Map Collector* 14 (1981) pp. 48-49.

Ajello, Michele. *Liguria marittima*, 1834.

Akademia Nauk St Petersburg [Academy of Sciences; Akademie der Wissenschaften]. Founded in 1726 by J.N. Delisle. *Atlas Russicus* [also French and German editions], 1745; *Plan Sanktpeterburga*, 1753 (engraved G.I. Unvertzagt); Russian Discoveries, 1775.

Akamatsu Kyūbei. Japanese publisher. Okada Gyokuzan's *Zōshū Kaisei Sesshū Ōsaka Chizu* [plan of the city of Osaka], 1844.

Akamidzu [Akamatsu] de Mito (*fl.*18th century). Japanese cartographer. Historical Atlas of China, 1789; World, 1796.

Akatsuki Kanenari. Japanese writer and painter of Osaka. Worked with the ukiyoe artist Matsukawa Hanzan on a guide map to the plum blossom viewing in part of Yamoto Province, Japan (colour printed woodblock), published by Kawachiya Kihei, Kawachiya Seicichi and Kajiya Heizō, 1858. **Ref.** YAMASHITA Kazumasa *Japanese maps of the Edo Period* (Japan 1998), pp.254-255.

Akbar-Ali. *Akbar Ali's Map of the Punjab*, 1882.

Akeleye, Jens Werner. *Söe-Kaart over Kattegattet*, 1770.

Åkerlund [Åkerland], Erik (1754-1832). Swedish chartmaker and engraver. Engraver for Johann Nordenankar's charts, 1787; *Petrografisk Karta öfver Skåne*, 1804 (with Samuel Andersson); *Karta öfver Gottland*, 1805; *Belägenheten omkring Åbo Stad...* [Turku] Finland, 1808; *Resakarta öfver Svea* *och Götha Riken*, Stockholm 1828; Ringsted, 1836.

Åkerman, Anders (1723-1778). Globemaker, cartographer and engraver of Uppsala. Globes, 1759, 1766, 1780, 1792. *Sinus Finnicus* [Finska Wiken], 1768; *Atlas hydrographica*, 1768; *Atlas juvenilis*, 1774.

Akersloot, W.O. (*fl.c.*1600-1640). Haarlem maps and plans, 1628.

Akhmatov, Ivan (*fl.*1840s). Russian cartographer of St Petersburg. Geographical, historical & chronological atlas of Russia, St Petersburg 1845.

Akhurst, Walter A. Printer of Sydney. *Lenehans Estate, Hunters Hill*, 1892.

Akin, James. *Sketch of the Santee Canal* [South Carolina], 1802; *A Plan of Charles Town* [South Carolina], 1809. **N.B.** May be same person as James Aikin listed earlier.

Akitaya Ryōsuke. Japanese publisher. Suigashi's *Dainihon Dōchū Kōtei Saiken Ki* [travel map of Japan] (colour printed woodblock), 1837 edition (with Hikawaya Zensaku, from an original by Kikuoka Ryōsen first published 1722).

Akiyama, Einen [Bokusen] (1805-1853). *Fujimi jusanshu yochi rotei no zenzu* [Map of the thirteen provinces surrounding Mount Fuji], 1841.

Akkeringa, A. *Kaart van Europa*, 1874.

Akkeringa, J.E. Map of the Netherlands East Indies, 1872-1873.

Akrel [Acrelius], father and son. Mapmakers of Stockholm.

• **Akrel** [Acrelius], Fredrik (1748-1804). Swedish engraver, studied under the cartographer A. Åkerman from 1768. Engraved Åkerman's *Mappa Hydrografica exhibens Sinum-Finnicum*, 1768; *Grundritning öfver Stockholms Stad...*, 1771; *Siöar och strömmer*, 1774; Sweden, 1778; *...Pater Noster Skaren*, 1795; *General Charta til Sveriges Sjö-Atlas*, 1795; *Trollhåttan med dess Nya Canal och Slussar*, 1800 (with C. Bergqvist).

•• **Akrel**, Carl Fredrik (1779-1868). Surveyor, cartographer and engraver of Stockholm. Son of Fredrik. Plan Stockholm, 1805; *Karta öfver Sverige*, 1811; *Res-karta öfver Sverige Södra delen*, 1830, 1853.

Akron Map & Atlas Co. *Summit County, Ohio*, 1891.

Alabern, family. Engravers of Barcelona, Spain.

• **Alabern i Moles**, Pablo (1804-1860). North America, 1830; maps for *Diccionario geografico universal*, 1830-1834.

•• **Alabern i Casas**, Ramón (*fl.*1835-1882). Son of preceding. Africa, 1845; *Oceania*, 1845; *Atlas universal*, 1857; *Deposito Hydrográfico*, 1865.

•• **Alabern i Casas**, Camilo (1825-1875). Engraver for Francisco Coello y Quesada's *Atlas de España*, 1848-1868.

Alagna, J. Giacomo. Hydrographer of Messina. *A Compleat Set of New Charts... Portugal and the Mediterranean Sea*, London, J. Mount & T. Page, 1764.

Alard. See **Allard** family.

Albach, Julius. Publisher. Maps of Austria, 1875-1879; *Umbegung von Wien*, *c.*1875; *Plan des Bricker Lager-Terrains im Maßstab* (3 sheets), 1877; *Das Salzkammergut*, *c.*1879.

Alban, E. *Mecklenburg-Schwerin*, 1891.

Albatenius. See **Battani-al** Abu Abdallah.

Albe, E.E.F. d'. See **Fournier** d'Albe.

Albe. See **Bacler** D'Albe, Baron Louis Albert Ghislain.

Albéca, Alexandre d' (1858-1896). French explorer and colonial administrator in Ouidah, West Africa; studied history and anthropology of the people of Bénin. *Côte occidentale d'Afrique*, 1889. **Ref.** BROC, N. *Dictionnaire illustré des explorateurs Français du XIXe siècle: I Afrique* (1988) pp.1-2 [in French].

Alberdingh [Alberdinck], Jan van (*fl.*1650-1687). Dutch land surveyor, in northeast Netherlands. *Desseijn van het graven eens canaals... loopende van Blijhamster... tot... Haseborgh*, 1687; *Verbaal... over het inunderen der moerassen ontrent de Bourtange,*

Coevorden, &c. (6 maps), 1681 MS. **Ref.** DONKERSLOOT-de VRIJ *Topografische karten van Nederland vóór 1750* (1981) N°750 p.147 [in Dutch].

Albernaz [Teixeira] family. **Ref.** CORTESÃO & TEIXEIRA DA MOTA *Portugaliae monumenta cartographica* (6 volumes) 1960 [Portuguese and English].
• [Teixeira], Luís. Father of João and Pedro. See **Teixeira.**
•• **Albernaz,** João Teixeira [I] (*c.*1602-1666). Son of Luís Teixeira, but used the name Albernaz. Portuguese cartographer and chart-maker, cosmographer to the King of Portugal. Produced many manuscript charts and atlases including: *Razão do Estado do Brasil* (18 charts), *c.*1616; portolan map of the Americas, Africa and Europe, *c.*1620; atlas of Brazil (19 maps), 1627; manuscript world atlases, *c.*1628-1643; atlas of Brazil (36 maps), 1631; atlas of the coast of Portugal, 1648; Atlas of the cities and fortresses of Oriental India, *c.*1648; and others.
••• **Albernaz,** João [II] (*fl.*1665-1681). Grandson of João I. Produced charts 1665-1681; atlases, 1665-1670, including: chart of the Atlantic, 1665; atlas of Africa (29 maps), 1665; atlas of Brazil (29 maps), 1666; Atlantic chart, 1667; atlas of Brasil (31 maps), 1670; map of Africa, 1677; and others.
•• **Albernaz,** Pedro Teixeira. Brother of João I. Chart of the Straits of Magellan, *c.*1619; plan of Madrid, 1656; map of Portugal, 1662.

Albers, Heinrich Christian (1773-1833). Naturalist and geographer of Lüneburg, Germany. He developed a conical projection, first used on a map of America by C.G. Reichard in 1809, he also wrote several cartographic monographs: *Beschreibung einer neuer Kegelprojection,* 1805; *Recension der Arrowsmithschen Charte von Ostindien,* 1807. **Ref.** KRETSCHMER, DÖRFLINGER & WAWRIK *Lexicon zur Geschichte der Kartographie* (Vienna 1986) p.9 [in German].

Albert I, Prince of Monaco (1848-1922). Founded Institut Océanographique, Monaco. Sponsored and supervised maps related to all aspects of his oceanographic and geographical work: currents of the North Atlantic, hydrographic and geographical surveys of Spitsbergen. Published *Carte générale bathymétrique des océans* (24 sheets), 1905.

Albert, A.M. *Mapa de los Ferrocarriles de España y Portugal,* 1905.

Albert, Joseph. Bird's eye view of Cologne entitled *Köln a./Rh.,* 1889.

Albert, R. *Der Harz. Strassen-Karte für Radfahrer,* 1898.

Albert, Th. *Rügen,* 1874.

Alberti, —. *Carta topografica Ducato Piacenzo,* 1816.

Alberti, A. Civil engineer. *Map of the Saginaw Valley, Michigan,* Cincinnati 1860.

Alberti, F. Leandro. Bressani, Padua, Verona, 1701; Mantoue, Modena, 1702.

Alberti, Gaspare. Print dealer and publisher of Rome. Successor to P. Palombi, re-issued two of his maps of Malta. *Nuovo et Verissimo Ritratto del Isola di Malta,* [n.d.] (first published 1565); *Nuovo dissegno della fortezza di.S. Ermo...,* [n.d.] (first published 1566).

Alberti [Albertus], Leandro (1479-1552). Dominican geographer and historian of Bologna. *Descrizione di tutta l'Italia,* 1550 and later editions (used by Abraham Ortelius as source material for the 'Catalogus auctorum'); *Isole appartenenti alla Italia,* 1567. **Ref.** KARROW, Robert W. *Mapmakers of the sixteenth century and their maps* (1993) p.278; MEURER, Peter *Fontes cartographici Orteliani...* (1991) p.101 [in German].

Alberti, Leon Battista (1404-1472). Italian geographer, wrote works on surveying and mapping. *Descriptio urbis Romae, c.*1445; *Ludi matematici, c.*1450.

Alberti, Lodewijk. Landdrost at the Cape of Good Hope. *Alagoa Bay,* 1810.

Albertin, Johann Heinrich (1713-1790). Swiss engineer and cartographer in Zurich and Fürstenberg [Germany]. *Grundriss über die Grenzen des Kelleramts und Fryamts...,* 1740; *...Landtgraffschaft Thurgeuw und der Graffschaft Kyburg...,* 1741; *Grundriss von Marchliken, Marthalen...,* 1746; *Geometrischer Entwurf der Situation v. Ausser Dettingen...,* 1750; *Grund-Riss der*

Gräntz Marchung...u. der Grafschaft Sargans, 1753; *Grund-Riss des Thurflusses und Bettes...*, 1754; *Geometrischer Grund-Riss einen Theils der Herrschaft Grüningen...*, 1760; *Geometrischer Plan von dem Tösstock...*, 1762. **Ref.** GROB, Dr.R. *Geschichte der Schweizerischen Kartographie* (1941) p.59.

Albertin de Virga. See **Virga**.

Albertis, L.M. d'. *Voyage of Beccari & Albertis to West New Guinea*, 1873; *Fly River, New Guinea*, Sydney 1876.

Alberts, Rutgert Christoffel. Publisher of The Hague. Plan Turin, 1706; *Geneva Antiqua et Nova*, 1725; published editions of the Blaeu Italian Townbooks, Savoy & Piedmont, 1724-1726 (Dutch, French or Latin text). **Ref.** KOEMAN, C. *Atlantes Neerlandici* Vol.II (1967-1971) (Bl) pp.95-112.

Albertsen, Dierick. Dutch engraver, printer. Nicolaas van Geelkercken's *Frisia occidentalis*, *c.*1616, 1618.

Albertus, Gaspar [Caspar]. Publisher of Rome. *Nuovo et Verissimo Ritratto del Isola di Malta*, after 1565 (first published by Pietro Paolo Palombo, *c.*1565).

Albi mappamundi (8th century). The oldest known geographical monument of Western Europe, from 8th-century manuscript of miscellanies.

Albin, John. *Map of the Isle of Wight* (includes an inset plan of Newport), Newport, L. Albin 1795, 1802, 1805, 1807, 1823.

Albini, *Rear-Admiral* Giuseppe (1782-1859). *Portolano della Sardegna*, 1842; *Liguria* (25 charts), 1854.

Albino de Canepa. See **Canepa**, Albino de.

Al Biruni. See **Biruni** Abu Rayhan.

Albizio, Antonio. *Princip. Christ. Stemmata*, Augsburg 1610.

Albrecht, —. *General Karte von Ungarn* (17 sheets), Vienna 1858.

Albrecht, von Bonstettin. Diagrammatic maps of Switzerland, 1479.

Albrecht, David. Bookseller and publisher of Breslau [Wroclaw]. Martin Helwig's *Silesia*, 1605.

Albrecht, Ignaz. Engraver. Switzerland, published F.J.J. von Reilly 1789; World, 1795.

Albrizzi, G. Father and son.
• **Albrizzi**, Girolamo (*fl.*late 17th century). Venetian cartographer and printer, father of Giovanni Battista.
•• **Albrizzi**, Giovanni Battista [Giovambattista] (*fl.*1734-1742). Publisher of Venice, son of Girolamo (above). *Distintissima descritionne della città, porto e isole di Scio...*, 1694; *Lo stato presente di tutti i paesi e Popoli del mondo* (3 volumes), *c.*1734-1766; published an edition of Guillaume Delisle's atlas as *Atlante novissimo* (2 volumes, 78 maps), 1740-1750.

Albu, Siegfried. *Flaggenkarte*, *c.*1870.

Albumasar. See **Abu** Mashar Jafar.

Alcedo y Henera, *Colonel* Antonio (1735-1812). Spanish geographer. *Dicionario geográfico historico*, 1786 (English edition, George Alexander Thompson 1812-1815).

Alcmarianus, Guilielmus [Guilhelmus] Janssonius. The name used by Willem Blaeu on his terrestrial globe of 1599. **N.B.** See **Blaeu**, Willem Jansz.

Alcuin of York (*c.*735-804). Concept of the quadrivium (as part of the seven liberal arts): music, arithmetic, geometry, and astronomy providing an outline of the natural sciences.

Alden, Edward C. *Plan of the City and University of Oxford*, 1888.

Alden, *Lieutenant* James, USN (*fl.*1837-1860). Hydrographer, worked with others on charts for the U.S. Coast Survey. *Chart of St George's Shoal & Bank*, 1837; *...Coatzacoalcos River to...Mina Tittan*, 1847; Canal de Haro & Strait of Rosario, 1854 (with G. Davidson); *Western Coast of the United States (Northern Sheet)...*, 1855 (with G. Davidson); *Canal d'Halteras*, 1856; *San Francisco Harbour*, 1856; and others.

Alden, John Berry (1847-1924). *Alden's Home Atlas of the World*, 1887.

Alden, Ogle & Co. Publisher of Chicago. *Plat Book of Marshall and Putnam Counties, Illinois*, Chicago 1890; *Plat Book of Christian County, Illinois*, Chicago 1891; *Plat Book of Macon County, Illinois*, Philadelphia 1891; *Plat Book of Hancock County, Illinois*, 1891; *Plat Book of Fayette County, Illinois*, Chicago 1891; *Plat Book of LaSalle County, Illinois*, Chicago 1892.

Aldenhoven, Ferdinand. *La Grèce* (8 sheets), Athens 1838.

Aldersey, William. Estate surveyor. Ditton Park [Buckinghamshire], 1695 MS.

Alderwereldt, Pieter (*fl.*1730s). Bookseller of Amsterdam, 'in de Wolvestraet op de hoek van de Keyzersgragt'. Sold Isaak le Long's *Historisch en Geographisch Berigt van Polen* (with maps), 1734.

Alderwerelt, R.P. Dutch cartographer. *IJsselmonde*, 1812.

Aldger, Cornelius. See **Adgiers**, Cornelis.

Aldous, M. Manitoba and North-West Territory, Winnipeg 1895.

Aldrich, *Lieutenant* Edward. *Jerusalem*, 1841 (with Symonds) published London 1857; Gaza, 1843.

Aldridge, Thomas. Script engraver. Battle Bridge, for James Rennell 1810.

Aldring, C. *fils*. Engraver. Engraved 47 maps for the Cassini *Carte de France*, from 1756; François Grasset's *Suisse*, 1769; for Jean, chevalier de Beaurain, 1782.

Aldwell, Thomas. Estate surveyor. *Cranborne Chace*, 1618 MS.

Aleau, Estienne. Bookseller of Beauvais, Rue S. Pierre. Guillaume Delisle's *Diocèse de Beauvais*, 1710.

Alefounder, Thomas, father and son. Estate surveyors.
• **Alefounder**, Thomas, *the elder* (*fl.*1757-1776). Essex estate surveyor. Surveys in Essex, 1762 MSS; *A Survey of the Estate of Abraham Reeve, Gent. lying in Shelly and Layham in the County of Suffolk*, 1776 MS (done by father and son).
•• **Alefounder**, Thomas, *junior* (*fl.*1776-1783). Essex estate surveyor. Manuscript surveys, 1776-1781.

Alegre, Francisco. Spanish pilot. Binan, 1745; de Cuyo Paraguay, 1750 MS.

Alekseyev, Fedor (*fl.*1815-1844). *Dvernyago Sveta*, 1841.

Aleman, N. *Evèché de Strasbourg*, 1659.

Alemand, Pierre. *Costes de l'Amérique septentrionale*, 1687.

Aleni, Giulio, *SJ* (1582-1642). Italian mathematician and astronomer in China, worked with Longobardi and Dias. World map *c.*1630. **Ref.** WALLIS, Helen *Imago Mundi* 19 (1965) pp.38-45.

Aleotti, Giovanni Battista [l'Argenta] (1546-1636). Architect, military surveyor, engineer of Argenta Ferrara, in service of Alfonso II d'Este (1571-1593). Thought to have worked in the Netherlands. *Northern Italy*, 1603; *Corografia dello stato di Ferrara*, 1603 (used in the Vrients editions of the Ortelius *Theatrum*) 1608, 1612. **Ref.** MEURER, P. *Fontes cartographici Orteliani...* (1991) p.101 [in German].

Aleph [*pseud.*] See **Harvey**, William.

Alès, Auguste François. 'Rue des Mathurins St Jacques N°. 1', Paris. Engraver for Charles V. Monin, 1840; for J.A. Fayard de la Brugère's *Atlas universel*, 1870; and for A.A. Vuillemin, 1873-1877.

Alessandri, Innocente (1741-1803). Engraver of Venice. Terrestrial and celestial globe gores, 1784 (with Pietro Scattaglia); also with Scattaglia engraved the plates for *Atlas Encyclopédique* (2 volumes), Padova 1789.

Alessandro, A. (*fl.* late 17th century). Italian engraver. Worked for Vincenzo Maria Coronelli.

Alessandro, d'Anna (*fl.* late 18th-early 19th centuries). Engraver of Naples.

Alexander, Andrew. *Headquarters of James River... Virginia*, Philadelphia 1814.

Alexander, C.H.M. *Etawah District Agra*, 1858.

Alexander, Daniel (*fl.*1794-1803). *The London Docks*, 1794; *Proposed London Docks* (2 sheets), 1796 published in: *Reports from Committees of the House of Commons* Vol.XIV, 1803.

Alexander, G. Publisher and engraver, 130 Strand, London (address also given as Clarks Place, Islington). William Darton's *An Entire New Plan of the Cities of London & Westminster*, 1817.

Alexander, *Captain Sir* James Edward. *Niagara Gore and Talbot Districts*, London 1842; *British Guiana*, 1832 MS; Missions Orinoco, 1835; profiles of Rideau Canal, Lake Ontario, St Lawrence, 1878.

Alexander, John Henry (*b.c.*1812). American mathematician and scholar. Surveyed parts of Maryland, 1833-1837; submitted a published report to the General Assembly: *Report on the New Map of Maryland* (with 2 maps), 1835; *A Geological Map of Maryland*, *c.*1840. **Ref.** PAPENFUSE, E.C. *Atlas of historical maps of Maryland, 1608-1908* (1982) pp.67 and 70.

Alexander, W.D. Superintendent, Hawaiian Government. *Map of the Hawaiian Islands*, 1876.

Alexander, *Sir* William, Earl of Stirling. Untitled map of New England, Nova Scotia and Newfoundland, published in *An encouragement to Colonies*, London, 1624, 1625, later republished with the title *The Mapp and Description of New England*, 1630; the map was also published in Samuel Purchas's *Purchas his Pilgrimes*, 1625. **Ref.** BURDEN, P.D. *The mapping of North America* (1996) pp.257-258.

Alexander, *General* William (1726-1783). Surveyor General of New York.

Alfani [Alphane], Domenico (*c.*1480-after 1553). Artist of Perugia. *Lago di Geneva*, 1590 [*sic.*].

Alfaro, Giovanni (1793-1837). Army officer, geodesist, astronomer and topographical engineer of Naples. Worked in the Officio Topografico di Napoli. Worked on the triangulation of the border between the Kingdom of Naples and the Papal States, 1816-1817; geodetic survey of the island of Ischia for the Accademia delle Scienze, 1827-1828. **Ref.** VALERIO, V. *Società uomini e istituzioni cartografiche nel mezzogiorno d'Italia* (1993) pp.432-433 [in Italian].

Alferden, Don Francisco de. See **Aefferden**.

Alferes, Joaquin Cardozo Xavier. *Campanha de Pilvens*, 1801 MS.

Alfiano, Epifanio de. Monk, and engraver of Giacomo Gastaldi's *Geografia moderna di tutta la Italia*, 1597.

Alfimov, Fyodor. Nobleman from Klin, teacher at Marine Guards Corps. Maps of the Caspian Sea, 1731 (with Farwharson); map of part of the Volga River.

Alfonse de Saintonge [Alfonce, Allefonse, Alphonse], Jean Fonteneau (*c.*1480-1550). French navigator who became an authority on the geography of Asia and the New World. In 1542 sailed to Canada as pilot to Jean-François Roberval's expedition. Notes of his voyages were later published as *Voyages adventureux du capitaine Jean Alfonse*. Map of the Saguenay River, *c.*1543; *Cosmographie*, 1544 MS; charts of the Gulf of St Lawrence, 1545 MSS.

Alford, Robert. Engraver. *Plan of Clink Liberty, Southwark*, 1827.

Alfraganus. See **Farghani**.

Alfred the Great (849-901). Collected material for geography, particularly of northern lands. The so-called Anglo-Saxon or Cotton manuscript world map [10th or 11th-century] may have had its origin in Alfred's acquired knowledge.

Alfthan, Georg von (1828-1896). Finland (30 maps), 1860.

Algarra, Agustín de. *Aduanas de España*, 1855.

Algate, A.J. Early twentieth-century maps of Canada, including: *Railway Guide Map, Ontario*, 1911; *Algate's Automobile, Motorist and Pedestrian Map, Toronto*, 1911; *Algate's map County of Brant*, 1913; *Map of ...Cobalt Silver Camp*, 1913.

Algermissen, Johann Ludwig (*b.*1847). Cartographer of Cologne. Maps of Alsace

Lorraine and Metz, from 1871; *Elsass-Lothringen*, 1878; *Karte des Königreichs Württemberg* (2 sheets), 1885; *Ümgebungs-Karte von Metz*, 1891; *Unter Elsass*, 1891; *Karte de Provinzen Rheinland und Westfalen*, 1898.

Algoet [Goethals; Panagathus], Liévin [Livinus] (*c*.1490-1547). Poet of Gent, worked as a teacher and calligrapher in Augsburg, Basle and Brussels, secretary to Erasmus (1519-1524). Chart of the world, 1530 (drawn for Johannes Dantiscus, Bishop of Culm, and possibly used later as the basis for a large terrestrial globe, *c*.1531); map of the northern lands, *c*.1530 (now lost), the first known published form appeared as *Terrarum septentrionalium exacta novissimaque descriptio...* (6 sheets), Antwerp, Gerard de Jode 1562 (the only known copy of De Jode's edition is to be found in the Bibliothèque Nationale, Paris), reissued on smaller scale by Gerard de Jode, 1570. **Ref.** KARROW, R.W. *Mapmakers of the sixteenth century and their maps* (1993) pp.34-36; MEURER, P. *Fontes cartographici Orteliani...* (1991) p.102-103 [in German].

Algué, José. *Atlas de Filipinas*, 1899.

Alhambra, D.G. *Atlas geográfico de la tierra*, Madrid 1849

Alibert, Jean (*d*.1787). Mapseller of Paris. *Nouveau plan routier de la ville et Faubourgs de Paris*, 1780.

Alibrandi, Giovanni Garti d'. See **Garti** D'Alibrandi.

Alioun Sal (*d*.1864). Senegalese explorer employed by French colonial service. *Relation abrégée du voyage*, 1864.

Alison, *Sir* Archibald (1792-1867). *History of Europe* (10 volumes), 1833-1842 (accompanying atlas volume produced by W. & A.K. Johnston).

Alix, Thierry (1530-1594). Historian and geographer of Lorraine, worked with Gerard Mercator. *Lorraine*, 1564; *Vosges*, 1579.

Alkemade, K. van. *Beschryving van de stad Brielle en den Lande van Voorn* (with P. van der Schelling), Rotterdam, Philippus Losel 1739.

Alingham, William. *A Short Account of the Nature and Use of Maps*, London 1703.

Allam, E.C., *RN*. Assisted with Admiralty surveys. Falkland Islands, 1860-1863.

Allan, A. (*fl*.1794-1820). Mysore, 1794; *Map of Ceylon* (2 sheets), 1803; *Map of India*, 1818.

Allan, A.C. Surveyor. *Beveridge Island... channels of the Murray River*, 1875; New South Wales plans, 1875; *Canterbury Park*, 1883 (with A. Tuxen).

Allan, D. Lithographer. *Chart of Tamar River, Van Diemen's Land*, 1833.

Allan, F. *Schoolkaart van Nederlandsch Indië*, *c*.1876.

Allan, Gordon. *A Plan of Belize, British Honduras* (14 sheets), 1885-1886, published 1887.

Allan, John (*fl*.1848-1857). Lithographer, 51 King St West (1847); 2 Bridge St (1851); 23 Bridge St (1855), all in Sydney, New South Wales. *47 Allotments being part of St James's Glebe*, 1848; *Plan of Kingston on the Newtown railway station*, 1854; *Map of the City of Sydney*, 1857.
• **Allan & Wigley**. Engravers and lithographers, 23 Bridge St (1857); 1 Bridge St (1858-1859), Sydney. *Plan of Kensington in the City of Sydney*, 1850; *Annandale estate north, section 1*, 1857.

Allan, T. *Environs of Nice*, 1818.

Allan, William. Worked with Jedediah Hotchkiss on *The Battlefields of Virginia. Chancellorsville*, 1867

Allard family. **Ref.** KOEMAN, C. *Atlantes Neerlandici* (1967-1971) Vol.I pp.31-45; KROGT, P. van der *Advertenties voor kaarten, atlassen, globes e.d. in Amsterdamse kranten 1621-1811* (1985) N°252 etc. pp.61 *et seq.* [in Dutch].
• **Allard**, Huych [Hugo] (1625-1691). Engraver and publisher, 'Calverstraet bij den Dam', Amsterdam. Father of Carel. World, 1650; East Indies, 1652; England, 1657; New York, 1660; Guinea, 1660 (engraved by Baptista van Doetecum); maps of the World, 1665; Leo Belgicus, 1665; Germany, 1666; New England, 1673.

•• **Allard** [Allardt], Carel [Carolus; Karel] (1648-*c.*1706). Engraver and publisher, 'op den Dam in de Kaertwinkel', Amsterdam. Son of Huych the first, took over business in 1691. Atlases included works of Janssonius, de Wit and others, but produced a number of original maps. *Totius Neobelgii Nova Et Accuratissima Tabula*, *c.*1674; *Orbis sive Americae Septentrionalis...*, 1690; issued undated composite atlases as *Atlas Major* or *Atlas Minor*, [the Vatican copy is dated 1697]; *Orbis Habitabilis Oppida et Vestitus*, 1710; *Magnum theatrum belli*, 1710 (100 plates of town views later reprinted by Pieter van der Aa, 1729); *...Africae...*, *c.*1720. Transferred some of the shop stock to his son Abraham in 1706.
••• **Allard**, Hugo, *the younger* (*b.*1673). Son of Carel (above). Republished some of his maps.
••• **Allard**, Abraham (1676-*c.*1730). Engraver and printseller of Amsterdam, son of Carel, above. *Sedes Belli in Polonia*, *c.*1700; *Spanje en Portugaal*, 1700; Holland, 1714.

Allard, Jacob. *Nova et accurata totius Europae tabula* (6 sheets), *c.*1685 (now in Leiden University Library).

Allardice, Samuel. Engraver. *East Indies*, 1793; *Ireland*, 1793; *Italy*, 1793; with Robert Scot engraved A.P. Folie's *...Plan of the City and Suburbs of Philadelphia*, 1794 (published in Benjamin Davies's *Some Account of the City of Philadelphia*); *Russia or Muscovy*, 1796; *Scotland*, 1796; C.P. Hauducoeur's *A map of the head of Chesapeake Bay*, 1799. **Ref.** See WHEAT & BRUN *Maps and charts published in America before 1800* (1978) for descriptions.

Allardt, G.F. Surveyor and engineer. *Tide lands in San Francisco*, 1869.

Allart, Johannes (*fl.*1781-1811). Book, map and print-seller of Amsterdam. *....Buitenstreeken van Parys*, *c.*1794; De Jongh and M. Sailleth's *Atlas van alle de Zeehavens der Bataafsche Republiek*, 1802.

Allason, Thomas. Architect. *Plans of the ruins of Olympia*, *c.*1814; *Site of Elis*, 1819; *Course of the Alpheus*, 1824.

Allday, J.L. *Dennis's Official Cab Radius Map... Birmingham*, 1886.

Alldridge, *Captain* George M., *RN*. Admiralty surveyor. *River Severn*, 1849; *West coast of England*, 1850-1864; *Survey of the River Usk*, 1867, *c.*1870.

Allefonse, Jean. Manuscript charts of the Gulf of St Lawrence, 1545.

Alleis, M. Publisher, associated with James Gilbert of Paternoster Row in the publication of the earlier maps from Fisher, Son & Co's *Fisher's County Atlas of England and Wales*, London *c.*1842-1845.

Alleman, L'. Engraver. Pierre du Val's *La Suisse*, 1654.

Allen. See **Black**, Parbury & Allen.

Allen. See **Lackington**, Allen & Co..

Allen, A.M. Surveyor. *Macamish...surveyed...* [Johnson County, Kansas], *c.*1860.

Allen, Adam (1757-1823). Lieutenant in the Queen's Rangers in North America. *Skirmish at Richmond Jan. 5th 1781* [Virginia], *c.*1781 (published in Simcoe's *Journal of the Operations of the Queen's Rangers*, 1787).

Allen, C.R. *Illustrated Historical Atlas of Pictou County, Nova Scotia*, 1879; *Harrison County, Iowa*, 1884; *Pottawattame County, Iowa*, 1885.

Allen, Daniel B. *Peoria Co., Illinois*, 1861; *Map of Peoria County* [Illinois], Chicago 1873; map of the city and county of Peoria, 1882.

Allen, Eleazer. British surveyor in North America. *Boundary of Lord Carteret's Grant* [North Carolina], 1746 MS.

Allen, Elias (*fl.*1601-*d.* by 1637). English estate surveyor. Manuscript surveys in the counties of Berkshire, Gloucestershire, Herefordshire, Kent, Oxfordshire, Surrey and Worcestershire.

Allen, F.D. *British Possessions in North America*, 1813.

Allen, Fawcett. *An Atlas of Commercial Geography*, 1913.

Allen, Fordyce A. *A Comprehensive Geography*, Philadelphia, J.B. Lippincott & Co. 1864 (with Benjamin F. Shaw).

Allen, George S. (*fl.*1790-1821). Engraver at No 4 Sadler's Wells Row, Islington (1791-1792), later at 19 Shoe Lane, Fleet Street, London. *North Wales* for John Harrison, 1791; *...Island of Barbados* for Bryan Edwards, 1794; *Pacific* for Thomas Butler, 1799; *Whale Fishery*, 1796; *Isle of Wight*, 1800; Aaron Arrowsmith's Asia and India, 1801, 1804; *A Survey of the River Thames...*, David Steel 1802; *East India Pilot*, 1804; William Faden's *English Channel*, 1804; Laurie and Whittle's *Chart of the Caribbee*, 1810.

Allen, Horatio (1802-1899). American civil engineer and inventor. Patented a terrestrial globe, 1879.

Allen, Ira (1751-1814). United States Surveyor General. *Vermont*, 1798.

Allen, J. [John?]. Engraver of 17 Poplar Row, New Kent Road, Surrey. Engraver of R.G. Baker's *Map of the County of Cambridge* (4 sheets), 1821.

Allen, [James]. Publisher and bookseller, 20 Warwick Lane, off Paternoster Row near St Paul's Cathedral. Publisher of the *Registered Exhibition Hand Guide to London* (glove map), 1851. **Ref.** HYDE, Ralph *The Map Collector* 35 (1986), p.47.

Allen, *Lieutenant* James. *Map of the Route & Country, passed over by an Expedition into the Indian Country*, 1832 MS, published by Congress *c.*1833 (later reduced and reprinted to illustrate Henry Rowe Schoolcraft's *Narrative of an Expedition... to Itasca Lake*, 1834); *St. Croix & Misacoda*, 1834; *Sources of the Mississippi River*, 1834.

Allen, James Baylis (1803-1876). Engraver for John Tallis's *Illustrated Atlas*, *c.*1851.

Allen, Joel (1755-1825). American artist and engraver. William Blodget's *A New and Correct Map of Connecticut...*, *c.*1791.

Allen, John. Engraver of *Europe* and *Africa* for J. Morse's *The American Universal Geography*, 1793.

Allen, John (*fl.*1820-1848). Surveyor of Oxford, produced manuscript estate, inclosure and tithe maps.

Allen, John William. Lithographer, 104 Fleet Street, London. *Liverpool Docks*, 1848.

Allen, Joseph. Survey of Wexford Harbour, published Scalé & Richards 1764.

Allen, Joseph. Publisher and engraver of 3 Hampden Str. Sommerstown, London. Engraver of [Robert] *Rowe's Plan of London, Westminster and Southwark...*, 1804 and later (Robert Rowe is also named as engraver); published a set of county maps in the form of cards entitled *A Geographical Game - Allen's English Atlas*, 1811 (R. Rowe's name also appears on some cards), these maps were amended and republished by John Thomson as *The New English Atlas*, 1823, and by Orlando Hodgson as *The Pocket Tourist and English Atlas*, 1827. **Ref.** BATTEN, K. 'Second edition of Allen cards' in: *IMCoS Journal* N°·58 (Autumn 1994) pp.25-27; BERESINER, Y. *The Map Collector* 30 (March 1985) pp.40-41. **N.B.** Compare with J. **Allen** above.

Allen, Mark. *Roads of Great Britain*, Dublin 1832.

Allen, Michael. Estate surveyor. *...Thorneborough in the County of Buck...* [Buckinghamshire], 1637 MS (Huntingdon Library).

Allen, Michael. Co-partner in the publishing company of Allen & West.
• **Allen & West**. Publishers, 15 Paternoster Row, London (1795-1798). Maps in John Malham's *Naval Gazette*, 1795.

Allen, Patrick. Plans of Irish Baronies, including Deece, 1657; Foore, 1657; Skreene, 1657; many were published in 1908.

Allen, Thomas. Publisher. *A new history of London, Westminster and the Borough of Southwark*, 1828, 1833, 1839; *Allen's New Plan of London from Actual Survey*, 1830.

Allen, W. Printseller of 'N°·88 Dame Street' (1783), 32 Dame Street (1810), Dublin. Co-publisher of W. Wilson's *The Post-Chaise Companion: or Traveller's Directory through Ireland...*, Dublin 1783; *Allen's New and Correct Pocket Map of Ireland...*, 1810.

• **Allen**, M.H. & J.W. (fl.1827-1831). Engravers, lithographers and publishers of 32 Dame Street, Dublin. *Irland*, 1828; *Map of Ireland, With the lines of Water Communication between Dublin and the Interior...*, 1830.

Allen, W.H. *Man River San Juan, Nicaragua*, 1851.

Allen, William. American engraver, of Philadelphia. Assisted H.S. Tanner, *United States of America*, by H.S. Tanner, 1829, 1830, 1832, 1839; *Plan of the City of Philadelphia...*, 1830.

Allen, *Captain* William, *RN*. Admiralty charts of the West Coast of Africa, 1830-1840; *Shores and Islands of the Mediterranean*, *c.*1838.

Allen, William [& Son] Publishers. Dane St, Dublin. Ireland, 1778, 1783; Carlow, 1798; *Pocket Map of Ireland*, 1815; Meath (4 sheets), 1817; *Allen and Son's Map of Dublin*, 1825; Wicklow (2 sheets), 1834.

Allen, William Houghton [& Co.] Publishers, 7 Leadenhall St, London. *New and Improved Map of India*, 1820, 1841; Punjab, 1846; Java, 1852; Route Map World, 1856.

Allen, William Thomas. *Proposed new Dock at Rainham* [Essex] (3 sheets), 1911.

Allen, Parbury & Co. See **Parbury**, Allen & Co.

Allertsz. [Allertszoon], Jan (*fl.*1630). Dutch mapmaker and surveyor, of Uitgeest. Map of area around Heemskerk, Uitgeest &c., 1630, MS.

Alley, William. *Journal of Voyages*, 1665-1678 (containing charts of Mauritius, Maldive Islands and Ceylon).

Alleyne, *Captain* (later *Major*) *Sir* James R.A. *Zululand*, for Quartermaster General's Department, 1879-1880.

Allezard, Jean Joseph. Hydrographer of Livorno. *Receuil des principaux plans des Ports et rades de la Mer Méditerranée...*, 1795 (revised edition of the maps first produced by Joseph Roux, 1764); *Nouveau recueil de...la Méditerranée*, 1800.

Allgood, *Captain* —. *Allied Armies in China*, 1860.

Allhusen, E.L. *Geological Map of Coolgardie*, 1898.

Alliaco, Petrus d'. See **Ailly**, Pierre d'.

Allin [Allen, Alleyn], Henry (*fl.*1599-*c.*1627). Surveyor of Tonbridge, produced plans of estates in Kent, including *...Hye Lese* [and] *Lowlese...* [Yaldinge, Kent], 1603.

Allin, Richard. Estate surveyor, of Robertsbridge, Kent. *...the Mannors of Eastdeane, Old Lavingeton and Grafham...* [Sussex], 1597.

Allioli, Joseph F. Biblical map (9 sheets), Landshut 1842-1844 (with L.G. Gratz).

Allman, Thomas Joseph (*fl.*1859-1879). Publisher of 42 Holborn Hill (1859-1862); 463 Oxford Street (1862-1879). D. Blair's *General School Atlas of Modern Geography*, 1850; D. Blair's *Classical Atlas on an entirely new plan*, 1835; *Atlas* (from Allman's Penny School Series), *c.*1871.

Allnut, H. *Record Map of the War between France and Germany* [Franco-Prussian war 1870-1871], *c.*1872.

Allnut, Zachary (*fl.*1789-1845). Surveyor to the Thames Navigation Commissioners. *Cookham, Berkshire*, 1812 MS.

Allodi, Pietro (*fl.*1859-1878). Draughtsman and engraver. *Carta postale, itineraria e amministrativa delle Prov[in]cie. Lombardo-Venete*, 1848; *Carta stradale dell'Italia*, 1859; *Carta Generale dell'Europa*, 1864; *Atlante universale*, 1867; *Carta nautica*, 1878; Pietro and Giuseppi Vallardi's *Oceania*, 1881.

Allouez, Claude, *SJ*. Missionary in the western Great Lakes area in the 17th century. Believed to be the supplier of information used by Claude Dablon in the making of *Lac Superieur...*, 1672.

Alloy, Luigi Sebastiano. *Diocesi di Milano*, 1821.

Alluaud, Charles (1861-1949). French naturalist and explorer in Africa. *Iles*

Canaries, 1891; *Iles Séchelles*, 1894; *Voyage au Soudan égyptien*, 1907; *Afrique orientale*, 1914 (with R. Jeannel).

Almada, Andreas d'. Professor of Theology. Revised Hessel Gerritsz's Spain, for Johannes Janssonius, 1631; maps in Nicolas Tassin's *France*, 1640-1643.

Almazan, —. *Departamento da Puebla*, 1855; *Mexicanische Stat Puebla*, *c*.1860; *Mexico...Veracruz*, 1862.

Almeda, Romão Eloi de. Portuguese cartographer and engraver. *Carta militar Portugal*, 1808.

Almeida, Candido Mendes d' (1818-1881). *Atlas do Imperio do Brasil*, 1868.

Almeida, Manuel de. Ethiopia, 1602 MS, printed 1640 (the first European printed map of Ethiopia), re-issued by Pieter van der Aa *c*.1710.

Almeida, Xavier d'. *Guarapuaia* [Brazil], 1770 MS.

Almera i Comas, Jaume (1845-1919). Catalan geologist. *Mapa geologico y topografico de la provincia de Barcelona*, 1888 (scale 1:100 000); re-issued at a scale of 1:40 000 (5 sheets), 1891-1913; *San Juan de Vilatorta*, 1905.

Almodovar. See **Gonzaga** y Luzon, Pedro.

Almon, Albert. *Plan of Sydney*, 1900.

Almon, John. Publisher, printer and bookseller, 178 Piccadilly (Autumn 1763-June 1781) 'opposite Burlington House'. Succeeded by John Debrett, 1781. One of many sellers of Gibson's edition of William Camden's *Britannia*, 1772 (with maps by Morden); *Environs of Boston* (in *Remembrancer*), 1775; *Middle British Colonies* (in Thomas Pownall's *America*), 1776; Thomas Hutchins's *Virginia*, 1778; Situation Fleet Sandy Hook, 1779.

Almond, James. *Plan of Chatham Street*, 1868.

Almonte y Muriel, Enrique d'. *Isla de Luzon* (4 sheets), Madrid 1883; *Provincia de Manila*, 1885; *Distrito de Leyte*, 1898; *Distrito de Samar*, 1898; *Isla de Negros*, 1898.

Aloja, family (*fl.* mid 18th century - mid 19th century). Draughtsmen and engravers of Naples, associated with the Officio Topografico de Napoli. Ref. VALERIO, V. *Società uomini e istituzioni cartografiche nel mezzogiorno d'Italia* (1993) pp.433-438.

• **Aloja**, Vincenzo (*c*.1768-1817). Master engraver of landscape, scientific texts and maps, professor in the school of landscape engraving at the Accademi di Belle Arti di Napoli. *Carta del Regno di Napoli...*, 1807; *Atlante del Regno di Napoli ridotto in VI Fogli*, 1808 (for Giuseppe Bonaparte).

•• **Aloja**, Luigi (1783-1837). Italian engraver. *Le Forche Caudine*, 1812; map of 5th-century Europe for *l'Atlante Universale e Portatile*, 1813; 3rd sheet of a 4 sheet map of Sicily, 1822-1826; 1st sheet of a map of the Adriatic Sea, 1824-1825; map of the neighbourhood of Naples, 1836-1837.

•• **Aloja**, Gennaro (1793-1868). Probably the son of Vincenzo. Engraver and lithographer specialising in mountain relief maps, views and illustrations of military uniforms. Engraved sheet 6 of a regional map of Naples, 1822; first sheet of a map of Sicily, 1824-1825.

Aloja, Giuseppe. *Villa di Partici*, *c*.1750; engraver for Giovanni Carafa's *Mappa topografia città di Napoli*, 1775; and *Città di Napoli*, 1788.

Alonzo de Santa Cruz. See **Santa** Cruz.

Alpenverein, Deutsche und Österreichische (*fl.* from 1862). Various topographical maps of German, Austrian, Swiss and French Alps. Ref. KRETSCHMER, DÖRFLINGER & WAWRIK *Lexicon zur Geschichte der Kartographie* (1986) Vol.C/I pp.11-16 [in German].

Alph, Johann. View of Münster, *c*.1628.

Alphand, Jean Charles Adolphe. *Plans et vues pittoresques*, 1868; *Les Promenades de Paris*, 1868; *Atlas des plans anciens de Paris*, 1880.

Alphane, D. See **Alfani**, Domenico.

Alpharabius. See **Farabi**

Alphen, Pieter van (*fl.*1690s). Son in law of Jacob Colom. Publisher of Rotterdam, 'bij de Roobrug inde Vijerige Colom'. *Zee-Atlas*, Rotterdam 1660 & 1682; *Kaert-Boek, waer in de XVII Nederlandse Provincie[n] door 47 curieuse kaerte, verthoont word*, 1691. **Ref.** *The Map Collector* 7 (1979), pp.56-57; SHIRLEY, R. *The Map Collector* 8 (1979) p.43 (letter); KOEMAN, C. *Atlantes Neerlandici* (1967-1971) Vol.I p.48.

Alphonse, Jean de Saintonge. See **Alfonse**, Jean.

al-Qazuini. See **Qazwini**.

al-Shafi. See **Shafi-al**.

Alsop, George. *A Land-skip of the Province of Mary Land...*, London, 1666, (rare promotional map published in *A Character of the Province of Mary-Land*).

Alsop, John. Map Friends Meeting, Rhode Island, 1782.

Alst, Peter G. van. *Map of the Hunter, Van Alst and De Bevoise Farms*, 1874.

Alstein, — Van. *Métropole commerciale de la Belgique*, 1872.

Alt, Franz (1821-1914). *Wien...*, 1873; *Aus den Alpen*, *c.*1877.

Alt, H. Engraver for Adolf Stieler's *Hand-Atlas*, 1863.

Alt, W. Engraver for Adolf Stieler, 1850; and K. Spruner von Merz, 1855.

Altena, Maas van. Draughtsman. Plan of Muiden, 1787 (engraved later by J.W. Strunck).

Altenburger, Peter. Map of Austria, 1843.

Altermatt, *Colonel* Johann Baptist (1764-1849). French born cartographer and draughtsman, worked in Solothurn, Switzerland. Map of the canton of Solothurn, 1795-1798 MS (State Archives of Solothurn); plan of the city of Solothurn, Basle 1822; *Plan der Stadt Solothurn*, Solothurn 1833.

Altheer, J. (*fl.*1791-1827). Publisher of Utrecht, with B. Wild. *Nederlandsche ontdekkingen*, 1827. **Ref.** KROGT, P. van der *Advertenties voor kaarten, atlassen, globes e.d. in Amsterdamse kranten 1621-1811* (1985) N° 1670 p.312 and other entries [mostly in Dutch].

Alting, Menso, *the younger* (1636-1712). Geographer and Burgemeester of Groningen. *Notitia Germaniae Inferioris*, Amsterdam 1697; *Uitbeelding der Heerlijkheit Friesland*, 1718. **Ref.** DONKERSLOOT de VRIJ *Topografische karten van Nederland vóór 1750* (1981) pp.146-147 [in Dutch].

Altischofen, P. von. See **Pfyffer** von Altischofen.

Altmann, J. *Chaussée Eisenbahn und Canal...Petersburg & Moskau*, 1844; *Neuer Schul- und Handatlas über alle Theile der Erde*, Berlin, Silber 1847.

Altmüller, H.W. *Relief plan von Jerusalem*, 1859; *Das Heilige Land und der Libanon*, 1860; *Sinai & Golgotha*, 1861.

Altmütter, Georg (1787-1858). Professor of mechanical engineering at Vienna. Wrote a treatise on the making of globes.

Altschüler, Jechiel Hillel. *Eretz Israel* [Land of Israel], 1775, 1782.

Álvares, Cristóvão (*fl.*1620-1656). Portuguese military engineer and cartographer. *Forte Reale, Pernambuco*, 1629; *Cidade de San Salvador*, 1638; *Baía de Todos os Santos*, 1638.

Alvares [Alvarez]. See **Seco**, Fernando Alvares.

Alvarez, —. See **Valverde** y Alvarez, Emilio.

Alvarez, Manuel. *Plano de la Ciudad de Mexico*, 1867.

Alvarez, Manuel Anibal. *Plano de las Ruinas de Numancia*, 1924.

Alvarez de Colmenar, Juan. *Les delices de l'Espagne & du Portugal*, Leiden, Van der Aa 1715.

Alvarez-Malgorey, Cárlos. *Atlas geográfico*, 1886.

Alves, A.F. *Plan of Victoria*, Hong Kong, 1862.

Alves, *Captain* Walter, *RN*. Charts of Malaya, surveyed 1763-1782. Pulo Pinang in the Strait of Malacca, 1763 published Dalrymple 1774; Point Romaina, 1763 published Dalrymple 1774; sketch of Maloza Bay, 1764 published by Dalrymple 1782; Island Cheduba, Dalrymple 1782.

Alvord, G.A. Printer. Charles G. Colby's Diamond Atlas, 1857.

Alwin, Richard. Estate surveyor. Quarrendon, Buckinghamshire, 1621 MS.

Alypius (*fl.*360-371). Geographer of Antioch.

Alzate y Ramirez, José Antonio de (1739-1790). Mexican astronomer and geographer. North America (4 sheets), 1768; *Nuevo Mapa geographico de la America septentrional...*, 1772; *Mapa de las Aguas...al Laguna de Tescuco*, 1786.

Amabile, V.M. *Carte générale de Cuba*, Paris 1896.

Amador, Alberto. *Limite entre Mexico y Guatemala*, 1931.

Amante, Fidele (1794-1851). Mathematician, geodesist, astronomer, teacher and scholar of Naples. Professor of geodesy in Naples, wrote many didactic and scientific works on geodesy and triangulation, and the use of contours in the representation of relief. A key figure at the Istituto Topografico Napoletano. **Ref.** VALERIO, V. *Società uomini e istituzioni cartografiche nel mezzogiorno d'Italia* (1993) pp.438-440 [in Italian].

Amari, Michele (1806-1889). Historian of Sicily. *Carte de la Sicile*, Paris, Lemercier 1859.

Amaseo, Gregorio [Grigorio] (1464-1541). Cartographer of Udine. Manuscript map of Friuli, now lost, but thought to have been a source for Abraham Ortelius, 1570. **Ref.** KARROW, R.W. *Mapmakers of the sixteenth century and their maps* (1993) pp.37-39; MEURER, P. *Fontes cartographici Orteliani...* (1991) p.103 [in German].

Amát de Tortosa, Andrés. *Yslas de Canaria*, 1779.

Amat di San Filippo, *conte* Pietro (1822-1895). Historian of cartography. *Bibliografia dei Viaggiatori Italiani*, 1874; *Studi bibliografii*, 1875.

Amati, Amato (1831-1904). Geographer. *Dizionario corografico dell'Italia*, 1864-1872.

Amati, Pietro (*fl.*1791-1834). Engraver working in Turin. *Nouvel Plan de Malte*, 1798; *Città di Torino*, 1800; maps for Carlo Antonio Barbiellini's *Atlante delle geografia antica e moderna*, 1807; also maps and plans for guide books.

Amato. See **Bellarmati**, Girolamo.

Amberger, Christoph (*c.*1505-1562). Painter in Augsburg. Plans of the River Lech, near Füssen, 1549 MSS.

Ambile, N.N. *Rios Colorado y Gila*, 1757 MS.

Ambridge, J. Bookseller and Stationer, 1 Lower St, Islington. *Australia*, 1852; *15 miles round London*, 1855; *Australia, fully exhibiting the scene of insurrection in Victoria in 1854*, 1855.

Ambrogio da Milano. See **Ambrosio** da Milano.

Ambrosini, Floriano. *Corografia... Contado di Bologna*, 1605.

Ambrosio [Ambrogio] da Milano. Italian military engineer and surveyor, worked in Antwerp *c.*1580s.

Ambrosius, *Doctor* Ernst. Publisher and editor of *Andree's Allgemeiner Handatlas...*, 1914 and editions to 1930.

Ambrosius, Marcus. See **Ambrozy**, Marek.

Ambrozy [Ambrosius], Marek [Marcus] (*c.*1530-post 1566). Silesian historian and humanist, from Neisse [Nysa]. Worked in Antwerp during early 1560s. *Arma regni Poloniae*, Antwerp, Gilles Coppens van Diest 1562; credited by Abraham Ortelius with a map of Livonia (no trace known, but before *c.*1566). **Ref.** KARROW, R.W. *Mapmakers of the sixteenth century and their maps* (1993) pp.40-41; MEURER, P. *Fontes cartographici Orteliani...* (1991) pp.103-104 [in German].

Ameline, F. ...*l'Arrondissement de Bernay*, 1926; *Département de l'Eure* (25 plans), 1927.

Amendale, *sieur d'*. See **Damme**, Jan van.

à Merica. See **Heyden**, Gaspard van der.

American Atlas Co. Plat books of counties in Illinois, Michigan, Indiana, 1893-1907.

American Express Co. *Tourist's Pocket Atlas*, 1881; Note Book & Atlas, 1889.

American Globe & School Supply Co. (*fl.*1891-1897). Made globes designed to carry gores produced by firms such as Rand McNally and W. & A.K. Johnston. **Ref.** WARNER, D. *Rittenhouse* Vol.2 (1) (1987), p.23.

American Photo-Lithographic Co. New York publisher. [Osborne's Process] Railway routes to Pacific, 1869; American Tract Society Bible Atlas, 1862.

American School Apparatus Co. School equipment manufacturer, 19 Murray St., New York. 5-inch terrestrial globe, 1867. **Ref.** WARNER, D. *Rittenhouse* Vol.2 (1) (1987), p.23-24.

Amerigo Vespucci. See **Vespucci**.

Amerigius, P. van der. See **Heyden**, Gaspard van der.

Amery, —. Antilles, published by L. Mondhare, 1782.

Ames, I. See **Amyce**, Israel.

Ames, J. Engraver. Robert Snell's *Map of the County of Monmouth* (4 sheets), 1785.

Ames, Joseph. *Town & Pier of Ramsgate*, 1736; *Thamesis Descriptio*, 1753 (taken from an original by Robert Adams, 1588).

Ames, N.S. Canadian surveyor. Westmoreland County, 1857; Bradford County, 1858; Lawrence and Beaver Counties, 1860.

Ameti, Giacomo Filippo (*fl.* late 17th century). *Il Lazio* (4 sheets), Rome, Giovanni Giacomo de Rossi 1693; *Parte meridionale del Latio*, Rome, Domenico de Rossi 1693; map of Papal Territories (4 sheets), Rome, de Rossi 1696; Tuscia, 1696.

Amettier, B. Engraver. *Plano... Havana*, 1798.

Amezaga. See **De Amezaga**.

Amherst, Edward. Surveyor. Peninsula in Bay of Fundy (with George Mitchell), 1735.

Amico, Fra Bernardino [called da Gallipoli]. Italian Franciscan friar. *Trattato delle piante & imagini di Terra Sancta* (34 maps and plans after Antonio Tempesta, including the plan of Jerusalem by Antonio de Angelis), Rome 1609; Jerusalem view, Rome 1612. **Ref.** MOLDOVAN, A. *The Map Collector* 24 (1983) p.24.

Amicus, J.B. *De motibus corporum coelest.*, Venice 1536.

Amirutzes, Georgius (*fl.*1465-1470). Cartographer of Trapezuntios [Trebizond, Turkey]. Worked in Constantinople. Maps for Ptolemy's *Geographia*.

Amman, Ignaz Ambrosius von (1753-1840). German surveyor of Mülheim. *Charte von Schwaben*, (56 sheets) 1798-1828 (with J.G.F. Bohnenberger).

Amman, Jost (1539-1591). Born in Zurich. Painter and wood engraver, worked in Zurich and Nuremberg. Cut the decorative border and ornaments for Philipp Apian's map of Bavaria, 1568; *Globus terrestris*, twin-hemispherical world map (*attributed*), Nuremberg *c.*1564. **Ref.** SHIRLEY, R. *The mapping of the World* (1983) N° 113 p.129.

Ammann, Johannes (1695-1751). Swiss engraver and mapmaker in Schaffhausen.

Amnelius, Andreas (*fl.*1704-1705). Swedish surveyor, worked in Skåne. *Geometrisk Karta öfwer Wallebiera och Löderup*, 1704-1705.

Amon, Anton (1761-1798). Engraver in Vienna. Poland and other maps for F.J.J. von Reilly, 1796.

Amouroux, L. *Plan de Nantes*, 1852.

Amrein, Kaspar Constantin (1845-1898). Swiss teacher of geography, born in Lucerne, worked in Hitzkirch and St Gallen.

Amschewitz, J.H. Cartoon map of the European war area, 1916.

Am Stein, *Major* Johann Rudolf (1777-1862). Swiss military cartographer and map collector in Holland and Malans.

Amthor, Eduard. Co-author (with Wilhelm Issleib) of *Volkes-Atlas über alle theile der erde*, Gera, Issleib & Rietzschel 1868; *Neuester Repetitions Atlas*, 1876.

Amundson, Albert. Post & telegraph map of Sweden, 1873.

Amyce [Ames], Israel (*fl.*1576-1607). Estate surveyor. Belchamp St. Pauls, 1576 MS; Manor of Plomborowe, 1579 MS; Castle Hedingham, 1592; surveyed Theobalds Park, 1607 (with Henry Cumings).

A. Myrica. See **Heyden**, Gaspard van der.

Anania, Giovanni Lorenzo d' (*c.*1525-1582?). Italian scholar. *L'Universale Fabrica del Mondo*, Naples 1573, Venice 1576 (no maps), Venice edition of 1582 included maps engraved by Girolamo Porro (maps were later used in G. Botero's *Le Relationi Universale*, from 1596). **Ref.** BURDEN, P. *The mapping of North America* (1996) N°.54 pp.68-69.

Ananias of Shiraz (*fl.*5th century). Armenian geographer.

Anaximander of Miletus (*c.*610-546 BC). Greek philosopher, said to have been first Greek mapmaker. Constructed a circular world map and has been credited with the introduction of the *gnomon* into Greece. **Ref.** KAHN, C.H. *Anaximander and the origins of Greek cosmology* (New York 1960).

Anbury, Thomas. *Travels through the interior parts of America*...[march of the army], 1789 (with 2 maps).

Ancelin [Ancelijn; Asselin], Pierre (*c.*1653-1720). Surveyor and cartographer of Rotterdam. One of the early users of the isobath line to indicate depth, such lines appeared on his map of the Meuse: *Kaart van de Maas bij Rotterdam*, 1697 (one of the earliest so published). **Ref.** *Tijdschrift voor kadaster en landmeetkunde* 40 (1924), pp.51-63, 133-144 [in Dutch]; see also under **Cruquius**.

Ancelin et P. Legrand. *Atlas de toutes les Russies*, Moscow and St Petersburg 1795.

Ancessi, Victor-Antoine (1844-1878). *Atlas Géographique et Archéologique...*, 1876; *Atlas de l'Ancien et Nouveau Testament*, 1885.

Anciau, G. *Map of Barbados*, 1887.

Andel, A. van. *Nederland en de provinciën* (12 maps), *c.*1930.

Andelfinger, Johann Jacob. Map of Constantinople, Matthäus Seutter 1735.

Anders, Friedrich Gottlieb Eduard. *Historischer Atlas der Evangelischen Kirchen im Schlesien*, 1856.

Andersen, Jürgen. *Orientalische Reise-Beschreibung...*, 1696.

Andersen, Karl Gustaf (1738-1808). Swedish land surveyor.

Andersen, S. Swedish engraver. Charts, 1837.

Andersin, Anders. *General Charta... Ingermannland*, 1704.

Anderson, *Lieutenant* —. Surveyor and draughtsman. *Plan of the Attack & Positions of the Allied Armies before Sebastopol*, 1858.

Anderson, A. Engraver for the 1814 London edition of *Atlas of Ancient Geography*, originally by J.B.B. d'Anville.

Anderson, A.H. *Plat Book of Redwood County, Minnesota*, 1898.

Anderson, Adam. *The World, with the latest discoveries...*, 1787 for *An historical and chronological deduction of the origin of commerce.*

Anderson, Alexander (1775-1870). American engraver. *An accurate map of the Holy Land*, 1792, 1795, 1796; *...the countries surrounding the Garden of Eden...*, 1792, 1795, 1796, 1799; *Jerusalem*, 1793; *An Accurate Map of the United States. According to the Treaty of Peace of 1783*, 1796; Kentucky, for William Winterbottom's Atlas, and John Reid's *American Atlas*, 1796; maps of New

Hampshire, Vermont, Massachusetts, Connecticut, Maryland and Delaware, Virginia and France for John Payne's *A New and Complete Universal Geography*, 1799. **Ref.** See WHEAT & BRUN *Maps and charts published in America before 1800* (1978) for descriptions; STAUFFER, D.McN. *American engravers* Vol.I pp.8-9, Vol.II pp.11-14.

Anderson, Alexander C. *Map of Fraser's River*, San Francisco, 1858.

Anderson, *Captain* Allen. New Mexico, 1864.

Anderson, Anders. See **Andersson**, Anders.

Anderson, Charles John. See **Andersson**, Charles John.

Anderson, Charles Wilgress. *Lower Essequibo and Cuyum Rivers* (2 sheets), 1905; *Northern portion of British Guiana*, 1908.

Anderson, D. American engraver. Engraved two maps for William Wintherbotham's *The American Atlas*, Philadelphia, John Reid 1796.

Anderson, David. *Map of the British Colonies and the United States...*, 1814.

Anderson, E.M. *Geological Survey of Scotland...*, 1906.

Anderson, *Captain* F.C. District of Shajehanpoor, 1862-1864, published 1866.

Anderson, George William. *A New Authentic and complete collection of Voyages round the World* [published in 80 weekly parts], A.Hogg 1784, 1785.

Anderson, Hugh (1782-1866). Born in Ireland, emigrated to Philadelphia, USA, 1809, later moved to Ohio. Engraver of maps and prints, also collaborated with F. Lucas Jr.. Luke Munsell's *A map of the State of Kentucky...* (3 sheets), Frankfort, 1818; John H. Robinson's *...Mexico*, 1819; John H. Robinson's *Louisiana and the Missouri Territory*, Philadelphia, 1819; B. Warner's *United States of America*, 1820; McCarthy and Davis's *United States of America*, 1824; [James] *Turnbull's map of the States of Ohio, Indiana and Illinois*, Ohio, 1835; and others.

Anderson, *Lieutenant* J. Punjab & Kashmir, 1846.

Anderson, J.A. *Map of the Railroads of New Jersey...*, 1871; *Map of the Railroads of Pennsylvania and parts of adjoining States*, 1872; *Map of the Railroads of Pennsylvania and New Jersey*, 1873.

Anderson, James. Hebrides, 1785; Aberdeen, 1794.

Anderson, James. Plan for a tunnel under the Forth, 1818; *Plans & sections of a bridge...over the Firth of Forth at Queensferry*, 1818; *line of the Union Canal*, 1818; *Plan of Callender Park...in the County of Stirling*, 1818.

Anderson, *Lieutenant* James. *...Chart of the Gulf-Passage or New Bahama Channel*, 1814.

Anderson, Johan [Johann]. Burgermeister of Hamburg. *Beschryving van Ysland, Groenland en de Straet Davids*, Amsterdam, S. van Esveld 1746 (Dutch edition), 1749; *Nova Gronlandiae Islandiae et Freti Davis...*, 1747.

Anderson, John. Map & Bookseller, 62 Holborn Hill, London (1787-1799). Demerary, 1795.

Anderson, *Reverend Doctor* John. *Geological Map of Fifeshire*, 1841.

Anderson, John [*the younger*]. Engraver for John Payne's *System of Universal Geography*, 1798.

Anderson, *Major* (later *Lieutenant-Colonel*) John. United States Army Topographical Engineers. Sketch map of Crown Point [New York], 1818 MS (with I. Roberdeau); Portland Harbor, 1833 MS.

Anderson, John Corbet. *...Commons of Croydon*, 1889; *Great North Wood, Norwood*, 1897.

Anderson, John George Clarke. *Asia Minor*, 1903.

Anderson, P. American civil engineer. *Map Exhibiting the Experimental and Located*

Lines for the New-York and New-Haven Rail-Road, New York, Snyder & Black 1845; Canal Map of Hartford, Connecticut, 1847.

Anderson, R.T. Mapmaker of Cadiz, Ohio. Map of Ohio, St. Clairsville, James Turnbull 1828-29; *Map of the Countries mentioned in the Holy Scriptures...*, St. Clairsville, 1845. **N.B.** See also Hugh **Anderson** (above), who was also at some time based in Ohio, and shared the same publisher.

Anderson, Thomas. Edgefield District, South Carolina, 1817; Barnwell District, South Carolina, 1818

Anderson, William. Chart Southern Hemisphere, 1775 (*attributed*).

Anderson, Young. District of Santo Tomas, 1880.

Andersson [Anderson], Anders (*d.*1695). Surveyor of Södermanland, Sweden. *Nykjöping Stad*, 1674 MS; *Hölbo häradt*, 1679.

Andersson [Anderson], Charles John (1827-1867). Ornithologist and explorer. Maps of Damaraland, 1866 MS.

Andersson, Samuel. Engraved charts for Nils Gustaf Werming, 1804; *Petrografisk Karta öfver Södra delarne af Swerige och Norrige*, 1804; *Petrografisk Karta öfver en del af Göta Rike*, 1804; *Petrografisk Karta öfver Skåne*, 1804, the last with E. Åkerlund.

Anderton, G. Engraver for Isaac Taylor's *Oxford*, 1750.

Andonaegui, *General* Joseph de. *Plano parte de la America meridional*, Buenos Aires 1755 MS.

Andouard, F. (*b.*1734). One of the engravers for Cassini's *Carte de France*: Cherbourg, 1758; Gien, 1759.

Andrade, B.M.F. de. *Oceano Atlantico Norte*, 1870-1873, published 1885.

Andrássy, István (1802-1890). Died in Gyulafehérvár [now Alba Julia, Romania]. Geodesist and astronomer at the bishop of Gyulafehérvár's observatory; assisted in surveys of Transylvania.

Andrau, K.F.R., *Lieutenant-ter-zee*, Royal Netherlands Navy. *Atlantische Oceaan*, 1856; *Maurij's Wind-Kaart*, 1856 (Dutch version of M.F. Maury's wind chart).

André, Aimé. Publisher of 'Quai des Grands Augustins 59' and 'Rue Christine', Paris; son-in-law of Jean-Baptiste Poirson. For H. Fournier, 1820; for Conrad Malte-Brun, 1837.

André, Gaspard. Letter engraver. Worked for Cassini (sheet 162), *c.*1768-1770; G.A. Rizzi-Zannoni, 1769-1772; for G.T. Raynal, 1780; for Rigobert Bonne, 1783-1787; Nicolas Desmarest, 1787; for Edme Mentelle's *Atlas universel*, 1797-1801.

André, H. *Paris imprenable et La France invincible*, *c.*1877.

André, *Major* John (1751-1780). Military engineer and surveyor, executed in New York as a British spy. Several manuscript maps and plans of towns, forts and counties in the United States, many of which are now in the Huntington Library. **Ref.** GUTHORN, Peter *The Map Collector* 2 (1978), pp.15-16; GUTHORN, P.J. *British maps of the American Revolution* (1972) pp.9-10.

Andre, Karl. North America, 1851.

André, Louis. *Plan...de Biarritz*, 1860.

André, Noel (1728-1808). Capucin geographer and astronomer. *Deux planisphères*, Paris 1778.

André, Peter (*fl.*1762-1777). Surveyor, of 7 Noel St Berwick St, London. Surveys of Essex with John Chapman, 1772-1774, published as *A Map of the County of Essex* (24 sheets), W. Keymer 1777, 1785, later edition (25 sheets) Chalk, Meggy & Chalk 1833.

Andre de Gy. See **Chrysologue**.

Andrea, Alvaro. *Rio Limpopo*, Lisbon 1896.

Andrea [Andreae], Johann Ludwig (*fl.*1717-1726). Cartographer of Nuremberg. Terrestrial and celestial globes.

Andrea, P. Publisher of J.U. Müller's *Neuaußgefertiger Kleiner Atlas...*, Frankfurt am Main 1702.

Andrea, T. *Plano do Porto e Cidade de Dilly* [Portuguese Timor], 1870.

Andreae, Johann Philipp (1700-1757). Copper engraver. Plates for a 3.5 cm globe used by the German globemaking firm of Klinger, Nuremberg 1790.

Andreas, Alfred Theodore (1839-1900). American compiler and publisher of county maps at 'Lakeside Building, Cor. Clark & Adams Sts.', Chicago (1874), partner in Andreas, Lyter & Co.. Began as agent for Thompson and Everts in the late 1860s, then from 1870 produced many large-scale county atlases for western Illinois and eastern Iowa, from *c.*1876 moved on to state atlases of Minnesota, Iowa and Indiana. *Atlas Map of Pike County, Illinois*, Davenport, Andreas, Lyter & Co. 1872; *Atlas Map of Tazewell County, Illinois*, Davenport, Andreas, Lyter and Co. 1873; *Atlas Map of Peoria County, Illinois*, Chicago 1873; *An Illustrated Historical Atlas of Hancock County, Illinois*, Chicago 1874; Iowa, 1875; Atlas of Dakota, 1884; and others. **Ref.** CONZEN, M. *Chicago mapmakers* (1984) pp.46-63; RISTOW, W.W., 'Alfred T. Andreas and his Minnesota Atlas' in: *Minnesota history* 40 (1966), pp.120-129.

• **Andreas, Lyter & Co.**. *Atlas Map of Knox County, Illinois*, Davenport 1870; *Atlas Map of Fulton County, Illinois*, Davenport, Iowa 1871; *Atlas Map of McDonough County, Illinois*, Davenport, Iowa 1871; *Atlas of Adams County, Illinois*, Davenport, Iowa 1872; *Atlas Map of Jersey County, Illinois*, Davenport 1872; *Atlas Map of Greene County, Illinois*, Davenport 1873; and others.

Andreas, Lambert (*fl.*1590-1598). Publisher of Cologne. Some of the individual maps used in the works which Andreas published have been attributed to him. Publisher of Matthias Quad's *Europa ... Descriptio*, Cologne, 1594, 1596; Giovanni Botero's *Tabulae geographicae*, 1596 (editions in Latin or German); *Theatrum principium*, 1596. **Ref.** BURDEN, P. *The Mapping of North America* (1996) N°90 p.115; MEURER, P. *Atlantes Coloniensis* (1988) p.78 *et seq.* [in German]; SHIRLEY, R. *The mapping of the World* (1983) N°190 p.210.

Andrée, Otto. *Königreich Sachsen*, 1862; *Sächsisch-Böhmischen Schweiz*, 1865.

Andree, *Doctor* Richard (1835-1912). Born

Alfred Theodore Andreas (1839-1900).

in Braunschweig, worked as a geographer and cartographer in Leipzig. *Volksschul-Atlas*, 1876; *Physikalischer-Statistisch Atlas des deutschen Reichs*, 1878 (with O. Peschel); *Allgemeiner Handatlas*, 1880; *Afrika*, 1885; *Historischer Handatlas*, 1886; *Afrika*, 1890.

Andreossy, François (1633-1688). Hydrological engineer. *Canal royal de Communications des Mers en Languedoc*, 1669.

Andrews. See **Thomas** and Andrews.

Andrews, Alfred H. [& Co]. Publishers of Chicago. School globes of various sizes from 3-inch to 30-inch diameter, 1866; nickel-plated brass orrery, *c.*1880s; 30-inch terrestrial globe, *c.*1890; *New Colored-Zone Terrurian Globe*, [n.d.]; *Deep Sea Globe*, 1897. Taken over by C.F. Weber & Co., *c.*1900; became Weber-Costello, 1907. **Ref.** WARNER, D. *Rittenhouse*, Vol 2 (1) (1987), p. 24-26.

Andrews, Arthur Westlake. *Diagram Atlas of South Africa*, 1901; *Macmillan's Orographical Map of Europe* (2 sheets), 1908; *Comparative Series of Physical Maps*, 1910.

Andrews, C.A. *River Pungue* [Mozambique], 1891.

Andrews, E.B. Gallia County [Ohio], 1874; Meigs County [Ohio], 1875.

Andrews, H. Engraver for B. Warner of Philadelphia. United States of America, 1820.

Andrews, Henry (*fl.*1838-1850). Estate surveyor of Modbury, Devonshire. Farms in Plymstock, Devon, 1850 MS.

Andrews, I. see **Andrews**, John.

Andrews, Israel D. *Straits of Florida*, 1852; *Railroads in the United States*, 1853.

Andrews, James. *Chart of Skerryvore Rocks*, 1836.

Andrews, John (1736-1809). Geographer, surveyor, engraver and mapseller at 'N°. 5 Fish Market, Westminster Bridge', at 'Mr. Branches, 40 Corner of Buckingham St. Strand' (1776-1777), '29 Long Acre at Mr. Blissets opp. Mercer St.' (1777-1782), and '211 facing Air St. Piccadilly' (1781 onwards). Also worked with Andrew Dury, Matthew Wren and others, see below. ...*Town of St Albans in Hertfordshire*, 1766 (with Matthew Wren); *Plan of Hartford*, 1766 (with Matthew Wren); *Topographical map of Hartfordshire* (9 sheets), 1766 published W. Faden 1782, single sheet version, A. Dury 1777, W. Faden 1782 (with Andrew Dury); ...*City of Canterbury*, 1768 (with Matthew Wren); *Blackheath and Lesnes Hundreds* [Kent], 1769; *A Topographical Map of the County of Kent* (25 sheets), 1769, published R. Sayer and J. Bennett 1779, Laurie and Whittle 1794 (surveyed with A. Dury and W. Herbert); Isle of Wight (4 sheets), London, A. Dury 1769 (surveyed with P. Bell and A. Dury, engraved by J. Andrews); ...*Plans of Capital Cities of Europe, Asia, Africa and America*, 1771; *A Plan of the City of Québec*, 1771; map of the county of Wiltshire (18 sheets), 1773 (surveyed with A. Dury, engraved by J. Andrews); *A New map of the British Colonies in North America...*, 1777 (with A. Dury); *Andrews's new and accurate map of the country thirty miles round London*, 1782; *Derbyshire*, 1786; *A new Physical, Historical and Political Map of England and Wales* (6 sheets), 1786; *A New Chart of New Holland*, 1787; *Canals and Rivers of England and Wales* (2 sheets), 1788; *Map of the Rivers...of England*, 1796; *Historical Atlas of England*, 1797; *Geographical Atlas of England*, 1809; and many other works.

Andrews, Norman S. Surveyor. Map of Constantine, 1874.

Andrews, Peter. Engraver of John Rocque's map of Surrey, 1754; Maerschalk's plan of New York City, 1762; of Mary Ann Rocque's *A Set of Plans and Forts in America...*, 1763-1765; Samuel Holland's Isle of St. John, 1765; *Town, Port and Citadel of Calais*, *c.*1765; John Montresor's *A Plan of the City of New-York & its Environs*, London, Mary Ann Rocque 1766; Sea of Marmara, 1770 published in *The American Pilot*, 1772; *Hydrographical Map of the Atlantic Ocean*, 1772.

Andrews, S.H. Surveyor, 28 King Street, Manchester. Salford, 1800.

Andrews, Samuel (*fl.* mid 18th century). Chart Cork Harbour, 1792.

Andrews, Thomas C. *The World; Bi-Conical projection*, 1895.

Andreyev, Stepan. Geodesy Sergeant. Surveyed and produced maps in the Medvezh'i Islands, the coast from Kolyma to the mouth of the Anadyr [Siberia], and the Bering Strait, 1762-1764.

Andries, Justin. *Carte Spéciale de la Province d'Anvers*, 1875; *Carte des Chemins de Fer de la Belgique*, 1868.

Andriessz. [Andriesszoon], Abraham (*fl.c.*1656). Bookseller and printer of Dordrecht. *Nieuwen Atlas ofte Beschryvinge van het noyt meer gevonden Eylant Bensalem*, 1656.

Andriveau [Andriveau-Goujon], firm and family. Mapsellers and publishers of Paris, variously located at 'rue du Bac n°.6' (1805-1820); 'rue du Bac 6' (1835); 'rue du Bac n°.17' (1841-1850); 'rue du Bac 21 près le Pont Royale', Paris (1854-1876); 'rue du Bac 4' (1876). **N.B.** With the exception of works naming J. Andriveau, works are often credited simply to

Andriveau-Goujon. The position of Gabriel-Gustave within the firm and family is unclear, and the work referred to has not been verified.

• [Goujon], J. (*fl.*1805-1824). Mapmaker and mapseller of Paris. Father in law and partner of J. Andriveau. See **Goujon.**

•• **[Goujon et Andriveau]** (*fl.*1825-*c.*1830). J. Goujon, and son-in-law J. Andriveau, cartographers and publishers of Paris. See **Goujon.**

•• **Andriveau** [Andriveau-Goujon], J. (*fl.*1831-1868). Son-in-law and partner of J. Goujon, succeeded Goujon and was joined by his son Gilbert-Gabriel *c.*1830. *Carte générale des routes de France*, Andriveau-Goujon 1831 and editions to 1846; *Carte physique et routière de la Suisse*, 1831; *Carte physique et routière de l'Italie*, *c.*1832; *Atlas classique et universel*, 1835; *Atlas Elémentaire simplifié de Géographie Ancienne et Moderne*, *c.*1840; *Carte routière et postale de la France*, 1846 and many later editions; *Carte générale des routes et des chemins de fer de la France*, 1852, 1858, 1862, 1873 (reduced version of the map of 1846); *Atlas ancienne et moderne*, 1856; *Carte physique et routière des Royaumes de Hollande et de Belgique*, 1867; *Atlas Elémentaire*, 1868.

••• **Andriveau-Goujon** (*fl.*from 1831). 'Géographe-éditeur', publishers of Paris, continued until the end of the 19th century.

••• **Andriveau**, Gilbert Gabriel Benjamin (1805-1884). Bookseller, publisher and geographer, joined his father in the family firm *c.*1830, passed the business to his son Eugène in 1858. *Océanie*, 1832, 1856; *Atlas classique et universel*, 1856.

•••• **Andriveau-Goujon**, Eugène (1832-1897). *Atlas de Choix*, 1862; *Carte Générale d'Allemagne*, 1865; Planisphère, 1872; *Atlas classique*, 1875; *Environs de Paris*, 1880; *Algérie*, 1890 (engraved by Gérin, Dalmont, Smith and Fontaine).

••• **Andriveau-Goujon**, Gabriel-Gustave (*b.c.*1808). Printer and geographer, born in Paris. ['Il a fait réduire et graver, avec un grand succès, une suite de cartes, qui pour la plupart, manquaient dans le commerce']. **Ref.** LERMINA, J. *Dictionnaire universel illustré biographique de la France contemporaine* (1884) [in French].

Androuet du Cerceau, Jacques. (1510-1584). Engraver of Macé Ogier's *Nova et Integra Cenomaniae Descriptio vulgo Le Mans* [Maine], 1539, 1565 (published in Maurice Bouguereau's *Théâtre françois*, 1594); *Plan de Paris*, 1555 (*attributed*); map of ancient Rome *Antiquae Urbis...* (6 sheets), 1578 (reduced version of Pirro Ligorio's map of 1561).

Andruzzi, E. Map of Sardegna, 1869.

Anemaat [Anemaet; Annemaet], family. Family of 17th and 18th-century Dutch land surveyors. **Ref.** DONKERSLOOT DE VRIJ *Topografische karten van Nederland vóór 1750* (1981) p.189 [list and dates in Dutch].

Anexarius or Anexerius. See **Buondelmonte.**

Anfora, Francisco Saverio (1833-1871). Officer and topographical engineer at the Officio Topografico di Napoli. His cartographic career was disrupted by military events and he went on to work on fortifications at Mola e di Gaeta, 1860. Frontier surveys in the region of Abruzzo, 1855-1858. **Ref.** VALERIO, V. *Società uomini e istituzioni cartografiche nel mezzogiorno d'Italia* (1993) p.441 [in Italian].

Angeli, Vittorio. Engraver. *Lombardia*, 1837; *Alessandria* (2 sheets), 1860.

Angelidev, Dmitri. Bulgarian cartographer. *Karta inostrana na Evropeiska Turiki'* [Extensive map of European Turkey], Plovdiv 1862 (first map printed in Bulgaria).

Angelieri, Giorgio. Printer of Venice. Tommaso Porcacchi da Castiglione's *L'isole più famose del mondo*, 1576 edition; published Giovanni Botero's *Le Relationi Universali*, 1596, 1600.

Angelino, *Don* Mariano Angel. *Mapa de Texas*, 1788.

Angelis, *Fr* Antonio de. Minorite monk, of Naples, artist and topographer. *Hierusalem* (copper engraving on two sheets), 1578, used by Bernadino Amico in his description of the Holy land, 1609. **Ref.** MOLDOVAN, A. *The Map Collector* 24 (1983), pp.17-24.

Angelis, Girolamo *SJ* (1568-1623). Italian Jesuit, travelled through Honshū and southern Hokkaidō, Japan, 1618-1623. Manuscript map of Japan and Ezo, *c.*1621 (now in the Archivam Historicum, Society of jesus, Rome. **Ref.** PAULY, U. in: LUTZ, W. *Japan: A cartographic vision* (1994) p.26.

Angelius, Vadius. See **Angelus**, Jacopo d'.

Angell, Samuel (1800-1866). Surveyor and architect of 26 Ely Place (1834), 18 Golder Street (1860), London, fellow of the Royal Institute of British Architects. *Cornhill Ward, London*, 1833 (Wm. Standidge lithog.); commissioned by the City of London to produce a detailed survey with Michael Meredith (31 sheets), *c*.1860.

Angelloti, P. *Pianta della valli di Conacchio*, 1658.

Angelo, M. de S. *Carte de Corse*, Paris, Jean Lattré 1768; maps for Roch-Joseph Julien's *Théatre du monde*, 1768.

Angelo, [Theodore] Gottfried Nicolai (1767-1816). Danish engraver and publisher, born Schlesvig, died Copenhagen. Aalborghuus, 1793; Anstrup, 1795; *Kort over Bornholm...*, 1805 (by T. Bugge & F. Wilster).

Angelocrater, *Fr* Daniel. German pastor of Frankfurt am Main. 'Novum orbis terrarum schema' (the world in four globe gores) in: *Doctrina de ponderibus, monetis et mensuris*, Frankfurt am Main 1628. **Ref.** SHIRLEY, R. *The mapping of the World* (1983) N°·320 p.344.

Angelos [Angelius, Angelus; Angiolo; Agnolo], Jacopo d'[Jacobo; Jacobus], of Scarperia, Tuscany. Translated Ptolemy into Latin, *c*.1409. His text was used as the basis of several printed editions from 1475 onwards.

Angelucci, Giovanni Battista. *Carta topografica Scandoglia*, 1788.

Angelus, Jacopo d'. See **Angelo**, Jacopo d'.

Angers, Jean-Baptiste (*fl*.1726-1732). 'Arpenteur du Roy', land surveyor worked in North America with Chaussegros de Léry. *Carte du Lac Champlain...*, 1732, published 1748.

Angeville, A. d' (1796-1856). Published thematic maps of France relating to morality and the incidence of crime. *Essai sur la statistique de la population française...*, Paris 1836. **Ref.** SMITH, David *The Map Collector* 30 (1985), pp.4-5; ROBINSON, A.H. *Early thematic mapping in the history of cartography* (1982), pp.162-164.

Anghiera, Peter Martyr d'. See **Martyr**, Peter.

Angieli, —. Publisher of Padua, for G.A. Rizzi-Zannoni, 1780.

Anginieur, Fernand (1868-1914). French traveller and explorer in Central Asia. *En Asie centrale. Turkestan, Thibet, Cachemire*, 1903, 1904.

Angiolini, Giorgio. Draughtsman. *Pianta topografia Veduta Generale della Città di Firenze*, 1826, 1836, 1837, 1840.

Angiolo, Jacopo d'. See **Angelos**, Jacopo d'.

Anglois, N. See **Langlois**.

Angove, W.H. *Western Australia* (4 sheets), 1887.

Anguissola, Leander (1652-1730). Italian cartographer and military engineer in the Imperial Austrian Army. Took part in relief of Buda 1683-1684. Danube, 1681; plan Vienna, 1683 (with J. Marinoni), engraved by Pfeffel and Engelbrecht, 1706; surveys of Hungarian forts to 1689.

Anguissola, M.A. *Pianta di Widino nova presente*, 1737.

Anhalt, Gisbertus (*fl*.1690s). Engineer, surveyor and schoolmaster of Emden, Germany. *Fondamentale onderwysinge in de Geometriae*, 1694.

Anhaltin, Christiaan Martini (*fl*.1640s). Bookseller of Amsterdam 'op de Zeedijck naest de Witte Roos'. *De lichtende fakkel of Schatkamer van de kunst der Stuyrluyden*, 1649.

Anias, Abraham. Examiner of the Chamber of Zeeland. Manuscript chart of the wrecks of the *Anna Catharina* and *'t Vliegend Hart* made for an enquiry, 1735. **Ref.** SCHILDER, Günter *The Map Collector* 27 (1984), pp.2-6.

Anich, José, *SJ*. Amazon, 1769.

Anich, Peter (1723-1766). Austrian charcoal burner of Oberperfusz, Tyrol, became cartographer, globe and instrument maker. Sundial, 1745; celestial globe, 1756; terrestrial globe, 1759; gores for *Deciphering the*

Chronogram, 1757; survey of the South Tyrol, 1759; and with Blasius Hueber *Atlas Tyrolensis* (21 sheets), Vienna 1774, reduction on 9 sheets, published Paris 1808. **Ref.** KRETSCHMER, DÖRFLINGER & WAWRIK *Lexicon zur Geschichte der Kartographie* (1986) Vol.C/1 pp.17-18 [in German]; BONACKER, W. 'On the track of the Tyrolean peasant cartographer Peter Anich', *Imago Mundi* 21 (1967) pp.111-112.

Anito, Nicolo. *Pianta...Città di Palermo*, 1777.

Ankarkkrona, —. *Plan de la Baie de Coquimbo*, 1858.

Anker, M.U. Madagaskar, 1875.

Anna, Alessandro d'. See **D'Anna**, Alessandro.

Annamaet. See **Anemaat**, family.

Annin, William B. (*d*.1829). American engraver of Boston. Engraver for *Map of the District of Maine*, 1815; Jacob Abbot Cummings's *School Atlas*, Boston 1815; State of Maine, 1822; engraved globes for Josiah Loring and Gilman Joslin, 1830s.

• **Annin & Smith** [Annin, Smith & Co.] (*fl*.1820-1831). Engravers of Boston, founded 1820. In 1828, under the name Senefelder Lithographic Company, William Annin and George Smith set up a printing operation using the relatively new process of lithography. The company was taken over by Pendleton's in 1831. Plan of Salem, 1820; amended maps for 1820 edition of *School Atlas to Adams' Geography* (first published 1814); *United States*, published in *Modern Atlas adapted to Morse's New School Geography*, Boston, Richardson & Lord 1822 and later atlases; James Hayward's *Plan of a Survey for the proposed Boston and Providence Rail-Way*, Boston, 1828 (one of Annin & Smith's first lithographed maps); Newton, 1831. **Ref.** RISTOW, Walter *The Map Collector* 21 (1982), pp.29-31.

Anquetil-Duperron, Abraham Hyacinthe (1731-1805). Orientalist, philologist and traveller. Wrote on history and geography of India and China.

Anquier, —. French engineer and draughtsman, contributed to the large scale Cassini *Carte de France*, 1765-1769 including: Valence, 1768 (engraved by Dupain-Triel and Bourgoin); Luçon, 1769 (with Saussaye de Blamecourt, engraved by Aldring); La Rochelle, 1769 (with others, engraved by Aldring).

Ansart, Felix. *Atlas historique des Etats Européens*, 1834, 1836; Amérique, 1838.

Anse [Anze], Luggert [Luggardus] van (*fl*.1690-1716). Engraver for the Visscher and Van Keulen firms at Amsterdam 'in de Langestraet' (1702-1716). *Tabula nova provinciae Ultrajectinae*, for Nicolaes Visscher, *c*.1696; *Nieuwe Kaart van de XVII Provincien in Neerland*, 1700; *'t Hooge Heemraedschap van Delflant* (25 sheets), for Nicolaas and Jacob Cruquius 1712. **Ref.** DONKERSLOOT-de VRIJ *Topografische karten van Nederland vóór 1750* (1981) N°806 p.159 [in Dutch].

Ansel, —. Map of Rochester, Massachusetts 1856.

Anselin, —. Printer of Paris. Odet Julien Le Boucher's Atlas, 1830; Laurent Gouvion Saint-Cyr's *Atlas à l'histoire militaire*, 1831; and Antoine Henri Jomini's *Atlas portatif*, 1840.

Anselm, C. *Telegraphenkarte von Baden*, 1870.

Anselmier, father and son.

• **Anselmier**, Claude-Henri-Jules (1815-1895). French engineer, cartographer and topographer. Worked in Geneva and Zug.

•• **Anselmier**, Gaston (1840-1909). Son of Claude-Henri-Jules, born in Geneva. Worked as an engineer and topographer in Einsiedeln and Bern. Stadt Bern, 1868.

Ansoleaga, Florencio. Pacific Ocean, 1743.

Anson, —. Engraver for Carl Ferdinand Weiland's *Europa*, 1849.

Anson, *Captain* George (1697-1762), later Admiral of the Fleet. Circumnavigator in 1739-44. *A Voyage round the World...*, 1744, published 1748, also many later editions and translations.

Anstatt, —. Environs of Berlin (10 sheets), 1816-1819.

Ansted, David Thomas (1814-1880). Geologist. *Atlas of Physical and Historical Geography*, 1852-1859.

Anstey, *Captain* Thomas Henry. *Battlefield of Isandhlwana*, 1881; *...the country around Isandhlwana*, 1881.

Anthoine, Edouard (1848-1919). *Environs de Paris*, 1881; *L'Atlas de géographie moderne*, Paris, Hachette *c.*1889, 1891, 1907 (with F. Schrader and F. Prudent); *La nouvelle France, paix de juin 1919*, Paris, Hachette 1920 (with Franz Schrader and Ferdinand Prudent).

Anthonisz. [Antoniszoon; Anthonij; Thonisz], Adriaan (1541-1620). Dutch land surveyor and mapmaker from Alkmaar, died Amsterdam. *Plattegrond van Alcmaer*, 1597 (engraved Cornelis Drebbel). **N.B.** Adriaan had three sons: Dirk Adriaensz. Schelven; Adriaan Metius (also known as Adriaan Adriansz. Schelven); and Antonius Metius (also known as Anthonie Adriaansz.), which has led to much confusion. **Ref.** DONKERSLOOT-de VRIJ *Topografische karten van Nederland vóór 1750* (1981) p.189 [in Dutch]; SCHILDER, Günter *The Map Collector* 17 (1981), pp.6-7.

Anthonisz. [Anthoniszoon; Antoniades; Teunisse; Thonisz.], Cornelis (*c.*1507-between 1553-1557). Dutch cartographer, painter, engraver and woodcutter, established himself in Amsterdam in 1533. Bird's-eye view of Amsterdam (pen and ink), *c.*1532, woodcut version by Jan Ewoutszoon (12 sheets), 1544, and later (used by Braun and Hogenberg *q.v.*); *De stadt van Aelgier*, 1542; *Caerte van Oostlant*, 1543 (basis of Abraham Ortelius's *Daniae regni typus*, 1570); pilot book *Onderwijsinge vander zee*, 1544; Weesp, 1551; Therouanne, 1553. **Ref.** KARROW, Robert W. *Mapmakers of the sixteenth century and their maps* (1993) pp.42-48; MEURER, P. *Fontes cartographici Orteliani...* (1991) p.104 [in German]; LANG, A.W. *Die 'Caerte van Oostland' des Cornelis Anthonisz., 1543* (Hamburg 1986) [in German].

Anthony, Daniel. American surveyor. *A Map of the Town of Providence*, 1803

Antillon y Marzo, Isidoro de (1778-1820). Spanish geographer. North America, Madrid 1803; Atlas, 1804.

Antipov, —. Stratigraphic map Donets Coalfield, St Petersburg 1872.

Antoine, Louis. Printer of Paris, 70 rue des Noyers. For Eugène Andriveau-Goujon, 1854-1874; for A.H. Dufour, 1864; *Grand atlas universel*, 1870.

Antoine, Pierre. *Arrondissement de St-Dié*, 1872; *Gerardmer et Bruyères* [Vosges], *c.*1876.

Antoli, Jacob (*fl.*12th-13th century). Jewish astronomer and geographer of Provence.

Anton, Manuel. Spanish Navy. Mindanao Harbour, 1875.

Antonelli, Giuseppe. Publisher of Venice. *Dizionario geografico*, 1833; America, 1834; Oceania, 1839.

Antonelli, Juan Battista. Map Mexico, 1590.

Antoniades. See **Anthonisz.**, Cornelis.

Antonine Itinerary. Roman land and sea itinerary (3rd century AD), perhaps originally designed for journeys of Caracalla (M. Aurelius Antoninus, AD 188-217) and re-used for troop movements.

Antonisz, Adriaan. See **Anthonisz.**, Adriaan.

Antonius [Antonio], Johannes [Giovanni]. Publisher. *Romae Descriptio*, 1600.

Antony, Carl. *Sanitäts-Carte des Königreiches Böhmen*, 1881.

Antropov, P.A. Financial-Statistical Atlas of Russia, 1898.

Anuchin [Anoutchine], D. Russian geographer. Directed geological review *Zemliebiedmie*.

Anville, H.F.B. d'. See **Gravelot**.

Anville, Jean Baptiste Bourguignon d' (1697-1782). Royal geographer and cartographer, 'Aux galéries du Louvre', Paris. Built up a collection of rare and contemporary maps, now in the Bibliothèque nationale. Elected to Académie des Sciences 1773. Compiled over 200 maps. France, 1719; *Carte de l'Ethiopie Orientale*, 1728; *Royaumes d'Angola de Matamba et de Benguela*, 1731; *Royaume de Congo*, 1731; *Nouvel atlas de la Chine*, 1737; *Atlas generale* from 1737; *Amérique*

Jean Baptiste Bourguignon d'Anville (1697-1782). (Taken from Map Collectors' Circle*)*

Peter (Petrus) Apian (1495-1552). (Taken from Map Collectors' Circle*)*

Septentrionale, 1746; Amérique Méridionale (3 sheets), 1748; *Côte Occidentale de l'Afrique*, 1751; *Carte d'Asie*, 1751; *Mer Caspienne*, 1754; *Louisiane et Terres Angloises (4 sheets)*, 1755; World, 1761; *Arabique ou Mer Rouge*, 1765; *Géographie Ancienne et abrégée*, 1775 (published by Laurie & Whittle as *A Complete body of Ancient Geography*, 1795 and later editions); etc. **Ref.** KOEMAN, C. *Atlantes Neerlandici* (1967-1971) Vol.I pp.48-49; CORDIER, H. 'Du Halde et d'Anville: Cartes de la Chine' in: *Recueil de memoires orientaux par les professeurs de l'ecole des langues orientales* (Paris 1905) pp.389-400 [in French].

ANWB [Algemeene Nederlandsche Wielrijdersbond]. *Wielrijders Atlas van Nederland* (Atlas of the Netherlands for cyclists), many editions from 1884, later published as a loose leaf atlas. **Ref.** KOEMAN C. *Atlantes Neerlandici* Vol.VI (1985) pp.67-70.

Anxerinus, C. See **Buondelmonte**.

Anze, Luggardus van. See **Anse**, Luggert van.

Anzenmaru Saikichi. Japanese chartmaker. *Dainihon Kairo Zu* [chart of the eastern sea], published by Suharaya Mohei and others 1842.

Aodo Denzen [Nagata, Zenkichi] (1748-1822). Japanese artist and copperplate engraver. Engraved a world map and a map of Japan for Kageyasu Takahashi, *c*.1816.

Aou Tokei (1764-1838). Japanese geographer. *Kokugun zenzu* [General atlas of the provinces of Japan] (2 volumes), 1828.

Apáczai Csere, János (1625-1659). Hungarian polymath, born Apácza [now Apata, Transylvania, Romania], died Kolozsvár [now Cluj, Transylvania, Romania]. Studied at Gyulafehérvár [Alba-Iulia, Romania] and Utrecht. On his return to Transylvania, he taught at Kolozsvár [Cluj] and Gyulafehérvár. Best known for his Encyclopaedia, 1653, which discusses geodesy and triangulation.

Aparici i Mercadal, Josep (1653-1731). Catalan cartographer. *Nueva descripción geographica del principado de Cataluña*, 1720, 2nd edition 1769.

Apens. See **Appeus**, Cornelis.

Apeus. See **Appeus**, Cornelis.

Apgar, E.A. Maps for *A Brief Course in Geography*, Philadelphia, Cowperthwait & Co. 1874.

Apian [Apianus; Bennewitz; Bienewitz; Panewitz], family.

• **Apian**, Peter [Petrus] (1495-1552). Born Leising, Saxony, and died at Ingolstadt. Geographer and Astronomer Royal at Ingolstadt; in 1528 set up printing press with his brothers Georg (below) and Martin. *Tipus orbis universalis*, woodcut world map on truncated cordiform projection, in Caius Julius Solinus's *Polyhistor*, Vienna 1520, also found in Pomponius Mela's *De situ orbis*, Basel 1522; Liber cosmographicus, Antwerp 1524, and for eighty years thereafter; engraved and printed Lázár, Ziegler and Collimitius's woodcut map of Hungary *Tabula Hungariae*, 1528; *Tabula orbis cogniti* [heart shaped world map], 1530; Sebastian Rotenham's *Das Francken Landt...*, 1533. **Ref.** KARROW, R. *Mapmakers of the sixteenth century and their maps* (1993) pp.49-63; MEURER, P. *Fontes cartographici Orteliani...* (1991) pp.105-106 [in German]; SHIRLEY, R. *The mapping of the World* (1983); ORTROY, Fernand G. van *Bibliographie de l'oeuvre de Pierre Apian* [in French]; STETTER, G. and A. FAUSER. (Introduction to facsimile edition of *Bairische Landtaflen*, 1568) (Munich 1979); WOODWARD, David *Imago Mundi* 24 (1970), pp.43-48; KISH, George *Imago Mundi* 19 (1965) pp.13-21.

•• **Apian** [Apianus], Philipp (1531-1589). German cartographer, mathematician and physician, son of Peter Apian, born Ingolstadt, and died at Tübingen. Succeeded his father as mathematician at Ingolstadt university in 1552. Known for his great survey of Bavaria on which he worked 1554-1561, aided by his brother Thimoteus and the painter Bartholomaus Refinger, *Ein neue Beschreibung des Fürstenthums Ober- und Nieder Bairn*, 1563 (original destroyed 1782, now known only from 40-sheet copy of 1756 by Franz Xaver Pitsch, itself destroyed in the Second World War); *Bairische Landtaflen XXIIII... Zu Ingolstadt MDLXVIII.* (24 sheets), 1568 engraved by Jost Amman, (another edition engraved by Peter Weiner, Munich 1579); Terrestrial globe, 1576; later printed and published his own works using his father's old press. **Ref.** KARROW, R.

Mapmakers of the sixteenth century and their maps (1993) pp. 64-70; MEURER, P. *Fontes cartographici Orteliani...* (1991) pp.106-107 (with extensive further references) [in German].

•• **Apian,** Thimoteus (*fl.*1554-1561). Son of Peter (above); assisted his brother Philipp (above) in the survey of Bavaria.

• **Apian,** Georg. Brother of Peter (above) and Martin, with whom he established the Apian printing house at 918 Haderstrasse, Ingolstadt *c.*1526. While working for Johann Weyssenburger in Landshut he cut the woodblock for Johannes Aventinus's *Obern und Nidern Bairn...*, 1523 and later version.

• **Apian,** Martin. Worked with Peter and Georg.

Apollonio, Levinio (*fl.*16th century). Born Bruges, died in the Canary Islands. *De Peruviae regionis inter Novi Orbis Provincias*, Antwerp, Jean Bellère 1567, German edition, Basle, Sebastian Henricpetri, 1583.

Appé, —. Département de l'Indre 1862.

Appeus [Apens; Apeus], Cornelis (*c.*1634/1635-*c.*1688). Publisher and mapmaker of Friesland, worked in Leeuwarden and Amsterdam. Engraved Frederik and Wilhelm Coenders van Helpen's *Prov. Groningae et Omlandiae tabula* (4 sheets), *c.*1677-1678.

Applegate, J. A Chart of the Bahama Islands, London 1739.

Applegate, H.S. & J. Publishers in Cincinnati, Ohio. George Conclin's River Guide, 1850.

Appleton, A. Map of Orfordness, 1568 MS.

Appleton, Daniel & Co. Printers and publishers of 346 & 348 Broadway (1865), 443 & 445 Broadway (1859), New York. Published a series of travel guides which included maps by W. Williams and others; also geographies by S.S. Cornell which included some maps. *Cornell's Companion Atlas to Cornell's High School Geography*, 1856, 1859, 1864; *Morse's General Atlas of the World*, 1856; *Railway map of the United States and the Canadas*, 1865; Atlas of the United States, New York 1888; Library Atlas, 1892, 1895; *The handy atlas of modern geography*, 1892.

Appleton, J.W. Bay of Bengal, 1876; *Windward Passages and Gulf of Mexico*, 1876; *Coast of England and Holland*, 1877; Mediterranean, 1877; *North Pacific Ocean*, 1878; *North Sea Fishing Banksevillette*, 1878; North America, 1880; *Baltic Sea* (6 sheets), 1882; Adriatic, 1884; charts for Charles Wilson.

Après de Mannevillette, Jean Baptiste Nicolas Denis d' (1707-1780). French hydrographer, navigational mathematician and nautical cartographer, native of Le Havre. 'Chevalier de l'ordre du roi, Capitain des vaisseaux de la Compagnie des Indes, Correspondant de l'Académie royale des Sciences, associat de l'Académie royale de la Marine'. Visited the Caribbean aboard *Le Marechal d'Estrées* in 1726. *Le Neptune Oriental ou routier general des côtes des Indes Orientales...* (22 charts), Paris, Jean-François Robustel 1745 and later editions; *Supplément au Neptune Oriental* (15 charts), Paris, Demonville 1781 and later. **Ref.** *Dictionnaire de biographie française*, Tome 3ème, 1939 [in French].

Arabia, Luigi (*b.*1823). Geographical engineer in the Officio Topografico di Napoli, and the Istituto Topografico Militare. Base surveys of Foggia, 1860-1864 (with Federico Schiavoni); surveys of the area around Naples, 1862; geodetic surveys of the area around Catania, 1865. **Ref.** VALERIO, V. *Società uomini e istituzioni cartografiche nel mezzogiorno d'Italia* (1993) p.442 [in Italian].

Aragon, Anne Alexandrine (*b.*1798). Atlas, 1843.

Aragon y Abollado, *Don* Yldefonso de. Maps and plans for Deposito Topografico de Manila. *Plano Chorografico de la Provincia de Pangasinan... Ena la Ysla de Luzon...* [Philippines], 1821 (engraved by Francisco de Herrera); *Plano Esferico de la Provincias del Norte y Sur de Ylocos...* [Philippines], 1821 (engraved by Juan de Sevilla).

Aragona, S. di. See **Sebastiano** di Aragona.

Arai Ikunosuke (1835-1909). Japanese surveyor. As director of the Office of Survey, Department of the Interior, established the trigonometrical survey of Japan *c.*1870s. *Hokkaido jissokuzu* [Map of Hokkaido], 1875 (surveyed by American engineer M.S. Dee and others).

Arai Yahei. Japanese publisher. One of the publishers of Oshima Takeyoshi's map for *Yamashiro Meishōshi* [A geography of Yamashiro Province], 1711.

Arana, Barros (*b*.1830). Chilean geographer and historian.

Arandia y Santistevan, Pedro Mañuel de. Plano Puerto Sisiran, 1757.

Arany, J. Surveyor. Hamburg Park, 1762.

Arason, Magnus (*c*.1680-1728). Icelandic map and chartmaker of Hági Bardaströnd. District maps of Iceland, 1721-1728.

Aratus of Soli [Aratos] (*c*.271-240/239 BC). Greek astronomer and poet from Cilicia. His *Phaenomena*, three times translated into Latin, described fixed stars, circles of celestial sphere, and weather signs.

Araujo, A.J. *Carta da Bahia*, 1837; *Rio de Janeiro*, 1858.

Arbousset, Thomas (1810-1877). French protestant missionary and explorer in southern Africa. *Relation d'un voyage d'exploration au nord-est de la colonie du Cap de Bonne-Espérance...*, 1836, published 1841; *Tahiti et les îles adjacentes*, 1867.

Arc, A. d' See **Gauttier** D'Arc.

Arce, Arinando da. *Mapa del Rio Mississipi*, 1699.

Arch, John. Publisher and bookseller. Cole & Roper's *The British Atlas*, 1810.

Archdale, John (*d*.1717). Governor of Carolina 1694-1695. *Papers relating to ye Province of Carolina with a Draught of ye Town* [Charles Towne] *Mapps of ye Forts, Rivers, Coasts &c.*, 1695 MS.

Archdall, Mervyn. *A Map of Ireland...*, 1786 (published in Archdall's *Monasticun Hibernicum or an History of the Abbies*, Dublin, Luke & White 1786).

Archdeacon, William E., *RN* (*fl*.1850s). Admiralty chart surveyor, he was involved in surveys of the coast around the Cape of Good Hope for the Hydrographic Office. *Table Bay to Donkin Bay*, Admiralty 1869; *Donkin Bay to Milkbosch Point*, 1869-1870, published for the Admiralty 1879; *Table Bay breakwater and Docks*, Admiralty 1870; *Saldanha Bay*, Admiralty 1870.

Archer, J. Bookseller of Dublin. *Plan of Dublin*, 1797.

Archer, J. Engraver for Robert Wilkinson's General Atlas, 1802.

Archer, *Captain* James. *Plan of the Greate Antiente and famous Citty of York*, *c*.1643 MS.

Archer, James, *junior*. Map of Alsatia, 1676 (for Moses Pitt's Atlas).

Archer, James. Portsea Island, 1773.

Archer, James. Engraver, for T.G. Bradford's *Illustrated Atlas*, 1838.

Archer, John Lee. Civil engineer, Sydney, Australia. Map of Stanley on the estate of Van Diemen's Land Company, Circular Head, 1843; plans of farms in Van Diemen's Land, 1843-1844.

Archer, Joshua. Draughtsman and engraver of Pentonville, London. County maps, 1833; Wales, 1833; *New Plan of London*, 1833-1837; *Cities of London and Westminster*, 1841; ecclesiastical maps of England and Wales and the Dioceses, published in *British Magazine* 1841-1843; English county maps for James Gilbert, later incorporated into *Fisher's County Atlas of England and Wales*, 1842-1845; *Colonial Church Atlas*, 1842-1850; maps of London, 1847-1861; *Ancient Atlas*, 1847; engraver of Thomas Dugdale's *Curiosities of Great Britain*, Tallis 1848; *The Emigrants' Map of the World*, *c*.1850; maps for William Hughes's Atlas, 1866.

Archer, Mark. *Map of the Principal Collieries in the County of Durham...*, 1890.

Archer, Thomas (*c*.1694-after 1747). Estate surveyor in Essex and Norfolk. Leyton, 1721 MS; Walthamstow, 1725 MS.

Archer, Thomas. *Presentation Map of Egypt* (2 sheets), 1885.

Archimedes of Syracuse (*c*.287-212 BC). Greek inventor and mathematician, who amongst many other achievements established mathematical formulae for the volumes of spheres and other solid figures. He was killed by a Roman soldier at the siege of Syracuse. Wrote *Sphairopoiia*, (now lost), in which he described the construction of a type of armillary sphere. He is also said to have constructed a celestial globe.

Arciszewski, Krzyzstof (1592-1656). Polish general in service Dutch West India Co. in Brazil.

Arçon, Jean-Claude Eléonore Le Michaud d' (1733-1800). 'Ingénieur ordinaire du roi', 1755. Mapped the frontier areas of the Jura and the Rhine Valley. *Plan de Gibralter*, Strasbourg, Weiss 1782; ...*Considerations militaires et politiques sur les fortifications*, 1795.

Arctowski, *Doctor* H. Chart Falkland Islands, 1898 MS.

Arculf. Saxon pilgrim who visited Jerusalem. Plan of the city of Jerusalem, AD 670 MS, later manuscript copies.

Ardagh, *Major General Sir* John Charles. *Map of the Geyser* [Iceland], 1867; Egypt, 1883; *Atlas of the boundary Guiana-Venezuela*, 1898.

Ardeiser, Johann. See **Ardüser**, Johannes.

Arden, Edward (*fl*.1796-1841). Estate surveyor. Produced manuscript surveys of Huntingdonshire parishes including: ...*the Parish of Alwalton...*, 1809; *Stibbington cum Wansford*, 1810; *Woodstone*, 1811.

Arderne, R. *Beveridge Island* [Murray River, South Australia], 1875.

Ardill, J. Roche. Surveyor, Goulburn, New South Wales. Plan of 'Garrorigong', the property of J.A. Broughton, 1859 MS; plan of Braidwood [New South Wales], 1859 MS.

Ardüser [Ardeiser], *Captain* Johannes (1585-1665). Fortifications engineer from Davos, Switzerland. Engineer to the town of Zurich. Vallistelina, 1620, 1625. **Ref.** REINLE, A. *Italienische und deutsche Architekturzeichnungen. 16. und 17. Jahrhundart* (Basle 1994) pp.165-186 [in German].

Arebault, J. See **Aerlebout**, Josias.

Arelebout, J. See **Aerlebout**, Josias.

Arenales, J. Valdivia, 1795; Carta general Patagonia, 1845 MS.

Arendts, *Professor Doktor* Carl (1815-1891). Austrian geographer and professor, founded Geographischer Gesellschaft München. Edited *Vollständiger Hand-Atlas...*, Regensburg 1858; *Preussen*, 1862; *Geographischer Schul-Atlas*, Regensburg 1868 and later editions; *Post und Reise-Karte von Mittel Europa*, 1869; *Schul-Atlas*, 1871; *Hydrographische Karte von Süd-Deutschland*, 1871; *Italien*, *c*.1876; *Bayern, Württemberg & Baden*, *c*.1877; *Frankreich*, *c*.1877; *Plan von Peking*, 1890.

Arendts, W. *Elsass & Deutsch-Lothringen*, 1871.

Arenhold, Gerard Justus. Hildesheim, 1727; maps for Homann's Heirs, 1753.

Arentsz., Symon. See **Bruningen**, S.A. van.

Aretin, *Baron* Johann Christoph Anton Maria von. *Baierns grösster Umfang unter den Agibolfingern, Carolingern, Welfen in Wittelsbachern in 4 geographischen Karten...*, 1809.

Aretin [Aretinus; Aretin von Ehrenfeld], Paul [Paulus] (*fl*.1608-1627). Bohemian surveyor of Klattaw, worked in Prague 1608-1627. *Regni Bohemiae nova et exacta descriptio*, 1619 and later (copied by W.P. Zimmerman); *Moraviae nova et post omnes priores...*, Amsterdam, 1627 (after Comenius).

Arfeville [Arfreville], *sieur de*. See **Nicolay**.

Arfwedson, C.D. *De colonia Nova Suecia* [map of Delaware Bay], Uppsala 1825.

Argaria, Casparo (*fl*.1538-67). Italian cartographer. Naples, 1538; Città di Messina, 1567.

Argelander, Friedrich Wilhelm August (1799-1875). Astronomer. *Uranometria Nova* (18 charts), 1843; *Atlas des Nordlichen Gestirnten Himmels*, 1863.

Argensola, Bartolomé Juan Leonardo de. See **Leonardo** y Argensola, Bartolomé Juan.

Argenson, M. d'. *Description de la ville et faubourgs de Paris*, Paris, Jean de La Caille 1714.

Argento, Carmelo (*fl.*late 18th century). Engraver of Sicily. Engraved charts in Fileti's *Carta Pianta del Mare Mediterraneo*, 1802.

Argentré, father and son.
• **Argentré**, Bertrand d' (1519-1590). Lawyer and chief magistrate of Rennes, born at Vitré [near Rennes], died at Château Tizé, Césson-Sevigné. 'Description du pays Armorique a present Bretaigne', in: *L'histoire de Bretaigne*, Paris, Jacques du Puys 1588 (used as basis for Abraham Ortelius's *Britanniae et Normandiae typus* 1595). **Ref.** MEURER, P. *Fontes cartographici Orteliani...* (1991) pp.107-108 [in German].
•• **Argentré**, Charles d' (*c*.1550-1625). Worked with his father Bertrand on *L'Histoire de Bretaigne*, negotiating publication of later editions. Prepared plan of *Rennes*, 1616 (engraved by G. Closche), published in 3rd edition of *L'Histoire*.... **Ref.** JONES, Ieuan E. *D'Argentré's history of Brittany and its maps* (University of Birmingham 1987).

Argoli, Andrea (*fl.* mid-17th century). Italian astronomer. *Pandosion Sphaericum*, Padua 1644, 1653.

Argonauti, Accademia cosmografica degli. See **Accademia** cosmografica degli Argonauti.

Arias Montanus, Benito [Benedictus] (1527-1598). Spanish philosopher and theologian of Frejenal de la Sierra, Estremadura. Edited Polyglot Bible (with maps of the World and Palestine), Antwerp, Christopher Plantin 1572. **Ref.** MEURER, P. *Fontes cartographici Orteliani...* (1991) pp.108-109 [in German]; SHIRLEY, R. *The mapping of the World* (1983) N° 125, p.147.

Arigoni[o], *Fra* Bono (*fl.*16th century). Venetian cartographer.

Arima Kisōta (1708-1769). Japanese cartographer, worked for the feudal lord of Hagi [now Yamaguchi Prefecture]. *Bōchu jige jōshin ezu* [Surveyed map of Suō and Nagato province], 1727-1753.

Arimaya Kyūbei. Japanese publisher. Woodblock map of the Nasu-Yumoto Hot Springs, [Edo Period].

Arisawa Nagasada.Japanese cartographer and strategist. Manuscript plan of Tsuwano Castle, Iwami Province, 1692.

Arista, *General* Manuelo. Movements United States Army in Mexico, 1846.

Aristagoras of Miletus (*fl.*495 BC). Greek administrator who went to Sparta (499-498 BC) to get help against the Persians, taking with him a bronze tablet engraved with a map of the World (described by Herodotus). First recorded example of a map being used for political purposes. **Ref.** DILKE, O.A.W. *Greek and Roman maps* (1985) pp.23-24.

Aristequi, E. de. *Croquis del Teatro de la Guerra Campaña del Norte*, 1874.

Aristotile, Nicolo d'. 16th-century Venetian printer and publisher. First edition of Benedetto Bordone's *Isolario*, 1528.

Aristotle of Stagira (384-322 BC). Greek encyclopaedic writer and tutor to Alexander the Great. In *Meteorologia* postulated 12 winds at 30° intervals (11 were named); either drew or intended to draw a wind-map (corrupt example known in MS in Madrid). Much translated into Latin and revered during Middle Ages in Europe.

Arkell, James (*b.*1829). Englishman, settled in New York State. Obtained various patents for globes, *c.*1879

Arlett, *Lieutenant* William, *RN* (*fl.*1844-1851). Hydrographer for the Admiralty. Survey of the West Coast of Africa, 1831-1834; *Costa Occidental de Africa*, 1857; *Gran Canaria*, 1834-1862.

Arluc, —. *Plan... de Cannes*, [1895].

Armani, Basilio. *Innsbrucke*, *c.*1870.

Armani, Domenico. Engraver. *Territorio Vicenza*, 1775.

Armann, —. Lithographer and publisher. *Plan Monumental von Cassel*, *c.*1869; *Plan Monumental von Wilhelmshoehe*, *c.*1870.

Armas, Duarte de (*fl.*1509-1510). Manuscript maps and views of Portuguese fortifications in D. Darmas's *Livro das Fortalezas do reino*, *c.*1509-1516 (two different copies exist, one with 71 maps [in Madrid] and another with 114 maps [in Lisbon]).

Armato. See **Bellarmati**, Girolamo.

Armendale, *sieur d'*. See **Damme**.

Armentaria, San Juan N. Bolivia, La Paz 1888.

Armitage Benjamin (*fl.*1770-1778). Estate surveyor in Essex and Yorkshire. Latton, 1778 MS.

Armour & Ramsay. Publishers in Montreal. Edward Staveley's *Canada*, 1846.

Armroyd, George. *A Connected View of the Whole Internal Navigation of the United States*, Philadelphia, Carey & Lea 1826.

Arms, C.J., *senior* and *junior*. Engravers. Joseph A. Caldwell's *Atlas of Harrison County, Ohio*, 1875.

Arms, Walter F. *Historical Atlas of Adams County, Ohio*, 1880.

Armsheim, Johannes. See **Schnitzer**, Johannes.

Armstrong, father and son. Surveyors and publishers of large-scale maps of the British counties. Originally based in the Scottish borders; in 1775 Andrew and Mostyn John moved to Norfolk where they continued their mapmaking.

• **Armstrong**, *Lieutenant* (later *Captain*) Andrew (1700-1794). Surveyor. *The County Palatine of Durham* (4 sheets), engraved and published T. Jefferys, 1768, 1791; *A Map of the County of Northumberland* (9 sheets), London, T. Kitchin 1769; *A New Map of Scotland...*, 1775; Pocket Map of Scotland, 1775; *Map of Lincoln-Shire* (8 sheets), 1779 (engraved by S. Pyle); Rutland, 1781 (engraved by J. Luffman). Jointly with his son Mostyn; *Map of the County of Berwick* (4 sheets), 1771 (engraved by A. Bell); *Map of the Three Lothians* (6 sheets), 1773 (engraved by T. Kitchin); *A new map of Ayrshire* (6 sheets), 1774, 1775 (engraved by A. Baillie); *Map of the County of Peebles or Tweedale* (2 sheets), 1775 (engraved by S. Pyle). **N.B.** See also Mostyn John, below.

•• **Armstrong**, Mostyn John (*fl.*1769-1791). Surveyor and publisher. Son of Andrew above. *Thirty Miles round Boston*, 1775; *An actual survey of the great Post-Roads between London and Edinburgh*, 1776; *Armstrong's actual survey of the Great Post Road between London and Dover*, 1777; *A Scotch Atlas...* (30 maps), 1777, 1787 and 1794; *Encampment of...the Norfolk Regt...*, 1779; corrected Nathaniel Kinderley's *Corporation of the Bedford Level. A Map of the Great Level of the Fens*, 1781. **Ref.** HODSON, D. *County atlases of the British Isles published after 1703* (1997) pp.96-100.

Armstrong, Arthur C. *Short spins round London*, 1903, more comprehensive edition 1906.

Armstrong, George. *Palestine* (21 sheets), 1890; *Raised Map of Palestine*, 1893; *Photo-Relief Map...of Palestine*, 1902; *Photo-Relief Map of Palestine*, 1921

Armstrong, J. Essex estate surveyor. Marks Tey Essex, 1809 MS.

Armstrong, J.B. Civil engineer. *Grand Island City* [Michigan], 1857.

Armstrong, *Colonel* John. Navigation King's Lynn-Cambridge (7 maps), 1725.

Armstrong, John. Surveyor-General of Ordinance. *A Plan of St Philips Castle...Minorca*, 1756.

Armstrong, John (1755-1816). American military cartographer. *Surveys of Road from Morristown, New Jersey, to Sussex Court House*, 1780 (for Robert Erskine); Illinois River, MS; Wabash River, 1790.

Armstrong, John. Surveyor of Sydney, New South Wales, based at Jamieson St (1833); George St (1842); 7 Macquarie St (1847); 114 Macquarie St (1855-1867). *The Bellambi estate Illawarra*, 1840s.

Armstrong, Marcus. *A New Map of Scotland or North Britain* (4 sheets), 1782.

Armstrong & Cia. *Plano de Mexico, 1890.*

Arnaud [Arnault], Henri (*fl.*1432-1466). Astrologer and astronomer of Zwolle. Astronomer to Charles VII of France, mathematician and physician to the duc de Bourgogne, died Dijon. Constructed a map of the Saône River, *c.*1446-1447.

Arnaud [Arnaud-Bey], Joseph Pons d' (1812-1884). French engineer, worked in Egypt and Sudan, contributed to numerous maps of the Nile valley, 1842-1881; surveys presented to the Société de Géographie in 1881. **Ref.** BROC, N. *Dictionnaire illustré des explorateurs Français du XIXe siècle: I Afrique* (1988) pp.4-5 [in French].

Arnaud, Th.J. (*fl.*1834-1884). French archaeologist and explorer in North Africa, Arabia. Bahrel-Abiad, 1843; *Relation d'un voyage à Mareb dans l'Arabie méridionale entrepris en 1843*, 1845; *La mer Rouge*, 1849.

Arnault, Henri. See **Arnaud,** Henri.

Arnaves, T. Engraver. Laurent Brémond's Marseille, 1730 edition.

Arnd, F.H. *Europäischen Russland*, *c.*1877; Frankreich &c., *c.*1880.

Arnd, J.J. Publisher of Leipzig. W. Koch and C. Opitz's *Eisenbahn-und Verkehrsatlas von Europa* (26 sheets), 1894, 1897.

Arndt, C.M. *City of Vancouver*, 1908.

Arngrimur-Jónsson, —. (1568-1648). Icelandic historian and mapmaker. *Liber de Gronlandia*, MS.

Arnheim, E.H. Surveyor. *Plan of Goulburn* [Australia], 1859.

Arnim, Hans von. *Garnison-Karte der Deutschen Armee*, 1888.

Arnold, *Brigadier-General* Benedict (1741-1801). Born in Connecticut, died in London; American turncoat who joined the British in the American Revolution, associate of Major John André *q.v.*. *Passes into Canada*, 1778; *Attack on Fort Trumbull*, 1781; *Road from Sleepy Hole by which Colo. Simcoe marched to Portsmouth* [North America], 1781

Arnold, Charles R. *Map of Hancock County, Illinois*, Buffalo 1859 (with J.W. Holmes); *Map of Pike County, Illinois*, 1860 (with J.W. Holmes); *Map of Madison County, Illinois*, Buffalo 1861 (with J.W. Holmes); *Map of Greene County, Illinois*, Buffalo 1861; and others.

Arnold, E.G. *Topographical Map of the original District of Columbia...* [Washington DC], New York, G.W. Colton 1862.

Arnold, Edward J. British publisher. ...*Yearbook of Memory Maps*, 1889; *School Atlas*, 1919.

Arnold, O.E. *Feld & Flötz-Karte des Zwickauer Steinkohlen-Reviers*, 1877; *Zwickau mit Umgebung*, 1900.

Arnold, Richard (*fl.*1702-1740). Essex estate surveyor. Wanstead, 1740 MS.

Arnold, *Captain* Stuart Amos. British hydrographer. *Coasts of Kent, Essex and Suffolk* (4 sheets), 1777; *Irish Sea or St George's Channel*, 1783, 1796, 1800; *Lewis and West Coast of Scotland*, 1783, 1796, 1800; *The West Coast of Ireland*, 1783, 1796; *Coast of Holland*, 1784, 1795, 1800; *Flanders and Walcheren*, 1784; Mull of Kintyre to Cape Wrath, 1800.

Arnold, *Doctor* Thomas. Headmaster of Rugby School, England. Revised Henry Teesdale and Co.'s *General Atlas*, 1856.

Arnold-Forster, *Right Honourable* Hugh Oakeley. *Cassell's Map Building Series* (72 maps), *c.*1890; *London School Atlas* (48 maps), 1900, 1911.

Arnoldi [Fiamengo], Arnoldo di [Aernout; Aernoutszoon] (*d.*1602). Flemish cartographer, worked with Giovanni Antonio Magini in Bologna 1595-1600, and Matteo Florimi in Siena, 1600-1602. *Descrittione universale della terra...* (10 sheets), 1600 and later, (also reduced edition *Universale descrittione del mondo* (2 sheets), Siena 1601); *America...*, Siena *c.*1600. **Ref.** BURDEN, P. *The mapping of North America* (1996) p.165; SHIRLEY, R. *The mapping of the World* (1983) N[os.]227-228 pp.242-243.

Arnoldo, Johann Gerhard. *Tabulae geographicae*, Frankfurt am Main 1698.

Arnott, —. *Arnott's Cycling and Tourist Map* [Dublin], 1896.

Aaron Arrowsmith Senior (1750-1823). Engraving by T.A.Dean after H.W.Pickersgill. (Taken from Map Collectors' Circle*)*

Arnoul, —. Letter engraver of 'rue du Petit Pont n°·26', Paris. Worked for J.A.C. Buchon, 1825; A.M. Perrot, 1827; E. Andriveau-Goujon, 1841-1862.

Arnould, —. *Le Royaume de France*, Paris 1705 (engraved by J.B. Nolin).

Arnoullet, Balthazar d'. Publisher of Lyon. Plan de Paris, 1530; Chorographia Europae, 1553.

Arnswaldt, B. von. *Umgegend von Eisenach*, 1857.

Arnz, A. [& Co.] (*fl.*1825-1851). Lithographers of Düsseldorf and Leiden, at Amsterdam after 1851, also lithographic plates for works published by E.J. Brill of Leiden, *e.g.* Siebold's Japan, *etc.*. *Kleiner Handatlas*, 1822; *Atlas der Alten Welt* (16 coloured charts), Düsseldorf 1829; *Schul-Atlas von Deutschland*, 1833; *Taschenatlas über alle Theile der Erde*, Düsseldorf 1836; *Zakatlas voor scholen*, Leiden and Düsseldorf [1836]-1846; *Allgemeiner Schul- und Hand-Atlas über alle Theile der Erde*, 1837.

Arnz, J. Publisher of Düsseldorf. *Schulatlas*, 1842, 1845; *Atlas van de oud wereld*, 1845. **N.B.** See also previous entry.

Aron & Abraham. Maps & diagrams of Godthaab District [Greenland], 1860.

Arosander, Peter (*d.*1696). Surveyor of Uppland, Sweden.

Arosenius, Jakob Fredrick Neikter (1809-1890). Swedish cartographer and surveyor. Globe, 1850-1853.

Arppe, Wilhelm. Finland, 1872 (with M. Wijkberg).

Arquer, Sigismund. Sardinia for Sebastian Münster, 1544.

Arquiza, Jacob de. *Maps of provinces of Philippine Islands*, 1832; Las Almas, 1845.

Årre, Olof. *Charta öfwer Slussarne wid Carlsgraf och Åkerstråm*, 1770.

Arredondo, Antonio de. Spanish diplomat, soldier and military engineer, involved in discussions about the Spanish-English borderlands in Florida. Plan San Augustín, Florida, 1737 MS; San Simon, 1737 MS; *Descriptio Geographica de la part que los Españoles poseen Actuelmente en el Continente de la Florida*, 1742 MS. **Ref.** CUMMING, W.P. *The Southeast in early maps* (North Carolina 1958) pp.211-212.

Arrenberg, Gerard Abraham (*fl.*1780s). Bookseller of Rotterdam. Sold maps, prints, advertised libraries for sale 'op de Zaal boven de Beurs'. **Ref.** KROGT, P. van der *Advertenties voor kaarten, atlassen, globes e.d. in Amsterdamse kranten 1621-1811* (1985) N°·1616 p.300 [in Dutch].

Arrian [Flavius Arrianus] of Bithynia (*fl.*2nd century AD). Roman governor of Cappadocia under Hadrian, AD134. Author of *Periplus of the Euxine Sea* [Black Sea] and *Indica* (account of India after Alexander's travels).

Arrioeta, Filippo de. Map of plague cordons and quarantine areas in Bari [Italy], 1694.

Arrigoni, *Captain* Ferdinando. *Carta postale della Italia*, 1846; *I contorni di Milano*, 1843; *Lombardia*, 1860.

Arrivet, J. (*fl.*1764-1786). French draughtsman and engraver (particularly title-pages and cartouches). For *Hydrographie françoise*, 1765; for Gilles et Didier Robert de Vaugondy's *Atlas universel*, 1760-1778; for Rigobert Bonne, 1771; *Plan de Varsovie*, 1772.

Arros, C.H.O.H. d'. See **Hallez** d'Arros.

Arrott, W. Miami County, Ohio, 1855.

Arrowsmith family.

• **Arrowsmith**, Aaron (1750-1823), *senior*. Prominent English cartographer, engraver and publisher from County Durham. Established in London *c.*1770 as land surveyor, working for Faden and Cary amongst others. Set up his own enterprise at Castle St, Long Acre (1790); moved to 5 Charles St, Soho Sq. (1794); to 24 Rathbone Place (1799); to 10 Soho Sq. (1808). In all he issued about 200 maps.He became Hydrographer to the Prince of Wales *c.*1810, and to the King 1820, and died in Soho, London. Arrowsmith's work includes: pedometer measurements and drawings for

THE ETON COMPARATIVE ATLAS of Ancient and Modern Geography, in 53 Plates imperial 4to., on a new plan; coloured, 2*l.* 2*s.*; full-coloured and half-bound, 2*l.* 15*s.*—A set of *Skeleton-outlines* to the above, 5*s.*—INDEX to the ETON ATLAS, containing more than 30,000 Ancient and Modern Names, with their Latitudes and Longitudes, and references to the Plates in which they may be found; 7*s.* boards.

A COMPARATIVE ATLAS of Ancient and Modern Geography, in 57 Plates medium 4to. reduced from the above, and including two complete Indices of the Ancient and Modern Names (the Ancient Names being accentuated); 30*s.* coloured and half bound.—A set of *Skeleton-outlines* to the above, 4*s.*

AN ATLAS of ANCIENT GEOGRAPHY, in 26 Plates medium 4to. with an accentuated Index of all the Names; 15*s.* coloured and half bound.—A set of *Skeleton-outlines* to the above, 4*s.*

AN ATLAS of MODERN GEOGRAPHY, in 30 Plates medium 4to. with an Index of all the Names; 16*s.* coloured and half bound.—A set of *Skeleton-outlines* to the above, 4*s.*

Compiled from original Authorities, and containing the latest Discoveries,

By A. ARROWSMITH,
Hydrographer to the King, Soho-square.

*** Shortly will be published a *Compendium of Ancient and Modern Geography*, with illustrations of the most interesting points in History, Poetry and Fable, accompanied by Plans of Athens, Rome, Syracuse &c. &c.

An advertisement for several atlases being offered by the Arrowsmith firm.

Cary's Great Post Roads between London & Falmouth, 1784; *Chart of the World...* (8 sheets), 1790 (Arrowsmith's first publication in his own right); *Survey of the harbour of Panama*, 1791; *Map of the World...*, 1794; *Chart of the Pacific Ocean* (9 sheets), 1798; Marquesas & Oregon, 1798; Asia, 1801 (these large maps were also available for sale as an atlas up to *c.*1809); Africa, 1802; America, 1802; Chart of the West Indies (4 sheets), 1803; untitled atlas containing maps and charts of Africa, 1803; Chart of the Cape of Good Hope (4 sheets), 1805; *Island of Ceylon*, 1805; *A Pilot from England to Canton*, in seven charts, 1806; *A New and Elegant General Atlas*, 1812 (with S. Lewis); *A Chart of the Cape of Good Hope*, 1812; *Atlas of Thompson's Alcedo; or dictionary of America and West Indies...* (5 maps on 19 sheets), 1816; *Improved Map of India* (9 sheets), 1816; *A new general Atlas...*, 1817; *Asia* (4 sheets), 1818; *Atlas of Europe* (10 maps on 54 sheets), 1821; *An Atlas of Ancient Geography*, 1828; A series of Diocesan maps of England & Wales printed for George III, and many other publications. **Ref.** HECKROTTE, Warren *The Map Collector* 39 (1987), pp.16-20; TOOLEY, R.V. *The Map Collector* 9 (1979), pp.19-22.

•• **Arrowsmith**, Aaron, *junior* (*fl.*1820-1830). Son of Aaron, above. Sometimes worked with his brother Samuel (below), sometimes solo. Denmark, for Archibald Constable, 1823; *Ceylon*, 1826 (with John Kershaw); *Orbis terrarum vetus descriptio*, 1828; *Outlines of the World*, 1828; *Chart of Van Diemen's Land and Bass Strait*, 1833.

•• **Arrowsmith**, Samuel (1805-1839). Son of Aaron, *senior*. Charts of New Guinea, 1832; Philippine Islands, 1832; Sandwich Islands, 1832; Ireland (4 sheets), 1832; Southern Italy, 1833; Syria, 1833; Bible Atlas, 1835; Africa, 1836. Maps published posthumously by John Arrowsmith and Stanfords up to 1889. Also worked with his brother Aaron, *junior*.

•• **Arrowsmith**, John (1790-1873). 10 Soho Square, nephew of Aaron. Founder member of the Royal Geographical Society. *British North America*, 1832; *London Atlas of Universal Geography*, 1834, continually revised and extended (later editions are valuable, particularly for new surveys in Australia and North America, Texas), subsequently published by Edward Stanford as *Stanford's London Atlas of Universal Geography: folio edition*, to 1920s; *Australia*, 1842; *Colony of Western Australia*, 1842; *Harbour and City of Auckland*, 1842; *Northern Asia*, 1842; *Eastern Provinces of Australia*, 1850; *Map of the Countries round the North Pole*, 1850; *Burmah and adjacent Countries*, 1853; *Map of South Africa*, 1857; *Discoveries in the Arctic Sea*, 1859; *Arctic shores of America*, 1859-1860; *Austrian Empire*, 1876. **Ref.** HERBERT, Francis *Imago Mundi* 41 (1989), pp.99-105.

Arrowsmith, James William. *Arrowsmith's Map of Bristol*, 1909.

Arseniev [Arsen'ev], Vlad (1872-1930). Russian military officer, explorer, ethnographer, naturalist and writer from St Petersburg, died in Vladivostok, 1930. Possibly best known for a colourful account of his adventures with the Nanai guide Dersu Uzala, which was published in many editions, but turned out to be fictitious. Led, and recorded, nine expeditions of exploration and mapping in the Russian Far East, 1906-1927. *Kratkii voenno-geograficheskii i voenno-statisticheskii ocherk Ussuriiskogo kraia, 1901-1911* (2 volumes), 1912. **Ref.** KHISAMUTDINOV, A.A. 'Vladimir Arsen'ev - Life under a lucky star' in: *The Russian Far East: historical essays* (University of Hawaii, Honolulu, 1993) pp.81-98; KHISAMUTDINOV, A.A. 'Yego zvali Arsen'ev-Ussuriiskii' ['He was called Arseniev from Ussuri'] in: *Zapiski obshchestva izucheniia amurskogo kraia* [Proceeding of the Society for the Study of the Amur Region] Vol.XXXI (Vladivostok, 1997).

Arsenius [Aertssens]. Flemish family of instrument makers and engravers of Antwerp and Leuven. The exact relationship between the two sets of brothers is unclear.

• **Arsenius**, Brothers [I]. Neohews of Gemma Frisius. At their workshops in Leuven they produced astronomical and geographical instruments including globes, astrolabes and quadrants. Gilt brass planispheric astrolabe, 1559, contains a world map derived from that in Apian's *Cosmographicus liber*, 1524. **Ref.** *The Map Collector* (1988) 45 p.44 (news).

• **Arsenius**, Gauthier [Walter] (*fl.*1555-1588). Brother of Remy and Regnier. Considered the leading craftsman in the family.

• **Arsenius**, Remy. Brother of Gauthier and Regnier.

• **Arsenius**, Regnier. Brother of Remy and Gauthier.

• **Arsenius**, brothers [II] (*fl.*1598-1615). Worked together as engravers of maps for

Christopher Plantin 1589-1628 (used by Abraham Ortelius from 1595); for Michel Coignet's *Epitome* [Ortelius], 1601; for J.B. Vrients, 1601-1612; for Franciscus Haraeus, 1614-1615.
• **Arsenius**, Ambrose. Brother of Ferdinand.
• **Arsenius**, Ferdinand. Brother of Ambrose.

Artaria & Co.. Print sellers, music and map publishers of Vienna, later cartographers and engravers. Founded by Italian cousins Francesco and Carlo in 1770, partners included Giovanni Cappi, continued publishing until 1920 when the company merged with Freytag & Berndt to form Freytag-Berndt & Artaria. *Grundriss der Kaiserl: Königl: Haupt und Residenz Stadt Wien...*, 1799; Hungary, 1801; Venice, 1805; Bohemia (4 sheets), 1808; Poland (9 sheets), 1813; Lombardy, 1815; *Atlas der neuesten Geographie für Jedermann und jede Schulanstalt*, 1822; *Venezia*, 1830; *Karte der Königreiche Ungarn und Galizien*, 1837; *General Post-und Strassen-Karte des Herzogthums Steyermark*, 1841; *Italia*, 1843; published many works by Steinhauser and Scheda 1869-1884, including: *Joseph Ritter von Scheda's Hand-Atlas des neuesten Geographie*, Vienna 1869 and later (edited by Anton Steinhauser); *Scheda und Steinhauser's Hand-Atlas der neuesten*, 1876-1877; and many maps of Europe. **Ref.** KRETSCHMER, DÖRFLINGER and WAWRIK *Lexicon zur Geschichte der Kartographie* (1986) pp.27-28 [in German]; DÖRFLINGER, Johannes *Imago Mundi* 35 (1983), pp.58-66.
• **Artaria**, Francesco (1744-1808). Co-founder with cousin Carlo of Artaria & Co.
• **Artaria**, Carlo (1747-1808). Publisher, art dealer, and co-founder of firm Artaria & Co. in Vienna 1770.

Artaria, Domenico (1775-1842). Viennese publisher and art dealer.

Artemidorus (*fl.*2nd century BC). Greek geographer.

Artero y González, Juan de la Gloria. *Atlas de la España*, 1879; *Atlas de Geografia*, 1895; *Atlas de historia universal*, *c.*1897; *Atlas histórico-geográfico de España*, *c.*1897.

Arthaud, Guy. Duché d'Angers, 1652.

Arthur, Pierse. *...trigonometrical map of Weymouth & Melcombe Regis*, 1857.

Arthur, T.W. *Arthur's Plan of Carlisle*, 1880.

Arthur, Thomas. *Map of the District where the Epidemic Fever prevailed in 1809, 1810, & 1811*, 1816.

Artopäeus [Becker], Petrus [Peter] (*c.*1500-1563). Theologian and grammarian at Stettin [Szczecin]. Map of Pomerania used by Sebastian Münster, 1550, in turn used by Abraham Ortelius, 1573; *Latinae, graecae et hebraicae linguae grammatica*, Basle, Heinrich Petri 1543. **Ref.** MEURER, P. *Fontes cartographici Orteliani...* (1991) pp.109-110 [in German].

Artus, —. Engraver for Adrien Hubert Brué, 1833.

Arundell, V.J. *Phrygia*, 1840 (with W.J. Hamilton).

Árvay, József (1823-1879). Hungarian cartographer, born Kisráska [now Male Raskovce, Slovakia], died at Sárospatak. Estate maps kept at Festetics collection, Keszthely. **Ref.** FODOR, F. *Az Institutum Geometricum* (Budapest 1954) [in Hungarian].

Arvidsson, G. *Helsinki*, *c.*1899.

Asaki, Gheorghe (1788-1869). Romanian cartographer.

Asano Yahei. Japanese publisher. Nagakubo Sekisui's *Kaisei Chikyū Bankoku Zenzu* [world map], *c.*1785; joint publisher (with Suharaya Mohei) of Nagakubo Sekisui's *Kaisei Nihon Yochi Rotei Zenzu* [map of Japan], 1811 edition (first published 1779).

Asaph, Judaeus. 11th-century *Mappamundi*, in Bibliothèque Nationale, Paris.

Asbóth, Mihály (1790-post 1853). Hungarian cartographer. Mapped Nagyág river at Óbuda, also maps in Hont County. **Ref.** FODOR, F. *A magyar térképirás* [Hungarian cartography] (Budapest 1952-1954) [in Hungarian].

Asbóth, Sándor (1811-1868). Born Keszthely [Hungary], died Buenos Aires. Hungarian engineer, studied cartography at Academy in Selmecbánya [now Banska Stiavnica, Slovakia]; participated in 1848 Revolution and went into exile in USA, assisted in mapping of New York City. **Ref.** BONA, G.

Tabornokok és törzstisztek [Generals and Staff Officers] (Budapest 1983) [in Hungarian].

Asbury, Samuel G. *Census Map of Texas*, 1850.

Ascensíon, Antonio de la *SJ* (*d.*1663). Accompanied Sebastian Vizcaíno's expedition to the south west coast of N. America in 1602. Believed to be the originator of the idea that California was an island. **Ref.** POLK, D.B. *The island of California* (1991) p.257.

Aschbach, A. *Pocket map showing the probable theatre of war...*, Philadelphia, M.H. Traubel 1861.

Aschbach, G.A. Pennsylvania surveyor. *City of Allentown*, 1871; *Lehigh Valley Rail Road*, 1871; Lookout Park, 1872.

Asche, A. *Karte Hannover und Linden*, 1894.

Ascheberg, *General* Rutger von. Governor-General of Skåne. Commissioned Gerhard von Buhrmann's survey of Skåne 1681.

Aschenburg-Halevi, Ha'Gaon Shimon. *Eretz Israel* [Land of Israel], 1788.

Aschke, O. Ems-Mündung, 1895.

Aschland, Arent. Coast of Iceland, 1820-1824.

Asensio Torres, José (1759-*c.*1820). Copper and letter engraver from Valencia, possibly son of Francisco (below). N.E. Atlantic, 1780; Island of Menorca, 1780; San Felipe, 1780; 4 maps in Vincente Tofiño de San Miguel's *Atlas marítimo de España*, 1789; map of Valencia for A.J. Cavanilles's History of Valencia, Madrid, Imprenta Real 1795-1797 (with T.L. Enguídanos); Fausto Martinez de la Torre's *Plano de la Villa y Corte de Madrid*, *c.*1800.

Asensio y Mejorada, Francisco (1725-1794). Engraver. *Plano de la ciudad de León*, Madrid 1792 (for M. Risco's *Historia de la Ciudad y Corte de León*); *Plano de Madrid. Plano de Paris* (pair of circular maps with Latin and Arabic text), [n.d.].

Aseo, Claudio. *Provincia di Venezia* (32 sheets), 1876.

Ash, John. Engraver of Adams Place, Borough [Southwark]. For sheet H1 of Richard Horwood's *Plan of the Cities of London and Westminster*, 1799.

Ash, Joshua W. Physician, member of the Delaware County Institute of Science. Delaware County, Pennsylvania, 1848 (lithographed by Robert Pearsall Smith).

Ash, S.I. Dean Forrest, 1821 MS.

Ashbridge, *Captain* John. *Plan of Quedah Road* [Malay Peninsula], published Dalrymple 1781.

Ashby, H. (1744-1815). Engraver for John Noorthouck's *A New Plan of London Westminster and Southwark*, 1772; for Mostyn John Armstrong's *A Scotch Atlas...*, 1777; Peter Crosthwaite's *Maps of the Lake District*, 1783 (with S. Neele and J. Ellis); Peter Fannin's Isle of Man, 1789.

Ashby, Robert. Cartographer. Lough Foyle, 1601.

Ashcroft, A.E. Survey of mineral claims in British Columbia, in *Stuarts' Maps of Boundary*, 1901.

Ashcroft, T.H. Nigeria, 1880.

Asher & Adams. Publishers of Washington, and 59 Beekman Street, New York. *The Historical War Map*, 1864; ... *Township Map of Indiana*, 1865; ...*State of Iowa*, 1868; ...*Atlas and Gazetteer of New York*, 1871; *New Commercial, Topographical and Statistical...Atlas and Gazetteer of the United States*, 1872; *Asher & Adams' Michigan*, 1873; *Asher & Adams' New Commercial, Topographical and Statistical Gazetteer of the United States*, 1873.

Ashley, *Sir* Anthony (1551-1625). Translated and edited Lucas Janszoon Waghenaer's *Speculum nauticum* as *Mariners Mirrour*, 1588.

Ashman, *Reverend* J. See **Ashmun**, Reverend J.

Ashmead, George C. Revised and completed John Plumley's *Plan of Bristol and its Suburbs*, 1829, 1833.

Ashmore, John. Estate surveyor. Stanton under Bardon, Leicestershire, 1780 MS.

Ashmore, *Captain* Samuel, *RN*. *West Coast of Sumatra*, 1822; *Tracks through the Barrier Reefs*, 1840.

Ashmun, *Reverend* Jehudi (1794-1828). American colonial agent and missionary. Surveyed the west coast of Africa. Map of Liberia, 1833; surveys used in *Colton's Map of Africa*, 1862

Ashpital, W.H. Surveyor. Estate of John Clark Powell, Whitechapel, 1805; St John Hackney, 1831.

Ashton, Henry G.G. Surveyor. *Liverpool Bay*, 1896-1897, and many other editions to 1908.

Ashwell, Samuel. Essex estate surveyor. Havering, Essex, 1662 MS.

Asiatic Lithographic Press, Calcutta. Press established 1825, the first of its kind in India. Road maps: Bengal, Madras and Bombay, 1828.

Aspheris, Petrus. Bronze celestial globe, 1586.

Aspin, Jehosophat (*fl.*1805-1840). British historian. Map of ancient Europe, Liverpool, Nuttall, Fisher & Co. 1816 (engraved by A. Keith); produced historical maps to accompany his own *Systematic Analysis of Universal History* (2 volumes), Liverpool, Leigh 1816-1820; *Chart of New South Wales*, 1816 [found occasionally in some editions of John Thomson's *New General Atlas* to 1828]; assisted C. Gros's in the revision of C.V. Lavoisne's *Genealogical, Historical and Chronological Atlas*, 1814 and later editions (the first edition of 1807 had no maps); Arctic Regions for John Thomson, 1822; maps for *A course of geography, by means of instructive games, invented by the Abbé Gaultier...by John Olding Butler*, 1822 (maps engraved by N.R. Hewitt); *An Improved moveable Planisphere*, 1844. **Ref.** GARDINER, R.A. in: WALLIS, H. & TYACKE, S. (Eds.) *My head is a map* (1973) pp.55 & 59.

Aspley, William. Bookseller of London. William Camden's *Britannia*, 1637 (English text, with William Kip and William Hole maps).

Asquith, Richard. Estate surveyor in Cumberland. *Park Estate in the Parish of Bewcastle*, 1862; *Scaleby Hall Estate*, 1862.

Assa, José Frederico d'. *Carto do Territorio Portuguez de Goa*, 1878.

Assaf. (*fl.*11th century). Jewish mathematician and geographer. Planisphere.

Asselen, C.F. von (*fl.*1720-1726). Ambt Apen, 1721 MS; Gegend von Sandfurth, [n.d.] MS.

Asselin, Pierre. See **Ancelin**, Pierre.

Asser, James. Essex estate surveyor. West Thurrock, 1798 MS; Bulphen, 1803 MS.

Assheton, J. *South America*, 1826; *Ireland*, London, T. Tegg 1826 (engraved by J. Shury).

Assheton, J.T. Involved in the production of Mathews & Leigh's *Scripture Atlas*, 1812; *Historical Map of Palestine*, 1825; *Geographical Companion England & Wales*, 1832.

Assiotti, Francis Aegidius. *Plan of the Island of Minorca* (2 sheets), 1780; *...Siege of St Philips Castle on the Island of Minorca*, 1782 (drawn by John Adams).

Assum, G. van. See **Gerritsz**, Hessel.

Aster, Ernst Ludwig (*d.*1778). German cartographer.

Aster, Friedrich Gottlieb (1788-1813). Military cartographer of Dresden.

Aster, Friedrich Ludwig (1732-1804). Military cartographer of Dresden. Triangulation of Saxony, 1792 published 1821.

Aster, H. *Plan von Angriff der Grossen Böhmischen Armée auf die Stadt Dresden, 26 Aug 1813*, [*n.d.*]; *Schlacht bei Dresden, 27 Aug 1813*, 1842.

Astley, Philip. *Astley's New and Correct Map of the Empire of Germany...*, 1800; New map France, 1800; *A New Map of Europe*, 1808.

Astley, Thomas. Bookseller, 'at the Rose over-against the North Door of St Paul's', London. One of the publishers of *The London Magazine*, 1735-1746; collection of Voyages and Travels, 1745-1747 [compiled by John Green (*pseud.* Braddock Mead)].

Astolfi, Gaetano (*fl.*18th century). Engineer of Bologna, worked in Rome. Pontine Marshes (4 sheets), [n.d.].

Aston, D.W. *Northern Lincolnshire Railway*, 1855.

Aston, John. Publisher of Coventry. *Warwickshire...*, 1816, republished by John Turner, 1830

Aston, Joseph. Publisher, printer and mapseller 84 Deansgate, Manchester, also at Exchange Herald Office St Ann's Street. Map of Manchester & Salford, 1804, 1816.

Astor, Diego de. *Descripcão do Reino de Bengallia*; *Descripcão do Reino de Guzarate*; *Descripcao da Ilha de Iaoa*, all in João de Barros's *Quarta Decada da Asia*, 1615.

Astruc, —. *Carte des vents...de la province du Languedoc*, 1737.

Asturlabi-al, Ali Ibn-Isa (*fl.*9th century). Instrument maker and astronomer. *Kitab al-Amal bi al-Asturlab* [The astrolabe and its use].

Aszalay de Szendrö, József (1798-1874). Hungarian cartographer, pioneer of full colour-lithographed maps 1830. *Mappa generalis regni Hungariae* (7 sheets), Vienna 1830-1838 (engraved David Burghart). **Ref.** KISS, L. 'Aszalay József, XIX. századi magyar térképszerkesztö' in *Geodézia és Kartográfia* [Geodesy and cartography] (1987-2) [in Hungarian]; STEGENA, L. *Térképtörténet* [History of maps] (Budapest 1981) [in Hungarian].

Atanasiu, A.D. *Roumainie et les Provinces Roumaines*, 1919; *Carte Ethnographique du Banat de Temesvar*, 1919.

Atchison, C.A. (*fl.*1880s). Surveyor, '141 Pitt Street, Opposite City Bank', Sydney, Australia. *Chisholm & Kenyon Estate, Fairfield*, *c.*1881; *Zahells Estate, Coogee*, 1884; *Mowbray Park Estate, Willoughby*, *c.*1880s.

Atchison & Schleicher (*fl.*1880-1890s). Surveyors, Eldon Chambers, 92 Pitt Street, Sydney. *Redfern, City of Sydney*, 1884; *Gladesville Park*, 1886; *Oaks Estate*, 1892.

Athbridge, *Captain* John. See **Ashbridge**, *Captain* John.

Atherstone, *Doctor* William Guybon (1814-1898). *Geological Sketch map of Cape Colony*, 1875.

Atienza, José de. Spanish navigator. Charts, 1830 MS.

Atienza y Cobos, Augusto. *España y Portugal*, 1888.

Atkinson, Alfred. *Levels within the Wapentakes of Manley...in the County of Lincoln*, 1875.

Atkinson, G.W.E. *Turkestan*, 1879; *Route map Tibet & Mongolia*, Gotha, Justus Perthes 1885.

Atkinson, George. Surveyor-General. *Map of Ceylon*, William Faden 1813.

Atkinson, George W. American land surveyor. Produced land ownership maps in Minnesota and South Dakota, all published by E. Frank Peterson of South Dakota, 1898-1900.

Atkinson, James. Mathematical instrument maker, '...on ye E. side of S. Saveris Dock over against the Griffen and at his shop at Cheney Garden Stairs on Rotherhithe Wall...', London. Joined with John Seller, John Thornton, John Colson and William Fisher in publishing or re-issuing works originated by Seller including *The English Pilot, Mediterranean*, 1677; and *Atlas Maritimus*, 1677. **Ref.** THROWER, N.J.W. *The compleat plattmaker* (1975) pp.149-150.

Atkinson, John. *...Forest of Dean*, 1845.

Atkinson, Joseph. Surveyor of Spanish Town, Jamaica. *An Hydrographicall Chart of the Virgin Islands*, London, Mount & Page 1739.

Atkinson, Thomas. Surveyor. Plan of the River Crake [Lancashire], 18th-century MS.

Atkinson, Thomas. *Plan of the Abbey of St Mary at Byland* [Yorkshire], 1806.

Atkinson, William G. Coal and Iron Fields of Alabama, 1872 MS.

Atlas. Mythical Greek giant thought to hold up the skies, since classical times depicted supporting a celestial globe on his shoulders. In the Renaissance the popularity of such representations of Atlas gave rise to the use of the term to describe a book of maps.

Atlas, Laureano. Engraver. For Pedro Mañuel de Arandia y Santistevan, Manila, 1757; Spanish Empire in form of woman, 1761.

Atlas Publishing Co. Universal Atlas, 1893; Erie County, Ohio, 1896.

Atlasov, Vladimir Vasil'evic (*c.*1661/1664-1711). Russian Cossack. First map of Kamchatka surveyed after 1697, used by Johann Baptist Homann *c.*1720.

Atthalin, *Baron* —. (1784-1836). Military cartographer.

Attinger frères. Publishers of Paris. O. Reclus's *Atlas pittoresque de la France*, 1910-1912; O. Reclus's *Atlas de la plus grande France*, 1913.

Attree, H.R. *Topography of Brighton*, 1809 (plan by J. Marchant): *Picture of the roads to the Metropolis*, 1809 (maps by M. Phillips).

Atwell, George. *The Faithfull Surveyor*, 1662.

Atwood, John M. (*fl.*1840-1883). New York engraver. *Michigan, Wisconsin & Iowa*, 1845, in: William Channing Woodbridge's *Modern Atlas, Physical, Political & Statistical...*, 1845; for Phelps's *National Map of the United States*, 1847; for J.H. Colton's *United States*, 1849; *Map of the Western States...*, Ensign & Thayer 1849, Thayer, Bridgman & Fanning 1853; *Map of the United States, Canada, Mexico, Central America and the West India Islands*, 1851; *Colton's Railroad & Township Map of the State of Ohio*, 1852; *Disturnell's New Map of the United States and Canada*, 1855; United States, 1865.

Atwood, Holmes & Read. *Pocket Atlas of the United States*, 1884.

Au, Herman. *Mapa de la República de Guatemala*, 1868, 1876.

Aub. See **Friend** & Aub.

Aubant, A. d'. See **D'Aubant**, A.

Aubel, O. Compiled geographical games: *Europe. Jeu des Voyages*, *c.*1883; *France. Chemins de Fer*.

Auber, H.P. In East India Company service. Survey Canton River, 1815-1816, published 1818.

Aubert, —. Printer of the Cassini *Carte de France* (181 sheets), 1764-1793.

Aubert, father and son. French script engravers. **N.B.** As they often signed themselves simply Aubert, the works cited below could be by either or both.

• **Aubert**, L., *père*. Script engraver, at 'Port St. Jacques n°·87', Paris. For J.F.G de La Pérouse, 1797; for Edme Mentelle, 1797-1801; for Philippe Buache, 1798; Charles Paultre's *Carte physique et politique de la Syrie...*, 1803 (with Blondeau); Cassini's *Carte générale de la France par Départments Servant à l'Assemblage...*, 1809 (key map showing the 182 sheets); for Alexander von Humboldt, 1811; P. Lapie's *Carte des Iles Britanniques ou Royaume-Uni...*, 1814 (with Blondeau).

•• **Aubert**, — [J-B-L?], *fils*. Script engraver of 'Rue et Porte St Jacques No161', Paris. J.B. Poirson's *Carte Itinéraire de la France*, 1798; J.B. Poirson's *Afrique*, 1809 (for Malte-Brun); for Jean-Baptiste-Marie Chamouin, 1812.

Aubert, André. *Carte du Théâtre de la Guerre Franco-Allemande*, 1914; *Carte Militaire, Département du Nord*, 1914; *Guerre Européenne*, 1914.

Aubert, Benoni d' (1768-1832). Born Copenhagen moved to Norway 1790. Worked with Carl Grove and Niels Wibe on a survey of the Norwegian coast, Paris, Dépot de la Marine 1793-1817.

Aubert, F. *Carte géologique...de Tunis*, 1892.

Aubert, J.M. *Plan of Pulo-Pontaugh* [Java], 1788.

Aubert de Gaspé, Ignace-Philippe (1714-1787). *Carte de Phis, St Frederick jusqu'à la fin de la Rivière de Corlac* [Canada], 1759.

Aubin, — d'. Script engraver. For Louis Bretez's *Plan de Paris*, 1739; Maraldi and Cassini's *Nouvelle carte Qui Comprend les principaux Triangles...de la France*, 1744, 1745; worked on some sheets of the Cassini *Carte de France*, published from 1756.

Aubonne, F. [von] Jenner von. Carte des Cantons, Bern 1820.

Aubrée, Armand. Publisher, rue de Vaugiraud 17, and rue Tavanne n°. 4, Paris. C.V. Monin's *Atlas classique et universel*, 1837; Dufour & Duvotenay's *La terre, atlas historique...*, 1840; A.-H. Dufour and T. Duvotenay's *Atlas de France et ses colonies*, 1860.

Aubrey, John (1627-1697). English antiquarian and topographer. Map of Surrey, 1673.

Aubry, Nicolas, *SJ*. French Jesuit active in Canada. *Canada*, 1713; *Limites entre France et Angleterre*, 1715.

Aubry, Peter. Engraver. Map and plans in German Bible, 1665.

Au Capitaine, *Baron* Henri (1832-1867). French geographer.

Aucher, Anthony (*fl.*1538-1558). Born in Appledore, England, died in Calais. Chart of Dover Harbour, *c.*1541 (with John Bartelot, Sir Richard Cavendish and John à Borough).

Aucher-Eloy, Pierre-Martin-Rémi (1793-1838). Pharmacist and botanist of Blois, traded as a bookseller and printer at '47 de la Grand-Rue', Blois (1817); '35 de la Grand-Rue' (1824); also '65 rue Saint-André-des-Arts', Paris (1826). Abandoned his ailing business in 1829 in favour of botanical and scientific explorations in Russia, the Middle East, and the Orient, an account of which was published in Paris in 1843. Sold Cassini's *Carte de France*. **Ref.** PACHA, B. in: PACHA, B. & MIRAN, L. *Cartes et plans imprimés de 1564 à 1815* (1996) p.43 [in French]; BROC, N. *Dictionnaire illustré des explorateurs Français du XIXe siècle: II Asie* (1992) pp.5-6 [in French].
• **Aucher-Eloy et Compagnie** (*fl.*1824-1829).

Aucuparius. See **Vogler**, Thomas Heinrich.

Audebeau, —. Basse-Egypte, 1897.

Audebert, Denys (*fl.*1735-1739). Clock and globe maker of Amsterdam 'agter de Beurs' (1735); 'op 't Rokkin by de Beurs' (1736). *De nieuwe Hemel en Aerd-globens, aenwyzende de uuren van dag en nagt*, 1735. **Ref.** KROGT, P. van der *Advertenties voor kaarten, atlassen, globes e.d. in Amsterdamse kranten 1621-1811* (1985) N°·684 p.131 [in Dutch].

Audemart, Louis (*b.*1865). French naval lieutenant and traveller in China. Carried out a navigational survey of the Yang-Tse and Yalong. *Exploration hydrographique du Yalong et du Yang-Tse supérieur*, 1911; Atlas of the Yangtse from Li-Kiang to Soui-Fou, the Yalong and the Min Ho (37 sheets), 1914. **Ref.** BROC, N. *Dictionnaire illustré des explorateurs Français du XIXe siècle: II Asie* (1992) pp.6-7 [in French].

Audet, P. *Ville de Toulon*, 1900.

Audibert, Jean Pierre (1680-1763). Military cartographer from Montpellier, France.

Audiffret, Jean Baptiste d' (1657-1733). French traveller and geographer. *La Géographie ancienne et moderne*, 1691.

Audin, —. Publisher, Paris. Carte routière d'Allemagne, 1836; Cours du Rhin, 1840.

Audoin, Antoine (1874-1932). French hydrographer. Surveyed the coasts of Gabon, Congo, also Lake Tchad and the River Niger. **Ref.** BROC, N. *Dictionnaire illustré des explorateurs Français du XIXe siècle: I Afrique* (1988) p.9 [in French].

Audry, C. *Carte des Communications postales, maritimes et des lignes télégraphiques du Globe*, Paris 1891.

Augouard, *père* Prosper (1852-1921). French missionary and explorer. *Carte du Congo*, 1882; *Le Congo de Brazzaville à Liranga*, 1906; *L'Oubangui de Liranga à Bangui*, 1908. **Ref.** BROC, N. *Dictionnaire illustré des explorateurs Français du XIXe siècle: I Afrique* (1988) p.10 [in French].

Augustin, Caspar (*d.*1643). *An Itinerary, giving views of the principal Cities in Europe, and their relative distances from Augsburg...*, *c.*1610.

Augustin, *Lieutenant-General* Vincze [Vincenz], báró [baron] (1780-1859). Hungarian military cartographer, professor at military academy Wiener Neustadt; established triangulation between Buda and Györ: *Bestimmung der Langedifferenz zwischen Raab und Wien*, Pest 1813. **Ref.** REGÖCZI, E. 'A pesti és a györi alapvonalmérés 1808 és 1810-ben' in *Geodézia és Kartográfia* 1955 (3) [in Hungarian].

Augustine, *Saint, Bishop of Hippo* (*fl.*AD 395-430). His geographical views influenced medieval mapmakers.

Augustini, J.L. Bookseller of Haarlem. *Plan van Zeist*, 1802.

Augustodunensis. See **Honorius**.

Augustus. Roman Emperor (63 BC-AD 14). Had world map constructed. **N.B.** See also **Agrippa**, Marcus Vipsanius.

Auhagen, E. *Karte des Harzgebirges*, 1867.

Auld & Litchfield. Sketch of Adelaide River, 1865.

Auliacus. See **Ailly**, Pierre d'.

Aulick, R. *River Jordan and the Dead Sea*, 1849.

Aumale, *sieur d'*. See **Damme**, Jan van.

Aunpekh, Georg (1423-1461). See **Peuerbach**, Georg von.

Auracher, Klemens (1748-1790). Born in the Netherlands, worked in Austria as a military cartographer.

Aurelius, Cornelius, of Gouda (*c*.1460-1523). Poet and historian; pupil of Desiderius Erasmus. Untitled woodcut world map in *Die Cronycke van Hollandt Zeelandt en Vrieslandt*, Amsterdam, Joannes Seversz., Amsterdam, 1514; later states in 1517, 1529, 1535. **Ref.** SHIRLEY, R. *The mapping of the World* (1983) N°37 p.42-43.

Aurelius. See **Veen**, Adriaen.

Aurelle de Paladines, Louis Jean Baptiste, Marquis d' (1804-1877). General in the French army in Africa and the Crimea, later commander of the Première Armée de la Loire. Prepared four maps of the Loire to accompany his book *La Première Armée de la Loire*, 1872.

Auret, William Henry. South Africa, *c*.1854 MS.

Aurigallus, Matthaeus. See **Auriogalus**, Matthaeus.

Aurigarius. See **Waghenaer**, Lucas Janszoon.

Auriogalus [Aurigallus], Matthaeus (*c*.1490-1543). Map of Saxony for Sebastian Münster.

Auroux, Nicolas. *Description de la Terre Universelle*, for Pierre Du Val's *Mémoires géographiques*, Lyon 1674, 1676 and 1688.

Ausfeld, Johann Carl. Engraver for C.E. Caspari, 1821; Adolf Stieler's Nord-Amerika, 1825; Wilhelm Ernst August von Schlieben, 1830; Fr. von Stulpnagel, 1832.

Auspitz, Jacob. Compiler of the first Hebrew atlas *Beer Haluchot*, 1817, 1818 (with 5 maps engraved by Gottfried Prixner).

Austen, E. *Road-Book from London to Naples*, 1835.

Austen, *Captain* F.W., *RN*. *Orfordness to Bawdsey*, 1812; *The Harbour of St. Helena*, 1847.

Austen, Henry. *Liverpool*, 1834; *Liverpool with part of River Mersey*, 1854.

Austen, John. Part of Is. of Enisherkin Co. Cork, 1709 MS.

Austen, Joseph Thomas. Survey of Fowey Harbour, Cornwall, 1813.

Austen, Robert Godwin. See **Godwin**-Austen.

Austen, Stephen T. Publisher of Newgate Street, London. Henry Popple's *A Map of the British Empire in North America...* (20 sheets and key), 1733 (engraved by William Henry Toms); T. Salmon's *Modern History*, 1744-1746; Austen's name appears on some copies of Willdey's *A Map of the King's roads...in the Highlands of Scotland*, 1746.

Austen, Thomas. Middlesex estate surveyor. Old Lock Acre, Enfield Chase, 1827 MS.

Austin, *Captain* (later *Admiral*) Horatio T., *RN*. *Arctic America*, 1850-1851; *Charts shewing the North-West Passage* 1853; *South Atlantic Ocean*, 1854.

Austin, *Lieutenant* Loring. Eastern Canada and New England, 1814 MS.

Austin, Robert. Assistant surveyor. Surveys in Western Australia, 1854.

Austin, Stephen Fuller (*c*.1790-1836). After whom state capital of Texas is named. *Map of Texas with Parts of the Adjoining States*, H.S.Tanner 1830, 1836, 1839. **Ref.** MARTIN, R.S. 'Austin's contribution to the cartographic history of Texas', *Bulletin*, Geography and map division, Special libraries association N°133 (1983), pp.2-13; MARTIN, R.S. 'Maps of an Impresario: Austin's contribution to the cartography of Texas', *Southwestern historical quarterly* 85, N°4 (1982), pp.371-400.

Autenrieth, E.L. Isthmus of Panama, 1851.

Auteroche, Chappé d'. See **Chappé** d'Auteroche, Jean.

Authes, Abraham. Sauria Staette Buch, Frankfurt am Main 1658.

Autolycus [Autolykos] (*c*.315-240 BC). Greek mathematician and astronomer of Pitane, Asia Minor.

Autrechy, —. Drew plan of Cazenovia, New York, 1794.

Auvergne, E.B. d'. *Gaze's New Map of Europe*, 1902.

Auvergne, Philip d' (1755-1816). *Plan of the Bay of Saldanha*, 1781; *Topographical Plan of the Isle Trinidada*, 1787.

Auvray, Alexis (*fl*.1763-1819). Print and map seller of 'rue Saint-Jacques' (1763); '11 quai Malaquais' (1782); '5 quai Voltaire' (1802); '7 quai Malaquais' (1817-1819), Paris. *Nouvelle carte d'Espagne*, 1763; sold sheets of the Cassini *Carte de France*. **Ref.** PACHA, B. & MIRAN, L. *Carte et plans imprimés de 1564 à 1815* (Bibliothèque Nationale de France, Paris 1996) p.43 [in French].

Auvray, F. *Plan de la ville et du Château de Caen*, 1718.

Auvray [Auvrai], Pierre-Laurent (1736-1781). French engraver, worked in Basle and Paris. Canton Basel, 1766; worked on G.A. Rizzi-Zannoni's *Carta della Sicilia Prima*, 1769.

Auvray, Th. L'Europe, 1630.

Auxerinus, C. See **Buondelmonte**.

Auzolles, Jacques d', *sieur de* la Peyre (1571-1642). Geographer.

Auzzani, C. Contorni di Firenze, 1870.

Avant, R.H. *City of Winnipeg* (33 sheets), 1920.

Avanzo, Dominique. Lithographer. *Plan de le Ville de Bruges*, 1855.

Aveele [Avelen], Jan van den (1655-1722). Cartographer, engraver, and mapmaker of Leiden. Worked in Amsterdam and Stockholm. *Zuid Holland*, 1682; *Oldenburg*, 1683; for Erik Jonsson Dahlberg's *Topographia sueciae*, 1700-1706; also numerous engraved titles and frontispieces for atlases and books.

Avelar, André do (1546-1622). Portuguese cosmographer. *Spherae utriasque tabella*, Coimbra, A. de Barreira, 1593; *Chronographia òl repertorio dos tempos*, Coimbra, Simão Lopez 1594.

Avelen. See **Aveele**, Jan van den.

Aveline family. Family of Parisian engravers and mapsellers flourishing in the seventeenth and eighteenth centuries. Engraved for J. Rocque, 1750; engraved 4 maps for the Cassini *Carte de France*, from 1756; for G.L. le Rouge, 1762; for L.C. Desnos, 1769. **N.B.** Sources are confused about the relationships between various members of the family. The following seems the most logical interpretation of secondary source material.

• **Aveline**, Pierre I (1656-1722). French cartographer and engraver, specialised in topographical views. 'Rue de la Juiverie, proce de la Madeleine, à l'Image St Nicolas', Paris. *Carte des environs de Maintenon*, 1684; Jerusalem, *c*.1700; also vues d'optique.

•• **Aveline**, Pierre II (1679-1722). Artist and surgeon, son of Pierre, above, brother of Antoine, below, father of Pierre Alexandre, below.
••• **Aveline**, Pierre-Alexandre (1702-1760). Son of Pierre the surgeon, grandson of Pierre, top. Draughtsman and engraver of Paris. D'Anville's *L'Italie*, 1745.
•• **Aveline**, Antoine II (1691-1743). Son of Pierre, top, brother of Pierre II, above. Publisher and print seller of the 'rue de Foin', Paris. Married Marie-Madeleine Taté. Sons François-Antoine, Jean Alexandre, Louis and Jean-Antoine continued as engravers.
•• **Aveline**, *Mme* Marie-Madeleine (*née* Taté), wife of Antoine II. 'Marchande d'estampes à la Reine de France', Paris. Sold vues d'optique.
The following are assumed to be the children of Antoine II and Marie-Madeleine:
••• **Aveline**, François-Antoine (1718-1762). Born in Paris, died in poverty in London. Engraver, cousin and pupil of Pierre Alexandre. Engraver for J.B.N.D. d'Après de Mannevillette, 1745; for J.N. Bellin, 1751; contributed to three sheets of Cassini's *Carte de France*, 1759-1761; engraved for *l'Histoire générale des voyages de l'abbé Prévost*, [n.d.].
••• **Aveline**, Jean Antoine (*b*.1720). Engraved four sheets of Cassini's *Carte de France*, 1744-1760
••• **Aveline**, Pierre Alexandre (*b*.1722).
••• **Aveline**, Louis (*fl*.1732-1753). Engraver.
••• **Aveline**, Jean Alexandre (*fl*.1739-1781). Engraver.
••• **Aveline**, Antoine III (*b*.1727).
••• **Aveline**, Marie-Anne (*fl*.1742-1781). Print and map-seller of Paris. 'Rue St Jacques, à l'enseigne de la Reine de France'. Issued vues d'optique from address 'Aux armes de la ville'.
••• **Aveline**, Madeleine Gabrielle (*b*.1729).
• **Aveline**, Antoine I [Antonio] (1662-1712). Engraver, rue St Jacques. Assumed to be brother of Pierre I, top. Views of cities, *e.g.* Köbenhavn; London; Malta, 1665; Basel; St. Cloud, 1680; Roma, 1700; Tangier, 1700.

Avellana, Miguel. Maps of Spain, 1858-1861.

Avena, brothers. Military topographers of Naples. **Ref.** VALERIO, V. *Società uomini e istituzioni cartografiche nel mezzogiorno d'Italia* (1993) pp.442-444 [in Italian].
• **Avena**, Carlo (*b*.1824). Conducted triangulation surveys on behalf of the Officio Topografico di Napoli. He relinquished an active career to take charge of the library at the Officio, and later became professor of mathematics at Naples university.
• **Avena**, Alfredo (1836-1880). Joined the Officio Topografico di Napoli in 1855. Conducted initial surveys for part of a map of the Kingdom of Naples.

Avenet, —. *L'Adriatique et le Monténégro* (7 maps), 1878.

Aventinus [Thurmair; Turmair], Johannes (1477-1534). Historian and cartographer from Abensberg, near Regensburg. *Obern und Nidern Bairn bey den alten im Latein und Griechischen Vindelicia etc.* (woodcut by Georg Apian), 1523 (only example destroyed at Munich 1945); another version *Abern unnd* [sic] *Nidern Bairn*, drawn 1533 and cut by Apian, 1535 (only example also lost in 1945), latter version copied by Abraham Ortelius 1570. **Ref.** KARROW, Robert W. *Mapmakers of the sixteenth century and their maps* (1993) pp. 71-77; MEURER, P. *Fontes cartographici Orteliani...* (1991) pp.110-111 [in German].

Averill, F.L. Corps of engineers. *Map of Washington, D.C. and suburbs...*, Washington, A.B. Graham 1892.

Averill, H.K., *junior*. Map of Decorah, Iowa, 1856.

Averill & Hagar. Clinton County, New York, 1879.

Aversano, Giovanni (*fl*.19th century). Engraver of Naples.

Avery, David. Publisher. Captain John Mitchell's Chart of the Thames, 1733.

Avery, Egbert. Surveyor. South Nebraska City, 1855.

Avery, J.C. Isthmus of Tehuantepec, 1851.

Avery, Joseph. Hydrographer. Chart south coast of England (3 sheets): *An exact draught from Arundel in Sussex to St Albans in Dorset*, 1721 used in Greenvile Collins' *Great Britain's Coasting Pilot*, 1723, later editions 1731, [1775] (revised by George Burn 1794); *The Coast, Rivers and Inlets of the Province of Georgia*, 1780, in Joseph Frederick Wallet

Pierre D'Avity (1573 - 1635). (By courtesy of Rodney Shirley)

Don Félix de Azara (1746-1811). Copperplate engraving by H. Lefévre after Antonio Rodrigues. (Taken from Map Collectors' Circle*)*

Des Barres's *Atlantic Neptune*. **Ref.** DE VORSEY, Louis *The Map Collector* 19 (1982), pp.30-31.

Avet, *General* Enrico (1826-1895). Italian army. Inventor of photographic process for reproducing engraving. *Carta delle Provincie Meridionali* (24 sheets), 1870-1874.

Avezac de Castera-Macaya, Marie Armand Pascal d' (1800-1875). Geographer and pioneering historian of cartography. *Etudes de géographie critique*, 1836; *Notice sur les découvertes faites au Moyen Age dans l'océan Atlantique*, 1845; *Les navigations terreneuviennes de Jean Sébastien Cabot*, 1869.

Avianello [Vianello], Baldassare (*fl.c.*1552). Italian military engineer, of Padua. Worked with Giovanni Maria Olgiatti in the Netherlands, 1552.

Avico, Joseph. Map of the area covered by Treaty of Turin, 1754 MS (published Geneva *c.*1955).

Avinea, Antonio. See **Wijngaerde**, Antonij van den.

Avity [D'Avity], Pierre de Montmartin, *sieur d'* (1573-1635). *b.*Tournon, *d.*Paris. *Les estats et empires du monde*, 1626, (later editions of which included world maps by makers such as M. Merian, 1638; H. le Roy, 1643; M. Tavernier, 1661); *Description générale de l'Amérique*, 1640 and later (from 1643 included map by Bertius). **Ref.** SHIRLEY, R. *The mapping of the World* (1983) N°341 p.365; N°345 p.369; N°349 p.373.

Avivas, A. Plan d'Angers, 1875.

Avramov, Il'ya (*fl.*1755-1760s). Russian student of geodesy. Globe engraved on copper, 1755 (St Petersburg); under Mikhail Vasilyevich Lomonosov prepared geographical extracts, data sets and map in 1760s.

Avril, Charles. Engraver for Feliks Petrovich Fonton's *Plan du siège et de l'assault d'Akhaltsikh*, 1830; *Atlas Russie dans Asie-Mineure*, 1840.

Avril, *père* Phillippe. *Nouvelle carte de la Sibérie*, 1692.

Awaya Bunzō. Japanese publisher. Imamura Yoshikage's route map to Kompira Shrine from land and sea (woodblock, colour map), 1778.

Ayala, Juan Manuel de (1745-1797). Surveys of the Spanish explorations of the west coast of America. *Coast of the Californias*, 1775-1778, .

Ayaseki Gyuko. Map of Osaka, Kochiya Kibei and Kochiya Gisuke, 1836.

Ayasse, *Captain* Joseph (*b.*1872). Captain in French colonial army in Tchad. *De Nguigni à Bilma*, 1908.

Ayer, Edward Everett (1841-1927). Bibliophile and collector who donated his collection of books, maps and drawings to the Newberry Library of Chicago in 1911.

Aylen, J.F.R. *Wolstenholme Sound*, 1850.

Ayling, Nicholas. Middlesex estate surveyor. Survey Manor of Langley in Ealing, 1692 MS.

Ayloffe, Benjamin. Essex estate surveyor. Dedham, 1573 MS.

Aymard, *Capitaine* A. (1870-*c.*1911). French explorer. *Le long du chemin de fer du Yunnan*, Paris 1911.

Aymès, Antoine (1836-1910). French Naval Officer, explorer in equatorial Africa. Surveyed the estuaries of Gabon and Como, 1870.

Aymonier, Etienne (1844-1929). French archaeologist, worked in Indochina. Géographie du Cambodge, 1876; *Notes sur l'Annam*, 1885-1886; *Voyage dans le Laos*, 1895-1897; *Le Cambodge*, 1900-1904.

Ayot, D'. See **Dayot**, Jean Marie.

Ayres, Jason C. *Map of Lee County, Illinois*, Chicago, Edward Mendel 1863 (with Joseph Crawford).

Ayrouard [Ayrouart], Jacques (*fl.*1730s-1740s). French hydrographer. *Recueil de plusiers plans des ports... Mediterranée*, *c.*1720, 1732-1746.

Aytoun, James. Engraver. Map for Bartholomew Sharp's *Journey over the Isthmus of Darien*, in William Hacke's *A Collection of original voyages*, 1699.

Azambuja, Carlos Augusto Nascientes d'. Chart Colonia [Brasil], 1839 MS.

Azara, *don* Felix de (1746-1811). Spanish traveller and cartographer. *Voyage dans l'Amérique méridionale*, 1809.

Azarquiol. See **Ibn** Zarqalah.

Azevedo, José da Costa. Charts Brazil, 1860; Rio Japura, 1871; Rio de Janeiro, 1871.

Azevedo, P. de. Corrected A.T.E. Vidal's *Madeira*, 1855.

Aznar, Pantaléon. Publisher, Madrid. Francesco Vázquez's *Atlas elementar*, 1786.

Azumaya Shinkichi. Japanese publisher. Ichikawa Yoshikazu's *Gokaikō Yokohama no Zu* [plan of Yokohama], 1860-1861.

Azurara [Zurara], Gomes Eannes de. Portuguese historian and geographer.

Azuz, Abd el. See **'Abd** al-'Aziz.

Azzerboni, Giuseppe (*b*.1764). Engraver at the Stamperia Reale, Naples, for G.A. Rizzi-Zannoni's *Atlante del regno di Napoli*, 1806-1808.

Azzi, Evangelista (1793-1848). Military topographer of Parma, also engraver. *Emisfero occidentale e orientale*, 1835.

B

B., —. *Karte vom Herzogthum Holstein*, 1801.

B., A. [possibly **Brambilla**, Ambrosio *q.v.*]. *La famosa citta d'Ancona*, *c.*1554; *Chales*, *c.*1565; *Genevra*, *c.*1565.

B., A. of Frankfurt am Main. *Regionis intra Mosellam, Rhenum & Mainem...*, *c.*1762.

B., A. Chart Tibee Inlet in Georgia, 1776.

B., B. de. *Atlas Geográphico, Historico y Estadístico de España...*, 1848.

B., D. See **Bertelli**, Donato.

B., F. See **Bertelli**, Ferrando.

B., G.A. *Geographische Beschreibung der Landt-Schafften welche die Aposteln und sonderlich Paulus durchraiset*, 1655.

B., H.F. *Plan von der in Nieder-Schlesien an der Oder Pass...*, *c.*1710; *Plan von der im Hertzogthum Schlesien berühmten Haupt- und Handels-Stadt Bresslau*, 1742.

B., J. *Piscataqua River*, 1700.

B., J.B.D. See **Bouge**, Jean Baptiste de.

B., J.G. *Carte topographique du Port et de la ville de Cherbourg et ses Environs...*, 1787.

B., N. See **Béatrizet**, Nicolas; **Bellin**, Jacques Nicolas; **Bonifacio**, Natale.

B., R. [Burton, Robert]. See **Crouch**, Nathaniel.

B., W. See **Bowes**, William.

B., W. Hudson River, 1757.

Baade, Heinrich. *...Mecklenburg*, 1894; *Grossherzogthum Mecklenburg-Strelitz*, 1895; *Grossherzogthümer Mecklenburg-Schwerin und Mecklenburg-Strelitz*, 1895; *Karte des nordöstlichen Mecklenburg*, 1899.

Baader, Friedrich (1802-1867). Swiss engineer, cartographer and superintendent of roads in Basle. Triangulation, survey and map of the canton of Basle, 1836-1845 MSS (Federal Office of Topography in Wabern); *Canton Basel Stadttheil*, Basle 1838 (taken from the manuscript survey); also produced local manuscript plans.

Baagøe, Peder Hansen (1791-1826). Danish draughtsman and engraver. *Kort over Siælland og Möen* [Denmark], 1813.

Baalde, Stephanus Jacobus (*fl.*1757-1789). Publisher and bookseller 'in de Beurssteeg', Amsterdam. *Nouvel atlas géographique & historique*, 1770, 1778 (also titled *Nieuwe Astronomische Geographische en Historische Zak en Reis Atlas*); *Reis- en Hand-Atlas van Vlaanderen, Braband...*, *c.*1770; *Atlas der VII Nederlandsche Provintien*, 1788; G. Brender à Brandis's *Nieuwe... Zak-en Reis-Atlas*, 1788 edition (first published by N.T. Gravius). **Ref.** KOEMAN, C. *Atlantes Neerlandici* (1967-1971) Vol.I pp.50-51.

Baarle, C. van. See **Barlaeus**, Caspar.

Baars, W.T. *Kaart van de Eilanden beoosten Java*, 1842 (with A. Edeling, L. de Vriese and C. van de Velde).

Baarsel family of Dutch engravers; 'C. van B. et fils'.
• **Baarsel**, Cornelis van (1761-1826). Dutch engraver for 'Ollefen en Bakker'. *Plan van Douai*, 1794; *Plan Nouveau & très exact de la Ville d'Amsteldam*, 1795; Dutch edition of Matthew Flinders's *Basses Straet*, 1798-1799; *Kaart van de Omstreeken der stad Haarlem*, 1805; *Pays Bas*, 1816.
•• **Baarsel**, Willem Cornelis van (1791-1854). Son of Cornelis, engraver. *Nouvelle Carte du Royaume de Hollande*, c.1828; *Nieuwe Kaart van het Koningrijk der Nederlanden...*, 1831.

Baas [Baes], Bastiaan de (*fl.*1720s). Dutch land surveyor and cartographer; lived in The Hague, worked in Zuid-Holland. *Heerlijkheid van Streefkerk*, 1723 (with P. van Rooyen).

Babbage, B.H. Explorer in Australia. *Rough sketch of Adelaide*, 1853.

Babel, *père* — (*fl.c.*1600). Traveller in Canada. *Routes suivies de Babel* (9 sheets), MS [n.d.].

Babel, Florian (1771-1811). Austrian cartographer.

Babelon, Ernest. *Atlas archéologique de la Tunisie*, 1893.

Baber, Edward Colborne (1843-1890). *Route Map of Explorations in Western China* (4 sheets), 1877-1879; Grosvenor's Mission (4 sheets), 1878; Chart Min River [China], 1880 MS.

Baberi, Giuseppe. See **Barberi**, Giuseppe.

Babin, Charles (1860-1932). French engineer and traveller, attached to archaeological expedition to Persia 1884 under Marcel Dieulafoy. *A travers la Perse méridionale*, 1892.

Babinet, Jacques (1794-1872). *Atlas universel de géographie physique, politique et historique*, c.1860; *Mappe-Monde*, c.1860; *Planisphère*, 1865.

Babinski, A. *Carte lithologique des Mers de l'Amérique-Nord*, 1869; *Carte lithologique des Mers de l'Europe*, c.1869.

Babylonians. Introduced sexagesimal system, dividing sky into 360 degrees, degrees into minutes, and minutes into seconds. Also invented the gnomon. Clay tablet world map, c.2000 BC. **Ref.** WALLIS & ROBINSON *Cartographical innovations* (1987) N°4.121 p.176; *ibid.* N°4.031 p.163; KRETSCHMER, DÖRFLINGER & WAWRIK *Lexicon zur Geschichte der Kartographie* (1986) Vol.C/I pp.60-63 [in German].

Bach, Heinrich (1812-1870). Geological mapmaker. Produced some notable geological maps of Germany.

Bach, John. *Philippine Islands*, Manila, Carmelo & Bauermann 1911 and later.

Bach, Joseph (1774-1853). Engraver. *Plan von der Stadt Warschau ...1808*, Dresden 1809, 1831.

Bach, Karl Philipp Heinrich (1812-1870). Geologist, topographer, draughtsman and calligrapher. *Das Konigreich Württemburg und die Hohenzollerschen Furstenthümer...*, 1850; *Geognostiche Karte von Württemburg, Baden und Hohenzollern*, 1850, 1870; *Geognostische Uebersichtskarte von Deutschland, der Schweiz, &c.*, 1856; *Karte von Württemburg, Baden und Hohenzollern...*, 1857, 1864, 1869; *Geologische Karte von Central Europa*, 1859, 1868; *Neuer Schul-Atlas über alle Theile der Erde*, 1862.

Bach, M. *Historischer Plan der ehemaligen Reichstadt Nüremburg*, 1882.

Bachácek, Martin (1539-1612). Bohemian astronomer, teacher, calligrapher, writer, theologian and cartographer.

Bache, Alexander Dallas. Superintendent, United States Coast Survey. *Coast Survey...* (58 charts and sketches of US coast), 1854, published Washington 1855; *Albemarle Sound, North Carolina. Eastern Part...*, 1860, 1895; *Iso-magnetic lines Pennsylvania*, 1862; *Southern Georgia and part of South Carolina*, 1865. **Ref.** GUTHORN, P.J. *United States coastal charts 1783-1861* (1984) pp.19 and other references.

Bache, *Lieutenant* George M. Hydrographer for the U.S. Navy and United States Coast Survey. *Little Egg Harbor* [New Jersey], 1846; *Map of Delaware Bay and River...*, 1848; *Patapsco River and the Approaches* [Chesapeake Bay], 1859.

Bache, *Major* Hartman. American cartographer and surveyor (Topographical Engineers). *Charleston Harbour and the Adjacent Coast and Country; South Carolina*, 1823-1825; Kennebec River [Maine], 1834; Cape May Roads, 1836; *That Part of the Headland of Sandy Hook, State of New Jersey...in 1835*, 1842; *...Atlantic Coast from Absecum Inlet to Penobscot River...*, 1851.

Bache, R.M. Topographer and Sub-Assistant on the U.S. Coast Survey (under A.D. Bache). *Preliminary Chart of the Entrance to St John's River, Florida*, 1853.

Bachelder, John Badger (1825-1894). Artist, draughtsman and lithographer. *Map of the Battlefield of Gettysburg, July 1st, 2nd, 3rd, 1863...*, 1876.

Bachelier, Jean. 'Géographe du roi', 1620.

Bachelot, C.M. *Carte statistique de France*, 1846.

Bacher, J. See **Bauer**, J.

Bachiene, Willem Albert (1712-1783). *b.* Leerdam, *d.* Maastricht. Army chaplain, appointed Professor of Astronomy at Maastricht in 1764. *...Lands Kanaan* (4 sheets), 1757-1758; *Kerkeljke Geographie*, Amsterdam 1768; *Nieuwe Kerkelijke Geographische Zak- en Reis-Atlas*, 1780; *Atlas tot opheldering der hedendaagsche Historie*, 1785 (maps engraved by J. van Jagen). **Ref.** KOEMAN, C. *Atlantes Neerlandici Vol.I* (1967) pp.53-55.

Bachiller, Doroteo (*d.*1866). Draughtsman and lithographer. *Atlas de España*, 1849-1851.

Bachmann, Franz. *Post-und Reisekarte von Deutschland, den Niederlanden, Belgien und der Schweiz*, 1865, 1867, 1874, 1893.

Bachmann, Isidor. *Carte géologique de la Suisse* (26 sheets), 1859-1887; *Karte der quartären Bildungen des Kandergebiets*, *c.*1874.

Bachmann, John. Bird's-eye view maps of the American Civil War. *...Seat of War around Richmond...*[Virginia], 1862; *...Birds Eye View of the Seat of War...*[Washington and the Potomac River], 1862; *Bird's eye View of Louisiana, Mississippi, Alabama and Part of Florida*, 1862.

Bachmeier [Bachmayer], Johann Wolfgang (1597-1685). Theologian. Ulm, 1653, used later as insert on Johann Baptist Homann's *Ulm*, 1715.

Bachofen, Johann Heinrich [Jean-Henri] (1821-1889). Swiss architect and engraver, born in Nänikon, Uster, worked in Aargau and Geneva, also for the Swiss Topographical Bureau. *Karte der Schweiz*, 1848; *Carte topographique du Canton de Genève*, *c.*1849.

Bachot, Jérôme (*c.*1588-1635). Engineer and engraver, 'géographe du roi'. *Carte du Pais d'Aunis, Ville et Gouvernement de La Rochelle*, 1627.

Back, —. *Plan du Siège du Château de Dilenburg*, 1760.

Bäck, Elias. See **Baeck**, Elias.

Back, *Sir* George, *Captain* (later *Admiral*), *RN* (1796-1878). Artist, draughtsman and Arctic explorer 1833-1837. Charts published in Sir John Franklin's *Narrative of a journey...*, London 1823; *Narrative of the Arctic land expedition...*, London, John Murray 1836; *Narrative of an expedition in HMS Terror...*, London, John Murray 1858.

Back, Gerbrand Nicolaas (*fl.c.*1694-1740s). Dutch land surveyor and cartographer. *Kaarte van de landen gelegen onder Oostveen*, *c.*1694 (now lost).

Back, Jean [Johan] Conrad. Engraver. *Plan du château de Harburg*, 1700; *Britisch* [sic] *Empire in America*, *c.*1740 (after Henry Popple).

Back, Joseph. Engraver. *Post und Reise-Karte von Deutschland, und den Nachbar-Staaten, bis London, Paris, Montpellier etc.*, Frankfurt am Main 1843, 1850, 1852, 1864.

Backenberg, Franz Heinrich (*c.*1750/1754-1813). Soldier and mathematician. *Situationscharte von einem Theile des Churfürstenthums Sachsen etc.*, *c.*1765.

Backer, George de. Mapseller in Bruxelles. *Plan de la ville de Brusselles... avec les Campemens de l'Armée...*, 1697.

Backer, J. de. *Champ de Bataille de l'Yser...*, 1919.

Backer, Jan Jansz. Dutch land surveyor and cartographer. Worked in Noord- Holland (1626-1642) and Drente (1634). *Caerte vande Wormer met Haere Weegen Wateringen en Waetermolens...*, 1626; *Caerte ven de Einghe-Wormer, met Eenighe Aengheleghen Plaetsen...*, 1638; *...de Nesse Landen Binnen den Ringdijck vande Purmer Gelegen...*, 1642.

Backer, Remmet Teunisse. *Sterre Kaert of Hemels Pleyn*, 1703; Map for Danckerts's Atlas, 1710. **Ref.** KOEMAN, C. *Atlantes Neerlandici* Vol.II (1969) p.5.

Backer, Theodorus (*fl.*1705-1726). Dutch surveyor and cartographer. Worked in Odijk (1710, 1717). Reduced Justus van Broeckhuysen's *Carte van den hoek-thiend competerende D'Hr Jacob Woertman, gelegen onder den geregte van Odijk*, 1717.

Backhoff, Anton Edvard Theodor (1840-1922). Lithographer and cartographer. *Res-Atlas öfver Sverige*, *c.*1876; *...Norra Sverige...*, *c.*1877; *Sveriges Iernväger*, *c.*1877.

Backhouse, James. Publisher in Darlington. *A Map of the* [Quaker] *Meetings...of Lancaster, Westmoreland, Cumberland, Northumberland, Durham & York*, 1773.

Backhouse, Thomas. Charts of Nova Scotia and the harbour of Halifax, 1798, published in Thomas Jefferys' *The first part of the North American Pilot*, London, Laurie & Whittle 1799.

Backofen, Johann Heinrich (1706-1788). Swiss clock-maker, mathematician and map-maker in Zurich.

Backwell, T. Publisher. *Asia*, 1742.

Baclé, —. Bookseller and publisher of 'rue de Dunkerque', St Omer. *Carte des environs de Saint-Omer*, 1815 (engraved by L. Huby); *Le Conducteur dans Saint-Omer et des environs*, 1827. **Ref.** PACHA, B. & MIRAN, L. *Carte et plans imprimés de 1564 à 1815* (Bibliothèque Nationale de France, Paris 1996) pp.44 [in French].

Bacler D'Albe, *Baron* Louis Albert Ghislain (1761-1824). 'Rue des Moulins au coin de St. Thérèse 542', Paris. Engineer, geographer, engraver, director of the Dépôt de la Guerre, 1814-1815, director Topographical Bureau of War in Italy. *Carte génèrale du Théâtre de la Guerre en Italie* (55 sheets), 1798-1802; *Charte von Süd-oder Unter-Italien...*, 1821; *Promenades pittoresques et lithographiques dans Paris et ses Environs*, 1822.

Bacon, George Washington [& Co.] (1830-1922). Publisher of 48 Paternoster Row (1862); 337 Strand (1868); 127 Strand (1874-1890s). British publisher for J.H. Colton; took over the Wyld firm in 1893, later re-used some of Cary's plates. *Bacon's Shilling Series of War Maps*, *c.*1862; *Colton's General Atlas*, 1863; *Bacon's Road Map of London with the Railways in Operation and Constructing*, 1865 (taken from Collins); *Bacon's County Atlas*, 1869 (based on maps originally published in the *Weekly Dispatch*, 1856-1862); *Bacon's Illustrated General Atlas*, *c.*1871; *Complete Atlas of Modern Geography*, *c.*1871; *Bacon's New Map of London*, *c.*1876; *Bacon's New Quarto County Atlas*, 1876 (based on Cassell); Frederick Rose's *Serio-Comic War Map for the year 1877* (Russia as an octopus); *New Large Scale Ordnance Atlas of London & Suburbs*, *c.*1885, 1886 (based on Cassell); *Ordnance Parish map of London and Suburbs*, 1885; *New Map of London & Suburbs, Showing Main Throughfares, Railway Stations &c.*, 1887; *Cycle Road Map of 150 Miles around London*, *c.*1890; *Bacon's New Large-Scale Atlas of London & Suburbs*, 1891 (based on Cassell); *Bacon's New Map of South Africa*, 1892 and later editions; *Commercial & Library Atlas of the British Isles*, 1895 and editions to 1901 (based on Cassell); War in the Philippines, 1898; *Portable Atlas of London and Suburbs*, *c.*1899; *Bacon's Cycling Road Map 30 Miles around London*, 1904; *Ward map of Lewisham*, 1910; *Bacon's Cloth Road Map*, 1912; *Bacon's New Map of the environs of*

London, 1913; *Bacon's library map of the environs of London* (4 sheets), 1917; and many more thematic, touring and comic maps. **Ref.** SMITH, D. *The Map Collector* 65 (1993) pp. 10-15; BURDEN, E. *The Map Collector* 53 (1990) pp.52 (letters); SMITH, D. *Victorian maps of the British Isles* (1985) many entries; NICHOLSON, T. *Wheels on the road: Road maps of Britain 1870-1940* (1983); HYDE, R. *Printed maps of Victorian London 1851-1900* (1975) pp.14-15 and many entries; BACON, G.W. *Catalogue of Bacon's maps, atlases and globes* (1912).

Bacon, J. Barnitz. Surveyor. *Pier Map of the City of New York*, 1864.

Bacon, Roger (1214-1294). Philosopher, geographer, and scientist. Map of travels of Rubroeck to Tartary, 1253-1255; world map made for the Pope. **Ref.** BURKE, Robert Belle. *The Opus Majus of Roger Bacon. A translation* (Philadelphia 1928); BRIDGES, John Henry (ed.) *The Opus Majus of Roger Bacon* (Oxford 1897).

Bacot, Jacques (1877-1965). French orientalist, explorer in Tibet, 1906-1908. *Dans les Marches Tibétaines*, Paris 1909; *Les Abords du Thibet Sud-Oriental* (4 sheets), 1912.

Bacot, P. Letter engraver for Francisco Coello de Portugal y Quesada's *Atlas de España...*, *c.*1851-1867; *Mar Mediterráneo. Costa oriental de España*, 1883.

Bacqueville de la Pothérie, Claude Charles Le Roy. Historian. Maps for *Histoire de l'Amérique Septentrionale...*, 1722, later edition in 1753.

Badder, John (*fl.c.*1730/5-1756). Surveyor and editor of the *Gentleman's Diary*. *A new Plan of the Town of Nottingham*, 1744 (with T. Peat).

Baddison, Thomas. *Chart of the Track of the Ship Princess Augusta*, 1788.

Badeigts de la Borde, Jean Jacques Hippolyte Amédée. *Plan de l'Entrée du Hyal Fiord*, 1855.

Badeslade [Bladeslade], Thomas (*fl.*1712-1741). Surveyor and mapmaker concerned with river improvements and fens reclamation. *Howland Great Dock near Deptford*, *c.*1720; *A Map of the Great Level of ye Fens*, 1723; River Ouse, 1723; drew the maps for Badeslade & Toms' *Chorographia Britanniae*, 1724-1727, engraved and published by W.H. Toms 1741-1742 and editions to 1749 (miniature county maps); *History of the Navigation Port of Kings Lyn* (containing many maps of fens and rivers), 1725; Thames, 1725; Wisbech River, 1735. **Ref.** HODSON, D. *County atlases of the British Isles published after 1703* Volume I (1984) pp.154-157.

Badge, T. Engraver. *A Survey for a Line of Navigation between Lough Erne and Lough Neagh*, 1815.

Badger, George Percy. *Map of Abyssinia*, 1869; *Oman*, 1871.

Badger, S.G. *A Map of the City of Concord, New Hampshire*, 1855.

Badlidge, A. Middlesex and Essex, 1768.

Badré, L. *Carte de l'Europe*, 1838.

Badslade, Thomas. See **Badeslade**, Thomas .

Baeck [Bäck], Elias (1681-1748). Publisher and engraver of Augsburg to the Prince of Saxe-Weimer. *Atlas Geographicus...*, 1710; Map of Switzerland, 1710; *Amsterdam*, 1720; *Augusta Vindelicorum* [Augsburg, *c.*Bavaria], 1729; *Ingolstadt*, 1729; *Cölln am Rhein* [Cologne], 1730; *...fluviis Maeno, Mosella et Mosa...*, *c.*1730; *...Philipsbourg...*, *c.*1734; *Accurater Grundriss von Moscau*, 1738; Prague, *c.*1745.

Baeda, Venerabilis. See **Bede**.

Baedeker [Baedecker], family. German publishers of Coblenz (1827-1872), Leipzig (1872-1959), Freiburg (1948-1987), also Stuttgart & Munich. *Hand-Atlas für preußische Volksschulen*, Essen 1862.
• **Baedeker**, Karl (1801-1859). German editor, geographer and publisher of the celebrated series of guide books. J.A. Klein's *Rheinreise*, 1832; the first French Baedeker guide was published 1846; the first English Baedeker guide was published in 1861.
•• **Baedeker**, Ernst (*b.*1833). Son of Karl, above, whose business he took over in 1859.
•• **Baedeker**, Karl (*b.*1837). Son of Karl, took over his father's business in 1861.
•• **Baedeker**, Fritz (*b.*1844). Son of Karl, took over his father's business in 1878.

••• **Baedeker**, Hans (*b.*1874). Son of Fritz, took over the company in 1925.
••• **Baedeker**, Ernst (*b.*1878). Son of Fritz, involved in the Baedeker guide publishing business.
••• **Baedeker**, Deitrich (*b.*1886). Son of Fritz, also involved in the family business.
•••• **Baedeker**, Karl Friedrich (1910-1979). Part of the same firm and family, but the exact relationship is not clear. Published in Freiburg and Stuttgart.
••••• **Baedeker**, Florian (*d.*1980). Son of Karl Friedrich. Took over the company in 1979.
•••• **Baedeker**, Eva. Widow of Karl Friedrich, ran the Baedeker company from 1980 until 1984.

Baedeker, Adolf. Possibly a brother of Karl Baedeker, above. Publisher of Rotterdam (1836-1844), then Cologne, taken over by Otto Petri *c.*1850. *School-Atlas* (22 maps), 1840, (24 maps) 1844, 1848, (25 maps) 1851, 1852, 1858; *Kleine Hand-Atlas*, 1842, 1851.

Baehr [Bähr; Bar], Johann Leopold von (1793-1893). Engraver of Leipzig. *Neuer Atlas...*, 1850; one of the engravers for C.G.D. Stein's *Neuer Atlas der ganzen Erde*, 1852 and later; *Spezial-Karte der Elb-Mündungen...*, 1858; *Spezial-Karte der Jade- und Weser-Mündungen*, 1858; *Spezial-Karte der Weser- und Elb Mündungen*, 1859.

Baer, *Rabbi* Dov ben Joseph Yospa. *Eretz Israel* [Land of Israel], 1881 (lithographed by Peretz Feingroit).

Baer [Bar], Johann Christian (*fl.*1834-1871). Engraver for Stieler, 1834; *Post und Reise-Karte von Deutschland...*, 1835; *Deutschland, Königreich. der Niederlande...*, 1836 and later editions; *Atlas über alle Theile der Erde...*, 1847; *Deutschland, Niederlande, Belgien und Schweiz*, 1848; *Eisenbahn-Atlas von Deutschland, Belgien, Elsass und den nördlichsten Theile von Italien*, 1848; *Post und Eisenbahn-Karte von Deutschland...*, 1848, 1853, 1858; *Karte von Europa und dem Mittelländischen Meere...*, 1871, 1874, 1877.

Baer [Bear; Beer], Johann Christoph (1673-1753). *Das Neu-Genharnschte Gross-Britannien*, Nuremberg 1690.

Baer, Joseph Christoph. See **Bär**, Joseph Christoph.

Baerentzen, Emilius & Co. (1799-1868). Lithographic institute founded in Copenhagen, 1837. *Nörrejylland* (9 sheets), 1841-1848; *Sleswig*, 1849; *Nyeste Skole-Atlas over alle Jordens Dele*, *c.*1854.

Baerle, Caspar. See **Barlaeus**, Caspar.

Baermann, —. *Königreich Bayern...* (5 sheets), 1892.

Baert, Charles Alexandre Balthazar François-de-Paul, *baron de* (1751-1825). Geographer, traveller, writer and politician. *Table de la Grande Bretagne et des Possessions anglaises*, 1800.

Baertls, Joseph Anton (*fl.*1750-1761). Engraver. *Plan de Mannheim*, 1758.

Baes, Bastiaan de. See **Baas**, Bastiaan de.

Baeyer, Johann Jacob (1794-1885). Geodesist and military cartographer. Numerous works on mathematical geography. Map of East Prussia, 1830s; *Über die Größe und Figur der Erde*, Berlin 1861. **Ref.** KRETSCHMER, DÖRFLINGER & WAWRIK *Lexicon zur Geschichte der Kartographie* (1986) *passim.* [in German].

Baffin, William (1584-1622). English navigator who went on 5 voyages 1612-1622. Charts of Hudson Bay, 1615 MS; North of Davis Straits, 1616; *A description of East India conteyning th' Empire of the Great Mogoll*, published in Samuel Purchas's *Purchas his pilgrimes*, London 1626 (first English map of this area).

Bagay, Nicolás de la Cruz. Engraver and geographer of Manila. *Carta Hydrographica y chorographica de las Ylas Filipinas* (4 sheets), 1734 (published on 1 sheet by Homann's Heirs among others, 1744); Murillo y Velarde's *Historia de la Provincia de Philipinas*, Manila 1749; *Aspecto geografico del Mundo Hispanico* [World in the form of a woman], 1761.

Bagetti, Giuseppe Pietro (1764-1831). Water colourist, topographer, architect and musician. *Vues des Champs de Bataille de Napoleon en Italie dans les années 1796, 1797 et 1800* (68 plates), 1835.

Bagg, Egbert. *Forest Hill Cemetery Utica* [New York], 1870.

Bagge, Arthur H. *Map of Tenasserim* [Siam] (4 sheets), 1868.

Bagge, C. *Afrique*, 1886.

Bagge, G. Draughtsman. L. Vivien de St. Martin's *Australie*, Hachette, 1889.

Bagge, Thomas (1753-1811). Map of the Fens, 1796.

Baggesen, A. von. ...*Holstein und Lauenburg*, 1827.

Baggi, Jacome Martins. *Provincia da Bahia* (4 sheets), 1886.

Baglivi, Giorgio (1668-1706). Clinician, medical scientist and earth scientist. Produced a map of Otranto, Italy, designed to illustrate the area affected by Tarantism, 1699 and later editions (demonstrates an early use of cartography in medicine). **Ref.** JARCHO, Saul, *Imago Mundi* 35 (1983) pp.14-19.

Bagnoni, Jacopo. Publisher of Florence. Co-publisher with Antonio Francesco Lucini of Sir Robert Dudley's *Dall'arcano del mare*, 1661 (2nd edition).

Bagot, —. *Carte topographique du Département du Rhône*, 1868.

Bagshaw, Samuel. Publisher, Wentworth Terrace, Sheffield. Gazetteer of Cheshire, 1850.

Bahadur, Iman Sharif Khan. Lake Darbat, Dhofar, 1894; Hadhramaut, 1894 MS; Dhofar, 1895.

Bahr, *Doctor* —. *Karte der Heilquellen Deutschlands und der Schweiz*, Berlin 1844.

Bahre, Hans Georg (*fl.*1628-1644). Engraver and draughtsman. Abraham Holtzwurm's map of Upper Austria, Regensburg 1628.

Baidon, father and son.
• **Baidon**, *Lieutenant* Afanasiy von [Athanasius]. Captive Prussian officer, later Russian General. Father of Fyodor (below). Karta öfwer Baikalsiön, 1667; Map of the land between the Amur and Yenisey, *c.*1685; worked in north-eastern Siberia, 1710-1711.
•• **Baidon** [Beton; Beyton], Fyodor. Son of Afanasiy (above). Map of the Selenga and its tributaries, *c.*1726.

Bailey, A.M. Draughtsman. *Parker's sectional & geological map of Iowa...*, 1856; *New Map of the Pike's Peak Gold Region*, in *The Illustrated Miners' Hand Book*, St Louis 1859.

Bailey, *Captain* F.M. *Sketch map of the route of Capt. F.M. Bailey through South Eastern Tibet*, 1912; *Part of North Eastern Frontier* [of India] *and Tibet*, 1914.

Bailey, Frank and Robert. Printers and publishers at 'Yorick's Head, N°·116 High Street, Philadelphia'. *F. Bailey's Pocket Almanac*, 1784; Joseph Scott's *The United States Gazetteer*, 1795; Joseph Scott's *An Atlas of the United States...*, 1796 (re-used maps from the *Gazetteer*).

Bailey, Franklin H. (*b.*1845). Science teacher of Michigan. Patented teaching aids advertised through the New England School Furniture Co. *Astral Lantern*, 1881; *Bailey's Cosmosphere*, 1886 (terrestrial globe inside a transparent celestial sphere). **Ref.** WARNER, *Rittenhouse* Vol. 2 (1) (1987), 26-27.

Bailey, J. Engraver for Valentine Gill's map of Wexford, 1811 (with E. Ramsey).

Bailey, John (1750-1819). Surveyor, agriculturalist, draughtsman and engraver of northern England. Produced estate and inclosure maps. Inclosure map of Woodborough, 1798.

Bailey, William Lee. *A Correct Chart of the Islands... of the Kingsmills Group* [Gilbert Islands, Pacific], 1833-1841.

Baillairge, G. *Le Canada Catholique*, 1893.

Baillard family. See **Baillieul** family.

Baillergeau, —. One of the contributors to the Cassini survey, worked mostly in areas of north-east France, *c.*1758-1759.

Bailleul family. See **Baillieul** family.

Bailleul, Nicolas [*le jeune*]. 'N. Bailleul,

Geographe Lyon, chez Daudet'. *Théatre de la Guerre en Savoye et en Piemont...*, 1747; ...*Plan de la Baye de Cartagene...* (2 maps and 4 plans on one sheet), Paris 1750; *Gouvernement générale de Languedoc*, Lyon, Daudet 1753. **N.B.** This may be Nicolas de Baillieul (below), the sources are unclear.

Bailleux family. See **Baillieul** family.

Bailley, F. & R. See **Bailey**, Frank & Robert.

Baillie, Alexander. Engraver, living in Rome in 1764, later in Edinburgh. *A General Map of Great Britain*, 1747; John Laurie's *A Plan of the County of Mid-Lothian or Shire of Edinburgh*, 1763; *A Map of Ayrshire*, 1774; *A new map of Lanark-shire...*, 1775; Loch Lomond, 1777.

Baillie, Charles William (1844-1899). Superintendent of the Meteorological Office 1888-1899. *Charts showing the surface Temperature of the Atlantic, Indian and Pacific Oceans etc.* (13 sheets), 1884; *Charts showing the Mean Barometrical Pressure over the Atlantic, Indian and Pacific Oceans*, 1887.

Baillie, Thomas. *Map of New Brunswick*, London, James Wyld 1832.

Baillie, W. *Plan of Calcutta*, 1794.

Bailliere [Bailiere], Ferdinand François (*c*.1836-1881). Publisher and bookseller of Melbourne. Gazetteers and road books: *Bailliere's Victorian Gazetteer and Road Guide*, 1865; *Bailliere's South Australian Gazetteer and Road Guide*, 1866; *Bailliere's Victorian Atlas*, 1866; *Bailliere's New South Wales Gazetteer and Road Guide*, 1870.

Bailliere, H. Thought to be the publisher of J.C. Prichard's ethnographical maps of Africa, Europe, America, Polynesia, 1843.

Baillieul [Baillard, Bailleul, de Bailleul, Bailleux, Baillieu] family. **N.B.** Information on the Baillieul family is unclear. Compare with Nicolas **Bailleul** of Lyons (above).

• **Baillieul**, Gaspard (fl.1700-1744). Publisher, editor, engineer and geographer. Paris, 'au bout du pont au change vis-à-vis l'orloge du Palais au Neptune François'. Some of his work was sold through the house of Daudet, in Lyons. *Plan de la Ville de Luxembourg*, *c*.1695/1700, 1708; *Ville et Citadelle de Turin*, 1700; ...*Seraglio de Mantoue...*, *c*.1703; *Brabant et Limbourg*, 1704, 1744; *Royaume d'Espagne*, 1705; *Le Théatre de la Guerre en Piémont et Savoye*, *c*.1705; Chastellerie de Lille, 1707; La Provence, 1707; l'Alsace, 1708; ...*Cours du Rhin...Basle jusqu'à Mayence*, 1708; *Plan et environs de Barcelonne*, 1708; ...*environs de Tournay*, 1709; *Carte Nouvelle de la Milanez...*, 1710; *Le Gouvernement de Dauphiné*, *c*.1710; *Plan de St Sebastian et de ses environs*, 1719; *Plan de Fontarabie*, 1719; *Nouveau Plan de la Ville...de Paris*, 1725; *Portugal et ses Frontières*, 1725; *Carte de la Forêt de Compiègne...*, 1728; *Plan de la Ville de Rouen*, 1731; diocesan maps, 1743; *Carte particulière des Duchez de Brabant et Limbourg* (2 sheets), 1744.

•• **Baillieul**, Marie. Daughter of Gaspard, worked on books and plans published by her father. Jean-Aymar Piganiol de la Force's *Le Nouveau voyage de France*, Théodor le Gras 1724; diocesan maps, 1743.

•• **Baillieul**, François (*c*.1690-*c*.1748). Geographer and engraver, eldest son of Gaspard. *Plan de Paris, ses Faubourgs et ses Environs*, 1722; *Carte de la Touraine*, 1722 (later published in Piganiol de la Force's *Nouvelle Description de la France*); *Carte du Berry...*, 1722 (published in the work of Piganiol de la Force); ...*Canal de St.-Denis à Paris...*, 1726; *Plan de Luxembourg*, in: l'Abbé Deidier's *Le parfait Ingenieur françois, ou La Fortification offensive et defensive...*, Paris, Charles-Antoine Jombert 1742 edition; *Plan des Vingt Quartiers de la Ville de Paris*, 1745; *Carte Générale de la Rivière du Sénégal*, 1750; draughtsman and engraver for Piganiol de la Force's 'Provinces de France' (24 sheets), in *Nouvelle Description de la France* (15 volumes), 1752-1754.

•• **Baillieul**, Nicolas. Engraver, youngest son of Gaspard, worked on the plan of Paris and the 'Provinces des France' with his brother, above. *Plan des Vingt Quartiers de la Ville de Paris*, 1745; *L'Amérique Septentrionale et Méridionale...*, Lyon 1748; Europe, Lyon 1778.

Baillieux, C. Plan de Maestricht, [1708].

Baillou, Giovanni de (1758-1819). Geographer of Tuscany. *Nuovo atlante generale* (96 maps as pack of *Tarot* playing cards), 1779 (with Agostino da Rabatta, engraved by Aniella Lamberti); survey of Tuscany, early 19th-century (now lost).

Bailly Brothers (*fl.*19th century). Printers and publishers of London, recorded at '3 Royal Exchange Buildings' (1846-1862). *Wilme's Symbolic Map of London, or Visitors Guide to the Sights and Amusements of the Metropolis*, 1851.

Bailly, Jean Sylvain (1736-1793). Politician and 'savant'. *Histoire de l'astronomie ancienne*, 1781.

Bailly, Robert [Robertus de] (*fl.*1525-1530). Terrestrial globe, 1530 (bronze globe known as the Verrazano globe).

Baily & Son. *Map of Bureau County...* [Illinois], Princeton 1903.

Baily, Francis (1774-1844). Astronomer, Vice-President Royal Geographical Society. Writer on longitude.

Baily, *Lt* John. Surveyor in Nicaragua 1837-1840. *Map of Central America*, 1850.

Bain, A.C. *Geological Sketch Map of Cape Colony*, *c.*1875; *Bain's Railway Map of South Africa*, 1904.

Bain, A.D.N. *Geological Survey of Nigeria*, 1924; *Geological Map of Bauchi Town...* [Nigeria], 1926; *Geological Map of the Tinfields of Zaria & Kano Provinces...*, 1927.

Bain, Nisbet. Contributed maps to Reginald Lane Poole's *Historical Atlas of Modern Europe*, 1896.

Bainbridge, —. *...Seat of the War in Florida*, 1839.

Bainbridge, John (1582-1643). Astronomer. *Cometae Anno 1618 descriptio.*

Bainbridge, Thomas (*fl.*1774-1819). Estate surveyor. *...Bampton Estate in the County of Oxford*, 1789; *...Brimstage in the County of Chester*, 1789; *...Burghfield in the County of Berks*, 1789; *...Dunham on the Hill ...in the County of Chester*, 1789; *...Inclosures and Part of the Common Land at Lew*, 1789; *...Huddington Estate...in the County of Worcester*, 1790; *Mickle Trafford in the County of Chester*, 1792; and many other estate and village plans.

Baines, —. Engraver. *Kentucky*, London 1792.

Baines, Edward. Publisher of Leeds. J. Talbot's *United States of America Exhibiting the Seat of War on the Canadian Frontier from 1812-1815*, 1816; directory of Lancashire with city maps, 1820s.

Baines, John [& Co.]. Publisher of Leeds. Joshua Thorp's *A map of the country extending ten miles round Leeds* (4 sheets), 1831

Baines, John Thomas (1820-1875). Landscape artist, draughtsman and writer. Artist to Livingstone's Zambezi expedition of 1858, later went to Victoria Falls region with James Chapman, 1861-1863. Map of the Zambesi (3 sheets), 1859 MS; *Explorations in South-West Africa*, 1864 (account of his travels with two maps); Route to the Transvaal Goldfields, 1874 MS; *Cape Colony*, 1876; *...Goldfields of South Eastern Africa*, 1876 (with A. Belville).

Bains, Jean de . See **Beins**, Jean de.

Baird, Edward. Proposed Railways in Texas, 1840.

Baird Smith, *Lieutenant-Colonel* R. Chief engineer for the British force at Delhi. Irrigation Canals Lombardy, 1855; *Siege of Delhi*, 1876.

Baisier [Beziers], Jean-Baptiste-Laurent (*d.*1768). Engraver of 'Rue St Jacques, vis-à-vis St Yves', Paris. Engraved frontispieces. L.C. Desnos and Michel's *L'Indicateur fidèle*, 1764; G.A. Rizzi-Zannoni's *Atlas historique de la France*, Paris 1764.

Baist, George William. American publisher. *Atlas of Camden* [New Jersey], 1886; *Town Estate and Property Atlas*, 1886-1909; Atlas of Philadelphia, 1888.

Bajdon, A. See **Baidon**, A.

Bajetto, G.F.M. *Topographische Kaart der Residentie Pekalongan...*, 1871, 1892, 1913.

Bakalowitz, Jan. Geographer to Stanisław II, King of Poland. Credited with *Regni Poloniae Magni Ducatus*, Königsberg [Kaliningrad], Jacob Kanter 1770 (with Charles Perthées); also plan of Crakow, Charles Perthées 1772.

Bake, J.D. (*fl.*1749). Dutch map copyist. Copied De Vigneau's *Kaarte van Staats-Vlaanderen...*, 1748.

Baker, family.
• **Baker**, *Sir* Samuel White (1821-1893). English traveller and explorer; discovered Lake Albert (1864); governor of equatorial regions of the Nile (1869-1874). Route Khartoum to Albert Nyanza, 1862-1865 MS; *A Map of the Albert N'yanza and of the Routes leading to its Discovery in 1864*, 1866; *Map of the Country annexed to Egypt*, 1872-1873 MS (drawn by his nephew Julian Alleyn Baker).
•• **Baker**, Julian Alleyn. Nephew of the traveller and explorer Sir Samuel White Baker, for whom he drew *Map of the Country annexed to Egypt*, 1872-1873 MS.

Baker, —. Lithographer in New York. *Juliana, Brown County, Wisconsin*, 1824; *Bouckville, Madison, Madison County*, 1836.

Baker, *Lieutenant* A. *Reconnoitering Survey of the Khyber Pass from Jumrood to Dhaka*, London 1842 (lithographed by L.J. Hebert).

Baker, Alfred E. American artist and lithographer. *Plano de Habana*, 1837.

Baker, Benjamin, (*fl.c.*1780-1824). Engraver and publisher in the High Street Islington, (1791), and Lower Street Islington, (1798). Engraver to the Board of Ordnance at the Tower of London. Also engraver for William Faden, 1779; *A Plan of the line of the proposed London and Western Canal* (2 sheets), 1790; for *The Universal Magazine*, 1791-1797, (including a series of county maps); *Plan of the City of Washington...*, W. Bent 1793; *Potamographia Gallica or A chart of the Coast and Rivers of France* and other charts in *Le Petit Neptune; or French Coasting Pilot*, London, W. Faden 1793, 1805 (3rd edition); *A Plan of the Warwick and Braunston intended Canal*, *c.*1793; J. Lindley and W. Crosley's *Surrey*, 1793, 1814; R. Wilkinson's *Atlas Classica*, 1796; engraved [Daniel] *Paterson's Twenty Four Miles round London...*, Bowles & Carver 1797 (with J. Ellis); Thomas Kitchin's *Universal Atlas*, 1799; Francis Owen's *Chart of Spithead*, 1801; maps from *The Universal Magazine* were republished by Laurie & Whittle as *New and Improved English Atlas*, 1807; *A Modern Plan of the City and Environs of Dublin*, 1808; *Plan of London*, *c.*1810. **N.B.** Compare with **Baker**, Edmund and Benjamin.

Baker, Benjamin Richard (*c.*1780-1840). Engraver and lithographer. *Map of the Course of the Nile*, 1845.

Baker, Charles. Cartographer of Massachusetts. *A Map of the County of Worcester* (with *Captain* John Peirce [*sic*], engraved by Joseph H. Seymour), in Peter Whitney's *The History of the County of Worcester* [Massachusetts], 1793.

Baker, Chas. [& Co.]. 271, 272, 273 & 274 High Holborn. *The A.B.C. Pictorial Plan of London*, 1892.

Baker, E. *France with its aggrandizements in the Netherlands and Italy etc...*, 1802; *South America*, 1806.

Baker, Edward (1780-1821). Maker of chronometers in Pentonville [London].

Baker, Edward and Benjamin. *Plan of the Town of Islington*, 1793.

Baker, Francis (*fl.*1765-1785). Surveyor. *Plan of the City of Rochester* [Kent], 1772.

Baker, G. Bookseller near the Royal Exchange, London. Sold Richard Blome's *A Geographicall description of the Four Parts of the World*, 1670 (containing maps after Nicolas Sanson, many engraved by Francis Lamb).

Baker, *Captain* George. *Chart of the South Coast of Borneo*, *c.*1760, 1786; *Bay of Ivanna on the Island Batanes*, 1774; *Plan of Musa Bay*, 1774; *Chart of the River Persaim*, 1775; *Plan of the Typa near Macao*, 1780; *Chart of part of the Coast of Choromandel*, 1784; *Chart of the Coast of Sumatra*, 1786.

Baker, George Holbrook (1827-1906). Lithographer and engraver of 522 Montgomery Street, San Francisco. *Direct*

Route to Washoe and the Mountain Town, 1863; *Horton's addition to San Diego*, 1868; *Map of the White Pine Mining District, Lander County, Nevada*, 1869; *Map of the Mining Region of California* 1855, 1656.

Baker, George Percival. *Sketch Map of the Canadian Rocky Mountains*, 1899 (with J. Collie).

Baker, J. Estate plans, 1792-1810.

Baker, James (*fl.*1798-1806). British road maps, views and sketches. *Fashionable Tours from London...*, 1802; *The Imperial Guide*, 1802; *The Intelligent Traveller, and Picturesque Guide, Throughout the Borders of the Thames, Essex, Suffolk, Norfolk, The Fens, and Isle of Ely*, c.1806; *Picturesque Guide through Wales and the Marches*, [n.d.]. **Ref.** SMITH, D. & WEBB, D. *The Map Collector* 42 (1988) pp.20-26.

Baker, James T. *Map of Chicago and Environs*, 1854; *Map of the Port Huron and Milwaukee, and the Detroit and Milwaukee Railways...*, New York, W. Felt & Co., 1854.

Baker, John. *Chart of the Harbour of Rio de Janeiro*, 1838.

Baker, *Lieutenant* Joseph (*d.*1853). Brother of Benjamin Baker. Prepared charts for George Vancouver's *A Voyage of Discovery*, London 1798.

Baker, Mathias (*fl.*1759-*c.*1791). Surveyor of Warwickshire, produced estate and inclosure maps in several English counties. Uffington Berkshire, 1785 MS.

Baker, R.W. *Map of the City of Lowell, Massachusetts*, 1868.

Baker, Richard Grey (*fl.*1821-1836). Surveyor and publisher. *Map of the County of Cambridge, and Isle of Ely* (4 sheets), surveyed 1816-1820, published 1821 (engraved John Allen); *Baker's New Map of the University and Town of Cambridge*, 1830.

Baker, Robert. Surveyor. *A New and Exact Map of the Island of Antigua, ...made...1746 by Robert Baker, Surveyor General of that Island. Engraved and improved by Thomas Jefferys...*, in Thomas Jefferys's *West-India Atlas*, London, 1775, 1777, 1794 and later editions including Laurie & Whittle, 1810.

Baker, Robert. District surgeon working in Leeds in the early 19th century. Map of the Leeds cholera epidemic of 1832, 1832; *Sanitary Map of the Town of Leeds*, 1842. **Ref.** ROBINSON, A.H. *Early thematic mapping in the history of cartography* (1982) pp.172, 186-187.

Baker, *Lieutenant* Samuel, *RN*. *...the island of St Christopher in America*, 1753, 1780.

Baker, Sidney. *Map of the Cumberland County, Maine*, 1857; *Map of Sagadahoc County, Maine*, 1858.

Baker, T. (*fl.*1765-1804). Bookseller and printer in Southampton. *Isle of Wight*, 1785 and later; *Twelve Miles round Southampton with The Isle of Wight*, 1801; *Plan of Southampton*, 1802.

Baker, T.W. *Map of the State of Ohio*, 1861 (with M. Clemens).

Baker, W.B. *Kickapoo City, Kansas Territory*, 1855.

Baker, W.E. Bengal Engineers. *Plan of the Battle of Sobraon*, 1846, 1851.

Baker, William. English estate surveyor. Estate at Brill, Buckinghamshire, 1768; Snells Estate, 1768.

Baker, William. *Map of Railways in England, Wales & Scotland*, c.1867.

Baker, William E. Publisher of Philadelphia. H.F. Walling's *Map of New London County Connecticut*, 1854 (engraved by E. Herrlein, printed by Wagner & McGuigan); Ashtabula County, Ohio, 1856.

Baker, William R. Engraver and lithographer of Sydney, New South Wales from 1835. 101 King St (1835); 19 King St (1838); 103 King St (1844); 119 King St (1851). Plan of the site for the new Government House, 1836; *Map of a portion of Australia... twenty...counties of New South Wales*, 1841; Chippendale, New South Wales, 1842; *Baker's Australian County Atlas*, 1843-1846; Overland Expedition to Port Essington, 1846.

Baker, William S.G. *Atchison, Kansas Territory*, 1854.

An advertisement for Thomas Bakewell's shop in Fleet Street, London.

Bakewell, Robert (1768-1843). Geologist. Introduction to Geology, 1813.

Bakewell, Thomas (*d.*1749). Printer, book, map and printseller, joined Timothy Jordan *c.*1729 and had succeeded to his business (including Richard Blome's atlas) by 1731. Like Thomas Taylor and Timothy Jordan before him he traded from 'Next ye Horn Tavern in Fleetstreet' and from 1746 'against Birchin-Lane in Cornhill'. Republished Thomas Taylor's *England Exactly Described...*, 1731, 1735 editions (this re-used the maps from Richard Blome's *Speed's Maps Epitomised*); *An exact Description of the Principality Of Wales*, edition *c.*1731; Europe, 1739; *America...*, 1740; Africa, 1742; Asia, 1742; *A New & Accurate Map of Scotland or North Britain...*, 1746; View of New York, 1746; his imprint appears on some maps in Bowen & Kitchin's *Royal English Atlas*, 1762 and later; perspective map of Shrewsbury with the arms of 23 guilds, [n.d.].

• **Bakewell**, Elizabeth (*fl.*1749-1762). Widow of, and successor to Thomas Bakewell from 1749. She was joined by Henry Parker before 1758, and the business later passed to him. Re-published *England Exactly Described...*, and Bowen and Kitchin's *Royal English Atlas*, 1762 (with others).

•• **Bakewell & Parker**. Partnership of Elizabeth Bakewell and Henry Parker. The imrpint appeared on some maps in the *Royal English Atlas*.

Bakhturin, Aleksandr Nikolaevich. Map of Russia, 1802.

Bakhuis, R.W.M. *Zakatlas van den sterrenhemel*, 1934, 1939.

Bakker, Frans de (*fl.*1736-1767). Map maker and engraver of Amsterdam. *Plan du Siège de la ville de Namur*, 1746; *Plan de la Ville de l'Ecluse* [Sluis] *en Flandre*, 1748; plates for Dutch edition of George Anson's *Voyage*, 1749.

Bakker, J.D. Schoolmaster of Groningen, worked with F. Deelstra. *Schoolatlas*, 1885; *Atlasje voor School en Huis...*, 1900; *Schoolatlas de Geheele Aarde*, 1901 and many later editions. **Ref.** KOEMAN, C. *Atlantes Neerlandici* Vol.VI (1985) pp.70-72.

Bakker, L.E. *Schul-Wandkarte van Ostfriesland*, 1876.

Bakker, M. *Map of the Surveyed Mineral Claims of the Trail Creek Mining District, West Kootenay, BC*, 1897.

Bakmeister, L. *Kratkaya Rossiiskoi imperii geografiya*, 1773.

Bakonyi, Károly [Carl]. Hungarian military cartographer. *Plan der Umgebung von Peterwardein* [Pétervárad, now Novi Sad, Serbia], 1872 (with A. Brosch).

al-Bakri. See **Abu** Ubayd Abdallah.

Baladhuri-al, Ahmad, Ibn-Yahya (*d.c.*892). Historian and geographer, lived most of his life in Baghdad. *Futuh al-Buldan* [Conquest of the Countries], translated into English as *The Origins of the Islamic States*, 1916.

Balandin, —. Map of the rivers discharging into the River Lena [Siberia], 1640.

Balassanian, Baruir H. *Map showing Mineral Lands...Nipissing District* [Ontario], 1908.

Balbi, father & son. Geographers.

• **Balbi**, Adriano [Adrien], *the elder* (1784-1848). Geographer of Venice and Vienna. *Compendio di geografia*, 1819; *Atlas ethnographique du Globe...*, Paris 1826; *L'Empire russe*, 1829; *Statistique comparée de l'état... des crimes... de France*, Paris 1829 (with André Michel Guerry); *Essai historique, géographique et statistique sur le Royaume des Pays Bas*, *c.*1831; *Abrège de géographie*, 1832. **Ref.** ROBINSON, A.H. *Early thematic mapping in the history of cartography* (1982) pp.158-159.

•• **Balbi**, Eugenio (1812-1884). Geographer, son of Adriano, above.

Balbi, Johann Friedrich (1700-1779). Engineer and cartographer in Berlin. *Karte der Mark Brandenburg* (30 sheets), 1748-1749.

Balch, Edwin Swift (1856-1927). *Antarctica*, 1902; *Chart of the Antarctic*, 1909; *Pole and Bradley Land*, 1913.

Balch, *Lieutenant Colonel* George Thatcher (1824-1894). U.S. military cartographer. *Reconnaissances in the Dacota County*, 1856.

Balch, S.W. *Map of the central portion of the town of Yonkers...* [New York], 1892.

Balch, Thomas Willing (1866-1927). *The Alaska frontier*, [n.d.].

Balch, Vistus (1799-1884). Engraver of New York, worked in his own right for publisher William Williams. Was a member of the engraving firm of Balch, Rawdon & Co. In New York, set up the firm of Balch & Stiles with Samuel Stiles, 1830. Engraver (with S. Stiles) of John Farmer's *A Map of the Surveyed Part of the Territory of Michigan...*, 1825; William Williams's *The Tourist's Map of the State of New York*, 1827.
• **Balch, Rawdon & Co.** *Map of a part of the Niagara River...*, *c.*1829.
• **Balch & Stiles**. Engravers of New York, formed by Vistus Balch and Samuel Stiles in 1830. Maps for John Farmer's *Atlas of Michigan*, 1831.

Bald, William. Irish cartographer. South Uist and Island of Benbecula, 1805-1836; *Map of the Bogs lying in the south west part of the County of Mayo* (4 sheets), 1814; *Map of the Maritime County of Mayo* (25 sheets), 1830; Western Isles, for John Thomson's *Atlas of Scotland*, 1832; *...Western Harbour and Dublin & Limerick Railway*, 1845.

Baldacci, Antonio. *Botanische Original-Karte von Mittelalbanien u. Epirus*, 1897.

Baldaeus, Philippus. *Naauwkeurige Beschryving van Malabar en Choromandel...*, Amsterdam, 1672, later published in English translation as *A Collection of Voyages and Travels... in Four Volumes...*, 1704.

Baldamus, Alfred (*fl.*1889-1923). Mapmaker and editor. *Wandkarte zur Geschichte der Völkerwanderung...* (6 sheets), 1922; F.W. Putzger's *Historischer Schul-Atlas...*, 1923.

Baldamus, C. Edited and corrected Freiherr von Schroetter's *Karte von Ost-Preussen...*, Berlin, S. Schropp u. Comp. 1853 (first published 1802).

Baldauf, F. *Territoire d'Erfurt*, 1813.

Balde, L. *Piano topografico della città di Trieste*, *c.*1820.

Baldelli, Francesco. Translator of vol.II of Zacharias Lilius's *Breve descrittione del mondo*, Venice, 1551; translated *I Commentari Di Caio Giulio Cesare...* (with 4 maps), Venice 1573.

Balding, R. *Map of the Town of Geelong...*, 1864.

Baldouino, G. See **Baudouin**, Gaspar.

Baldrick, J. *A Plan of the City and Bay of Cadiz*, 1739.

Balducci, L. *Carta geologica dell' isola d'Ischia*, 1884.

Baldwein, Eberhard (*d.*1593). Instrument maker.

Baldwin, family. Civil engineers of Charlestown, Massachusetts, responsible for many roads, canals, tunnels and water systems, and the maps, plans and surveys to go with them. Their large collection of manuscript maps and papers, many their own, is held by the William L. Clements Library.
• **Baldwin**, Loammi [I] (1740-1807). Civil engineer and soldier. A number of plans and surveys of Charles River, Mill Creek, and Massachusetts.
•• **Baldwin**, Loammi [II] (1780-1838). US engineer and surveyor concerned with the supply of water to Boston. Many manuscript surveys including maps for Pendleton's Lithography; *Plan and profile of Chelmsford River*, 1812; *...from Farm Pond in Framingham to Long Pond in Natick for supplying water to the city of Boston*, 1834; *Plan and profiles of surveys for feeders to Jamaica Pond for Boston Aqueduct Corp.*, 1834; *Plan and profiles...for a canal from the Altamaha River to Brunswick Harbour in Georgia*, 1836.
•• **Baldwin**, George Rumford (*b.*1798). American surveyor and engraver. Manuscript plans of aqueducts for *Quebec Water Works*, 1847; *...Occupation ...of Boston Harbour*, 1866; plans of Charlestown and other cities and towns in Massachusetts.

Baldwin, family. Publishers and booksellers of London.
• **Baldwin**, Richard, *junior* (*fl.*1746-1770). Publisher and bookseller 'at the Rose in Paternoster Row'. *The London Magazine*,

1746 and later; Thomas Salmon's *Universal traveller*, 1752-1753; *Virginia*, 1755. Succeeded by his cousin Robert Baldwin, below.

• **Baldwin**, Robert. Bookseller. Continued the business of Richard Baldwin 1770 to 1810, at Union Street, New Bridge Street (1805-1810). *A View of the Rivers Kenebec and Chaudière...*, 1776; *Baldwin's New and Universal System of Geography*, 1786.

• **Baldwin**, Robert. Assumed to be related to Robert and Richard, above. Bookseller and printer at 47, Paternoster Row (1810-1855); 16, Essex Street, Strand (1856-1861). Publisher for the Society for the Diffusion of Useful Knowledge [SDUK]. Entered into partnership with Cradock & Joy, then Cradock alone. From 1842 he traded alone.

•• **Baldwin, Cradock & Joy** (*fl.*1816-*c.*1828). Partnership of Robert Baldwin and Cradock & Joy, publishers of 47 Paternoster Row, London, one of the successors to Benjamin Crosby. James Wyld's *General Atlas*, 1822 (quarto edition); maps for the SDUK, 1827.

••• **Baldwin & Cradock** (*fl.c.*1829-1837). Publishers for the SDUK, pioneers of the production of maps for the mass market until thier insolvency in 1837. Firm taken over by Simkin, Marshall & Co in 1841. *Baldwin & Cradock's Terrestrial Globe* in 6 maps and other maps and plans for the SDUK; *Atlas*, 1831 and later; *North America*, SDUK 1832.

Ref. CAIN, M.T. 'The maps of the Society for the Diffusion of Useful Knowledge: A publishing history' in: *Imago Mundi* 46 (1994) pp.151-163; SMITH, D.,'John Heywood and others' in: *IMCoS Journal* 69 (1997) pp.23-30.

Baldwin, B.F. *...proposed canal from Souhegan River...to the mouth of Nashua River* [N. America], 1802 MS.

Baldwin, C. & R. Charles and Robert Baldwin, partners from 1805-1811, publishers of Bridge St., Blackfriars, London. *Map of Honduras*, 1809.

Baldwin, George. *Plan of Sevastopol*, 1857; *Plan of the Attacks and Positions of the Allied Armies before Sebastopol in 1854-1855*, 1858.

Baldwin, James Fowle (1782-1862). U.S. engineer and surveyor. In addition to railway plans he produced many manuscript charts from coastal and river surveys, including the Saco River. *Plan...for a rail-road from Boston to Albany, NY*, 1828; *A Plan and profile of the Boston and Lowell railroad*, Boston 1836.

Baldwin, Oran. U.S. Civil engineer. *Map of the Town of Montrose, Lee County, Iowa*, 1853.

Baldwin, Richard (*fl.*1681-1698). Bookseller of London, 'Bail Court near the Black Bull, Great Old Bailey' (1681), 'Near the Oxford Arms Inn Warwick Lane' (1690). *Plan of Mons*, 1692.

Baldwin, T. *A Complete Pronouncing gazetteer...of the World*, Philadelphia, J.B. Lippincott 1856 (with J. Thomas).

Baldwin, Cradock & Joy. See **Baldwin** family of London.

Bale, Jan. Dutch surveyor and cartographer. Worked in Zeeland (1656, 1658); styled as 'landmeter van de stad Gent' (1656).

Bale and Woodward [George]. Makers of miniature globes. *New Celestial Globe*, 1845.

Baleato, *Doctor* Andres. *Cuzco*, 1792; *Plano Intendencia de Truxillo*, 1792; *Montanas Peru*, 1795; *Plano del Virreynato del Peru*, 1795.

Balen, C. van. *Kleine Atlas der Havensteden*, The Hague 1924.

Balen, M. Artist. Contributed to François Valentijn's *Oud en Nieuw Oost-Indien*, 1724-1726.

Balestrier, Louis de. *Cartes commerciales du Mexique*, 1889.

Balfe, James. *Map of Mount Wellington*, 1860.

Balfour, —. Canadian lithographer in partnership with Hugh Scobie as Scobie & Balfour, lithographers of Toronto, 1846-*c.*1850. Succeeded by Hugh Scobie *q.v.*.

Balfour, *Commander* A.F. Royal Naval officer of HM Surveying Ship *Stork*. *Approaches to Port Victoria*, 1890; *Mahe Islands*, 1890; *Simon's Bay* [South Africa], 1890.

Balfour, *Professor* Isaac Bayley. *Map of Socotra...*, 1880, published Edinburgh 1888.

Balicki, K. *Historyczny Atlas Polski*, *c.*1900, 1904.

Balinski, —. *Mapa Rzeczypospolitej Polskiej...*, 1864.

Balkhi, Abu Zayd, Ahmad, Ibn Sahl (850-934). Born near Balkh [Greater Khurasan]. Persian scholar, known for his geographical work *Suwar al-Agalim* [Shapes of the Countries], called also Taqwim al-Buldan [Assessment of the Countries], originally with maps. This was the basis of the second classical school of Islamic geography that flourished during the 10th century, known as the Balkhi School (the first school being the one influenced by Ptolemy).

Ball, Allen Edward. *Chart of the South-East Coast of Arabia*, 1839; *Kooria Mooria Bay and Islands*, 1840; *Trigonometrical Survey of the North-East Coast of Africa*, 1848.

Ball, Carl Johannes. German historian. *Plan von Jerusalem*, 1843.

Ball, *Lieutenant* H.L. Chart of Lord Howe Island, 1788 MS.

Ball, J. *Schizza geologica degli Alpi*, 1877.

Ball, John (1818-1889). English geographer and naturalist. Treatises on physical and geographical science.

Ball, John. *Map of Baharia Oasis*, 1916.

Ball, Richard. *Plan of Algiers*, 1776.

Ball, *Sir* Robert Stawell (1840-1913). *Atlas of Astronomy*, 1892.

Ball, Sheldon. Map of the village and view of the harbour of Buffalo in: *Buffalo, in 1825...*, [n.d.].

Ball [Balle], William (*d.*1690). Astronomer.

Balla, Antal [Antonius] (1739-1815). Hungarian mapmaker, born Nagykörös. Studied at Nagyszombat [now Trnava, Slovakia] and Szempc [now Senec, Slovakia]. Prolific output of some 50 printed maps, died at Nyáregyháza. Szeged, 1764; *Nova tabula geographica*, 1765; *Delineatio praediorum Kis Telek*, 1775; Mirko, 1779; Pest, 1785; *Mappa specialis regionis coeli*, 1793; *Planum L.R. Civitatis Pesth*, 1803. **Ref.** STEGENA, L. *Térképtörténet* [History of maps] (Budapest 1981) [in Hungarian]; FODOR, F. *Balla Antal élete és müszaki munkássága* (Budapest 1954) [in Hungarian].

Ballanti, G. Publisher. Carta del Gran Chaco, Faenza 1763.

Ballantine, —. Lithographer. *Plan of Dundee*, 1827; *Plan of Paisley*, 1828, for John Wood's *Town Atlas of Scotland*, 1828.

Ballard, Christophe. Publisher of Paris. Philippe Buache's *Considérations géographiques*, 1753; Nicolas Louis La Caille's *Entretiennes géographiques*, 1760.

Ballard, Samuel (*fl.*1704-1733, *d.*1761). Publisher at the Blue Ball in Little Britain, London. Wars of Europe, 1704.

Ballard, *Captain* V.V. Gorontalo River, 1798; *Survey of Kema Roads*, 1801; Chart of the Moluccas, 1811.

Ballast Office, Dublin. *Lights on the Coast of Ireland*, 1830.

Ballay, Noël Eugène (1847-1902). French explorer and traveller in the Congo. *Expédition sur les cours supérieurs de l'Ogooué, de l'Alima et de la Licona*, Paris 1879.

Balleis, —. Engraver. *Grundriss der Stadt Stuttgart*, 1780; *Siège de Schweidnitz* [siege of 1762], Frankfurt am Main, *c.*1790.

Ballendine, John (*d.*1781). *A Map of Potomack and James Rivers...Shewing their Several Communications with the Navigable Waters of...the River Ohio*, 1773 MS. **Ref.** CHURCH, R.W. 'John Ballendine, unsuccessful entrepreneur of the 18th century' in *Virginia Cavalcade* 8 (Spring 1959) p.46.

Ballester i Mas, Juliá (1750-1800). Mallorcan cartographer and engraver. *Mapa de la Isla de Mallorca*, 1785.

Balling, Friedrich. *Übersichts-Karte...Saazer Kreise Böhmens befindlichen Bergwerksmasse...*, 1870.

Ballino, Giulio (*fl.*1560-*d.c.*1592). Italian geographer. *De' disegni delle più illustri città...* (50 maps and views), Venice, Bolognini Zaltieri 1569.

Ballintyne & Byworth. Publishers of London. Captain Thomas Hurd's *The Channel Atlas*, 1811.

Ballisat, G. *Plan...of ground granted from the Crown in New South Wales*, 1814.

Ballivian, José. *Mapa corográfica de la República de Bolivia*, 1845.

Ballivian, Manuel Vicente (*b.*1849). Bolivian geographer. Various works on South American geography. *Diccionario geográfico de la República de Bolivia*, 1890-1901.

Ballot, Marie Paul Victor (1853-1939). Administrator and explorer. *Carte des Etablissements français du Golfe de Bénin*, 1889.

Balluseck, *Lieutenant* F.V. *Topographische Kaart van Krawang* [Netherlands East Indies], 1877.

Balmain & Menzies. Surveyors. Royles' Chambers, Bond St, Sydney, New South Wales (1885). *Plan shewing Albert Road and allotments near Homebush*, *c.*1880s.

Balsamo-Crivelli, G. *Una carta geologica d'Italia*, 1847.

Balthasar, Floris. See **Berckenrode**, Floris Balthasar.

Balthasarbrys. See **Berckenrode** family.

Baltimore, *2nd Baron*, Cecil Calvert. Founder of Maryland. *A Relation of Maryland*, 1635, containing first separately published map of the colony: *Nova Terrae-mariae tabula* (the maker is unknown, but the map is referred to as Lord Baltimore's Map), used later by John Ogilby in his *America* 1671.

Baltimore, *6th Baron*, Charles Calvert. Parts of Pennsylvania, Maryland and Virginia, 1763.

Baltzer, A. *Carte géologique de la Suisse* (26 sheets), 1859-1887; *Der Glärnisch*, 1873; *Geologische Karte des Kanton Bern*, 1889.

Baltzer, C.H. Celestial wall map, 1854.

Baluarte, Rafael E. *Aya cúch...*, 1901; *Rios Apurimac, Ene y Tambo*, 1901; *Mapa del Perú...*, Lima 1903.

Balugoli, Alberto (*d.*1579). Cartographer of Modena, Italy. Map of Modena, 1571.

Bambe, W. Engraver for Heinrich C. Kiepert, 1878.

Bamberg, Karl. Specialised in wall maps for schools. *Europa*, 1877; *Deutschland*, *c.*1878; *Klimatologische Schulwandkarte von Europa* [climatological wall map], 1888 (with G. Coordes); *Schul-Wandkarte von Deutschland*, 1889; *Süd-Amerika*, 1889; *Schul- Wandkarte*, 1894-1895.

Bamford, T. *Map of part of Slocan Mining Camp, West Kootenay BC*, 1897.

Bampton, *Captain* William. Marine surveyor. *Plan of part of the Strait of Sunda*, 1787, published Alexander Dalrymple 1792; Chart of Torres Strait, published Aaron Arrowsmith, 1798 (inset chart of Bampton Shoal).

Banaré, *Lieutenant* A. French navy. *Japan...channels between Misima Nada and Iyo Nado*, 1872, Admiralty Chart 1877; *Japan...Gulf of Yedo*, Admiralty Chart 1875.

Banchem, Jacob van (*fl.*1590-*c.*1608). Dutch surveyor and cartographer of Leiden. Worked in Rijnland (1590-1599). *Chaerte...van seekere landen leggende inden Ambachte van Warmont*, 1595.

Banchero, Giuseppe. *Nuova Pianta di Genova*, 1868.

Bancker, Evert. *The State of New Jersey*, 1754, 1784, 1785.

Bancker, Gerard. American public surveyor of Dutch extraction; treasurer of the State of New York, 1784. Worked on Faden's map of New Jersey, 1770-1771; *A Map of the Province of New York...*, 1772; *A Plan of Fort George in the City of New York*, [n.d.]; *Plan of the Fort*, 1774; *The Province of New Jersey...*, William Faden 1777, 1778 (with Bernard Ratzer).

Bancroft, A.L. [& Co.]. Lithographers and publishers of San Francisco, probably successor to H.H. Bancroft, below. *Bancroft's Map of California, Nevada, Utah and Arizona*, 1871 edition; *Bancroft's New Map of Central California*, 1871; printers of *Froiseth's New Sectional & Mineral Map of Utah*, Salt Lake City, B.A.M. Froiseth 1871, 1875, 1878, 1879 editions; E.S. Glover's *Bird's eye view of Logan City, Utah Territory* [Salt Lake City], 1875.

Bancroft, *Colonel* Hubert Howe [& Co] (1832-1918). Historian, bookseller, stationer and publisher of 609 Montgomery St. San Francisco. Founder of the firm H.H. Bancroft & Co.. D.M. Warren's *The Common-School Geography*, 1860 edition; D.M. Warren's *A First Book of Geography*, 1860, 1864; *Bancroft's Map of the Colorado Mines*, 1863; *Bancroft's Map of the Pacific States*, 1863, 1864, 1867, 1868, 1870 (compiled by W.H. Knight); reduced versions sold as *Bancroft's Map of California, Nevada, Utah and Arizona*, 1864, 1868, 1871, 1873, 1876, 1878 (the 1871 edition of this map was published by A.L. Bancroft, above); *Clarke's New School Geography*, 1866; *Map of the Southern Pacific States*, 1866, 1868; *Bancroft's Map of the Rocky Mountain States and the Pacific Coast*, 1866 (compiled by W.H. Knight); *Clark's New School Geography*, 1866; *Oregon, Washington, Idaho, Montana & British Columbia*, 1868.

Band, Otto. *Wandkarte des Kreises Colmar*, 1874.

Banduri, Anselmo. *Urbis Constantinopolitanae...*, 1711; Tabula Geografia [Asia Minor and Greece], 1718.

Banfield, J.A. Surveyor and geologist. *...map of the iron, silver, lead and gold region of Lake Superior, Michigan*, 1864.

Banga, Jarich (*fl.*1669-*post* 1679). Dutch surveyor, official surveyor of Friesland 1679. No maps extant.

Banister, John. Engraver of *A New Map of Leicestershire* for John Throsby's *The Memoirs of the Town and County of Leicester*, 1777.

Banister, T.C. Cartographer. *Warwickshire*, 1825.

Banister, W. (*fl.*1732-1736). Estate surveyor. Estates in Sussex, 1732-1736.

Bank, John. *Plan of the Roads West of Hastings*, 1836.

Bankes, Henry. Surveyor. *A Mapp of the lands in North Meols* [Lancashire], 1736.

Bankes, Richard (*fl.*1606-1642). Surveyed estates in several English counties. Sherwood Forest, 1609 MS; Manor of Hardmead, Buckinghamshire, 1638 MS; and others.

Bankes, Thomas. *A New Royal Authentic and Complete System of Universal Geography*, 1787-1810 (with Edward W. Blake and Alexander Cook); *New System of Geography*, 1790.

Banks, —. *Ruins discovered at the Corner of Whitcomb Street near Charing Cross*, 1821, 1822.

Banks, John Henry & Co., 4 Little Queen Street, Holborn. *A Balloon View of London*, 1851, later re-issued by Edward Stanford 1862.

Banks, *Sir* Joseph, FRS (1743-1820). Accompanied Captain James Cook on his voyage round the World, 1768-1771.*The Map of the Great Pacific Ocean*, 1769 (based on a map by A. Dalrymple, engraved by John Bayly).

Banks, T.B. *Banks' Map of the Island of Guernsey*, 1897.

Banks, William. Engraver. Lakes of Cumberland, 1860.

Banning, Emile (1836-1898). Belgian geographer. *L'Afrique et la Conférence Géographique de Bruxelles*, 1878.

Bannister, John. Surveyor. *Map of the Town of Two Rivers*, 1831.

Banse, Ewald. *Ewald Banse's Reise im Nördl. Mesopotamien 1908*, 1911; *Ewald Banse's Reise am Mittleren Euphrat 1908*, 1913.

Bansemer, Jan Marcin (1802-1840). Atlas of Poland, James Wyld 1837 (with Piotr Falkenhagen-Zaleski).

Bante, Hans. Dutch military engineer and cartographer. Commissioned 1586 'ingenieur pour fortifier la ville de Heusden'.

Baptist Missionary Society. Cameroons, 1884; Central Africa, 1884.

Baptista, Pedro João. *Lands of Cazembe*, 1873.

Baptiste, J. Mayo, 1585 MS; Mayo and Sligo, 1587 MS (with John Browne).

Baquet, H. *Plan de Sfax* [Tunisia], 1881.

Baquol, Jacques (1813-1856). Publisher and historian of Strasbourg. *Atlas historique* (2 volumes), 1860.

Baquoy, Jean Charles (1721-1777). Engraver of the title page of Gilles Robert and Didier Robert de Vaugondy's *Atlas universel*, 1757 and later.

Bär [Baer], Joseph Christoph (1789-1848). Trained with Adolf Stieler, then worked for Justus Perthes from 1826. Worked with Stieler on the *Hand-Atlas über alle Theile der Erde...*, Gotha, J. Perthes 1817-1823 (Perthes's first large cartographic publication); *Karte von Deutschland in 25 Blättern*, Perthes 1829-1836 and editions to 1876 (prepared by Bär, Stülpnagel, and Stieler to accompany Stieler's *Hand-Atlas*); together with Stülpnagel, assisted Emil von Sydow in the preparation of wall maps for school geography lessons, published as *Wandatlas über Theile de Erde* (24 maps), 1833; *Eisenbahn Atlas von Deutschland, Belgien, Elsass und nördlichsen Theile von Italien*, Gotha, Perthes 1847 (with F. von Stülpnagel).

Baracio, Francisco. *Cosmographia*, 1585.

Baradelle, brothers. Globemakers of Paris.
• **Baradelle**, Jacques-Nicolas (1701-*c*.1770). Instrument and globemaker of Paris, 'Quai de l'Horloge au Palais à l'enseigne de l'Observatoire'. *Globe Terrestre*, 1740; *Globe Celeste*, 1750.
• **Baradelle**, Jean-Jacques (*fl*.1732-1744). Brother of Jacques (above).

Baraga, Franz [Francesco] Xaver [Xaviero]. *Alpkarte*, Ljubljana 1778; Istria 1778.

Baraize, Edouard (1855-1888). Churchman and historian. *Plan des Nécropoles Thébaines...*, 1904.

Baralle, Alphonse. French cartographer. *Atlas universel*, Paris 1877 (with J.A. Fayard de la Brugère).

Baranov, Aleksandr. Map of the Naval Discoveries in the Pacific Ocean, St. Petersburg 1802.

Baranovski, Stepan Ivanovic (1817-*c*.1890). *Kratkiy geograficheskiy atlas sovremennago mira*, St Petersburg 1845; Physical map of the World, St Petersburg 1847.

Barateri, Marco Antonio [Marc Antonio]. *Nuova Descrittione del Stato di Milano* (4 sheets), 1636; *Novo et accurato disegna della Lombardia*, 1701.

Baratier, Albert (1864-1917). French explorer and traveller in Algeria and the Congo. *Carte du haut Oubangui*, Paris 1910. **Ref.** BROC, N. *Dictionnaire illustré des explorateurs Français du XIXe siècle: I Afrique* (1988) p. 15 [in French].

Baratta, Allessandro. ...*Urbis Neapolitanae... delineatio*, 1627.

Baratta, *Professor* Mario. *Grande Atlante geografico*, 1922.

Baratti, A.L.J. Map battle of Stockach, 1800.

Baratti, Domenico. Engraver of Venice, for Pietro Santini's *Atlas universel*, 1775-1784.

Barbalan, *Fra* Michelli. See **Barbolano**, *Fra* Michelli.

Barbari, Jacopo de' (*c*.1440-1515). Engraver. *Venetie* (6 sheets), 1500 (woodcut perspective plan). **Ref.** SCHULZ, J. 'Jacopo de Barbari's view of Venice' in: *Art Bulletin* 60 (College Art Association of America 1978) pp.425-74; HARVEY, P.D.A. *History of topographical maps* (1980) p.75.

Barbat, Louis (1795-1870). *Histoire de la Ville de Châlons sur Marne...*, 1855-1860.

Barbatini, J. *Pianta topografica di Lucca* (4 sheets), 1804.

Barbatinti, Tommaso. Engraver. *Carta*

topografica ... Bologna Ferrara e Romagna, 1790, additions 1816, 1825, 1836.

Barbé-Marbois, François *marquis de* (1745-1837). Diplomat and politician. *Carte de la Louisiane*, in *Histoire de la Louisiane*, 1829.

Barbeau de la Bruyère, Jean Louis (1710-1781). Geographer. *Mappemonde Historique...*, 1749.

Barber, E. Geography of England and Wales, 1877.

Barber, J. 'Gentleman from that city' [New York]. *Twenty Five Miles Round New York*, London, W. Hawkes 1776.

Barber, J.W. American engraver. S.G. Goodrich's *United States*, 1826 (with A. Willard).

Barber, John. *A map of Ireland...*, 1776 (engraved by William Watts for R. Twiss's *A Tour in Ireland in 1775*). **N.B.** Compare with John **Barber** (below).

Barber, John (*fl.*1770-1800). Engraver. John Chandler's *Greece*, 1776; worked on George Taylor and Andrew Skinner's *Roads of North Britain or Scotland*, 1776; also their *Roads of Ireland*, 1777; *Map of North Wales*, 1777 (published in Joseph Cradock's *An Account of some of the most Romantic Parts of North Wales*).

Barber, John W. (1798-1885). Antiquarian and historian from New Haven, Connecticut. *Plan of the City of New Haven*, 1825.

Barber, John Warner. *Our Whole Country: or, the Past and Present of the United States...* (2 volumes with maps), Cincinnati, Henry Howe 1861. **N.B.** Although this could well be John W. Barber, and/or J.W. Barber (above), the editors have not assumed this.

Barberet, Charles. *Atlas de géographie ancienne et moderne*, Paris 1865; *Carte d'Allemagne et des autres états de l'Europe Centrale*, 1870.

Barberi [Baberi], Giuseppe (1746-1809). Draughtsman of Venice, active in Rome. Designed frontispiece for Rome edition of C. Cellarius's *Geographia antiqua*, 1774.

Barberi, Rafael. *Plano de la Ciudad de Mexico*, 1867 (with Luis and Manuel Espinosa).

Barbey, Antonio (*fl.*1684-1714). Engraver, for Giovanni Giacomo de Rossi's *Dalmatia*, Rome 1689; *Piemonte* (4 sheets), 1691; *L'Irelanda o'uero Hibernia*, 1692; *Governo Generale d'Orleans...*(2 sheets), 1692; *Italia* (10 maps), 1693; ...*Citta di Roma*, 1697; *Liguria*, 1697; *Romagna*, 1699; and others.

Barbié du Bocage, family.

• **Barbié du Bocage**, Jean-Denis (1760-1825). French geographer and cosmographer, pupil of J.B.B. d'Anville. Librarian of the French Royal Library, prepared inventory of d'Anville's map collection. One of the founders of Société géographique de Paris, géographe du roi, 1782. *Atlas du voyage du jeune Anacharsis* [l'abbé Barthélemy], 1788; *Plan des environs d'Athenes*, 1785; corrected d'Anville's *Hémisphères*, 1786; *Recueil de cartes géographiques...de l'Ancienne Grèce*, 1791; *Carte réduite de la Mer des Indes*, 1800; *L'Ecosse avec ses Isles...*, 1800; *Cartes des campagnes d'Alexandre*, G. de Ste-Croix 1804; *Carte générale de la Grèce*, 1811; *Mer des Indes*, 1811; *Atlas d'histoire ancienne*, 1816; *Carte de la Grèce Moderne*, 1821; *Plan du Constantinople*, 1822. **Ref.** TOLIAS, G. 'The cartographer Barbié de Bocage and the approach to the Greek world in the late 18th and early 19th century' in: *Cartography of the shores and islands of Greece* [Proceedings of the IMCoS Seventh International Symposium, Athens 1989] (Athens: Society for Hellenic Cartography, 1994); POINDRON, P. 'Les cartes géographiques du ministère des Affaires Etrangères, 1780-1789, Jean-Denis Barbié du Bocage et la collection d'Anville' in: *Sources* (1943) pp.46-72.

•• **Barbié du Bocage**, Jean-Guillaume, *le jeune* (1795-1848). Geographer, son of Jean-Denis. School maps including: *Carte de la Province d'Alger...*, 1830; *Carte statistique de la France*, 1834; *Atlas illustré*, 1842, 1852, 1867; and others.

•• **Barbié du Bocage**, Alexandre-François (1798-1835). Second son of Jean-Denis. Worked for La Société de Géographie. *Gallia Ulterior vel Transalpina sive Galliae*, 1818; *Carte du Province d'Alger*, 1830.

Barbiellini, Carlo Antonio. Publisher in Milan. *Descrizione geografica dell'Italia*, 1806; *Atlante delle geografia antica e moderna* (49 maps), 1807.

Barbier, Joseph-Victor. (1840-1898). Author of several maps and geographical memoirs. *Carte Spéciale des Colonies*, 1883.

Barbier, Nicolas. Publisher, with Thomas Courteau, of the Bible of 1559 containing maps known as the Geneva 'set of four'.

Barbier de Meynard, Charles Adrien Casimir (*b.*1826). Orientalist. *Dictionnaire géographique, historique et littéraire de la Perse*, 1861.

Barbieri, C. Italian engineer and geographer. *Direzione de' viaggiatori in Italia*, 1771, 1773.

Barbin, Alexandre Fréderic (1798-1837). *Trait de géographie générale*, 1832.

Barbin, Claude. Publisher of Paris. Jean-Baptiste Tavernier's *Les six Voyages*, 1676; *Relation de la Chine*, 1688.

Barbinais, Le Gentil de la (*fl.*1714-1731). Navigator. *Nouveau Voyage autour du Monde...* (3 volumes with charts and plates), published Paris 1727, and Amsterdam 1728.

Barbolano, *Fra* Michiel. *Universale Orbe della Tera*, Venice 1514 MS. **N.B.** David Woodward has raised doubts as to the authenticity of this and other manuscript world maps of the period. WOODWARD, D. *The Map Collector* 67 (1994) pp.2-10.

Barbolano, *Fra* Hieronymo. *Universale charta de navigatione...*, Venice 1525 MS. **N.B.** Now believed to be a fake. See note at Barbolano, above. WOODWARD, D. *The Map Collector* 67 (1994) pp.2-10.

Barbot, Alphonse. *Track of the U.S. Surveying Brig, Dolphin*, *c.*1852.

Barbot, Louis J. Civil Engineer. *Historical Map of Charleston 1670-1883*, 1883.

Barbot, R. *Atlas nationale*, 1874; *Carte Celeste...*, *c.*1874; *Nouveau Plan de Lille*, 1874; *Plan de Bordeaux*, 1874; *Plan de Lyon*, *c.*1874; *Brest*, 1875; *Ville de Nantes*, 1875; *Mappemonde...*, 1878.

Barbou, J. Publisher. Louis Brion de la Tour's *Atlas et tables élémentaires*, 1777, 1783.

Barbuda, Luís Jorge de [Ludovicus Georgius; Georgio; Giorgius], *SJ* (*fl.*1575-1599). Portuguese Jesuit cartographer. *Chinae... descriptio* used by Abraham Ortelius in his *Theatrum*, 1584. **Ref.** KARROW, R.W. *Mapmakers of the sixteenth century and their maps* (1993) entry 1/152 p.18; MEURER, P. *Fontes cartographici Orteliani...* (1991) pp.111-112 [in German].

Barcaiztergui, Ventura de. *Carta...de la Isla de Cuba*, [1793], 1821.

Barcelona, Fr. Carlos de. *...Provincia de Guayana...*, 1771; *Ferras perenecientes a la conquista de los R.P. Capuchines Cathalanes de Guiana...*, 1779.

Barcheston, Warwick. Tapestry maps woven at Barcheston, Warwickshire in 1588 for Ralph Sheldon . See **Sheldon**, Ralph.

Barclay, —. Dictionary of North America, 1806 (printed by C. Brightly & E. Kinnersley).

Barclay, Anthony W. Boundary Commissioner of the 19th century, surveyed and signed maps, also commissioned many surveys of the Great Lakes area to accompany his reports to the Boundary Commission. St Lawrence, 1817; mouth of the River St Louis, 1825; Lake Superior, 1826 (with Peter Porter and David Thompson); South West End of Lake Erie, Pendleton's Lithography 1828; and many others.

Barclay, C. Engraver. Isle of Wight, 1832.

Barclay, Charles William (1759-1832). *Plan d'un partie de la côte de l'Amerique*, 1786.

Barclay, Henry. *Plan shewing Explorations between Alice Springs and the Eastern Boundary of the Province* [of South Australia], 1878.

Barclay, *Reverend* James A. *A Complete and Universal Dictionary* (54 maps and plans), George Virtue 1840-1852 (some editions have Thomas Moule's county maps, others have maps engraved by E.P. Becker's Patent Omnigraph, yet others have maps engraved by Sidney Hall); *Universal History*, *c.*1850.

Barclay, Malcolm de. *Map of the Island of Montreal*, 1914.

Bard, —. *Carte des Communications Postales, Télégraphiques et des Chemins de Fer du Royaume de Serbie*, 1893.

Bard, I.L. *Pianta...della Città di Firenze*, 1739.

Bardelle, —, *junior* (fl.*c.*1804). 'Fabricant des globes et sphères, Rue du Faubourg-St Jacques n°·7', Paris.

Bardet di Villanova, family. Family of Naples, involved in various capacities as map-makers. **Ref.** VALERIO, V. *Società uomini e istituzioni cartografiche nel mezzogiorno d'Italia* (1993) pp.395-397, 446-447.

• **Bardet di Villanova**, Pietro. Archaeologist, born Paris, worked in Naples, father of Luigi. Directed the archaeological excavations at Pompeii and Herculaneum, 1741-1745, from which he prepared a number of manuscript maps.

•• **Bardet di Villanova**, Luigi (1758-1834). Topographer from Pescara, professor of the military academy; in 1800, he found himself in charge of the fortifications at Gaeta, for which he built a topographical model; director of the Officio Topografico de Napoli, 1807. Map of the city and harbour of Toulon, 1793 (engraved by Giuseppe Guerra); planned the defences for the town of Trapani, 1808-1810; supervised the verification of Schmettau's 18th-century map of Sicily, 1809-1810; manuscript plan of the island of Ischia, 1814.

••• **Bardet di Villanova**, Federico (1796-1868). Probably son of Luigi, but there is no supporting evidence. Wrote the first Italian treatise on lithography, 1830, and brought lithography to Naples through the Litografia Militaria, 1823-1839. Worked for the Officio Topografico di Napoli. Sailed on board Captain William Smyth's ship *Aid* and surveyed parts of the Albanian coast, 1819; Census maps for the district of Castelvolturno, 1819 (used by the land registry); surveys in Terra di Lavoro [Kingdom of Naples], 1819-1820.

Bardey, Alfred Xavier (1854-1934). French explorer in Arabia and East Africa (1880-1897), *b.* Besançon. *Somal, Harar et pays de des Gallas*, Algiers 1883; *Rapport sur El-Yemen et partie du pays de l'Hadramaut (Arabie)*, Paris 1899. **Ref.** BROC, N. *Dictionnaire illustré des explorateurs Français du XIXe siècle: I Afrique* (1988) p.16 [in French].

Bardi, Lorenzo. Publisher in Florence. *Pianta topografia, veduta Generale della città di Firenze*, 1837.

Bardin, A.G. *Map of the City of Binghampton, Brown County, New York* (6 sheets), Philadelphia 1872; *Map of the Corporation of Ithaca, Tomkins County, New York*, 1872.

Bardin, Libre-Irmand (1794-1867). Mathematician and politician. *La Topographie...par des plans-reliefs et des desseins*, 1855.

Bardin, William & Son. English firm of globe and scientific instrument makers. *The Celestial Globe*, 1785; *Wrights New Improved Terrestrial Globe...*, 1790; *Bardin's new British Globes*, 1798 (18-inch terrestrial and celestial globes); *The new Twelve inch British Celestial globe*, 1800 (made for William and Samuel Jones); *A New Terrestrial Globe*, 1807.

• **Bardin**, William (*c.*1740-1798). First began making globes with Gabriel Wright in about 1780. When Wright died in 1783, William's son Thomas (below) came into the business.

•• **Bardin**, Thomas Marriott (1768-1819). Son of William Bardin. Globemaker of 'Hind Ct, Fleet St.' (to *c.*1795), then 16 Salisbury Square, London.

Bardsley, D.W. *The Monarch Series of Uniform Memory Maps*, 1895.

Bardwell, William. *...the proposed new Street from Westminster Abbey to Grosvenor Place*, *c.*1835.

Baren, Jodocus van der. *Louvain*, *c.*1620; *Bruxella*, [n.d.].

Barendszoon, Willem. See **Barentsz.**, Willem.

Barengi, Giovanni. Italian astronomer. Considerazioni di sistemi ptolemaico e copernicano, Pisa 1638.

Barentin de Montchal, Charles A. Louis (*fl.*1781-1807). *Atlas de la Géographie ancienne et historique...d'après les cartes de d'Anville...*, 1807.

Barentsen, Dirk. *Il vero disegno dell'Isola di Malta*, Venezia 1550.

Barentsz. [Barendszoon; Barentszoon; Bernard; Bernardi; Bernardus], Willem (1550-1597). Dutch navigator and cartographer from Amsterdam, died Novaya Zemlya. Attempted a Northeast passage to China 1594-1597 (2 voyages). *Nieuwe Beschryvinghe ende Caertboeck vande Midlandtsche Zee*, 1593-1595, and later editions (said to be the earliest printed sea atlas of the whole of the Mediterranean), French editions 1599, 1607, and later editions to 1626; some charts in later editions of Waghenaer's *Spieghel der Zeevaert*, and Linschoten, 1599. **Ref.** MEURER, P. *Fontes cartographici Orteliani...* (1991) p.112 [in German]; KOEMAN, C. *Atlantes Neerlandici* (1967-1971) Vol.IV, pp.22-25.

Bareri, Alfonse di. Teatro delle città del mondo, Venezia, Rasciotti 1600.

Baretti, M. Carte geologiche della provincia di Torino, 1893.

Barfield, John. Publisher of 91 Wardour Street, London (1796-1840). *Lavoisne's complete genealogical, historical, chronological and geographical Atlas...A new edition enlarged by C. Gros and J. Aspin*, 1814 and later.

Bargas, *Lieutenant —. Plano de la Bahia de Cartagena de las Yndias...*, 1735.

Bariolle, —. Text engraver for charts of d'Entrecasteaux, 1807.

Bar Kepha, Moses (*d.*982). Map of southern Syria.

Barker, family. Family of artists specialising in panoramas. Founded and ran the Rotunda in Leicester Square for the exhibition of Panoramas. It remained open from 1793 to 1863 during which period it displayed about 126 panoramas, of which many were by the Barkers. **Ref.** HYDE, R. *Panoramania* (1988).

• **Barker**, Robert (1739-1806). Irish born artist who invented the technique of the 360 degree painting *c.*1785, patented it in 1787, and coined the name Panorama in 1791. *A Panoramic View of the City of Edinburgh*, displayed in Edinburgh and Glasgow in 1788, moved to London 1789.

•• **Barker**, Henry Aston (1774-1861). Son of Robert Barker, also an artist working almost exclusively on panoramas from the age of 12. Took over the family business on the death of his father, on his own retirement in 1822 the Rotunda passed to John Burford and Robert Burford. *London from the roof of the Albion Mills*, exhibited 1791-1794, published 1792; *...Panorama of the City of Constantinople...*, 1813; *Panorama of Paris...*, 1802; *Rock of Gibralter*, 1804.

•• **Barker**, Thomas Edward. Son of Robert, above. Began in business with his father and brother, then set up a rival panorama in the Strand with Ramsey Richard Reinagh.

Barker, Ebenezer. *Plan of the City of Charlestown...*, 1848 (with others).

Barker, Elihu. *A Map of the State of Kentucky*, Philadelphia, Matthew Carey 1794; re-issued smaller scale 1795 and 1797.

Barker, *Sir* George Digby. *Lucknow and its environs*, 1857.

Barker, J.N., Surveyor. Route to Shoa [Abyssinia], 1841.

Barker, James K. Surveyor. *Map of the City of Lawrence, Mass.*, 1853.

Barker, *Captain* John. Marine surveyor. Draught Cockle and St. Nicholas Gatts [off Great Yarmouth], 1753.

Barker, Robert (1570-1645). Northumberland House, Aldersgate St., London. King's Printer, licensed to print charts and maps. Sea chart English coast, 1604.

Barker, W.P. Surveyor. Montgomery, Alabama, 1870.

Barker, William. Prolific American map and script engraver, based in Philadelphia. Maps and diagrams for Thomas Dobson's *Encyclopedia*, 1792; *A chart of the West Indies*, 1794; one of the engravers of maps for Mathew Carey, published in different atlases, starting with *A General Atlas for the Present War*, 1794; *Carey's American Atlas*, 1795 and later editions; Carey's editions of *Guthrie's Geography Improved*, 1795; *Carey's General Atlas*, 1796; and *Carey's American Pocket Atlas*, 1796 and later editions; *Plat of the Seven Ranges of*

Townships...N.W. of the River Ohio, Carey 1796; *Map of the United States exhibiting the Post-Roads*, 1796; map for Jonathan Condy's *A Description of the River Susquehanna*, 1796; *Plan of Beula, Pennsylvania*, *c*.1797; *Egypt, Nubia and Abissinia*, and *France*, for John Gibson's *Atlas Minimus*, 1798; maps for John Payne's *A New and Complete Universal Geography*, 1799; Jonathan Price and John Strother's *A Map of Cape Fear...*, before 1800; plan of the city of Philadelphia, 1800; *Map of Tennessee*, 1805; *Map of Kentucky*, 1805; and others.

Barker, William Charles. *Trigonometrical Survey of the African Coast*, 1844.

Barker, William J. Surveyor. *Barker's Map of Washington County, Pennsylvania*, 1856; *Clark's map of Fairfield County, Connecticut*, 1858.

Barklay [Barkley], *Captain* C., Commander HMS *Imperial Eagle*. Discovered Barkley Sound, Vancouver Island, Canada (1787). *A Plan of Port Effingham in Berkley's Sound*, London, J. Walter 1790. **N.B.** Compare with Charles William **Barclay**.

Barlacchi, Tommaso. Editor in Rome. Ingoldstatt, 1541; *Templum Fortunae Virilis ad ripas Tiberis in Foro Piscario nunc Mariae Aegiptiacae Sacratum*, 1550.

Barlaeus [van Baarle, Baerle], Caspar (1584-1648). Historian, philosopher and theologian of Amsterdam. Oversaw the maps in Antonio de Herrera's *Descripción de la Indias Occidentales*, 1622; *Rerum per octennium in Brasiliam Gestarum...Historia* (with Georg Marcgraf's large scale map of Brazil), Amsterdam, J. Blaeu 1647, and other editions. **Ref.** WHITEHEAD, Peter J. *The Map Collector* 40 (1987) pp.17-20.

Barlow, brothers.

• **Barlow**, Peter William (1809-1885). Civil engineer and Fellow of the Royal Society, brother of William, below. Worked on canals, docks and railways. *South Eastern Railway*, 1835.

• **Barlow**, *Lieutenant* William Henry (1812-1902). Civil engineer, Fellow of the Royal Society. Maps of docks and railways. *Ordnance Survey of the Town of St. Helens*, 1851.

Barlow, —. Engraver. *A Plan of the City and Suburbs of Canterbury*, 1815.

Barlow, A.E. *Geological Map of Portions of Hastings...and Peterborough Counties*, (with F.D. Adams) 1909.

Barlow, Edward David. Lithographer, of Sydney, New South Wales. 'Barlow's Lithographic Establishment, Bridge St' (1838-1840). Map of part of Keelogues at Illawarra..., 1839; Plan of extension to the town of Yass [known as O'Connell Town], 1840.

Barlow, Francis (1626-1702). Engraver of frontispiece title of Ogilby's *Britannia*, 1675.

Barlow, Frederick. *Plan of the Foundling Estate*, Middlesex 1844.

Barlow, Inigo. Engraver. *St. Albans Abbey*, 1790.

Barlow, James. Celestial Hemispheres, 1790; *The north part of Great Britain called Scotland*, *c*.1790; T. Telford and M. Downie's Road map of Scotland, 1803, 1804.

Barlow, James S. Engraver for George Grierson of Dublin. *The South Part of Great Britain called England and Wales*, Dublin *c*.1733 (copied from Herman Moll's map); charts for Grierson's *The English Pilot*, 1749; *coast of Newfoundland...*, 1767.

Barlow, Peter, *the elder* and *the younger*. *A Chart of Magnetic Curves...*, 1833; *A Chart of Magnetic Curves...in the Northern Regions*, 1833.

Barlow, Robert. Draughtsman of Montreal. *Map of the Province of Canada*, 1832, 1855; *Geological Map of Canada*, 1865; *Geological Map of Newfoundland*, [n.d.]; and others.

Barlow, Roger (*d*.1554). Cartographer and navigator. Sailed with Sebastian Cabot. *A brief summe of geographie*, *c*.1541.

Barlow, Thomas Harris (1789-1865). Credited with making about 14 planetaria in America, each about 13 feet in diameter, *c*.1849-1853. **Ref.** WARNER, *Rittenhouse* Vol 2 (1) (1987) p.27; LAGEMANN, R. 'Thomas H. Barlow and his planetariums', *Rittenhouse* Vol 1 (1986) pp.7-9.

Barlow, William. *The Solar System representing the Sun and planets in their natural proportions*, 1750.

Barmston, A.S. Vancouver Island, 1864 MS.

Barnard, C.W. Emur Bay, Tasmania, 1893.

Barnard, *Captain* Charles H. *Chart of the Falkland Islands*, 1829.

Barnard, Chauncey. American surveyor and engineer. *Thompsonville* [Connecticut], 1849; *Tariffville* [Connecticut], 1850.

Barnard, George D. [& Co.]. Lithographers and engravers of St. Louis, Missouri. Maps of Texas, 1892.

Barnard, *Brigadier General* John Gross. *Map of the Isthmus of Tehuantepec*, 1851; *Tehuantepec Survey*, 1852; *Official Plan of the Siege of Yorktown*, 1862.

Barnard, Joseph H. District Surveyor. Goliad Land District [Texas], 1856.

Barnard, Thomas. *...The East India Company's Lands on the Coast of Coromandel*, 1778.

Barnard, W.S. Engraver of New York. *Railroad Map of the United States...*, 1856.

Barnes, A.S. & Burr. Publishers of 51 & 53 John Street, New York. Francis McNally's *An Improved System of Geography...for schools, academies, and seminaries...*, 1856.

Barnes, John. *St. Helena Island*, Admiralty Chart 1846; French edition 1851.

Barnes, Joseph. Publisher of Oxford. Publisher of John Smith's *Virginia*, 1612. **Ref.** BURDEN, P. *The mapping of North America* (1996) N°164 pp.202-205.

Barnes, M.C. *Topographical Map of Danville... Pennsylvania*, 1870.

Barnes, Oliver W. *New York and New England Showing the Routes of the Dutchess and Columbia & Connecticut Western Railroad*, 1868.

Barnes, R.L. *A Township Map of the State of Iowa*, 1855; *Barnes' Map of Philadelphia*, 1859; *State of Pennsylvania*, 1860; *Barnes' Township Map of Western Pennsylvania*, 1862; *Barnes' Railroad, Canal & County Map of Pennsylvania, New Jersey and adjoining States*, 1865; *Barnes' Driving Map of Philadelphia and Surroundings*, 1867; *Barnes' Township Map of Pennsylvania*, 1867; *Map of Forest County, Pennsylvania*, 1868.

Barnet, James. *International Exhibition, Sydney* (9 plans), 1880.

Barney, *Lieutenant* Joshua. American army surveyor and mapmaker. *Map of the...routes surveyed for the Balt. and Ohio Rail-Road*, *c.*1829; *Profiles of Routes for the proposed Boston and Providence Rail-Road*, 1831.

Barney, W.J. *Map of North East Iowa and South East Minnesota*, 1863.

Barnickel [Barnicelius], Johann [Johannes] Christoph (*d.*1746). Architect. Le Duché de Courlande, 1747, for Homann's *Grosser Atlas*.

Barnout, Hippolyte. *Atlas Astronomique de la Poche*, *c.*1875.

Barnsley, *Lieutenant*, later *Captain*, Henry (*fl.*1742-1767). *Draught of the Island of Rattan*, 1742, used by Jefferys; Chart of New England, 1760; *...chart of the sea coast of New England from Cape Codd to Casco Bay*, 1767; *Ruatan or Rattan Surveyed...with improvements by Thomas Jefferys*, 1775.

Barnstad, Hieronymus Emanuel (van) (*fl.*1752-1789). Dutch military engineer and surveyor, worked at Sluis in Zeeuws-Vlaanderen, 3 maps recorded.

Barnston, *Brevet Major* —. Surveyor. South West Crimea (28 sheets), 1856.

Barnum, H.L. American surveyor and topographer. *A Map of the City of Middletown* [Connecticut], 1825 (engraved by J.T.Porter).

Barnum, Wallace. *Ashland, Lancaster County, Nebraska*, 1856; *Tecumseh, Johnson County, Nebraska*, 1856; *Clinton, Otoe County, Nebraska*, 1857; *Plat of Syracuse, Otoe County, Nebraska*, 1860.

Barnwell, *Colonel* John ['Tuscarora Jack']. Engineer to the Province of Carolina; Agent for the South Carolina Commons House; built Fort King George. *The Northern B[ra]nch of Alatama River...*, 1721 MS; Southeastern North America from Cape Charles in Virginia southwards to Cape Canaveral in Florida, *c.*1720-1722 MS (used as the basis for many later maps of the region, including those of Catesby, 1731; Popple, 1733; Bull, 1738; and Mitchell, 1755). **Ref.** CUMMING, W.P. *The Southeast in early maps* (1958) pp.42-47.

Barocci, Giovanni Maria (*fl.*1560-1570). Italian globemaker.

Baron, Bernard (1696-1766). Engraver for Henry Popple, 1733.

Baron, Paul (*fl.*1896-1901). Regional wall maps of Germany including *Wandkarte des Kreises Bunzlau*, 1896; *Kreis Gostyn*, 1898; *Kreis Rawitsch*, 1898; *Wandkarte des Kreises Fraustadt*, 1898; *Wandkarte des Kreises Neutomischel*, 1898; *Kreis Greifenberg*, 1899; *...Wandkarte des Kreises Bonst*, 1899; *Wandkarte des Kreises Schmiegel*, 1899; *Kreis Krotoschin*, 1901.

Baron, W.H. *Map of the City of Sydney*, Woolcott & Clarke 1854.

Baroncelli, Adrien de. *Plan vélo des environs de Paris...*, 1892.

Baroncelli, Bernardo. *Plan vélo des environs de Paris*, 1892. **N.B.** Compare with **Baroncelli**, Adrien de.

Barotte, J. *Carte géologique du Département de la Haute-Marne*, 1859-1865.

Barozzi, Properzio (*fl.*1579-1590). Italian military engineer and surveyor. Plan for rebuilding citadel of Antwerp, 1585.

Barraga, Francesco Xaviero. Istria, 1778.

Barral, Jean Augustin (1819-1884). Chemist and agronomist. *Atlas du Cosmos contenant les cartes...relatives aux oeuvres d'A. Humboldt et de F. Arago*, 1861-1867.

Barral, L.M. (died after 1850). 'Lieutenant de Vaisseau... Au Dépot-général de la Marine en 1829'. *Plan de la Baie de Rio-Janeiro*, 1829, 1866.

Barralier, Emmanuel. *Carte de l'Afrique Occidental Française*, 1903; *Carte du Congo Française*, 1906; *Afrique Equatoriale Française*, 1911.

Barrallier [Barralier], *Ensign*, later *Captain*, Francis Luis (1773-1853). Engineer, explorer, marine surveyor on Matthew Flinders's expedition to Bass Strait 1798-1803. *Coal Harbour & Hunter River* [N.S.W., Australia], 1801; *Chart of Bass's Straits...*, 1803; *Chart of Western Port and Coast to Wilson's Promontory* [Victoria, Australia], 1803; *Barbados...Trigonometrical survey* (6 sheets), 1825, published (2 sheets), 1827. **Ref.** SCURFIELD, G. & J.M. *The Map Collector* 62 (1993) pp.14-17.

Barran [Baran], —. One of the surveyors working on Cassini's *Carte de France*, *c.*1772-1778, mainly in south west France including: Cahors, 1776 (with others); Agen, 1778 (with others); Roquefort, 1778 (with others).

Barrande, Th.. French geographer. *Carte de l'Asie Centrale*, 1875.

Barras de la Penne, Jean Antoine (*d.c.*1750). *...Détroit de Gibraltar*, 1756.

Barrat, Maurice (1868-1896). French explorer and prospector in the French Congo. *La géologie du Congo français*, Paris 1895.

Barraud, Charles D. *New Zealand graphic & descriptive*, 1877.

Barre, Jean François Le Febvre de la (*fl.*1747-1766). *Carte Nouvelle de la France équinoctiale...*, 1766.

Barrechia [Barrecchia], *Professor* B. *Carte du théâtre sud-occidental de la guerre européenne* [Italy], Paris 1917; *Carte kilométrique de la Nouvelle France*, 1919; *Carte des Canaux et Rivières navigables de la France*, 1920; *Carte commerciale et routière de la Nouvelle Belgique et du Luxembourg*, 1921; *Carte commerciale et routière du Département du Nord*, 1921; *Carte commerciale et routière du Département de Seine et Marne*, 1921; *Carte de la Nouvelle Europe*, 1921; *Carte des Chemins de Fer de la Nouvelle France*, 1921.

Barreira, A. de. Publisher of Coimbra, Portugal. For André do Avelar, 1593-1594.

Barreiro, Francisco Alvarez. ...*Nueva Mexico*, 1727 MS; ...*Extremadura*, 1729; ...*Nueva Mexico*, 1770; *Plano corographico de el nuevo reyno de Toledo...*, [n.d.]; ...*provincias de Sonora, Ostimuri y Sinaloa...*, [n.d.].

Barren, A. *A Map of the Collectorate of Trichinopoly* [India], 1835.

Barreo, —. Maps for Abraham Wolfgang's *Atlas Minor*, Amsterdam *c.*1680.

Barrera, Daniel. *Map of Peru*, 1871.

Barrère, Henri. *Afrique*, 1893; *Plan Géométral de Paris* (2 sheets), 1894; *Mapa geográfico-historico de la república del Ecuador*, Quito, A. Gentil 1906; *Carte du Congo Française*, 1912; *Carte pour suivre les opérations du Général Maison en 1814*, 1914.

Barrère, Pierre (*c.*1690-1755). Coast French Guiana and Brazil, 1743.

Barres, J.F.W. Des. See **Des Barres**, Joseph Frederick Wallet.

Barreto, Don Manuel. *Carta nautica del Mar, Costas y Islas de las Yndias occidentales*, 1680.

Barrett, Lucas (1837-1862). Geologist and Director of the Geological Survey of Jamaica, 1859. *Geological map of the neighbourhood of Cambridge*, 1859.

Barretté, *Lieutenant* Thomas George Leonard, 23rd Royal Welch Fusiliers. Battle plans American War of Independence, 1780; *Plan of the Battle of Gum Swamp* [Camden, South Carolina], 1780 MS. **Ref.** GUTHORN, P. *British maps of the American Revolution* (1972) p.11.

Barri, Gerald de. See **Giraldus**.

Barrière frères. Engravers for C.F. Delamarche's Atlas, 1820.
• **Barrière**, —. Engraver, brother of Alexis. *Carte de la France divisée en 83 départements...*, 1791 (for Pierre de Belleyme); *Carte générale de la Campagne de l'Armée de Réserve en Italie*, 1799; *Carte de la République Italienne*, 1803.
• **Barrière**, Alexis. Engraver. *Plan du siège de Tarragone par l'armée française d'Aragon en 1811*, 1811.

Barrière, H. *Carte tourist de France* (4 sheets), 1914-1917.

Barrière, Marcellin (*fl.*1868-1875). *Système Planetaire-Barrière Cosmographie*, 1872.

Barrois, Charles. *Carte géologique de la France* (4 sheets), 1889, 1905.

Barrois, Théophile, *l'ainé*. Publisher of 'quai des Augustins', Paris (1783), also 'Rue de Seine', Paris. E. Mentelle's *Géographie comparée ou analyse*, *c.*1778; J. Crozet's *Nouveau Voyage à la Mer du Sud, Commencé sous les ordres de M. Marion* [-Dufresne]... (5 charts on 2 sheets), 1783; *Atlas militaire*, 1809.

Barron, Andrew. Australian surveyor. *South West Lake Eyre*, 1877-1878, published Adelaide 1879.

Barron, *Captain* Edward. Officer in the King's Own Regiment in North America. Prepared many manuscript maps of Nova Scotia and South Carolina, 1776-1780; *Fort Edward in Nova Scotia*, 1779; ...*Fort Howe on the River St John, Nova Scotia*, 1779; ...*Fort of Annapolis, Nova Scotia*, 1779; ...*intrenchment of...Citadel Hill, Halifax*, 1779. **Ref.** GUTHORN, P.J. *British maps of the American Revolution* (1972) pp.11-12.

Barros, João de (1496-1570). Portuguese traveller, geographer, historian and possibly cartographer. His accounts of parts of Asia were used as a source by several cartographers including Gastaldi, *c.*1553; Ramusio, *c.*1587; Van der Aa, *c.*1727; D'Anville, *c.*1753.

Barros Gomes, B. Cartas elémentarias de Portugal, 1878.

Barrot, Adolphe (1801-1870). French orientalist and diplomat, at Whampoa and Canton, China. *Un voyage en Chine*, 1839.

Barrow, John. Navigatio Britannica, 1750; History of Discoveries, 1756.

Barrow, *Sir* John (1764-1848). English explorer, Private Secretary to Lord Macartney when he was ambassador to China, also accompanied Macartney to the Cape Colony in 1797. Secretary to the Admiralty, 1804; founder member and vice-president of the Royal Geographical Society, 1830; encouraged the exploration and mapping of the Arctic. Worked with William Alexander and H.W. Parish on maps and views representing Lord Macartney's Embassy to China, 1793-1794; *General Chart of the Colony of the Cape of Good Hope*, Cadell & Davies 1800 in: *An account of travels into the Interior of Southern Africa*, 1801.

Barrow, John. Surveyor. *Lady Bay Warrnambool* [Victoria], 1853; *Map of the Town of Mount Gambier* [South Australia], 1869.

Barrow, T.T. *A Plan of Tunbridge Wells*, 1808.

Barry, Frederick. Engineer. *A General Map of the Grand Junction Railway of Ireland*, 1853.

Bartelot [Bartlet], John (*fl.*1530-1541). Navigator and ship owner active in Dover and English Calais. Chart of Dover harbour, 1541 MS (with Anthony Aucher, John à Borough and Sir Richard Cavendish).

Bartels, Wilhelm. Wildbad Gastein, *c.*1830 MS.

Barten, J.B. *Nederland en Indië*, 1933; *De Gehele Wereld in Kaarten en Lessen...*, 1934; *Nederland*, 1936.

Barth, B. *Map of the Siege of Vicksburg* [Mississippi], 1863.

Barth, F. *Wien, Graz, Komorn*, 1855?.

Barth, *Dr* Heinrich (1821-1865). German explorer from Hamburg. One of several who travelled in North and central Africa in the service of the UK Foreign Office, 1850-1853. He produced nine manuscript maps, which were used by A. Petermann as the basis for three coloured, lithographed maps in *An account of the progress of the expedition to Central Africa... under messrs Richardson, Barth, Overweg and vogel*, London, E. Stanford 1854, also Gotha, J. Perthes 1854.

Bartha, István (1719-1785). Hungarian map-maker, cleric and astronomer. Produced instruments for 1776 survey of Hungary, preserved at Eger.

Barthel, Otto. Atlas of Atlantic City, New Jersey, 1896 (with E. Kiser).

Barthélemier, J. Publisher of Paris, 'Quai des Grands-Augustins 39, ci-devant Quai Conti'. Successor to Charles Picquet, proprietor of the maps and stock of Adrien Hubert Brué. *Atlas universel de géographie*, 1816, 1865, 1872; *Nouvelle Carte Physique et Politique des Etats-Unis*, 1867; *Nuevo Mapa de las Repúblicas de Chile, Rio de la Plata, Paraguay y Uruguay*, 1869; *France et Prusse, Théâtre de la Guerre*, 1870.

Barthélemy, *abbé* Jean Jacques (1716-1795). Learned man and numismatist. *Recueil de Cartes géographiques*, 1787, 1790, 1799; *Voyage du Jeune Anacharsis en Gréce*, 1788, 1789 (with Barbié de Bocage); *Atlas Oeuvres complètes*, 1822.

Barthélemy, *Lieutenant* Pierre, *comte* de (1870-after 1909). French explorer and colonial officer in Indochina from 1894. *Carte provisoire de Cambodge*, *c.*1903.

Barthelet [Bartlett], Richard (*fl.*1601-1603). Cartographer, killed while surveying in Donegal in 1603. Produced at least eighteen manuscript maps of Ulster, 1600-1603 including: Fort of Mount Norris, *c.* 1601 MS (National Library of Ireland); Armagh, *c.*1602 MS (National Library of Ireland); Dundalk to the River Blackwater, 1602 MS (British Library); and others. **Ref.** HAYES-McCOY, G.A. *Imago Mundi* 19 (1965) pp.33-36; HAYES-McCOY, G.A. *Ulster and other Irish maps c.1600* (Dublin 1964).

Bartholomaeus, Fr. Atlas von dem Preussischen Staate, 1859.

Bartholomaides, László (1754-1825). Hungarian cartographer, historian, teacher and cleric, born Klenócz, died Ochtina. Produced maps for his own works on history. Cave of Baradla, 1806, reprinted 1976.

Bartholomew, family. Firm of publishers based in Edinburgh. Founded by George B. Bartholomew *c.*1826 as a firm of engravers

John George Bartholomew (1860-1920). (By courtesy of the Royal Geographical Society, London)

working with W.H. Lizars on Thomson's *Atlas of Scotland*, 1831; and on Lizars' own *General Atlas of the World*, 1836; with Sidney Hall and W. Hughes on Black's *Atlas of Australia*, 1853; with George Swanston on Fullarton's *Royal Illustrated Atlas*, 1860. Began printing and publishing in 1860, achieving a phenomenal output of general and specialist maps, atlases and plans. A great deal of their output was published by other companies such as Nelson, Macmillan, A. & C. Black, Oxford University Press, and reproduced in guides and directories. The firm was taken over by Reader's Digest Publications in 1990. **Ref.** SMITH, D. *IMCoS Journal* (Winter 1998); SMITH, D. *Victorian maps of the British Isles* (1985); *The Map Collector* 27 (1984) (Cartographical curiosities) p.49; GARDINER, Leslie *Bartholomew 150 years* (1976) plus family archives.

• **Bartholomew**, George B. (1784-1871). Engraver of Edinburgh, apprenticed to Daniel Lizars, 1797, later worked with W.H. Lizars. *Lothian's Plan of the City of Edinburgh..*, 1825; *Lothian's Plan of the Town of Leith...*,1826 (in Wood's *Town Atlas of Scotland*); and others.

•• **Bartholomew**, John (*Senior*) (1805-1861). Apprenticed to W.H. Lizars in 1820, became freelance map engraver for Lizars and other publishers in 1826. Important works include *Post Office Directory Plan of Edinburgh (for W.H. Lizars)*, 1832; contributed to *Lizars Edinburgh General Atlas of the World*, 1836, Black's General Atlas, 1846, and Fullarton's Royal Illustrated Atlas; and others.

••• **Bartholomew**, John (*Junior*) (1831-1893). Geographer and engraver. A. & C. Black's General Atlas: Solar System, 1856; A. & C. Black's Large Tourist Map of Scotland, 1862; Six-inch Directory Plan of Edinburgh and Leith, 1868; District maps of Scotland, 1876; Edinburgh, Large City Plan, 1885; and others.

•••• **Bartholomew**, John George B. (1860-1920). Geographer and cartographic publisher who developed contour-layer colouring on maps. From 1888 he worked in partnership with Thos. Nelson as map publisher. Reduced Ordnance Survey maps of Scotland (29 sheets), 1890; RSGS Atlas of Scotland, 1895 (1st national atlas); England & Wales (37 sheets), from 1897-1903; Citizen's Atlas, 1898; Botanical Survey of Scotland, 1901-1907; John Murray's Bathymetrical Survey of Scottish Fresh Water Lochs, 1901-1907; Climatological Atlas of India, 1906; *The comparative atlas, physical & political...*, 11th edition, London, Meiklejohn & Son Ltd 1914; *Australasian school atlas...* (58 maps), Melbourne, H. Milford 1915 (with K.R. Cramp); Times World Survey Atlas, 1922; and others.

••••• **Bartholomew**, John B. (1890-1960). Geographer and cartographic publisher.

Bartle, H. (*b.*1840). Works on mathematical geography.

Bartlet, John. See **Bartelot**, John.

Bartlett, F.A. *Topographical Survey of the Borough of St. Marylebone*, 1832, 1834 (published at the scale of 600 ft to one inch).

Bartlett, G. *Index Geologicus*, 1841.

Bartlett, John Russell (1805-1886). Antiquarian and bibliographer, born Providence R.I., moved to New York in 1836. In partnership as a bookdealer with Charles Welford as Bartlett & Welford; later became one of the founders (with John Carter Brown) of the book collection and catalogue which was to form the basis of the John Carter Brown Library, Providence, Rhode Island. In 1850 became the U.S. Commissioner for the boundary between the United States and Mexico. His two volume narrative included: *General map showing the countries explored and surveyed by the United States & Mexican boundary commission*, New York, J.H. Colton 1854.

Bartlett, Richard. See **Barthelet**, Richard.

Bartlett, S.M. *County Atlas of Monroe Michigan*, New York, F.W. Beers & Co.1876 (engraved by L.E. Neumann).

Bartlett, William. Ordnance draughtsman. Dover Castle, 1781.

Bartman, George. *Plan of Fort Edward*, *c.*1755 (with Henry Gordon); *...the lines at Saratoga*, 1757.

Bartoli, father and son. **Ref.** VALERIO, V. *Società uomini e istituzioni cartografiche nel mezzogiorno d'Italia* (1993) pp.450-453.

• **Bartoli**, Gennaro (1770-1837). Engraver of the Stamperia Reale, and the Officio Topografico di Napoli (1809-1835), also

associated with publications arising from the excavations at Herculaneum. Maps for *Atlante universale e portatile di geografia...* 1813, 1817; *Provincia di Napoli...*, 1814; *l'Atlante delle quindici provincie al di quà del Faro del Regno delle Due Sicilie*, *c.*1816; amended 6-sheet map of the Kingdom of Naples, *c.*1818; outline map of the island of Capri, 1818 (part of a 9-sheet map of the Kingdom of Naples); 3rd sheet of a map of the neighbourhood of Naples, 1821-1822.
•• **Bartoli**, Leopoldo (1806-1865). One of the sons of Gennaro. Geographical engineer in the Officio Topografico di Napoli. Worked on several sheets of a large scale survey of the Kingdom of Naples at intervals between 1839-1860; map of the straits of Medina, 1840-1841.

Bartoli, Daniello [also known as Langobardi]. *Geographia moralibus et politicis* (7 maps), Constance 1674.

Bartolí, Manuel Escudé. *Atlas Geográfico Ibero-Americano España*, 1901-1903.

Bartoli, Papirus and Simone. *Civitas Nova in Piceno*, 1613.

Bartolommeo dalli Sonetti [Zamberti]. Venetian sailor and cartographer. *Isolario* (49 woodcut maps of Aegean islands, with descriptions in verse printed on facing pages), 1485 MS, *c.*1485 (printed), reprinted 1532.
Ref. WALLIS, H. & ROBINSON, A.H. *Cartographical innovations* (1987) p. 321.

Bartolozzi, F. Artist and engraver. Best known for the so-called 'Death of Cook' plate added to copies of the atlas to the Cook voyages. Designed and engraved vignettes on title pages of some atlases and contributed to Wilkinson's Atlas, 1802.

Barton, James. Surveyor on the sloop *Endeavour*. *Chart of the Island Palawan*, 1774, published Laurie & Whittle *c.*1792; *A Chart of the Northwest coast of Borneo...*, 1780.

Barton, *Reverend* John. *India*, 1873.

Barton, T.A. *Chart of Owls Head Harbour* [Maine], 1837 (with A.R. Flint).

Barton, W. *City of Troy, New York*, 1858, 1869.

Bartonec, Franz. *Geognostische Übersichts Karte...*, 1894.

Bartram, John. *Town of Oswego* [New York], 1751.

Bartram, William. American naturalist. *A Map of the Coast of East Florida*, 1791.

Bartruff, Friedrich Johann Jakob (1754-1833). Military cartographer.

Bartsch, Jakob (1599/1600-1633). Professor of Mathematics at Strasbourg, astronomer. Son-in-law of Johann Kepler. *Planisphaerium Stellatum*, 1661 (derived from earlier globe by Isaac Habrecht, 1621, itself based on a celestial globe by Plancius and Van den Keere, *c.*1612).

Bartsch, Johann Gottfried (*d.*1684). Engraver in Berlin.

Bartsz, Maarten. Meseen, 1754 (engraved by Thomas Jefferys).

Baruffaldi, Ambrogio. Engraver. *Corografia del ducato di Ferrara*, 1758.

Baruffi, *abbé* Giuseppe Filippo (*b.*1800). Italian savant. Peregrinazioni auturnali, 1841-1843.

Barvalet, J.J. Engraver. Map of Ireland, 1793.

Barvicius, Johannes Antonius. Globi geographicae et astronomicae, Cologne 1577.

Bary, Hendrik (1640-1707). Copper engraver. Map of Palestine in Bible, 1672.

Bas, F. de. *Waterwegen in Nederland*, 1880.

Basanta, Joseph. Surveyor-General. *Island of Trinidad*, Arrowsmith 1838.

Bascarini, Nicolo (*fl.*1547-1548). Venetian printer of a pocket edition of Ptolemy with maps by Gastaldi, *La Geografia di Claudio Ptolemeo* (60 copperplate maps), Giovanni Baptista Pedrozano 1548 (with additions and amendments by Sebastian Münster).

Basch, A. & Co. Publishers of Sydney. *Atlas of New South Wales*, 1874.

Basch, L. *Atlas of the Settled Counties of New South Wales*, 1874; *New Victorian Counties Atlas*, 1874.

Baschin, Adolph Carl Otto (*b*.1865). Based in Berlin. *Bibliotheca Geographica*, 1895-1914.

Baseggio, Giacomo. *Nuova Carta Maritima del Mar Negro*, 1793.

Basil [Basyll, Basill, Basilius], Simon (*fl*.1590-1615). Surveyor of HM Buildings, 1608. Plans of Hertford Castle; Hatfield House; Ostend, 1590; Sherborne Lodge, MS 1600; and others.

Basilicata, Francesco (*fl*.1612-1638). Cartographer who lived and worked in Crete, producing manuscript maps of the island. Atlas of Candia, 1612 MS (British Library); *Pervetusti atque nobilissimi cretensis regni urbes...* (map and 42 plans and drawings), 1618-1619 MS (Museo Correr); *Relatione di tutto il Regno di Candia...* [map of Crete], 1629 MS (Biblioteca Nazionale Marciana). **Ref.** CLUTTON, Elizabeth *Imago Mundi* 34 (1982) pp.48-65.

Basilius, Simon. See **Basil**, Simon.

Basire, family (*fl*.1720-1855). English engravers, produced work for the Royal Society of Antiquaries. **N.B.** As three members of the family had the same name, their work cannot be individually attributed with accuracy. **Ref.** GARDINER, R.A. in: WALLIS, H. & TYACKE, S. *My head is a map* (1973) pp.55-57.

• **Basire**, Isaac (1704-1768). Engraver of Clerkenwell, London. Credited with engraving Francis Drake's *Plan of the City of York*, 1736, 1744 (to accompany his *History*); Kirby's map of Suffolk, 1736-1737; Lemprière's *Cartagena*, 1741; engraved *Town and Fortification of Gibraltar, Town and Harbour of Mahon* for Nicholas Tindal's continuation of Rapin de Thoyras's *History of England*, *c*.1745; for John Rocque's *Plan of the Cities of London and Westminster*, 1749 (the lettering was engraved by R.W. Seale).

•• **Basire**, James [I] (1730-1802). Son of Isaac, picture engraver of Great Queen Street, Lincoln's Inn, London. Succeeded George Vertue as engraver to the Society of Antiquaries. *Plan of the Tower of London*, *c*.1750; *Siege of Pontefract Castle*, 1750; *Plan of the battle of Saragossa*, *c*.1751; *Plan of the Country and Camps of Almanar*, 1751; *Le Champ de Drap D'Or*, 1778; engraved part of Hereford Cathedral's *Mappa Mundi*, Richard Gough *c*.1780; *Mappa Britannica* for William Roy's *Antiquities*, 1793.

••• **Basire**, James [II] (1769-1835). Son of James Basire [I], also engraver to the Society of Antiquaries, specialising in maps and plans. Maps and plans for House of Commons Reports, 1796-1803; *...Coal District in the Counties of Northumberland and Durham*, 1800; *Plan and Elevation of London Bridge*, 1803; *A Plan of the District supplied with water from Shadwell Waterworks*, 1803; *...proposed docks at the Isle of Dogs*, 1803 (and many other plans of the Thames and proposed docks); ancient territorial districts of Scotland, 1805; *Stonehenge*, 1810; engraved several maps for the Irish Bogs Commissioners (66 sheets), 1810-1814 including William Bald's map of the county of Mayo (2 sheets), 1812-1814; *Plan of improvements on the Crown's Estate in Hamilton Place and Piccadilly...*, 1812; *...street proposed from Charing Cross to Portland Place*, 1813; *Red River Settlement* [Canada] (3 maps), 1819; *Mary-bone Park Estate now called Regents Park*, 1824; Swan River, 1827.

•••• **Basire**, James [III] (1796-1869). Engraver, son of James Basire II with whom some of his work is confused, the third of the name to be engraver for the Royal Society of Antiquaries. *Map of London, shewing the Boundaries of the General and Two Penny Post Deliveries...*, James & Luke Hansard 1830; Roman Norfolk, 1831; *Plan of a Roman Villa... in Sussex*, 1832; *A Chart of Magnetic Curves*, 1833; maps for S. Woodward's *History of Norwich Castle*, 1836-1847; *Exeter Canal in the reign of Queen Elizabeth*, 1839; map of Westminster, 1846; *Skeleton Map of Scotland*, 1846 (for the Board of Trade).

Bashmakov, D. (*fl*. 1650-1688) Russian surveyor and cartographer. Plan of the main road from Vologda to Moskow, 1665 MS; Map of the course of the river Zapadnaya Dvina and its mouth, 1666 MS; and several others.

Baskin, O.L. Partner in the firm of Everts, Baskin & Stewart, county atlas publishers of Chicago. *Ogle Co. Illinois*, 1872;

Combination Atlas Map of McHenry County, 1872.
• **Baskin, Forster & Co.** Publishers of Chicago. *Illustrated historical atlas of the State of Indiana*, 1876.

Basola, *Rabbi* Moses ben Mordechai (1480-1560). Italian rabbi, scholar and traveller. *Eretz Israel* [Land of Israel].

Bass, George (*d*.1812). Explored coast of New South Wales and circumnavigated Tasmania. Twofold Bay, 1798; Pacific Isles, 1802; Straits of Malacca, 1805.

Bassac, Edmond. *Carte Hydrographique, Topographique et Archéologique du Golfe de Morbihan*, 1878.

Bassac, N. *Plan de la Ville et des environs de Vannes*, 1898.

Bassaglia, G. Printer of Venezia. *Nuovo atlante portatile*, 1785.

Bassani [Bassanus], Bernardino. Engraver. Milano (4 sheets), 1636; *Novo et accurato Disegno della Lombardia*, 1701.

Bassantin, James (*d*.1568). Scottish astronomer. *Discours astronomique*, Lyon 1557.

Bassanus, Bernardino. See **Bassani**, Bernardino.

Basset family. Publishers of 'Rue St. Jacques', Paris. C.T. Chaumier's *Carte Routière de la France*, 1792; H. Jaillot's *Carte routière du Théâtre de la guerre Dans la partie Septentrionale de l'Italie...*, 1800; *Carte des environs de Paris*, 1815; Hérisson's *Carte routière du royaume de France...*, 1816, 1841, 1842; Noël's *Carte routière de la France...*, 1822 and later editions; *Carte de l'Amerique Septentrionale et Meridionale*, 1833; A.H. Dufour's *Atlas départmental de la France* (36 maps), 1835; and many others. **N.B.** The relationships between members of this family are uncertain, as are the correct attributions for the maps.
• **Basset**, Antoine *l'ainé* (*d*.1760). 'Rue St. Jacques à la Comète d'Or', Paris. *Veue de la ville de Chartres...*, after 1724.
• **Basset**, Catherine Françoise Dalendon (*fl*.1740-1767). Widow of Antoine, continued business as 'imagère'.
•• **Basset**, André *dit le jeune* (*fl*.1750-1787). Engraver and publisher, 'Rue St Jacques au coin de celle des Mathurins à Ste Genevieve', Paris, thought to be the son of Antoine, and the father of Paul-André. *Vue de la Ville de Richelieu en Poitou*, *c*.1763; Carte de la Baye de Gibraltar, 1781.
••• **Basset**, Geneviève. Daughter of André. Married publisher Jacques-Simon Chéreau in 1787.
••• **Basset**, Paul-André *le jeune* (*fl*.1755-1839). Parisian publisher, son of André. 'Rue St Jacques 64 au coin de celle des Mathurins à l'Image S.e. Genevieve' (1782). Lorraine, 1762; Théâtre de la Guerre en Amérique, 1779; Golphe du Mexique, 1780; L. Denis's *Carte du theatre de la guerre dans l'Inde...*, 1781; *La Mer Mediterranée...*, 1782; for L. Denis, 1793; Phelipeau's *Carte de la France*, 1801; Bonne's *Carte Générale de l'Europe...*, 1806; Asie, 1809; for Alexandre Emil Lapie 1817-1819; for Eustache Hérisson, 1829; *Carte de l'Amerique Septentrionale et Meridionale*, 1833.

Basset, Thomas (*fl*.1659-1693). Bookseller 'at the George in Fleet St.', later at Westminster Hall in London. Basset was declared bankrupt in 1696. *Cosmography*, 1669; Richard Blome's *...description of the World*, 1670 (with N. Brooks and E. Brewster); Speed's *Theatre of the Empire of Great Britaine* [with *Prospect of the most famous parts of the World*], 1676 (with Richard Chiswell); Richard Adams' *A New Large Map of England*, 1677; *...Islands in the Archipelago, Sea of Constantinople and Gulph of Smyrna...*, 1687; numerous titles of travel literature to the 1690s.

Bassett, A. Surveyor. *Boundaries of the parish of Ealing*, 1777.

Bassett, George (*fl*.1796-1846). Surveyor, 35 Arundel Street, London. Broadwater, 1806 MS; The Coast from Dymchurch to Sandgate, 1809 MS; *A Plan of the intended New North Road from London to Highbury*, 1812 (with C.T. Christian); Kinsale, *c*.1815 MS; Parish St. Mary Kensington, 1822 (with C.T. Christian and Hodskinson).

Bassett, Nathaniel. Surveyor. ...*Woodgatesen Farme*, 1705.

Basso [Bassus], Francesco [Franciscus]. Cartographer and instrument maker of Milano. Globe, 1570.

Bassot, *General* L.A. Léon (*b.*1840). Director of 'Service Géographique'. Mathematical geography meridian France, 1870-1892.

Bassus, Franciscus. See **Basso**, Francesco.

Bast [Bastius], Pieter [Petrus] Jacobsz. (1570?-1605). Publisher, engraver and geographer from Antwerp, worked in Emden and Leiden. Engraver of topographical drawings mainly of the Low Countries. Perspective view of Amsterdam (4 sheets), 1597; two other profiles of the city, 1599 and 1601; numerous similar plan-views including Leeuwarden, 1603; Middelburg, c.1596. **Ref.** KEYES, G. *Pieter Bast* (Alphen aan den Rijn, 1981).

Bastard, Brothers. Family of surveyors and architects from Dorset, All undertook work involving surveys within the county.
• **Bastard**, John (*c.*1688-1770). Architect and surveyor. Produced at least one estate survey.
• **Bastard**, William (*c.*1689-1766). *Plan of the Town of Blandford*, 1731.
• **Bastard**, Benjamin (*c.*1698-1772). Mason and land agent of Dorset. Map of the manor of Oborne, 1766.

Bastard, Eugène (1865-1910). French naturalist, traveller and colonial administrator in Madagascar (1895-1910). Map of Onilahy, 1899.

Bastert, A. *Map of the city of Washington* (10 sheets), Boston, A. Petersen & J. Enthoffer 1872 (with P.H. Donegan and J. Enthoffer).

Bastian, A. *Die Indochinesischen Reiche Birma, Siam, Kambodja und Annam*, 1867; *Uebersichtskarte der ethnologischer Culturkreise...*, 1868.

Bastide, *Captain* Henry. ...*City and fortress of Louisbourg*, 1745 MS; Plan Harbour & Fortifications of Louisbourg, 1745; *Prospect of Annapolis Royal in Nova Scotia*, 1751; Harbours of Nova Scotia, 1761.

Bastide, John Henry. Engineer. Elizabeth Castle, 1737 MS; Mount Orgueil Castle Jersey, 1741 MS; *Prospect of St Peter's Port and Town*, 1745; *Prospect of the Fort and Town of St Aubins*, 1750; *Prospect of St Helier's*, *c.*1750.

Bastingius, Jeremias. Dutch land surveyor and cartographer. Lived in Strijen (1637), worked in Noord-Brabant (1658-1660). *Afmeetinge van...de stad s'Hertogenbosch* 1658, reprinted 1757.

Bastius, Petrus. See **Bast**, Pieter Jacobsz.

Basyll, Simon. See **Basil**, Simon.

Batalha, Caetano María (*fl.*1855-1878). Portuguese cartographer. *Carta do Berlenga, Farihõese Enscada de Peuiche*, *c.*1855; *Plano hydrographico da Barra do Porto de Lisboa*, 1857; *Carta topographica do Pinhal Nacional de Leira*, 1859; *Plano hydrographico da Barra do Porto*, 1871.

Batatzi [Battatzi], Basil. Mapmaker of Constantinople. *Charta...pars Asiae*, 1732.

Batchelor, Thomas. Publisher, 'At the Globe near Charing Cross', London, in conjunction with Christopher Browne, 1690-1691. Map of Piedmont and Savoy, 1690; A New and Large Map of Ireland, 1691.

Bate, H.C. *Underground survey principal mines at Stawell* [Victoria] (6 sheets), 1879.

Bate, Robert Brettell (*d.*1847). Map seller of 21 Poultry, London. Agent for Admiralty Charts, 1814-1849; sold Edward Smith's map of Syria, 1840.

Bate, William Thornton. *Mer de Chine*, 1863.

Bate & Kennish. Surveyors and publishers of New York. *Map of the Village of Kent Litchfield County Connecticut*, 1855.

Batelli & Fanfani. Publishers in Milan. *Mappemonde*, 1820.

Bateman, Henry William. *Plan of the District around Shanghai*, 1864.

Bateman, J.F. Maps of Wales for Water Supply, 1870.

Bateman, John. Engraver of some of James Horsburgh's charts for the East India Company, 1813-1831; *Survey of the North*

& South Sands in the Straits of Malacca, 1820.

Bateman, Richard. Engraver. *Plan of the Harbour near Singapore*, 1829; *A Survey of Singapore Harbour*, 1830; *A Survey of Kyoak Phyoo Harbour*, 1833, 1843; *A Survey of the Passages leading to the Lymoon*, 1833.

Bateman, Thomas (*fl.*1746-1783). Estate surveyor, globemaker and engraver of London. Apprenticed to Nathaniel Hill, and succeeded Elizabeth Hill at 'the Globe and Sun in Fleet Street', 1773. Succeeded by John Newton and William Palmer. Anderson Estate St Neots, 1757; Old Priory and Priory Farm, 1757; surveys of many other English estates.

Batenburg, D.C.J. van. *Teekenatlas de Geheele Aarde*, 1901; *Atlas ten gebruike bij het Onderwijs in de Natuurkundige Aardrijkskunde*, 1907.

Batenham, George, *junior*. Engraver. *A Plan of Chester*, 1821.

Bates, Charles Ellison. *Route Map of the Western Himalayas, Kashmir etc.*, 1893.

Bates, Elias. *Plan of Merjee...*, 1725; *Comptee on the Malabar Coast*, 1775; *Plan of Onore on the Malabar Coast*, 1775.

Bates, H.S. New Plymouth New Zealand, 1863.

Bates, Henry B. *Guide Map of Walks and Drives about Buxton*, 1850.

Bates, Waldron. *Map of Mount Desert Island, Maine*, 1896 and later editions (with Edward L. Rand and Herbert Jaques).

Bath, Neville (*fl.*1776-1806). Irish cartographer. *The County of Cork surveyed...* (6 sheets), published 1811.

Bathalos, Manoel. Portuguese cartographer, 17th century.

Báthory, István (*fl.*1821-1850s). Hungarian cartographer. Received engineering diploma at Pest 1821, in 1830s participated in mapping of Temesköz region [now Romania]. In 1850s worked with László Vörös mapping Danube and Pest.

Bathurst, J. *A Map of Part of Lower Egypt...*, 1801.

Batley, R. Engraver. *Plan of the River and Sound of Dawfoskie* [South Carolina], 1773; East Coast England, by J. Chandler 1780; cartouche on Mount and Page's chart, Caribee Islands, *c.*1794.

Batonnet, —. One of the surveyors for Cassini's *Carte de France*, worked with others on three sheets (44, 45, 80) to the north and east of Paris, 1750-1758.

Batres, Leopoldo. *Plano de la Ciudad de Tenochtitlan en la Ano de 1519*, 1892.

Batson & Co. Lithographers of Sydney. *Redbank via Picton*, *c.*1880s.

Battani-al [Albatenius], Abu-Abdallah, Muhammad Ibn-Jabir (858-929). Astronomer and mathematician; born in Asia Minor, died in Mesopotamia. His book *Astronomical treatises and tables* was translated into Latin *c.*1143, published in Nuremberg, 1537.

Battatzi, Basil. See **Batatzi**, Basil.

Batten, J. Lithographers, 79 Queen Street, Melbourne. *Canterbury Park*, 1883.

Battenlen, C.N. van. *Kleine Atlas...*, 1924.

Battersby [Battersbey], William. A Travelling Map of England, 1680.

Battey, Jeremiah. Bookseller, 'At the Dove in Paternoster Row', London. Guillaume Delisle's South America, 1725.

Batthyáni, Ignác *SJ* (1741-1788). Bishop of Transylvania from 1780, founded observatory and library at Gyulafehérvár [now Alba Julia, Romania]. Initiated accurate mapping of Transylvania. **Ref.** BARTHA, L. *Régi magyarországi csillagászok* [Hungarian astronomers] in *Technikatörténeti Szemle* [History of Technology Review] 1979 [in Hungarian].

Battista Genovese [= of Genoa]. Portolan, 1514.

Baturin, Boris. (*fl.*1721-1730). Russian topographer. Numerous district [zemstvo] and town maps in central Russia, 1721-1730.

Bauchop, Robert (*fl*.1793-1835). Scottish land surveyor. Oversaw maps of Bute, Linlithgow and Lanarkshire, for J. Thomson's *Atlas of Scotland*, 1832.

Baud, Joseph Marie Louis (1864-1904). French explorer in Asia, 1889-1891, and West Africa, 1893-1895, respected for his work on the mapping of the River Niger. *Carte des Missions...dans la Bouche de Niger*, 1895. **Ref.** BROC, N. *Dictionnaire illustré des explorateurs Français du XIXe siècle: I Afrique* (1988) p.18 [in French].

Baudesson, *Capitaine* Henry. French army officer, topographer and surveyor. Surveyed and mapped route for railway across Vietnam, *c*.1902; topographical surveys in Indochina, 1902-1903. **Ref.** BROC, N. *Dictionnaire illustré des explorateurs Français du XIXe siècle: II Asie* (1992) p.17 [in French].

Baudin, *Captain* Nicolas (1754-1803). French mariner, in command of *le Géographe* and *le Naturaliste* on an expedition to explore the coasts of Australia 1802-1804, with Louis-Claude de Freycinet and Henri-Louis de Freycinet. Surveys used by J.W. Norie for his *Chart of the North, West and South, coasts of Australia...*, 1834; used by John Arrowsmith for his map of Australia, 1838; and for other charts. **Ref.** MARCHANT, Leslie R. *Imago Mundi* 40 (1988) pp.47-56; TOOLEY, R.V. *The mapping of Australia* (1979) p.82.

Baudoin, —. Some contributions to Melchior Tavernier's *Théâtre géographique de France*, 1634; and Nicolas Tassin's *Cartes générales de toutes les provinces de France*, 1640-1643.

Baudouin [Baldouino], Gaspar. *Charte de la Suisse*, Milan, P. Stella 1625.

Baudouin, H. *Plan der Stadt Stettin*, 1828.

Baudous, Robert Willemszoon de (1574-1659). Engraver & publisher from Brussels, worked in Amsterdam.

Baudrand, *abbé* Michel Antoine (1633-1700). 'Géographe du roi'. *Descrittione delle stato della chiesa e della Toscana*, 1667; *Geographia ordine litterarum deposita*, 1682, 1705; *Mercurio Geografico*, *c*.1688; *...Postes de l'Italie*, 1695; later editions of Desgranges' 1689 map *La Carte des Royaumes d'Angleterre, d'Ecosse et d'Irlande*, were attributed to Baudrand, published J. Besson, Paris 1695, 1699, 1701; *La Principauté de Catalogne et le Comté de Roussillon*, 1703.

Baudrand, P.F. Lexicon geographicum, 1674.

Baudwin, Gabriel S. *St. Sebastian*, 1740; *A Draught of the coast of Biscay*, 1742.

Bauer, family. Engravers and globemakers of Nuremberg.
• **Bauer**, Johann Bernard (1752-1839). Cartographer and globemaker. Made globes on his own account, and in association with Johann Georg Klinger.
•• **Bauer**, Carl Johann Sigmund (1780-1857). Made globes for Friedrich Campe, 1825; miniature globe, *c*.1825.
•• **Bauer**, Peter (1783-1847). Worked for J.G. Klinger.

Bauer, Antonius. *Ichnographia Universarum...*, 1827.

Bauer, F.A. Engraver. *Ratisbona....*, Frankfurt am Main 1634-1635, Vienna 1672.

Bauer [Baur; Bawr], Friedrich Wilhelm von [Frederic Guillaume de] (1731-1783). Military cartographer in Prussian and Russian service. Bataille de Créfeld, 1758, published 1765; Théâtre de la guerre (4 sheets), 1759; *Plans pour servir à l'histoire de la guerre de Sept ans*, 1763-1766; *Théâtre de la Guerre en Allemagne entre la Grande-Bretagne et la France depuis l'an 1757 jusqu'à l'an 1762* (6 sheets), 1769; *Carte de la Moldavie...* (6 sheets), *c*.1772, Amsterdam 1775.

Bauer [Bacher], J. Globe, *c*.1800; revised F.J. Maire's map of Silesia-Moldova-Bosnia-Switzerland, published with the twin title *Carte Hydrographique et de Poste / Hydrographische Karte*, Vienna, Tranquillo Mollo, 1808, 1809 (engraved by Mansfeld).

Bauer, Norbert (1812-1881). Geodesist, in charge of Sopron triangulation district from 1864 [Hungary]; first director of Magyar Kataszteri Térképnyomda [Hungarian cadastral lithographic establishment], 1868-1872.

Bauer, Robert. *Reise- und Gebirgs-Karte vom Lande Salzburg*, 1890.

Bauerkeller, Georg (*fl.*1830-1870). Relief cartographer and globe maker. *Post und Routen-Karte van Deutschland*, 1832; *Nouveau Plan de Paris en Relief*, 1839; embossed plan of London, 1841; *Bauerkeller's Handatlas der allgemeinen Erdkunde*, 1846 (many later editions lacked the Bauerkeller imprint); embossed plan of Paris, 1848; *Neuester Plan...Stadt Hamburg*, *c.*1850; *La Suisse...*, *c.*1850; *Carte en relief de l'Empire Roman*, *c.*1864; *La Russie Européenne*, *c.*1864; *Deutschland und die Niederlande*, *c.*1865; *Frankreich und Belgien*, *c.*1865.

Bauernfeind, Karl Marx. Railway map of Bavaria, 1845.

Baugh, Robert. Engraver. John [1] Evan's Six Counties of North Wales, 1795; ...*Shropshire* (9 sheets), 1808, reduced version 1811.

Bauland, --. Special Karte von dem Odenwald, 1808.

Baum, Johan Christoph. Map of Mansfeld, 1732, used later by Homann's Heirs.

Bauman, *Major* Sebastian (1739-1803). German, emigrated to New York, served in the French and Indian Wars; Captain in New York Regiment of Artillery. *Plan of Yorktown in Virginia...*, 1781, published Philadelphia, H.S. Tanner 1825; *Plan of the Investment of York and Gloucester*, 1782 (engraved by R. Scot). **Ref.** GUTHORN, P.J. *American maps and map makers of the American Revolution* (1966), pp.8-9.

Baumann, H. *Wanderkarte der Luneburger Heide*, [n.d.]

Baumann, Joseph Martin (1767-1837). Swiss relief model designer of Dussnang, worked in Brunnen.

Baumann, *Doctor* Oscar (1864-1899). ...*Insel Fernando Poo*, 1887; Mittlere Congo, 1888-1890; maps for Petermann's *Geographische Mittelheilungen*, 1889; ...*Deutsch Ost-Afrika* (4 sheets), 1894; *Karte des Forschungsgebietes der Massai...*, 1894; *Der Unterlauf des Pangani Flusses...*, 1896; *Sansibar Archipel*, 1897.

Baumann, Victor. *Karte von Elsass-Lothringen*, 1873; *Karte der Rench-Bader...*, *c.*1877.

Baumann, William. *Pulaski County*, 1871.

Baumeister, I. Maps for Homann's Heirs. Stadt Atlas, 1762.

Baumeister, Johann Jacob. *Ducatus Mantuani*, 1720; *Nouveau Plan de Carolsruhe*, Homann's Heirs 1737.

Baumgarten, —. Karte der Schweiz, 1799.

Baumgarten, Max von. *Herzogthum Bukowina*, *c.*1877.

Baumgartner, A. *Historischer Atlas von Sachsen* (25 maps), Leipzig 1816 (engraved by Leutemann); Hypsometric maps of Austria, 1827.

Baumgartner, Johann Jacob (*fl.* 18th century). Cartographer of Baden. *Augsburg...*, *c.*1700.

Baur, Carl Friedrich. German cartographer, numerous maps of Europe, France, Germany etc., Cairo, 1841; *Neuer Schul-Atlas...*, 1862; *Neuester Atlas*, *c.*1868; World maps, 1871, 1893; *Neueste Karte von America*, 1872; *Rakousko Uherske*, *c.*1874; *Elementar-Schul-Atlas*, *c.*1874; *Monarchie Oesterreich-Ungarn*, *c.*1874; *Oesterreichische-Ungarische Monarchie*, *c.*1874; *Böhmen*, 1876.

Baur, F.W. See **Bauer**, Friedrich Wilhelm von.

Bausa, Felipe. See **Bauzá**, Felipe.

Bausemer, J.M. Atlas of Poland, Wyld 1837 (with Zaleski).

Bauser, G.W. ...*Württemberg u. Hohenzollern*, 1868.

Bauzá [Bausa], *Don* Felipe. Chartmaker. Accompanied the Malaspina Expedition, 1798. *Atlas Maritimo de España*, 1789; *Manila Bay*, 1792, used by Norie in *Complete East India Pilot, 1838;* Antillas, 1802; *Carta Esferica de la parte Interior de la America Meridional*, 1810; *Seno Mejicana*, 1829; Colombia, 1830, 1841; *Carte de la Côte Occidentale du Mexique...*, 1836; Madrid, 1841; *Mapa de una parte de Territorio de Colombia eu La America Meridional*, 1841. **Ref.** MARTIN-MERAS, L. 'Felipe Bauza: sus Trabajos sobre el Mapo de Espana' in *Revista de Historia Naval* Vol. VII, N° 27 (1989) pp. 33-47 [in Spanish].

Bavier, Simon. Swiss engineer. *Die Strassen der Schweiz*, 1878.

Bavio, E.A. *Atlas escolar de la república Argentina...*, Buenos Aires, A. Estrada & Cia. *c.*1910.

Bawr, F.G. de. See **Bauer**, Friedrich Wilhelm von.

Bawr, F.W. See **Bauer**, Friedrich Wilhelm von.

Baxter, family. **Ref.** KINGSLEY, D. *Printed maps of Sussex 1575-1900* (1982) pp.175-6.
• **Baxter**, John (1781-1858). Bookseller, printer and publisher in Lewes, Sussex. Many publications including Horsfield's histories of Lewes and Sussex.
•• **Baxter**, William Edwin (1808-1873). Son of John. He and his father founded the *Sussex Agricultural Express* in 1837. Re-issued Greenwood's *Sussex*, *c.*1860s.
•• **Baxter**, George (1804-1867). Son of John and brother of William Edwin. Engraver and artist. Assisted with illustrations for Horsfield's histories of Lewes and Sussex.

Baxter, F. *A Plan of the Parish of Elton*, 1784.

Baxter, *Captain* John. Map of Donegal Bay & Part of Fermanagh, finished by Baptista Boazio, 1599 MS.

Baxter & MacGowan. Publisher of Washington. H.W. Newby & Co.'s *Map showing the proximity of Carlin Springs to Washington and Alexandria*, 1887.

Baye, *baron* Joseph de (1853-1931). French archaeologist and traveller in Central Asia. *En Géorgie*, 1898; *Au nord de la chaine du Caucase*, 1899.

Bayer, Fritz. *Neuester Hand-Atlas*, 1894.

Bayer [Bayr], Johann (1572-1625). Astronomer of Ulm, first to use Greek letters to denote magnitude of stars. *Uranometria omnium asterismorum* (51 maps), Augsburg, 1603 and later editions.

Bayer, Joseph (1772-1841). *Karte des Maehr. Schles. Gouvernements...* (4 sheets), 1817-1818.

Bayfield, *Captain* (later *Admiral*) Henry Wolsey, *RN* (1795-1885). Admiralty surveyor, took over the survey of the Great Lakes in 1817 and completed it in 1825. Gulf and River of St Lawrence (110 charts & plans), 1828-1865; *Lake Superior*, 1828; *Chart of Part of the North Coast of Lake Superior*, 1828; *A Survey of Lake Erie...*, 1828; *Map of Prince Edward Island*, 1852; Nova Scotia Pilot, 1856.

Bayle, —. Engineer and draughtsman for sheet 93 [north-east part of the Cherbourg peninsula] of the Cassini *Carte de France*, 1757.

Bayles, Frederick Henry. *The Race Courses Atlas of Great Britain and Ireland*, 1903.

Bayley, *Lieutenant* T. *Environs of Batavia*, 1812.

Baylis, Thomas. *Plan of the intended Rail or Tram Road from Stratford upon Avon to Moreton in the Marsh* [England], 1820.

Bayliss, Edward Percival. *Antarctica*, 1939.

Baylor, H.B. Atlas of Atlanta, Georgia, 1893 (with E.B. Latham).

Bayly, John Abraham (*fl.*1755-1794). Engraver, at St. Thomas Apostle, London. *The City Guide or a Pocket Plan of London, Westminster and Southwark*, 1761; Sheffield's Jamaica, 1770; Matthew Stobie's *Roxburghshire*, 1770; Captain Collet's *A Compleat map of North Carolina*, London, S. Hooper 1770; maps and charts for Joseph Smith Speer's *West India Pilot*, 1771; Dorset, 1773; Pitcairn, 1773; engraver for John Byres's maps of islands of the Caribbean, 1776; Pennant's Scotland, 1777; many maps for W.F. Martyn's *Geographical Magazine*, 1782; *A New Map of the World in three sections*, 1782; and numerous others.

Baylye, Jeremie. *Plott of the manor of Shingle Hall*, 1618 MS.

Bayne, J. Printer and publisher, 3 Bridge Street Row, Chester. *Plan of the City and Liberties of Chester*, *c.*1855.

Baynes, *Lieutenant* Charles Edward Stuart (d.1855). *Plan of the Siege Works thrown up by the British Troops for the Attack of Sevastopol*, 1855 (with J.F.D. Donnelly).

Baynton, Edward. Surveyor General of the Duchy of Cornwall. Plan of Sutton Pool [Plymouth], 1757.

Bayol, Jean-Marie (1849-1905). French naval surgeon, served in West Africa. Records of his travels included notes on geography and ethnography as well as some maps; *Voyage en Sénégambie, Haut-Niger, Bambouk, Fouta-Djalon et Grand Bélédougou 1880-1885*, published 1888.

Bayot, —. Artist for Hyacinthe-Yves-Phillippe Florentin de Bougainville's Atlas, 1837.

Bayr. Johann. See **Bayer**, Johann.

Baz, Gustavo. *Historia del Ferrocarril Mexicano*, 1874.

Bazalgette, *Sir* Joseph William (1819-1891). Engineer and surveyor; drainage engineer for London's Metropolitan Board of Works. Central Kent Co. Railway, 1836; *Map Shewing the...Sewers, Constructed for the Improved Drainage of the Metropolis, and the Purification of the River Thames*, 1865.

Bazeley, C.W. *The New Juvenile Atlas*, Philadelphia 1815.

Bazelius, Nicolaus. Celestial planisphere, published in Prognostication nouvelle, 1578.

Bazendijk, P.M. *The Handy Atlas of the World*, Rotterdam 1913.

Bazendorff, J.F. von (*fl.* 18th century). Bavarian cartographer.

Bazewicz, Jósef Michat. *Mapa Królestwa Polskiego*, 1900; *Polska i Rosja*, 1920; *Atlas historyczny...*, 1921.

Bazilevski, Ivan and Feodor. Orenburg Province, Russia, 1880.

Bazin, —. *Carte de Champagne et de Brie* (2 sheets), 1790.

Bazin, François. *Atlas...de géographie... de la France*, 1855; *L'Europe économique*, 1887.

Bazley, Thomas Sebastian. *Stars in their courses*, 1878.

Beach, Harlan P. *World Atlas of Christian Missions*, 1911.

Beake, —. Booksellers, King Street, Sydney. *Map of New South Wales...*, 1871.

Beal, John [& Son]. Booksellers and publishers of 55 East Street, Brighton, Sussex. *John Beal & Son's Hunting and Tourists' map of Sussex...*, Brighton 1885 (based on the Ordnance Survey map); *John Beal & Son's Hunting and Tourists' map of Sussex & c...*, Brighton 1889 (different scale, based on Cary).

Beale, E.F. *Map of Public Surveys in California & Nevada*, 1858-1863.

Beall, George W. *...Ann Arundel County, Maryland*, 1860; *Prince George's County Maryland*, 1861.

Bean, Charles. *Elementary Atlas of Modern Geography*, 1854; *Introductory School Atlas*, 1854; *Map showing the overland route to India*, 1854.

Bean, *Reverend* James Philip. *A New Atlas of Classical Geography*, 1835.

Bean, John. *Charts of the Coasts & Harbours of Great Britain and Ireland*, 1781-1802.

Bear, J. C. See **Baer**, Johann Christoph.

Beard, Charles L. *Map showing the Niagara and Detroit Rivers Railway...*, 1858 (with H. Gregory, lithographed by C.W. Hamilton).

Beard, Ithamar A. *A rough plan of the survey of a route for a canal from Shirley to Concord River* [Massachusetts], 1825.

Beards, Thomas (*fl.*1832-1869). Surveyor of Hampshire. Manuscript estate plans in Buckinghamshire, Hampshire and Somerset, 1848-1858.

Beardsley, Charles. *Map of Elkhart*, 1871.

Beare, James. Sailed with Martin Frobisher on three voyages as surveyor, 1576-1578, credited with maps in George Best's *A True Discourse of the Late Voyages... of Martin Frobisher*, 1578. **Ref.** BURDEN, P. *The mapping of North America* (1996) N°51 pp.64-65; SHIRLEY, R. *The mapping of the World* (1983) N°138 p.160.

Sir Francis Beaufort (1774-1857). (By courtesy of the National Portrait Gallery, London)

Bearhop, Andrew. *Map of the Baronry of Stobbo in the Sherreffdom of Peebles*, 1740 (for A. Murray).

Beato [Beatus], San [Saint] Liebanensis [San Beato de Liébana] (*d.*798). Benedictine monk born in Valcorado, Spain. Biblical commentator. World maps to illustrate his text *Commentary on the Apocalypse of Saint John*, *c.*776 (several variants known).

Béatrizet [Beatrice; Beatricetto; Beatricius; Beautrizet], Nicolas (*fl.*1540-*c.*1577). Painter, engraver, printer and publisher, native of Lorraine. Worked in Rome, initially for others such as Salamanca, Barlacchi, Tramezini and Lafreri, and on his own behalf from 1558. Engraved Michele Tramezini's *Nova Germaniae descriptio*, 1553; *Thionville*, 1553, 1558; ...*Romae topographia*, 1557; *Melita Nunc Malta*, 1565. **Ref.** GANADO, A. & AGIUS-VADALÀ, M. *A study in depth of 143 maps representing the Great Siege of Malta 1565* (1994) pp.37-44

Beatson, *Captain* A. *The marches of the British Armies in the Peninsula of India 1790-1791*, 1792.

Beatson, R. *Tor Bay* [Devon], *c.*1785.

Beattie, J. Lithographers of Cleveland, Ohio. *Proposed Route for a Rail Road From Copper Harbor to Fond du Lac Winnebago*, 1855.

Beatty, John. *Survey of the Manor of Livingston*, 1714.

Beatty, Walter (1836-1911). Surveyor in Canada, specialised in township plans. Athens, 1860; Seelys Bay, 1863; surveys of Gananoque, 1863-1867; Map of Sherbrooke, 1865; survey of Delta, 1865; Lyndhurst, 1876; township of Armour [Ontario], 1877; Township of Strong, 1878.

Beatus. See **Beato**.

Beaublé, —. Script engraver for de Belleyme's *Carte de la France*, 1791.

Beauchamp, —. One of the surveyors for sheets 1, 26 and 27 [Paris area] of the Cassini *Carte de France*, 1756-1757.

Beauchamp, —. *Charte vom Schwarzen Meer*, 1798.

Beauchamp, Adolphe [Alphonse] de. *Nouvelle carte du Brésil*, 1815.

Beauchesne de Gauin, —. Sailed in the southern seas, 1698-*c.*1704. *Carte du Paraguay, du Chili, du détroit de Magellan*, 1710 (maps of South America based on his information were published by Homann, 1733; Bellin, *c.*1764; J. Callander, 1766-1768; and J. Burney, 1803-1817).

Beaudouin, *Major* —. *Carte de l'Empire de Maroc* (2 sheets), Paris 1848.

Beauford, William (*d.*1819)....*Map of Antient Ireland*, (2 sheets), 1801; *Plan of the City and Suburbs of Cork*, 1801; maps for ...*Irish Atlas*, Grierson *c.*1818.

Beaufort, father & son. **Ref.** ALLPRESS, Peter, 'The Beauforts - father and son' in *IMCoS Journal* (Summer 1995).

• **Beaufort**, *Reverend* Daniel Augustus (1739-1821). Founder of the Royal Irish Academy, father of Francis. *A New Map of Ireland-Civil and Ecclesiastical*, London, W. Faden, Dublin, W. Allen 1792 and later (engraved by S.I. Neele); *The Diocese of Meath*, 1797.

•• **Beaufort**, *Captain* (later *Admiral*) *Sir* Francis (1774-1857). Son of Daniel, above. Hydrographer for the Navy 1829-1855, supervised the work of Admiralty hydrographers such as Belcher, the Beechey brothers, Collinson, FitzRoy, Graves, Spratt and many others. Well known for his 'Beaufort scale', devised to indicate wind velocities. *Sketch of the islands of Papadoula*, 1812 MS; prepared atlas plates used by the Society for the Diffusion of Useful Knowledge, 1828-1843, and later; *North Pacific Ocean*, 1861; *Charts of the North Atlantic Ocean*, 1863 (these charts were published after his death). **Ref.** RITCHIE, G.S. *The Admiralty Chart: British naval hydrography in the nineteenth century* (1995) pp.208-218, 259-270; CAIN, M.T. *Imago Mundi* 56 (1994) pp.151-167; FISHER, Susanna *The Map Collector* 54 (1991) pp.18-23.

Beaufort, Henri Ernest Grout de (1798-1825). French marine officer, explorer in West Africa (1818-1821). Surveyed cataracts of the Félou and Gouina rivers, 1824-1825.

Beaufoy, Mark. *Routes in Mexico*, 1828.

Beaujour, Felix *Le Chevalier* de. *Sketch of the United States of North America*, 1814.

Beaulieu, G. *Nouvel Atlas Portatif de la France*, Amsterdam, Covens & Mortier 1795.

Beaulieu, family. **Ref.** PASTOUREAU, M. *Les atlas français XVI-XVII siècles* (1984) [in French]; PASTOREAU, M. *Imago Mundi* 32 (1980) pp.59-61 [in French].

• **Beaulieu**, Sébastian de Pontault, *sieur de* ['le Donjon'] (*c.*1612-1674). French military engineer, 'Ingenieur du Roy'. Best known for his battle plans, especially *Les glorieuses conquestes de Louis le Grand* [the so-called 'Grand Beaulieu'], begun in 1642, probably completed and published by his niece *c.*1676; *Carte de L'Isle de Malthe*, 1646 (drawn by Della Bella, engraved by Nicolas Cochin); *Plans et Profils des principales villes, et lieux considerables de la principauté de Catalogne*, 1668; *Villes de France* [known as the 'Petit Beaulieu'] (14 small volumes), [*n.d.*].

•• **Beaulieu**, Reine-Michèle de (*fl.*1674-1695). Niece of Sébastien (above). Together with her husband Jean Baptiste Hamont, sieur Des Roches, she succeeded Beaulieu in 1674, completing and publishing his *Les glorieuses conquêtes de Louis le Grand*, 1676.

Beauman, Guy. Maps of Illinois counties inbcluding: Jackson County, 1896; Johnson County, 1898; Gallatin County, Vienna 1899; *Map of Franklin County, Illinois*, Vienna 1900; *Map of Jefferson County, Illinois*, 1900.

Beaumarchey, L. *Carte Celeste*, 1868.

Beaumier, Auguste (1823-1876). French interpreter at Mogador, Tangier, Tunis and Rabat, 1847-1865. *Le Maroc*, 1867; *Itinéraire de Mogador à Maroc et de Maroc à Saffy*, 1868; *Itinéraire de Tanger à Maroc*, 1876.

Beaumont, Albanis. *Charts of...the route from Lyons...to Turin...*, 1806.

Beaumont, Carmine de. *Mapa geometrica... città di Genova*, [n.d.].

Beaumont, Elie de. See **Beaumont**, Jean Baptiste Armand Louis Léonce Elie de.

Beaumont [Bouthillier de Beaumont], Henri (1819-1889). Swiss geographer. Founder and president of Société Géographique de Genève.

Beaumont, Jean Baptiste Armand Louis Léonce Elie de (1794-1874). Studied geological mapping in England in the 1820s. Worked with P.A. Dufrénoy on a geological survey of France, 1825-1836. *Geognostische Ubersichts-Karte von Deutschland, Frankreich, England...*, 1838; *La Carte Géologique de la France* (6 sheets), 1841 (from his surveys with Dufrénoy); *Carte Géologique de l'Europe*, 1875.

Beaumont, Fr. Pablo. *Descripcion de la parte de la America Septentrional que se llama Nueva España*, *c.*1691.

Beauplan. See **Levasseur** de Beauplan.

Beaupré, B. de. Engraver. Rue de Vaugirard, Paris. *Territoire de Michigan*, 1825.

Beaupré, Charles François. See **Beautemps**-Beaupré, Charles François.

Beaurain, Jean de. Father and son. Cartographers of Paris.

• **Beaurain**, Jean, *chevalier de* (1696-1771). Engineer and 'Géographe du roi' (1721), studied under Pierre Moulart-Sanson. 'Quai des Augustins au coin de la rue Git-le Coeur quartier St. André aux Arts,' Paris. *La ville d'Orleans avec ses Environs...*, 1720 (engraved by Incelin); *...Ducheé de Mantoue*, 1734 (engraved by de Favanne); *Plan de Luxembourg*, *c.*1740 (engraved by C. Inselin); *...Ville et Faubourg de Cartagene...* (3 maps and 4 plans on one sheet), Paris 1741; *Carte...de la Guerre d'Allemagne...* (6 sheets), 1750; *Atlas de géographie ancienne et moderne* (14 volumes), 1751; Brabant, 1756; *Histoire Militaire de Flandre...* (6 volumes), 1756, 1776 (this atlas contains works by Homann, Delisle, Moll, Rocque, Bellin, de Fer, Le Rouge, Hondius, Jaillot etc. in addition to those of Beaurain); *Plan géométral de Gibralter Avec les nouveaux Ouvrages de cette Ville faits par les Anglois...*, 1756; *Carte Générale des Isles Grenesy, Jersey, Aurigny etc.*, 1757; *Carte Topographique de l'Isle de Jersey*, 1757; *Carte Topographique des Isles d'Aurigny...*, 1757; *Carte de la Manche*, 1760; *Carte Hidrographique...*, 1760; Cours du Rhin, 1760; *Carte Topographique... de l'Isle de Belle Isle*, 1761; *Plan de... Hanovre...*, 1761; *Carte du Détroit de Gibralter*, 1762; Allemagne, 1765; and others.

•• **Beaurain**, Jean de. Son of the above. *Carte du Port et havre de Boston*, 1776 (based on Des Barres' *Map of the port of Boston with Nautical Remarks*, 1775).

Beausard [Beausardus], Pierre [Petrus] (*fl.*1535-1577). Physician and mathematician of Louvain. *Annuli Astronomici Instrumenti...*, 1553; Universalis cosmographici, 1556.

Beautemps-Beaupré, Charles François (1766-1854). Engineer and hydrographer; 'Sous-chef du Dépôt-général de la Marine, contre Amiral de France'. Neptune de la Baltique, 1785; Atlas de la Mer Baltique, 1796; *Plan de la Baie de l'Aventure*, *c.*1800; manuscript atlas of Ports of the Gulf of Venice compiled for Napoléon, 1806; *Carte de l'Extrémité Septentrionale de la Nouvelle Zéeland*, 1807; charts for *Atlas du voyage de Bruny-d'Entrecasteaux*, 1792-1793, published 1807; published *Méthodes pour la levée et la construction des cartes et plans hydrographiques*, 1808; coasts of France, 1824-1829; *A New Chart of the Bay of Biscay*, 1852.

Beauvais, Charles-Nicolas-Dauphin de (1730-1785). Script engraver for some sheets of the Cassini *Carte de France*, published from 1756.

Beauvau, Henri, *Baron* de. Soldier and traveller in the Levant. 47 maps and plans including Jerusalem, 1615; Jaffa, 1615; *Carte des lieux saincts*, 1615; *Ramma*, 1615.

Beauvilliers [Bauvillier], —, *sieur de*. (*d.*1730). *Carte nouvelle de la...Louisiane*, 1719; *Partie ouest de la Province de la Louisiane*, 1720 MS; *Coste de la Louisiane*, 1721.

Beavan, *Captain* R. Plan Cantonments Kandehar, 1879; Marri Hills [India], 1881.

Beavan, T. Jubilee and Philip Rivers, New Guinea, 1887.

Beaver, *Captain* Philip. *Nautical...West Coast of Africa*, 1805.

Beawes, William. Archipel des Isles Philippines, 1750 MS.

Becard de Grandville, Charles (*fl.*1673-1703). *Carte...lac des Ilinois*, 1673; *Carte Generale du Grand fleuve de s. Laurent*, [n.d.].

Beccario [Becharius], Battista (*fl.*1426-1435). Genovese portolan maker, untitled portolan chart of Europe, Atlantic and west African coast (attributed), 1434 MS.

Beccaro [Becaria, Becharius], Francesco. Genovese portolan maker. Chart of the Mediterranean and Black Seas, 1403. **Ref.** KELLEY, J.E. *Imago Mundi* 31 (1979) pp.18-23; SKELTON, R.A. *Imago Mundi* 22 (1968) pp.107-108.

Becharius, Battista. See **Beccario**, Battista.

Becharius, Francesco. See **Beccaro**, Francesco.

Beche, *Sir* Henry Thomas de la (1796-1855). Geologist of London. First Director of the Geological Survey, 1832; founder of the Geological Museum, 1851. *Geology of the environs of Bath* (colour lithograph), Bath, J. Barratt 1823 (with William D. Conybeare); also wrote several manuals and papers on geology.

Bechein, J.L. Surveyor. *Lower reach of Saginaw River* [Michigan], 1856.

Becher, *Commander* A.B., *RN*. Hydrographer. Surveys used in the compilation of *A General Chart of South America*, Admiralty 1824 (engraved by J. Walker); Orkney, 1847-1848.

Becher, Arthur. *Plan of Simla*, 1864.

Becher, J.R. *Map of the District of Jalindhur*, 1852.

Becher, S.J. *Mining Map of the Boulder Belt, East Coolgardie*, 1900; *Topographical Map of Kalgoorlie*, 1900.

Bèchet, Bernadin. *Nouvelle carte d'Europe*, 1878.

Bechler, Gustavus R. Allegany County, New York, 1856; ...*Washtenaw County, Michigan*, 1856; *Lenawee County, Michigan*, 1857; *New Illustrated & Historical copper-plate Map of the World*, 1862; *Atlas...Campaigns of the Armies of the Potomac in Virginia*, 1864; ...*Counties of Washtenaw and Lenawee, Michigan*, 1864; *Atlas showing battles, engagements, and important localities connected with the campaigns in Virginia*, Philadelphia 1864.

Bechstatt, I.C. *Situations-Carte von Darmstadt und der umliegenden Gegend*, *c.*1820.

Bechtold, Christian von. *Worms und Umgegend*, 1860.

Beck, Anton August. Engraver. *Plan de la ville de Bronsvic*, 1759; *Plan der Stadt Inverness*, 1760.

Beck, Bastiaan. Dutch surveyor working in Groningen. *Caerte van de nieuws ingedijckte landerijen de stadt Groningen*, 1696.

Beck, E.H. Italia con le sue poste (for Georg Matthäus Seutter), *c.*1740.

Beck, Eduard (1820-1900). Lithographer and topographer from Bruchsal, worked in Bern. Best known as a maker of relief maps. *Carte geologique de la Suisse*, 1855; *Relief de la Terre Sainte*, 1856; *Carte pour le Relief de l'Oberland Bernois*, 1858; *Karte der Umgebungen von Bern*, 1858.

Beck, George. Surveyor. Maps of Mahé [Seychelles], 1783-1785.

Beck, Henry C. [Harry] (*b.*1902). Draughtsman with London Underground Railways, later taught at the London School of Printing. Beck devised the original schematic map of the London Underground. The first draft was presented in 1931, it was then developed for publication as a card folder for free distribution, 1933; the map has since been published in countless updated editions. **Ref.** GARLAND, Ken *Mr Beck's Underground map* (1994).

Beck, Hugo. Central de Bolivia, London, James Wyld, 1862.

Beck, Jens Michelsen. Danish cartographer. *Tilforladelige Kort over Eylandet de St. Croix udi America*, 1754. **Ref.** HOPKINS, Daniel *Imago Mundi* 41 (1989) pp. 44-58.

Beck, Joseph. *Historische-Geographische Atlas für Schule Freiburg*, 1870.

Beck, Lewis Caleb (1798-1853). *A Gazetteer of the States of Illinois and Missouri*, 1823 (maps & plans).

Beck, Thomas George. *A Map of the Parish of St. Paul, Covent Garden* [London], 1851.

Becker, *Captain* —. *Neue topographische Charte der Umgegend von Dresden*, 1821.

Becker, B. *Plan von dem Curorte Grafenberg und Umgebung*, 1876.

Becker, Carl Friedrich. *Historisch-geographischer Atlas*, 1839-1842; historical work used in Rudolph von Wedell's *Historisch-geographischer Hand-Atlas*, Berlin, Duncker 1843-1849, 1856, 1862.

Becker, F. Quarter Master General Staff Darmstadt. *Karte von dem Grossherzogthume Hessen*, *c.*1874.

Becker, Francis Paul. Steel-engraver of Stationer's Court, London, and inventor of Becker's Patent Omnigraph process. Engraver for *A New Map of China*, *c.*1842; *Becker's Omnigraph Atlas of Modern Geography*, London, Smith, Elder & Co.1843 and other editions; *The Visitors Map of London...*, Charles Cradock & Co. 1851; *Cab Fare and Guide Map of London*, William S. Orr *c.*1853; *Map of Africa from the Equator to the Southern Tropic*, 1853; *Norfolk*, 1854; *Post Office Directory Atlas of England and Wales*, 1860; *Post Office Map of Lincolnshire*, 1861.
• **Becker,** F.P. [& Co.]. Omnigraph engravers. Prepared maps for some editions of James A. Barclay's *Universal Dictionary*, 1840-1852; maps for *Fisher's County Atlas of England and Wales*, 1842-1845.

Becker, François. Lithographer. Carte du District industriel de Liège, 1875.

Becker, *Professor*, *Colonel* Fridolin (1854-1922). Swiss topographer and cartographer-from Linthal. Worked in Zurich, and in Bern for the Swiss Topographical Bureau, 1876-1884. Becker had a great influence on the development of cartography in Switzerland, through contributions to the so-called 'Siegfriedkarte', and his many maps, articles and books relating to relief cartography. *Relief-Karte des Kantons Glarus*, Winterthur, 1889; *Karte der Ferwallgruppe*, 1899; *Säntis- und Churfirstengruppe*, Winterthur 1903; *Karte von Bodensee, Rhein & Bregenzerwald*, 1911; *...Schweiz...*, 1921. **Ref.** SCHERTENLEIB, U. *Cartographica Helvetica* 15 (1997) 3-10 [in German].

Becker, George Ferdinand (1847-1919). Geologist. *Atlas...Quicksilver Deposits of the Pacific Slope* (14 sheets), 1877-1888; *Geology of the Comstock Lode and Washoe District* [Nevada] (21 sheets), 1882.

Becker, Johann. *d.*1756. Engraver of Linz.

Becker, John. Engraver for A. Taylor's *Topographical Map of the Island of Barbados*, 1859 (from W. Mayo's survey of 1721).

Becker, Lewis M. Draughtsman and engraver of 3 Dyers Buildings, Holborn. Map of Cheshire, Sheffield, Francis White & Co. 1860; *The Picture Map of London*, William Tweedie *c.*1861.

Becker, Matthias. Publisher and engraver of Frankfurt am Main. Published individual volumes from Levinius Hulsius's 'Voyages' series between 1599-1660; printed editions of Bertius' *Tabularum Geographicarum Contractarum*, *c.*1612 (for Henry Laurentsz).

Becker, Moritz Alois R. von (1812-1887). Austrian geographer. General works on geography and topography. Edited S. Steinhauser's map of Lower Austria, 1853.

Becker, Peter. See **Artopäeus**, Petrus.

Becker, Wilhelm. *Karte der Eifel*, *c.*1890.

Beckeringh, Theodorus (*fl.c.*1750-1781). Dutch lawyer and cartographer. *Topographisch kaart van Groningen en Ommelanden* [wallmap of Groningen] (4 sheets), 1768, published 1781 (engraved J. van Jagen).

Becket, Thomas (*fl.*1730-1767). Estate surveyor active in Foulsham, Norfolk. Also conducted an estate survey in Barking, Essex, 1734.

Beckett, W.R.D. *Map of Siam*, 1893.

Beckit, H.O. Editor of the Oxford series of wall maps including *The World*, 1910; *Asia* (3 sheets), 1911; *The Western War Area*, 1915; *New Oxford Wall Atlas of Australia* (11 sheets), 1929.

Beckit, Robert. Map of Arabia for Linschoten's *Discours of Voyages into ye Easte & West Indies*, *c.*1598 (English edition); A.F. van Langren's *Delineatio Orarum maritimarum Terrae vulgo indigitatae Terra do Natal item Sofala Mozambicae & Melindae...*, London, John Wolfe 1598.

Beckman, C. Swedish chartmaker. *Charta öfver Åland*, 1789.

Beckman[n], *Captain* (later *Sir*) Martin. Swedish engineer in the service of Charles II of England. Fortifications of Hull, 1681-1682; north-east coast of Guernsey, 1694; Jersey, 1694; plans of Tweed, Humber, and Portsmouth; *NewCastle upon Tine*, [n.d.].

Beckmann, Otto. *...Post und Eisenbahn-Karte des Deutschen Reichs*, 1899; *...Plan der Stadt Worms*, 1902; *...Plan von Stuttgart*, 1902; *Beckmann's neuester Plan der Haupt-und Residenz-Stadt Karlsruhe*, 1903; *...Plan von Heidelberg*, 1903; *Plan von Landau i. Pfalz*, 1903; *Plan von Augsburg*, 1904; *Beckmann-plan von Basel*, 1913.

Beckwith, Edward (*fl.*1726-1727). Yorkshire estate surveyor. Sleningford Hall & Rudding's Farm, 1726 MS.

Beckwith, *Captain* Edward Griffin. *Explorations and Surveys for a Rail Road from the Mississippi River to the Pacific Ocean* (4 maps), 1855.

Becq-Rouger, —. *Plan de la ville de Tours*, 1898.

Bede [Beda, Baeda], 'Venerable' (637-735). Theologian, historian, and cosmographer. Climate and zonal maps; Tractatus de situ. al de locis sanctis, plans Holy Sepulchre and Mount Sion.

Bedekovich, Lörinc (1751-1839). Hungarian cartographer. Mapped numerous villages of the Alföld region. *Nagykunság*, 1795 (of special interest is the cartouche showing surveying instruments of the period). **Ref.** FODOR F. 'Bedekovich Lörinc gazdasági és települési térképei' [The economic and settlement maps by Lörinc Bedekovich] in: *Térképészeti Közlöny* [Cartographic Journal] 1934-III [in Hungarian].

Bedford, E.I. *Approaches to Liverpool*, 1834; *Survey of the coasts of North Wales*, 1837.

Bedford, F. *University Town of Cambridge*, *c.*1840.

Bedford, George Augustus. *Ascension Island*, 1838.

Bedford, Captain J. *Map of the District of Budaoon*, 1848.

Bedford, T.W. *Map of the Nemaha Land District, Nebraska Territory*, 1858.

Bédigis, —. French engineer of the 'Services des Ponts et Chaussées' in Corsica, from 1770, worked with Testevuide as co-director of the cadastral survey of the island. *Carte topographique de l'Ile de Corse*, 1824 (with Testevuide). **Ref.** HUGUENIN, M. *Imago Mundi* XXIV (1970) pp.129-131.

Bedingfeld, *Commander* Norman B. *RN*. Took part in Charles Livingstone's surveying expedition to Central Africa. *River Ogun* [West Africa], Admiralty 1861.

Bedö, Albert (*fl.*1870-1884). Hungarian cartographer. Known for his county maps, 1870-1884; forestry map of Hungary (12 sheets), 1884, reissued 1896. **Ref.** HRENKÓ, P. and PAPP-VÁRY, A. *Magyarország régi térképeken* [Hungary on old maps] (Budapest 1990) p.226 [in Hungarian].

Bedolina map. One of a series of Bronze age maps carved on rock in the Oglio Valley at Valcamonica [Brescia, northern Italy] dated to between 2000-1500 BC. **Ref.** DELANO SMITH, Catherine 'Prehistoric maps and the history of cartography' in *The History of Cartography* Volume I (1987) pp.78-79; DELANO-SMITH, Catherine *Imago Mundi* 34 (1982) pp.9-25; BLUMER, Walter *Imago Mundi* 18 (1964) pp.9-11.

Bee, Stephen. *A Survey and Plan of the Road from...Lincoln...to Peterborough...*, 1756 (engraved by William Bower of Boston).

Beebe, A.N. *Map of Kendall County, Illinois*, Yorkville 1890.

Beechey, *Captain* E.A. River Severn Old Passage surveyed, 1845-1846.

Beechey, *Captain* (later *Rear Admiral*) Frederick William, *RN* (1796-1856). President Royal Geographical Society, 1855. Was involved in explorations of the Arctic regions in the ship *Blossom*, later surveyed the Irish Sea, and the River Severn from Worcester to Minehead. Voyage of discovery towards the North Pole, 1818; survey of the coasts of North Africa, 1821-1823; surveys of the Irish sea, 1829; *Voyage to the Pacific & Bering's Strait*, 1831; South America, 1835; Ireland, 1837; *Chart of the Hawaii archipelago*, 1840; *Plan of the Skerry Islands*, 1850; *Chart of the Tidal Streams in the English Channel & North Sea*, 1851.

Beechey, *Commander* Richard B. Admiralty surveyor, younger brother of Frederick, above. Surveys of the western coasts of Ireland from 1835; Irish Sea, 1838; Holyhead, Irish Channel, River Severn; Black Sod Bay, 1851.

Beeck [Beek], Anna (*b.*1678, *fl.*1693-1717). Engraver, publisher, and colourist of The Hague. About 60 plans and prints, many of them of fortifications, battlefields and sieges, were published at The Hague with her imprint, 1697-1717. Pierre Husson took over her work when she left The Hague in 1717.

Beeck Calkoen, J.F. van (1772-1811). Astronomer of Leiden. Issued celestial charts in 1797.

Beecq, Jan Karel van. *Veuë du Siège...de la ville et château d'Auguste en Sicile*, *c.*1678; *Veuë de la Ville de Gènes*, *c.*1685.

Beecroft, Robert, *Master RN*, HMS *Crocodile*. Jervis Bay Terra Australis, 1830.

Beek, Anna. See **Beeck**, Anna.

Beek, James (*fl.*1816-1843). Surveyor. *Plan of the Tower Hamlets*, London 1843.

Beekel, Petrus. See **Boeckel**, Pieter.

Beekman, Anton Albert (1854-1947). Geographer, hydrologist, historian and military engineer in the Netherlands. Studied the history and development of the Dutch polders, and published several works on the subject. *Schoolkaart van Nederland*, 1888; *Schoolatlas van Nederland*, 1889; *Atlas van Nederland*, 1892 (with R. Schuiling); *Kleinste Atlas van Nederland*, 1892; *Schoolatlas van De Geheele Aarde*, 1893 (with R. Schuiling); *Kleine Schoolatlas van Nederland*, *c.*1911; *Geschiedkundige Atlas*, 1912-1938. **Ref.** KOEMAN, C. *Atlantes Neerlandici* Vol.VI (1985) pp.74-85.

Beekman, Barnard (1721-1797). American surveyor.

Silas Beers (b.1837).

Beeldsnijder [Beeldsnyder; Bilhamer], Joost Jansz. de [also called Joost Janszoon] (*c.*1541-1590). Architect, engraver, surveyor and cartographer from Amsterdam. *Land caerte ende water caerte van Noort Hollandt ende West-Vrieslandt* (6 sheets), Amsterdam 1608 (printed by H.A. van Warmenhuysen, based on original woodcut of 1575, no copies now known).

Been, father and son.
• **Been** [van Wena], Matthijs Janszoon [de] (*fl.*1584-1620). Dutch land surveyor 'Gezworen landmeter van Rijnland, Delfland, Schieland ende den Lande van Putten' (admitted 1584). Delfland, 1606 MS, later engraved by Floris Balthasar van Berckenrode.
•• **Been** [van Wena], Jacob Matthijs [Matijsz.] de (*fl.* post 1620). Dutch surveyor, son of Matthijs Janszoon de Been, took over from his father in 1620.

Beer, Johann Christoph. See **Baer**, Johann Christoph.

Daniel Beers (d.1913).

Beer, Josef. *Eisenbahn-Karte der Oesterreich-ungarischen Monarchie* (2 sheets), 1884; *Eisenbahn-Karte der Oesterreichisch-ungarischen Monarchie*, 1897; *Eisenbahn-Karte von Böhmen, Mähren und Schlesien...*, 1897.

Beer, Julio. *Division Civil de la Ciudad de Buenos Aires*, 1859.

Beer, Wilhelm. *Mappa Selenographica...*, *c.*1834, 1870 (with Johann H. von Maedler); *General-Karte...*, 1837.

Beers, family. Surveyors, cartographers and publishers originating from Newtown, Connecticut. Originally associated with John Homer French's surveying enterprises of the mid-19th century, the family made a great contribution to the publication of county and state atlases of North America. Their family and business relationships are complicated and not clarified here. **Ref.** CONZEN, M.P. in: WOLTER, J.A. & GRIM, R.E. *Images of the World* (1997) pp.338-344; RISTOW, W. *American maps and mapmakers* (1985) pp.408-409; RISTOW, W. *The Map Collector* 36 (1980) p.22-26.

• **Beers**, James Botsford (*fl.*1870s-1880s).

Atlas publisher of New York associated with his son, F.W. Beers.

•• **Beers**, Frederick W. (*fl.*1858-1929). Son of James (above), and cousin of Silas and Daniel Beers. Joined John Homer French's surveying team in 1857, preparing county surveys of New York. Later set up on his own as a publisher, becoming a partner in Beers, Ellis & Soule. In 1870 the firm became F.W. Beers & Company based in New York City, where for a while he was associated with his father, James Botsford Beers. Oneida County, 1858 (with Silas N. Beers and D. Jackson Lake); *Atlas of the Oil Region of Pennsylvania*, Beers, Ellis & Soule 1865; four county atlases of Ohio, 1866; county atlases of Pennsylvania, New York, Michigan, Connecticut and Vermont, 1866-1870; *Atlas of Otsego Co. New York*, New York, F.W. Beers, A.D. Ellis & G.G. Soule 1868; and others.

••• **Beers, Ellis & Soule**. (*fl.c.*1865-1868). See F.W. Beers, above.

••• **Beers**, F.W. & Co. American publishers of 36 Vesey Street, New York (1873), founded by Frederick Beers, often worked with Charles Hart (printer) and L.E. Neuman (engraver) both also of 36 Vesey Street, New York. Numerous county atlases for North American states, 1860-1890. *Beers' Fire Insurance Atlas...of the City of Boston*, 1874; *The Atlas of New York and Vicinity*, 1867; and many others. **Ref.** McCORKLE, Barbara B. 'The Strange Case of F.W. Beers and The Atlas of New York and Vicinity' *Meridian* N°5 (1990) pp.39-43.

•• **Beers**, Silas N. (*b.*1837, *fl.*1856-1867). Cousin of Frederick (above), brother of Daniel (below). Surveyor of Connecticut. Began working in 1856 for his former headmaster, John Homer French, on surveys of New York counties, gave up mapmaking in 1867. Schuyler County, 1857; Oneida County, 1857 (with Frederick W. Beers and D. Jackson Lake); *Map of the Vicinity of Philadelphia*, Philadelphia, John E. Gillette and C.K. Stone 1860 (with D.J. Lake, L.B. Lake, and D.G. Beers); *New Topographical Atlas of Jefferson County, New York*, Philadelphia, C.K. Stone 1864 (with Daniel Beers); and others.

•• **Beers**, Daniel G. (*d.*1913). Cousin of Frederick (above), brother of Silas (above), began working as a surveyor with Silas and Frederick Beers, and D. Jackson Lake on county surveys of New York state for John Homer French, and Robert Pearsall Smith. He later began producing county atlases of New York and Pennsylvania with his brother Silas, 1864-1867. After Silas retired from mapmaking in 1867, Daniel compiled his own atlases in New York (1868), and later in Philadelphia (*c.*1870); he produced few, if any, maps between 1878 and 1905. *Map of the Vicinity of Philadelphia*, Philadelphia, John E. Gillette and C.K. Stone 1860 (with D.J. Lake, S.N. Lake, and S.N. Beers); *New Topographical Atlas of Jefferson County, New York*, Philadelphia, C.K. Stone 1864 (with Silas Beers); *Boughman, Thomas & Co.'s map of the peninsula, embracing Delaware, and the Eastern Shores of Maryland and Virginia*, Wilmington, Boughman, Thomas & Co. 1868; *Atlas of the State of Delaware*, 1868; wall maps of Kentucky and Tennessee counties, 1876-1878; *Bi-Centennial Map of the Town of Newtown, Connecticut*, 1905; other atlases of counties in New York, Pennsylvania and Massachusetts.

Beers, John Hobart. A distant cousin of the Beers family, above, and like them a native of Newtown, Connecticut. Joined J. Silliman Higgins and Augustus Warner (Warner & Higgins) to form the publishing firm of Warner, Higgins and Beers of Chicago, and similarly named off-shoots. Succeeded those companies *c.*1870, and as J.H. Beers & Co. published state and county atlases until 1884.

• **Warner, Higgins & Beers**. (*q.v.*)

• **Warner, Beers & Co.** (*q.v.*)

• **Warner & Beers**. (*q.v.*)

• **Beers**, J.H. [& Co.]. (fl.1870-1884). Historical atlas of Johnson county, Indiana, 1881; *Atlas of Kankakee County, Illinois*, Chicago 1883; *Historical Atlas of Ford County, Illinois*, Chicago 1884.

Beest, Johan van (*fl.*1632-1645). Dutch land surveyor and cartographer.

Beetjies, *Lieutenant* P.I. Dutch engineer. *Chart of Defence for the Island Banda Neira* [Dutch East Indies], 1801.

Beeton, Samuel Orchart. Publisher. *Beeton's Map of Paris, its fortifications and its environs*, 1870; *Beeton's Map of the Seat of War* [in Europe], 1870.

Bégat, Pierre (1800-1882). French hydrographer. Géographie mathématique, 1834; Traité de géodesie, 1839; Atlas des côtes méridionales de France, 1850; *Carte Particulière des Côtes d'Italie...*, 1854.

Begbie, P. Engraver. J. Craig's *Plan of the new streets and squares intended for the City of Edinburgh*, 1768; *Plan of a new designed Square in Edinburgh*, 1773; W. Garden's *A map of Kincardineshire*, 1776; for W. Day & C.H. Master's *County of Somerset*, 1782.

Beggiato, F. Carta geologica della provincia di Vicenza, 1866.

Begule, L.C. Engraver for Nicolas de Fer, 1705.

Behaim [Bohemus], Martin (1459-1507). Geographer, globemaker and mariner, pupil of Regiomontanus. Worked in Nuremberg and Lisbon. Behaim made the oldest surviving terrestial globe (his 'Erdapfel'), 1492 (now in Nuremberg City Museum); contributed to Hartmann Schedel's Nuremberg Chronicle, 1493. **Ref.** RAVENSTEIN. E.G. *Translations and commentary on Martin Behaim's 'Erdapfel'* [earth apple], London, Greaves & Thomas (1992); MOSELEY, C.W.R.D. *Imago Mundi* 33 (1981) pp.89-91.

Beham, Hans Sebald. Engraver. Map of the eastern Mediterranean showing the travels of St. Paul and the Apostles (4 sheet woodcut), Ingolstadt, Peter Apian *c.*1528 (part now lost). **Ref.** KARROW, R.W. *Mapmakers of the sixteenth century and their maps* (1993) pp.55-56.

Behm, Ernst (1830-1884). Geographer of Gotha. Contributor to Petermann's *Geographische Mittheilungen* and founder of *Geographisches Jahrbuch*.

Behn, Charles. *Profiles of Routes proposed for a Pacific Railroad*, 1855.

Behr, Guillermo. *Los saladeros y Graserias del Rio de la Plata, Paraná y Uruguay*, 1883.

Behrens, C.A. Chartmaker. *Jadekarte*, 1799; *Heiligeland*, 1801.

Behrens, G. *Topographische Karte des Gebiets der freien Hanse-Stadt Lübeck*, 1827; *Topographisch-Statistische...Lübeck*, 1843.

Behrens, H.C. *Karte von dem Herzogthum Oldenburg*, 1804.

Ernst Behm (1830-1884). (Taken from Gothaer Geographen und Cartographen, *Haack 1985)*

Behrens, Herman (1854-1902). Engraver for Justus Perthes.

Behrens, Tankred Tunstall. *Anglo-German Boundary, East Africa 1904 & 1905*, 1906.

Behring, Vitus. See **Bering**, Vitus Jonassen.

Behrisch, D. *Plan von Dresden*, 1865.

Behrmann, Walter Emmerich (1882-1955). Wrote many theoretical works on cartography. Topographical maps of the Sepik Basin, New Guinea, 1912-1913.

Behrns, Bernhard. *Zak-Atlas van Friesland in 36 Kaarten*, Franeker 1861, 1863.

Behsel, Anton (1780-1838). Town surveyor of Vienna. He undertook a survey of the city and its suburbs, 1818-1824, later used as a basis for a cadastral map of Vienna, 1829.

Beich, D. See **Beuch**, Daniel.

Beiger, Hans Conrad. Engraver. *Helvetia cum confiniis*, 1637.

Beighton, family.
• **Beighton**, Henry (1687-1743). Surveyor and engineer, editor of the *Ladies' Diary*. Survey of Bedworth Field, 1707 MS; Inclosure surveys, 1729-1731; *A Map of Warwickshire* (4 sheets), 1728, published Elizabeth Beighton 1750, and in Bowen and Kitchin's *Large English Atlas*, various editions.
• **Beighton**, Elizabeth. Widow of Henry, published his *A Map of Warwickshire*, 1750.

Beijerinck. See **Beyerinck**.

Beilby, R. Plan of Newcastle and Gateshead, 1788.

Beiling, Carl. *Karte von Palästina*, 1846.

Beilly, —. Engraver for James Murray's *Impartial History*, 1780.

Bein, Julius. Lithographer and printer of 16 & 18 Park Place, New York. C. Roeser's *State of Michigan*, 1878.

Beins [Bains], Jean de (1577-1651). French military surveyor, engineer and mathematician of Paris. 'Géographe du roi' from 1607. Compiled many manuscript maps and plans of frontier regions in southern France, contained in J. Leclerc's the *Théâtre géographique du Royaume de France*, 1621, and used in the so-called 'Richelieu Atlas' *c*.1624-1642; *Plan...de la Ville de Suze*, 1629; *Carte et Description Générale de Dauphiné*, 1630; other maps used by Blaeu, 1631, Hondius, 1633, and later. **Ref.** KRETSCHMER, DÖRFLINGER & WAWRIK *Lexicon zur Geschichte der Kartographie* (1986) Vol.C/I p.75 [in German]; PELLETIER, M. 'La representation du relief au XVIIe siècle: L'oeuvre de Jean de Beins (1577-1651)' in *Bulletin de la Bibliothèque Nationale* N°3 (1980), pp.104-106 [in French]; DAINVILLE, François de *Le Dauphine et ses confins vus par l'ingénieur d'Henri IV, Jean de Bains* (Paris 1968) [in French].

Beiton, Athanasius [Afanasy] von (*fl.c.*1685). Captive Prussian Officer, later Russian General. Made a manuscript map of the territory between the Amur and the Yenisei.

Beiton, Thomas. Carte de la Selenga, *c*.1726 MS.

Beitzke, H. *Uebersichts-Charte der Mittel und Ost Alps*, 1840.

Beke, Charles T. The lands of Cazembe, 1873.

Beke, Daniel van (*fl.*1657-1668). Dutch surveyor and cartographer in Zuid-Holland. Dircxland in Overflackee, 1668 (with Gerrit van Rest).

Beke, Johan van (*fl.*1632-1645). Dutch land surveyor and cartographer. 'Kaert-Bouck' of lands in Heerhugowaard and Schermer districts, *c*.1635.

Beke, Pieter van der (*d.*1567). Flemish cartographer of Ghent. *Charta Flandriae* (woodcut, 4 sheets), 1538.

Bekovich-Cherkassky, *Prince* Aleksandr. Russian explorer. Map of the Caspian coast, 1714-1715 MS, first published 1952.

Bél, Mátyás [Matthias] (1684-1749). Historian and polymath, born Ocsova, died Pozsony [now Bratislava, Slovakia], encouraged use of maps in teaching of history. *Notitia Hungariae novae historico-geographica* (5 volumes), 1735-1742 (with maps by Sámuel Mikoviny); Parvus Atlas Hungariae, 1750-1751. **Ref.** DOBOS, J. *Gondolatok Bél Mátyás születésének 300. évfordulóján* [Commemorating the 300th Anniversary of the birth of Mátyás Bél], *Hidrológiai Tájékoztató* [Hydrological Bulletin] (Budapest 1980-X) [in Hungarian].

Bel, Pierre (1742-1813). Swiss engineer and topographer, worked in Bern.

Belam, Henry. Surveyor. *Liverpool Bay*, 1892-1893, many editions to *c*.1910.

Bélanger, Charles (1805-1881). French naturalist and explorer in Asia and the East Indies. Atlas for *Voyages aux Indes orientales...pendant les années 1825, 1826, 1827, 1828 et 1829* (7 volumes, of which 3 form the atlas), Paris 1834-1838. **Ref.** BROC, N. *Dictionnaire illustré des explorateurs Français du XIXe siècle: II Asie* (1992) pp.22-23 [in French].

Bel Armato, Girolamo. See **Bellarmati**, Girolamo.

Belbridge, William Charles (1829-1892). Government printer, newspaper proprietor and surveyor general in Sydney and Brisbane. *Atlas of the Colony of Queensland*, 1865.

Belch, William. Publisher of 1 Staverton Row, Newington Butts, London (1820); 258 High Street, Borough (1828). Until 1820 he worked closely with Edward Langley as Langley & Belch. Co-publisher of *Langley's New County Atlas of England and Wales*, 1818; republished the 1812 Langley & Belch map of London as *William Belch's plan of London*, 1820. **N.B.** See also **Langley**, Edward and **Langley & Belch**.

Belcher, *Commander* (later *Admiral*) *Sir* Edward (1799-1877). Coastal surveys used by the Hydrographic Office. Africa, 1830-1833; Irish Channel, 1834; Pacific Ocean, 1835-1842; China Seas, 1842-1847; Simon's Bay [Cape Province], 1846 published Admiralty 1849; Table Bay, 1846; *Chart of the Coast of China*, 1851; *Chart of the Coast of Oregon and Columbia*, 1851; *Chart of the West Coast of South America*, 1851; *Chart of the Western Coast of America*, 1851-1853; *Chart showing the North West Passage*, 1852-1853; *Nicaragua*, 1860.

Belcher, G.R. Surveys used by W.H. Holmes for his relief map of the Yellowstone National Park, USA, 1876.

Belcher, R. *Lake Nyassa*, 1889, Admiralty Chart 1893.

Belcikowski, K. *Mappa Wol. Miasta Krakówa*, 1828.

Belden, Henry [& Co.]. *Illustrated Historical Atlas of Kent County, Michigan*, 1876; *Illustrated Historical Atlas of Milwaukee County, Wisconsin*, 1876; *Illustrated Historical Atlas of Wayne County, Michigan*, 1876; and others.

Beleke, —. *A Map of the Environs of Berlin in 10 Sheets*, 1816.

Beleski, J.L. Dantzig, 1822.

Belga, Aegidius Bulionius. See **Boileau** de Bouillon, Gilles.

Belgrand, Marie François Eugène (1810-1878). Hydrographer and geologist. *Carte Géologique des Environs de Paris*, 1865.

Belgrano, Giovanni Maria. Italian engraver. Giovanni Tommaso Borgonio's *Piedmont*, 1680; *Carta Generale di Stati di Sua Altezza Reale* (6 sheets), 1680.

Belgrano, Luigi Tommaso. *Atlante idrografico*, 1867.

Belgrave, William. Surveyor. Dagenham and Barking [Essex], 1653.

Belidar, —. Maps and charts for maritime ports of France, Faden and Jefferys 1774.

Belin De Launay, J.H. Robert (1814-1883). French geographer and historian.

Belkmeer, *Doctor* Jan. Engineer. Banken, dieptens voor de stad Enkhuyzen, 1714; Nieuwe Afteekening...Stad Enkhuysen inde Zuyder Zee, Gerard van Keulen 1734.

Belknap, Jeremy. American cartographer. *A New Map of New Hampshire* in *History of New Hampshire*, 1791 (map engraved by Samuel Hill).

Belknap & Hamersley. Publishers of Hartford, Connecticut. William C. Woodbridge's *School Atlas to accompany the Modern School Geography*, 1843; Woodbridge's *Modern Atlas, Physical, Political & Statistical...*, 1845.

Belkum Azn., P. van. *Zak-Atlas...van Alle Provincien en van Nederland* (12 maps), Amsterdam *c.*1917.

Bell, —. *A Map of the District of Kantyre*, 1793.

Bell, *Lieutenant Colonel* —. *...Movements, Battles and Sieges...during the War in the Spanish Peninsula* (38 plates), 1840.

Bell, Alexander. Engraver. Plan of Edinburgh, 1759; *...County of Berwick*, 1771.

Bell, Alfred. *Sub-surface contour Map of the Illinois Basin*, 1937; *Oil and Gas map of Illinois*, 1939.

Bell, Allan [& Co]. Publishers of Pancras Lane, Warwick Square, London. J. Dower's *The Stranger's Companion Through London and its Suburbs*, 1835 (published with *Diamond Gazetteer of Great Britain and Ireland*); *General Atlas of the World*, 1837-1839; *New General Atlas*, 1837.

Bell, Andrew. Bookseller 'at the Cross-Keys and Bible in Cornhill', London. One of a group of booksellers associated with the publication of works by Robert Morden. Morden's *The New Description and State of England...*, 1704; *Fifty Six New and Acurate* [sic] *Maps of Great Britain, Ireland and Wales...*, 1708 (Herman Moll's reworking of *The New Description...*); involved with others in the publication of early volumes of the Morden partwork *Atlas geographus*, 1708.
• **Bell**, Elizabeth. Widow of Andrew. Succeeded Elizabeth Nutt, M. Nutt, and Elizabeth Morphew as owner and publisher of the partwork *Magna Britannia*, 1723-1724 (based on the work of Robert Morden). In 1724 ownership passed to Thomas Cox. **Ref.** For a full breakdown of the history of the Morden maps see HODSON, D. *County atlases of the British Isles published after 1703* Vol.I (1984) pp.3-31.

Bell, Andrew (1726-1809). Engraver. Northumberland, 1753; Martinico, 1759; *A New Map of the World*, 1759; *A Map of the River of Northesk*, *c.*1760; Berwick, 1771; Armstrong's *Thirty miles round Boston*, 1775; Scotland, 1797.

Bell, C.G.H. Basutoland [Berea], 1881.

Bell, Charles Davidson (1813-1882). Artist and Surveyor-General, *Cape of Good Hope*, 1848.

Bell, Clarence. Surveyor. *Stony Lake District*, 1897.

Bell, Euphemia Young. *The Bermuda Islands*, 1937 and later.

Bell, Henry. *The Ground plat Kings Lyn*, 1561, engraved 1670.

Bell, J. Publisher of Oxford St, London. *La Plata*, 1807; *Monte Video*, 1807.

Bell, James (1769-1836). Geographical and topographical author and editor. Edited Rollin's *Ancient History*, 1828; *System of Geography*, Glasgow 1830; *A New and Comprehensive Gazetteer of England & Wales* (4 volumes), Glasgow, A. Fullarton & Co. 1833, 1836, 1843.

Bell, Jared W. *Bell's Statistical Sheet Atlas of the United States of America*, 1833.

Bell, John. *Map of the Route between Mosco & Pekin...*, London, 1764 (published in Bell's 2 volume work *Travels from St Petersburg in Russia...*).

Bell, John (*fl.*1792-1833). Surveyor and engraver of Edinburgh, particularly for estates in Scotland. *Kinross-shire*, 1796; Fife and Kinross, for Thomson's *Atlas of Scotland*, 1832.

Bell, John Thomas William (*fl.*1831-1861). Surveyor. *Plan and Section of the Hartlepool Railway*, 1831; *A Series of Plans of the Great Northern Coalfield*, 1843-1861.

Bell, *Colonel* M.S. *Route from Peking to Si-Nan-Fu* (5 sheets), 1889.

Bell, Peter. Geographer and engraver. *A Plan of the Harbour of Lynn*, 1762; Black Sea, 1765 (with Andrew Dury); *A New and Accurate Map of North America*, Bowles 1768; *Environs of London*, Dury 1769; *A Map of the British Dominions in North America...*, 1772; Andrews's Isle of Wight, 1775; *Theatre of War in the Provinces of Massachusetts Bay, Connecticut &c.*, 1776; *A new and accurate map of the Seat of War...Morea, the Islands of the Archipelago and...Turkey in Asia*, *c.*1785; British Isles, 1787.

Bell, R.B. *The Harbour of Belfast...*, 1862.

Bell, Robert. 'The present state of Europe and America, or The Man in the Moon taking a view of the English Armada', in *Plea of the Colonies*, 1777.

Bell, *Doctor* Robert. Maps of Canada, 1875-1897; *Map showing the limits of various trees in Canada*, 1897; *Preliminary map of the South Coast of Baffinland*, 1901.

Bell, Thomas (1785-1860). Antiquary and surveyor of Newcastle upon Tyne. Many manuscript maps of northern England.

Bell, W. Lithographer. *Map of Venango County*, 1865.

Bell, *Doctor* W.A. Map of the South West portion of the United States, 1869.

Bell, William. Teacher. *A New General Atlas*, Brown & Nelson 1834, 1837.

Bella, Stefano della (1610-1664). Artist and engraver of Florence, teacher of the engraver Antonio Francesco Lucini. *Schlacht bei Freiburg*, 1644; [Siege of]...*Posto de Longone*, 1650, published *c.*1680.

Bellairs, E. Australian surveyor. *Subdivision of Land on & about Moorabool Creek...Victoria*, 1855; *Plan of Red Bluff Estate, Victoria*, 1858; *Plan of 6 Government Allotments in Latrobe St. West, Melbourne...*, *c.*1863.

Bell Amato, Girolamo. See **Bellarmati**, Girolamo.

Bellamy, Charles Vincent. *A Geological Map of Cyprus*, 1905.

Bellanger, *Mademoiselle* —. Engraver of Denis's *Golphe du Mexique*, 1780; *Carte du theatre de la guerre dans l'Inde*, 1781 (by L. Denis); *Carte Chorographique de la Belgique...* (69 sheets), 1797.

Bellarmati [Amato; Bell Amato; Bel Armato; Bell'Armato], Girolamo [Hieronimo] (1493-1555). *b.* Siena, *d.* Châlons-sur-Soâne. *Chorographia Tusciae* (4 sheets), Rome 1536 and 1554, used later by Abraham Ortelius, 1589.

Bellefonds, A. Linant de. See **Linant** de Bellefonds, Louis.

Belleforest [Belleforêt], François de (1530-1583). Poet, historian and editor of Gascony. *La Cosmographie universelle*, 1575 (a French edition of Münster's *Cosmographia*). **Ref.** PASTOUREAU, Mireille *Imago Mundi* 32 (1980) pp.51-54 [in French].

Bellère [Bellero; Bellerus; Bollero], Jean [Ioan, Joan, Joannis] (*fl.*1550-1595). Printer, publisher, and printseller of Antwerp. López de Gómara's *Historia general de las Indias*, Antwerp, 1554 (with a woodcut map of Peru, *Brevis exactaq. totius novi orbis...*, which was used in several other books of the period including Boileau de Bouillon's *La sphère des deux mondes*, 1555); *De Peruviae regionis inter Novi Orbis Provincias*, Antwerp, 1567; republished Peter Apian's *Cosmographicus Liber...*, 1575 (in Spanish), 1581 (in French), 1584 (in Latin) (the first edition had been published by Gemma Frisius in 1529); edited French edition of Waghenaer's, *Miroir de la navigation*, 1590.

Bellero [Bellerus], Joannis. See **Bellère,** Jean.

Bellery Desfontaines, —. Engraved 9 sheets for the Cassini *Carte de France*, from 1756.

Bellery-Desmarets, —. One of the engravers for sheet 82 of the Cassini *Carte de France*, 1759.

Bellessort, André (1866-1942). French orientalist, travelled in Japan and Korea (between 1898 and 1920). *Voyage au Japon*, 1902.

Bellevue, L.B.F. See **Fleurieu** de Bellevue, L.B.

Belleyme, Pierre de (1747-1819). 'Ingénieur Géographe du roi. Rue du Paon N°·1, Faubourg Saint Germain', Paris. Worked with others on four sheets (in south-west France) of the Cassini *Carte de France*, 1767-1775; map of Aquitaine (4 sheets), 1787, 1791; *Carte de Guyenne* (54 sheets), 1789; *Carte de la France divisée en 83 départements*, 1791 (engraved by Beaublé and Barriere), amended edition published 1793; *Carte de l'Empire français divisée...*, 1812; *Carte des Pays Bas, de la Hollande et du Pays sur le Rhin* (2 sheets), 1812; *Carte itinéraire de la France*, *c.*1820.

Bellier, George. Publisher of 'Rue Saint Jacques à l'ensigne du gril de fer', Douai, France. Cornelis van Wytfliet's *Histoire universelle*, 1611.

Bellin, Jacques Nicolas.
• **Bellin**, Jacques Nicolas, *the elder* (1703-1772). 'Hydrographe du roi', first chief hydrographic engineer of the 'Dépôt des cartes, plans et journaux du Ministère de la Marine'. Maps of Japan for Pierre François Xavier de Charlevoix's *Histoire et Description Générale de Japon*, 1736; *Carte des lacs du Canada*, 1744; P.F.X. de Charlevoix's *Nouvelle France*, 1744; maps for Abbé Prévost's *Histoire générale des voyages*, 1746 and later; *Carte réduite des terres Australes*, 1753; *Le Neptune François*, 1753; *Carte de la Louisiane et des Pays Voisins...*, 1755; *Hydrographie françoise* (2 volumes), 1756-1765, and later editions; *Isles Antilles*, Paris 1758; *Essais géographiques sur les Isles Britanniques*, 1759; *Déscription...de la*

Guyane (20 charts and plans), Paris 1763; *Petit Atlas français* (5 volumes), 1763, re-issued as *Le Petit atlas maritime* (5 volumes), 1764; *Atlas de Corse* (33 maps and plans), 1769; *Carte Hydrographique du Golphe de Venise*, 1770; *Description géographique des Débouquements qui sont au nord de l'isle de Saint Domingue...*, 1773.
•• **Bellin**, Jacques Nicolas, *the younger* (1745-1785). Son of Jacques Nicolas the elder. Engraver of Paris.

Bellingen [Bellingan], Floris (*fl.*1660 and later). Dutch surveyor and cartographer. Worked in Gelderland. *Eijlandt van Nederhemert...voor de stad Heusden*, 1719.

Bellingero, Giorgio. Genoese portolan maker, *c.*1687.

Bellingshausen, Fabian Gottlieb Benjamin (1770-1852). Russian admiral and polar explorer, whose sketch map of Princess Martha Coast is reputedly the earliest map of the Antarctic continent. *Karta Yuzhnago Polushariya*, St Petersburg 1821; Atlas, 1822.

Bellmond, F.B. *Plan des Bromberges dem Tit. Bürger-verein zu Bern...*, 1819; *Plan über das sogenante Kapelle dem Tit. Bürger-verein zu Bern...*, 1819.

Bellnato, F. Discorso di Cosmographia, Venice 1573.

Belloc, Hilaire. *The Land and Water Map of the War*, 1915.

Bellot, —. Map of Port Mahé, Seychelles, 1846 (with Leblanc and Cloué).

Bellot, Joseph René. Chart of the Northwest Passage, 1852-1853.

Bellue, Pierre. Atlas ou Neptune Mediterranée 1830, *c.*1834.

Bellus, Nicolaus. Author of *Laurea austriaca*, Frankfurt am Main, 1627 with David Custodis's map *Angliae Scotiae et Hiberniae sive Britannicarum Insularum*, re-issued in Bellus' *Ostreichischer Lorberkrantz*, 1627-1628 (German translation); re-issued 1644.

Belly, Felix. *Canal Interocéanique de Nicaragua*, 1858.

Belly, Léon (1827-1877). French archaeologist and surveyor in Syria and Palestine (1850-1851) with the Louis-Félicien Caignart de Saulcy expedition. Provided maps for the atlas to de Saulcy's *Voyage autour de la mer Morte*, 1853. **Ref.** BROC, N. *Dictionnaire illustré des explorateurs Français du XIXe siècle: II Asie* (1992) pp.22-23 [in French].

Belmas, Jacques Vital (1792-1864). Maps of Peninsular War campaigns, 1807-1814; *Atlas* (24 plates), 1836.

Belmont, *abbé* -- de (1645-1732). *Description généralle de l'Isle de Montréal...*, 1702.

Belmont [à Bello Monte], Simon [Somon]. Trajectum ad Mosam Maastricht, used by Braun and Hogenberg *q.v.*, 1574.

Belmotus. See **Gervase** of Tilbury.

Belon du Mans, Pierre (1517-1564). Born at La Soultière, near Le Mans, died in Paris. *Observations de plusieurs singularitez et choses memorables trouvez en Grèce, Asie, Indée, Égypte, Arabie et autres pays étrangers*, Paris 1553; inset map of Limnos on Abraham Ortelius's Cyprus, 1573. **Ref.** MEURER, P. *Fontes cartographici Orteliani...* (1991) pp.113-114 [in German].

Belpaire, Alphonse (1817-1854). Engineer of the Belgian 'Ponts et Chausées', who experimented with coloured flow maps. These showed the comparative density of movements of goods and people by road, river, canal and railway. *Carte du Mouvements des Transports en Belgique... Année 1834*, *c.*1845; *Carte du Mouvements des Transports en Belgique... Année 1844*, *c.*1845; another undated map with the same title referring to the transport in the year 1843 was later published as an *Extrait...*, showing the region of Brussels, Malines and Louvain. **Ref.** ROBINSON, A.H. *Early thematic mapping in the history of cartography* (1982), 149-150.

Belt, Thomas (1832-1878). Geologist. Australian Gold Diggings, 1852-1862.

Beltrami, Giacomo Costantino. Manuscript map of the sources of the Mississippi. **Ref.** WOLF, Eric W. *Imago et mensura mundi...* (Rome 1985) pp.127-132.

Beltran, Francisco de P. *Plano Topográfico de la Ciudad de Merida*, 1864-1865.

Belville, Alfred. *A Map of the Gold Fields of South Eastern Africa by T. Baines, assisted A. Belville*, 1876.

Belwe, C.F.G. *Grundriss der Stadt Aschersleben*, 1798.

Bembo, Pietro (1470-1547). Venetian Cardinal. Collector of globes and maps, said to have made a terrestrial globe, 1547.

Bement, Alberto (*b*.1862). *Peabody Atlas: shipping, mines and coal railroads in the central commercial district of the United States*, 1906.

Bemi, Peter Lewis. Surveyor of Elizabeth St. Cattle Market, Sydney (1847). *Map of part of the Keelogues at Illawarra...*, 1839; *Kensington...*, *c*.1840s; *Map of 84 allotments of land situate on the Surry Hills in...Sydney*, 1852.

Bemmelen, W. van. *Uitkomsten der Regenwaarnemingen op Java* (7 maps), Batavia, 1915.

Bemrose, William [& Sons.]. Engravers, printers and publishers of Derby, England. *Map of the Borough of Derby*, 1854, 1867, 1871, 1876.

Ben, Jon. Descriptio insularum Orchadiarum, 1525 MS.

Bénard, Jacques-François (*d*.1751). Publisher 'sur le Quai de l'Orloge à la Sphère Royale', Paris. Son-in-law and successor (with William Danet) to Nicolas de Fer (who died in 1720). *Mappemonde*, 1705, 1717; *Le Berri et le Nivernois, la Beauce et la Sologne*, *c*.1720.

Bénard, Robert. Engraver. *Amérique Méridionale*, in André Prévost's *Histoire générale des voyages*, 1754; French edition of Cook's voyages, 1774-1785.

Benard & Frey. Lithographers, for H.Y.P.F. Bougainville's Atlas, 1837.

Benazech, P. Engraver, for Hervey Smyth's *A View of the City of Québec*, 1760.

Benci, Carlo (1616-1676). Pair of Globes, 1671 MS.

Bendeffy [Benda], László (1904-1977). Hungarian geographer and geologist, born Vasvár, died Budapest. Specialised in cartographic history and hydrographic cartography. Worked with the Hungarian State Survey Office at Szeged, 1929-1930, and at the Triangulation Office from 1931, Associate Research Fellow of the Hungarian Hydrology Reasearch Institute, 1959-1971. Hungarian Land Survey, 1890-1920; Geodetic Bibliography, Vols I & II. **Ref.** RAUM, F. *Bendeffy László geodéziai munkássága* [The geodetic work of László Bendeffy] in: *Vasi Szemle* [Review of Vas county] (1984) issue 4 [in Hungarian].

Bendorp, K.F. Dutch waterways surveyor and engraver worked on Rhine surveys, e.g. *Het Afkomen van een Houtviot op den Rhyn*, 1785.

Benecke, E.W. *Geognostische Karte der Umgegend v. Heidelberg*, 1874-1877; Geologische Karte des Griqua-Gebirges, 1885.

Benedetti, Giuseppe. Engraver. ...*Contorni di Genova*, *c*.1760.

Benedetti, Ignazio. *La Topografia di Roma*, 1773.

Benedetto, Giovanni. Italian cosmographer in service of France, 1535-1540. Portolan, 1543 (attributed).

Benedict [Benedicti], Anton (1767-1854). Viennese engraver for F.J.J. von Reilly and for Artaria & Co.

Benedict [Benedicht], Lorenz [Laurentz] (*d*.*c*.1604). Printer and publisher of Copenhagen. Sea charts of Baltic and North Sea, 1568: *Søkartet offver Øster oc Vester Søen*, 1568.

Benedicti, Anton. See **Benedict**, Anton.

Benedicti, Hieronymus (*c*.1756-1809). Engraver for F.A. Schraembl, 1787-1788. *Post Karte von der Halbinsel Taurien oder Krim*, 1787; for J.M. von Liechtenstern, 1790-1806; *Carte von West-Galizien* (12 sheets), Vienna 1808; *Grundriss des Lustschlosses und des Gartens von Schönbrunn*, 1810.

Beneke, Wilhelm. Elbe, 1868; *Plan der vier Städte Hamburg, Altona, Ottensen und Wandsbeck*, 1872; Heidelberg, 1874-1877.

Benese, Richard. Wrote an early manual on the principals of surveying and the use of surveying instruments. *The Boke sheweth the maner of measurynge all maner of Lande*, 1537, and editions to 1565.

Beneventanus, Marco. Neapolitan cosmographer. Editor of Ptolemy's *Geographia*, Rome 1507-1508.

Benham, Charles E. *Benham's Map of Colchester*, 1893.

Benicken, Friedrich Wilhelm. *Handatlas der allgemeinen Weltgeschichte*, Weimar, Geographischaes Institut 1812-1824; *Historischer Schulatlas oder Übersicht der allgemeinen Weltgeschichte*, Weimar 1820; *Historischer Hand-Atlas...*, Weimar, 1820; *Orbis Terrarum Antiquus*, 1829.

Benincasa, father & son.
• **Benincasa**, Grazioso (*c*.1420-1482). Born at Montesino near Ancona. Navigator and portolan chart maker from 1461, moved to Venice 1462, returned to Ancona 1470, worked at Hungarian royal court at same time as Regiomontanus. Several portolan chart atlases known (mostly of 5 charts: 3 of the Mediterranean and 2 of the African coast); British Isles, 1467; Black Sea, 1474; Europe, 1482. **Ref.** SZATHMÁRY, T. *Descriptio Hungariae* (1987); KRETSCHMER, DÖRFLINGER & WAWRIK *Lexicon zur Geschichte der Kartographie* (1986) vol. C/1, p.78 [in German].
•• **Benincasa**, Andrea (*c*.1440-1508. Geographer and portolan chart maker of Ancona, son of Grazioso (above). Portolan chart atlas 1476; portolan chart 1508.

Benitez y Parode, Mañuel (1845-1904). Spanish military cartographer.

Benjamin of Tudela (12th century). Jewish rabbi, traveller and writer on geography. *Itinerarium Beniamini Tudelensis*, 1543.

Benjamin, *Lieutenant* —, *USN*. Coastal chart: *Monterey & its Approaches*, 1846.

Bennefeld, *Lieutenant* —. *Hannover*, 1770? (with Pentz).

Bennens, Bontko (*fl*.1656-1690). Dutch surveyor and cartographer. Worked in Groningen. De landen van Midwolda..., 1656 MS.

Benner, J. *Karte der farbigen Wegebezeichnungen von Wallmerod u. Umgegend*, 1899.

Bennet, Roelof Gabriel (1774-1829). *Atlas, behoorende tot de Verhandeling ... wegens de Nederlandsche Ontdekkingen*, Dordrecht, J. de Vos 1829 (with J. van Wijk).

Bennett, I.R. Burlton. *Post Office map of the Provinces of Bengal, Behar, Orissa and Arracan*, 1853.

Bennett, James. Engraver of charts for John Seller's *Atlas Maritimus*, 1675; *English Pilot*, 1677, and later editions.

Bennett, John (*d*.1787). In 1774, Bennett joined Robert Sayer in partnership as publishers Sayer & Bennett [*q.v.*] at 53 Fleet Street, London. When Bennett left the partnership *c*.1784, Sayer reverted to trading alone. **Ref.** HODSON, D. *County atlases of the British Isles published after 1703* Vol.II (1989) pp.72-77, 186-187.

Bennett, Lyman G. *Map of Kendall County, Illinois*, 1859 (with J. McWilliams); *Map of Du Page County, Illinois*, 1861; *Map of Ramsey and Manomin Counties and Hennopin East, Minnesota*, 1867; [atlas of] *Winona County, Minnesota*, 1867.

Bennett, Richard. Engineer, draughtsman and engraver. For Solomon Bolton's London editions of J.B.B. d'Anville's maps, 1750, 1752, 1772, 1775; *An Accurate Plan of Cherbourg*, 1758; *Europe*, 1758; *A New and Accurate Map of the English Empire in North America*, 1758; *A New Map of the Coast of Africa*, 1758; *America*, 1759; *Asia*, 1759, *A map of the Caribbe Islands...*, 1759; *A Plan and View of the Town and Fortifications of Montreal in Canada*, 1759; *A New...Post Map of the Great Roads and Cross Roads through England and Wales*, *c*.1765.

Bennett, T. Downes. *Geology simplified*, 1845.

Bennett & Walton. Publishers of Philadelphia, for Robinson, 1807; for H.S. Tanner, 1832.

Bennewitz. See **Apian**, family.

Benning, Richard. *Plan of Bristol*, 1780.

Benning, Robert. Engraver for Peter Martell's *Nouveau Plan de la Ville de Luxembourg*, *c*.1745; engraver for John Rocque, 1746-1761; Edward Oakley's map of Calais, 1750; *City of Winchester*, 1750; *Plan of Rome*, 1750; *Plan of Madras and Fort St George*, 1751; Isaac Taylor's *New Map of the County of Hereford*, 1754; *Plan of the Cities of London and Westminster* (8 sheets), 1755; Edward Oakley's *Plan des Ports et Château de Brest*, 1757; Isaac Taylor's *Hampshire* (6 sheets), 1759; *Plan of New Orleans*, 1761; *Topographical survey of the County of Berks*, 1761; Samuel Loup's *Occident de l'Oberland*, 1766.

Bennison, Jonathan. Liverpool, 1835.

Benoist, —. Engraver. *Plan de Bruxelles*, 1825.

Benoist, Michel *SJ* [Chiang Yu-Jen]. French Jesuit cartographer in China. Engraved Asia and Europe [Qianlong atlas] (104 sheets), Peking (wood-cut) 1769, (copper engraving) 1775, (based on surveys by Felix de Rocha *SJ* and Joseph d'Espinha *SJ*, 1756-1759).

Benoist, Nicolas. Map of the Island of Tobago, *c*.1760 MS.

Benoit, François Chrétien de (1729-1812). Military cartographer. *Osnabrugensis Episcopatus delineata, 1775.*

Benoni, Karol. *...Atlas geograficzny...Cześć I Galicya*, 1873.

Benson, Charles. *A Statistical Atlas of the Madras Presidency* (23 maps), 1895.

Benston, John. Chartmaker, 'Dwelling over against New Grauell Lane in Radcliffe Highway', East London. Portolan chart, 1640.

Bent, James T. Hadhramaut, 1894; Dhofar [Arabia], 1895; Fadhli Country, 1897.

Bent, Theodore. Travels in Arabia and East Africa, 1890-1892.

Bent, William. Publisher at the King's Arms Paternoster Row. *Universal Magazine*, 1784-1799; Benjamin Baker's *Plan of the City of Washington...*, 1793.

Benthuysen, C. van. Lithographer. Map of the Cities of New York and Brooklyn, 1855.

Bentinck, William (1764-1813). Liscomb Harbour, 1784 MS.

Benton, Henry. States of Pennsylvania, New Jersey & Delaware, Hartford 1833.

Bentze, Hans. *Oost Zee*, Copenhagen 1700.

Bentzen, Lauritz Christian von (1803-1881). Danish military cartographer.

Benzinger, J. Plans of Jerusalem, 1896-1897.

Benzoni [Benzo], Girolamo [Jerome] (b.1519). Italian geographer and historian. His account of voyages to America was used in T. de Bry's *Grands Voyages*, parts IV-VI, Frankfurt 1594-1596. **Ref.** ALEXANDER, M. (ed.) *Discovering the New World* pp.122.-151.

Beotio, Johannes Baptista. See **Boazio**, Giovanni Battista.

Ber., Fer. See **Bertelli**, Ferrando.

Berain, N. Engraver, frontispiece to *Neptune François*, 1693.

Beranek, Karl (1779-1842). Austrian military cartographer.

Beranger, Jean. *Plan de la Baye de St. Joseph*, 1718.

Bérard, Antoine. Timor, 1818; *Carte de la Côte septentrionale d'Afrique*, 1836; *Carte Générale de la Mer Méditerranée et de la Mer Noire*, 1859-1860.

Bérard, Auguste. One of the officers on Duperrey's expedition to New Zealand 1822-1825. *Plan du Port Manawa*, 1824; *Carte de la Baie des Isles* [New Zealand], 1824 published in the atlas to accompany Duperrey's *Voyage autour du monde...*, 1827.

Beraud, *Major* —. *Carte des parties centrales du Sahara*, 1862.

Berbrugger, Adrien (1801-1869). French Arabist and traveller in Algeria from 1835, founded Société Historique Algérienne. L'Algérie historique, pittoresque et monumentale, 1843; Antiquités d'Alger, 1845.

Berch, Jan Rutgersz. van den. See **Bergh**, Jan Rutgersz. van den.

Berchem, Jacob Pierre van (1763-1832). Cartographer of French descent working in Lausanne. *Carte pétrographique du Saint Gothard*, 1791.

Berchem, Nicolaes Pieterszoon (1620-1683). Printer of Haarlem and Amsterdam. Nicolaas Janszoon Visscher's *Orbis Terrarum*, 1658 (engraved J. de Visscher), in Janssonius' *Novus Atlas*, *c*.1660; various Bible maps, 1660-1669.

Bercheyck, Laurens Lodewyck van. *Carte van de Rivier Demarary*, 1759, English edition 1781.

Berchtold, Joseph Anton (1780-1859). Swiss theologian, born Greich, dean in Leukerbad and Sion. Worked with J.A. Müller on a survey of the canton of Wallis, 1831-1844 (surveys used in a topographical map of Switzerland). **Ref.** GATTLEN, A. *Walliser Jahrbuch* 29 (1960) pp.22-32; GATTLEN, A. *Vallesia* 14 (1959) pp.117-166 [in German].

Berck, Cornelis van. Dutch land surveyor and cartographer. Worked in Utrecht and de Betuwe. *Landt van Altenae...*, 1563; *Chaerte ende beschrijvinghe vande Middelweert...*, 1571 (with Evert van Schaick Jansz.).

Berck, Jan van. See **Bergh**.

Berckenrode [Balthasar, Balthasarbrys], family, of Delft.

• **Berckenrode**, Floris Balthasar [Balthsars] van (1562/1563-1616). Surveyor, goldsmith and engraver of Delft. Ostenda, 1604 (used by Braun and Hogenberg, *q.v.*); *Africae pars* [Saldanha Bay to beyond Cape Agulhas], 1605 (used by De Bry, 1605, and Commelin); engraved M.J. Been's 1606 survey of Delfland, 1611, (used by Hondius, Janssonius and Blaeu); *Kaart van Rijnland*, 1610-1615 (used by Hondius, Janssonius and Blaeu).

•• **Berckenrode**, Balthasar Florisz. [Baltesar Florissen; Floriszoon] van (1591-1644). Son of Floris Balthasar. Surveyor, engraver and publisher in Amsterdam (1618-*c*.1636) and The Hague. *Expugnatio Sylvae Ducis* (6 sheets), 1629 (with Solomon Savry); *Ware afbeeldinge vande bedyckte Beemster-landen inden jare M.D.C.XLIII* (6 sheets), 1644.

•• **Berckenrode**, Cornelis Florisz. van (1607/1608-1635). Son of Floris Balthasar. Cartographer in Amsterdam. *La Siège de la ville de Boldue*, 1629; engraved the animals on C.J. Visscher's *Ysel-stroom*, *c*.1630.

Berckmann, Arnold. Publisher of Antwerp. Apian's *Cosmographia*, 1539, 1540, later by his heirs, 1574.

Berecz, Antal (1836-1908). Hungarian cartographer. Founding member and first secretary of Magyar Földrajzi Társaság [Hungarian Geographical Society], leader of the cartographic section formed in 1891. **Ref.** FARKASFALVI, J. 'Berecz Antal emlékezete' [Remembering Antal Berecz] in: *Földrajzi Közlöny* [Geographical Journal] 1909-X [in Hungarian].

Bereghy, Károly (1829-1900). Hungarian cartographer. Began work on a cadastral survey of Tirol; from 1856 in charge of survey of Hungary; map librarian at Kassa [now Košice, Slovakia], 1869; at Györ, 1870; central inspector of mapping,1880-1889.

Berendsen, A.W.H.M. *Atlas behorende bij Leven en Streven*, Groningen and Batavia, J.B. Wolters 1935-1938.

Berendt, C. Hermann. *Mapa de la peninsula de Yucatan...*, 1878 and other editions; Crao Para, 1888.

Berendt, Gottlieb. *Geologische Übersichtskarte der Umgegend von Berlin*, Berlin, *c*.1887.

Berendts, F. Campen, 1625 MS.

Berentsz., Thomas (*c*.1576-1625). Dutch surveyor and cartographer from Emmerich. Admitted as surveyor for Overijssel 1611, died at Kampen. *Caerte der Over-IJsselsche limiten boven Deventer...*, 1612.

Beresford, Arthur de la Poër. *Map showing route of exploring party under command of Col. P.E. Warburton from the Centre of the Continent to Roeburne, Western Australia*, 1874.

Berey, father and sons.
• **Berey**, Nicolas I (*c*.1606-1665). 'Marchand d'estampes et enlumineur de la reine, au bout du Pont Neuf proche les Augustins aux Deux Globes'. Father-in-law of Alexis-Hubert Jaillot. Plan de Paris, 1645; *Carte générale de l'Isle de France*, 1648; *Carte de tous les Pars de Mer du Monde*, *c*.1650; *Nouvelle carte marine de tous les ports de l'Europe*, 1650; *Plans et profils de la ville et Fauxbourgs de Paris*, 1650; *Carte de la Suisse*, 1654; *Provinces de France et Espagne*, 1655.
•• **Berey**, Nicolas II (1640-1667). 'Graveur, enlumineur et marchand d'estampes, Île du Palais, rue de Harley pres St Barthelemy', son of Nicolas I (above).
•• **Berey**, Claude-Auguste (*c*.1651-1732). 'Graveur de l'Académie royale des Sciences', son of Nicolas I, brother-in-law of Alexis-Hubert Jaillot. Address at 'Rue St Jacques a la vieille Poste', Paris. Charts for *Le Neptune François*, *c*.1696; Henri Liebaux' *Carte de Thrace*, 1727; also engraved for Guillaume Delisle.

Berey, Antoine. 'Graveur en taille-douce' [copper], 'Faubourg St Antoine', Paris. Plates and maps in Amédée Frézier's *Voyage*, 1716.

Berey, Charles Amadeus de. Engraver in 'Rue St. Jacques', Paris. Engraved '...Irlande..' for *Le Neptune François*, 1693; for Guillaume Delisle, 1700; *Le Cours du Po* (5 sheets), 1702; *Cours du Danube*, 1703; Diocese de Rouen, 1715; *Carte de les Isles de Malte*, 1720.

Berey, Charles Amadeus de, *le jeune*. Les villes, Forts et Châteaux de Malte, 1724.

Berg, family. Dutch military cartographers in the 18th century.
• **Berg** [Bergh], Ernst Willem (1712-1777). *b*. Maastricht, *d*. Breda. Father of Frans Jacob Alexander Berg; brother of Frans Jacob Berg. Plans of Hulst 1744; Nijmegen *c*.1750.
•• **Berg**, Frans Jacob Alexander (1755-1793). *b*. 's-Hertogenbosch, *d*. Lincelles. Son of Ernst Willem (above). Fortification plan of Breda, 1774.
• **Berg**, Frans Jacob (1726-1774). Brother of Ernst Willem (above).

Berg, Alfred. *Universal-Atlas und illustriertes geographisches Handbuch*, Leipzig 1911.

Berg, J. van den. Schoolatlas voor Bijbelsche Geschiedenis, 1875; *De geheele Aarde. Kleine Atlas van de Geheele Wereld*, 1881, 1885; *Atlas der geheele aarde in 24 Kaarten...*, 1885.

Berg, L.W.C. van den. *Hadhramaut*, 1886.

Berg, Pieter van den. See **Montanus**, Petrus.

Berg, *Lieutenant* Staffan Wolter (1671-1745). Swedish military surveyor under Eric Dahlbergh, for *Suecia antiqua et hodierna* (completed 1715).

Bergamo, Antonio da (*fl*.16th century). Manuscript plan of the fortifications at Berwick, with alterations proposed by him, *c*.1570.

Berganus [Bergano], Georgius Jodocus. Italian Benedictine monk of Verona. Description of Lake Garda, before 1550 (used by Leandro Alberti, and by Ortelius).

Berge, Pieter van den (*c*.1665-1737). Dutch draughtsman, engraver and publisher, of Leiden and Amsterdam 'te Heiligenweg'. *Theatrum Hispaniae...*, 1700-1705; *Watergraafs of Diemer-meer*, 1719.

Bergelin, Walter. *Map of the lake district between Sydney and Newcastle*, 1906; *Jervis Bay City estate...*, 1917.

Bergeman, J.C. *Casteel St. George Del Mina & het Foort Couraadsburg de Berg St. Jago*, 1770 MS.

Bergen, Ernst Gottlieb (1649-1722). Mapmaker of Berlin. Ukraine, *c*.1680 MS.

Bergen, J.H. *Great Circle Chart of the North Atlantic*, 1880.

Bergen, Johannes van. Chart of South West Luzon, 1746 MS.

Bergenhold, Johann Ferdinand Schmidt von. *Geognostisch-montanistische Geschäfts- und communicationen-Karte des Konigreichs Böhmen*, 1873.

Berger, C. Heilquellen im Nordwesten Böhmen, 1840.

Heinrich Berghaus (1797-1884). (Taken from Gothaer Geographen und Kartographen, *Haack 1985)*

Berger, Daniel. Engraver of Berlin. *Carte de la Crimee*, 1776; *Die Stadt Hirschberg im Schlesien*, 1810.

Berger, Franz. Wien, 1873; Plan de Dôle, 1874.

Berger, Friedrich Gottlieb (1713-*c*.1800). Engraver of Berlin. For J.C. Rhode, *Theatrum Belli in America Septentrionali*, *c*.1755; Prussia, 1765; *Accurater Plan der Stadt London...*, 1770; *Neuer Geometrischer Plan der... stadt Berlin*, 1772.

Berger, G. Worked with Engelmann on *Portefeuille géographique et ethnographique*, 1820.

Berger, George. Military Map of India, 1854.

Berger, Johann Christian. *Delineationem Liberae in Silesia Dynastiae Drachenberg*, *c*.1676 (re-engraved by P. Schenck in Amsterdam).

Berger, Johann Otto [Giovanni Ottone di] (*fl*.1748-1799). Cartographic draughtsman from Austria, headed the school of cartography in Southern Italy. Also worked with Rizzi-Zannoni in Naples, Paris and Bayerne. *Pianta della Citta di Taranto*, 1757; *Plan du Fort Philippe...*, 1759; *Carta delle Reali Cacce*, 1784; *Atlante piccolo marittimo des Regno di Napoli*, 1785; map of Calabria, 1788; *Carta Topografica dell'Isola di Malta*, *c*.1791; map of the Eolie Islands, 1793. **Ref.** VALERIO, V. *Società uomini e istituzioni cartografiche nel mezzogiorno d'Italia* (1993) pp.453-455.

Hermann Berghaus (1828-1890). (Taken from Gothaer Geographen und Kartographen, *Haack 1985)*

Berger, John George. War Chart, *c.*1854; War Map [Northwest and Central Europe], *c.*1870.

Berger Levrault, *veuve* [et fils]. *Carte Topographique du Département Bas Rhin*, 1854; Haut-Rhin, 1857; *Plan Topographique de la Ville de Strasbourg*, 1857.

Berger-Levrault. Photo-lithographers of Nancy. *L'atlas-index de tous les théâtres de la guerre*, 1914-1916; *Les opérations russo-turques. Mer Noire-Crimée. Caucase-Dardanelles*, 1916.

Berggreen, V.F.A. Maps of Denmark, 1870-1886 including; *Kort over Aarhus Kjøbstad*, 1870; *Kjøbenhavn med naermeste omegn*, 1879; *Carre-Planer af Kjøbenhavn*, 1886.

Bergh, Ernst Willem. See **Berg**, family.

Bergh [Berch; Berck], Jan Rutgersz. van den (*fl.*1594-1609). Dutch surveyor and cartographer of Utrecht. Landerijen... onder Amerongen, 1608.

Bergh, Th. Danish engraver. *Sinerissap Kavdlanâkarfiligtâ...* [Greenland], 1870; *Vulkaner paa Island*, 1882.

Berghaus, family. **Ref.** KRETSCHMER, DORFLINGER and WAWRIK *Lexicon zur Geschichte der Kartographie* (1986) vol. C/1, pp.80-81 [in German].

• **Berghaus**, *Doctor* Heinrich Carl Wilhelm (1797-1884). Engineer and geographer from Cleve, in French service from 1811, studied at Marburg 1814-1815 and Berlin 1816-1817, worked in the Prussian Land Survey Office.

Teacher of geodesy and pioneer of scientific thematic cartography. Author of many theoretical works on geography and cartography. Worked for the Geographical Institute in Weimar, for S. Schropp & Co., and for Georg Reimer. From 1829 worked in association with Justus Perthes. *d.* Stettin [Szczeczin]. Afrika, 1826; Atlas von Asien, 1832 (the first part of a larger work for Perthes); *Berghaus' Physikalischer Atlas*, from 1836, published Gotha, J. Perthes 1837-1848; maps for Stieler's *Schul Atlas*, 1839; *Royal Prussian Maritime Atlas* (10 sheets), Berghaus 1839-1847; thematic maps for A.K. Johnston's *National Atlas*, 1843; revised English edition of the Berghaus atlas published by A.K. Johnston as *The Physical Atlas*, 1848; editor of *Vollständiger Hand-Atlas*, from 1850. **Ref.** DREYER-EIMBCKE, O. 'Heinrich Berghaus and August Petermann' in: *IMCoS Journal* 70 (Autumn 1997) pp.19-27; ESPENHORST, J. *Andree, Stieler Meyer & Co: Handatlanten des deutschen Sprachraums (1800-1945)* (1994) pp.138-144 [in German]; ROBINSON, A.H. *Early thematic mapping in the history of cartography* (1982), 65-67, 114-115, 134-137; ENGELMANN, G. *Heinrich Berghaus: der Kartograph von Potsdam* (Halle 1977) [in German].

•• **Berghaus**, *Doctor* Hermann (1828-1890). Gotha. Nephew and pupil of Heinrich (above). Austrian Atlas, 1855; *Welt-Karte* (4 sheets), 1859; *Welt-Karte* (8 sheets), 1863; issued third edition of Heinrich Berghaus's *Physikalischer Handatlas*, Gotha 1886-1892.

Berghes, Carl de. *Cartá Geognostica de los Principales Distritos Minerales... de Mexico*, 1827.

Bergier, Nicolas. *Histoire des grands chemins de l'empire romain*, 1726.

Bergius, E.W. *Fiedlitz bei Marienwerder*, *c.*1835; *Erinnerungs-Blattan Marienwerder*, 1840.

Bergman, Carl Johan. *Gotland och Wisby*, 1875.

Bergman, Johan. Swedish cartographer. Plan of Lund, 1704 MS.

Bergman, P.C. *Plankarta öfver Staden Jönköping*, 1868; *Jönköping*, *c.*1876.

Bergman and McGonigale. American surveyors. *Plots of Land in the Vicinity Louisville* [Kentucky], 1860; Preston, 1865.

Bergmans, J. *Kleine Atlas van Nederland*, 1917.

Bergmiller, —. Engraver. *St.-Eustache Insel*, *c.*1785.

Bergqvist, Carl. Swedish engraver. *Grundritning öfver Stockholms Stad*, 1771; *...Siöar och Strömmar*, 1774 (with F. Akrel).

Bergslien, Knud. *Kart over Dr. F. Nansen's Polar Expedition*, 1893.

Bergstrand, P.E. *Bohus Län*, 1867.

Bering [Behring], Vitus Jonassen (1680-1741). Danish navigator in Russian service. Northern Siberia, 1729 MS, first printed in J.B.B. d'Anville's *Nouvel atlas de la Chine*, 1737; also in J.N. Delisle's *Atlas russicus*, 1745 **Ref.** MULLER, G. F. *Bering's voyages: The reports from Russia* (1987).

Beringer, David (1756-1821). Globemaker of Nuremberg.

Beringer, O.L. *A Map of the Shire Highlands* (2 sheets), 1897; *North Eastern Rhodesia*, 1907.

Berjaques, Louis de. *Menina vulgo Meenen*, published in Blaeu's Townbooks of the Netherlands, 1649-1653.

Berjon, Jean. Publisher of Paris. Samuel de Champlain's *Voyages*, 1613.

Berk, Gidion van. *Curieuse Aftekening van de glorieuse actie voorgevallen by Donawaart*, *c.*1704.

Berka, Johann (1758-1820). Engraver. Plan of Prague, 1790.

Berken, Johann [János] von (1765-1822). Viennese engraver. Maps for H.W. von Blum von Kempen, *c.*1796; Demeter Goerög, *c.*1802; maps of Hungary for J.M. Korabinsky, 1804.

Berkenmeyer, P.L. *Le curieux antiquaire ou recueil géographique et historique*, Leiden, Pieter van der Aa, 1729.

Berkum, A. van. *Atlas der oude geschiedenis*, 1901.

Berlepsch, K. *Göttingen und Umgegend*, 1897.

Berlinghieri, Francesco di Nicolo (1440-1501). *Geographia di Francesco Berlinghieri Fiorentino in terza rima* [a metrical rhyming paraphrase, and the first edition of Ptolemy in Italian], 1482 (printed by Nicolo Todescho).

Berlinguero, Nicolas. Graduate of Spanish School of Navigation. *Plano de la Bahia de Panzacola* [Florida], 1774 MS; *Plano de la Bahia de Sn. Joseph Uno...*, 1774.

Bermann. Publisher in Vienna.
• **Bermann**, — [and son]. *General-Postkarte des Kaiserthumes Oesterreich...*, 1839; *Karte der Markgrafschaft Mähren und Schlesien*, 1841 (both the above maps were by F. Raffelsperger).
•? **Bermann**, Joseph. *Nord-Deutschland*, 1848.

Bermingham, T. *Oporto & Environs, including Villa Nova*, 1840.

Bermont, P.L. *Plan de la Ville de Brest*, 1779.

Bermudes, —. *Nouvelle Carte de Mexique...*, Paris, A. Vuillemin, 1862.

Bernadin-Bechet, —. *Nouveau Plan de Paris*, 1878.

Bernard, A. Publisher of 'N°1 rue des Grands Augustins', Paris. Plan of Paris, 1850 (with P. Maire); *Théâtre de la Guerre en Europe*, 1866, 1870.

Bernard, Augustin (1865-1947). Professor of Geography at the Sorbonne, Paris. *De Adamo Bremensi Geographo*, 1895; *Atlas d'Algérie et du Tunisie*, 1923.

Bernard, D.H. *The Navigator's Sky Pilot*, *c.*1907.

Bernard, Frédéric (1851-1927). French explorer in the Sahara, Indochina and Sumatra. *Carte des régions parcourues par la mission Flatters* [Trans-Saharan journey], 1892; *Délimitation de la frontiére franco-siamoise*, 1909.

Bernard, Guillaume. See **Barentsz.**, Willem.

Bernard, John A. Associated with Edmond Bohun's *A Geographical Dictionary...* by Edmund Bohun, London, 1694 (4th edition).

Bernard, Jean (*fl.*16th-century). Historian. *La Guide des Chemins D'Angleterre*, Paris, Gervais Mallot 1578.

Bernard, Jean Frédéric. Publisher in Amsterdam. *Recueil De Voiages Au Nord...* (volumes I-IV), 1715-1718, (volumes I-X) 1731-1738; N. de Fer's *Mississippi*, 1718; French edition of the voyages of Nicolas de Graaf published as *Voyages De Nicolas De Graaf Aux Indes Orientales...*, 1719 (first published in Dutch in 1701); *Voyages de François Coreal aux Indes Occidentales... depuis 1666 jusqu'en 1697*, 1722; second edition of Constantin de Renneville's *Recueil Des Voyages...*, 1725 (first published by Estienne Roger, 1702-1706); Garcilaso de la Vega's *Histoire des Incas*, 1733-1737.

Bernard, L. Publisher of Paris. L. Bouffard's *Carte des Etats-Unis d'Amérique*, 1836.

Bernard, Pierre. Mediterranean, 1623.

Bernard, *General* S. *Map of a reconnaissance between Baltimore and Philadelphia exhibiting several routes of the Mail-road*, 1827.

Bernardes Branco, Manoel. *Diccionario geográfico abreviado de Portugal e suas possessoes ultramarinas*, 1862 edition.

Bernardi, Willem. See. **Barentsz.**, Willem.

Bernardin-Bechet, --. *Nouveau Plan de Paris*, Paris, 1878.

Bernardino di Firenze. Woodcut view of Florence, published in his *le bellezze et chasati di Firenze*, Florence *c.*1500.

Bernardinus. See **Bordeaux**, Bernardin de.

Bernards, Christoffel. Dutch surveyor and cartographer. Worked in Zeeland. Kaarte en afteijkeninge van het eijland Walcheren, 1643 MS (destroyed in 1940, known from 18th-century copies by A. Hattinga).

Bernardus, Willem. See **Barentsz.**, Willem.

Bernasconi, G. Italian engraver. *Vedute della coste d'Italia*, 1897.

Bernat, Josef. *Přehledna mapa o vyrobe piva v kralovství českem za obcansky rok 1873*, *c.*1874; *Übersichts-Karte der Rübenzucker-Production im Königreiche Böhmen...* 1875-1876.

Berndt, Gustav. *Übersichtskarte des schweizerischen Fohngebietes*, 1886?

Berndt, Johann Christoph, *the elder* (1707-1798). Military engineer and engraver of Nuremberg. Globe, compiled earlier by Jean Pigeon, 1756; War in America, 1776.

Berndt, Johann Christian, *the younger* (1748-1812). German engraver and publisher of Nuremberg. Engraved for Jaeger, *Iles Etats de la Saxe* (12 sheets), 1779.

Berndt, P. Swedish engraver. *General Karta öfver Sverige, Norge och Danmark* (6 sheets), 1880, 1887, 1892.

Berndt, Wilhelm (1851-1920). Co-founder, with Gustav Freytag, of the Austrian publishing house of Freytag & Berndt *q.v.*, which later merged with Artaria to form Freytag-Berndt & Artaria in 1920. **Ref.** ESPENHORST, J. *Andree, Stieler Meyer & Co: Handatlanten des deutschen Sprachraums (1800-1945)* (1994) pp.338-344 [in German].

Bernegger, Matthias. Austrian professor at Strasbourg, edited Wolfgang Lazius's maps of Austria, before they were republished as *Austriae chorographia*, J. van der Heyden 1620 (the maps by Lazius were first published in 1563).

Bernet, *Professor* Johann Jakob [Jacob] (1800-1851). Theologian and librarian of St. Gallen, Switzerland. *Karte der Cantone St. Gallen und Appenzell*, 1840.

Bernewitz, F.A. von. *Zwickau mit seinen Umgebungen*, 1845.

Bernhard, C. Map of Bayern, 1866; *Karte vom Chiem See*, 1866; *Schul-Atlas von Bayern*, 1872; *Schul-Handkarte von Bayern*, 1876; *Umgebungskarte von München*, 1897.

Bernhardi, Karl [Carl] (1799-1874). *Sprachkarte von Deutschland* [dialect map of Germany], 1843, 1849.

Bernhardt, E.F. *Carte du Caucase, Géorgie et Perse*, 1810; *Grundriss der Freien Stadt Hamburg*, 1819, revised by R. Kener and F.E. Schuback, Hamburg 1839.

Bernhardt, L. Engraver. Maps for Wieland's *Hand-Atlas*, 1848; *Karta öfver Medlersta och södra Sverige* (4 sheets), 1852; *Norra delen af Elfsborgs Lan*, 1856; *Södra delen af Elfsborgs Lan*, 1860.

Bernier, —. Engraver, maps and plans for Louis Claude Desaulces de Freycinet, 1802.

Bernier, François. French physician of the Faculté de Montpellier. Travelled extensively in Asia. *Voyages de François Bernier...*, Amsterdam, Paul Marret 1711 (included the following maps: *L'Empire du Grand Mogol*; *Carte Nouvelle du Royaume de Kachemire*; *Sources du Nil*).

Bernière, Henry. See **De Bernière**, Henry.

Bernizet, Sébastien. For J.F.G. de La Pérouse, *Baie de la Conception*, 1796; *Plan de la Baie de Castries*, 1800.

Berno, Pierantonio. *Pianta e dissegno della fabbrica fatta ad' uso delle fiere di Verona nell' anno 1722*, *c.*1722.

Bernou, Claude. *Carte de la decouverte du Sr. Jolliet où l'on voit la communication du Fleuve St Laurens*, 1676; *Carte de l'Amérique Septentrionale et partie de la Méridionale*, 1682 MS (this map was for some time attributed to Jean Baptist Louis Franquelin).

Bernoulli, *Professor* Daniel (1700-1782). Dutch physician, physicist and mathematician, active in St. Petersburg and Switzerland. Produced works relating to the theories of geodesy.

Bernoulli, Johann [Jean] (1742-1807). Astronomer and mathematician in Basle and Berlin. Published a work about India which included the geographical notes and maps of India by Joseph Tieffenthaler, and the research of Anquetil du Perron. *Description historique et géographique de l'Inde* (3 volumes), Paris and Berlin, 1786-1789 [Berlin title *Historisch-Geographische Beschreibung von Hindustan*]. **Ref.** GOLE, S. *India within the Ganges* (New Delhi 1983) pp.212-214.

Bernuset, Pierre [& Co.]. Publisher of Lyon. Bertholon's *Atlas Moderne Portatif...*, Lyon 1799 (also published in Paris by Louis Perisse).

Béro, D. *Plan de la Ville de Bordeaux...*, 1823, revised 1831, 1840, 1866.

Beroaldo [Beroaldus], Filippo [Phillipus], *the elder* (1453-1505). Cosmographer and print seller of Bologna. Bologna edition of Ptolemy, 1477 (first edition with engraved maps).

Berquin-Duvallon, —. *Carte détaillée de la Basse Louisiane et Floride Occidentale*, 1802.

Berre, —. *Carte particulière des Côtes de France*, 1848.

Berrien, *Lieutenant* J.M. *Survey of the Mouth of Trail Creek, Indiana*, 1834; *Survey of the Mouth of Galien River*, 1835.

Berriman & Co.. Publishers of Philadelphia. Maps used variously in the works of Flavius Josephus, and editions of the *Holy Bible*. Thomas Bowen's *An Accurate Description of the Holy Land*, 1792, 1795, 1796 (engraved by Alexander Anderson); *A Correct Map of the Countries Surrounding the Garden of Eden or Paradise*, 1792, 1795, 1796, 1799 (engraved by Alexander Anderson).

Berry, family. See also **Newton**, family and company.
• **Berry**, Miles. (*fl.*1830s). Globemaker, in partnership with members of the Newton family of Chancery Lane, London.
• **Berry**, W. (*fl.*1816-1849). Globemaker, also in association with the Newton family.

Berry, Edward. New York publisher. *A Map of General Dumouriez's Campaign on the Meuse...1792*, 1794.

Berry, Edwin S. *Map of route travelled and discoveries made by the South Australia Government Central and Western Exploring Expedition...*, 1873; *Map showing routes travelled... by Thomas Elder... under command of Ernest Giles*, 1876.

Berry, Henry C. 15 High Street, St Albans. *Sketch Map of St Albans*, *c.*1880.

Berry, J. *A Plan of the Towns of Manchester and Salford* (2 sheets), 1741 (with Russel Casson).

Berry, R.W. Produced a series of relief maps of the continents, *c.*1928.

Berry, William (1639-1718). Engraver, bookseller, globemaker and publisher 'at the Signe of the Globe between Charing Cross and White-Hall', and elsewhere. Member of the Weavers' Company of London. Sometimes known as 'the English Sanson'; collaborated with Robert Morden. England, 1671; *A New Map of Scotland...*, 1679 (with Robert Greene); playing cards depicting English county maps, 1676 (with Robert Morden); untitled world atlas after Sanson, 1680-1689; 14-inch terrestrial globe, 1683 (with Morden and Lea); *A General Map of Ireland, Surveyed by Sr William Petty. Divided into its Four Provinces and its 32 Counties...*, 1689 (republished by Willdey, *c.*1720); Gascoyne's Cornwall, 1700 (with Thornton & Mount).
Ref. BANNISTER, D. *The Map Collector* 21 (1982) pp. 22-24; TYACKE, Sarah *London map-sellers 1660-1720* (Tring 1978) pp. 109-110.

Bersenev [Bersenieff], *Lieutenant* (later *Captain*) Ivan. Russian Admiralty Surveyor, produced manuscript atlases including: Black Sea and Sea of Azov (11 sheets), 1755-1785; two atlases of the Dnepr river, 1775 and 1786; chart of Black Sea, 1777 (with Lieut. Lyubim Pustoshkin); Arkhipelago [Aegean], 1788 and 1798.

Bert, Paul. Henri Mager's *Atlas colonial* [atlas of the French colonies] (20 maps), 1885.

Bert[s], Pierre. See **Bertius**, Petrus.

Bertan, Giàcomo. *Lucemburg*, in *Raguaglio historico...*, Venice 1699 (copied from the plan of the same name by Demetrio Degni, 1698).

Bertarelli, L.V. *Carta d'Italia del Touring Club Italiano...* (51 sheets), 1907-1914; *Carta d'Italia del Touring club Italiano...la regione Sicilia* (7 sheets), 1916.

Bertaux, Emile (fl.1880-1900). Globemaker of Paris. Produced many terrestrial and celestial globes for the schoolroom.

Bertelli, Andrea. *Disagno del Territorio Padoano*, 1568.

Bertelli, Donato (*fl.*1558-1584). Venetian publisher and mapseller 'Alla libreria del segno del S. Marco in Mezzaria'. Re-issued some of P. Forlani's work; also acquired many of Camocio's plates, which later passed to Stefano [Mozzi] Scolari. *Italia Nuova*, 1558; *Germania*, 1559; *Austria e Ungaria*, 1559; Tripoli, 1567; *Universale descrittione di tutta la terra conosciuta*, 1568; Greece, 1569.

Bertelli, Ferrando [Ferando; Ferdinando] (*fl.*1556-1572). Venetian publisher and engraver, sold printed and composite atlases which he assembled to order; associated at times with Camocio, Forlani and Zenoi. *Mapa de España y Portugal*, 1559; *Britannia Insula...continet Angliam et Scotiam cum Hibernia adjacente*, 1562; *Germaniae... descriptio*, 1562; *Descrittione del Ducato di Savoia*, 1562; World map, 1562, 1565; Belgium, 1565; Dalmatia, 1565; Great Italy, 1565; Russia, 1566; Santo Domingo, 1566; *Civitatum Aliquot Insigniorum...*, Venice 1568.

Bertelli, Francesco (1594-1629). Publisher, editor, and engraver of Padua. *Theatro delle Città d'Italia* (75 plates), 1629.

Bertelli, Lucca. Italian publisher. Algiers, 1565; Rome, 1589; took over the plates for Ferrando Bertelli's World map, *c.*1571; etc.

Bertelli, Nicolo. *Isola de Palmosa*, *c.*1570; *Milo*, *c.*1570; *Nicosia*, *c.*1570; *Negroponte*, *c.*1570.

Bertelli[us], Pietro [Petrus] (*fl.*1594-1616). *Theatrum Urbium Italicarum*, Venice 1599, 1616, 1629; *Italia*, *c.*1620.

Bertera, —. *Carte géologique du Département de la Nièvre*, 1864.

Berthault, M. (*fl.*18th century). French engraver, worked for Jean Baptiste Louis Clouet. Engraved title for Clouet's *Geographie Moderne*, 1780.

Berthe, —. Engraver and editor. 'Rue St. Jacques N°·66', Paris. Historical plans of Paris, 1823; *Carte des Côtes de Terre Firme*, 1828; *L'Océan Atlantique Septentrional*, 1828.

Berthe, L.H. *Carte de la Mer Baltique...*, 1870; *Carte géné du Royaume de Prusse*, 1870; *Nouvelle Carte de L'Empire de Russie*, 1877. **N.B.** Compare with **Berthe** of 'Rue St. Jacques', above.

Berthelemier, —. Engraver for Conrad Malte-Brun. Océanie, 1842.

Berthelot, *Captain* —. Régiment de'Isle de France. Issued printed portolans and charts. Isles Seychelles et Ste. Anne, 1782.

Berthelot, François. *Nouvelle Carte de l'Archipelle*, 1695; *Carte des Costes de la Mer Méditerranée* (2 charts), *c.*1695.

Berthelot, P. French geographer. Madagascar, 1635.

Berthelot, Sabin (1794-1880). Geographer, naturalist and diplomat from Marseilles. French consul in Canary Islands. *Carte topographique de l'Île de Teneriffe*, 1825-1827, published 1834; *Histoire naturelle des Îsles Canaries* (9 volumes), 1836-1850 (with P.B. Webb); *Carte topographique de l'Îsles de Canaria*, 1838.

Berthelsen, Rasmus. Cartographer and lithographer. Map of Greenland, Godthaab 1858.

Berthelt, A.J. *Handatlas für Schüler beim Unterrichte in der Geographie und Geschichte*, Leipzig, Klinkhardt 1851

Berthier, Jean Baptiste (1721-1804). Military cartographer. *Carte des chasses royales*, 1764.

Berthier, Louis-Alexandre (1753-1815). French general, engineer and cartographer. Manuscript maps of the movements of Rochambeau's army in Connecticut, New York, New Jersey and Massachusetts, 1781-1782; *Ville...de Baltimore*, *c.*1781; *Carte générale des États Prussiens ou théâtre de la guerre*, 1806.

Bertholon, *Citoyen* —. *Atlas Moderne Portatif...*, Paris, Louis Perisse, 1799, Lyon, Pierre Bernuset & Co. 1799, B. Cormon & Blanc 1804.

Berthoud, Edward L. Civil Engineer. *Map of Township 8... and of Leavenworth City*,

Petrus Bertius (1565-1629). (By courtesy of Rodney Shirley)

1857; *Map of ... Routes to the Gold Region of Western Kansas*, 1860.

Berthoud, Henri. *Carte des Districts de Zoutpansberg (Transvaal)*, 1886.

Bertin, I. Engraver in Padua. *Tabulae Geographicae* of Typographica Seminary, 1699.

Bertius [Bert, Berts], Petrus [Pierre, Pieter] (1565-1629). Born in Flanders, settled first in Amsterdam, then became professor of mathematics and librarian at Leiden; appointed Cosmographer to Louis XIII from 1618, died in Paris. Brother-in-law of both Jodocus Hondius and Pieter van den Keere, with whom he collaborated. Wrote a geographical treatise in Latin published as the pocket atlas *Tabularum geographicarum contractarum...*, Amsterdam, Cornelis Claesz 1600, and many editions to *c.*1650 in Latin, French and German (accompanied by miniature maps engraved by J. Hondius and P. van den Keere and first published by Barent Langenes in *Caert-Thresoor*, 1598); an edition of Ptolemy's *Geographia*, published as *Theatrum geographiae veteris*, 1619 and other editions (with maps by Mercator and Ortelius). **Ref.** KOEMAN, C. *Atlantes Neerlandici* Vol.I (1967) pp. 60-66; *ibid.* Vol.VI p.xiii.

Bertolotti, — (*fl.* late-18th century). Engraved maps for Barbiellini's atlas, 1807.

Berton, Thomas. Map of Selenga and its affluents, 1746 Russian MS.

Bertou, *comte* Jules de. French surveyor and explorer in the Middle East. Undertook a topographical study of the Dead Sea and valley of the River Jordan, 1839.

Bertram, Carl Julius (1723-1765). Teacher in Copenhagen. Forger of so-called Roman Map of Britain by Richard of Cirencester, *Mappa Brittaniae Faciei Romanae...*, 1755. **Ref.** SHIRLEY, R.W. *The Map Collector* 53 (1990) pp. 8-13.

Bertran, C. Draughtsman and cartographer. *Plan von Swinemünde* [Świnoujście, Poland], Berlin 1839.

Bertran [Bertram], Jacobus [Giacomo]. Catalan portolan maker. Portolan chart, Barcelona 1456 (with Berenguer Ripoll).

Bertrand [Betrand], Alejandro. *Carta de la Hoya de Lago Titicaca*, 1879; *Map of the Deserts of Tarapaca and Atacuma*, 1879.

Bertrand, Alexandre J.F. (1795-1831). French geographer and physician. Revolutions du globe, 1824; Lettres sur la physique, 1825.

Bertrand, Alfred. Swiss explorer, army captain and President of the Geographical Society of Geneva. Explored the Zambezi basin in 1895. *Au Pays des Ba-Rotsi Haut-Zambèze*, 1898

Bertrand, Arthus. Lithographer and publisher of Paris. Issued many fine lithographed books. *Plano comparativo... de la extremidad occidental de la isla de Cuba*, *c.*1840; *Carte de la Côte de l'Amérique sur l'Ocean Pacifique Septentrionale...*, 1844.

Bertrand, Jean. *Paskaart van West-Indien van Rio Oronoque tot Carthagena* (for the Van Keulens), *c.*1728.

Bertrand, L. *Carte géologique de la Rance...*, *c.*1889; Carte géologique de la Tarentaise, 1894.

Bertrand, P. Publisher of 'Rue de l'Arbre Sec N°·22', Paris. Francis de la Porte Castelnau's *Géographie des parties centrales de l'Amérique du Sud...* (30 maps), 1854 (relating to an expedition of 1843-1847).

Bertrand-Gonlin, Charles. *Carte de la terraine de transport dans le Valdamo*, 1833.

Bertres, Felipe. *Mapa Corografico de la Républica de Bolivia*, 1845.

Bertuch, Friedrich Justin (1747-1822). Engraver, printer and publisher of Weimar, founder of the Landes-Industrie-Comptoir which developed into the Geographisches Institut Weimar. F.L. Güssefeld's *Carte itinéraire de l'Europe*, 1793; publisher of the periodical *Allgemeine Geographische Ephemeriden*, 1800-1804 (with maps by A. Stieler).

Bertulius, Sigismund. Map of the river Danube (3 sheets), Jacob Sandrart 1664.

Berty, Adolphe. *Topographie historique de Vieux Paris*, *c.*1897.

Bes, A. *Carte des Pays-Bas*, *c*.1869.

Bes, N. *Magni Amazoni fluvii in America Meridionali nova delineatio*, 1655, 1661.

Besançon, —. (*fl*.1798-1835). Engraver for the Dépôt de la Marine. Golfe du Mexique, 1800; Bahama, 1801; charts for Bruny-d'Entrecasteaux, 1807; Côtes du Brésil, 1822-1826; *Carte de False Bay et la Baie de la Table*, 1824 (in *Neptune François*, 1824); for Dumont d'Urville, 1833.

Beseglio, Giovanni (*fl*. mid-17th century). Venetian. Author of manuscript Isolario, 1645.

Beseke, C.J. *Der Nord-Ost-See Kanal*, 1886.

Besier, J.A. *Waterstaatskaart van Nederland*, 1865.

Besley, E. *Topographical Map of the State of Colorado*, Denver 1887.

Besley, Henry. Route maps of Devon, 1850; *Plan of Ilfracombe and Lynton*, 1892.

Besme, Victor. *Projets des quartiers nouveaux... Bruxelloise* (2 sheets), 1863.

Besoet, Iven van (1720-1769). Engraver and publisher of The Hague.

Besold, Christoph. *De nove orbe conjectana*, 1619.

Bessel, Friedrich Wilhelm (1784-1846). German astronomer. Devised a system of measurement for accurate surveying. Star catalogue, 1830.

Besseling, H. Kaerte van de Reize der Kinderen van Israel, *c*.1750.

Bessels, Emile (1847-1888). German geographer and explorer. US Arctic expedition *Scientific Results*, 1876.

Besserer, Wilhelm (*fl*.16th century). Painter and mapmaker from Speyer. Made a large scale chart of the Rhine, 1580-1590 (the map was 12 metres in length); also credited with a wall chart of the river: *Rhein Abrisz von Linckenheim... bis vnder Vdenheim* (watercolour, 4.58 metres long), 1590.

Besson, J. Publisher, '...sur le Quay de l'Horloge du Palays...', Paris. Duval's *Les Isles Britaniques...*, 1680; Placide's Siam, 1686; abbé Baudrand's version of Desgranges' *La Carte des Royaumes d'Angleterre, d'Ecosse et d'Irlande*, 1695, 1699, 1701; *Royaume de Portugal* (4 sheets), 1704; *L'État du Duc de Savoie* (6 sheets), 1704.

Besson, Jacques. *Le cosmolabe ... un instrument universel*, Paris, Rouille 1567.

Best, Edward. Geologist. *Geological Map of the British Isles* in *Stanford's atlas of universal geography*, 1887, 1894, 1904 editions.

Best, George (*d*.1584). Navigator with Martin Frobisher 1576-1578. *A True Discourse of the Late Voyages ... of Martin Frobisher* (with woodcut world map), 1578.

Beste, J.R. Digby. *Pianta dei Contorni di Roma*, 1858.

Bestehorn, C.B. Map for Homann's Heirs' *Stadt Atlas*, 1762.

Bestehorn, L.S. *Perspectivische Vorstellung des berühmten Blocken oder Blokenbergs...* 1732, 1749 (for Homann Heirs' *Atlas Selectus*).

Beszédes, Ferenc (*b*.1789). Hungarian cartographer. Map of Tolna County, 1836. **Ref.** FODOR, F. *A magyar térképirás* [Hungarian Cartography] (Budapest 1952-1954) [in Hungarian].

Beszédes, József (1787-1852). Hungarian cartographer, engineer and cleric. Made hydrographical and estate surveys. **Ref.** FEJÉR, L. 'Beszédes József maga által leírt élete' [Autobiography of József Beszédes] in *Vizügyi Közlemények* [Hydrographic Journal] (1987 :4) [in Hungarian]; 'Emlékbeszéd Beszédes József felett' [Obituary of József Beszédes] *Akadémiai értesítö* [Transactions of the Academy] (1852) [in Hungarian].

Betagh. William. *The World in Planisphere...*, 1728.

Bétemps, Adolphe-Marie-François (1813-1888). Born in Evian, France. Worked as a military cartographer and topographer for the Swiss Topographical Bureau in Geneva, 1838-1861, 1869-1874.

Betgen, Johann Friedrich. *Plan von der in Lithauen [...] Stadt Gumbinnen* [Gusev], 1733 published in: *Hancce Lithuaniam Borussicam...corographice exhibentur...*, Nuremberg, Homanns' Heirs 1735.

Bethune, *Captain* —. Chart of the Brunei River, 1845.

Beton, Fyodor. See **Baidon**, Fyodor.

Betrand. See **Bertrand**, Alejandro.

Bettencourt, Emiliano Augusto de. *Atlas Pecuario de Portugal*, Lisbon 1870 (lithographed by C. Maigne).

Bettison, Samuel. *Map of the Country Twenty miles round Cheltenham*, *c.*1830; *Plan of the Town of Cheltenham*, *c.*1830.

Betts, A.C. *Beveridge Island* [Australia], 1875.

Betts, Henry D. Civil engineer and surveyor. *Map of Corington...*, Nebraska, 1857.

Betts, John (*fl.*1844-1863). Publisher of London, 7 Compton St., Brunswick Sq. and 115 Strand. *Itinerant and commercial map of England*, 1839; Family Atlas, 1848, 1863 (with W. Carson); Sixpenny Maps, 1846; *...Betts's New Portable Terrestrial Globe*, London 1850; numerous educational maps 1852-1872.

Betts, W. Publisher. Walter Hoxton's *Chesapeake Bay*, 1735 (with E. Baldwin).

Beuch [Beich, Beuech], Daniel (*d.*1670). Painter in Ravenstein and Munich. *Vera totius Marchionatus Badensis delineatio*, *c.*1635.

Beudant, François Sulpice. (1787-1850). French physician & geologist. *Voyage minéralogique et géologique en Hongrie*, 1822.

Beudeker, Christoffel (1675-1756). Dutch collector and antiquarian. Compiled a personal atlas *Germania Inferior, sive XVIII. Provinciarum Geographicae...* (24 volumes), now in the British Library. **Ref.** SIMONI, Anna, 'The Beudeker Collection in the Map Library of the British Library' *British Library Journal* 11 (1985), pp. 143-175.

Beuech, Daniel. See **Beuch**, Daniel.

Beughem, Cornelis van. *Polimetria Britannica...*, *c.*1695.

Beugo, John. *Scotland with the roads*, 1789.

Beuker, Ernts [Erents] (*fl.*1745-1762). Dutch surveyor and cartographer. Worked in Drente. Untitled manuscript map of Drente, 1748.

Beumer, P.J. *Der Kleine Geograph oder Hand-Atlas für Elementarschulen...*, Wesel, Bagel 1841 and later editions; *Neuer Atlas über alle Theile der Erde*, Wesel, Bagel 1847.

Beusch, Th. *Plan von Stettin*, *c.*1879.

Beust, *Second lieutenant* Friedrich von (1816-1899). German teacher and cartographer in Prussia and Zurich. *Historischer Atlas des Kantons Zürich*, 1873.

Beutel, Tobias (*d.*1690). Estate surveyor of Saxony and Thuringia.

Beutler, Joseph. Engraver. *Australia...*, *c.*1840; *Oceanien*, 1840.

Beuvelot, J.B. *Environs de Paris*, 1817.

Bevan, Benjamin, father and son.
• **Bevan**, Benjamin *the elder* (1772-1833). Architect, civil engineer and instrument maker. Father of Benjamin, below. ... *Grand Union Canal*, 1809.
•• **Bevan**, Benjamin *the younger* (*fl.*1819-1839). Estate surveyor and engineer. Manuscript surveys of Bedfordshire, Buckinghamshire and Northamptonshire.

Bevan, George Philips (1829-1889). Royal relief atlas, 1880; Statistical atlas of British Isles, 1881-1882; *A Guide Map of Kent*, E. Stanford 1891.

Bevan, John. American surveyor and civil engineer, produced U.S. county maps, *c.*1850-1870 including: *Map of the Borough of Stonington, Connecticut*, 1851 (printed by Saront & Major of New York).

Bevan, William Lattan. *Mediaeval Geography: An Essay in Illustration of the Hereford Mappa*, 1873 (with H.W. Phillott).

Bever, Th. van. *Ville d'Anvers*, 1865, 1872.

Beveridge, —. Engraver for Ensign, Bridgman & Fanning. *Township Map of Michigan*, New York 1855.

Beverly, Henry. *Plan of Tappahanock*, 1706 MS.

Beverly, T. *Reflections upon the Theory of the Earth*, 1699.

Bevers, A. *Kaart der Omstreken van Scheveningen*, *c*.1862.

Bevers, Fendol. American surveyor. *Map of Wake County*, 1871.

Beville, *Lieutenant-Colonel* [of Dragoons], *Sieur de* —. Plan to accompany a reconnaissance made in England, 1768 MS.

Bevis, John (1693-1771). English astronomer. *Uranographia Britannica* (35 sheets), 1750, published as *Atlas Celeste*, 1786.

Bew, John (*d*.1793). Bookseller and publisher of 29 Paternoster Row, London; publisher of *Political Magazine*... which appeared from 1780-1791 and included many maps of all parts of the World engraved by John Lodge.
• **Bew & Murray**. Published the magazines jointly for a brief period in 1785, publication then passed to J. Murray. **Ref.** Full listing of the maps can be found in JOLLY, D. *Maps in British periodicals Part I* (1990) pp.145-172.

Bewlay, Robert (*fl*.1739-1781). Surveyor of York. Monastery of St Mary York, 1770 MS; Haworth Grange, 1780 MS.

Bewsher, *Lieutenant* J.B. Surveyor in Iraq 1862-1868. *Survey of Part of Mesopotamia*, 1868.

Beyer, Carlos. *Mapa de la República del Paraguay*, 1886; Atlas General de la República Argentina, 1889.

Beyer, Edward. *Album of Virginia*, 1857.

Beyer, F. Norwegian publisher. *Route Map of Norway*, *c*.1895; *Turist-og Kontor-Kart over Norge*, *c*.1898.

Beyer, Herman. *Schul-Atlas der neuesten Erdbeschreibung*, *c*.1868.

Beyer, I.C. *Kaart van het Koningrijk der Nederlanden... Luxemburg*, 1866.

Beyer, Johannes (1673-1751). Instrument maker of Hamburg. Globes, 1718, 1720.

Beyer, Leopold. Engraver for Wieland, 1848.

Beyer, Max (1894-1982). *Sternatlas*, Hamburg 1925 (with K. Graff).

Beyer, Th. *Plan der Stadt Liegnitz*, 1877, 1891.

Beyerinck [Beijerinck], Frederik (*c*.1720-1783). Dutch surveyor and cartographer. Worked in Gelderland. *Kaart...van den Boven Rhyn, van Emmerik, langs het Spyk*, 1730-1772; *Kaart waar opis afgebeeld ... de Boven Rhijn*, 1772; and numerous other river surveys.

Beyerinck [Beijerinck], L.W. *Kaart van het Gouvernement Sumatras West-Kust*, 1852; *Topographische Kaart van de Residentie Banjoemaas*, 1868, 1892; *Topographische Kaart Soerakarta*, 1873, 1894; *Topographische Kaart van de Residentie Semarang* (6 sheets), 1873, 1898; *Topographische Kaart van de Residentie Cheribon* (6 sheets), 1894, 1909.

Beyerinck [Beijerinck], Martinus. *Kaart van de Rivier de Whaal*, 1778; *Plan van het Nieuwe Kanaal door den eer tyds Bylandschen nu Hollandschen Waard*, 1778.

Beyrich, Ernst (1815-1896). Worked for the Journal of the German Geological Society. *Geognostische Karte des nördlichen Harzlandes*, 1851; with W. Hauchecorne initiated the *Geologische Karte von Europa* (49 sheets), from 1878 (compiled from the findings of the international geological conference of 1878, completed by F. Beyschlag).

Beyschlag, Franz (1856-1935). Completed the *Geologischen Karte von Europa* (49 sheets), begun by Beyrich and Hauchecorne in 1878; *Höhenschichtenkarte des Thüinger Waldes*, *c*.1893.

Beyton, Fyodor. See **Baidon**, Fyodor.

Bezard, Johann von Ritter (1871-1954). Austrian military cartographer. Constructed surveying instruments, including a compass, designed specifically for cartographic use. **Ref.** BONACKER, W. *Kartenmacher aller Länder und Zeiten* (Stuttgart 1966) [in German].

Bezaure, Gaston de (1851-*c.*1901). French traveller and diplomat in China, particularly Szechuan (1875), later consul-général at Shanghai. *Le Fleuve bleu. Voyage dans la Chine occidentale*, 1879; *Le pays du Fleuve bleu*, 1899.

Bèze, Theodore de (1519-1605). French writer and theologian in Paris, Lausanne and Geneva. *Mappemonde Nouvelle Papistique*, *c.*1566 (not a geographical map).

Beziers, [Jean-Baptiste-Laurent]. See **Baisier**, Jean-Baptiste-Laurent.

Biancani [Blancano; Blancanus], Giuseppe [Josephus], *S.J.* (1566-1624). Italian mathematician, astronomer and cosmographer. *Sphaera Mundi, Sev Cosmographia...*, 1620, 1653.

Bianchi, Giulio Cesare. *Piano della Città di Milano*, *c.*1820.

Bianchini, Francesco (1662-1729). Astronomer and archaeologist of Verona and Rome.

Bianco, Andrea. Venetian cartographer, chart maker and 'comito' (captain of a galley). Atlas, 1436; world map, 1436; chart of the Atlantic, London 1438; chart West Africa, 1448 MS; assisted Fra Mauro on his *Mappa mundi*, 1457-1459.

Bianco, Noe (16th century). Italian author and traveller. Edited *Viaggio al Santo Sepolcro da Venetia* (with a view of Jerusalem), Venice 1521 and later editions.

Bianconi, F. Various *Cartes commerciales*, 1885-1900: Bulgarie, 1887; Venezuela, 1888; Guatemale, 1890.

Biandrate di San Giorgio, *conte* Guido (*d.*1579). Italian military engineer from Monferrato, in service of Alessandro Farnese at Maastricht, where he planned the defences, 1579.

Biart, Alphonse (1825-1898). Belgian geologist.

Biasini, Gregorio. Publisher of Cesena, Italy. Francisco Clavigero's *Storia Antica del Messico...* (4 volumes, 2 maps, 1 view), 1780.

Bibb, *Captain —*. *Plan of the Harbour of Messaua*, 1763, published 1784.

Bibbs, Thomas Franklin. Australian cartographer. *County of Mornington, Victoria*, 1838; *County of Bourke, Victoria*, 1857, 1866; *Excursionists Map...*, *c.*1860.

Biberger, Ulrich Johann (*fl.*1700). Austrian engraver.

Biberstein, H. Surveyor, Confederate States Marine Corps. *Fort Mason*, 1863.

Bibliographisches Institut. Cartographic publishing house of Gotha, founded by Joseph Meyer, 1828, moved to Leipzig 1874. Issued numerous atlases.

Bickham, family. Printers, publishers, engravers and booksellers of London. **Ref.** HODSON, D. *County atlases of the British Isles published after 1703* Vol.II (1989) pp.81-93; LYON, David, 'A bird's-eye view of the Bickhams' in: *The Map Collector* 2 (1978) pp. 30-33.

• **Bickham**, George, *senior* (1684-1758). Author, engraver, and calligrapher of James Street, Bunhill Fields (1743), later Featherstone Street, Bunhill Fields. *Universal Penman*, 1733-1741 (partwork); *The British Monarchy* (partwork), 1743-1749 (includes curious bird's-eye views of English counties, American colonies and settlements in India and Africa), re-issued with county maps, 1749-1751, and later; the bird's-eye views were re-issued by Laurie and Whittle 1796.

•• **Bickham**, George, *junior* (*d.*1771). Son of George senior. Engraver, publisher and art master at 'Ye corner of Bedford Bury New St. Covent Garden', and later 'Mays Buildings Bedford Court Covent Garden'. *A Map Twenty Miles Round London*, 1740; *A New and Accurate Chart of the Sea-Coasts of Great Britain, Ireland with a view of Dunkirk...*, 1743; American Colonies, 1743; Counties of South Britain, 1750; *Bird's-Eye Views of the Several Counties in England And Wales*, *c.*1753; St. George's Parish Hanover Sq., 1761; Topographical Survey of the Parish of Kensington, 1766.

• **Bickham**, John. A relative of the two Georges who worked as engraver and publisher at 'Seven Stars in King St. Covent Garden'.

George Bickham, the elder (1684-1758).

Biddle, Mary. Bookseller of Philadelphia. Named as editor and seller (with Matthew Clarkson) of a 1762 issue of Scull's map of Philadelphia.

Biddulph, John Burnett (1796-1837). Plan mouth of Kowie River [South Africa], 1820 MS.

Biddulph, *Lieutenant Colonol* Michael Antony Shrapnel. *Siege of Sevastopol*, 1854; *Sketch of the Battle of Alma*, 1854; *Field of Inkermann...*, 1855; *Map of Europe...* [showing the telegraph communications in Turkey, and between India and Europe], 1860.

Bidwell, Oliver Beckworth. Missionary maps. *China*, 1850; *Africa*, 1860; *Asia*, 1865.

Biedermann, E. Engraver for *A new map of the State of California...*, 1852 (published in *Meyer's Hand-Atlas*).

Biedermann, Henry Augustus (1743-1816). Surveyor, born Blankenburg [Germany]. Many manuscript maps of parts of England including: Bettswells & Hollybush Hall, 1772; *Plan of Elmham Park in the County of Norfolk*, 1778.

Biedma, J.J. *Atlas Histórico de la República Argentina*, Buenos Aires 1909.

Bielck, Johannes. Publisher of Cellarius's *Nucleus Geographicae Antiquae et Nova*, 1692.

Bieling, Johann Heinrich Gottfried. Printer of Homann's Heirs' Atlas, 1737.

Bielitz, Robert. *Principatul Romaniei. Das Fürstenthum Walachei* (2 sheets), 1854.

Bielski, M. (1596-1673). Compiled a work on Polish cartography, *Kronika Wszytkiego świata*, [n.d.].

Bielz, A. *Charta Românii...*, *c.*1865.

Bielz, E.A. *Karte der geognostischen Verhältnisse*, Hermannstadt 1854.

Bien, Hans (1591-1632). Plans of Nuremberg, from *c.*1621; Pflegamt Lauf, *c.*1628 MS.

Bien, Joseph R. *Atlas of the State of New York*, Julius Bien 1895.

Bien, Julius [later Bien & Co.] (1826-1909). Born and educated in Germany, moved to America in 1849. Addresses given as 24 Vasey Street, NYC (1866); 139 Duane Street, New York (1884, 1889); also Fulton Street, New York. In 1849 Bien started his business as an engraver and lithographer, later trading as a publisher. He built up a leading map-making establishment employing more than two hundred people and several lithographic and copperplate presses. Worked extensively for US Government. Engraved A. Boschke's *Map of Washington City*, 1857; G.K. Warren's *Map of Routes for a Pacific Railroad*, 1857; *Map of the territory of the United States*, 1858 (drawn by E. Freyhold); *Map of the State of Florida*, 1859; engraved the second, updated edition of Warren's *Map of the territory of the United States from the Mississippi River to the Pacific Ocean*, 1863; *Map of the United States Military Rail Roads*, 1866; *American Central Railway*, 1866; Warren's Military maps, 1869; *Statistical Atlas of the United States* (54 plates), 1874; Clarence King's *Geological and Topographical Atlas Accompanying the Report of the Geological Exploration of the Fortieth Parallel*, 1876; C. Roeser's *Territory of Utah*, 1876; F.V. Hayden's *Geological and Geographical Atlas of Colorado*, 1877; C.E. Dutton's *Topographical and Geological Atlas of the District of the High Plateaus of Utah...*, 1879; *Map of the Rio Grande and Pecos Railway*, 1882; *State of Kansas*, 1884; *Topographical map of the District of Columbia and a portion of Virginia*, 1889; *Atlas of the Metropolitan District* [New York], 1891; *Atlas to Accompany Official Records of the Union and Confederate Armies*, 1891-1895; *Statistical Atlas of the United States, Based Upon the Results of the Eleventh Census* (with many analytical population maps), 1898; and others.
• **Bien & Co.** New York publishing company founded by Julius Bien.

Bienewitz. See **Apian**, family.

Biens, Cornelis. *Topographia Enchusae delineata*, 1649.

Bierens de Haan, *Doctor* D. *Kaart van de Province Overijssel*, 1889.

Biermann, —. *Berlin und seine Umgebungen*, 1833.

Biermann, J.P. *Plan de la Ville de Luxembourg*, 1878.

Biernatzki, H. *Nationalitäten-und Sprachenkarte des Herzogthums Schleswig*, 1850.

Biet, *père* Félix (1838-1901). French missionary and explorer in China and Tibet, best known for his meteorological and geological observations. *Observations météorologiques*, 1884.

Bifezzi, Giuseppe (1798-1881). Geographical engineer of the Officio Topografico di Napoli (from 1826). Surveys, including the city of Messina, for the Officio's map of the Kingdom of Naples, from 1830; *Nuovo atlante corografico, storico, statistico ed idrografico del Regno delle Due Sicilie*, Naples 1837, 1845, 1847; surveys for a map of the frontier between the Kingdom of Naples and the Papal States, 1844-1847; survey of Palermo, from 1849; survey of Brindisi, 1863. **Ref.** VALERIO, V. *Società uomini e istituzioni cartografiche nel mezzogiorno d'Italia* (1993) pp.455-458.

Bigges, *Captain* Walter. *A Summarie and True Discourse of Sir Francis Drake's West Indian Voyage*, Leiden 1588, London 1589 (maps by Giovanni Battista Boazio).

Biggs, Benjamin (*fl.*1809-1834). Surveyor. East Smithfield, *c.*1834 MS.

Bigrel, —. *Carte Générale de la Cochinchine française*, 1872-1873.

Bije [Bye], Cornelis de. Dutch surveyor, engraver and copyist, worked in Zeeland (1691-1719). *De Stad Goes*, 1719 (for Pieter van der Aa); copied A. Braams's Generaele beschrijvinge ofte Caerte des Marquisaets van Bergen op Zoom, *c.*1575, 1719.

Bilberg, Johan. *Refractis solis in Occidui* (2 maps), 1695.

Bilhamer. See **Beeldsnijder**, Joost Janszoon de.

Bill, John (*d.*1630). Printer of Northumberland House, St Martins Lane, Aldersgate St., and Hunsdon House, Blackfriars, London. Apprenticed to printer John Norton 1592; associated with the King's Printing House from *c.*1617. Involved with John Norton in the production of *Theatre of the Whole World*, 1606 (English language edition of the Ortelius atlas); abridgement of Camden's *Britannia*, 1626 (small county maps).

Billaine, Louis. Publisher, 'St Augustin rue St Jacques', Paris. Davity, 1660; Voyage de Laval, 1679.

Biller family. Engravers of Vienna. Workshop engraved maps of Károly Nagy; Demeter Görög and Joseph Schede. **Ref.** DÖRFLINGER, Johannes *Imago Mundi* 35 (1983) pp. 58-64.
• **Biller**, Bernhard, *the elder* (1778-1855). Viennese engraver and globemaker, 1820-1830; plan of Vienna, Artaria & Co. 1830.
•• **Biller**, Bernhard, *the younger* (1802-1840). Engraver and globemaker, son of preceding.
•• **Biller**, Dominick (1812-1884). Copper engraver, son of Bernhard Biller *the elder*; took over the business after 1840. Engraver for Artaria & Co.: Franz Fried's *Karte der Königreiche Ungarn und Galizien...*, 1837; *Wein*, 1838; R.A. Schulz's *General-Post und Straßen Karte des Herzogthums Steyermark*, 1841; R.A. Schulz's Illyria, 1841; R.A. Schulz's *Karte der Gegend von Maria-Zell bis zur Schneealpe*, 1843; *Italia*, 1843; *Neuester Plan von Wien...*, 1848; engraver for Joseph Schede, 1879.

Billès, *Captain* —. (*fl.*early 1900s). French surveyor and explorer in China and Indochina, carried out trigonometrical surveys in Mekong valley, *c.*1904.

Billig, G.A. *Situations-Plan und Grundriss von Berlin*, *c.*1874.

Billinge, Thomas. Engraver. *A Chart of the Harbour of Liverpool*, 1771; P.P. Burdett's *County Palatine of Chester*, 1777; William Yates's *County Palatine of Lancaster*, 1786.

Billings, Joseph. English sailor, accompanied Cook on his Third Voyage as cabin boy. Later entered service of Russia as a naval captain engaged in a survey of the Bering Strait. Expedition to eastern Siberia, 1787-1788.

Billings, T. Lincolnshire, 1790 (with W. Yates).

Billington, William. Engineer and surveyor. *Town of Bertrand*, 1859; *Masonic Map of Illinois*, 1862; *Skeleton Map of Illinois*, 1867.

Billschier, Pehr (*fl.*1755). Swedish surveyor. Surveyed roads in Västmanlands län, 1755 MS.

Billy, Edouard de. *Carte géologique du département des Vosges*, Paris 1848.

Billyard, William. Civil engineer. *Map of the Western District in the Province of Canada...*, Toronto, Scobie & Balfour 1847 (with Richard Parr).

Bilsing, G.G. *Der Stadt Magdeburg*, *c.*1600.

Binder, —. Engraver. *Provincia Moxos* [South America], 1791.

Binder, S. Fülöp (1735-1811). Hungarian engraver, worked in Buda from 1760. Plan of Buda-Ofen, 1761.

Binder, Henri (1855-*c.*1888). French explorer in Central Asia. Map of the valley of the Grand Zab, Kurdistan, 1885.

Bindon, John. Surveyor. *Map of Cullompton Court* [Devon], 1820 MS.

Bindoni, Augustino, and Francesco di. Venetian booksellers and printers. Pietro Coppo's Portolano, 1528; Istria, 1540.

Bineteau, P. Geographer, printer and engraver. *Carte de l'An-Nam, de la Cochin-Chine*, 1860-1863; *Europe*, 1876; Wall maps of France, Belgium and Switzerland, 1877; *Carte du Kurdistan*, 1888.

Binfield, Richard (*fl.*1769-1826). Surveyor of Eton, England. Various manuscript surveys including *A Survey of the Parish of Great Greenford... Middlesex*, 1775.

Binger, Louis Gustave (1856-1936). French explorer and surveyor in West Africa. Maps of Niger and Sénégal regions, 1886-1898; *Carte du Haut-Niger au Golfe de Guinée* (4 sheets) 1889; *Carte du Haut-Niger au Golfe de Guinée par le pays de Kong et le Mossi* (4 sheets), 1892. **Ref.** BROC, N. *Dictionnaire illustré des explorateurs Français du XIXe siècle: I Afrique* (1988) pp.27-29 [in French].

Bingham, *Lieutenant* H.M. Crimea, London, Wyld 1855.

Bingle, J. *Entrance to Port Hunter New South Wales*, 1828.

Bingley, James. Engraver of Goswell Road (1823), later Sidney St., Islington, London. For Henry Teesdale's map of Yorkshire, 1828; *Environs of London*, Cowie & Strange 1828; G. Hennet's *Map of the County Palatine of Lancashire*, 1830; Thomas Moule's *The English Counties Delineated*, 1837.

Binkerd, J.S. *Map of the City of Dayton* [Ohio], 1862.

Binneman, Walter. Engraver. *Middlesex... by J. Ogilby*, 1672, 1673; Ogilby's London, 1677; R. Daniel's *A Map of ye English Empire in... America*, Robert Morden 1679; William Berry's ... *New England and Annapolis*, re-issued Christopher Browne *c.*1690.

Binney, Edward William (1812-1881). Geologist. Founded the Manchester Geographical Society.

Binns, Jonathan. Surveyor and publisher. Surveys in northern England, 1820-1821, 1825; *Map of the Town & Castle of Lancaster*, 1825; *Lancaster & Preston Junction Railway*, 1836.

Binstead, W.H. Surveyor of 88 Pitt Street, Sydney. *Woodford Park, Blue Mountains*, 1880; *Auburn-Vale* [South Granville], 1881; *Mossvale*, 1881; *Hill End Estate, Doonside*, 1885.

Bion, father & son. French globemakers.
• **Bion**, Nicolas (1652-1733). 'Ingenieur du Roi pour les instruments de mathématique', Quai de l'Orloge du Palais au Soleil d'Or', Paris. *L'usage des globes...*, 1699 (and later editions); terrestrial and celestial globes ranging from 5cm to 32cm produced from 1700.
•• **Bion**, Nicolas (*d.*1775). Globemaker, son and successor of Nicolas, above. Married the daughter of Jean-Baptiste Delure, another globemaker.

Biot, Jean Baptiste (1774-1862). French geographer and astronomer.

Birago, Karl, *Freiherr von* (1792-1845). Military cartographer.

Birancourt, —. One of the surveyors for the first sheet [Paris] of the Cassini *Carte de France*, 1745-1755.

Birch, E. ...*East & West of England Junction Railway*, 1845; ...*Southampton and Salisbury Junction Railway*, 1845; also worked with J.B. Birch.

Birch, J.B. Railway maps, 1844-1845 (with E. Birch).

Birch, J.W. Part of Malay Peninsula, 1876.

Birch, William Russell (1755-1834). Atlas of Philadelphia, 1800; The Country Seats of the United States of America (19 maps), 1808.

Birchall, James. *The Student's Atlas of English History*, *c*.1859.

Birckmann [Birckmann], Arnold. Bookseller and publisher of Antwerp. Gemma Frisius's revision of Peter Apian's *Cosmographicus Liber*, 1533, 1539.
• **Birckmann**, Heirs [Hae redes]. Booksellers of Cologne. Ptolemy's *Geographia*, 1540 edition; P. Apian's *Cosmographicus Liber*, 1574.
•• **Officina Birckmannica**. C. van Adrichom's *Theatrum Terrae Sanctae et Biblicarum*, 1590.

Bird, Caspar. Plans of Berlin, 1848-1868; Sans Soucci, 1848; Gradnetz Atlas, 1868.

Bird, Charles. ...*Twelve Maps for Map Drawing*, 1875; Geological map of Yorkshire, 1881.

Bird, J.S. Civil Engineer. *Albion*, 1871; *Jackson*, 1871; *Monroe*, 1871; *Ionia*, 1872.

Bird, John (1815-1896). Government surveyor in Natal. Surveys of Pietermaritzburg, 1849.

Bird, Thomas. Architect and surveyor, Kent St (1847), N°·5 Exchange (1858-1859), Sydney, New South Wales. *Township of Brecon*, *c*.1841; *Chippendale...*, 1842.

Bird, W.A. *Map of a part of the Niagara River*, Balch, Rawdon & Co. *c*.1829.

Bird, William A. Maps of the boundary between Upper Canada and the United States for the Treaty of Ghent, 1817-1822 (with David Thompson, and S. Thompson). **N.B.** Compare with W.A. **Bird**, above.

Bird, William Wilberforce (1758-1836). Civil servant at Cape Town. Plan of Cape Town & harbour published in *State of the Cape of Good Hope in 1822*, 1823.

Bird & Chalmers. Surveyors of Australia. *Plan of 16 agricultural farms...Huskisson*, [post 1840].

Birdwood, W.S. *Plan of Attack... City of Canton*, London 1841 (lithographed by L.J. Hebert).

Birk, Caspar (*fl*.1848-1868). Lithographer and publisher in Berlin.

Birken, Sigismund von (1626-1681). *Danubius Fluviorum Europae Princeps*, 1664, Nuremberg 1683.

Birman et fils. Publishers of Basle. Aquatint views, panoramas, etc., to *c*.1850.

Birmingham, F.W. Surveyor. *Catlets Farm* [North Rocks], *c*.1881; *Parramatta* [New South Wales], 1881; *Calder Estate* [Dundas, New South Wales, [n.d.].

Birmingham, John (1816-1884). Astronomer.

Bironius Gallus. See **Brion** de la Tour, Louis.

Birt, William Radcliff. *The Mare Serenitatis...*, 1869; *Outline Lunar Map Zone IV*, 1870.

Biruni, Abu Rayhan Muhammad Ibn-Ahmad (973-1050). Persian mathematician, astronomer and geographer; one of the greatest intellectual figures of his time in the Muslim world. The titles of some 155 works by Biruni are known. *Al-Qanun al-Masudi* [Masudic Canon of Astronomy] - an encyclopaedia of astronomy.

Bischof, D.D. Goldfass. *Fichtelberg*, Nuremberg 1816.

Bischoff, C. (*fl*.1878-1885). Engraver. H. Kiepart's *New Original Map of the Island of Cyprus...*, Berlin, D. Reimer 1878 (also published in London by Williams & Norgate);

H. Kiepart's *Generalkarte der Südost-europäischen Halbinsel*, D. Reimer, Berlin 1885.

Bischoff, Max. *Flaggen Karte der seefahrenden Nationen*, 1869.

Biscoe, John. Ice Chart of Southern Hemisphere, 1866.

Bishop, F.A. ...*Pacific Wagon Roads Map of...Fort Kearney, South Pass & Honey Lake Road*, 1857.

Bishop, George (*c*.1562-1611). Printer and bookseller at the Bell, St. Paul's Churchyard, London. Succeeded by his cousin Thomas Adams. Camden's *Britannia*, 1590, 1594, 1600, 1607 (with John Norton), 1610; Richard Hakluyt's *Principall Navigations*, 1598.

Bishop, George. Surveyor, Sydney, New South Wales: 80 Cooper St (1858-1859), 112 Pitt St (1881), 130 Pitt St. (1885). *Northumberland New South Wales*, 1864; *Bishop's County of Cumberland in the Colony of N.S.W.*, *c*.1870s; *East Brighton*, 1888; *Clarke's Estate, Burwood*, [n.d.]; *Hunters Hill*, [n.d.]; *Parramatta Park*, [n.d.].

Bishop, H.G. *Atlas of Keokuk County Iowa*, 1895.

Bishop, John. Printer of Denmark Street, Camberwell, Surrey. Maps for the Society for Promoting Christian Knowledge [SPCK], *c*.1850.

Bishop, *Captain* John S. *War map of the Southern States* [Confederacy], 1863.

Bishop, *Captain* Robert. Hydrographer. *Gulf & Windward Passages from Jamaica*, 1761; *Old Straits Bahamas* (6 charts), 1794-1796.

Bishop, S.E. Hawaiian Islands, 1886.

Bisselius [Bisselii, Bissel], Johann. German theologian. ...*Argonauticon americanorum*, Munich 1647, 1698 (Latin translation of Pedro Goveo de Victoria's *Relacion de su Naufragio*..., Seville 1610); Two maps of Palestine in *Palaestinae typothaesia*, 1659.

Bissell, E. American surveyor. *Map of Manchester* [Connecticut], Cheney Brothers 1849.

Bissuel, H. French West Africa, 1888.

Biston, V.J. Lille, 1822.

Bitsch, Johann. Publisher of Georg Matthäus Vischer's *Topographia Stiriae*, Vienna *c*.1681.

Bittner, A. *Geologische Übersichtskarte von Bosnien-Herzegovina*, 1880.

Bittorf, —. *Wandkarte von Unter-Franken*..., *c*.1847.

Biurman, Georg (*d*.1755). Swedish mapmaker and engraver. *Charta öfwer Nerike*, 1715; *Charta öfwer Stockholms Stads*, 1720; *Finland*, 1740; *Charta öwfer Upland*, *c*.1742; *Charta öwfer Westmanland*, *c*.1742; Nerike, *c*.1742; Skåne, *c*.1745; *Svea*, 1747.

Bizemont, Henri Louis-Gabriel de (1839-1899). French traveller and geographer in Central America and Indochina. Les grandes entreprises géographiques depuis 1870, 1877; L'Amérique centrale et le canal de Panama, 1881; L'Indochine française, 1884.

Bizet, A. Letter and text engraver. *Mer des Indes: Golfe d'Aden*, 1882 (for Dépôt des Cartes et Plans de la Marine); for Migeon, 1891.

Björkbom, Karl Niklas (1742-1809). Surveyor of Åbo [Turku], Finland.

Björnlind, Per (*fl*.1781). Swedish surveyor. Kvarnsjö, Jämtland 1781 MS.

Blachford. Chart publishers of 116 Minories and elsewhere in London.

• **Blachford**, Robert (*fl*.1801-1840). Hydrographer, chart publisher and seller, son-in-law of J.H. Moore, probably related to Michael (below) and William, (below). In partnership with Moore 1803-1804 at Little Tower Hill, subsequently alone at 137 Minories (1805-1808), 114 Minories (1809-1819), 79 Leadenhall Street (*c*.1821-1827), then with William Blachford at 116 Minories (1828-*c*.1836). *An improved Chart of the Baltic or East Sea*..., 1803; ...*this New Chart of the Cattegat*, 1803; *A new and correct chart of the North Sea*, 1809; *Blachford's New Chart of the Sleeve*, 1810; *Blachford's enlarged chart of the Firth of Clyde*, R. & W. Blachford 1830; *Blachford's new chart of the* *west coast and islands of Scotland...*, 1835.

• **Blachford**, William (*fl.*1828-1840). Hydrographer, in partnership with Robert Blachford (to whom he was presumably related) at 116 Minories (1828-*c.*1836). *Blachford's new chart of the coasts of Holland and Germany from Amsterdam to Hamburg*, London, R. & W. Blachford 1827? (drawn by William Blachford); *A chart of the east coasts of England and Scotland from Yarmouth Roads to Duncansby Head*, 1830; *Blachford's new chart of Liverpool Bay with the entrances to the rivers Dee and Mersey*, 1830; *St George's, British Channel and North Sea Pilot*, 1838-1845.

• **Blachford**, Michael (*fl.*1836-1844). Probably related to Robert and William (above). Although he wrote sailing directions, 1836-1842, there is no evidence that he prepared any charts himself.

•• **Blachford & Imray**. Partnership of Michael Blachford and James Imray, from 1836. Sold charts and instruments, and published charts at 116 Minories, London, possibly later at 7 Stilbon Terrace, Commercial Road, London (1848-1850).

Black family [A. & C. Black]. Publishing and bookselling business established by Adam Black at 57 South Bridge, Edinburgh (1807-1822); at 27 North Bridge, Edinburgh (1822-1851), and finally sharing premises with Bartholomew at 6 North Bridge from (1851), before moving to London in 1889.

• **Black**, Adam (1784-1874). Publisher of Edinburgh, elected to Parliament in 1856, and served twice as Lord Provost of the City of Edinburgh. Apprenticed to Lackington, Allen & Co. in London, set up on his own in Edinburgh, 1807. Adam Black acquired copyright to the *Encyclopaedia Britannica* in 1827, to illustrate which he commissioned maps and plans from both John Bartholomew and W. & A.K. Johnston. Formed the company A. & C. Black with his nephew Charles in 1834, later to be joined by his sons, who took over management of the business when he retired in 1865.

• **Black**, Adam & Charles [A. & C.]. Initially formed by Adam Black and his nephew Charles in 1834, and passing to his sons in 1865. *Encyclopaedia Britannica*, 1827-1899; *Black's General Atlas of the World*, 1840 and editions to 1900; *Black's County Atlas of Scotland*, 1848 (from John Lothian's atlas of *c.*1828); *Black's School Atlas of Modern Geography*, 1851 and later; *Black's Pocket Atlas*, 1852; *Atlas of Australia*, 1853; *Black's School Atlas of Physical, Modern, Ancient and Scripture Geography*, 1853 and later; *Black's Atlas of North America*, Edinburgh 1856; *Black's School Atlas for Beginners*, 1859 and later; *Black's Modern Atlas*, 1874; *Black's Handy Atlas of England and Wales*, 1892; and many others including numerous railway and tourist guides. **Ref.** SMITH, David. 'A & C Black, 1807-c.1900' *IMCoS Journal* 71 (1997) pp.15-21.

•• **Black**, James Tait (1826-1911). Son of Adam.

•• **Black**, Francis (1830-1892). Son of Adam.

•• **Black**, Adam William (1836-1898). Son of Adam. Joined his father in A. & C. Black in 1861 after serving an apprenticeship in London. Moved business to London in 1889 and retired 1891.

•• **Black**, Charles (1807-1854). Nephew of Adam Black, whom he joined in partnership in 1834. Company traded as A. & C. Black even after the death of Charles.

Black, George. *Contour Map of Melbourne and Suburbs*, 1888.

Black, Kingsbury, Parbury and Allen. Publishers and booksellers. Horsburgh's East India Pilot, 1806-1836; Arrowsmith's Delhi to Constantinople, 1814-1816; Mahratta War, 1817-1819; Blomfield's *Philosophy of History*, 1819; India, 1820. **N.B.** See also **Kingsbury**, Parbury & Allen, and **Parbury** & Allen.

Black, Parry & Co. Publishers of Leadenhall Street, London. East India Pilot, 1885.

Blackadder, John (*fl.*1793-1830). Surveyor and agriculturalist, produced manuscript surveys in Scotland. *Berwickshire* (2 sheets), 1797; Part of Skye, Thomson 1832.

Blackburn, George. *The Map of South Carolina*, 1822.

Blackburn, William (1750-1790). Surveyor and architect of London, County Surveyor of Surrey.

Blacker, *Lieutenant Colonel* Valentine. Surveyor General for India. *Maps and Plans illustrating...the Mahratta War*, 1821; Hindostan, 1824.

Blackett, T.O. *Plan of Burnham Salt Marshes*, 1822 MS; *A plan & section of an intended railway or tram road from the Town & County of Newcastle upon Tyne to the City of Carlisle* (12 sheets), London, 1828 (with B. Thompson and J. Studholme).

Blackford, William Willis. Engineer. *Map & profile of the Virginia & Tennessee Railroad*, 1856 (lithographed by Ritchie & Dunnavant of Richmond).

Blackie, Walter Graham. Geographer and publisher. *The Imperial Atlas of Modern Geography*, London 1860.
• **Blackie & Son**. Publishers of Glasgow, Edinburgh, and London (49 & 50 Old Bailey). *General Atlas*, 1840; *Blackie and Son's Imperial Atlas*, 1859; *The Comprehensive Atlas and Geography of the World*, 1880-1883; Descriptive Atlas of the World, 1893; and others.

Blackmore, E. Published John Smith's *The Generall Historie of Virginia*, London, 1632 edition.

Blackmore, Nathaniel (*d.*1716). ...*Chart of the Coasts... of Nova Scotia*, 1715 MS; ...*Bay of Fundy*, *c.*1720; Annapolis Royal, 1729. **Ref.** ROBINSON, Arthur H. *Imago Mundi* 28 (1976) pp. 137-141.

Blackney, W., *RN*. Hydrographer, 1850-1855.

Blackwell, E.R. Civil engineer. *Trigonometrical Survey of the falls of Niagara; Executed for the Geological Report of the Fourth District*, *c.*1838 (with W.S. Haines).

Blackwell, Thomas E. *The London, Newbury and Bath Direct Railway*, 1845; *Severn & Wye Railway & Canal*, 1852.

Blackwood, *Captain* Francis P., *RN*. Hydrographer. South West Pacific, 1841-1846; *Map of the Eastern Provinces of Australia*, 1850.

Blackwood, James. *Shilling Atlas*, 1858, 1870; *Map of the Seat of War* [in Europe], 1872.

Blackwood, William [and Son]. Publishers in Edinburgh and London. British North America, 1833; *Blackwood's County Atlas of Scotland*, 1838, and later editions to 1853 (engraved W.H. Lizars); Atlas to A. Alison's *History of Europe*, *c.*1833-1842; Berwick, 1850; Atlas of Physical Geography, 1850; A.K. Johnston's *Royal Atlas*, 1861.

Blades, Robson. *The map of the Surrey New Roads*, 1826.

Blades, R. [& Co.]. Lithographers, 11 Abchurch Lane, London (1825-1838). *Van Diemans Land*, 1828.

Bladeslade, T. See **Badeslade**, Thomas.

Bladie, John. *Castricom's Bay on Celebes*, 1666, in *Chart of part of the East Coast of Celebes*, 1781.

Blaettner, Hans Samuel. Territorium Episcopatus Spirensis, 1753.

Blaeu [Blaeuw] family. One of the most famous 16th-century publishers of maps, globes and atlases, in addition to poetry, theology and mathematical works. Also cartographers, instrument and globe makers, and booksellers. Flourished in Amsterdam for more than 40 years until a fire destroyed their premises on the Gravenstraat in 1672, when the loss of all equipment, plates and stock effectively ruined the firm. **Ref.** GOSS, J. *Blaeu's The Grand Atlas of the 17th-century World* (1990); KROGT, P. van der *Old globes in the Netherlands* (1984) pp.22-25; KEUNIG, J. / DONKERSLOOT de-VRIJ, M. *Willem Janszoon Blaeu. A biography and history of his work as a cartographer and publisher* (1973); KOEMAN, C. *Joan Blaeu and his Grand Atlas* (1970); KOEMAN, C. *Atlantes Neerlandici* Vol.I (1967) pp. 67-377.
• **Blaeu**, Willem [Guilielmus, Guilhelmus] Jansz. [Alcmarianus; Janssonius; Janszoon] (1571-1638). Born in Uitgeest, near Alkmaar and studied under Tycho Brahe. Founded his company in 1596, initially as globe and instrument makers but soon expanded into publishing maps, topographical works and sea charts. From *c.*1621 Willem adopted the name Blaeu to differentiate himself, his firm and his family from that of Johannes Janssonius. Appointed chartmaker to the East India Company (VOC) in 1633. On the death of Blaeu in Amsterdam in 1638, the business passed to his sons Joan and Cornelis. Globes, 1599, 1602, 1603; numerous single maps

Willem Blaeu (1571-1638). (By courtesy of Rodney Shirley)

including *Hollandia*, 1604; *Nova Regni Hispaniae descriptio* [Spain and Portugal], 1605; *Nova Totius Terrarum Orbis...*, 1606; map of continents, 1608; Blaeu's first sea atlas *Het Licht der Zeevaert*, 1608; wall map of Germany, 1612 (copy in Staatsarchiv, Nuremberg); globe, 1616; in 1629 he bought 36 plates from Hondius (*q.v.*) and issued his first land atlas as *Atlantis Appendix...* (60 maps without text), 1630 (eight copies recorded to 1991); *Theatrum Orbis Terrarum* or *Novus Atlas*, 1634 and later editions; and many other works. **Ref.** STEVENSON, E.L. *Willem Janszoon Blaeu 1571-1638*; CAMPBELL, Tony *The Map Collector* 30 (1985) pp.36-38; SCHILDER, G. *Imago Mundi* 31 (1979) pp.36-54; SCHILDER, G. *The Map Collector* 3 (1978) pp.2-10; CAMPBELL, T. *Imago Mundi* 28 (1976) pp.21-50; SCHILDER, G. *Imago Mundi* 28 (1976) pp.129-136; DONKERSLOOT-de VRIJ, M. *Willem Jansz. Blaeu* in: KEUNING, J. (Ed.) (1973); KOEMAN, C. *Imago Mundi* 26 (1972) pp.9-16; WARNER, D.J. *Imago Mundi* 25 (1971) pp.29-38.

•• **Blaeu,** *Doctor* Joan I [Johannes; Jan] (*c.*1599-1673). Born and died in Amsterdam. Collaborated with his father Willem Janszoon and with his brother Cornelis, and ran the business after the death of Willem. Appointed official cartographer to the East India Company (VOC) in 1638 in succession to his father. What was left of the company after the fire of 1672 and the death of Joan in 1673 passed to his son Joan [II], who sold the globemaking part of the business to J. van Keulen. Worked in partnership with his father on *Appendix Theatri Ortelii et Atlantis G. Mercatoris*, 1631 and *Novus Atlas*, 1634; wall maps, 1647-1649; *Tooneel der Steden van der Vereenighde Nederlanden* [Town books of the Netherlands], 1649-1653; expanded his father's atlas *Theatrum Orbis Terrarum*, publishing six volumes up to 1655; *Atlas Maior* (11 volumes, known as the Grand Atlas), 1662-1672 in editions with French, Latin, Dutch and Spanish text (often described as the most magnificent work of its kind ever produced); town plans of Italy, 1663; *Regni Hungariae nova et exactissima delineatio*, 1664. **Ref.** WIERINGA F. et al. *De Wereld Folgens Blaeu...1646* (Rotterdam 1986) [in Dutch]; STONE, Jeffrey C. *The Map Collector* 10 (1980) pp.25-29, 50-52; KOEMAN, C. *Joan Blaeu and his Grand Atlas* (1970).

••• **Blaeu,** Willem (1635-1700). Son of Joan [I], above, briefly involved with brothers Joan II and Pieter in the management of the Blaeu business.

••• **Blaeu,** Pieter (1637-1706). Briefly involved with brothers Joan II and Willem in the management of the Blaeu business.

••• **Blaeu,** Joan II (1650-1712). Succeeded to the remnants of the Blaeu publishing house on the death of his father in 1673, much of the stock was auctioned soon after. The globemaking part was sold to J. van Keulen in 1682.

•• **Blaeu,** Cornelis (1610-1644). In partnership with his older brother Joan [I] for about six years. His name appears with Willem and Joan on the imprint of the German edition of *Novus Atlas*, 1634.

Blagrave, John (*fl.*1560-1611). Mathematician of Reading. Estate map of Harpsden, Oxfordshire, 1586; *Forest and Manor of Feckenham*, 1591; *Nova Orbis Terrarum Descriptio*, 1596 (engraved by Benjamin Wright and published, together with a celestial map, in *Astrolabium Uranicum Generale*); Abbey of Marksby, 1608.

Blagrave [Blagrove], Oliver (*fl.*1609-1617). Surveyor. *Sonning*, 1617 MS.

Blagrave, *Lieutenant* T.C. *Map of the District of Jalindhur* (2 sheets), 1852; *Map of the District of Hoshiyarpoor* (4 sheets), 1853.

Blair, *Lieutenant* Archibald. Manhora River, 1777; *An Eye-Draught of Dewgur Harbour*, 1784; *Chart of the Andaman Islands* (4 sheets), 1793, 1795.

Blair, *Reverend* David. *General School Atlas of Modern Geography*, London, T. Allman 1850; *A Classical Atlas on an entirely new plan*, London, T. Allman 1853.

Blair, J.A. Map of the West Indies and Middle Continent of America, 1768.

Blair, *Reverend* John (1722-1782). Chronologist and geographer, Fellow of the Royal Society, Fellow of the Society of Antiquaries. *Chronology & History of the World*, 1754, later edition with *Fourteen maps of ancient and modern geography*, 1768 (maps by Delarochette), further editions to 1834; North America, 1768; *East Indies*, 1773; India, 1773; World, 1779; *The Kingdom of France*, retitled *The Republic of France*, 1814, 1834.

Blaisot, —. *Nouvelle Carte d'Espagne et de Portugal*, *c.*1870.

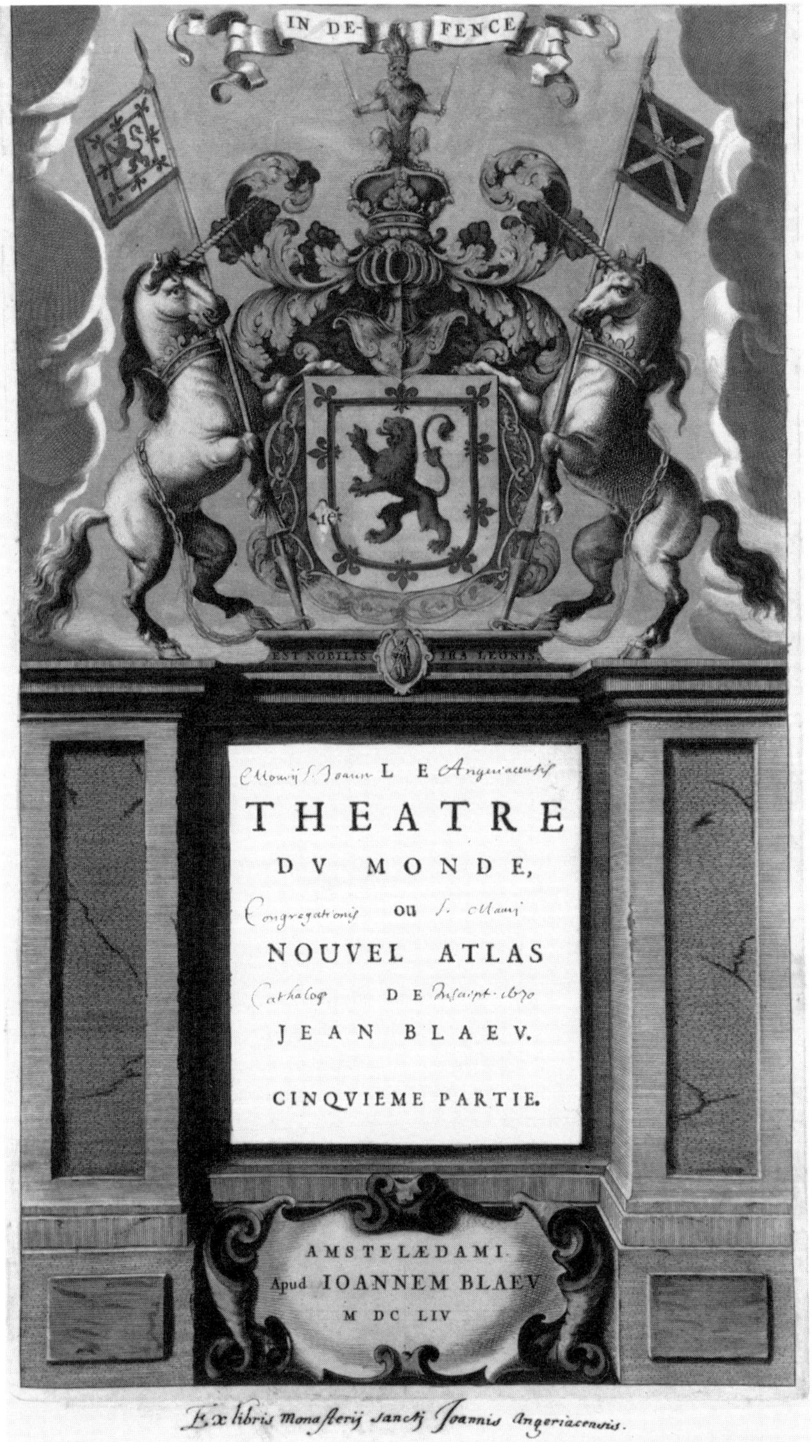

The title page of Joan Blaeu's Atlas of Scotland. *(By courtesy of The Bodleian Library, Oxford)*

Blake, E.J. *Map of North Warwickshire, South Staffordshire, and East Worcestershire Mineral Districts*, 1888.

Blake, Edward Warren. *A New Royal Authentic and Complete System of Universal Geography* (2 volumes), London 1787-1810 (with Thomas Bankes and Alexander Cook).

Blake, *Lieutenant* George S. Worked on the United States Coastal Survey. *Narragansett Bay, Rhode Island*, 1832 (with Alexander S. Wadsworth, Thomas R. Gedney, & Charles Wilkes Jr.); ...*Harbour of Bridgeport, Connecticut*, 1835 (with Alex D. Mackey, Oliver Tod, T.A. Budd, T.A.M. Craven, & B.I. Moeller).

Blake, J. Engraver. *A Plan of the N.W. Harbour in... Mauritius*, 1738 MS.

Blake, J. *An Accurate Plan of the Camp on Coxheath* [Kent], 1778.

Blake, *Lieutenant* J. Edmund. *Part of the Boundary between the United States and Texas...*, 1838, 1841; *Map of the Seat of War in Florida*, 1839.

Blake, *Reverend* J.L. Geographical Atlas, 1816; *A Text Book: or, First Lessons in Modern Geography*, Boston, Richardson and Lord 1821; *American Universal Geography for Schools and Academies*, Boston, Russell, Odiorne & Co. 1833, Lilly, Wait & Co. 1834.

Blake, William P. Expert on the geology of the western United States. Stikine River, 1863; *Geological Map of the United States*, 1874 (with Charles H. Hitchcock) for Francis A. Walker's *Statistical Atlas of the United States*, 1874.

Blakelock, D.H. *Theatre of Military Operations in Luzon*, 1899.

Blakely, W.R. ...*Track through the Gillolo Passage*, 1823; *A Survey of the Passages leading to the Lymoon*, 1830.

Blakiston, —. See **Lindsay** & Blakiston.

Blakiston, *Captain* Thomas, R.A. *China. The Yang-tse-Kiang from Han-Kow to Ping-Shan...from the survey of Captn. T. Blakiston, R.A.*, John Arrowsmith, 1862.

Blamey, Jacob. *A Plan of Amelia Harbour...in East Florida*, 1775, in *North America Pilot*, Sayer & Bennett 1776; revised 1779 for republication in *Neptune François*, 1800.

Blamey, R.H. Surveyor. Port of Alexandria, Admiralty chart 1857 (with E.W. Brooker and F. Skead).

Blanc, Edouard (1858-1923). French geographer, engineer and explorer in Central Asia and North Africa. *Carte du bassin de l'ancien Oxus*, 1892. **Ref.** BROC, N. *Dictionnaire illustré des explorateurs Français du XIXe siècle: II Asie* (1992) pp.35-37 [in French].

Blancano, Joseph. See **Biancani**, Giuseppe.

Blancardus, Nicolas. See **Blankaart**, Nicolas.

Blanchard, A. (*fl.*1832-1860). Script engraver for Dépôt de la Marine. *Baie Saldanha*, 1832.

Blanchard, *Colonel* Joseph. Worked with the Reverend Samuel Langdon on *An accurate map of His Majesty's Province of New Hampshire* (engraved by Thomas Jefferys), 1761 (appeared in Thomas Jeffery's *A General Topography of North America and the West Indies*), additions and amendments were made by Abel Sawyer *Jr.* and the map was republished in Boston, 1784. **Ref.** WHEAT & BRUN *Maps and charts published in America before 1800* (1978) N° 181 p.39.

Blanchard, Rufus. (1821-1904). American publisher and bookseller operating in New York and then at 146 Lake St., Chicago, also noted at 52 LaSalle St., Chicago (1858), and 132 Clark Street, Chicago (1875). Worked with George Franklin Cram for a period. Map of Chicago, 1857; *Johnson's New Copper plate Map of Michigan and the Great Lakes*, 1858 (printed by D.& J. McLellan); *Blanchard's Map of Wisconsin and Northern Michigan*, 1866; *Colton's Guide Map of Michigan*, 1870; *Map of Cook and DuPage Counties*, Chicago 1871; *Commercial Atlas of the United States*, 1874; *Blanchard's Guide Map of Michigan*, 1875; Historical Atlas of the United States, 1895; *Atlas of Chicago* 1895-96, and many single maps. **Ref.** SELMER, Marsha L. in: CONZEN (Ed.) *Chicago mapmakers* (1984) pp.23-31.

Blanchard, W.B. *Map of the Salt Lakes*, 1877.

Blanchin, Jean. Engraver. *Potamographie de Garone*, 1628, *c.*1650; Hondius's Palestine, 1630.

Blanchini, Francesco (1662-1729). Italian astronomer. *Observationes circa planetam Veneris*, 1728 containing gores for first globe of Venus, (said to be the first non-terrestrial globe).

Blanco [Blancus], Christophoro [Christophorus]. Reworked and re-issued Forlani's world map, Rome 1599; *Iaponia...*, 1617 (based on Ignacio Moreira's manuscript map). **Ref.** HUBBARD, Jason C. 'The map of Japan engraved by Christopher Blancus, Rome 1617' in: *Imago Mundi* 46 (1994) pp.84-99.

Blanco, Manuel. Philippines, Manila 1845.

Blanco, Placido. Lithographer. *Mapa General del Estados de Mexico*, 1851-1852.

Blanco y Crespo, Miguel. Missiones de Moxos y Chiquitos, 1770.

Bland, A. Surveyor. Plan of Clapham, 1849.

Bland, Edward. *A Mapp of Virginia*, (engraved John Goddard), in *Discovery of New Brittaine*, London 1651.

Bland, Joseph (*fl.*1714-1760). Worked with Payler Smyth on surveys for John Warburton. *A new and correct Map of the County of York*, 1720; Essex, 1726; Little Ilford, 1737 MS; *A new and correct Mapp of Middlesex, Essex and Hertford-shire*, London 1724 (engraved by S. Parker).

Blandford, W.E. Geologische Karte von Indien, 1879.

Blandowski, —. *Plan von Eckernförde*, 1849.

Blaney, George A. Middlesex Canal in Charleston, 1804 MS.

Blanford, Henry Francis (1834-1893). Geological Survey of India, 1855; Rainfall chart of India, 1883.

Blankaart [Blanckaert, Blancardus], Nicolas (1629-1703). Professor of Leiden. Ancient Europe, Asia, Africa, for Janssonius 1652.

Blanken, J.W. *Kaart van de Provincie Groningen*, 1860.

Blanken, Jan Antonius (1790-1837). Dutch surveyor, worked in Utrecht.

Blanken, Willem (Jansz.) (1769-1831). Dutch military cartographer. Worked in Delfland.

Blankensteiner, I.G. Danish publisher and mapseller, 'Östergaden 36', Copenhagen. Christian Jochum Pontoppidan's *Sydlige Norge*, 1785; *Nordlige Norge*, 1785; *Post-Kaart over Dannemark*, 1796.

Blanmont de Thierville, —. Comté de Lippe, 1761.

Blaquière, G. *Sketch of the action at Kioge*, *c.*1807.

Blaschnek, Sámuel Benjámin (*fl.*1830-1840). Hungarian cartographer. Magyarország, Horvát, Tót, Dalmát és Erdély országok ... Földképe [Wall-map of Hungary] (9 sheets), German edition published L.J. Schedius, 1830 (engraved J. Kiss), Hungarian edition 1836, 1847; miniature maps, 1840.

Blasco, Michel Angelo de. *Stato di Milano e sue Provincie*, *c.*1710.

Blaskowitz, Charles (*fl.*1760-1823). Numerous manuscript surveys in North America, notably Rhode Island and Canada. Many of these surveys were later engraved for publication by others, for example Faden's *North American Atlas*, 1777. **Ref.** GUTHORN, P.J. *British maps of the American Revolution* (1972) pp.13-14.

Blasquez, Antonio. Terrestrial globe, 1530.

Blass, —. *Plan der Stadt Worms*, 1891.

Blatchford, Torrington. *Geological map of Coolgardie* (4 sheets), 1899.

Blathwayt, William (*fl.* 1651-1683). First Secretary for the Lords of Trade and Plantations. *Atlas of the Overseas Possessions of England* (13 manuscript maps and 38 printed maps), 1683. Now at John Carter

Brown Library, Providence, Rhode Island. Facsimile published in 1975. **Ref.** BLACK, Jeannette *Imago Mundi* 22 (1968) pp.20-22.

Blau, Otto. *Karte der Herzegowina im Jahre 1861*, 1861.

Blaufuß, I.A.W. *Charte über die sämtlichen fürstlich sächsischen Länder* (33 sheets), 1798-1800 (with J.V. Blaufuß).

Blaufuß, Johann Valentin (1796-1850). Worked with I.A.W. Blaufuß on the map of Saxony (see above).

Blaurer, Thomas. Manuscript map of Lake Constance, used by S. Münster 1540, 1542 and 1545 (with Johann Zwick).

Blaxland, George. *Sketch of part of the East Coast of China*, 1838.

Bleeker, G.H. *Schoolatlas Voor Bijbelsche Geschiedenis...*, 1894.

Blegen, John. Veileder og verdens atlas, *c.*1892.

Bleiswijk, F. Designed and engraved frontispiece title to Pieter van der Aa's *Le nouveau théâtre du monde*, 1713.

Bleuel, —. Publisher of Nuremberg. Ertl's *Des Chur-Bayerischen Atlantis*, 1705.

Bleuse, Abram. See **Bluzé,** Abram.

Bleysweyck [Bleysweijck], D.E. van. Map of Delft (4 sheets), 1703.

Bligh, *Captain* (later *Admiral*) William (1754-1817). Captain of *The Bounty*, Fellow of the Royal Society. Sailed with Captain James Cook. Governor of New South Wales 1806-1808. Chart of Van Diemen's Land, 1777 MS; Voyage to the South Sea, 1792; Humber, Faden 1797; *Survey of the Bay of Dublin*, 1800, 1803; Dungeness, 1803 MS; *Laurie and Whittle's new Chart of the North Sea*, 1811.

Blijdenzin, Nicolaas [Silvester?]. Dutch surveyor and cartographer. Worked in Noord-Holland. Afcontrafeijtinge van t' Landt van den Zijpe, bedijckt inden jaere 1597, 1600 (engraved by Baptista van Deutecum).

Blink, H. (1851-1931). *Atlas der Natuurkundige Aardrijkskunde, behoorende bij Onze Aarde...*, 1885; De Bewoners der vreemde werelddeelen, Amsterdam 1898-1899; *Onze Planeet Atlas der Wis- en Natuurkundige Aardrijkskunde*, 1899.

Bliss, E. & E. White. Publishers of New York. *Atlas to Morse's New School Geography*, New York edition of 1824.

Bliss, Henry I. Maps of counties of the United States, 1857-1874.

Bliss, Nathaniel (1700-1764). British Astronomer Royal.

Bliss, Sylvester. *Bliss's Outline Globe*, 1851 (made by Gilman Joslin of Boston).

Blith, Charles. Bookseller 'at the White Hart in Launceston', [Cornwall]. Quoted in the cartouche as one of the sellers of Joel Gascoyne's *A Map of the County of Cornwall*, 1699.

Block, Adriaen. Dutch merchant and explorer. With Cornelis Doetsz, produced an untitled manuscript map of the area of New England from Chesapeake Bay to Penobscot, 1614 (Block's name is commemorated in Block Island off Long Island). **Ref.** COHEN, P.E. & AUGUSTYN, R.T. *Manhattan in maps 1527-1995* (1997) pp.22-23.

Block, Joan. *Theatrum Belli...*, 1745; *Verus Chersonesi Tavricae...*, 1745.

Block, *Doctor* Maurice (1816-1901). With the help of A. Petermann, Block created a series of coloured economic and population density maps to accompany and illustrate his studies of Europe. *Bevölkerung des Französischen Kaiserreichs...*, Gotha, J. Perthes 1861; *Puissance comparé des divers états de l'Europe - Atlas*, 1862. **Ref.** ROBINSON, A.H. *Early thematic maps in the history of cartography* (1982), pp.130-131 and 146-148.

Blodget, Lorin. Climatological maps of the United States, 1856-1886.

Blodget, Samuel (1724-1806). *A Prospective Plan of the Battle fought near Lake George...8th September 1755*, Boston 1755 (engraved by Thos. Johnston).

Richard Blome (1641-1705). (By courtesy of Ashley Baynton-Williams)

Blodgett, William. *Topographical map of... Vermont*, 1789; *A New and Accurate Map of Connecticut*, 1792 (engraved by Joel Allen).

Blödner, Cyriak (1672-1733). General Staff Austrian Army. Best known for his coloured manuscript military maps of campaigns in Flanders and the Rhine valley, 1702-1713: *Theatrum Belli Rhenani* (20 sheets), 1713-1715.

Bloemen, Jan. Map-copyist, worked in Kleve. G. Passavant's *Caerte van eenige landerijen omtrent Wageningen gelegen*, *c.*1676, copied 1759.

Bloemswaerdt, C.C. van. Dutch cartographic draughtsman, worked in Utrecht. *Nieuwe kaart van Loenen*, Amsterdam, Johannes Covens and Cornelis Mortier 1730 or 1737, and later editions.

Blohn J.T. *Karte der Insel Yap*, 1873.

Bloijs [Bloys], Cornelis de (*fl.*1656-1680). Dutch military surveyor and cartographer, worked in and around Den Briel.

Blois, Théodore Julien de la Calande de. Worked with August Bérard and Jules de Blosseville on charts of New Zealand. *Carte de la Baise des Iles*, 1827; *Plan du Port Manawa*, 1827.

Blokhuysen, Rutger Christoffel. Plans published in Antonius Sanderus's *Verheerlykt Vlaandre* (2 volumes), The Hague, Christiaan van Lom 1732.

Blom, Adam. See **Blum**, Adam.

Blom [Bloom], Bernt Anker (1796-1858). Norwegian cartographer.

Blom, Gustav Peter (1785-1869). *Kongeriget Norge*, 1843.

Blomberg, Carl Johann. *Carte de la Livonie*, 1705.

Blome [Bloome], Richard (1641-1705). Map and bookseller of London, and Freeman of the Company of Stationers (1660). Principal work *Britannia* (an atlas of county maps based on Speed). *London*, 1666 (engraved by W. Hollar); *A Generall mapp of the Isles of Great Brittain designd ... N. Sanson*, 1669; *A new Mapp of Africa*, 1669 (taken from Sanson's map of 1650); *A Geographical Description of the Four Parts of the World...*, 1670 (with maps after the Sanson family, published jointly with Thomas Basset, Edward Brewster and Nathaniel Brooks); *A Description of the Island of Jamaica*, 1672; Camden's *Britannia...*, 1673; *Speed's maps Epitomis'd*, 1681 (later republished by Thomas Taylor and successors as *England Exactly Described...*, 1715 and later); Bernhard Varenius' *Cosmography and Geography*, 1682, 1683, 1693 (used some of the same maps as the *Geographical Description*); *The Present State of his Majesties Isles & Territories in America*, 1687, french edition *L'Amérique Angloise*, 1688. **Ref.** TYACKE Sarah *London map sellers 1660-1720* (Tring 1978) pp.110-111.

Blomefield, *Reverend* Francis. Cleric and local historian of Fersfield, Norfolk. *A Plan of the City of Norwich* (2 sheets), 1746, in his *Topographical History of the County of Norfolk*.

Blomfield, *Reverend* Ezekiel. *General View of the World*, Bungay 1807; *Philosophy of History*, London 1818.

Blomiley, Thomas. Surveyor. Lands near Wigan, 1639.

Blommaert, Samuel. Basse Ethiopie, N. Sanson 1655.

Blommaerts, Jan. *Frisia orientalis*, Antwerpen, J.B. Vrients 1602 (in Ortelius's *Theatrum orbis terrarum*).

Blommendal, A.R. Dutch hydrographer. *Hydrographische Kaart der Zee-Gaten van Goeree en Maas*, 1858, 1869; *Hydrographische Kaart der Monden van de Eems*, 1859; *Hydrographische Kaart van de Zuiderzee*, 1870.

Blondeau, A. Engraver of *Central States* for *Guyot's Grammer-School Geography*, New York, Scribner, Armstrong & Company 1874 (with W. Speller).

Blondeau, Alexandre. Royal engraver of 'Rue des Frances Bourgeois N°·131 près la place Michel', later 'Rue des Noyers N°·24', Paris.

First engraver to Dépôt de la guerre, then for Dépôt de la marine. Maps in *Atlas Universel pour la Géographie de Guthrie*, Paris, Hyacinth Langlois 1798-1799; engraved for French edition of George Vancouver's voyages entitled *Voyage de Découvertes à l'Ocean Pacifique du Nord*, 1800; Charles Paultré's *Carte physique et politique de la Syrie...*, Paris 1803 (with L. Aubert); Lapie's *Map of the United States. Carte des Etats-Unis.*, London 1814, published in Le Chevalier Felix de Beaujour's *Sketch of the United States of North America*; P. Lapie's *Carte des Iles Britanniques ou Royaume-Uni de la Grande Bretagne & d'Irlande...*, 1814; for Alexandre-Emile Lapie, 1817; Jean-Baptiste Poirson, 1817; C.F. Delamarche, 1820; Port du Croisic, 1827; St. Nazaire, 1828.

Blondel, *Colonel*, —. Atlas de la guerre d'Orient, 1858.

Blondel, François (1618-1686). 'Maréchal de camp..., directeur de l'Academie royale d'architecture, maître de mathématiques du dauphin...'. *Plan de Paris* (12 sheets), 1676 (with Pierre Bullet).

Blondel la Rougery, E. Publisher of Paris. City plans of France (20 plans), 1909 (engraved by G. Peltier); *La France en 15 régions à l'usage des automobilistes, cyclistes et touristes*, 1911-1915; *Paris en 9 cartes (Extrait des cartes Campbell)*, *c.*1914; *Carte générale des installations aéronautiques diverses militaires et civiles de l'Allemagne...*, *c.*1914; *Italia*, *c.*1915 (Carte-guide Campbell); War map of Europe, 1915; *Dardanelles, mer de Marmara et Bosphore*, *c.*1915; *Nouveau plan de Paris...*, 1916 (one of the so-called Cartes Campbell prepared by Blondel la Rougery); *Carte du grand-duché de Luxembourg*, *c.*1916; *Collection des cartes du front...* (13 sheets), Paris *c.*1917; *Carte de l'Aéro-club de France*, Paris, 1917-1918.

Blondiaux, *Lieutenant* Paul (1867-*c.*1899). Surveyed frontiers of Liberia and French West African colonies from 1892-1897.

Blood, —. *...Forest County, Pennsylvania*, *c.*1868.

Bloodworth, Charles James. Surveyor. Estate plans, Huntingdonshire, 1778-1859; e.g. *A Plan of the Hamlet of Wornditch*, 1795.

Bloom, Bernt Anker. See **Blom**, Bernt Anker.

Bloom, Richard. See **Blome**, Richard.

Blooteling [Blotelingh], Adriaan (1640-1690). Dutch engraver of maps, views, frontispieces and title pages. Six views of Amsterdam after Jacob van Ruysdael, [n.d.]; *Situation du Paradise et du Païs de Canaan*, *c.*1669.

Bloss, *Lieutenant* — von. *Gefecht bey Épinal 1814*, *c.*1820.

Blosseville, Jules Alphonse René Poret de. Junior officer on Duperrey's voyage to the Pacific, 1824. Compiled charts of North and South Island, New Zealand, including the mythical Taranaki Bay and surveys of the Bay of Islands, published in *Nouvelles Annales des Voyages...*, Paris 1826, and in Duperrey's *Atlas hydrographie et physique*, 1827. **Ref.** HOOKER, B. *The Map Collector* 44 (1988) pp.33-34; *The Map Collector* 43 (1988) pp.20-21.

Bloteling [Blotelingh], Gijsbert (*fl.*1676-after 1714). Dutch surveyor, cartographer and hydrological engineer. Admitted as surveyor 1676, worked mostly in Zuid-Holland. *Caerte figurative van 't Slaeck ende inkomen vande riviere d'Eendracht... tussen het landt van Tolen en Philipslandt...*, 1682.

Blottière, François de la. See **La Blottière**, François de.

Bloud, Charles. Instrument maker of Dieppe. Pocket sundial, *c.*1700.

Blouzet, Jules. *Carte Général du Théâtre de la Guerre en Asie*, *c.*1860.

Blowe, Daniel. Publisher of London. *A Geographical, Historical, Commercial, and Agricultural View of the United States of America*, London 1820.

Bloys, Cornelis de. See **Bloijs**, Cornelis de.

Bloys van Treslong Prins, J.A.P. *Kaart van Straat Sunda*, *c.*1850; *Kaart van de Lampong Baai*, 1857.

Blücher, Ephraim I. *Karte von Palästina...*, 1858.

Bluhme, I.P. *Grönland*, 1866, 1871.

Blum [Blom], Adam (*fl.*1733-1749). Dutch surveyor and cartographer. Worked in Limburg. Caerte van het Dorp Bree, 1749 MS.

Blum von Kempen, Heinrich Wilhelm von (*c.*1756-1796). *Natur und Kunst Producten Atlas der Kaernten*, Vienna 1795; *Natur und Kunst Producten Atlas der Oestreichischen deutschen Staaten*, Vienna, 1796; *Natur und Kunst Producten Atlas der Böhmen*, Vienna 1796; and other similarly named works.

Blume, *Lieutenant* I. *Atlas des Königreichs Preussen* (27 sheets), 1831; *Die Anhaltischen Herzogthümer Bernberg, Dessau, Cölter...*, 1839.

Blume, J. *Allgemeiner Atlas über alle Theile der Erde*, 1828, 1832; *Atlas über alle Theile Europa's*, 1832-1834; *Atlas über alle Theile der Erde*, 1833.

Blume, K. *Mwambani-Bai*, 1863; *Pangani Mündung*, 1893; *Kamerun Mündung*, 1895; *Pemba Kanal*, 1895.

Blume, Wilhelm. *Die Operationen der deutschen Heere*, 1871.

Blumenau, Hermann. *Übersichtskarte der deutschen Ansiedlungen... in Brasilien*, 1867.

Blumenauer, —. *Plan der Residenzstadt Cassel* (9 sheets), *c.*1896.

Blunck, Otto. Lithographer. *Plan der Stadt Kiel und Umgegend*, 1876.

Blundell, H. *The City of Bristol* (panorama), 1717 (engraved Johannes Kip).

Blundell, James J. & Co. Publishers, 44 Collins St., West, Melbourne (1856). *Map of the settled districts around Melbourne*, 1854; *Electoral plan of Melbourne*, *c.*1856; *General Map of Australia*, 1857; *Squatting map of Victoria*, 1861.

Blundeville, Thomas (*c.*1560-1603). English mathematician. A Briefe Description of Universal Mappes & Cardes, 1589; *Exercises*, 1594 (a work describing the Plancius World map of 1594).

Blunt, father & sons [E. & G.W. Blunt]. 'Nautical Book & Chart Publishers, at the sign of the Bible and quadrant', Newburyport, Massachusetts; 391 Broadway, New York (1810); 202 Water Street, New York (1811); other addresses included 133 Maiden Lane, 154 Water Street and 179 Water Street. Became 'E. & G.W. Blunt' when Edmund March retired, flourishing until 1872, when plates and copyrights were transferred to the government. Florida Bank, 1846; *Coast of North America*, 1847, 1848, 1852, 1860 (engraved by Charles Copley); *North East Coast of North America*, 1853; Charles Heyne's *Map of part of Virginia, Maryland and Delaware*, New York 1861; United States coastal surveys, 1860-1864. **Ref.** GUTHORN, P. *United States coastal charts 1783-1861* (1984) pp.9-11 *et passim*; GUTHORN, P. *The Map Collector* 23 (1983) pp.14-16.

• **Blunt**, Edmund March (1770-1862). American publisher, born New Hampshire, began as a bookseller, moved on to publishing charts *c.*1796, he retired to concentrate on revising his *Coast Pilot.* Captain Lawrence Furlong's *The American Coast Pilot*, 1796, issued in 23 editions up to 1867; Paul Pinkham's *...George's Bank*, 1797; *Blunt's New and Correct Chart of Long Island Sound...*, 1813 (from an 1805 survey by J. Cahoone and N. Fosdick, engraved by Peter Maverick); *Blunt's New Chart of the North-Eastern Coast of North America*, 1813; Charles Morris's *Chart of the harbor of New London...*, 1815 (engraved by Hooker); *Plan of the City of New York*, Hooker 1817, 1830; *...Coast of the United States of North America...*, 1827; *Blunt's Charts of the North and South Atlantic Oceans...* (29 sheets), 1830.

•• **Blunt**, Joseph (1792-1860). Lawyer, son of Edmund March Blunt. Contributed legal material for the *Coast Pilot*, which was later expanded to form *The Merchants and Shipmaster's Assistant*, then *The Shipmaster's Assistant and Commercial Digest*, 1837 and later.

•• **Blunt**, Edmund (1799-1866). Hydrographic surveyor, son and associate of Edmund March Blunt, above. Sailed with his brother George William aboard the *Orbit*, and helped with a survey of Bahama Bank, 1819-1820, also carried out coastal surveys for his father, 1823-1830. In 1833 became assistant to the Director of the Coast Survey. *Long Island Sound From New York to Montock Point*, New York, E. & G.W. Blunt 1830; *The Bahama Banks and Gulf of*

Samuel Bochard (1599-1667). (Taken from Map Collectors' Circle*)*

Florida, 1834, 1841, 1844, 1848, 1853, 1860.
••• **Blunt**, Edmund *Jr*. Son of Edmund, above. Took charge of E. & G.W. Blunt in 1866, together with F.M. Nash and John H. Nichols. See below.
•• **Blunt**, George William (1802-1878). Son of Edmund March, above. Left a career at sea to join his father's business in 1822. First Assistant to the U.S. Coast Survey, 1833-1878. Marine surveys of the Bahama Banks and New York Harbour were used to update Blunt charts. Produced later editions of *Bowditch's Navigator*, and *Blunt's Coast Pilot*.
•• **Blunt**, E. & G.W. (*fl.c.*1830-1866). Edmund and George William Blunt trading together. Edmund seems to have remained a practical hydrographic surveyor, while George William managed the editorial, publication and sales parts of the business. The business passed to Edmund Blunt Junior and associates (below).
••• **Blunt**, Edmund *Jr*., F.M. Nash & John H. Nichols. Successors to E. & G.W. Blunt. On the departure of Nash it was succeeded by Blunt & Nichols.
•••• **Blunt & Nichols**. Last successors to E. & G.W. Blunt. The company closed in 1872.

Blunt, Charles John. Railway Maps, 1843-1845.

Bluzé [Bleuse], Abram (*fl.*1676-1688). Dutch surveyor and cartographer. *Caarte van t Eyerland, ende West Vlieland...*, 1688 MS (with Isaac Haringhuijsen and Adriaan van Twuyver).

Bo da Venecia, Gioane. Said to have drawn or engraved a world map or terrestrial gores, *c.*1520 (not traced). **Ref.** SHIRLEY, R. *The mapping of the World* (1983) N°44 p.51.

Board, Stephen. Engraver for John Seller's chart of Russia in *English Pilot, Part I*, 1671.

Board, W. *Chart of Exmouth Harbour*, 1830.

Boardman, Harvey. ...*White Mountains, New Hampshire*, 1858.

Boas, Franz. Baffin Land, 1885; *Sprachenkarte von Britisch-Columbien*, 1896.

Boase, Henry Samuel. *A Geological Map of Cornwall* (4 sheets), 1828.

Boazio [Beotio], Giovanni Battista [Johannes Baptista; John Baptist]. Italian cartographer working in England (1585-1606). Compiled 5 charts for Walter Bigges's *A Summarie and True Discourse of Sir Francis Drake's West Indian Voyage*, 1588 and later; Isle of Wight, 1591; Raid on Cadiz, 1596; Chart of the Azores, 1597; Great Bay of Falmouth, 1597 MS; *Irelande*, London, John Sudbury 1599 (using material by Robert Lythe, engraved by Reynold Elstracke, 3 examples known; later used in reduced form in J.B. Vrients's 1606 edition of Ortelius); manuscript plans of Calais, 1596; coast of Ulster, *c.*1600 MS; Oostende, 1604 MS. **Ref.** BURDEN, P. *The mapping of North America* (1996) pp.81-83; MEURER, P. *Fontes cartographici Orteliani...* (1991) pp.114-115 [in German].

Bobé, *père* Jean (1654-1735). Worked with Guillaume Delisle, 1710-1721. *Carte des Mers et des Pays qui sont a l'Ouest, au Nord du Lac Superieur et du Mississippi*, 1718. **Ref.** LAGARDE, Lucie *Imago Mundi* 41 (1989) pp. 27-30 [in French].

Bobillier, Charles. *Genève et ses environs*, 1899.

Bobrik, Hermann (1814-1845). Geographer. Atlas, Königsberg [Kaliningrad] 1838.

Bocage. See **Barbié** du Bocage, family.

Bocage, Georges. See **Boissaye** du Bocage.

Boccard, Ludwig. *Plan-Karte von Coblenz*, 1854; *Karte des Rheins*, 1858, 1865.

Bochard [Bochart; Bochartus], Samuel (1599-1667). Theologian and geographer. *Geographiae sacrae*, 1646 (with numerous maps including *Paradisi terrestris situs Syria et Aegypta*).

Bock, —. *Treffen bei Warre... 1815*, *c.*1821.

Bock, A. *Plan showing Streets and Houses within the Limits of the Wynberg Municipality* [Cape of Good Hope] (6 sheets), 1888.

Bock, Arnoldus Joannes de (*fl.*1760-1795). Dutch military engineer, surveyor and cartographer. Worked mostly in Brabant, at Breda and Bergen op Zoom.

Bock und Pollach [Polach], Joseph Ferdinand von. Economic map of Bohemia: *Totius Regni Bohemiae...* (4 sheets), Vienna, Artaria 1808.

Bockelberg, F. von. *Wildbad-Gastein*, *c.*1860.

Bod, Peter (1712-1769). Hungarian linguist, historian and writer, born in Transylvania, died at Magyarigen, Romania. His maps of the four continents were published in *Az Isten vitezkedo anyaszentgyhaza allapotjanak historiaja* [History of God's Valiant Mother Church], Basle 1760 (engraved by C. Burckhardt).

Boddington, Henry. Strangeway's Brewery, Manchester. *Proposed Manchester Ship Canal*, 1882.

Bode, Johann Ehlert (1747-1826). Astronomer of Hamburg. Astres, 1782; *Stereographischer Entwurf des gestirnten Himmels...*, 1787; Celestial globe, 1790; Atlas céleste, 1801; *Uranographia...*, Berlin 1801; 31-cm celestial globe, 1804.

Bodega y Cuadra, Juan Francisco de la. Charts, west coast of America, 1766-1791: the coast of northern California from Bodega Head around the mouth of San Francisco Bay..., 1775 MS; Bodega Head, Bodega Bay etc..., 1775 MS (copied by Alexander Dalrymple and others).

Bodenehr, family. Publishers and engravers of Augsburg. **N.B.** The editors have been unable to establish reliably the relationships between mapmakers of this name, or distinguish between the work of Gabriel the elder and younger. The following is inferred. Several individuals of the name of Bodenehr were involved in some capacity with the publication of work by the Stridbeck father and son.

• **Bodenehr**, Johann [Hans] Georg (1631-1703). Engraver and publisher of Augsburg. Sold the earlier works of Stridbeck. *Totius Germaniae Itinerarum*, 1660; *Sueviae descriptio*, 1670 and later editions; *Teutschland zu bequemen Gebrauch...* (32 sheets), 1677; *Provinciarum Europæ Geographica Descriptio* (34 maps), 1679.

•• **Bodenehr**, Georg Conrad (1663-1710). Engraver and publisher of Augsburg. *Augusti Vindelicorū*, *c.*1685; engraved H.C. Gyger's *Helvetia, Rhaetia, Valesia...*, H.L. Muoss 1698; engraved map of Japan for J.U. Müller's *Kleiner Atlas*, 1702; *Atlas Minor Sive Orbis Terrae...*, [n.d.].

•• **Bodenehr**, Moritz (1665-1749). Engraver and publisher of Dresden. *Delineatio Norwegiae novissima*, 1719; A.F. Zürner's *Neue Chur-Saechsische Post Charte*, 1736.

• **Bodenehr**, Gabriel *the elder*. Engraver and printseller of Augsburg. Some of the work published under the name of Gabriel Bodenehr was that of the Stridbeck father and son, whose plates Gabriel bought in 1716. The Stridbeck plates later passed to Georg Christoph Kilian, who likewise continued publishing them under his own name. *Atlas curieux*, 1704 and later editions (after Nicolas de Fer, first published Stridbeck, 1677); *Curioses Staats- und Kriegs- Theatrum Dermaligen Begebenheiten (in Spanisch Niederland)...*, c.1716; *Force d'Europe...* (with 200 town and fortification plans), 1720 (after de Fer and others); *Curiose Staats und Kriegs Theatri am Rhein...*, c.1720.

•• **Bodenehr**, Gabriel *the younger*. J.M. Hase's *Tabula expeditionis Alexandri M. Macedonis*, 1716; *Europens Pracht und Macht* (200 pages), 1720.

Bodener, Jakob. *...Bregenz und Umgegend*, *c.*1876.

Bodger, John (*fl.*1784-1815). East Anglian land surveyor, publisher and mapseller at Stilton, England. *Newmarket Heath*, 1787; *...Whittlesea Mere*, 1786.

Bodley, *Sir* Josiah (*c.*1550-1617). English military surveyor. North of Ireland 1609; County of Wexford, 1617.

Bodmer, Carl. *...Reise durch Nord-America*, *c.*1841.

Bodmer, *Lieutenant* Samuel (1652-1724). Swiss surveyor and engineer, born in Bern, worked in Amsoldingen.

Bodoki [Henter], Mihály (1782-1838). Hungarian cartographer. Surveyor Békés County from 1832; mapped territory east of Tisza river, also flood plain of Berettyó and Hortobágy rivers, 1831. **Ref.** FODOR, F. *A magyar térképirás* [Hungarian Cartography] (Budapest 1952-1954) [in Hungarian].

Bodola, Lajos [Louis] (1859-1936). Hungarian cartographer and academic. Parents emigrated to Italy after abortive War of Independence 1848-1849. Studied at Catana, later at Budapest; Professor at Technical University 1894-1912. As a member of 'Comité International des Poids et Mesures', he was influential in the adoption and use of the metric system in cartography. **Ref.** OLTAY, K. 'Louis Bodola', in: *Bulletin géodesique* 57 (1938); OLTAY, K. 'Bodola Lajos élete és müvei' in: *Geodéziai Közlöny* [Geodetic Bulletin] 1938/1 [in Hungarian].

Boeck, Caesar Lesur (1746-1832). Danish cartographer.

Boecke, H. ...*City of St. Louis and Vicinity*, 1859.

Boeckel [Beekel; Boekel; Bokel], Pieter [Petrus] (*c.*1530/1535-1599). *b.* Antwerp, *d.* Wismar, Germany. *Bescribung vom landt zu Ditmers*, Antwerp 1559 (later used by Ortelius, 1570).

Boeckel, Peter August Gottlieb (1719-1797). Member of Homann firm 1744-1746.

Boeckh [Böckh], Richard (1824-1907). *Sprachkarte vom Preussischen Staat* (2 sheets), 1864; *Historische Karte vom Elsass*, 1871; *Verbreitung der Deutschen in Europa* (8 sheets), *c.*1891.

Boeckler [Böckler], Georg Andreas. ...*Beschreibung der Kinder Israel Auszug...*, *c.*1655; *Geographische Beschreibung dess Gelobten Landes*, *c.*1655; ...*Die Gegendt dess Paradyss*, *c.*1655, *c.*1660.

Boecop, Hendrik Justus Michiel van (*c.*1744-1812). Belgian military engineer and cartographer from Namur. Worked in southern Netherlands and Brabant. He died in Breda.

Boeddicker, Otto. *The Milky Way*, 1892.

Boehler, F.L. *Geometrischer Plan von Frankfurt*, 1874.

Boehm, A. *Cours-Karte von Bayern*, *c.*1866.

Boehm [Böhm], F. *Grundriss von Berlin*, 1850; *Plan von Berlin*, 1861, 1865.

Boehm [Böhm], Josef Georg (1807-1868). Astronomer of Prague. **Ref.** MUCHA, L. *Der Globusfreund* 43/44 (1995) pp.227-236.

Boehme, August Gottlieb (1719-1797). Mathematician and military cartographer of Dresden, worked for Homann's Heirs. *Carte de l'Asie*, 1744; worked on the maps of J.M. Hase after his death, 1744; Homann's Heirs' *Americae Mappa Generalis...*, 1746, 1752; *Scandinavia*, 1776.

Boekel, Pieter. See **Boeckel**, Pieter.

Boelen, J.J. Surveyor in Netherlands East Indies. *Straat Lagoendi*, 1855; ...*Lampong Baai*, 1857.

Boelens, Broer (*fl.*1649-1686). Dutch surveyor and cartographer. Worked in Friesland. Worked on C. Schotanus à Sterringa's maps of Friesland, *c.*1664.

Boelhouwer, F.K. Kaart van Gooiland, 1709, (engraved by Ottens, 1740).

Boell, William (*fl.*1860-1879). Lithographer. ...*Licking County, Ohio*, 1854; ...*Baltic*, *c.*1862; ...*Bergen Heights, New Jersey*, 1867; ...*North Adams* [Massachusetts], *c.*1870.

Boem, J. Recueil de diverses histoires, Paris 1543.

Boener, Johann A. *Territorium Norimbergense*, *c.*1650.

Boerio, C. Maps of Italian regions. Mincio, 1802; Olona, 1802; Po, 1802.

Boersma, H.L. *Kaart van Europa*, *c.*1874.

Boese, K.G. *Karte vom Herzogthume Oldenburg*, 1865; *Grossherzogthum Oldenburg*, 1872.

Boethme, —. *Plan von Tharand*, *c.*1835.

Boetius, Christian Friedrich. Engraver. *Prospect der... Elb Brücken zwischen Neustadt und Dresden*, 1762.

Boettcher, Carl. *Handkarte von Mitteleuropa*, *c.*1889.

Boettger, Hermann. *Eisenbahn-Karte von Deutschland und Frankreich*, 1864.

Boetticher, Wilhelm. *Carthage*, 1835.

Boettus, J. Associated with the 1682 edition of Joan Blaeu's Town books of Italy.

Bofinger, G. *Karte des nördlichen Theils der Schwäbische Alb*, 1880.

Bogaerd, Jan. Printer of Louvain. Printed Cornelis van Wytfliet's American supplement to Ptolemy - *Descriptionis Ptolemaicae Augmentum...* (19 copperplate maps of America), 1597.

Bogaerts, Adrian J. van (1813-1891). Printer, lithographer, publisher and artist. Maps of Germany, Netherlands and Italy, 1858-1873.

Bogár, Francisco. *Episcopatus Diocesis Javrinensis*, 1821.

Bogardus, Joannes. See **Bogaerd**, Jan.

Bogdanich, Imre Dániel (1762-1802). Hungarian cartographer and astronomer from Veröce [now Vitrovitica, Croatia], died in Pest. Director of the Buda Observatory, 1796; provided astrofixes for János Lipszky, 1798 onwards. **Ref.** BENDEFFY, L. 'Emlékezés Bogdanich Imrére' in: *Geodézia és Kartográfia* [Geodesy and Cartography] (1972/5) [in Hungarian]; BARTHA, L. *Bogdanich Imre Dániel, évfordulók (1987)* [in Hungarian].

Bogert, M.L. Maps of Wisconsin including *...Fond du Lac County, Wisconsin*, *c.*1862.

Bogerts, Cornelis (1745-1817). Dutch engraver of Amsterdam. Titlepage for W.A. Bachiene's *Atlas tot opheldering der hedendaagsche historie*, 1785.

Bohemio, A.G. See **Boehme**, August Gottlieb.

Bohemus, Martinus. See **Behaim**, Martin.

Böhm, F. See **Boehm**, F.

Böhm [Cseh], Ferenc [Franz] (1736-1799). Hungarian cartographer. *b.* Vienna, *d.* Tata. Worked for count Eszterházy, made maps of Transdanubia, 1757; directed mapping of Sárviz river, 1767. **Ref.** REVESZ, V. *Magyar müszaki alkotók* [Hungarian technical engineers] (Budapest 1964) [in Hungarian].

Bohme, August Gottlieb. See **Boehme**, August Gottlieb.

Bohn, Anton *S.J.* Map of Paraguay, 1697 (with A. Sepp von Rochegg).

Bohn, Carl Ernst. Publisher in Hamburg. A. Eschels-Kroon's *Beschreibung der Insel Sumatra...*, 1781; C.D. Ebeling's *Erdbeschreibung und Geschichte von America*, 1793-1816 (the first edition contained no maps, later editions had maps by Sotzmann); D.F. Sotzmann's *Atlas von den Preußischer Staaten*, Bohn 1794.

Bohn, Casimir. Publisher of Washington, produced many editions of *Bohn's hand-book of Washington*, 1849-1866 most of which contained a map of the city entitled *Map of the city of Washington in the District of Columbia*, 1854, and later.

Bohn, François. Publisher of Haarlem. Asia, 1807; *Nieuwe en beknopte vezameling der noodigste landkarten...*, *c.*1820; *Hand-en Schoolatlas van het Konigrijk der Nederlanden*, 1830; *Nieuwe en Beknopte Hand-en School-Atlas der Nederlanden*, 1830, 1840.

Bohn, Henry George (1796-1884). Publisher. *Seat of War in North Italy*, 1859; Atlas of Classical Geography, 1861.

Bohn, Oscar. *Geological Sketch Map of the Wrangel Mountains*, 1899.

Bohn, P.D. *...Sea of Marmora*, 1770.

Bohnenberger, *Professor* Johann Gottlieb Friedrich (1765-1831). Astronomer, mathematician and surveyor. *Charte von Wirtemberg*, Tübingen, Cotta 1797; *Charte von Schwaben* (62 sheets), Tübingen 1798-1828 (with I.A. Amman, later used by J.V. Moreau).

Bohnenberger, Wilhelm. Orbis terrarum antiquus, 1847.

Bohnert, Friedrich. Tourist maps of Germany, 1870-1893.

Bohun, Edmund. *A Geographical Dictionary*, 1688 and later; Theatrum geographicae, 1695; contributed to Heylin's *Cosmography...*, 1703 edition (first published 1652).

Boileau de Bouillon, Gilles [Aegidius Bullonius Belga; Boleavus; Darinel, pasteur des Amadis; Egidius Boleavus Bulionus; Gielis Bouloigne] (*c.*1510-1563). Flemish writer, poet, diplomat, cartographer and printer, Governor of Cambrai, 1541, in service of Charles V. Map of Germany after Sebastian Münster, in *Commentaire du Seigneur Don Loys d'Avila, contenant la guerre d'Allemagne*, Paris 1551 (only 2nd edition has map); *La sphère des deux mondes*, Antwerp 1555 (with maps by Johannes Honter); *Nova et exactissima Sabaudiae ducat. descriptio*, 1556 (used by Ortelius 1570); *Gallia belgica*, 1557; *Campaigne de Romme*, *c.*1563; and others. **Ref.** KARROW, R. *Mapmakers of the sixteenth century and their maps* (1993) pp.83-88.

Boilly, C. Engraver for Rigobert Bonne, 1787.

Boim, Michal Piotr. See **Boym**, Michal Piotr.

Bois, Abraham. *La géographie moderne* (4 volumes), The Hague 1736.

Boisgelin, Louis de. *A New Sea and Land Chart of Malta*, 1804.

Boisquenay, *Captain* M. de. Revised plan of Port Louis, 1775 (in Après de Mannevillette's *Le Neptune Oriental*).

Boissaye du Bocage, father and son.
• **Boissaye du Bocage**, Georges *the elder* (1626-1696). Hydrographer at Le Havre 'Idrographe et Profeseur Roial en la navigation au havre de Grace'. *Carte ronde ou Reduite Fort* [North Atlantic Ocean], 1669 and later editions (taken from Blaeu's *West Indische Paskaert* of *c.1630*). **Ref.** BURDEN, P. *The mapping of North America* (1996) pp.511-512.
•• **Boissaye du Bocage**, Georges *the younger* (*c.*1661-1717). Son of above, hydrographer of Le Havre. Assisted father in observations on tides.

Boisse, Adrien. *Carte géologique... du Département de l'Aveyron*, 1858.

Boisseau, Jean (*fl.*1637-1658). French geographer, topographer, colourist and genealogist. 'Enlumineur du roi pour les cartes géographiques. Paris, l'Isle du Paris sur le quai qui regarde la Mégisserie à la fontaine Jouvence'. Terrestrial globe, 1635 (copied from Blaeu, 1616); Nouvelle et exacte description de la terre universelle, 1636; Canada, 1641; published the surveys and views of C. de Chastillon as *Topographie françoise* (532 plates), 1641 (sold to L. Boissevin); re-edited the work of Bouguereau and Leclerc as *Théâtre géographique du royaume de France* (103 plates), 1641 (with copies of Ortelius, Mercator, Hondius and Blaeu maps); *Trésor des cartes géographiques...*, 1643; *Description de la Nouvelle France...*, 1643; *La Clef de la géographie générale*, 1645 (by I. Dumas de Fores); *Tableau géographiques des Gaules*, 1645; republished P. Bertius's *Geographia Vetus Ex Antiquis...*, Paris 1645; *Theatre des citez*, 1648; Ville d'Anvers, 1650; Paris, 1652. **Ref.** BURDEN, P. *The mapping of North America* (1996) pp.331-335.

Boissevin, Louis (*c.*1610-1685). 'Marchand et enlumineur, Paris, rue St. Jacques près Sévérin à l'image Sainte Geneviève'. *Trésor des cartes*, 1653; Claude de Chastillon's *Topographie françoise*, 1655 (first published by Jean Boisseau in 1641); Tableau historique, 1658.

Boissière, Gilles de la (*fl.*1660-1671). Tables géographiques, 1669; Jeu de France, 1671.

Boisson, —. *Plan général du Canal de Paris*, 1729.

Boiste, Pierre Claude Victoire (1765-1824). French cartographer. Carte de l'Amérique, 1795; *Atlas Portatif*, Paris 1806; Dictionnaire de géographie universelle, 1806.

Boitet, Reiner. Het Eiland van Kaap Breton, 1746; Noord America, 1746.

Boitout, L. Chart of the River Medway, *c.*1675 MS.

Böjtös, Imre (*fl.*1816-1822). Hungarian cartographer and military surveyor. Mapped Temes Bánát region, 1822. **Ref.** FODOR, F. *Magyar kéziratos térképek 1867-ig* [Hungarian manuscript maps to 1867] (Budapest 1956) [in Hungarian].

Bokel, P. See **Boeckel**, Pieter.

Bokma, R.W. (*fl.*1729-1742). Dutch surveyor and cartographer working in Friesland. Ingedijkte landen bij het Doccumer Diept, 1729 (with C. Tadema).

Bokushin, Eami. See **Eami** Bokushin.

Bolam, Robert George. *Map of Norham and Islandshires*, 1844 (after Robert Rule).

Bolcher, *General* André (*b*.1828). Imperial Russian War Office, Topographical Section.

Bold, Edward H. Surveyor. *Township of Ross* [New Zealand], 1867.

Boldoduc, Eduard. ...*Arrondissement de Béthune*, 1867.

Boleavus Bulionus, Egidius. See **Boileau** de Bouillon, Gilles.

Boles, Ary van. Dutch map copyist. Worked in Zuid-Holland. Het eylandt West-Voorn, 1741 (copied from the work of Nicolas Cruquius).

Bolger, F. Bridgeland, 1878; Township of Whitman, Toronto 1878.

Bolia, C. Lithographer. *Karte des Grossherzogthums Baden*, 1850; *Plan der Stadt Freiburg*, *c*.1860.

Bolifon, Antonio. See **Bulifon**, Antonio.

Boling, C. Dutch cartographer. Kaart van Texel, 1795.

Bollaart, David (*fl*.1650s-1679). Dutch surveyor and cartographer, copyist of Bergen op Zoom. Vertoogh van de Steden... in Vlaenderen, 1664 (copied 1679).

Bollaert, Abraham (*fl*.1643-after 1649). Dutch surveyor and cartographer of Bergen op Zoom. Polder surveys in Zeeland, 1649.

Bollain, —. *Carte de la population des départements du Nord et du Pas-du-Calais*..., 1844.

Bollan, William. Folding maps of Cape Breton in *The importance and advantage of Cape Breton*..., London, 1746.

Bolland, *Captain* Richard. *A Draught of the Streights of Gibralter with Some Observations upon the Currents*..., 1675.

Bolle, D. *Touristen Zakatlas...van Nederland*, 1898.

Bollero. See **Bellère**, Joan.

Bollin, J.R.F. ...*Sumatras Westkust*, 1852.

Bollin, Rudolf Jacob (1770-1844). Swiss surveyor in Büren an der Aare. *Plan von der Stadt... Bern*, 1808, 1811.

Bollmann, Hermann (*b*.1911). Twentieth-century German cartographer who produced bird's-eye views (Bildkarten) of historic towns and cities. *Staedte: 25 Bildpläne von Hermann Bollmann*, 1968. **Ref.** KAY, Terry *The Map Collector* (letters) 34 (1986) p.56.

Bollmann, L. von. *Reise Karte der Schweiz*, 1839.

Bolotov [Bolotoff], *General* Aleksei Paulovich (1803-1853). *Carte Géographique de l'Asie Mineure*, 1853.

Bollo, Luis Cincinato. Atlas geografico... del Uruguay, 1896.

Bolstra, B.A. (1580-1633). Views in Amsterdam, 1609.

Bolstra, Melchior (1704-1779). Dutch surveyor, map copyist and cartographer of Leiden. Ordinaris landmeter for Rijnland (1732-1776). Chart of the River Merwede, and map of Rhynlands, both published in Hendrik de Leth's *Nieuwe Geographische en Historische Atlas*, 1766; revised Jan Jansz. Dou and Steven van Broeckhuysen's *'t Hooge Heemraad-schap van Rhijnland* (13 sheets), 1642-1646, 1746; also Jan Pietersz. Dou's Haarlemmermeer, *c*.1636, 1740.

Bolton, *Mr*.—. 18th-century cartographer. Set of 5 maps, after J.B.B. d'Anville, for later editions of Malachy Postlethwayt's *The Universal Dictionary of Trade and Commerce*, 1752, 1755, 1766, 1774 (one map is also marked S. Bolton). **N.B.** Mr. Bolton may be Solomon **Boulton**. **Ref.** GARDINER, R.A. in: WALLIS, H. & TYACKE, S. (eds.) *My head is a map* (1973) p.57-58.

Bolton, *Colonel Sir* Francis. *Map Shewing the Districts Supplied by the London Water Companies*, 1884.

Bolton, John (1791-1874). Tithe surveyor and geologist. ...*Parish of Aldingham in the County of Lancaster*, 1848.

Bolton, S. Map of Africa, Robert Sayer 1787 and editions to 1800 (after d'Anville). **N.B.** Compare with Mr **Bolton** above, and/or Solomon **Boulton**, below.

Bolton, Solomon. Publisher. *Geographia Antiqua Delineata*, 1755. **Ref.** GARDINER, R.A. in: WALLIS, H. & TYACKE, S. (eds.) *My head is a map* (1973) p.57-58.

Bolts, William. *Carte du Bengale*, *c.*1770; *The Kingdoms of Bengal, Bahar and ...Orissa*, 1773, 1775.

Boltze, Julius. *...Saarburg*, 1877; *Wand-Karte des Stadt und Landkreises Metz*, 1877.

Bolus, Harry (1834-1911). Sketch map Eastern Province Cape Colony, *c.*1870s MS.

Bolzini, Giovanni Andrea. *Novo ed accurato Disegno del Territorio Cremasco* (2 sheets), 1741.

Bolzoni, Andrea. Engraver. *Stati del ...Duca di Modena*, 1746; *Nuova Pianta... della Città di Ferrara*, 1747.

Bombicci, L. *Carta geologica della Montagana Bolognese*, 1881.

Bommer, Cornelis Claesz. (*fl.*1630-1667). Dutch surveyor and cartographer, worked in Noord-Holland and Drente (1640-1643). Caerte van de Belckmeerdijck, *c.*1631 MS.

Bompar [Bompare; Bompario; Bomparius], Pierre-Jean de [Petrus Ioannes de]. *Provinciae, Regionis Galliae....* [Provence], published by Ortelius, 1595, also by Jodocus Hondius in his edition of Mercator, 1606 (and later), and by Willem Blaeu, 1631 (and later).

Bomsdorf, R. von. Mecklenburg-Schwerin (2 sheets), 1895; Mecklenburg-Strelitz, 1895.

Bomsdorff, Oscar von. Maps of German states and regions, Hannover, Sachsen, Vogtland, etc., 1870-1898.

Bomsdorff, Theodor von. Magdeburg, 1853; Austria, 1867; Nord-Amerika, 1882; Saxony (4 sheets), 1889; Railway and telegraph maps &c.

Bonaccina, Giovanni Battista. *Disegno particolare del Piemonte et Monteferrato*, 1630.

Bonacina, Cesare. *Disegno di Pavia*, 1654.

Bonaldo, Dolfin (*fl.*16th century). World map: *Totius cogniti orbis typus universalis*, Venice 1541 MS. **N.B.** David Woodward has cast doubt on the authenticity of this and other manuscript world maps of the period. WOODWARD, D. *The Map Collector* 67 (1994) pp.2-10.

Bonanni, Jacobi. Syracusae antiquae delineatio (for Pieter van der Aa), 1728.

Bonaparte, *Prince* Louis-Lucien. *Carte des Sept Provinces Basques*, 1863; also ethno-linguistic maps.

Bonar, William. *A Draught of the Creek Nation*, 1757.

Bonardo, G.M. La grandezza... di tutte le sfere, Venice 1584.

Bonarelli, F. Carta geologica di Cammago, 1894.

Bonatti, Frères. Engravers of Milan. Worked for François Gandini, 1821.

Bonatti, C. Engraver. *Atlante geografico universale*, Turin 1854; also for J.-G. Barbié du Bocage, 1854. N.B. Compare with others of the same name.

Bonatti, Domenico. *Plan de la Ville de Gènes*, 1829.

Bonatti, Enrico. *Nuova Pianta della Città di Torino*, 1869; *Nuova Piano... della Città... di Genova*, 1875; *Pianta della Città di Perugia*, *c.*1875; *Pianta della Città di Messina*, *c.*1877.

Bonatti, G. Draughstman and engraver of Turin. *Carta degli Stati di S.M. Sarda...*, 1854; *Carta del Ducato di Savoia*, 1854; *Carta del contado di Nizza*, 1855; *Carta del Ducato di Genova*, 1855; *Carta del Piedmonte...*, 1855; *Carta Corografica de Stati di S.M. Sarda...*, 1859; *Carta delle Alpi Giulie*, 1864.

Bonatti, Giuseppe (*b.*1769). Engraver of Venice. Map of Ireland for Pazzini Carli, 1788; some maps for C.A. Barbiellini, *Atlante della geografia antica e moderna*, 1807.

Bonatti, Marco (*fl.* 1810-1830). Possibly son of Giuseppe (above), and possibly one of the Frères Bonatti (see above). Engraver working in Milan. Worked with Giovanni Antonio Sasso on maps of the continents and the Americas for Luigi Rossi's *Nuovo atlante di geografia universale*, 1820.

Boncompagni, Hieronymo (16th century). Italian cartographer.

Bond, —. *Martenet and Bond's Map of Montgomery County* [Maryland], 1865 (with Simon Martenet).

Bond, Frank and Frederick. *New Map of Wyoming*, 1885.

Bond, G.H. *Plan of Gravesend and Milton* [Kent], 1890.

Bond, H. An engraver for R. Montgomery Martin's *Illustrated Atlas of the World*, John Tallis 1851.

Bond, Isaak. ...*Frederick County*, 1858.

Bond, Rowland. Lecturer in geography. Popular Geography, London, Joseph Thomas 1836; *Fenner's Pocket Atlas of Modern and Antient Geography*, 1836.

Bondelmonte. See **Buondelmonti**, Christoforo.

Bone, J. Agar. Estate surveyor. *Map of the Town and environs of Maidstone*, 1860.

Bone, W. Population maps based on census reports. Map of Cheshire and Lancashire, 1851; *Division 1. London Registration Districts*, 1852.

Bonefaes [Bonefaas; Bonifacius], Claesz. [Nicolaas] Huybertsz. [Hubertsz.]. Dutch surveyor and cartographer. Worked in Noord-Holland. Caerte vande Naerder ofte Uytermeerse Meer..., 1629.

Bonetti, A. ...*Linee ferroviarie e di navigazione in Italia*, 1878.

Bonfa, Jean *S.J.* (1638-1724). Professor of mathematics at Avignon. *Le Comté Venaissin*...[S.E.France], 1696 (dedication re-engraved 1762). **Ref.** STROUP, A. *Imago Mundi* 47 (1995) pp.118-137.

Bonfadini, Giuseppe Tommaso. *Geografia... per dimonstrare l'Inondazione del Ferrarese*, 1705.

Bonfrère, Jacques (1573-1643). Flemish theologian. *Tabula geographica Terrae Sanctae*, 1631, re-issued by François Halma, Utrecht 1700.

Bongars, Johann. Printed 5 maps derived from 14th-century manuscripts, including an untitled circular world map credited to Marino Sanuto, and a map of the Holy Land taken from an original by Marino Sanuto and Petrus Vesconte. The maps were published as Volume 2 of *Gesta Dei per Francos*, Hanover 1611. **Ref.** SHIRLEY, R. *The mapping of the World* (1983) N°276 p.297.

Bonheur, Raimond. Painter. Designer of decorative vignettes for Victor Levasseur's *Atlas national*..., Paris 1840, and later editions.

Bonhomme, —. *Carte du Plan et des Environs de la ville d'Alger*, 1830.

Bonière, Hilarion. Europe [showing the houses of the Carthusian Order], 1785.

Bonifac, P. *Mappa geographica... Nigrae Silvae* [Black Forest], 1788.

Bonifacio, Giovanni (1547-1635). Lawyer of Rovigo (1573). He later became an official of the Venetian Republic (1593-1623), died in Padua. *Rhodiginiae peninsulae descriptio*, inset on Magino's *Romagna* map in Ortelius's *Theatrum orbis terrarum*, 1608 edition.

Bonifacio [Bonifacius; Dalmatino], Natale [Natalis Bonifacius Sebenicensis] (1538-1592). Painter and engraver, worked in Venice, Rome and Antwerp. *Cyprus Insula*, 1570; Bishop John Leslie's *Scotiae Regni*, 1578; *Citta de Mastrih*, 1579; plan of Palermo, Duchetti 1580; *La Provincia Ulteriore D'Abruzzo*, 1587 (used by Ortelius 1590); Palestine, 1590; *Citta di Gallipoli*, 1591; *Parigi*, 1591; *Calabria*, 1592; also plans and miniature views in various travel books such as Jean Zuallart's *Il Devotissimo Viaggio di Gerusalemme*, 1583, and later editions.

Bonifacius, Nicolaas. See **Bonefaes**, Claesz.

Bonin, Charles-Eudes (1865-1929). French diplomat and traveller in China, Indochina, Netherlands East Indies and Central Asia. *Le Leang-Shan*, 1898; *Le chemin de fer du Hedjaz*, 1909. **Ref.** BROC, N. *Dictionnaire illustré des explorateurs Français du XIXe siècle: II Asie* (1992) pp.37-41 [in French].

Boninsegni, Domenico di Leonardo (1384-1465). Drew maps for Latin version of Ptolemy's *Geographia*, 1415 MS.

Bonissel, N. *Carte de l'empire français...*, Paris, Jean 1811.

Bonk, R. Dutch navigator. Corrected Johannes van Keulen's Straat Davids, *c.*1730.

Bon Maison, Simon de. *Nieuwe en aldereeste Afteekening van 't Eyland St. Thomas*, Gerard van Keulen, 1718, 1728.

Bonn, Otto. *Plan... Würzburg*, 1894.

Bonnange, Ferdinand. Atlas graphique... du commerce de la France, 1878.

Bonnani, Jacobis. *Syracusae Antiquae Delineatae*, *c.*1700.

Bonne, father & son. **Ref.** KRETSCHMER, DÖRFLINGER & WAWRIK *Lexicon zur Geschichte der Kartographie* (1986) Vol.I p.101 [in German].

• **Bonne**, Rigobert (1727-1795). Engineer, mathematician and cartographer from Raucourt, died in Paris. Succeeded Jacques-Nicolas Bellin as hydrographer at the Depôt de la Marine in 1773. *Atlas maritime...*, Paris, Lattré 1762; maps for Jean Lattré's *Atlas moderne...*, 1762; Neptune Amerique septentrionale, 1762; *Carte de la Méditerranée*, 1763; *Petit tableau de la France...*, published Lattré, Paris 1764; Globe, 1771; G.T. Raynal's Histoire philosophique du commerce des Indes, 1774; maps for Guillaume Raynal's *Atlas de toutes les parties connues du Globe Terrestre*, Geneva 1780; *Royaumes d'Aragon et de Navarre avec la principauté de Catalogne*, *c.*1780; maps for Grenet's *Atlas portatif*, 1779-1782; *Recüeil de cartes sur la géographie ancienne...*, 1783 (with Nicolas Demarest); *Gouvernements d'Anjou, de Saumurois et de Touraine*, 1784 (drawn by Hérisson, engraved by Perrier and André); *Atlas Portatif*, Lattré 1785; *Atlas encyclopédique...* (2 volumes), 1787-1788 (with Nicolas Demarest and Bory de St. Vincent); *Atlas de France*, 1790; and others.

•• **Bonne**, — *fils*. Maps for Rigobert Bonne and Nicolas Demarest 1787.

Bonnecamp, Joseph Pierre, *SJ*. Voyage fait dans la belle rivière en la Nouvelle France, 1749.

Bonnefont, Louis. *Atlas de géographie ancienne*, 1868; *Carte Murale de la France*, 1873; *Atlas de géographie contemporaine*, Paris, Lanée 1875.

Bonner, —. Draughtsman and engraver. *Yorkshire. The Topographical Dictionary*, 1822.

Bonner, *Captain* John (1643-1726). English seaman and cartographer. *The Town of Boston in New England...*, 1722, 1725 (engraved by Francis Dewing). **N.B.** This map was re-issued after Bonner's death by William **Price**. **Ref.** REPS, John W. 'Boston by Bostonians: the printed plans and views of the colonial city by its artists, cartographers, engravers, and publishers' in: *Boston prints and printmakers 1670-1775* (The Colonial Society of Massachusetts, 1973).

Bonner, William G. Civil engineer. *Map of the State of Georgia*, 1857.

Bonnet, Charles. *Mappa Geographico da Provincia do Alemtejo e do Reimo do Algarve*, 1851.

Bonnet, Edouard. *Carte routière... du Département des Bouches du Rhône*, 1864.

Bonnet, V. Engraver. *Atlas Générale de la Ville... de Paris*, 1836.

Bonneville, *Major* Benjamin L.E. Map of the sources of the Rivers Colorado, Missouri and Snake, 1837; map of the territory west of the Rocky Mountains, 1837 (both maps published by Washington Irving in *The Rocky Mountains...*, 1837).

Bonney, E.A. Architect and surveyor, Scott Chambers, 96 Pitt Street, Sydney. *Mildura Estate, Campsie*, 1892.

Bonney, James. *Map of St. Joseph Co.*, 1863.

Bonnier, Adolf. Publisher of Stockholm. *Skol-Atlas öfver alla verldens delar...*, 1843, 1845; Stockholm edition of *Stieler's Hand-Atlas öfver alla delar af jorden och verldsbyggnaden*, 1851; *Stieler's Skol-Atlas öfver alla verldens delar...*, 1853 and editions to 1875; [Richard] *Andree's Handatlas*, 1881 (later published as *Bonniers Världsatlas*, 1924).

Bonnière, *Dom.* Hilarion. *Carte géographique des maisons de l'ordre des chartreux...*, 1785.

Bonnisel, J. *Plan de la ville et faubourgs de Paris*, 1814, 1819.

Bonnisselle, J.G. Designer and engraver. *Plan of London Westminister and the Borough of Southwark...*, *c.*1790.

Bonniver, J.F.M. Cartographer. Limburg (4 sheets), 1849.

Bonnor, F. Engraver for a map of Jamaica, published in E. Long's *History of Jamaica*, London 1774.

Bonomi, Joseph. *...Ancient Egypt*, 1848; *A Relief Map of Jerusalem*, 1858.

Bonomini, Bartolomeo. Chartmaker of Ancona. Mediterranean portolan chart, 1570.

Bonpland, Aimé. French botanist and physician. Travelled to and explored Spanish America with Alexander Humbolt, 1799-1803.

Bonsall, Joseph H. Civil engineer. *Map of the City of Philadelphia*, 1862; *...Denver, Colorado*, 1871; *...Colorado*, 1872.

Bonsignori, Stefano (*d.*1589). Completed the painted maps begun by Ignazio Dante in the *Sala de carte geographia*, Palazzo Vecchio, Florence; also did the maps in the *Loggia di Carte* of the Uffizi, 1589.

Bonson, E.W. *...Western Province of the Gold Coast*, 1884.

Bonstetten, Albrecht von *OSB* (*c.*1442-*c.*1504). Swiss historian, born Uster, worked in Einsiedeln. Wrote a geographical description: Superioris Germanie Confoederationis descriptio, 1479. **Ref.** SIEBER-LEHMANN, C. *Cartographica Helvetica* 16 (1997) pp.39-46 [in German].

Bonstetten, *Baron* Wilhelm von. *Carte Archéologique du Canton de Berne*, 1876; *Carte Archéologique du Canton de Friburg*, 1878.

Bont, A.L. de. *Geschiedkundinge Schoolatlas de Algemeene en Vaderlandsche Geschiedenis*, 1893; *Kleine Schoolatlas der Algemeene en Vaderlandsche Geschiedenis*, 1912 and later editions; *Kleine Schoolatlas der Vaderl. en Algem. gesch.*, 1914

Bontein, Archibald. Chief engineer in Jamaica. *An Exact Plan of Carthagena...*, (2 sheets), 1744; *A new and correct Map of the Caribee Islands*, 1751; *A Map of the Island of Jamaica*, 1753.

Bontein, William. Annapolis-Royal, 1754; Hudson River Albany to Fort Edward, 1757; Survey of Louisbourg, *c.*1758.

Bontems [Bontemps], *Colonel* Auguste-François (1782-1864). Swiss military cartographer. Map of Geneva, 1815.

Bontz, Geo. F. *Strategic map of central Europe*, Washington 1915 (with A.B. Williams).

Bonus, Claudius. *Chorographica descriptio Conventuum Hispaniae*, *c.*1700.

Bonvalet, H. Engraver, for Jean Mallat de Bassilan, 1846.

Bonven, Maurice. Engraver. Seven United Provinces, 1772.

Bonwick, James. *Victoria with parts of New South Wales and South Australia*, 1855.

Bonwicke, J. Publisher of Edward Wells's *A New Sett of Maps*, 1722, 1738.

Boog, F.A. Chart Moulmein Harbour [Burma], 1809 MS; Salween River, 1869 MS.

Boom, family. Printers and booksellers in Amsterdam. Atlas françois, Amsterdam 1667.
- **Boom**, Hendrik.
- **Boom**, Dirk.

Boom, E.H. *De Reede van Samarang*, 1855; *Straat Sunda en Westkust van Java*, 1855; Nederlandsch Oost-Indië, 1863.

Boonacker, Sijmon [Symon] Willemszoon (*fl.*1615-1627). Dutch surveyor and cartographer, of Edam. Provided data for Lucas Jansz. Sinck's *Caerte van de Purmer...*, 1622; *Kaarte vande Buyck-slooter, Broecker ende Belmer Meeren in Water-land*, 1627 MS (published by Blaeu, 1662).

Boonen, J. Staats. Bought the Van Keulen family business in 1823, sold on to Jacob Swart in 1840.

Boosen, C.B. Norge nordlige [- sydlige] del (2 sheets), Oslo 1862.

Boot, Alfred [& Son.]. 24, Old Bailey, London E.C. *The District Railway Miniature Map of London*, *c.*1881.

Booth, Charles (1840-1916). Used coloured maps to show the distribution and extent of poverty in London, published in *Life and Labour* [Vol.1 East London], 1889, [Vol.2 London], 1891. **Ref.** *Charles Booth's descriptive map of London poverty 1889* (London Topographical Society 1984) [facsimile with an introduction by D.A. Reeder].

Booth, George. Publisher of 1 Paragon Place, New Kent Road, London. *London*, 1846 (with J. Reynolds).

Booth, John (*c.*1688-1733). Surveyed estates in various midland counties of England, including manor of Ditton [Buckinghamshire], 1718 MS.

Booth, John. Publisher of Duke St., Portland Place, London. Settlements in Australia, 1810; P.J. Martin's *A geological memoir on a part of Western Sussex...*, 1828 (with a map engraved by Sidney Hall).

Booth, John C. *Geological and topographical map of the mineral district of Lake Superior, Michigan*, E. Warner & E.J. Hulbert 1855, 1864 (with E.J. Hulbert, engraved by H. Colton).

Booth, W. Chichester to Rye, 1704.

Booth, W.J. Plans of Newington Green, London 1826.

Booth, William. Draughtsman. *Plan of the Attack of the High Allies against the French near the Village of Fontenay... 1745*, 1745 MS.

Booth, *Lieutenant* William. *Plan of the Harbour, Forts, Town & Environs of Fort Royal in Martinique*, 1793, 1795.

Boothby, Frederick. *Map of Sussex*, Great Tower Street Tea Co. 1893 (this map also appeared in *A Handbook of Sussex for the pocket, the library, or the tea table...*, 1893).

Boottats, Gaspar. See **Bouttats**, Gaspar.

Booty, C. Publisher of 16 Kings Road, Brighton, England. *Hunting Map & Environs of Brighton*, 1860.

Boraigne, R. (*fl.*19th century). Lithographer of Naples.

Borchers, E.N.W. *General Gang-Churte des nordwestlichen Harzgebirges*, 1866.

Borcht [Borchius, Verborcht], Pieter van der (1545-1608). Painter and engraver of Antwerp, for Ortelius's *Theatrum orbis terrarum*, 1570 and later editions.

Borck, —. Belgrade, Berlin 1726.

Borckeloo [Borculo; Borculoo; Borculus], Harmen van (*c.*1505-1578). Engraver, printer and publisher. Plan of Jerusalem (woodcut, 8 sheets), 1538. **Ref.** RUBIN, R. 'The Map of Jerusalem (1538) by Herman Borculus and its copies - a cartogenealogical study' *Cartographical Journal* 27 (1) (1990) pp. 31-39.

Borda, Jean Charles, *chevalier de* (1739-1799). Mathematician and engineer. Voyage Flore, 1771-1772 (with J.R.A. Verdun de la Crenne and A.G. Pingré); *Carte des Îles Canaries*, 1780; *A new chart of the Coast of Africa*, 1794; *A General Chart of the Atlantic*, 1804; *Laurie & Whittle's New Chart of the Coast of Greece*, 1812.

Borde, —. See **La Borde**, Jean-Benjamin de.

Bordeaux, Bernadin de [Burdigalensis, Bernardinus]. Monk and cartographer. G. Montecalerio's *Chorographica descriptio provinciarum... S. Francisci Capucinorum*, Rome 1643.

Bordeaux, Paul René. French naval surveyor in Indochina, 1893-1896, and Tchad, 1906-1907.

Bordeaux Itinerary. *Itinerarium Burdigalense* [Latin itinerary for pilgrims from the west to Jerusalem], AD 333.

Bordee, —. Maps for Laurie & Whittle's *Complete East India Pilot*, 1800.

Borden, H.D. Surveyor and engineer. *City of Syracuse*, 1868.

Borden, Simeon. *Topographical map of... Massachusetts* (6 sheets), 1844.

Borden, William G. *Map of the Town of Tiverton*, 1854.

Bordiga, brothers [Fratelli Bordiga]. Engravers of Milan. *Carte Générale Du Théâtre de la Guerre en Italie...* (30 sheets), Milan, Bacler d'Albe 1798; also Mappamondo, *c.*1798; maps of England, France and South Africa for Barbiellini's *Atlante delle geografia antica e moderna*, 1807; published *Modelli di topografia* (5 sheets), 1810-1812; engraved all six maps in Formiggini's atlas, 1814. **Ref.** VALERIO, V. *The Map Collector* 45 (1988) p.14.
• **Bordiga**, Benedetto (1768-1847). *Carta topografica del Milanese*, 1788-1796.
• **Bordiga**, Gaudenzio (1773-1837). Brother of Benedetto.
•• **Bordiga**, Giovambattista (*b.*1799). Nephew of the Bordiga brothers, above. Engraver. The World for Luigi Bossi's *Nuovo atlante universale*, 1824; A. Litta's map of the State of the Church (6 sheets), 1824.

Bordiga, Giovanni Battista. Engraver and draughtsman. *Carte militare del Regno d'Etruria*, 1806; *Carta Geografica Postale dell'Italia*, 1813; *Carta della Grecia*, 1828; *Carta...dell'Europa*, 1832.

Bordone [Bordon: Bordonius; Burdonius], Benedetto (*b.*disputed 1445-1460-*d.c.*1530/1539). Miniaturist, illuminator, cartographer and engraver from Padua, worked and died in Venice. World map, 1508 (possibly in the form of globe gores, now lost); Italy, *c.*1508 (now lost); maps in Pliny's Natural History, 1513 (attributed); best known for island book, *Libro de Benedetto Bordone, nel qual si ragiona de tutte l'isole del mondo* (111 woodcut maps), Venice 1528, reprinted as *Isolario...*, 1534, 1537, 1547, etc. **Ref.** ARMSTRONG, L. *Imago Mundi* 48 (1996) pp.65-92; SKELTON R.A. Biographical note in facsimile edition of *Theatrum Orbis Terrarum* Third Ser., Vol.I. (Amsterdam 1966); KARROW. R.W. *Mapmakers of the sixteenth century and their maps* (1993) pp.89-93; MEURER, P. *Fontes cartographici Orteliani...* (1991) p.119 [in German].

Bordoni [Bordone], Girolamo (*c.*1510/1520-1615). Franciscan monk from Sermoneta, near Rome, *d.*Genoa. In the service of Cardinal Filonardo (to 1549), from 1564 he was an official of the Genoan Republic. Credited with a map of Corsica (now lost). **Ref.** MEURER, P. *Fontes cartographici Orteliani...* (1991) p.119 [in German].

Borel, Petrus (*fl.*17th century). Dutch astronomer.

Borelli, Jules (1853-1941). French geographer, surveyor and photographer in north-east Africa, 1885-1889. Surveys in southern Ethiopia, 1885-1888.

Borenfield, D.L. Plan of Ngatapa Pa [Poverty Bay, New Zealand], 1869.

Boret, A. de. *Camp de Châlons*, 1862.

Boreux, I.G. Evenements militaires, 1799.

Borg, Antonio. Famagusta, Larnaca 1770 MS; Plano del Isola di Cipro, 1770 MS.

Borgen, *Dr* —. Süd Georgien, Berlin 1886.

Borghi, Bartolomeo (1750-1821). Italian cartographer and engraver. Editor of *Atlante Geografico* (102 maps), Siena, Pazzini Carli 1788-1800; *Topographia della Città di Firenze*, 1817; *Atlante Generale*, Florence, Giarrè 1819 (some of the maps were first published in *Atlante Geografico*).

Borgia world map. Anonymous circular map on 2 iron plates (63-cm diameter), made in southern Germany, early 15th century. Acquired by Cardinal Stefano Borgia 1794 in Portugal for his museum at Velletri near Rome. Printed copies entitled *Apographon descriptionis orbis terrae*, issued 1797. **Ref.** KRETSCHMER, DÖRFLINGER & WAWRIK *Lexicon zur Geschichte der Kartographie* (1986) vol. C/1, pp.101-102.

Borgnis-Desbordes, Gustave (1839-1900). French marine officer. Led an expedition through Sénégal and Niger, which produced maps of the region.

Borgonio, Giovanni Tommaso (1620-*c.*1683). Military engineer and cartographer from Piedmont. *Tabula Generalis Sabaudiae*, 1670 (also published in Joan Blaeu's Italian Townbooks); Turin, 1672; *Piedmont* (16 sheets), 1680; *Carta...degli Stati di S. M. Re di Sardegna* (25 sheets), 1683.

Borland, Francis. Map of this Neck & Harbour [Darien], 1775.

Borlase, William. *Map of the County of Cornwall*, 1754.

Bormann, E. *Übersichtskarte des Bezirkes Tetscahen*, *c.*1872.

Bormeester, Joachim. Engraver, printer and publisher, 'in de Warmoes Straadt By de Niesel', Amsterdam. *Orbis terrarum nova accurata tabula* (4 sheets), *c.*1685 (drawn by F. Tieman, engraved by Gilliam van Gouwen); America, 1700; Europe, G. van Keulen1712; *Novissima et Accuratissima XVII Provinciarum Tabula*, [n.d.].

Born, C.L. von. *Kort over Foeroerne*, 1789, English edition 1806; *Carte Reduite de l'Island et des Îles Feröe,...1795*, 1836.

Born, Ignaz Edler von (1742-1791). Geologist and mineralogist of Prague and Vienna.

Born, Moritz L. Surveyed parts of the coast of Iceland, 1815-1819 (with A. Aschland).

Borner, —. Lithographer. Plan von Karlsruhe, 1817.

Bornhaupt, C. *Karte von Liv-, Est- & Kurland*, 1875.

Borodavkin, Korney. Russian topographer. Numerous district [zemstvo] and town maps in central Russia, 1720-1728.

Borodovsky, L. Map of Manchuria, St Petersburg 1897.

Boronat y Satorre, Francisco. Spanish publisher in Madrid. *Carta geográfica de España* (49 sheets), *c.*1875.

Boros, Frigyes [Fritz] (1824-1892). Hungarian engineer, mapped Tisza river 1848 onwards. **Ref.** SÁRKÖZY, I. *Régibb vizimérnökeink* [Our former hydrographers] (Budapest 1895) [in Hungarian].

Borough, family of mariners.
• **Borough**, John à (*fl.*1533-1542). Seaman and ship owner, father of Stephen and William Borough. Dover harbour, 1541 MS (with Anthony Aucher, John Bartelot and Sir Richard Cavendish).
•• **Borough** [Burrough], Stephen (1525-1584). Navigator, served under Richard Chancellor. Narratives of the Arctic voyages in search of the Northeast Passage with Chancellor, 1553-1554, 1556, 1557, published in Richard Hakluyt's *Principall Navigations*, 1589; Russian voyage, 1561; voyage to the Caspian Sea and Persia, 1579-1581 published by Samuel Purchas in *Purchas His Pilgrimes*, 1625.
•• **Borough** [Burrough], William (1536-1599). Younger brother of Stephen (above); navigator, hydrographer and writer on nautical subjects. Served under the Moscovy Company. Comptroller of the Navy. Chart of the Northern Navigation, 1560; Russian Arctic voyages, 1570-1575 published Hakluyt, 1589; drew chart of the Atlantic for Martin Frobisher's voyage of 1576 MS (with Frobisher's amendments to 1578, became the first chart to show magnetic variation); Inward Parts of Russia, 1578 MS; Outer Thames Estuary, 1588; chart of the Thames estuary and south-east England, 1596 MS.

Borowsky, Georg J. Viennese publisher of the second half of the 18th century. Also worked as copper engraver. *Prospekt der heutigen Stadt Jerusalem* [Plan of the present city of Jerusalem], 1865.

Borre, Pieter van. Dutch cartographer, worked in Zeeland. Oranjepolder, 1618 MS.

Bors, Johann Jakob von. *Neue und vollständige Postkarte durch ganz Deutschland* (16 sheets), 1764 (with F.J. Hegers, later available as a pocket atlas).

Borsano, Ambrosio (*fl.*1652-1687). Italian military surveyor in Spanish service in Barcelona. *El principado de Cattaluña y condados de Rosellon y Cerdaña*, 1687 MS.

Borven, M. Engraver for Robert Sayer. China, 1790.

Bory de Saint-Vincent, Jean Baptiste Marcellin (1778-1846). *Carte des Isles Canaries*, 1801-1803; *Teneriffe*, 1803; *Nouvel Atlas pour servir à l'histoire des Îles Ioniennes*, 1823; *Nouvelle Carte d'Espagne et de Portugal*, 1824; *Atlas encyclopédique...*, 1827 (with N. Desmarest & R. Bonne); *Île de Tine*, 1833; *Atlas géographique... de la France*, *c.*1845.

Bos, Albertus. See **Busius**, Albertus.

Bos, *Lieutenant* H. French explorer West and North Africa. Exploration de la rivière Kotto, 1901-1903.

Bos [Bossius Belga; Busius; Bussius], Jacob [Jacobus] (*fl.*1551-1577). Belgian lettering engraver who worked in Italy in the 16th century, engraver for Tramezini and Lafreri. Flanders, 1555; Switzerland, 1555; *Brabantiae Belgarum*, Rome 1556; *Hollandiae...*, 1556; *Gelriae...*, 1556; *Frisiae...*, 1558; *Septemtrionalium regionum Svetiae Gothiae Norvegiae Daniae...*, 1558; *Anteiquae urbis imago...*[Rome], 1561; *La descrittione della Prima Parte dell' Asia*, 1561 (Gastaldi map in a Lafreri Atlas).

Bos, Jacobus (*fl.*1695-1706). Dutch surveyor and cartographer. Worked in Friesland to 1706.

Bos, Pieter Roelof (1847-1902). Dutch schoolteacher, best known for his ubiquitous and enduring *School-atlas*. Atlases of various titles were originally published in Groningen by J.B. Wolters followed by many editions published over more than a century. *School-atlas der geheele aarde*, 1877 and later; *Atlas voor de Volksschool...*, 1878 and later editions; *Kleine Atlas voor de Volksschool*, 1881 and later editions; *Natuur-en Staatkindige Atlas*, 1886 and later editions; *Teekenatlas* [blind atlas], 1891 and later editions; and many other similar titles. **Ref.** KOEMAN, C. *Atlantes Neerlandici* vol.VI (1985) pp.99-126.

Bos, R. Produced school atlases and blind atlases, all published by P. Noordhoff of Groningen. *Voorlooper van den Eersten Teekenatlas*, 1890 (blind atlas); *Schoolatlas van Nederland*, 1892 and later editions; *Atlas der Geheele Aarde in Kaarte en Beeld*, 1893 and later editions; *R. Bos' Volledige Schoolatlas der Geheele Aarde*, 1909 and later editions; *R. Bos' Nieuwe Schoolatlas...*, 1910; *Eerste Atlas voor Schipperskiassen*, 1920. **Ref.** KOEMAN, C. *Atlantes Neerladici* Vol.VI (1985) pp.126-136.

Bosari, Bonifacius (*fl.*18th century). Italian instrument maker.

Bosatsu. See **Gyoki** Bosatsu.

Bosch, Hendrik. Bookseller of Amsterdam. Issued Dutch edition of Herman Moll's maps, 1727.

Bosch, J. (*fl.*1572). Engraver and publisher of Haarlem.

Bosch, Jacob (*fl.*1660-1694). Dutch surveyor and cartographer ['stadsingenieur en landmeter'] of Amsterdam. Revised Cornelis Dankertsz. de Rij's *Caarte daar in vertoont wert der over Amstels polder zo die nu lest bepoldert is...*, 1639, 1694.

Bosch, *baron* Jan van den (1780-1844). Of The Hague and Batavia [Jakarta]. *Atlas der Overzeesche Bezittingen* (12 maps), Amsterdam, Gebroeders van Cleef 1817.

Bosche, Philip van den. *Praga*, 1606.

Boschini, Marco (1613-1678). Venetian artist, printer, engraver and mapseller. Candia, 1645; Dalmatia, 1646; *Il Regno tutto di Candia* (2 maps, 60 plans & views), Venice, Nicolini 1651; *L'Arcipelago con tutte le Isole*, Venice, Nicolini 1658. **Ref.** CLUTTON, Elizabeth *Imago Mundi* 34 (1982) pp.59-65.

Boschke, A. American surveyor and publisher. *Map of Washington City*, 1857 (printed by J. Bien); *Topographical map of the District of Columbia, surveyed in the years 1856-'59*, Washington, D. M'Clelland 1861.

Boscovich, Ruggiero Giuseppe, *SJ* (1711-1787). Mathematician and astronomer from Ragusa [Dubrovnik], member of the Institute of Sciences, Bologna, from 1746; Royal Academy of Sciences, Paris, from 1748; and Accademia degli Arcadi, Rome. Worked with fellow Jesuit Christopher Maire on a geodesic survey of the Rome-Rimini meridian. An account, with a 3-sheet map by Maire, was published as *De Litteraria expeditione*, Rome 1755. **Ref.** PEDLEY, M. *Imago Mundi* 45 (1993) pp.59-76.

Bose, Hugo von (*d.*1856). Geographer of Dresden. *Karte vom Koenigreiche Sachsen*, 1845; *Special-Atlas der Oesterreichischen Monarchie*, 1856.

Boskovviz, Lorentz Slisansky de. Born in Moravia in the early 17th century, in 1660 made a pilgrimage to the Holy Land. On his return he published the work *Newe Reissebeschreibung macher Jerusalem* [New travel report on Jerusalem], Vienna 1662; *Wahre abzeichnung der alten H. Statt Jerusalem* [small plan of Jerusalem], [n.d.].

Bosma, Benjamin. De Geographische Onderwyzer, Amsterdam, Jacobus Loveringh 1765.

Bosman, H.W. *Atlas van de Nederlandsche Bezittingen in Oost-Indië*, 1908, Leiden *c.*1919.

Bossange, Adolphe. *Atlas des mémoires du maréchal Suchet, duc d'Albufera, sur ses campagnes en Espagne...1808-1814*, Paris, Firmin Didot 1828.

Bosse, Abraham. Engraver, worked with Melchior Tavernier. Untitled twin-hemispherical world map in shape of a shield, Paris, Melchior Tavernier *c.*1630.

Bosse, Paul. *Das Grossherzogthum Hessen*, *c.*1877.

Bosset, *Lieutenant-Colonel* — de. Harbour of Cephalonia, 1814.

Bossi, Giacomo. *Carte de la Route du Capitaine Clapperton*, 1851; *Carte des deux voyages de Mungo Park*, *c.*1851; *Cours du Quorra*, *c.*1851.

Bossi, Luigi (1758-1835). Italian cartographer, geographer and historian. *Nuovo atlante universale*, Milan 1824.

Bossius, Albertus. See **Busius**.

Bossius Belga, Jacobus. See **Bos**, Jacob.

Bossler, Frederick. American engraver. James Madison's map of Virginia, *c.*1805.

Bossoli, Carlo. *The Crimea, a bird's-eye view*, 1854; *The Kingdom of Sardinia*, 1859.

Bossong, Franz. *Touristen-Karte...von Ems*, *c.*1894.

Bostwick, Henry (1787-*c.*1836). Historical & Classical Atlas, New York 1828.

Bostwicke, Homer T. One of the surveyors for *Map Exhibiting the Route of the Norwich & Worcester Rail-road*, Boston, T. Moore's Lithography 1835.

Boswell, Henry. Credited on the title page of Alexander Hogg's partwork *Historical Descriptions of New and Elegant Picturesque Views of the Antiquities of England and Wales*, *c.*1787-1789 (re-using Thomas Kitchin's maps from the *London Magazine*). It is quite probable that Boswell was an invention of Alexander Hogg. **Ref.** HODSON, D. *County atlases of the British Isles published after 1703* Vol.III (1997) pp.165-170.

Boswijk, D. *Thieme's tweede teekenatlas de Europa & wereld*, 1894; *Atlas v.d. Nederlandsche Bezitt. in Oost-Indië*, 1908, 1919.

Boswijk, Johan (*fl.c.*1700). Dutch map copyist, worked in Zeeland. Made copy of Claes Janssen's ...*West Craeyert...*, *c.*1676, 1700 (with G. Poly).

Boteler, Thomas, *RN*. Admiralty surveyor. *Survey of the Cape of Good Hope*, 1822, Admiralty, 1828 (with Gibbons, Lechmere and Vidal of HMS *Leven*).

Botella y de Hornos, Frederico de. *Descripcion geológica-minera de las Provincias Marcia y Albacete*, 1868; *Mapa geológica de España y Portugal*, 1879.

Botero, Giovanni (1540-1617). Piedmontese priest and geographer, Secretary of the Duke of Savoy. *Le relationi universali...*, Rome 1591 (this edition had no maps), re-issued in Venice by Giorgio Angelieri 1596 (with maps from Giovanni Lorenzo d'Anania's *L'Universale Fabrica del Mondo*, 1582), many later editions with varying maps were published in Rome, Bergamo, Venice, and elsewhere until 1640; Tabulae geographicae, 1596.

Both, *Major* — von. Battle plans in Germany, 1807-1810; *Plan von Berlin*, 1811.

Both, L. Scandinavian cartographer. Maps of Denmark, France and Flanders, 1861-1877.

Botham, W.R. *Map of the Railways, Canals, Navigable Rivers of Great Britain*, *c.*1852; *Map of London and the Suburbs, Shewing the Existing and Proposed Railways*, W.H.Smith & Son 1853.

Bothams, J.C. *Plan of the Borough of Salisbury*, 1860.

Botley, Thomas. *Map of the County of Surrey* (8 sheets), 1765-1769.

Botta, C. Carta militaria d'Italia, 1800.

Bottats, Gaspar. See **Bouttats**, Gaspar.

Bottin, —. One of the surveyors for the Cassini *Carte de France*. Worked with others on sheets 30, 65 & 109, *c.*1753-1766.

Bottius, Adrianus. Credited by Tooley with the publication of Mercator's *Atlas minor*, 1610. **N.B.** See **Hondius** family.

Bottle, father and son.
• **Bottle**, Alexander I (*fl.*1722-1764). Surveyor of Harrietsham, Kent. *Kingsnorth*, 1729 MS; *Manor of Tremworth*, 1730; *Sutton Valence*, 1733.
•• **Bottle**, Alexander II (*fl.c.*1737-1771). Kent estate surveyor.

Bottle, Robert Thomas (*fl.*1792-1822). Surveyor. Estate plans of Hoo, Kent, 1808; *Farm in Smarden*, 1810 MS.

Botz, H. *Plan der Universitätsstadt Jena*, *c.*1858.

Bouchard, E. *Carta del Teatro della Guerra in Italia* (7 sheets), 1799.

Bouchard, Giuseppe. Pianta della città di Firenze, 1755; *Nuova Pianta di Gibilterra*, 1762.

Bouchenroeder, *Baron* B. Friedrich von. Explored parts of South Africa. *Reize in de Binnelanden van Zuid-Afrika dedaan in den jare 1803*, Amsterdam, Mortier, Covens & Zoon 1806 (included a map of the Colony of the Cape of Good Hope).

Boucheporn, F. de. *Atlas...du département de la Corrèze*, *c.*1875.

Boucher, F. Engraver. Frontispiece for Après de Mannevillette's *Le Neptune Oriental*, 1775.

Boucher, M. Engineer. *Plan du Port Dauphin*, 1733; Neptune François, 1778; Île Royale, 1780.

Bouchet, Joan Venantio *SJ*. *Sud-Theil von Indien Herwerts des Ganges...*, 1728 (for Joseph Stöcklein).

Bouchette, *Lieutenant-Colonel* Joseph Francis (1774-1841). Nephew of Samuel Holland, whom he succeeded as Surveyor General of Canada in 1804.*Plan of Toronto Harbour With the Rocks, Shoals & Soundings Thereof...*, 1792; *A Plan of York Harbour in Upper Canada...*, 1794; *Plan of Upper Canada*, *c.*1795; *A Plan of Lake Ontario*, 1799; *Connected plan of the several Townships...southward of the River St. Lawrence...*, Québec 1805; *Topographical Description of the Province of Lower Canada* (with 10 maps), London, Faden 1815; Town of Three Rivers, 1815; *Map of the Provinces of Upper & Lower Canada...*, Faden 1815 (engraved by J. Walker); *Plan of the District of Gaspé*, Faden 1815 (engraved by John Walker); *British Dominions in North America*, London, Colburn & Bentley 1831; *...District of Montreal, Lower Canada...*, James Wyld 1831 (engraved J & C Walker); Districts of Montreal and Quebec, James Wyld 1831; and many other maps of Canada. **Ref.** WINEARLS, J. *Mapping Upper Canada 1780-1867* (1991) many references.

Bouchette, Joseph, *junior* (*b.*1800). Deputy Surveyor General of the Province of Lower Canada. *Provinces of Lower & Upper Canada*, London, James Wyld 1831 (engraved J. & C. Walker); *Map of the Provinces of Canada, New Brunswick, Nova Scotia...*, 1846; *Crown Land Survey of Canada*, 1858. **Ref.** BOUDREAU, C. *L'analyse de la carte ancienne: essai methodologique: la Carte du Bas-Canada de 1831 de Joseph Bouchette* (Quebec 1986) [in French].

Bouchette, Robert Shore Milnes (1805-1879). Boundary surveys for the Treaty of Ghent from Cornwall to Lake George [Canada] (17 plans), 1824; *City of Quebec*, *c.*1850.

Lieutenant Colonel Joseph Bouchette (1774-1841). (By courtesy of the New Brunswick Museum)

Bouchier, M. *Plan de la ville de Troyes*, 1839.

Bouchier, Thomas. Buckland River, 1849; Eschacholtz Bay, 1850.

Bouclet, —. (*fl.*1793-1820). One of the engravers for *Atlas du Voyage de La Pérouse*, 1797; engraver for Bruny-d'Entrecasteaux, 1807; *Environs de Versailles* (12 sheets), 1808; for Freycinet, 1811; for Dépôt de la Marine.

Boudart, A. *Carte Générale des Chemins de Fer Belges*, 1876.

Boudène, François. *Plan de la Ville de Saint Pons de Thomières en 1699*, 1900 (with Joseph Sahuc).

Boudet, Antoine (*fl.*1745-1787). Publisher and printer in partnership with his stepfather Jean-Baptiste Coignard, and with his brother-in-law Pierre Gilles Le Mercier. 'Imprimeur du roi, rue St Jacques près St Yves' Paris. *Mappemonde*, 1752; Robert de Vaugondy's *Atlas universel*, 1757 and later editions; new edition of Antoine Bruzen de la Martinière's *Le Grand Dictionnaire Géographique*, 1766; Courtalon's *Atlas élémentaire de l'Empire d'Allemagne*, 1774 (with R-J Julien); *États-Unis de l'Amerique septentrionale*, 1785; Indes orientales, 1786. **Ref.** PEDLEY, M. *Bel et utile* (1992) pp.51-58.

Boudin, J.Ch.M. *Carte Physique et Météorologique du Globe*, 1851.

Boudinot, E.S. *Atlas of Vermilion County, Illinois*, Danville 1907.

Boué, *Doctor* Ami (1794-1881). Naturalist and physician of Hamburg, active in Paris and Vienna. *Carte géologique de l'Écosse*, 1820; Carte géologique de l'Europe, 1832; *Carte de la Turquie d'Europe*, *c.*1840; *Carte géologique du globe terrestre*, 1845, 1848.

Boue, H. Map of the Baltic, 1815.

Bouet-Willaumez, *Amiral* Louis Edouard (1808-1871). French colonial administrator in Ivory Coast, Dahomey and Gabon, governor of Sénégal 1843-1845. *Esquisse commerciale de la côte occidentale d'Afrique*, 1843; *Description nautique des côtes d'Afrique occidentale... entre le Sénégal et l'Equateur*, 1846.

Bouffard, L. Engraver. *Carte des Etats-Unis d'Amérique*, Paris, L. Bernard 1836; *Carte de la Province d'Alger*, 1838; *Carte...de l'Ile de Canaria*, 1838; *Carte de l'Ile de Cuba*, 1841; *Carte geologique...de la Russi meridionale*, 1844.

Bougainville, Hyacinthe-Yves-Philippe Florentin, *baron* de (1781-1846). French marine officer from Brest. *Journale de la navigation autour du globe*, 1837 (with accompanying atlas).

Bougainville, Louis-Antoine, *comte* de (1729-1811). French navigator, explorer and hydrographer. *Isles Malouines* [Falkland Islands], 1763-1764, Thomas Jefferys 1771; *Voyage autour du monde... En 1766, 1767, 1768 & 1769*, Paris 1771-1772.

Bougard, Réné. French hydrographer of Le Havre. *Le Petit Flambeau de la Mer...*, Le Havre, Jacques Gruchet 1684, second edition published by J. Hubault, 1694, reprinted 1742; 1709 edition published Guillaume Gruchet; English edition, translated and edited by John Thomas Serres, published as *The Little Sea Torch* (24 maps), London, J. Debrett 1801.

Bouge, Jean Baptiste de (1757-1833). Surveyor, engraver, geographer and publisher, 'Rue de Lille n$^{o.}$ 57', Paris (1812), also worked in The Hague. *Carte Itinéraire des Pays Bas* (4 sheets), 1786; *Carte du Diocese de Tournai* (4 sheets), 1789; *Plan de Luxembourg*, 1790 (engraved by Jeanne Cne. Maillart); Flandre, 1793; Europe (50 sheets), 1796-1798; *Nouvelle Carte...des États du Roi de Sardaigne*, 1800; *Nouvelle Carte...des États du Roi de Naples*, 1801; *Carte de l'Europe* (45 sheets), *c.*1820; *Petit Atlas Administratif du Royaume des Pays-Bas*, Ghent *c.*1820; *Carte chorographique du royaume des Pays Bas* (20 sheets), 1823; *Carte de Postes du Royaume de Pays-Bas...*, *c.*1827.

Boughman, Thomas & Co. Publishers of Wilmington, U.S.A.. *Boughman, Thomas & Co.'s map of the peninsula, embracing Delaware and the Eastern Shores of Maryland and Virginia*, 1868 (compiled by D.G. Beers).

Bouguer, Pierre (1689-1758). Mathematician of Paris. Joined Louis Godin, Charles Marie de la Condamine and Joseph Jussieu on an expedition to Peru for the French Académie Royale, 1735-1744. He conducted a survey by triangulation and measured the arc of the meridian. The results, combined with those of a similar party in the northern hemisphere, helped establish the shape of the Earth. *Carte de triangles de la Meridien de Quito*, published in Bouguer's *La Figure de la terre*, Paris 1749.

Bouguereau, Maurice (*d.c.*1595). Master printer and map publisher, 'à la Petite Fontaine du Carroy de Beaulne', Tours. *Le Théâtre françoys* (20 maps), 1594 (the first printed atlas of the provinces of France). **Ref.** PASTOUREAU, M. *Les atlas français, XVIe-XVIIe siècles* (Paris 1984) pp.81-84 [in French]; DAINVILLE, F. de *Le Théatre Francoys*, introduction to facsimile edition (Amsterdam 1966); DAINVILLE, F. de 'Le premier atlas de France. Le Theatre Francoys de M. Bouguereau' in: *Actes du Quatre-Vingt-Cinquieme Congress national des Societes Savantes* (Paris 1960) [in French].

Bouillet, —. Plan de la conquête de l'Isle de la Dominique, 1778.

Bouillet, Marie-Nicolas (1798-1864). *Dictionnaire et Atlas Universel d'Histoire et de Géographie*, Paris, Hachette 1865, 1872, 1877.

Bouillevaux, *père* Charles-Emile (1823-1913). French missionary and traveller in Indochina. First detailed description and survey of Angkor, 1850. *Voyage dans l'Indo-Chine 1848-1850* (with a map of Cambodia), Paris 1858.

Bouillon, A.M. Mining maps of British Columbia, 1897 (with R. Buck).

Bouillon, Gilles. See **Boileau** de Bouillon, Gilles.

Bouinais, Albert M. Aristide (1851-1895). French mariner and geographer, boundary commissioner in the Far East.

Boulangier, Edgar (1850-1899). French engineer, specialist in the railways of Russia and Indochina. *Études sur le relief terrestre*, 1886; *Chemin de fer transcaspien*, 1887; *Le chemin de fer transsibérien et la Chine*, 1891.

Boulengier [Bolengier, Boulanger], Louis [Ludovicus]. French astronomer and geographer. *Cosmographie introductio*, Lyon, Jean de la Place, 1514; attributed set of globe gores *Universalis cosmographie descriptio*, *c.*1514 (known from one copy).

Boullanger, —. Hydrographer on Freycinet's expedition. Charts of coast of Australia, 1801-1803, published 1811-1812.

Bouloigne, Gilles. See **Boileau** de Bouillon, Gilles.

Boulter, B.C. *Railways of Kent*, 1903, 1915.

Boulton, John G. Second Master, *RN*. Admiralty surveyor. Newcastle Harbour, New South Wales, 1867; Chart Table Bay to Donkin Bay, 1871.

Boulton, Solomon. *Geographia Antiqua Delineata*, 1755; map of Africa, Robert Sayer 1787 (derived from J.B.B. d'Anville). **N.B.** Possibly S. **Bolton**, see above.

Boulze, Auguste. *Carte Routière Industrielle et Minière de l'Arrondissement d'Alais*, 1867.

Bouman, Jacobsz. Publisher of Amsterdam. Schouten's *Journal...* (with woodblock world map), 1658, 1661.

Bouquet, *Colonel* Henry (1719-1765). Led the British army party into the Indian territory west of the Ohio River in 1764. The expedition was mapped by Thomas Hutchins *q.v.*.

Bourbon, Petit. See **Petit**, Pierre.

Bourcet [Bourcet la Saigne; Bourset], Pierre Joseph de (1700-1780).Directed surveys of the Alpine frontier between France and Piedmont, 1748-1760. *Nouvelle Carte de l'Isle de Corse...*, 1740; *Carte Géométrique du Haut-Dauphiné et du comté de Nice* (9 sheets), 1758; *Plan de la ville du Cap de Bonne Esperance*, 1770 (possibly used by J.N. Bellin in *Hydrographie françoise*, 1772); *Carte des Alpes*, 1800.

Bourdan, Jean. Map of the Seigneural Settlement in Canada, 1641 MS.

Bourdelin, Emile. *Plan de Paris*, *c.*1878.

Bourdin, Ernest. Editor and publisher of Paris. Maps to illustrate the travels of A. Demidoff. *Carte du voyage dans la Russie Meridionale...*, 1841; *Carte géologique de la Crimée...*, 1853 (both maps engraved by P. Tardieu).

Bourdon, Jean [Jehan] (1602-1668). *Vray plan du haut & bas de Quebec...*, 1660

Bourdonnaye,— de la. Carte de la Mer des Indes, 1750 MS.

Bourg, Marie François Cromot, *baron* du (1756-1836). Plans of French Army at Philipsburg, 1781; Siège d'York, 1781.

Bourgeois, Jaques. Published a map of the Garden of Eden in a French Bible, Geneva 1560 (with F. Jaquy and A. Davodeau); Eden in a Polyglot Bible, Geneva 1568.

Bourgoin, A. *Plan de la ville de Chartres*, 1889.

Bourgoin, Gustavo (1821-1863). Topographical engineer in the Officio Topografico di Napoli. During the 1840's worked as one of the surveyors for the Officio's map of the Kingdom of Naples. **Ref.** VALERIO, V. *Società uomini e istituzioni cartografiche nel mezzogiorno d'Italia* (1993) pp.459-460.

Bourgoin, Jean François (1750-1811). *Atlas...de l'Espagne moderne*, 1807, 1808.

Bourgoin, Pierre. Engraver and publisher of 'Rue de la Harpe vis à vis le passage des Jacobins', Paris. Asie, 1741; engraved for J.B.B. d'Anville, Italie, 1743; *Carte des Postes de France*, 1748-1763; engraved lettering on many sheets of the Cassini *Carte de France*, from 1756; *Carte itinéraire de la France...*, 1773 and editions to 1787; Amérique méridionale, 1774.

Bourguignon, Mr —. Bookseller of Liège 'Au Livre d'or' (1748). Sold Guillaume Dheulland's Carte topographique des environs de la ville de Maastricht, 1748.

Bourguignon, Hubert-François. See **Gravelot**.

Bourguignon d'Anville, Jean-Baptiste. See **Anville**, Jean Baptiste Bourguignon d'.

Bourjolly, —. *Carte politique de St Domingue*, 1803.

Bourjot, A. *Carte de la Géogénie du Sahel d'Alger*, *c*.1887.

Bourne, A. *Map of the State of Ohio from Actual Survey*, 1814 (with B. Hough, engraved by H.S. Tanner); *Map of the State of Ohio...in 1820*, Connecticut, E. Windsor 1820 (engraved by A. Reed).

Bourne, A. Engraver. *Street Map of Reading*, 1889.

Bourne, Ebenezer (*fl*.1797-1823). Engraver of City Road, Islington, London. Text engraver for the Ordnance Survey. Engraver for *8th Chart of the Coast of France...1st Part* (plate 13 in *Le Petit Neptune Français)*, W. Faden 1793; for Robert Wilkinson, 1797, 1800; John Fairburn's *...Country Twelve miles round London*, 1798; *The World on Mercator's Projection*, 1805; North America, 1823.

Bourne, Nicolas (*fl*.1601-1657). Publisher and bookseller of South Entrance Royal Exchange, Cornhill, London. Drake's *The World Encompassed*, 1628.

Bourne, Richard. Surveyor. The Manor of Great Linford [Buckinghamshire], 1678.

Bourne, Simon A.G. Mexico and Guatemala, Sidney Hall 1823.

Bournial, — de. 'Ingénieur des Ponts et Chaussées'. Assisted Dupain-Triel in the updating of his map *Navigation intérieure du royaume de France...*, 1789 (first published 1781).

Bourquin, Frederick & Co. Lithographer and printer of 602 Chestnut Street, Philadelphia (1864); '29 & 31 So. 6th St. Philadelphia' (1873). Printed many maps and county atlases for American publishers including C.O. Titus. Also worked with engravers Worley & Bracher. J.L. Hazzard's *Map of Lake Superior with its Railroad & Steamboat Connection...*, Charles Desilver 1857; S. Geil's *Map of the State of Michigan*, 1864; People's map of Michigan, 1864; *Plan of ...Land in ...the City of Auburn, New York*, *c*.1866; D.J. Lake's *Atlas of Allegan County Michigan*, Philadelphia, C.O. Titus 1873; and many other U.S. county maps and atlases.

Bourset, —. Worked with Foncet on a map of France and Savoie (14 sheets), Paris 1760. **N.B.** Compare with Pierre Joseph **Bourcet**, above.

Bourset, M. *A Plan of the Town of the Cape of Good Hope*, 1770, William Faden 1795.

Bourtruche, A. French cartographer. Atlas chronologique, Paris 1837.

Bourz von Seethal, Ferdinand (*d.*1827). Surveyor.

Bousquel, Auguste. *Plan de la Ville de Castres*, 1867.

Bousquet, J.G. *Atlas général des houillères* [Europe & USA], *c.*1909-1911 (with E. Gruner).

Bousquez, J. *Carte de l'Italie Centrale et des États Pontificaux*, *c.*1868.

Boussard, L. Engraver. Maps for Francis de la Porte, comte de Castelnau, 1853-1854.

Boutats, Gaspar. See **Bouttats**, Gaspar.

Boutflower, *Captain* Thomas, *RN* (*fl.*1752-1775). Series of manuscript maps of the Falkland Islands, 1768. **Ref.** BARBER, Peter *The Map Collector* 20 (1982) pp.50-51.

Bouthillier de Beaumont, Henry (1819-1898). French mathematician and geographer. *Carte des possessions Anglaises et Françaises sur le Golfe de Guinée*, 1884.

Boutigny, H. Surveyor and architect. Rouen, 1817 (with J. Heliot).

Boutin, F. Maps of Colombia, 1825.

Boutin, Vincent (1772-1815). French traveller and secret agent, assisted at French colonisation of Algeria, worked also in Egypt, Syria and Lebanon. Plan of Algiers, in: *Aperçu historique, statistique et topographique sur l'état d'Alger*, 1810; *Piano di Algeri e Contorni...1808*, 1830.

Boutourlin, *Colonel* —. Atlas de la campagne en Russie en 1812, Paris 1824; Théâtre de la guerre dans le Caucase, la Perse et la Turquie, 1846.

Boutrache, A. Atlas chronologique et synchronique de l'histoire universel, 1844.

Boutrois, -- Engraver of paris. Chantreau's *Carte de L'Irlande Pour le Voyages dans les trois Royaumes*, Paris, Briand 1792.

Boutroux, E. French naval hydrographer. *Plan du Golfe de la Spezia Côtes d'Italie...*, 1846 (with others), published 1849; worked with M.C.A. Vencendon-Dumoulin and others on charts of the Strait of Gibralter, 1854-1855, published 1859.

Bouttats [Boottats; Bottats; Boutats; Bouttate], Gaspar (*b.*1634). Engraver and artist in Vienna. *Lutzenburg*, Michael Thurnmayer 1673; Ostend, 1676; Esquemeling's *Carte de l'Isthme de Panama*, 1687 published in his *Histoire des Avanturiers*, 1688; Atlas of Hungary and Dalmatia, 1690; Nijmegen, 1718.

Bouttats, Philibert, father and son.
• **Bouttats**, Philibert, *the elder* (1635-*c.*1707). Engraver and draughtsman, of Antwerp, later in Amsterdam and The Hague. T. van der Heyden's *Ein gedeelte van de Heerlyckheyt van Maerseveen*, *c.*1690.
•• **Bouttats**, Philibert, *the younger* (1663-*c.*1717). Son of Philibert above. Engraver and draughtsman of Antwerp..

Bouvé, E.W. Lithographer and publisher of Boston. Nantucket by S.A. Mitchell, 1834; *Map of the Alton & Springfield Railroad*, *c.*1860.

Bouvet, L. *Atlas de la Confederacion de Argentina*, 1873.

Bouvet de Lozier, J.B.Ch. (1705-1786). French circumnavigator. Discovered Bouvet's Island, 1739. Carte de l'hémisphère méridionale, Guillaume Delisle *c.*1725 and later.

Bouville, J.B. *Chemin de Fer Indo-Européen*, 1881.

Bovinet, Giraldon. Engraver for L. Vivien's *Carte générale de l'Afrique Méridionale...*, 1826.

Bowden, James. *A Map of the Meetings of Friends in England and Wales*, *c.*1850, *c.*1855.

Bowden, Robert. *Plan of the ship canal and harbour... River Yarra Yarra* [Victoria, Australia], 1860.

Bowden, Thomas. *A map of the Colony of New Zealand*, 1870 (wall map); ...*Oceania*, 1874; *New Zealand*, *c.*1879.

Bowdich, Thomas Edward. *Esquisse de l'intérieur de Mozambique et de Benguela*, 1822.

Bowditch, father and son.
• **Bowditch**, Nathaniel (1773-1838). American mathematician, astronomer and hydrographer, born at Salem, Massachusetts. *New American Practical Navigator*, 1799 (40 editions up to 1867); *Chart of the Harbours of Salem, Marblehead, Beverly and Manchester...*, 1804-1806, 1834.
•• **Bowditch**, Jonathan Ingersoll. Son of Nathaniel, above.

Bowdoin, James (1752-1811). *Plan of Halifax Harbour*, 1780 MS.

Bowen, father and son. Engravers of London.
Ref. HODSON, D. *County atlases of the British Isles published after 1703* Vol.III (1997) pp.61-75; *ibid.* Vol.II (1989) pp.97-147; HARLEY, J.B. and HODSON, D. introduction to facsimile atlas *The Royal English Atlas 18th Century County Maps of England and Wales by Bowen E. and Kitchin T.*, (Newton Abbot 1971); HARLEY, J.B. introduction to facsimile atlas *Britannia Depicta or Ogilby Improv'd by Emanuel Bowen* (Newcastle upon Tyne 1970).
• **Bowen**, Emanuel (*c.*1693/1694-1767). One of the leading 18th-century map and printsellers and engravers in London, 'Next ye King of Spain in St. Katherines' (1720); 'the corner of Bolt-Court near the Leg-Tavern in Fleet Street' (1722); 'Near ye Stairs in St. Katherines'; 'Opposite to the Tunn Inn in Fleet Street'. Appointed Engraver of Maps to George II of England and possibly Louis XV of France. His apprentices included Thomas Kitchin (who necame his son-in-law) and Thomas Jefferys. Collaborated with other leading mapmakers of his time, including the Bowles family and John Owen. Asia, 1714; World map, 1717; *Britannia depicta or Ogilby Improv'd*, 1720 (with John Owen) and many later editions; map of the city of Cuzco [Peru], 1720; *A New and Accurate Map of South Wales...* (6 sheets), 1729; *A New and Accurate Map of England and Wales*, 1734; *A New and Accurate Map of America...*, 1744; *A Complete System of Geography* (70 maps), 1744-1747 (maps later used in *A Complete Atlas*, 1752); *A new and accurate map of Louisiana, with part of Florida & Canada*, 1747; maps for *The Universal Magazine of Knowledge and Pleasure*, John Hinton 1747-1766; *A complete Atlas...*, 1747, 1752; *A New Map of Georgia, with part of Carolina, Florida and Louisiana*, 1748; with Thomas Kitchin prepared the large scale county maps commissioned by John Hinton, published as individual maps from 1749 (the plates were sold to John Tinney *c.*1762, who was joined in partnership in the venture *c.*1755 by Thomas Bowles, John Bowles & Son and Robert Sayer, the set of maps was completed *c.*1760, and published as *The Large English Atlas...*); *A Map of the British American Plantations*, 1754 (engraved by T. Bowen); maps in Benjamin Martin's *The Natural History of England*, 1755 and later editions; *Atlas Minimus* (52 pocket maps), 1758; *Royal English Atlas*, 1762 (with Thomas Kitchin) and later; *North America describing... the British, Spanish and French Dominions* (4 sheet wall map), 1763 and many later editions; *Atlas Anglicanus*, T. Kitchin 1767, 1777 (completed by Thomas Bowen), later republished by Carington Bowles as *Bowles's New Medium English Atlas*, 1785; and many others.
•• **Bowen**, Thomas (*c.*1732/1733-1790). Son of Emanuel (above), engraver of 9, Charterhouse Lane, Clerkenwell, died in Clerkenwell workhouse. Engraver for his father's ...*American Plantations*, 1749; *A map of the British and French Settlements in North America*, 1755 and later; *A Correct Map of the Ottoman Empire*, *c.*1758; *Nth America from the Best Authorities*, *c.*1760; *Atlas Anglicanus*, 1767; Philip Morant's Essex, 1768; *A New and Correct map of the countries twenty miles round London*, 1770; maps for the *Gentleman's Magazine*, 1773-1779; *A Plan of the Black River on the Mosquito Shore...*, 1773; *Taylor & Skinner's Survey and Maps of the Roads of North Britain or Scotland*, 1776; Charles Middleton's *A New and Complete System of Geography*, 1777-1779; for J. Carver's *The New Universal Traveller*, 1779; for Alexander Hogg, 1784; *A New and Complete chart of the World...*, in *Bankes's New System of Geography*, 1790.

Bowen, brothers.
• **Bowen**, Abel (1790-1850). Wood and copper engraver, printer and publisher of Boston, Massachusetts. *New Hampshire*, 1823; *History of Boston*, 1825 (with maps); *Boston and Environs*, 1828.
• **Bowen**, Henry. Printer, brother of Abel.

Bowen, *Captains*. Pilots on the Mississippi River. Together updated and amended James T. Lloyd's map of the lower Mississippi River (5 sheets), New York 1862.
• **Bowen**, *Captain* Bartholomew.
• **Bowen**, *Captain* William.

Bowen, J.M. American publisher. Map of the United States, 1827 (from surveys by Amos Lay).

Bowen, John S. *Map of Chester County, Pennsylvania*, 1847 (with Samuel M. Painter).

Bowen, John T. [Bowen & Co.; J.T. Bowen & Co.] (1801-1856). Map publisher of New York and Philadelphia. Noted at 56 Liberty St. New York (1834-1835); 118 Nassau St. New York (1835-1836); 59 Cedar St. New York (1836-1838); 94 Walnut St. Philadelphia (1838-*c*.1844). *Map showing the lands assigned to emigrant Indians west of Arkansas & Missouri*, [n.d.]; *Plan of the Settlement at Prairie des Chiens*, 1820; *Map of the Route Passed over by an Expedition into the Indian Country in 1821 to the Source of the Mississippi*, [n.d.]; Plan of Detroit, 1830; R.C. Taylor's mining maps of Pennsylvania, *c*.1839; J.A. Sheaff's *Map of the projected railway from Harrisburg to Pittsburg*, 1840; Charles Cramer's *Map exhibiting that portion of the State of Pennsylvania...from Harrisburg to Pittsburg...*, *c*.1840.
• **Bowen & Co.** Name sometimes used by J.T. Bowen & Co. when at 94 Walnut Street, Philadelphia. A.C. Moreton's *Map of the European and North American Railway...*, 1850; map of Kansas and Nebraska, 1865; *Extract of military map of northeast Virginia...*, 1865; *Region embraced in the operations of the armies against Richmond and Petersburg*, 1865 (both prepared by the U.S. War Department, Corps of Engineers); State of Alabama, 1866; *Map showing the line of Boundary between the United States & British Possessions...*, 1871.

Bowen, Maurice. Geographer and engraver. Map of China, for R. Sayer and J. Bennett *c*.1780; *A Topographical Survey of the Country from Thirty five to Forty Miles Round London*, Sayer & Bennett, London 1775, 1787.

Bowen, N. American cartographer, worked on maps of the American civil war, 1862.

Bowen, Thomas. *An Accurate Map of the Holy Land with the Adjacent Countries*, Philadelphia, Berriman 1792 and later editions (engraved by Alexander Anderson).

Bowen & Son. Booksellers in Barbados. Edward Weller's Topographical Map of the Island of Barbados (after William Mayo 1721), 1870 (only one copy known).

Bower, A. *Plan of... Kingston upon Hull*, 1787, 1791; *A Plan of the River Witham*, 1803.

Bower, *Captain* H. Tibet and Western China, 1892-1893.

Bower, John (*fl*.1810-1819). Engraver of Philadelphia. Maps in [Mathew] *Carey's American Pocket Atlas*, 1814.

Bower, Robert A. New Handy Atlas of the United States and Canada, 1884.

Bower, Thomas Bowyer. ...*Southern Transvaal and Northern Orange River*, *c*.1900.

Bower, William. Engraver of Boston, Lincolnshire. S. Bee's *A Survey and Plan of the Road... from Lincoln... to Peterborough..*, 1756.

Bowes, Joseph. American engraver. *Plan of the City of Philadelphia*, in Cornelius W. Stafford's *The Philadelphia Directory for 1797*.

Bowes, William. Believed to have produced the earliest pack of geographical playing cards showing English county maps. The untitled set of cards was known as the 'W.B.' cards until a second later set was attributed to Bowes. 60 cards with 52 county maps divided into 4 regional suits, and eight text cards, 1590, *c*.1605. **Ref.** BERESINER, Y. *British county maps* (1984) pp.251-252; SKELTON, R.A. *County atlases of the British Isles 1579-1703* (1970) pp.16-18.

Bowler, L.P. Northern Goldfields Matabele & Mashonaland, Pretoria 1889.

Bowler, Thomas William (1812-1869). Bird's-eye view of the breakwater and dock works at Table Bay, in *...guide...to the Cape of Good Hope...*, 1869 (drawn on stone, lithographed by Saul Solomon & Co.).

Bowles family. Map publishers and sellers in London. Collaborated with others on atlas publication, including Emanuel Bowen's *Britannia Depicta*, 1720 onwards; and with Herman Moll, many of whose atlas plates were bought by the Bowles family. **Ref.** HODSON, D. *County atlases of the British Isles published after 1703* Vol.III (1997) pp.72-75; *ibid.* Vol II (1989) pp.97-147 *et passim*; *ibid.* Vol I (1984) pp. 186-191; TYACKE, S. *London map sellers* (1978) pp. 111-112; HOWGEGO, J. *Printed maps of London c.1553-1850* (1978) *passim*; MAXTED, I. *The London book trades, 1775-1800* (Folkestone 1977) pp.25-26; PLOMER, H.R. *Dictionary of booksellers & printers 1726-1775* (Oxford 1932) pp. 31-32.

• **Bowles**, Thomas I (*fl.*1683-*c.*1714, *d.*1721). Father of Thomas II and John. Member of the Joiners' Company of London, map and printseller in St. Paul's Churchyard, London.

•• **Bowles**, Thomas II (before 1695-1767). Engraver, publisher and bookseller, son of Thomas I, to whom he was apprenticed; took over his father's business *c.*1714, trading from 'next the Chapter House in St. Paul's church yard' (*c.*1714-*c.*1762). After 1724 many works were published in co-operation with his brother John, and in 1730 they jointly purchased stock from Anne Lea, they also bought Herman Moll's copperplates after his death in 1732. Thomas retired in 1763, and was succeeded by his nephew Carington Bowles. *Oxford Shire Actualy Survey'd &c.*, 1715 (with Philip Overton; *The Travellers Guide...South Britain*, 1715; Bowen's *Britannia Depicta*, 1720 and later; republished Francis Lamb's *A Geographical Description of the Kingdom of Ireland*, 1720 edition (later published by John Bowles); one of the booksellers for Herman Moll's *A New Description of England and Wales...*, 1724, 1733; bookseller for Herman Moll's *A Set of Fifty New and Correct Maps of England and Wales*, 1724, 1739 (this used the maps from *A New Description...*, and published them as an atlas without text), other editions were published as *The Geography of England and Wales...*, 1747 and *Moll's British Atlas...*, 1753 (all with John Bowles, or John Bowles & Son); *A Set of Thirty Six New and Correct Maps of Scotland...*, 1725 (with John Bowles, republished as *Scotland Delineated...*, 1745 and later); with Philip Overton published a map of *The Counties of Essex Middlesex & Hertfordshire* (2 sheets), 1726; *A New Map of the County of York* (2 sheets), 1728 (with Philip Overton); Moll's *Atlas Minor*, *c.*1732 (with J. Bowles); large scale county maps by Bowen & Kitchin (originally commissioned by John Hinton in 1749), published as *The Large English Atlas*, 1760 and later (with J. Tinney, John Bowles and Son, and Robert Sayer), *c.*1762 (with John Bowles and Son, and Robert Sayer), *c.*1763 (with John Bowles and Robert Sayer); and many others.

••• **Bowles**, Thomas III (*c.*1712-1762). Son of Thomas Bowles II, worked with, but predeceased his father.

•• **Bowles**, John (*c.*1701-1779). Younger son of Thomas I, apprenticed to his elder brother Thomas II in the Joiners' Company. Set up on his own *c.*1723 as map and printseller, publisher 'over-against Stocks Market', later trading from a variety of addresses in and around Cheapside and Cornhill in London. In addition, he frequently co-operated in publication with his brother. His son Carington Bowles joined him in partnership, 1752-1763; on John's death the business was taken over by Robert Wilkinson. *Britannia Depicta*, 1724 edition (with Thomas Bowles); republished Francis Lamb's *A Geographical Description of the Kingdom of Ireland*, 1728, 1732 editions; An actual Survey of all the Principal Roads of England and Wales, 1757 and later; The Roads through England delineated, 1757 and later; Bowen & Kitchin's *The Large English Atlas*, *c.*1763 edition (with T. Bowles and R. Sayer); Emanuel Bowen and Thomas Kitchin's *The Royal English Atlas*, *c.*1764 (with Carington Bowles and others); *A Traveller's Companion...*, *c.*1766 (re-using Moll's English county maps); *Kitchin's Post-Chaise Companion, through England and Wales*, 1767 (with Carington Bowles and Robert Sayer); and many others.

••• **Bowles**, John & Son. Company formed when Carington joined his father in business *c.*1752, and lasted until Carington bought the stock of his uncle, Thomas II *c.*1763. Bowen & Kitchin's *The Large English Atlas*, 1760 (with T. Bowles, R. Sayer and J. Tinney), *c.*1762 (with T. Bowles and R. Sayer).

••• **Bowles**, Carington (1724-1793).

Publisher at 'N°·69 in St. Pauls Church Yard', London. Son of John Bowles and nephew of Thomas II, like them a member of the Joiners' Company of London. He worked in partnership with his father from *c.*1752 to 1763, then took over his uncle's business. Continued Bowles family publications until succeeded by son Henry. *Britannia depicta*, 1764 and later; *The Large English Atlas*, *c.*1764 (with R. Sayer), 1767; Emanuel Bowen and Thomas Kitchin's *The Royal English Atlas*, *c.*1764 (with John Bowles and others); *Ellis' English Atlas*, 1766 (with Robert Sayer); *Kitchin's Post-Chaise Companion, through England and Wales*, 1767, 1770 and later (with John Bowles and Robert Sayer, maps by John Senex); *Post-Chaise Companion*, 1782 (with maps based on those of John Senex); *New Pocket Map of the United States of America*, 1783 and later; *Pocket Atlas of...England and Wales*, 1785; republished Emanuel and Thomas Bowen's *Atlas Anglicanus* (first published 1767-1768) as *Bowles' New Medium English Atlas*, 1785; *Bowles' Pocket Atlas of the Counties of South Britain*, 1785.

•••• **Bowles**, Henry Carington (1763-1830). Took over from his father Carington, and worked in partnership with Samuel Carver as the firm of Bowles & Carver.

•••• **Bowles & Carver** (*fl.*1793-1832). Successor to Carington Bowles, 1793. Publishers in London, '69 St. Pauls Churchyard'. Mainly printed from inherited copperplates. *Post-Chaise Companion*, 1793(?) and later; *The Large English Atlas*, *c.*1794; *The Royal English Atlas*, 1794 and later; *New Medium English Atlas*, *c.*1795; *New One-Sheet Map of America*, *c.*1796 and later.

Bowles, Charles. *Map of Europe*, 1871-*c.*1872; *Map of the World*, 1871.

Bowles, John. English estate surveyor, produced various manuscript surveys including Huntsmoor, Thurney [Iver, Buckinghamshire], 1736 MS.

Bowman, Alexander. *The City of Urbana*, *c.*1855; *Map of Champaign County, Illinois*, Chicago 1863; *Map of Vermilion County, Illinois*, Danville 1867.

Bowman, Amos. Dawson's Explorations in British Columbia, 1888; *Cariboo Mining District*, 1889; *Geological and Natural History Survey of Canada* (9 sheets), 1895.

Bowra, John (1714-1780). Surveyor of Groombridge, Kent. *A Survey of Tunbridge Wells*, 1738; *An exact Survey of the River Medway*, 1739; Estate plan of *Seven Oakes Park, Sevenoaks*, 1780; and many others.

Bowrey, Thomas. Chart of Ceylon, 1681 MS.

Bowyer, William. British Isles, 1567 MS.

Boxhorn, Marcus Zuerius. Leiden. *Theatrum sive Hollandiae comitatus et urbium nova descriptio*, Amsterdam, H. Hondius 1632.

Boxtart, Hinrich zum. Gräntze zwischen Ostfrieslandt Oldenburg und Munsterland, 1630.

Boycote, Thomas (*fl.*1652-1659). Estate surveyor of Kent, possibly son of William, below. Estate maps in Kent and Sussex.

Boycote, William (*fl.*1615-1648). Estate surveyor of Kent, drew manuscript surveys of various parts of England and Wales including *Buckland, Kent*, 1623; *Marden*, 1636; *Manor of Sturry, Kent*, 1643.

Boyd, —. See **Oliver** & Boyd.

Boyd, Andrew. *Map of the city of Washington*, Boyd & Waite 1866.

Boyd, Charles. Surveyor. *Chester District, South Carolina*, 1818; for Robert Mills, 1825.

Boyd, Hugh. *A Plan of Ballycastle Harbour*, 1743.

Boyd, J. Engraver for Mathew Carey's *American Atlas*, 1822 (maps of Maryland; Ohio; Windward Islands).

Boyd, W.H. *Boyd's map of the city of Washington and suburbs*, Washington 1884.

Boydell, John & Josiah. Engravers and publishers 'at the Shakespeare Gallery & at N°·90, Cheapside', London (1798). *A South Prospect of the City of Chester*, 1749; *Plan of the Town and Harbour of Mahon*, *c.*1750; *A Map of the Basque Road*, *c.*1758; John Hill's *Plan of... Philadelphia*, 1796.

Böyë, Herman. German engineer living in Virginia. *A Map of the State of Virginia...* (9 sheets), H.S. Tanner 1826, reduced edition 1827.

Boyens, C.J.R. *Plan communal de la Ville de Liège*, *c.*1820.

Boyer, Abel (1667-1729). *The Draughts of the most remarkable Fortified Towns of Europe* (44 plates), 1701; *A Map and Plan of the Town of Preston*, *c.*1715.

Boyer, M. French Marine. Plan du Havre Peraki [New Zealand], 1840.

Boyer-Peyreleau, *Colonel* E.E. Large folding map of French West Indies in Les Antilles françaises, 1825.

Boykin, J. Two maps for Robert Mills's *Atlas of the state of South Carolina*, 1825.

Boylan, J.D. *The North Coast of the Gulf of Mexico*, 1846, 1864.

Boyle, Augustin J. Draughtsman at the Hydrographic Office, London. Admiralty charts, 1860-1870. *La Poile Bay*, [Newfoundland], 1864, 1886; *Newfoundland - North West Coast*, 1869, 1883; *Table Bay Breakwater & Docks*, 1870, 1880; *The Bosporus*, 1871.

Boyle, Charles B. Artist of New York. Built a lunar globe, *c.*1869.

Boyle, G.E. *Southern suburbs of Cape Town*, 1885.

Boyle, George. Publisher of '290 Regent Street Langham Place', London. *Boyles Thames Guide* (8 sheets), 1839 (printed by J. Bradley).

Boyle, John Roberts. *Actual Survey of Newcastle-upon-Tyne*, 1723.

Boym [Boim], Michal Piotr, *SJ* (1612-1659). Polish Jesuit in China. Atlas of China, 1653-1655 MS; *Chung kuo t'u* [map of the Middle Kingdom], *c.*1653; Mappa Imperii Sinarum, *c.*1670. **Ref.** SZCZESNIAK, B. *Imago Mundi* 19 (1965) pp.113-115.

Boynton, George W. American engraver. Worked for S.G. Goodrich, 1830, 1835; for J.P. Edwards, 1832; for Thomas Bradford, 1834; worked on Thomas Bradford's *An Illustrated Atlas, Geographical, Statistical and Historical of the United States and Adjacent Countries*, 1838; T.F. Gordon's New Jersey, 1838; *City of Lowell, Massachusetts*, 1845; *Connecticut*, *c.*1845; *City of Boston*, 1846; *Jerusalem and its environs*, *c.*1855.

Boys, John. *A map of the soil of Kent*, 1805.

Bozzoli, fratelli. Publishers of Brescia, together they republished de Marchi's *Della architettura militare del Capitano Francesco de' Marchi*, Brescia 1609 (first published by Dall'Oglio 1599).

• **Bozzola**, Gio.Battista. Agostino Gallo's *Le Dieci Giornate Della Vera Agricoltura, e Piaceri Della Villa*, 1564.

• **Bozzola**, Antonio.

Braakensiek, Albert (1811-1883). Dutch cartographer, engraver, printer and publisher. Scandinavia, 1835; *School-Atlas van alle deelen der aarde*, Amsterdam 1853; *Atlas van het Koningrijk der Nederlanden*, *c.*1857, 1869; *Het Noordzeekanaal*, *c.*1872; *Kaart der Noordelijke Isjzee met reisroute...*, Amsterdam 1879 (given away to readers of *Het Nieuws van den Dag*).

Braakman, Adriaan (1664-1720). Dutch publisher and cartographer of Amsterdam, worked with P. Schenk. *Atlas minor...* (117 maps by different cartographers), Amsterdam 1706; *Schouw-Tonneel des Oorlogs*, 1706; *Le théâtre de mars...*, Amsterdam 1706 (with P. Schenck); *Le Théâtre de Bellone*, Amsterdam 1706-1708 (with P. Schenk): *Suite du Theatre de Mars en Europe*, 1710.

Braam, Jan [Johannes] van [de] (*fl.*1702-1744). Bookseller of Dordrecht. With Gerard onder de Linden (of Amsterdam) published François Valentijn's *Oud en Nieuw Oost-Indiën* (8 volumes), 1724-1726; *Kaart van het Eyland Borneo*, 1750; *Kaart van het Eyland Mauritius*, 1750; *Kaart van het Eyland Formosa*, *c.*1760; *Tabula Indiae Orientalis*, *c.*1770.

Braams [Braems], A. Dutch cartographer, possibly also surveyor. Worked in Zeeland. Generaele beschrijvinge ofte Caerte des Marquisaets van Bergen op Zoom..., 1575 (copy made by C. de Bije 1719).

Brabazon, *Captain* Luke. *Sebastopol. Plan & section of the Redan*, 1856; Route to China, 1860.

Brabner, J.H.F. Comprehensive Gazetteer of England and Wales, 1894.

Brabo [Bravo], Francisco Xavier [Javier]. Cartas geograficas de la America Meridional, Madrid 1872.

Brachel, P. von (*d*.1650). Publisher of Cologne. Published a Latin edition of Braun & Hogenberg's *Civitates Orbis Terrarum*, *c*.1606-1618.

Brachelli, *Professor* H.F. *Neuer Atlas der ganzen Erde*, 1863 (with others). **N.B.** This is possibly C.G.D. Stein's *Neuer Atlas*, which dates back to 1819.

Brackebusch, *Doctor* Ludwig (*b*.1849). Argentine geographer. Maps of the Republic, 1881, 1885, 1891.

Brackell, T. *Map of Lancashire*, 1883.

Brackenberg, —. Karte von Sachsen (10 sheets), *c*.1740.

Brackenridge, C.S. *Map of... Fort Wayne* [Indiana], 1871.

Braconnier, M.A. *Carte Géologique...du Département de Meurthe-et-Moselle*, 1882.

Bradbury, John. Map of the United States, in his Travels, 1819.

Braddock, David Cutler. *The Coast of Florida*, 1756 MS.

Braddock, *General* Edward. Leader of campaigns in Pennsylvania and Maryland, 1755. Associated plans and maps were used by Thomas Jefferys in *A General Topography of North America and the West Indies*, London 1768.

Braddock, J.H. *South Australia. Plan of the County of Eyre*, 1860.

Braden & Burford. Lithographers of Indianapolis. *Map of the Great Saginaw Route*, 1871.

Bradford, H. Plans of the Rivers Trent and Tame [England], 1759.

Bradford, J.L. Surveyor. *Olathe, Kansas*, *c*.1860.

Bradford, J.R. Post & Telegraph map of Queensland, Brisbane, 1892.

Bradford, L.H. [& Co.]. Lithographer of Boston. *Map of Part of the State of Michigan Showing the Pine Lands of the St. Mary's Ship Canal Company*, 1858

Bradford, Samuel (1725-1751). Warwickshire surveyor. *A Plan of... Coventry*, 1750; *A Plan of Birmingham*, 1751.

Bradford, Thomas Gamaliel (1802-1887). Publisher of Boston and New York. *Atlas designed to illustrate the abridgment of universal geography, modern & ancient...*, Boston, W.D. Ticknor 1835 (also published in New York by F. Hunt & Co.); *A Comprehensive Atlas, Geographical, Historical and Commercial*, Philadelphia 1835, Boston, W.D. Ticknor 1835; *An Illustrated Atlas; Geographical, Statistical and Historical of the United States and Adjacent Countries*, Boston, Weeks, Jordan & Company 1838, 1842; *A Universal Illustrated Atlas...*, Boston, Charles D. Strong 1842 (with Samuel Griswold Goodrich).

Bradford, William (1663-1752). New York printer and publisher, said to have set up New York's first printing press in 1693, and published its first newspaper, the *New York Gazette*, in 1725. James Lyne's *A Plan of the City of New York...*, 1730 (said to be the first copperplate plan of New York); *A New Map of the Harbour of New York*, *c*.1735. **Ref.** COHEN, P.E. & AUGUSTYN, R.T. *Manhattan in maps 1527-1995* (1997) pp.54-56.

Bradford & Inskeep. Publishers of Philadelphia. Nicholas Biddle's *History of the expedition under the command of Captains Lewis and Clark*, 1814.

Bradley, Abraham *junior*. United States Post Office Department, Philadelphia. *Map of the United States, Exhibiting the Post-Roads and Distances*, 1796, 1819; *Northern Part of the USA*, *c*.1797; *Southern Part of the USA*, *c.*1797; United States, 1804; Indiana, published John Melish 1817; *Map of the United States Intending Chiefly to Exhibit the Post Roads & Distances...*, Georgetown 1819 (engraved and printed by W. Harrison).

Bradley, J[ohn?]. *Bradley's Universal Atlas for the Use of Schools*, Liverpool 1818; *New General Atlas*, 1820.

Bradley, J.H. *Plan of the Isthmus of Tehuantepec*, 1851.

Bradley, T. Engraver. Naples, 1835.

Bradley, William. *A Chart of Norfolk Island*, 1794.

Bradley, William M. [& Bros]. Publisher of Philadelphia. Later editions of *Mitchell's New General Atlas*, 1880, 1882, 1884 (first published *c.*1865); *County and Township Map of Utah and Nevada*, 1884; *Bradley's Atlas of the World*, 1886, 1893; *Atlas of the Western Hemisphere*, 1887; *Pocket Atlas of the United States*, 1888.

Bradshaw, George (1801-1853). Engraver, draughtsman and publisher, of St Mary's Gate and Market Street, Manchester. Produced many maps in addition to his railway guides. *G. Bradshaw's Map of Canals situated in Lancashire, Yorkshire, Derbyshire and Cheshire*, *c.*1830; *G. Bradshaw's Map of Canals, navigable rivers, rail roads &c. in the Midland counties of England...*, 1830; *G. Bradshaw's Map of canals...in the southern counties of England...*, 1832; Railway Companion, 1844; *Bradshaw's Map of the Rhine...*, Manchester and London, Bradshaw & Blacklock 1848; *Bradshaw's Plan of Paris and Map of the Environs*, Manchester, Bradshaw & Blacklock 1850, 1860; *Bradshaw's map of Great Britain showing the railways*, *c.*1857 from *Bradshaw's Shareholders' Guide. Railway manual*; *Bradshaw's Indian Guide*, Manchester, Bradshaw & Blacklock 1860; *Bradshaw's Continental Railway Guide*, 1861 and many later editions.
• **Bradshaw & Blacklock** (*fl.c.*1848-1860). Publishers of Manchester and London. Published many of the *Bradshaw* series of guides.

Bradshaw, R., *Mate*, HMS Acheron. Survey of South Island, New Zealand, 1884.

Bradshaw, Richard. *Geological map Terre Haute and adjacent coalfields* [Indiana], 1875.

Brady J.H. *Shilling Atlas for South Africa*, 1892.

Brady, P. Trigonometrical survey of Algoa Bay, South Africa, *c.*1820 MS.

Braems, A. See **Braams**, A.

Bragg, B. *The Island of Jamaica*, 1705.

Brahc, Ernest, W. *Atlas of the City of Quincy*, Massachusetts, 1907, 1923.

Brahe, Tycho [Tyge] (1546-1601). Danish astronomer and teacher of mathematics (Willem Jansz. Blaeu was a pupil), built large observatory on Hven in Öresund between Denmark and Sweden, 1576. Used the principal of triangulation to establish the position of Van Island, Denmark, 1578-1579. When exiled in 1596 he went to Prague, where from 1599 he worked with Johann Kepler. Pair of globes, 1584; *Astronomiae instauratae mechanica*, Nuremberg, Hulsius 1602; Sphaera stellifera, 1606.

Brahm, See **de Brahm**

Brambilla, Ambrosio. Engraver and cartographer of Rome. *Genova*, 1581; *Antiquae Urbis* [Rome], 1582, 1590; plan of Pozzuoli, Apud Haeredes Claudij Ducheti 1586; Augsburg, 1602.

Brambilla, Valerio. North Italy (6 sheets), 1840; *Carta degli Stati di Sua Maestá Sarda in Terra Firma*, 1841; *I Contorni di Torino*, 1854.

Bramham, James. Engineer. Plan of St. John's Harbour in Newfoundland, 1751 MS (with Edmond Scott Hylton).

Bramston, W. *...City and Harbour of Macao*, 1840; *A Plan of... Canton*, 1840.

Brancaccio, Giulio Cesare. Italian military engineer and surveyor of Naples, worked in the Netherlands from 1542. Della nuova disciplina & vera arte militare, Venice 1582.

Brancas-Villeneuve, *abbé* André (*d.*1758). French cosmographer. Système de cosmographie, The Hague 1747.

Tycho Brahe (1546-1601). (By courtesy of Rodney Shirley)

Branch, William [Harper] (1761-1799). Surveyor of London. *A Mapp...of the Mannors of Butters...* [Kent], 1786 MS.

Branco, Alves. *Plano hydrogràphico da Enseada do Quicembo*, 1888.

Branco, W. Carta geologica della Valle del Sacco, 1877.

Brandard, E.P. Steel engraver for *The Indexed Bible Atlas*, *c.*1878.

Brandenburg, Lorens Wilhelm (1794-1850). Swedish engineer, worked on the official survey entitled *Karta öfver Sverige och Norige*, 1843.

Brander-Dunbar, *Lieutenant* —. Maps of Somaliland, 1896-1897.

Brandes, H.H. (1798-1874). German geographer. Wien, 1863, 1873; geographical manuals of Europe, 1872, 1885.

Brandes, Heinrich Wilhelm (1777-1834). German physicist. *Carte synoptique du 6 mars 1783* [air pressure map of Europe], 1820 (based on meteorological records of a storm of 1783).

Brandis, G.B. à. See **Brender** à Brandis, Gerrit.

Brandis, Lucas. Printer of Lübeck. Woodcut map of Palestine in *Rudimentum novitiorum* (woodcut), Lübeck 1475 (the first printed map which is not merely a diagram); also credited with designing the mappamundi in the same volume. **Ref.** NEBENZAHL, K. *Maps of the Bible Lands* (1986) pp.60-62.

Brandreth, Timothy. London publisher 'over against ye Royal Exchange in Cornhill'. Joint ventures with other publishers, including Charles Price's *A Correct Map shewing all the Towns, Villages, Roads, the Seats of Ye Nobility and Gentry wth whatever else is remarkable within 30 Miles of London*, 1712.

Brandt, C. *Karte des Kreises Ruppin*, c.1893.

Brandt, F.F. *Plan voor den Veldslag van Waterloo*, c.1895.

Brandt, John. *Map...of Albany*, 1843.

Brandt, T.J. Dutch cartographer. Kaart van Zeeland, 1860.

Brandt, Wouter Goderts van den (*fl.c.*1599-1600 and later). Dutch military engineer and cartographer, worked in and around Breda.

Brannan, J. Publisher of Washington. F.C. de Krafft's *Map of the city of Washington*, 1828 (engraved W.I. Stone).

Brannon, George. Engraver and publisher. ...*Isle of Wight*, 1831.

Bransby, John (*c.*1762-1837). Surveyor, astronomer & globeseller. *A Map of the Liberties of Ipswich*, 1812; *Plan of the River Orwell*, 1814.

Bransfield, Edward. *Master, RN*. *Plan of St. George's Bay, South Shetland*, 1820.

Branston, C. Engraver. *Pullan's Pictorial Plan of Chester*, 1851.

Branth, *Lieutenant* T. *Kaart over Kjöbenhavn*, 1854.

Brantz, Lewis (1766-1838). Merchant and sea captain of Baltimore. ...*Survey of the River Patapsco and part of Chesapeake Bay*... (2 sheets), Baltimore, F. Lucas Jr 1819; *Plan of the City of Baltimore*, 1822.

Braouzec, Jules Edmond (*b.*1828). French explorer in Gabon. *Exploration dans le Fouta et le Danga*, 1864; *Rivière Manéah et monts du Soumbouyah*, 1867.

Brasch, B.M. *Plan der Stadt Neu-Ruppin*, 1789.

Brascuglia, F. Topographical map of Friuli, 1739.

Braselmann, J.E. *Bibel Atlas*, 1868.

Brasier [Brassier], William Furness (1689-1775). Draughtsman and surveyor of the Tower of London, served with the army in North America. *Land...in Paddington... belonging to Sr. John Frederick*, 1742; *View of English Harbour... Antigua*, 1756; *Plan of the Fortifications of Louisbourg*, 1758; Fort Ticonderoga, 1759 MS (with Bream); Plan of Oswego, 1759; *A Survey of Lake Champlain...*, 1762, published Sayer & Bennett 1776, Laurie & Whittle 1794 (this map appeared in most editions of Jeffery's *American Atlas*); *A Map of the Country from the Western Lakes to the Eastern parts...*, 1765; *The Rapids of the Ohio*, 1766 MS; *A Plan of New Orleans*, 1769 (from a drawing by Lieut. P. Pittman); ...*Harbour of Pensacola...*, 1771; and many other manuscript maps. **Ref.** GUTHORN, P.J. *British maps of the American Revolution* (1972) p.14.

Brassier, William. See **Brasier**, William Furness.

Brath, James. *The Balaklava Railway Map*, 1855.

Brathuhn O. *Specialkarte von den beiden Mansfelder Kreisen*, c.1864.

Brau [Brauws], Johan de (1559-1619). Dutch military cartographer, born in Geersbergen, worked in Axel, Ter Neuzen and Biervliet, died in The Hague.

Brauchitsch, — von. *Karte der Umgegend von Braunschweig*, *c.*1895.

Braud, E. *Carte du Canton de la Chataigneraie*, 1877.

Brauer, *Doctor* A. Die Seychellen, 1896.

Brault, Louis Dexie Léon (1839-1885). Meteorologist and marine cartographer.

Braun, Ernst Eberhard. Engineer. *Carte von der Gegende um der Stadt Hannover*, 1750-1757.

Braun, Friedrich. *Himmels Atlas in Transparenten Karten*, *c.*1855.

Braun [Bruin] Georg [Joris] (1541-1622). Topographer, geographer and publisher. Best known for *Civitates orbis terrarum* (6 volumes), 1572-1617, on which he worked with the engraver Frans Hogenberg. It was published in many editions in Latin, French and German at Cologne. The first atlas of town plans and views embracing the known world, it was often found in conjunction with Ortelius's *Theatrum orbis terrarum*. **Ref.** GOSS, J.J.S. *The city maps of Europe* (London 1991); POINTER, S. *The Map Collector* 4 (1978) pp.7-14; KOEMAN, C. *Atlantes Neerlandici* Vol.II (1969) pp.10-25; SKELTON, R.A., introduction to facsimile edition (Amsterdam 1966).

Braun, Hieronymus (*d.*1620). Plan of Nuremberg, 1620.

Braun, Jacob (*fl.*1712). Swedish surveyor, worked in Uppsala. *Wendel Prästgarden*, 1712 MS.

Braun, W. *Eisenbahn-und Strassen Atlas von Kurhessen*, *c.*1855.

Brauns, J.C. Dutch cartographer. Kaart van Drenthe, 1840.

Braunschweig, L. *Telegraphen Karte von Deutschland*, 1869.

Brauws, Johan de. See **Brau**, Johan de.

Brave, Heinrich. *Wien und Vororte*, 1874.

Bravo, Francisco Javier. See **Brabo**, Francisco Xavier.

Bray, I. Master of missionary vessel *Morning Star*. Corrected Admiralty Chart of Truk or Hogolu Is., 1880.

Bray, J. & A. *New Map of Hastings and St. Leonards*, *c.*1898.

Bray, Salomon. *Kaarte van... de Stadt Haerlem*, 1644.

Brayer, Lucien, *comte* de. *Carte de la République du Paraguay*, 1863.

Brayley, Edward Wedlake. With John Britton compiled the text for the partwork *The Beauties of England and Wales* (18 volumes), *c.*1804-1810, (a separate set of maps to accompany the work was drawn by G. Cole and engraved by J. Roper).

Brazier, Edward John Richard Butler. *Survey of the Coast of Sindh and Cutch*, 1853 (with Lieutenant Grieve).

Brdičzka, Leopold. *Mapa Království Ceského*, 1861; *Übersichtskarte... für das Königreich Böhmen*, 1861.

Breaks, Thomas (*fl.*1767-1790). Estate surveyor. *A Complete System of Land Surveying*, 1771.

Bream, *Lieutenant* —. Assisted William Brasier with a plan of the Fort at Ticonderoga, 1759.

Brecher, Adolph. *Darstellung... des Brandenburgisch-Preussischen Staates*, 1868-1869, later editions to 1880.

Bredda, Bartolomeo. *Agri Patavini chorographia*, 1650, 1687.

Bredsdorff, Jacob Hornemann (1790-1844). Danish cartographer, worked with Oluf Nikolay Olsen on the production of a hypsometric map of Europe. *Esquisse Orographique de l'Europe...*, 1824 MS, revised Olsen 1830, published Olsen 1833; later copied by Heinrich Berghaus in *Physikalischer Atlas*, 1842.

Bree, A. & J. de. Charts for *Pas-caart van Hispangien*, Amsterdam, Hendrick Doncker 1658, Pieter Goos, *c.*1673.

Breen [Brenius], Daniël van (*c.*1599-1665). Engraver from Middelburg, worked in Amsterdam 1630-1650. Beverwijk, 1648-1649; engraved Jan Pietersz. Dou's *Haarlemmermeer en omstreken c.*1636; also B.F. van Berckenrode's *Ware afbeeldinge vande bedyckte Beemster-landen inden jare M.D.C.XLIII* (6 sheets), 1644.

Breen, H.H. St. Lucia, in *Breen's St. Lucia Historical Statistical and Descriptive*, 1844.

Breen, James. *Stanford's map of the Paths of Comets*, 1858.

Breese, Samuel. American map compiler, worked with the wax engraver S.E. Morse. Compiled maps for *Morse's North American Atlas*, 1842; *The Cerographic Atlas of the United States*, New York, 1842-1845 (Harper & Brothers are given as publishers for 1845); large wall map of the United States and Canada, New York, Harper and Brothers 1847 (wax engraved by Sidney Morse); maps for *Morse's General Atlas of the World*, D. Appleton & Co. 1856.

Bregante y Martinez, Salvador (*fl.*1869-1914). Spanish letter engraver. Maps published by the Dirección de Hidrografía, Madrid 1883-1892.

Bréhant de Galinée, René de (1645-1678). *Carte du Canada et des terres découvertes vers le lac d'Erié*, 1669; *Carte du Lac Ontario et des habitations qui l'environnent...*, 1670, published Rouen 1895.

Breidenbach, Bernhard von. See **Breydenbach**, Bernhard von

Breislak, Scipione (1748-1826). *Topografia fiscia della Campania*, 1797; Plan physique de la ville de Rome, 1801.

Breitenbach, G.A.V. Welt nach... Völkerschaften, 1786.

Breitenstein, W. *Carte topographique de la Mairie de Stolberg*, 1811.

Breithaupt, Johann Christian (1736-1799). Instrument maker of Kassel.

Breitinger, *Lieutenant-Colonel* David (1763-1834). Swiss precision instrument maker of Zurich and London. *Plan der Stadt Zürich*, 1804, 1814.

Breitkopf, Johann Gottlob Immanuel (1719-1794). Publisher in Leipzig. Numerous geographical texts and works on surveying and engineering. *Das Reich der Liebe*, 1777.

Bremden, D. van (*fl.*1650s). Engraved title page for some editions of Joannes Janssonius's *Atlas major*, 1657.

Bremicker, E. *Karte von der Hauptstrassen im Deutschen Zollverein*, 1848.

Brémond, Laurent. 'Hydrographe du Roy et de la Ville...', 'sur le Port a Marseille au coin de Reboul'. Worked with Henri Michelot on *Recüeil de plusiers Plans des Ports et Rades de la Mer Mediterranée* (21 maps), Marseilles 1727 (engraved by P. Starck-man), later edition (38 maps), 1730, English edition, 1802.

Brendel, Lucas (*d.*1657). Map of the Thurnau area, 1639.

Brender à Brandis, Gerrit (1751-1802). *...Handelskundige Atlas*, 1788; *Nieuwe Natuur-Geschied- en Handelkundige Zak-en Reis-Atlas*, Amsterdam, N.T. Gravius *c.*1787 (also published by S.J. Baalde).

Brenier, Henri (1867-1962). French merchant and explorer in China and Indochina. *Géographie élémentaire de l'Indochine*, 1909; *Essai d'atlas statistique de l'Indochine française*, 1914.

Brenius, Daniël van. See **Breen**, Daniël van.

Brenna, Giovanni. Engineer and geographer. *Carta geograficae postale del Regno Lombardo-Veneto*, 1831, 1837; *Carta topografica dei contorni di Milano...*, 1841-1851; *Pianta di Milano*, 1858.

Brennecke, L. *Officielle Karte vom Nord-Ostsee-Kanal*, 1890.

Brenner, Johann Jakob (1712-1774). Swiss surveyor in Bern. Manuscript plans of the city of Bern, 1752-1769 (State Archives of Bern); surveys for J.B. Micheli du Crest, 1755; plan of Lake Thun, Basle 1771.

Brenner, Sebastian. *Continuator temporis quinquennalis...*, Frankfurt am Main, 1598-1599 (contains the decorative untitled World map engraved by Adam Elsheimer).

Brentán, Károly (1694-1752). Hungarian cartographer, *b.* Komárom [now Komarno, Slovakia]. Studied in Nagyszombat [now Tirnava, Slovakia], posted to Seville, from where he was sent to Quito in South America, 1722. He mapped the course of the Upper Amazon and Marañón rivers; returned to Europe in 1751 and died in Rome. *Provincia Quitensis*, 1751. **Ref.** RAUM, F. *Magyar térképészek Délamerikában* [Hungarian cartographers in South America] in: *Geodézia és Kartográfia* [Geodesy and cartography] (1983/3).

Brentel, Frederic. *Urbis Nancei [Nancy] delineatio*, 1611 (used by Braun and Hogenberg, 1617).

Brentel, Georg. Instrument maker in Lauingen, Bavaria. Astronomical chart, 1573.

Brenton, E.P. American draughtsman. The Narrows [New York], 1798 MS; St. John's, Newfoundland, 1798 MS.

Breon, Fossey & Co. Insurance Atlas of the City of Reading, Pennsylvania, 1884.

Brereton, Robert. *Plan of Prince Rupert's & Douglas Bays...*, Dominica, 1758 MS.

Brereton, *Captain* W. Batangas Bay, Philippines, 1763.

Bressani, Francesco Giuseppe, *SJ* (1612-1672). *Novae Franciae accurata delineatio*, 1657. **Ref.** DAHL, E.H. *The Map Collector* 63 (1993) pp. 42-43.

Bressani, Genesio (*c.*1525-*c.*1610). Italian military engineer and surveyor, from Firenzuola d'Arda. Summoned to Brussels 1589 to provide information on fortification at Parma; also worked at Lucca and Palmanova.

Bressanini, Rinaldo (1803-1864). Born in Riva di Trente (then Austria, now Italy). Engraver in Milan, Zurich and Geneva. Worked for the Swiss Topographical Bureau, 1839-1864. *Thurgau*, *c.*1870.

Bresson, Jean Antoine. *Terroir, ville, port et rade de Marseille et ses environs*, 1772, 1773.

Bresson, Jean Pierre, *fils*. *Plan géometral de la Ville de Marseille*, 1773.

Bretez, Louis. Sculptor, painter and perspective specialist. Created a bird's-eye view of Paris at the request of Michel-Etienne Turgot in 1734, known as the 'plan de Turgot', it was published as *Plan de Paris...* (20 sheets and index), 1739 (engraved by Claude Lucas and Aubin).

Bretonnière, Coudre la. Calais and Somme areas, 1792-1797; charts for Dépôt de la marine, 1830.

Bretschneider, C.A. *Historisch-Geographischer Wand-Atlas*, Gotha, Perthes 1856, 1876.

Bretschneider, Emil W. (1833-1901). *Map of China* (4 sheets), 1896.

Brett, *Admiral Sir* Piercy. *A Plan of the intended Harbour at Ramsgate*, 1755 (with J.P. Desmaretz); Coast of Kent and Sussex, 1759; *A Map of the Downs*, *c.*1775.

Brettel, Joseph. Mining engineer in service of Mohammed 'Ali, Viceroy of Egypt, 1842-1845. *Bombardment and capture of Acre*, *c.*1841.

Brettingham, Matthew. *Plans ...of Holkham in Norfolk*, 1773.

Breugnot, —. Pays Bas (60 sheets), Paris, R.C. Piquet 1822.

Breusing, Arthur (1818-1892). German geographer. Hydrographical works and a history of geography.

Breval, J. Historian and geographer. Remarks on Several Parts of Europe, 1726.

Breventano, Angelo. *Mons Circaeivs* used by Abraham Ortelius, 1595.

Brewer, *Reverend* John Sheeren. *An Elementary Atlas*, *c.*1854, 1860, 1865.

Brewer, W.H. Engraver. *Twelve Miles Round St Pauls*, 1846.

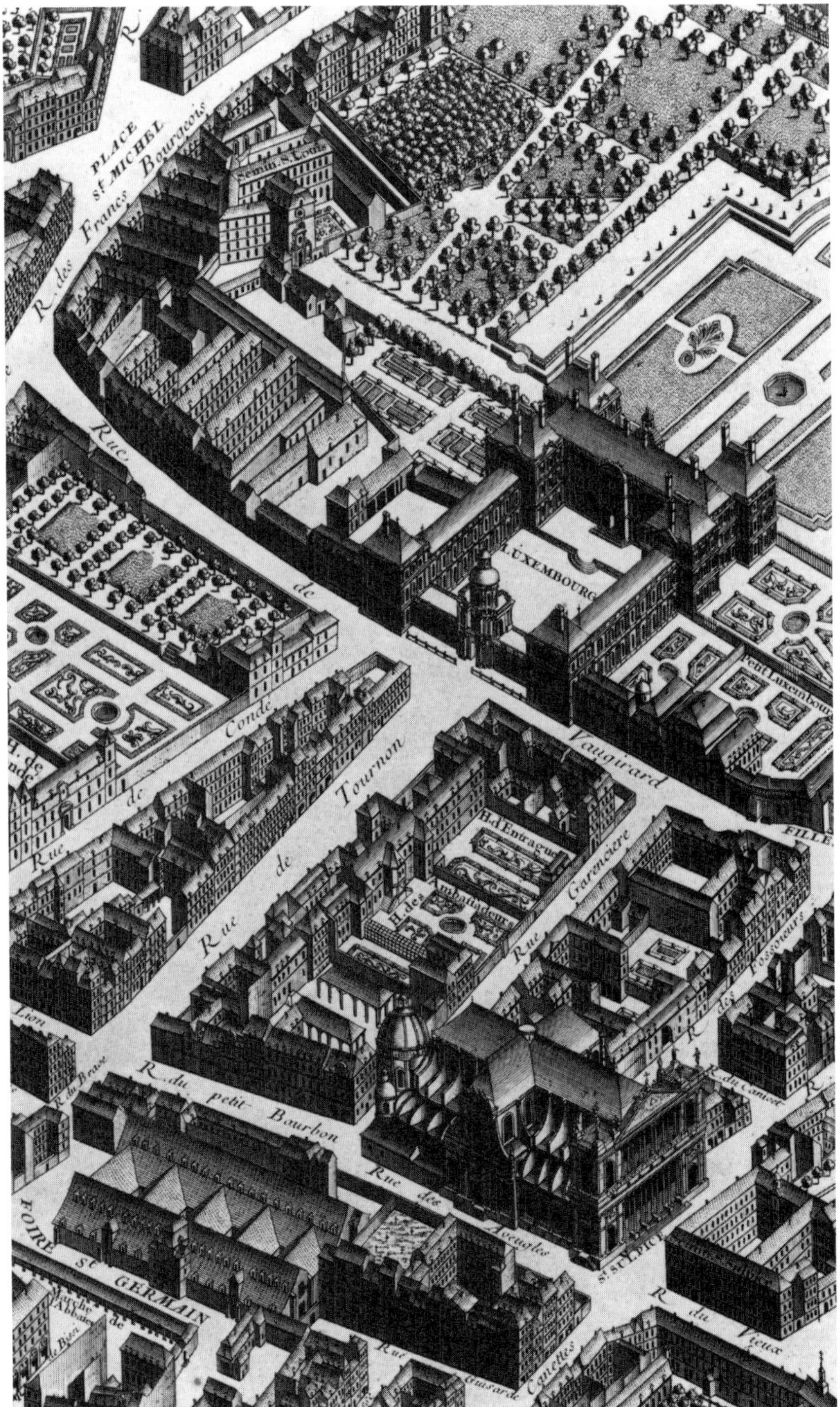

Plan of Paris by Louis Bretez.

Brewse, John. Plan of Halifax, Nova Scotia, 1749 MS.

Brewster, *Sir* David. Collaborated on A.K. Johnston's *Physical Atlas*, 1848.

Brewster, Edward. Publisher in London 'at the Crane in St. Paul's Churchyard'. Richard Blome's *...description of the World*, 1670 (with Nathaniel Brooks and Thomas Basset).

Breydenbach [Breidenbach], Bernhard von (*c*.1440-1497). Traveller and writer of Mainz. *Peregrinatio in Terram Sanctam*, Mainz, Erhard Reuwich 1486, several editions. Said to be the first printed book with folding plates, first printed book of town plan views by Reuwich, and first illustrated travel book. **Ref.** TOOLEY, R.V. *The Map Collector* 3 (1978) pp. 48-49.

Brialmont, Alexis Henri. *Les Régions fortifiées*, 1890.

Briard, P. Worked with others on sheets 94 and 95 [Normandy] of the Cassini *Carte de France*, 1754-1767.

Briaticho, Cola de. Atlas of 3 manuscript portolan charts: Atlantic Coast, Adriatic, Black Sea, 1430.

Brickell, John. *...North Carolina*, 1737.

Bridgens, Henry F. Publisher of Pennsylvania. Map of Elmira, New York, 1852; Cumberland County, Pennsylvania, 1858; Lawrence Fagan's *Map of Berks County, Pennsylvania* [wall map], 1860; Fagan's *Map of Berks County, Pennsylvania* [atlas], 1861, 1862; Lancaster County, Pennsylvania, 1864; Chester County, Pennsylvania, 1865.

Bridges, E.W. *Map of... a Canal from Baltimore to the Potomac*, *c*.1830.

Bridges, *Lieutenant-Colonel* George. *Military plan of the Cape Peninsula* [Cape of Good Hope], 1803, for Sir John Barrow's *Travels*, 1806.

Bridges, William. Surveyor of New York city. Copied the 1803 city plan of Mangin & Goerck, 1807; published John Randel's survey as *This map of the city of New York and island of Manhattan...*, 1811. **Ref.** COHEN, P.E. & AUGUSTYN, R.T. *Manhattan in maps 1527-1995* (1997) pp.100-105.

Bridgman, E.C. Publisher of 5 Barclay Street, New York. *Bridgman's Railroad, County & Township Map of Michigan*, 1874.

Briesmeister, William. *Ethnographic map of the Balkan peninsular*, *c*.1918.

Briet [Brietus], Philippe, *SJ*. (1601-1668). Jesuit priest of Abbeville. *Palestinae Delineatio...*, Paris, Michael van Lochom 1641 (engraved by Henri Le Roy); historical atlas *Parallela geographia veteris et novae*, 1648, 1653; *Royaume de Iapon...*, 1650. **Ref.** MOSKOPP, Jakob 'Alt und Neue Geographie am S. Buch des Ersten Bandes, Teil I, der Parallela Geographicae Veteris et Novae von Phillipe Briet (1601-1668)...' in: *International Yearbook of Cartography* Vol.26 (1986) pp. 61-76.

Briffaut, Etienne. Publisher in Vienna. L'Etat de la guerre en Italie, Vienna 1734; Théâtre de la guerre sur le Rhein, 1735.

Briggs, *Major* —. *...Tea Countries of Assam & Cachar*, *c*.1865.

Briggs, Carl R. [& Co.]. Shasta County, California, 1891.

Briggs, Henry. Mathematician. *The North Part of America*, in Samuel Purchas' *Purchas His Pilgrimes*, 1625 (the map, engraved by Reynold Estracke, shows California as an Island).

Briggs, Joseph. *Plan of... Monte Video*, 1823.

Brigham, H.G. *...Counties of Clinton and Gratiot, Michigan*, 1864.

Brigham, John C. *Nuevo sistema de geografia*, New York 1827-1828.

Brigham, W.T. Surveyor. *...Crater of Kilavea*, Hawaii, 1865; *An index to the islands of the Pacific Ocean* (24 maps), Honolulu 1900.

Bright, Frederick John. *...Bournemouth*, *c*.1885.

Brightly, C. Printers of Bungay, Suffolk. Barclay's Dictionary of North America, 1806 (with E. Kinnersley).

Brightly, H.A. *Western States*, in: R.M. Smith's *New Common-School Geography*, Philadelphia, Grigg, Elliott & Co. 1848.

Brigniel, Joseph *SJ* (*b*.1699). Austrian missionary in Paraguay from 1728. Contributed to a map of the Jesuit province of Paraguay, published as *Paraquaria*, Vienna, Johann Winnkler *c*.1750.

Brigonci, Pietro Antonio. Published last edition of Tomaso Porcacchi's *L'Isole più famose del mondo*, 1686.

Brillet, L.C.D. *Plan des Fontaines...de Paris*, 1738.

Brinck, Ernest. The first person to translate Mercator's Atlas into Dutch, published as *Atlas Minor Ofte Een korte doch grondige...*, Amsterdam, Joannes Janssonius 1630.

Brindley, James (1716-1772). Canal engineer, produced many maps and plans of the new canal network. Plan of the Bridgewater Canal, 1759 MS; Navigable Canals, 1769; *A Plan of the River Thames and of the intended canal from Monkey Island to Isleworth*, 1770.

Brindley, John. Bookseller at 'the Feathers in New Bond Street'. *The Theatre of the present War in the Netherlands and upon the Rhine* (map and 68 plans), London 1745.

Brine, *Lieutenant* Frederic, *RE*. Plans of battles of Crimea, 1857-1859; *Map of Sevastapol and surrounding country...*, London, Ackermann & Co. 1857 (engraved by B.R. Davies); *Map of Valentia*, 1859.

Bringas, Diego. Spanish missionary cartographer. California, 1795 MS.

Brink, C.F. Map of Southern Africa, 1761 MS. **N.B.** Compare with Ten **Brink** (below).

Brink, H. Ten. *Ten Brink's Touristenatlas van Nederland*, 1908; *Ten Brink's Zak-Atlasje van Nederland*, 1908 and later editions; *Ten Brink's Reisatlas van Nederland...*, 1910 and later editions; *Repetitie-Atlas van Nederland*, 1916; *Ten Brink's Wegenatlas van Nederland*, 1925.

Brink, M.J. *Map of the Colony of the Cape of Good Hope*, 1880.

Brink, Peter Henry. American surveyor. *Map of Ulster County, New York...*, 1853 (with Jay Gould and Oliver J. Tillson).

Brink, Ten. Surveyor. Part West Coast Africa, 1761 MS. **N.B.** Compare with C.F. **Brink** (above).

Brink, Wesley R. [& Co.]. American publisher. Left the firm of Alfred Andreas to produce his own atlases of southern Illinois and Missouri counties. *Illustrated Atlas Map of Mason County, Illinois*, 1874; *Illustrated Atlas Map of Cass County, Illinois*, 1874; *Atlas of Lawrence County, Illinois*, Philadelphia 1875; *Illustrated Atlas Map of St. Charles County, Missouri*, 1875; *An Illustrated Historical Atlas Map of Douglas County, Illinois*, Edwardsville 1875; *Illustrated Atlas Map of Iroquois County, Illinois*, Edwardsville 1884.

Brink & Co. Publishing firm probably founded by Wesley R. Brink *c*.1872. *Illustrated Atlas of Mercer County, Illinois*, 1875.

Brink, McCormick & Co. Publishing firm of St. Louis. *Illustrated Encyclopedia and Atlas Map of Madison County, Illinois*, St. Louis, Brink, McCormick & Co. 1873; *Illustrated Atlas Map of Sangamon County, Ill.* (27 Maps), St Louis 1874.

Brinkley, John (1763-1835). Bishop and astronomer.

Brinkman, Bernardus. *Sterren Kaart*, *c*.1873.

Brinkman, C.L. *Atlas van het Koningrijk der Nederlanden*, *c*.1852; *Geschiedkundige Atlas van Nederland...*, 1890.

Brinton, D.G. Worked with A. von Steinwehr on *An Intermediate Geography*, Cincinnati and New York, Wilson, Hinkle & Co. 1870, 1877; *A Primary Geography*, Wilson, Hinkle & Co. 1870 (also with A. von Steinwehr).

Brion de la Tour [Brion], father and son. French geographers. *Voyages dans les départements de la France*, 1793, 1800 (text by Lavallée and Breton, geography by Brion père, maps drawn by Brion fils); *Atlas géographique et statistique de la France en 108 départements*, *c*.1802 (this re-used maps from the *Voyages...*).

• **Brion de la Tour** [Bironius Gallus], Louis (*fl.*1756-1803). 'Ingénieur-géographe du roi. Rue du Petit Pont près la fontaine St. Severin maison de M. Landes. Libraire au 1er' (1793). *Cartes des Places Fortes et des Principaux Ports des Isles Britanniques*, 1756; *Côtes Maritimes de France*, 1757; *L'Amérique...*, 1764; *Atlas général, civil et ecclésiastique*, 1766 and later editions; *Atlas itinéraire portatif de l'Europe*, 1766; *Atlas Générale méthodique et élémentaire...*, Desnos 1768; *Nouvelle carte de la France divisée en ses 40 Gouvernements Généraux*, 1776; *l'Atlas itinéraire portatif de l'Europe*, 1776; *Atlas et tables élémentaires*, J. Barbou 1777; *Carte du Théatre de la Guerre entre les Anglais et les Américains...*, 1778; *Presqu'isle des Indes Orientales en deça du Gange, Comprenant l'Indostan ou Empire du Pogol...*, 1780; *Carte des Etats Unis de l'Amérique Septentrionale...*, 1781; *Tableau Général de l'Amérique*, 1783; *Carte des Etats Unis d'Amérique et du Cours du Mississippi*, 1784; *Carte du Canada*, Paris 1784; *Mappe-Monde*, 1786; *Carte de la partie de la Virginie*, 1787; *Itinéraire de l'Europe...*, 1788; *Carte du Département du Cher...*, 1789 in: Prudhomme and Tournon's *Révolutions de Paris*, 1790; *Guayane Terre Ferme...*, 1790; departmental maps in E. Mentelle's *La République française...*, 1793, 1794; *Description générale de l'Europe...*, 1795; *Mappemonde philosophique et politique où sont tracés les voyages de Cook et de la Pérouse*, 1800; *Carte élémentaire et statistique de l'Allemagne*, 1803 (with N.M. Maire).
•• **Brion** [Brion de la Tour], Louis, *fils*. *Atlas Portatif de la France en départements*, 1789; *Département de l'Eure et du Loire*, 1793; *Départment du Var*, 1800; *Tableau de l'Empire français*, 1812 (with Perrier).

Brion, Benjamin. Maps for P.M. Gourné's *Atlas abrégé*, 1763.

Brion, Hypacio de. *Plano hydrographico da Enseada do Quicembo*, 1888.

Brion, John [& Sons]. Relief maps. Brighton, 1858; Isle of Wight 1858; Jerusalem, 1858; England & Wales, 1859.

Brion, Martin de [Martinus Brionaeus Parisiensis]. *Totius Terrae Sanctae... elaborata descriptio*, Paris, Guillaume de Bossozel 1540 (but no surviving map). **Ref.** KARROW, R.W. *Mapmakers of the sixteenth century and their maps* (1993) pp.94-95.

Briot, B.J. Composite atlas, *c.*1660.

Briot, brothers. French engravers, amongst those employed by Jean Boisseau to prepare the numerous surveys and views by Claude de Chastillon for publication. Published as *Topographie françoise*, 1641, 1648, 1655.
• **Briot**, Nicolas.
• **Briot**, Isaac.

Briscoe, J.C. *Plan of the Battle of Gettysburg*, *c.*1863.

Brissart, P. Engraver for Pierre Du Val, 1669.

Bristow, *Captain* Abraham. Additions to Thomas Butler's chart of the Western Pacific, 1822 (original date 1799).

Bristow, William. *A Compleat Map of the County of Kent...*, Canterbury 1801.

Brito, Pedro Torquado Xavier de. *Nova Carta Corographica do Imperio do Brasil*, 1867.

Britto, Capello H.C. de. Maps of Portuguese Africa, 1883-1887.

Britto, Giovanni. *Africa*, 1541; woodcut map of Boulogne, 1550.

Britton, John. Writer, antiquary and topographer. Compiler, with Edward Brayley, of the text for the partwork *The Beauties of England and Wales* (18 volumes), *c.*1804-1810 (a separate set of maps to accompany the work was drawn by G. Cole and engraved by J. Roper); re-edited J. Feltham's work *Picture of London* as *The Original Picture of London enlarged and improved* (with 2 new maps), 1826, 1829, 1832.

Brives, Abel (1865-1929). French mineralogist and surveyor in Morocco. Les terrains tertiaires du bassin du Chélif, 1897; Voyages en zig-zag dans l'Atlas marocain, 1904; Voyages au Maroc (1901-1907), 1909.

Brix, C.C. *Skole-Atlas*, Copenhagen 1867 (with J.L. Jensen); *Haand og Skole Atlas*, *c.*1876 (with J. Hoffensberg).

Brjus, Yakov Vilimovich. See **Bruce**, Yakov Vilimovich.

Broaders, Edward R. *A New Universal Atlas of sixty maps...*, Boston 1835.

Broadie, Joseph. See **Brodie**, Joseph.

Broc, E. *Cartes commerciales, Colombie et Équateur*, 1887; *République Argentine*, 1887; *États-Unis de Vénézuela*, 1888.

Brocardus de Monte Sion. See **Burchard.**

Brocas, S.F. Engraver. *A Map of the Lakes of Killarney*, 1815.

Brocchi, Giovanni Battista. Carta fisica del suolo di Roma, 1820.

Broch, Ábrahám (1834-1914). Austrian cartographer, worked on 19th-century triangulation of Hungary, in charge of Carpathian triangulation, 1857; Director Triangulation Service at Vienna from 1891. **Ref.** LEGO, K. *Geschichte des ö. Grundkatasters* (Vienna 1980) [in German].

Broch, Lars. *Kart over den Norske Kyst*, 1839; *Oversigtskart til Lomme Rejserovce*, *c.*1876.

Broch, Ole Jacob (1818-1889). Norwegian geographer. Royaume de Norvège, 1878.

Brochant de Villiers, André Jean Marie François. Studied geological mapping in England in the 1820s. *Carte Géologique de la France*, 1840.

Brochard [Brochart], Bonaventure (*b.c.*1500). French Franciscan friar of Normandy. Travelled to Jerusalem and Mount Sinai and produced a description of this journey: *Le voyage de Hierusalem et du mont de Sinay*, 1535 MS. **N.B.** Not to be confused with the 13th-century German monk **Burchard** of Monte Sion. **Ref.** KARROW, R. *Mapmakers of the sixteenth century...* p.96.

Brochet des Roches, P. *Carta Geografica de la Provincia de Corrientes* (6 sheets), 1877; *Mappa... de la America del Sur*, 1880.

Brock, *Major General Sir* Isaac. *A Map of the American Lakes and Adjoining Country*, London, Luffman 1813.

Brockbank W.H. *Greater Bolton*, 1899.

Brockes, C. *Mappa do Imperio do Brasil*, 1878.

Brockett, John. Plan of New Haven [Connecticut], 1641.

Brockhaus, Friedrich Arnold (1772-1823). Bookseller of Leipzig. Founder of the publishing house F.A. Brockhaus.

• **Brockhaus**, F.A. [Verlag]. Publisher of Leipzig, produced many maps to support the Brockhaus encyclopaedia and travel literature. Henry Lange's *Vollständiger Hand-Atlas über alle Theile der Erde*, 1847; *Bilder-Atlas zum Conversations-Lexikon*, 1849; *Neuester Grundriss von Hamburg*, 1855; E. Leeder and T. Schade's *Illustrierter Hand-Atlas*, 1863; Henry Lange's *Geographischer Handatlas über alle Theile der Erde*, 1866; *Russland*, *c.*1870; *Australien und Polynesien*, 1870. **Ref.** ESPENHORST, J. *Bibliographie der Handatlanten* (1994) pp.378-380 [in German].

Brocklesby, John. *Elements of Physical Geography*, 1869.

Brocktorff, F. Arabic atlas, 1835.

Brodarics, István (1490-1539). Humanist, chancellor to Lajos II of Hungary, later bishop of Vác. Map of Hungary [now lost] possibly used by Lazarus Secretarius [Lázár]; wrote eye-witness account of Battle of Mohács, 1526 (illustrated with map). **Ref.** HRENKÓ, P. 'Magyarország legrégibb nyomtatott térképe' [The earliest printed map of Hungary], in: *Geodézia és Kartográfia* [Geodesy and Cartography] (1974/6) [in Hungarian]; SZATHMÁRY, T. *Descriptio Hungariae* (1987).

Brodhead, Charles. *A Map of a Tract of Land in the State of New York called Macoomb's Purchase*, *c.*1796.

Brodie [Broadie] *Lieutenant* Joseph. *A Chart of... the North Sea from Texel to the Naze of Norway*, 1798; *A New General Chart of the North Sea*, 1812.

Brodie, Paul. *...Original Cessions of Lands made to the United States by Treaty or Agreement with the Indian Tribes of the United States and Territories*, 1879; *Indian Reservations in the United States*, 1883.

Brodin, Rudolf. *Karta öfver Stockholm*, 1870, *c.*1874.

Broditzky, Joseph. *Post- und Eisenbahn-Karte der Oesterreichischen-Ungarischen Monarchie*, 1874-1876.

Broeck [Broek], Abraham van den (*c.*1616/1617-1688). Engraver, worked in Amsterdam (1641-1665). Zürich, Joannes Janssonius 1641; Transilvania, Janssonius 1645; Valois, Janssonius 1658; Germany, Nicolaes Visscher 1680.

Broeckhuysen, G. van. Surveys in Indonesia, *c.*1708, used by F. Valentijn, 1726.

Broeckhuysen, Justus van (*fl.*1696-after 1723). Dutch surveyor and prolific cartographer. Breukelerweerdt, 1701; Moersbergen, 1706; Gooiland, 1709 and many others.

Broeckhuysen, Steven [Pietersz.] van (*fl.*1627-1661). Dutch cartographer working in Leiden from 1627, in Drente from 1637. Kaart van Rijnland, 1647 (with Jan Jansz. Dou).

Broedelet, Guillaume [Wilhelm]. Publisher of Utrecht. L. Hennepin's *Carte d'un tres-grand Pays...*[North America], 1697; A. Reland's *Imperium Japonicum*, 1715; later edition of Clüver's *Introductio in universam geographiam*.

Broedelet, Johannes. Publisher of Utrecht. J. Haas's *Imperii Russici*, 1743

Broek, Abraham van den. See **Broeck**, Abraham van den.

Broelmann, Stephan (1551-1622). Historian and cartographer. *Epideigma sive specimen historiae*, Cologne 1608.

Broen, Joannes de (1659-1730). Engraver of Amsterdam. G.T. Borgonio's *Lac Leman, Pedemontium et...Italiae Regiae*, *c.*1682; O. Dapper's *Nova et Accurata Moreae*, 1688; De Wit's *Galliae Tabula*, *c.*1688; *Tabula Generalis Sabaudiae*, *c.*1700; *Conspectus Libari atque Antilibani*, *c.*1700 (after Borgonio in later eds of Blaeu's Italian Townbooks). Also worked for Cornelis Danckerts, Pieter Schenk and Johannes de Ram.

Broenner, Heinrich Ludwig. Bookseller and publisher of Frankfurt am Main. *Part. Sept. du Landraviat de Hesse Cassel*, 1760; War in Germany, 1760-1761; Samuel Holland's *Provinces of New York and New Jersey*, 1777.

Brognoli [Brognolio], Bernardo. See **Brugnoli**, Bernardino.

Brohez, L.J. Publisher of Brussels. Brussels edition of G.B. Depping's *L'Angleterre ou Description Historique et Topographique du Royaume Uni de la Grande-Bretagne* (6 volumes), 1828 (first published Paris, Ledoux 1823).

Broichmann, J. *Neuer Volksschul Atlas*, 1880.

Brolan, C. Publisher. *Reskarta öfver Skäne*, 1868.

Brolin, Jonas. Uppsala Stad, 1770; *Grundritning öfver Stockholms Stad*, 1771; Norrköping, *c.*1780, 1816.

Broling, G, Engraver. *Geographiske Chartor öfver Swerige* (18 maps), Stockholm, S.G. Hermelin 1797 and later (with F. Akrel and S.J. Neele).

Bromfield, Davenport. Map of San Mateo County, California, 1894.

Bromley, brothers. Atlas publishers of Philadelphia. *Atlas of the City of Baltimore*, 1885, 1887, 1896, 1906; Atlas Boston, 1888; Atlas of Philadelphia, 1889; Atlas of New York, 1891.
- **Bromley**, George W.
- **Bromley**, Walter S.

Bromme, Simon Traugott (1802-1866). Bookseller of Stuttgart. Edited *Atlas zu Alexander von Humboldt's Kosmos*, 1851/1854, 1861; second edition of *Dr Carl Glaser's topisch-physikalischer Atlas*, Stuttgart 1855; Carl Glaser's *Schul-Atlas*, 1858; *Illustrirter Hand-Atlas der Geographie und Statistik*, *c.*1858/1862, 1868/1874; Arminius Rheinhard's *Atlas Orbis antiqui*, 1860; *Deutschland*, *c.*1867; *Illustrirter Volks Atlas*, 1875.

Bronckhorst, J.G. van. Map of Amsterdam (9 sheets), 1625; *Caerte van de Belegeringe van Breda*, 1637.

Brongniart, Alexandre (1770-1847). French geologist and mineralogist. *Carte géognostique des Environs de Paris*, 1810 (with G.L.G.F.D. Cuvier); *Carte géologique des Environs de Paris*, 1811, 1821 (with G.L.G.F.D. Cuvier).

Broniowski, Marcin. Description of the Crimea, 1579; published 1595 edition of Georg Reichersdorff's geography of Moldavia and Transilvania, which included a copy of the Honter map of Transilvania.

Bronn, Heinrich G. *...Mineralogische Beschreibung der Gegend von Heidelberg*, 1830.

Bronner, H. Ludwig. Plan von Mainz, 1735.

Brook, R. Engraver of Gray's Inn Place, London. *Northern Hemisphere. Southern Hemisphere*, 1799; Synopsis of the Universe, 1799.

Brooke, *Sir* Arthur de Capell. *Plan of Sweden, Norway, and Lapland*, *c.*1850.

Brooke, B.O. Plan of the Potomack River, 1737.

Brooker, *Lieutenant* Edward W., *RN*. Second master and assistant surveyor aboard HMS *Spitfire*, which undertook surveys in the Mediterranean, the Dardanelles and Crimea, *1852-1855* (surveys were used as the basis of many maps associated with the Crimean war). Port of Alexandria, Admiralty Chart 1857 (with F. Skead and R.H. Blamey); Port Derwent, Hobart, 1861.

Brookes, P. *Travelling Companion through England and Wales*, 1812.

Brookes, *Doctor* R. Map of the World, 1817.

Brookes, Richard. *Brookes's General Gazetteer*, 1762 and later editions, first American edition published Philadelphia 1806; *Brookes's General Gazetteer Improved*, Philadelphia, Jacob Johnson 1806, Johnson & Warner 1812; and other gazetteers.

Brooking, Charles. *Map of the City and Suburbs of Dublin*, 1728 and later editions.

Brooks, J. Rapids of the Ohio River, Frankfort, Kentucky, 1806.

Brooks, J. Louisville Canal, Ohio River, 1825-1843. **N.B.** Compare with J. **Brooks**, above.

Brooks, Nathaniel. Publisher at the Angel in Cornhill. Blome's *...description of the World*, 1670 (published jointly with Edward Brewster and Thomas Basset).

Brooks, Nicholas. North American printer. Bernard Romans's *Seat of the Civil War in America*, 1775.

Brooks, T.B. Worked with C. Rominger and R. Pumpelly on the geological survey of Michigan, 1869-1873. The maps were published as *Atlas accompanying reports on Upper Peninsula...*, New York, Julius Bien 1869-1873; the maps resulting from Brooks's own work on iron-bearing rocks were published as *Geological Syrvey of Michigan. Atlas accompanying report on Iron Bearing Rocks* (14 maps), Julius Bien 1873.

Brooks, Vincent. Lithographer. Major William Kersteman's *Saldanha Bay* [Cape of Good Hope], 1803.

Brooks, Vincent. Lithographer of London. *Map of London and its Environs*, *c.*1868.

Brooks, W. *Plan of the Division of the Metropolis into Districts for Auxiliary Bible Societies* [London], *c.*1812.

Brooks, W.A. Civil engineer. Charts of the River Tees, 1833-1840.

Brooks, William. *The United States of America*, *c.*1820.

Broom, Jacob (1752-1810). American surveyor. Map of the Brandywine Campaign, 1777 MS.

Brosch, August. *Plan der Umgebung von Peterwardein*, 1872.

Brose, Heinrich, *the elder* (1805-1868). Engraver. Battle plans of Europe, 1821-1854; Alpine maps, 1822-1850; for Heinrich Kiepert, 1860.

Brose, W. Engraver for Tanner, 1833 and later.

Brossa i Trullàs, Eduardo (1848-1924). Catalan cartographer and engraver. *Mapa de la provincia de Barcelona*, 1881; *Cataluña*, 1883 and later editions; *Mapa geológico y topográfico de la provincia de Barcelona*, 1888, 1900 (with Jaume Almera); *Cataluña y país lindante*, 1904; *Carta topográfica de San Julian de Vilatorta...*, 1905; *Mapa de Carreteras para ciclistas y automóviles*, 1921.

Brossard de Corbigny, Charles-Paul (1822-1900). French naval surveyor in West Africa, later in Indochina, finally at the Service hydrographique de la Marine, Paris (from 1882). *Carte de la Sénégambie*, 1860; *Carte du Sénégal*, 1861.

Brosselard-Faidherbe, *capitaine* Henri-François (1855-1893). French officer in West Africa. *Voyage de la mission Flatters au pays des Touaregs Azdjers*, 1883; *La Guinée portugaise et les possessions françaises voisines*, 1889. **Ref.** BROC, N. *Dictionnaire illustré des explorateurs Français du XIXe siècle: I Afrique* (1988) pp.58-60 [in French].

Brosses, Charles de, of Dijon. *Histoire des navigations aux Terres Australes*, 1756 (with 7 maps by Didier Robert de Vaugondy, engraved by Guillaume Delahaye).

Brossier, Lafitte de. Cartes marines des Isles Seychelles, 1777 MS.

Brostenhuysen, Jan van [Johann] (*c.*1596-1650). Engraver and printer of Leiden. Worked for Huych Allard, 1646.

Brothers, Alfred. *Star Atlas*, 1870.

Broughton, Hugh (1549-1612). Puritan rabbinical scholar. Prepared *Concent of Scripture*, *c.*1588, containing two world maps *A Map of the earth...* and *A mapp of the north part of the equinoctial....*

Broughton, S.H. *Geological Map Rockland and Pewabic Township, Michigan*, 1863.

Broughton, W.E. Delves. See **Delves**.

Broughton, *Captain* William Robert (1762-1821). English seafarer, in command of the *Chatham* 1791-1792 (for part of George Vancouver's voyage to N.W. coast of America), then the *Providence*, 1793-1797. Surveyed N.W. Pacific coast of America, the N.E. coast of Asia, and the central Pacific. *Plan of Indermo Harbour in the Isld of Matsmay*, 1797 MS; *A voyage of discovery to the North Pacific Ocean* (engraved plates and 3 charts), London, 1804.

Brouscon, Guillaume. French hydrographer of Le Conquet, Brittany [Dieppe School]. First tidal charts and woodcut portolans in nautical guide and almanac, *Manuel de pilotage à l'usage des marins breton*, 1548. **Ref.** DUJARDIN-TROADEC, L. *Les cartographes breton du Conquet: la navigation en images 1543-1650* (Brest 1966) [in French].

Brouwer, Hendrick (1521-1643). Maps of Brazil and Chile, 1640.

Brouwer [Brouwers], Hendrik. Dutch surveyor and cartographer, working at Aardenburg, Zeeland. Landmeter for het Vrije van Sluis, 1733.

Brouwer, Jan Janssen. *Caert en Stedeboexken van Nederlandt...* (67 maps), Amsterdam 1672.

Brouwer, Meinte [Andries] (*fl.* post 1732). Dutch land surveyor and cartographer, worked at Harlingen.

Brown, --. See **Tulloch** & Brown.

Brown, *Captain* —. Manuscript maps and sketches of the area around Boston on the eve of the Revolution, 1775 (with Henry De Berniere).

Brown, *Major* —. Bantry Harbour, 1797 MS.

Brown, A.L. *...Dubuque County, Iowa*, 1866.

Brown, Charles. *A Plan of Man of War Bay, Island of Aegina*, *c.*1830.

Brown, Charles Barrington. *Geological Map of Jamaica*, 1865; *Geological map of British Guiana*, 1873.

Brown, Charles F. Engineer. *...Rock Island & Alton Railroad*, 1857.

Brown, F.A. Publisher of Hartford, Connecticut. H.E. Rockwell's *Map of*

Connecticut For Schools, Families, and General Use, 3rd edition 1855 (first published by Kellogg & Comstock, 1848). **N.B.** Compare with **Brown & Gross**, and **Brown & Parsons.**

Brown, George (1747-1816). Surveyor and engineer of Elgin, prepared various manuscript surveys in Scotland including ...*Moray and Nairn Shires*, 1810, 1813.

Brown, Henry Yorke Lyell (1844-1928). Part West Australia, *c.*1873; *Geological map of South Australia*, 1886, 1899; *Geological map of the Northern Territory of South Australia*, 1898; *Geological map South Australia* (4 sheets), 1899.

Brown, I. Engraver. *The Phoenix Park, Dublin*, 1789.

Brown, *Lieutenant* J. Four charts of the West Indies, 1778-1784.

Brown, J. Publisher and cartographer. Seymour Place, Bryanston Square, London. *Ireland, exhibiting the principal Roads*, 1842; *New General Atlas*, *c.*1852.

Brown, J.D. Engraver. *Geological Survey of Victoria*, *c.*1868.

Brown, J.W. Surveyor. *Street map of the Manor of Aston*, 1883.

Brown, John. Publisher in New York. *An Accurate Map of the Holy Land*, 1790.

Brown, John. *Northumberland and Lancashire Junction Railway*, 1846; *Wear Dock Railway*, 1846.

Brown, John Carter (1797-1874). Collector whose books formed the basis of the Collection at the John Carter Brown Library, Providence, Rhode Island.

Brown, Joseph. Bookseller of Boston. 3-inch and 5-inch globes, *c.*1836.

Brown, Joseph. Plan of New Haven [Connecticut], 1724 (now lost).

Brown, Joseph C. Surveyor. Survey and manuscript map of the Santa Fe trail, 1825-1827; Missouri, Illinois, 1841.

Brown, Lancelot ['Capability'] (1715-1783). Landscape gardener who produced many maps in connection with landscape projects in England and Wales.

Brown, Martin. Map of Alexander County, Illinois, 1897.

Brown, Milton R. Publisher of Philadelphia. *The continental atlas*, 1889.

Brown, Nicholas (*fl.*1790-*c.*1839). Surveyor, produced many manuscript surveys including *Plan of the Canal between Huddersfield in the County of York and Ashton-under-Lyne in the County of Lancaster*, 1793 (with Benjamin Outram).

Brown, Peter. Publisher of Edinburgh [with Thomas Nelson], used lithography as early as *c.*1831.
• **Brown & Nelson**. William Bell's *A New General Atlas*, 1834, 1837 (with maps by J. Clark).

Brown, R. Architect. *Plan of Plymouth, Stonehouse and Devonport*, *c.*1830.

Brown, S. Howell. *Map of Jefferson County, Virginia*, Staunton 1852.

Brown, *Lieutenant* T.S. *Map of Dunkirk Harbour Chautauque County ... New York*, 1835; *Map of... Presqu'ile Bay*, 1835.

Brown, Thomas (1786-1820). Bookseller, publisher and engraver at N°·1 North Bridge Street, Edinburgh (1786-1820). *A New & Accurate Travelling Map of Scotland*, *c.*1790; Plans of Edinburgh, 1793, 1820; *A General Atlas*, 1801; *Atlas of Scotland being a new set of county maps...*, 1802-1807 (in parts); and a smaller version, *Taylor & Skinner's Survey of the Roads of Scotland on an Improved Plan...*, *c.*1805, *c.*1813; *New Edinburgh General Atlas*, *c.*1818 (with William & Daniel Lizars). **Ref.** MOIR, D.G. *The early maps of Scotland* (Edinburgh 1973) Vol.I, p. 137.

Brown, Thomas [& Sons] (*fl.*1785-1841). Surveyor in Kent, produced various manuscript maps. *A Map of the Town of Maidstone*, 1821.

Brown, *Captain* W. Revenue Surveyor in India. *Map of the District of Mozufurnurgur*, 1827-1832 (with G.J. Fraser), published Agra

1858; *District of Boolundshurer*, 1828-1840, published Calcutta 1846, Allahabad 1859 (3rd edition); *District of Hissar*, 1832-1838 (with J.H. Simmonds) published 1848; *District of Hurriana*, 1832-1838 (with J.H. Simmonds) published 1848; *District of Gurgaon*, Calcutta 1849 (with T. Oliver and J.H. Simmonds).

Brown, W.S. Plat Books of Brown, Dunn and Wanpaca counties, Wisconsin, 1889-1890 (with C.M. Foote).

Brown, William. Deputy Surveyor, Northern District, North America. Cape Breton & Prince Edward Island, 1764; Hillsborough Bay, 1787.

Brown, William (*fl.*1823-1862). Surveyor of Tring [Hertfordshire], produced various manuscript maps including one of an estate at Great Missenden, *c.*1840.

Brown, William & Co. Engravers & printers of 40 & 41 Old Broad St. London E.C. *London Shewing the Postal Districts*, 1860.

Brown & Parsons. Publishers of Hartford, Connecticut. One of the later publishers of A. Daggett's *Map of Connecticut*, *c.*1847-1854 (first published by A. Daggett *c.*1827). **N.B.** Compare with F.A. **Brown**, above. Although there may be some connection between these publishers, the Rockwell map of Connecticut republished by F.A. Brown was much larger than the earlier Daggett map mentioned here.

•**Brown & Gross**. Publishers of Hartford, Connecticut. Succeeded Brown and Parsons as publishers of A. Daggett's *Map of Connecticut*, 1858 (first published *c.*1827).

Brown & Slight. Engravers of Clarendon St. Emerald Hill, Melbourne, Victoria. *Map of New South Wales and Victoria*, 1861; *Map of Melbourne & Suburbs*, *c.*1870. **N.B.** Compare with **Tulloch** & Brown.

Browne, *Captain* B. Revenue Surveyor. Districts of India, 1846-1863.

Browne, Christopher (*fl.*1684-1712). English cartographer, globe maker, printer and publisher 'at the Globe and Compass at the West end of St. Pauls Churchyard', London. Apprenticed to Robert Walton in 1683, to whose business and premises he succeeded when he was made a freeman of the Stationers' Company in 1688. After his retirement in *c.*1712, the business passed to George Willdey. Sir Jonas Moore's map of the Fens, *c.*1684; *A New Map of Virginia, Mary-land, Pennsylvania & New Jersey*, 1685, 1700; *A new Map of the Sea-Coasts of England, France and Holland*, 1688 (with Robert Morden); Flanders, 1700; *Nova totius Angliae tabula* (15 sheets), 1700; F. de Wit's *Atlas Major* (2 volumes), 1706; *Geographia classica*, 1712 (still credited on later editions to 1736). **Ref.** TYACKE, Sarah *London map sellers 1660-1720* (Tring 1978) pp. 113.

Browne, D. Jay. *River Niagara and Vicinity*, Boston 1838.

Browne, Daniel (*d.*1762). Publisher and mapseller without Temple Bar. John Senex's *A New General Atlas*, 1721; Thomas Hutchinson's *Geographia Magnae Britanniae*, 1748, 1756 (with several others); Thomas Kitchin's *Geographia Scotiae*, 1749, 1756 (with others); E. Gibson's edition of Camden's *Britannia*, 1753 (with several others; maps by R. Morden).

Browne, Henry. Parish, Town and Harbour of West Point, Hollymount, Mayo, *c.*1810.

Browne, I. Co-publisher of M. Drayton's *Poly-Olbion*, 1612, 1613 (with M. Lownes, J Helme & I. Busbie).

Browne, I. *Plan of the Island Socotra*, 1615, published Alexander Dalrymple 1774.

Browne, J.L. Surveyor [with F.H. Reuss]. *Map of... Sydney*, *c.*1856; *Map of New South Wales*, 1862.

Browne, J.O. *A Map of the Municipal Borough of Thetford*, 1837; Chelmsford-Norwich Railway, 1846 MS (with G.E. Hadden).

Browne, James. *A Map of the Country of Sikhs*, 1787; *A Map of Scotland*, *c.*1845.

Browne, John.
• **Browne**, John I (*fl.*1583-1589). Irish surveyor, uncle of John Browne II, below. Galway, 1583; Parte of the County of Mayo, 1584 MS; Mayo and Sligo, 1587 MS (with J. Baptiste).
•• **Browne**, John II. Irish surveyor. Connaught, 1592.

Browne, *Lord* John Thomas. *The Islands of Milo &c.*, 1851; *Cyprus South Coast. Limasol*, Admiralty chart 1878.

Browne, Joseph. Engraver. ...*This Map of Staffordshire*, 1682, for Robert Plot, in *Staffordshire*, 1686.

Browne, P.J. ...*Monroe County, New York*, 1852.

Browne, Patrick, *MD*. *A New Map of Jamaica*, *c*.1755.

Browne, Richard (*fl*.1658-1690). Surveyor, produced maps of parts of several English counties.

Browne, Robert. (*fl*.1705-31). British astronomer, instrument maker and carpenter. Produced a map predicting the track of the 1724 eclipse. *A Projection of the Passage of the Moon over the Sun....*, 1724. **Ref.** ARMITAGE, G. *The shadow of the moon: British solar eclipse mapping in the 18th century* (Tring 1998).

Browne, T. Lithographer in Hobart Town, Tasmania. *Wabbs Harbour*, *c*.1860.

Browne, Thomas (1702-1780). Land surveyor and herald (Blanch lyon pursuivant of arms). Manuscript plans and surveys of parts of several English counties including Essex plans, 1726-1733; plans of Micheldever, Hampshire, 1730.

Browne, W.F. Lithographer. *Norwich Corporation*, 1885; *Map of the Hundreds of Norfolk*, 1895.

Browne, William George. *Map of the Route of the Soudan Caravan*, 1799, in his *Travels in Africa*.

Browne & Brooks. Surveyors of Salt Lake City. *Map of Salt Lake City and Vicinity, Utah*, New York, G.W. & C.B. Colton & Co. 1888, (later used by Rollins & Sons, New York, 1893).

Browning, J.S. ...*Province of Canterbury, New Zealand*, 1863.

Brownlee, J.H. District Land Surveyor, Canada. *Brownlee's Railway & Guide Map of Manitoba*, Winnipeg 1888; Winnipeg District, 1888 (with R.E. Young); *Map of the Province of British Columbia*, 1893.

Brownrigg, John (*c*.1748-1838). Surveyor of Dublin, produced many manuscript plans of various parts of Ireland including *Grand Canal* [Dublin], 1788; *A Map of Antient Ireland...*, 1789 (engraved by C. Henecy, published in W. Seward's *The Hibernian Gazetteer*, Dublin 1789).

Brownrigg, William Meadows. Land surveyor, Sydney, New South Wales, at Jamison St (1847), 6 Grimes Bldgs, Argyle St. West (1851), 38 Pitt St (1855), 24 O'Connell St (1864). *Plan of the Curramore estate, Jamberoo, Illawarra...*, 1850; *Gold Country*, New South Wales, 1851 (used by John Tallis 1851); *Plan of the town and part of the parish of Nimmitabel*, 1858.

Brownrigg & Roe. Surveyors, Sydney, New South Wales. *Plan of Feltham, Fig Tree Bay, Parramatta River*, 1857; *Plan of valuable building allotments at Newtown...*, *c*.1850s.

Bruce, Alexander. Surveyor. *A Plan of Loch Sunart*, 1733.

Bruce, C. Engraver. Plan of Launceston, Van Diemens Land, 1832; Furneaux Islands, 1833.

Bruce, C.A. *Map of Upper Assam*, 1839.

Bruce, E. & L. Introduction to Geography and Astronomy, 1820.

Bruce, F.G. Pocket Map of Edinburgh, 1830.

Bruce, James (1730-1794). Scottish explorer in Africa. Charts and plans to accompany his *Travels to Discover the Source of the Nile* (5 volumes), Edinburgh and London, G.G.J. & J. Robinson 1790.

Bruce, John. *Nieuwe Paskaard van Hitland*, *c*.1711, published by Reinier and Josua Ottens, Amsterdam 1745.

Bruce, William Downing. *Kennet, Clackmannanshire*, 1855.

Bruce [Brjus; Bryus], *Field-Marshall* Yakov Vilimovich [Jakob] (1670-1737). Russian count of Scots descent, collaborator of Tsar

Peter I. Maps of southern Russia, 1699 (with Georgy Mengden), printed in Amsterdam (Russian and Latin editions).

Brucioli, Antonio. Trattata della sphera, Venice 1546.

Bruckner, D. *Canton Basel*, 1766 (with E. Büchel).

Bruckner, G. Terrestrial and celestial globes, 1842.

Bruckner [Brouckner], Isaak, *the elder* (1686-1762). Born in Diegten, Switzerland, worked as a mechanic, seal engraver, globe-maker, geographer and cartographer in Paris, St Petersburg, London, Berlin and Basle. Geographer to Louis XV of France. Globe for Empress of Russia, 1735; *Nouvel atlas de la marine*, Berlin 1749; Africa, 1749; Globe, 1752 (engraved by J. Handmann); Nieuwe atlas, 1759. **Ref.** SCHNEEWIND, Wolfgang *Der Globusfreund* 2 (1953) pp.22-29 [in German].

Brucks, *Lieutenant* (later *Captain*) George Barnes. Charts of the coasts of Arabia, Bushire, Bahrein and Persian Gulf, Admiralty charts 1826-1872.

Bruder, C. Lithographer. *Canton Schaffhausen*, *c.*1855.

Brué, Adrien Hubert (1786-1832). French geographer and publisher, 'rue des Macons-Sorbonne'. 'Géographe du roy', and 'Géographe de son Altesse royale Monsieur le comte d'Artois'. Succeeded by Charles Picquet, then J. Barthelemier. *L'Empire français*, 1813; *Grand Atlas universel...*, Paris, Desray 1815; *Atlas Universel...*, 1816 and later; *Atlas universel de géographie physique, politique et historique, ancienne et moderne...*, 1822, edition by Humboldt 1830, edition by E. Levasseur, published C. Delagrave 1875; Atlas de France, 1820-1828; ...*Colombia*.., 1825 (used by Humboldt *c.*1834); *Atlas géographique, historique, politique et administratif de la France* (24 maps), Paris, Veuve Desray 1828; Atlas classique, 1830; and others.

Bruechman, —. *Plan de la Ville & Citadelle de Gand*, *c.*1709; *Plan de la Bataille de Malplaquet*, 1709; *Plan des attaques de la Ville de Lille*, 1709.

Brueck, P.A. *Plan de la ville d'Arlon*, 1874.

Brueckner, Daniel. *Canton Basel*, 1766.

Brueckner, *Lieutenant* Johann. *Die Gegend Dresden-Freiberg-Meissen*, 1868; *Umgegend von Dresden*, 1868.

Bruegner, C. *Die Schlacht bei La Belle Alliance*, 1821; *Die Schlacht bei Ligny*, 1821; *Die Schlacht von Bautzen*, 1821;*Die Schlacht von Laon*, 1821; *Plan von Algier*, 1830; *Plan von Oporto*, 1833.

Bruegnor, C. Engraver. *Plan des Hafens von Navarin*, 1827. **N.B.** Compare with C. **Bruegner**, above.

Bruehl, H. ...*Plan de l'ancien Magontiacum*, *c.*1800; *Carte von Wiesbaden*, 1819.

Bruehl, Johann Benjamin. Engraver. *Prospect der Stadt Halle*, *c.*1710.

Bruel, Georges (1871-1944). French historian and geographer, explored large parts of west and central Africa. *L'Afrique équatoriale française*, 1918, 1930. **Ref.** BROC, N. *Dictionnaire illustré des explorateurs Français du XIXe siècle: I Afrique* (1988) pp.60-62 [in French].

Bruellow, Friedrich. *Geognostische Wandkarte*, 1856; *Wandkarte... von Berlin*, 1869.

Bruff, Joseph Goldsborough. Architect and mapmaker. *Map Exhibiting the position of the Lands occupied by Indian Tribes in Amity with the United States...*, 1839 (superimposed upon H.S. Tanner's map of the United States of the same year); *The State of Florida*, 1846; Mexican War, 1847; *Plan of the Isthmus of Tehuantepec*, 1851; *Army Map of the Seat of the War in Virginia*, New York, J. Disturnell 1861; *New map of the seat of war in Virginia and Maryland*, New York, J. Disturnell 1863.

Bruff, Peter. Engineer. Plans of the Quay of Ipswich 1850-1852.

Bruggen, Johann van der (1695-1740). Engraver in Prague and Vienna. *Grund-Riss von der... Stadt Wien*, *c.*1720; Autriche (11 sheets), 1737.

Brugisser, Ferdinando (1816-1840). Topographical engineer and geodesist for the Officio Topografico di Napoli. Conducted triangulation surveys of the border between the Kingdom of Naples and the Papal States. **Ref.** VALERIO, V. *Società uomini e istituzioni cartografiche nel mezzogiorno d'Italia* (1993) p.462.

Brugmans, S.J.H. *Historische Atlas...*, 1893.

Brugnoli [Brognoli; Brognolio], Bernardino [Bernardo] (1538/1539-1583) *b.* Verona, *d.* Mantua. *Veronae Urbis Territorium*, Venice, Paolo Forlani 1574 (used by Abraham Ortelius 1579).

Brugsch, Heinrich. *Karte des Alten Aegypten*, 1857; *Aegyptus Antiqua*, 1875.

Brugsma, F.C. *Atlas van het Koningrijk der Nederlanden...*, Groningen, J.B. Wolters 1856 and later editions; *Atlas der geheele aarde*, 1864; *Nederland en zijne Overzeesche Bezittingen*, Groningen, J.B. Wolters *c.*1866, 1869.

Brugueirolle, Ate. *Géographie du Département du Gard*, *c.*1874.

Bruhns, *Doctor* Carl Christian (1830-1881). Associated with the Weimar Geographisches Institut. One of the editors of *Hand-Atlas der Erde und des Himmels*, 1859 (with H. Kiepart, C. Graef and A. Graef); *Großer Hand-Atlas des Himmels und der Erde*, 1871 and later editions (with H. Kiepart, C.F. Weiland, A. & C. Graef, O. Delitsch); *Atlas der Astronomie*, 1872.

Bruijn, Adriaan de. See **Bruyn**, Adriaan de.

Bruin[s], Cornelis. See **Bruyn**, Cornelis de.

Bruin, Joris. See **Braun**, Georg.

Bruins, F. Geography teacher. *Het Beeld der Aarde* (30 maps), Groningen, P. Noordhoff 1877, (43 maps) 1878; *Kleine Schoolatlas* (16 maps), Groningen, P. Noordhoff *c.*1880; *Nederland en Insulinde...*, 1889; *Kleine Schoolatlas*, 1891; *Atlas van Nederland*, 1895.

Bruinsz., Pieter. See **Bruynsz.**, Pieter.

Bruist, Gilles. Dutch surveyor and cartographer, worked in Overijssel. Schets van de Heerlijkheit Kuinder..., 1701.

Bruist, Henrick. (*fl.c.*1698). Dutch surveyor and cartographer, worked in Overijssel. 'Het gebied grenzend aan de Rivier de Linde van Kuinre tot voorbij "Hillen Verlaat..."'.

Bruletout de Préfontaine, —. Carte géographique de l'Isle de Cayenne, 1762 (with Louis-Charles Buache).

Bruls, Michael. See **De Bruls**, Michael Godhart.

Brummer, V.R. *Karta öfver Åland*, 1838.

Brun, Giovanni. Engraver. *Roma antica*, 1801.

Brun, *Captain* Jayme. *Mapa de la Republica de la Nueva Granada*, 1843-1847; Port de Sabanilla, 1843-1856.

Brun, V.A.M.. See **Malte-Brun**, Victor Adolphe.

Brunacci, —. *Teorica e la practica del piloto*, Livorno 1819.

Brunacci [Brunacius], Baldo. *Mapa Mvndi*, 1516 MS. **N.B.** David Woodward has raised doubts as to the authenticity of this and other manuscript world maps of the period. WOODWARD, D. *The Map Collector* 67 (1994) pp.2-10.

Brunacci, Francesco. Italian astronomer. *Planisfero del globo celeste*, Rome 1687.

Brunckow, Berthold. *Grundriss von Berlin*, 1858; *Karte des Preussischen Staates*, *c.*1867.

Brunclík, Josef (1850-1929). Bohemian [Czech] cartographer and teacher. School atlases, 1904 onwards.

Bruneau, Capitaine —. *Carte du Sud Oranais* (4 sheets), 1884.

Bruneau, Robert. Engraver and printer of Antwerp. Additamentum to Abraham Ortelius's *Theatrum orbis terrarum*, 1603; Ottavio Pisani's *Globus terrestris planisphericus* (12 sheets), 1612. **N.B.** This information provided by Tooley is not supported by R. KARROW in *Mapmakers of the sixteenth century and their maps* (1993), or R. SHIRLEY in *The mapping of the World* (1983).

Brunel, Isambard Kingdom (1806-1859). Civil engineer and inventor. *Cheltenham and Great Western Railway*, 1834-1845; *Northumberland Railway*, 1844; *Worcester and Oxford Railway*, *c.*1845.

Brunelo, Felice. Italian engraver. Isola di Cipro, published R. Savonarola 1713.

Bruneseau, — (*fl.*18th century). French engraver.

Brunet, Roch. French engraver. Involved in the preparation of 13 sheets of the Cassini *Carte de France*, *c.*1751-1767.

Brunfaut, J. Carte sulfurifère de la Sicile, 1874.

Bruni-d'Entrecasteaux [Bruny d'Entrecasteaux], Antoine Raymond Joseph (1739-1793). Governor of Mauritius and Île de Bourbon. In charge of an expedition in search of the explorer Jean-François Galaup de La Pérouse (1791-1794); surveyed coast of Australia. *Carte générale de la partie Méridionale de la Nouvelle Hollande*, *c.*1800; *Atlas du Voyage de Bruny d'Entrecasteaux, 1791-1793*, published by Beautemps-Beaupré, Paris 1807. **Ref.** HOOKER, B. *The Map Collector* 43 (1988) pp.19-20.

Bruningen [Buningen], Symon Arentsz. [Aartsz.; Aerntsz.] van. Dutch land surveyor and cartographer, worked in Zuid-Holland; Rijnland (1586-1596).

Brunnetiann, Max. *Specialkarte der Provinz Hessen-Nassau*, *c.*1899.

Brunner, Christoph (1696-1733). Swiss draughtsman of Basle. *Territorium Basiliense*, 1729.

Bruno, Colao (19th century). Lithographer of Naples.

Bruno, J. Descriptio orbis terrarum veteribus, 1708.

Brun-Rollet, Antoine (1807-1858). French explorer in the Nile valley. Le *Nil blanc et le Soudan...*, 1855-1863.

Brunton, J. Maps of railways in India, 1863.

Brunton, R. Henry. *Nippon* [Japan] (4 sheets), 1876, 1880.

Bruny-d'Entrecasteaux, Joseph. See **Bruni-d'Entrecasteaux**, Antoine Raymond Joseph.

Brusch [Bruschius; Brusthius], Kaspar [Caspar] (1518-1557). Born at Schlaggenwald im Egerland, died in Steinach. Travelled in Germany, Austria, Italy and Switzerland, became Poet Laureate to Charles V (1542.) Fichtelgebirge, *c.*1541 MS.

Bruschi, Giacomo. *Topografia del Porto e Città di Genova*, 1789.

Bruschius. See **Brusch**, Kaspar.

Brusthius, Caspar. See **Brusch**, Kaspar.

Brutt, Peter. *Plan ... for Harwich Docks*, 1852 (printed by Cowell's Anastatic Press).

Bruun, Ph. J. (1804-1880). Finnish geographer. Works on Eastern Mediterranean.

Bruyn [Bruijn], Adriaan de. Dutch land surveyor and cartographer, worked in Leiden. Admitted as surveyor in 1599; worked in Rijnland, 1600-1631.

Bruyn [Bruin; Bruins; Le Brun], Cornelis de. *Cornelis de Bruins Reizen over Moskovie, door Persie en Indie*, Amsterdam, Willem en David Goeree 1711; *Voyage au Levant...*, Paris, Guillaume Cavelier 1714.

Bruyn, D.L.J. de. *Omstreken van Utrecht*, 1871.

Bruyn, Martinus de. *Bybelsche Huis en Zak-Atlas...* (15 maps), Amsterdam *c.*1780.

Bruynsz. [Bruinsz.], Pieter. Dutch land surveyor and cartographer, 'landmeter van Rijnland, Kennemerland en Waterland', lived in Haarlem from 1596. Map of the River Spaarne, 1584.

Bruzen de la Martinière, Antoine Augustine (1662-1746). Geographer. *Le Grand Dictionnaire Géographique* (10 volumes), 1726-1739, republished by Boudet, 1766; *Introduction a l'Histoire de l'Asie, de l'Afrique, et de l'Amerique*, Amsterdam, Zacharie Châtelain 1735 (with maps by N. Sanson d'Abbeville); wrote a preface for Marsigli's *La Hongrie et le Danube*, 1741.

Theodore de Bry (1528-1598). (By courtesy of Rodney Shirley)

Bry family. Compiled and published sets of voyages in many languages between 1590 and 1634, commonly referred to as 'De Bry', and based on work by John White and Jacques le Moyne de Morgues. **Ref.** FAUPEL, W.J. *A brief and true report of the new found land of Virginia - a study of the de Bry engravings* (1989); GARRATT, John G., 'The maps in De Bry' *The Map Collector* 9 (1979) pp. 2-11.

• **Bry** [Brij], Theodore de (1528-1598). Goldsmith, engraver and publisher, born in Liège, worked in England 1586-1588. Engraved charts for the English edition of Waghenaer entitled *The Mariner's Mirrour* 1588; *Grands Voyages* [to America, containing many famous engravings of the New World] (14 parts), begun in 1590 and completed by his sons in 1601 (many later editions), so-called because it was larger in size than the later version; re-issued Caspar Vopel's map of the Rhine, 1594. **Ref.** FAUPEL, W.J. *The Map Collector* 52 (1990) pp. 33-36.

•• **Bry**, Johann [Joannes] Theodore de (1561-1623).

•• **Bry**, Johann Israel de.

Sons of Theodore, above. *Petits Voyages* [to the east and north, covering Africa and the East Indies] (13 parts), 1598-1628; Filippo Pigafetta and Durarte Lopes's Africa, 1598, 1624.

•• [Bry] - **Merian**, Matthaüs. Son-in-law of Theodore. Involved in publishing later versions of the Voyages. See **Merian**, Matthaüs.

Bryan, *Captain* —. Corrected John Thornton's *A Large Chart of... Antegua*, 1701.

Bryan, Francis T. *Map of Routes to the Gold Region of Western Kansas*, *c.*1860.

Bryan, *Lieutenant* H. *Niger Territories*, 1899.

Bryan, Hugh. Surveyor. Surveys used as one of the sources for De Brahm's *A Map of South Carolina...*, 1757.

Bryan, Margaret (*fl.*1797-1815). British teacher and astronomer. Star charts in System of Astronomy, 1797 and later.

Bryant, Andrew (*fl.*1822-1835). English surveyor and publisher of 27 Great Ormond St., London. Prepared a series of large-scale English county maps as part of a proposed *British Atlas* of which only thirteen maps were completed: *Map of the County of Hertford* (4 sheets), 1822; Suffolk, 1823; Surrey, 1823; Norfolk, 1824; Oxfordshire, 1824; *...the County of Gloucester*, 1824; *Map of the County of Buckingham* (2 sheets), 1824; Bedfordshire, 1826; Northamptonshire, 1827; Lincolnshire, 1828; Yorkshire East Riding, 1829; *Map of the County Palatine of Chester...*, 1831; *Map of the county of Hereford...*, 1835.

Bryant, Jacob. *Map of Malta*, 1767.

Bryce, *Major* —. *Chart of the Western Harbour of Alexandria*, 1802-1804; Lower Egypt, A. Arrowsmith 1807.

Bryce, *Reverend* Alexander. Clergyman of Kirknewton, active in the Established Church of Scotland. With the encouragement of the Philosophical Society of Edinburgh he undertook a significant survey of the north coast of Scotland, 1740, published as *A Map of the North Coast of Britain from Row Stoir of Assynt to Wick in Caithness*, 1744 (engraved by R. Cooper).

Bryce, David. *Pearl Atlas of the World*, 1893; *Vest Pocket map of Glasgow*, *c.*1895.

Bryce, J. Annan. *Burma*, 1886; *Burma Siam-China Railway*, *c.*1888.

Bryce, James. *A Cyclopedia of Geography*, 1856; *The Family Gazetteer*, *c.*1862; *The Library Atlas*, 1875-1876; Comprehensive Atlas, *c.*1886.

Bryce, James (1806-1877). Geologist. Papers on Northern Ireland and Scotland.

Brydone, J. *Plan of... Edinburgh*, *c.*1858; *Tourist Map of Scotland*, *c.*1858.

Bryer, F. Engraver. *Post Office Map of Hertfordshire*, 1874 and later editions to 1898; *Post Office Map of Surrey*, 1874, 1878.

Bryus, Jakob. See **Bruce**, Yakov Vilimovich.

Buache [de la Neuville], family. Geographers and publishers of Paris, great rivals to the Robert de Vaugondy family. **Ref.** PEDLEY, M.S. *Bel et utile: the work of the Robert de Vaugondy family of mapmakers* (Tring 1992) pp.9 *et passim*; KONVITZ, J.W. *Cartography in France 1660-1848* (Chicago 1987) *passim*.

• **Buache**, Philippe (1700-1773). One of the foremost French geographers of the 18th century, noted for his theoretical cartography. Married daughter of Guillaume Delisle in 1720; entered Dépôt des cartes et plans de la marine, 1721; 'premier géographe du roi', 1729; member of the Académie des sciences, 1730. Addresses: 'Paris Sur le Quay de la Mégisserie près le Pont Neuf' (1737-1741); 'Quai de l'Horloge' (1745-1763). Re-issued maps of Guillaume Delisle and Alexis-Hubert Jaillot. Isles de l'Amérique, 1724 MS; Martinique, 1732; *Carte de France levée géometriquement* [outline map of France in 116 segments], 1735 MS (Bibliothèque Nationale); *Carte des Lieux où les differentes longuers du Pendule à Secondes ont eté Observées*, 1740; ...*Carte minéralogique sur la nature du terrain...de l'Europe*, 1746, 1751; *Carte physique et profil du Canal de la Manche...*, 1752, 1756; Nouvelles découvertes au Nord, 1753; Atlas physique, 1754; Atlas géographique et universel, 1762; map of the Seine, 1766; and many others. **Ref.** PEDLEY, M.S. *Imago Mundi* 36 (1984) pp. 48-63; KISH, G. *Imago Mundi* 28 (1976) pp. 129-136.

•• **Buache de la Neuville**, Jean Nicolas (1741-1825). Nephew of Philippe Buache, from whose widow he bought Buache's geographical collection 1773; collection sold on to J.A. Dezauche, 1780. 'Conservateur du Dépôt des cartes et plans de la marine'; 'adjoint géographe Académie des sciences'; 'premier géographe du roi', 1782. Dominica, 1778; *Nouvelle Carte de la Partie Septentrionale du Globe comprise entre le Kamchatka et la Californie*, 1781; maps for *La vie de Washington*, 1807; charts for La Pérouse.

Buache, Louis-Charles. *Carte géographique de l'Isle de Cayenne*, 1762 (with Bruletout de Préfontaine).

Buasso, V. *Pianta della Città di Torino*, 1800.

Bucelin, Gabriel (1599-1681). Swiss genealogist and map author, produced many manuscript maps of towns and regions in southern Germany and Austria. *Historiae universalis*, Ulm 1654; *Praecipuarum Universi Terrarum Orbis*, Ulm 1658, 1659, 1664; *Auctarium sive nuclei historici*, Augsburg 1664. **Ref.** KULLEN, S. *Im Oberland* (Ravensburg) 8, 1 (1997) pp.31-44 [in German].

Buch [Buck], Leopold, *Freiherr* von (1774-1853). German geographer and geologist. Süd-Tyrol, 1823; *Atlas zur physicalischen Beschreibung der Canarischen Inseln*, 1825; Deutschland, 1838.

Bucha, Zacharias. Armillary sphere, 1593.

Buchan, *Captain* David. Leader of expedition to North Pole (with Rear-Admiral F.W. Beechey) in 1818. *Chart... of the Expedition... 1818*, 1843.

Buchanan, George. Southern Hemisphere, for John Thomson's *A New General Atlas of the World*, 1814 and later; *Plan of the Estate of Corehouse*, 1841.

Buchanan, J. Chart of Cork Harbour, 1808.

Buchanan, Thomas. *A New Chart of the Northern Ocean*, 1820.

Buchard, —. *Fleuve Sénégal* (13 sheets), 1893.

Büchel, Emanuel (1705-1775). Swiss draughtsman of Basle, specialising in village and local maps of the region. Map of the canton of Basle, Basle 1766 (with D. Bruckner); contributed plans and drawings to D. Herrliberger's *Helvetische Topographie*, 1754 and later.

Bucher, A. *Plan de la Forteresse de Gironne*, 1812.

Buchnerd, Mrs —. *Plan of the City of New York in the Year 1735*, 1735 MS **Ref.** COHEN, P.E. & AUGUSTYN, R.T. *Manhattan in maps 1527-1995* (1997) pp.60-61.

Bucholtz, György (1688-1737). Hungarian cartographer of Késmárk [now Kezmarok, Slovakia]. Mapped caves at Deményfalva [now Demanova, Slovakia], 1719 (engraved by Sámuel Mikoviny).

Bucholtz, Leopold von. *Map of Virginia*, Richmond, Ritchie & Dunnavant 1858; ...*State of Virginia*, 1859.

Buchon, Jean Alexandre C. (1791-1846). Publisher of Paris. *Atlas Géographique, Statistique, Historique et Chronologique des deux Amériques*, Paris, 1825 (French edition of Carey and Lea's American atlas of 1822); *Notice d'un Atlas en langue catalane*, 1843.

Buchwalder, *Colonel* Antoine-Joseph (1792-1883). Swiss engineer and topographer in Bernese Jura and Bern. *L'ancien Évêché de Bâle*, *c.*1820. **Ref.** SCHWAB, S. *Le Colonel Buchwalder (1792-1883)* (Neuveville, 1893) [in French].

Buck, brothers. Topographic artists and engravers from Yorkshire. Worked together from *c.*1724 onwards, sharing premises in London from 1731. 'At the Golden Buck, Great Russell Street, Bloomsbury', (1731-1735); at the Middle Temple (from 1735); 'N°·1 Garden Court, Middle Temple' (from 1749); also at the 'Green Canister by the Crown and Anchor Tavern, near St. Clement's Church in the Strand'. They produced over 400 engravings of monasteries, abbeys, castles and other ruins [*Buck's Antiquities*] and several series of views of English and Welsh towns [*Town Prospects*]. **Ref.** HYDE, Ralph *A Prospect of Britain: the town panoramas of Samuel and Nathaniel Buck* (1994).

• **Buck**, Samuel (1696-1779). Began work on drawing and engraving views of towns and historic buildings of Yorkshire for John Warburton, 1719. With the encouragement of members of the Society of Antiquaries he began working on his topographical views *c.*1720, a work which would continue for most of his life. He set himself up at the 'White Swan in Brownlow Street, Off High Holborn', London, in 1724.

• **Buck**, Nathaniel (*d.c.*1753/1759). Younger brother of Samuel. Artist and engraver 'at the Golden Buck in Warwick Street near Golden Square, Soho' (1724-1731). He worked closely with his brother from around 1724.

Buck, James. *A Plan of an Estate in... St. Marylebone*, 1780.

Buck, L. von. See **Buch**, Leopold von.

Buck, Pieter de. Dutch surveyor and cartographer. *Caarte ende discriptie figuratief van de zeedijcken van Groot Merschmont...*, 1579 (with Jan Simonse, Dircks Hendrikz., and François Horenbault).

Buck, Rufus. Mining maps of British Columbia, 1897 (with A.M. Bouillon).

Buck, Walter M. *Map of... New Brunswick*, *c.*1874.

Bucker, Johan (*fl.*1738). Dutch surveyor and cartographer, worked in Noord-Brabant.

Buckert, Frederick. *General Chart of the Cattegat*, 1810-1811, published 1825, 1827 and later.

Buckhout, Leavitt. ...*Obey River Oil Region*, *c.*1863.

Buckinck, Arnold[us]. Printer of Cologne, partner and successor to Conrad Sweynheym, from whom he took over work on the Rome edition of Ptolemy, the first to contain maps, 1478 (text edited by Domitius Calderinus). **N.B.** Compare with Arnold **Pannartz**. **Ref.** KARROW, R.W. *Mapmakers of the sixteenth century and their maps* (1993) p.134.

Buckingham, D.W. County surveyor for New Haven, Connecticut. *Map of the City of New Haven*, Jocelyn, Darling & Co. 1830 (engraved by N. & S.S. Jocelyn).

Buckingham, John Silk. Map of Syria and Palestine, 1848.

Buckland, J.W. *Map of the Hop-growing District of Worcestershire and Herefordshire*, 1890.

Buckland, William Thomas (*fl.* 1838-1853). Surveyor, worked on estate, road and tithe maps in Buckinghamshire, including the parish of Datchet, 1839 MS; Dorney, 1844 MS; Parish of Horton, 1838 MS; Parish of Upton, 1848 MS.

Buckle, *Commander* C.H.M. *San Aleiro Island*, 1844; *San Joao Islands*, 1844.

Buckley, Samuel. Publisher 'at the Dolphin in Fleet Street', London. Draught of the City of Carthagena, 1699.

Buckley, William. Instrument maker. Quadrant, 1551.

Bucknall, Thomas (*fl.*1722-1753). Surveyor, produced estate maps of Hampshire, the Isle of Wight and Sussex including *Chillerton Farm*, 1722; *Mannour of Chillerton*, 1724; *Parnells Farm*, 1724.

Bucknor, Royston. Plan of the Town of Fredericksburg, Virginia, 1721 MS.

Bucknor, William. North American surveyor. Town of Marlborough, Virginia, 1691.

PROPOSALS

By Way of

SUBSCRIPTION,

FOR

Making a new Survey, and publiſhing a moſt correct MAP, of the County of *Suſſex*.

By *RICHARD BUDGEN*, Surveyor at *Frant* near *Tunbridge Wells*.

IT is propoſed, I. That an actual Survey or Admeaſurement be made of the whole County, divided into its Rapes and Hundreds; in which ſhall be deſcribed the Towns, Villages, Churches, Chapels, *&c.* according to their Forms and Situations; as likewiſe all Gentlemens Seats, Caſtles, Foreſts, Parks, Chaces, Creeks, Bays, Rocks, Shoals, Sands, Bridges, Lakes, Soil, (whether Woody, Heathy, Mountainous, or Champaign.) Alſo what belongs to the Iron Works, *viz.* Chief Mines, Furnaces, Forges, *&c.* Likewiſe Windmills, Watermills, whether for Corn, Gunpowder, *&c.*

II. That an actual Survey or Meaſurement be made, not only of the Poſt Roads that croſs the County, but likewiſe all the principal Roads that are of Note or Uſe to Travellers, inſerting the Names of all Commons, Downs, Foreſts, Greens, Croſs-ways, Villages, Streets, *&c.* that ſhall lye in or near the Roads. Deſcribing all Lanes and By-ways, that ſhall turn out of the Roads, to what Places they lead. To delineate and lay down the Roads thus ſurvey'd exactly, according to their ſeveral Angles, Turnings, Bearings and Situations, noting the Miles between Town and Town, whereby the Diſtance, not only of the Towns, but alſo of other interjacent Places will appear by Inſpection.

III. That every thing relating to Natural Hiſtory of the County, whether Mines, Iron, Coal, or other Minerals, ſhall be inſerted; as alſo the Manufactories; how far any River is navigable; Antiquities, where met with, whether *Roman*, *Saxon*, or *Daniſh*, to be noted.

IV. To give greater Advantages to this Map than any other yet publiſhed, not only as to the Roads, the true Longitude and Latitude of Places, the Value of Livings in the King's Books, whether Rectories or Vicaridges, *&c.* But alſo, and that more eſpecially, by giving remarkable Facts and Circumſtances to Places in eaſy Characters to be explain'd by themſelves, that thereby it may, in ſome meaſure give, an Hiſtorical as well as a Geographical Account of the County.

Engraved by JOHN SENEX, and printed on the fineſt Imperial Paper, containing in length about five Foot, and in breadth about four; which will make the Scale ſo large as to contain a Mile in an Inch.

Thoſe who are willing to have their Arms engrav'd are to pay half a Guinea in Hand, and half a Guinea more on the Delivery of a Map illuminated, paſted on Canvas with Rolls. Of this ſort no more will be made up than are ſubſcrib'd for: And thoſe who are not willing to be at that Expence may have one in Sheets for half a Guinea.

SUBSCRIPTIONS will be taken in by RICHARD BUDGEN, at *Frant* near *Tunbridge Wells*; BERNARD LINTOT, at the *Croſs Keys* between the *Temple Gates*, and by JOHN SENEX in *Saliſbury Court*; by at *Chicheſter*; by at *Petworth*; by at *Horſham*; by at *Eaſt Grinſted*; by at *Arundel*; by at *Lewes*; and by at *Eaſtborne* in *Suſſex*.

N. B. This Map is intended to be deliver'd at *Chriſtmas* next.

A published proposal for a survey and map of the county of Sussex, England, by Richard Budgen (1695-1731).

Bucquoy, Jacob [Jakob] de (1693-1772). Dutch surveyor and traveller. Baay Fals, 1752; *De Waterwereld*, 1752; Reize naar de Indien, Haarlem 1757.

Budai, Ésaiás (1766-1841). Hungarian cartographer, born Pér, studied at Oxford and Göttingen, later professor at Debrecen college. Died in Debrecen. *Oskolai Új Átlás* [New School Atlas], 1800 (first Hungarian school atlas); also made globes, 1791. **Ref.** STEGENA, L. *Térképtörténet* [Map history] (Budapest 1981) [in Hungarian].

Budd, T.A. *Harbor of Bridgeport, Connecticut*, c.1835 (with other surveyors of the United States Coast Survey); one of the surveyors on the Charles Wilkes survey of the Pacific Coast, 1841.

Buddingh, D. *Zak-Atlas van het Koningrijk der Nederlanden* (17 maps), 1842.

Budgen, family. Surveyors of Sussex. **Ref.** KINGSLEY, D. *Printed maps of Sussex 1575-1900* (Lewes 1982) pp.57-59.

• **Budgen**, Richard I (1695-1731). Estate surveyor and meteorologist in Sussex. Produced the first large scale map of Sussex, *An actual Survey of... Sussex* (6 sheets), 1723 and later (engraved John Senex); estate plans, 1724-1728; *An exact plan*, 1730 (accompanied his pamphlet *The Passage of the Hurricane from the Sea-side at Bexhill in Sussex to Newington-level... May 1729*).

•• **Budgen**, Richard II (1730-1789). Son of the above, credited on the 1779 edition of his father's map of Sussex. Also worked as an estate surveyor and plans by him are known dated 1757-1783; *A scetch of the roads within fourteen miles of Tunbridge Wells*, Jasper Sprange 1774, 1802, 1809, 1811, J. Clifford 1820.

••• **Budgen**, Thomas (1766-1832). Son of Richard II. Estate surveyor of Lewes, produced a large number of estate plans, 1788-1826; *A new and correct Plan of Brighthelmstone* [Brighton], 1788 (engraved by S. Neele); commissioned to undertake a survey of the River Ouse, 1791.

••• **Budgen**, Charles (1769-1838). Son of Richard II, reported as working for the Ordnance Survey, 1798-1824.

•••• **Budgen**, William (*b.*1790, *fl.*1809-1830). Grandson of Richard II, worked with his brother John (below) in Tunbridge Wells.

•••• **Budgen**, John (*b.*1794). Worked as a surveyor 1826-1840, often in collaboration with brother William.

Budington, A. Co-publisher (with R. Whiteford) of Whiteford's *Map of the County of New Haven, Connecticut*, 1852

Budker, A. *Plan de la ville de Mâcon*, 1893.

Buechel, Emanuel. *Canton Basel*, 1766.

Buechel, J. Military plans of Germany, mainly Rhineland, 1855-1856.

Buecking, H. *Geologische Karte von Attika*, 1891; *Geologische Skizze von Nord-Celebes*, 1899.

Buehler [Bühler], Adolph. *Geognostische Karte von Reichenhall...*, 1867; *Reichenhall*, 1874; *Berchtesgaden*, 1874, 1897; Salzkammergut, c.1879.

Buehler [Bühler], Edouard. *Karte von Curland*, 1848, 1874.

Buehler, James A. (1813-1870). Copperplate engraver of Nuremberg. 15-cm terrestrial globe, 1840-1850.

Buel, John. Publisher of New York. Geographical Ledger, 1794; New York, 1794.

Buell, Abel (1742-1825). Silversmith, type-founder and metal engraver of Connecticut. From c.1774 moved to New Haven and began working with Bernard Romans. Moses Park's *Plan of the Colony of Connecticut in North-America*, 1766; *Capt. Parkers. Chart, of Saybrook Barr* [Connecticut], c.1771; *Part of the Province of East Florida*, 1774 (engraved by Abel Buell and Paul Revere for Bernard Romans's *A Concise Natural History of East and West Florida*); *A New and Correct Map of the United States of North America*, 1784 (said to be the first map of the United States by an American publisher, and the first to show an American flag). **Ref.** WROTH, L.C. *Abel Buell of Connecticut silversmith type founder & engraver* (New Haven 1926, revised edition Middletown 1958).

Buell, Wyllys. Lawyer of Ohio. County map of Upper Muskingham and Scioto Valleys of Central Ohio, 1833.

Buenau, *Count* Heinrich von. *Atlas Historique*, 1763-1767.

Bueninck, Johan [Jan] (*fl.*1666-1707). Dutch land surveyor and cartographer, worked in Gelderland and Overijssel.

Bueno, F.A.P. Mappa de Amazonas, 1865.

Buenting, Heinrich. See **Bünting**, Heinrich.

Buerchner, L. *Die Insel Nikaria*, 1894.

Buergin, J. *Carte du Massif principal des Vosges*, *c.*1868.

Buesching, H.O.C. *Schauplatz der fünf Theile der Welt*, von Reilly 1789-1791.

Buesser, E. *Plan von Strassburg*, 1873.

Buettner, J. *Haupt- und Handelsstadt Frankfurt*, 1870.

Bufalini, Leonardo. *Roma*, 1551, second edition (woodcut, 24 sheets), 1560, used later by J.B. Nolin 1748.

Buffa, Frans. *Noord Zee Kanaal*, *c.*1876.

Buffier, Claude *SJ*. Theologian and geographer from Warsaw, working in Paris. *Géographie universelle* (*c.* 20 maps), Paris, Pierre François Giffart *c.*1749, 1754 (engraved by Desbruslins), Italian edition publishes as *Geografia Universale Del P. Buffier*, Rome, G. Petroschi 1775, and other editions.

Buffon, George Louis le Clerc, *comte* de. (1707-1788). Wrote *Histoire Naturelle*, 1749-1804 (with 2 maps by Didier Robert de Vaugondy, engraved by Guillaume Delahaye).

Bufford, J.H. Lithographer of 260, Washington Street, Boston. Ebenezer Barker's *Plan of the City of Charlestown...*, 1848; *Plan of the Michigan Central R.R.*, 1855; *Plan Portland*, 1858; *Map showing the war operations in Virginia and Maryland*, 1861.

Bugarsky, Jean. *Carte de la Principauté de Serbie...*, Belgrade, 1845.

Bugge, Thomas (1740-1815). Danish geographer and astronomer. Supervised the first triangulation of Denmark from 1762. Faeroe Islands, 1770; *Kort over Bornholm...*, Gottfried Angelo 1805 (with F. Wilster).

Bühler. See **Buehler**.

Buhrmann, Gerhard (*fl.*1670s-1684). Swedish military surveyor and engineer, worked in Skåne with Rutger von Ascheberg. Schoonenkarta, 1681-1684 MS.

Buhsel, —. *Plan von Danzig*, 1869.

Buisson, —. *Carte du Département de Haute Saône*, 1874.

Bula, —. Publisher of Paris, for A. Lorrain, 1831-1834.

Bulem, L. Lithographic artist specialising in hill and relief shading. For P.M.G. Vandermaelen, 1829-1833.

Bulifon [Bolifon], Antonio (*b.*1649). French bookseller, cartographer and engraver in Naples. Napoli (21 sheets), 1692; Pianta di Pozzuoli, 1696.

Bulionis Belga, Aegidius. See **Boileau** de Bouillon, Gilles.

Bull, Adolf (*fl.*1850-1880). Danish cartographer. *Kongeriget Danmark* (4 sheets), 1851; *Bornholm*, 1869; Skoleatlas, 1870; Støre Skoleatlas, 1876.

Bull, Fenwick. Mapseller. Re-issued G. Foster's *A New and exact Plan of the Cities of London and Westminster...*, 1754 (first published 1738).

Bull, G. Engraver for William Blachford's *New Chart of the Coasts of Holland and Germany*, 1827.

Bull, *Colonel* William (*d.*1755). Surveyor, Lieutenant-Governor of South Carolina. Manuscript map of Southeastern North America, 1738 (taken from the map by John Barnwell), one of the surveys used by William Gerard de Brahm as a basis for *A Map of South Carolina and a part of Georgia*, 1757 and many later editions.

Buller, William. *Plan of the Anchorage at Blewfield Jamaica* (with John Leard), 1793; *Plan of the anchoring places at Old Harbour*, 1793.

Bullet, Jean Baptiste. French cartographer. *Plan de Paris* (12 sheets), 1710 [not verified].

Bullet, Pierre (1632-1716). 'L'architect du roi et de la ville.' *Plan de Paris* (12 sheets), 1676 (with François Blondel).

Bullionis. See **Boileau** de Bouillon.

Bullock, *Lieutenant* (later *Commander*) Frederick. Admiralty Surveyor, 1820-1854. Newfoundland, 1821-1826; survey of the River Thames and Thames Estuary, 1827-1847; River Medway, 1840; Approaches to Harwich, 1847; Dungeness, 1847-1850; *A Survey of the Coast and Inland Waters between Dover and Worthing*, 1848-1854; *Estuary of the Thames and Medway*, 1849; Gravesend Reach, 1852; *The Downs*, 1857.
Ref. ROBINSON, A.H.W. *Marine cartography in Britain* (Leicester 1962) pp.134-140, 196-197.

Bullock, Henry. Map of disputed land on the borders between Scotland and England, *c.*1552 MS (Public Record Office, Kew).

Bullock, Thomas. Survey of Great Salt Lake City, London, Longman & Co 1860 (engraved by H. Adlard); *Sketch of Post Roads and Post Offices in Utah Territory...*, 1863 MS.

Bullock, William. Atlas historique pour servir au Mexique, 1823.

Bulman, W.J. Bird's-eye view of Winnipeg, 1900.

Bulow, *Baron* A. von. Nicaragua and Panama, 1847, published 1851.

Bumble, Otto. Karte von Schlesien, *c.*1876.

Bumpus, John. Cartographer and publisher of 23 Skinner Street, London. One of the publishers of *Crosby's Complete Pocket Gazetteer of England and Wales...*, *c.*1818, 1822; *A New Plan of London and Its Environs*, 1827, 1828, 1829, 1830 (engraved by S.R. Cooke); *Milton's Improved Plan of London and its Environs...*, 1828.

Buna, Wilhelm C. Bavaria, 1760; *Carte d'une partie des cercles du Haut et du Bas Rhin* (6 sheets), 1760; *Totius Ducatus Bavariae Tabula* (9 sheets), *c.*1760.

Bunau, *Count* Henri de. German cartographer. *Atlas Historique, Géographique, Topographique et Militaire*, *c.*1760.

Bundaiken Uhei. Kyoto publisher. Woodcut world map, 1710.

Buningen, Symon Arentsz. See **Bruningen**, Symon Arentsz.

Bunjeimon Ohata. Mapmaker and publisher of Nagasaki. Hishu Nagasaki no zu, *c.*1778.

Bunjiemon Toshimayo (*fl.*18th century). Map and print publisher of Nagasaki. Map of Kyushu province, *c.*1783; Plan of Nagasaki.

Bunker, *Captain* E. South End of New Zealand, 1808-1809 MS.

Bunney & Gold. Publishers. *Chart of the Lines of Magnetic Variation in the Seas around Africa*, 1799

Bunnik, H. ...*Karte von dem Furstenthum Ostfries*, 1804.

Bunning, J.B. Architect. *Proposed Improvements in the neighbourhood of Smithfield*, 1851.

Buno, Conrad. ...*Statt Hannover*, for J.L. Gottfried's *Archontologia Cosmica*, 1695.

Bunonis [Buno; Bunone], Johannes. Professor of Classics at Leiden, editor of 1661 edition of Philipp Clüver's *Introductionis in universam geographicam*, (first published 1624), for which he produced *America* and a world map *Typus Orbis terrarum...*, Wolfenbüttel 1661 and later editions; world map *Orbis terrarum...*, 1664, published 1677 (for Johannes Strubius).

Bunou, Philippe, *SJ* (1680-1759). Abrégé de géographie, 1716.

Bünting [Buenting], Heinrich (1545-1606). Professor of theology at Hanover. *Itinerarium sacrae scripturae*, 1581 and later editions in various languages, contained woodcut maps including: *Cosmographia Universalis* [world map]; *Die Ganze Welt in einem Kleberblat* [World in the form of a clover leaf]; Europe as the queen of the World; and Asia as a winged horse.

Bunyan, C. Stephen Holloway. *A Plan of South Leaga Bay* [Sumatra], Alexander Dalrymple, 1774.

Buondelmonti [Anexerius; Anxerinus; Auxerinus; Bondelmonte; Ensenius], Christoforo (*c.*1385-*c.*1430). Florentine humanist and cartographer, travelled through the Mediterranean and Aegean. *Liber insularum Archipelagi* [book of islands] (79 maps and views), 1420 MS (perhaps the earliest *Isolario*, many later manuscript copies made); used by Bartolommeo dalli Sonetti in a printed *Isolario*, *c.*1485; by Benedetto Bordone, 1528; and Marco Boschini. **Ref.** *Dizionario Biografico* vol.15 pp.198-200 [in Italian]; TURNER, H.L. 'Christopher Buondelmonti and the Isolario' in: *Terrae Incognitae* vol.19 (1987) pp.11-28.

Buono, Florino dal. *Città di Bologna*, 1636.

Buonopane, Camillo (1797-1862). Geographical engineer, worked for the Officio Topografico di Napoli. *Pianta dell'Isola di Favignana*, *c.*1820; worked on the surveys for the Officio's map of the Kingdom of Naples, from 1835; survey of the coast of Faro and the Strait of Messina, 1840-1843; contour and land use map of Messina, 1843-1844. **Ref.** VALERIO, V. *Società uomini e istituzioni cartografiche nel mezzogiorno d'Italia* (1993) pp.463-464.

Buonsignori, Stefano [Stephanus Monachus Montis Oliveti] (*d.*1589). Cartographer of Florence, Benedictine monk of Montoliveto Maggiore near Siena. Cosmographer to the Duke of Tuscany. *Dominii Florentini locorumq. adiacentium descriptio*, 1584; *Senarum locorumq. adiacentium descriptio*, *c.*1584 (both used by Abraham Ortelius 1595); also series of wall maps for the Palazzo Vecchio, Florence.

Buot, Jacques (*d.c.*1695). French astronomer. Star chart, Paris 1657.

Buraeus, Anders. See **Bureus**, Andreas.

Burbank, C.W. *Map of the City of Worcester*, 1872.

Burch, Bartholomeus van der (*fl.*1640-1643). Dutch land surveyor and cartographer, worked in Drente.

Burch, U. *Plan of the Battle of Busaco*, *c.*1810.

Burchard 'of Monte Sion' [Brocard; Brocardus] (*fl.*13th century). German monk. His description of the Holy Land was published in the 16th century. **N.B.** Not to be confused with the 16th-century French Franciscan Bonaventure **Brochard**. **Ref.** KARROW, R. *Mapmakers of the sixteenth century...* (1993) p.96.

Burchardi, *Lieutenant* —. Engineer. Colberg & Environs (16 sheets), 1857-1860 MS.

Burchell, William John (1782-1863). Explorer and naturalist. *A Map of the Extratropical part of Southern Africa*, 1822.

Burchgrave, Claas (*fl.*1564). Dutch cartographer, worked in Gelderland and Noord-Brabant.

Burcht, W. van den. Dutch cartographer. Bois le Duc ['s-Hertogenbosch], Hondius 1630.

Burck, A. Montana and Wyoming, 1870.

Burckhardt, Johann Ludwig [Ibrahim] (*b.*1784). Born in Lausanne, travelled through Egypt and Arabia. *l'Arabie Petrée*, 1834; *Karte der Sinaï-Halbinsel*, 1859; *The Imperial Map of Palestine*, *c.*1864.

Burdett, Peter Perez (*c.*1734-1793). Surveyor, mapmaker, draughtsman and engraver. Worked first in Derbyshire and then moved to Liverpool. In 1771 he produced the first aquatint to be published in England and sold the process to Paul Sandby. Moved to Germany in 1775 to conduct an official survey of the Margravate. Founded the surveying school which became the Karlsruhe Technical School. *...Survey of Derbyshire...* (copper engraving on 6 sheets), 1767, 2nd edition 1791; Liverpool Harbour, 1771 (engraved by Billinge); *...the County Palatine of Chester...* (4 sheets), 1777, 2nd edition 1794. **Ref.** LAXTON, P. *Dictionary of national biography: Missing persons* (Oxford 1993) p.102.

Burdigalensis, Bernardinus. See **Bordeaux**, Bernardin de.

Burdonius. See **Bordone**, Benedetto.

Bure, Anders . See **Bureus**, Andreas.

Bureau, E. *Atlas de Géographie Militaire*, 1871.

Bureau, Jacques. Carte du Mississippi, 1700.

Buret, Daniel (*fl.c.*1674). Dutch surveyor and cartographer, working in Zeeland.

Bureus [Buraeus; Bure; Bureus de Boo], Andreas [Anders] (1571-1646). Swedish cartographer and secretary to King Karl IX, later to Gustaf II of Sweden. Lapponiae, Bothniae &c nova delineatio, 1611 MS; *Orbis Arctoi nova et accurata delineatio* (wall map of Scandinavia in 6 sheets), 1626 (engraved by Valentin Thrautman of Stockholm), re-issued by Pieter Mortier *c.*1710, reduced edition engraved by Abraham Goos and published by Visscher, 1628, also used in later atlases including those of Hondius, 1633, and Blaeu. **Ref.** ÅKESSON, L.. *Andreas Bureus - Father of Swedish cartography*, in *IMCoS Journal* 75 (Autumn 1998) pp.49-54; BRATT, E. *En kronika om kartor över Sverige*, pp.47-63 [in Swedish].

Burford, father and son. Successors to the panorama business of Henry Aston Barker, Leicester Square and Strand, London, 1822. Many panoramas were displayed until 1863.
• **Burford**, John. In partnership with Henry Aston Barker, bought out the Panorama in the Strand, 1817. Retired in 1826, leaving the business in the hands of his son.
•• **Burford**, Robert (*d.*1863). Son of John. Worked with his father, and with Henry Aston Barker. **Ref.** HYDE, R. *Panoramania!* (London 1988) *passim*.

Burg, Michael. See **Burghers**, Michael.

Burgartz, F. *Höhenkarte von Vorarlberg*, *c.*1864.

Burgdorfer, Johann Jakob. Publisher in Bern. Plan Bern, 1811; Berner Oberland, 1816; *Le Canton de Berne*, 1830.

Burges, Bartholomew. The Solar System displayed, Boston 1789.

Burges, John (*fl.*1697-1702). Estate surveyor of Stanton St. John (Oxfordshire). Manuscript maps of estates in various English counties including the manor of Boarstall [Buckinghamshire], 1697 MS.

Burgess, Daniel [& Co.]. Publisher of Hartford, Connecticut (1835) and 60 John Street, New York (1854), associated with Paine & Burgess, and later, Cady & Burgess. Roswell C. Smith's *Smith's Atlas for Schools, Academies, and Families*, 1835; *Smith's Atlas of Modern and Ancient Geography...*, 1854; *A Concise and practical system of Geography*, 1855.

Burgess, F.I. *Topographical survey of... Dihlee district* (8 sheets), 1852.

Burgess, R.W. *West End of Lake Erie*, 1852; Chart of the head of navigation of the Potomac River, 1857.

Burgess, William. See **Burgiss**, William.

Burgett, H.W. & Co. Atlas of the State of Vermont, 1876.

Burgh, Michael van de. See **Burghers**, Michael.

Burgh, P. van. Dutch map copyist, worked in Noord-Holland. Copied Jan van Scorel's Noord-Holland, *c.*1552, 1650.

Burgh, Thomas. *...Bay and Harbour of Dublin*, 1728 (with J. Perry).

Burgh, Willebordus van der. See **Burght**, Willebord van der.

Burghart, David. Viennese engraver. For József Aszalay's *Mappa generalis regni Hungariae*, Vienna 1830-1838.

Burghers [Burg; Burght; Van de Burgh], Michael (*fl.*1670-1720). Born Amsterdam, came to England 1672. Engraver to Oxford University. Engraved the map in Dr Robert Plot's *Natural History of Oxfordshire*, 1677; several maps in Dr Edward Wells's *A New Sett of Maps*, 1700 and later editions; Devizes, 1714; some maps for E. Nutt's *Magna Britannia*, 1720.

Burghley, William Cecil, *Lord*. See **Cecil**, *Sir* William, *First Baron* Burghley.

Burght, Michael. See **Burghers**, Michael.

Burght [Burgh], Willebord [Willebordus] van der. Flemish cartographer and military engineer, worked in Noord-Brabant. *Quarta pars Brabantiae cujus Caput Sylva ducis*, *c.*1630 (used by Willem and Joan Blaeu, 1634 and later).

Burgi, Josef. Relief map of Palestine, 1845.

Bürgi, Jost (1552-1632). Swiss clock-maker, instrument maker, mathematician and globe-maker, born Lichtensteig, worked in Kassel and Prague. Celestial globe, *c.*1585, 1592. **Ref.** KRETSCHMER, DÖRFLINGER & WAWRIK *Lexicon zur Geschichte der Kartographie* (1986) pp.125-126.

Burgiss [Burgess, Burgis], William. Draughtsman, painter, designer, publisher and possibly engraver. Arrived in New York from England c.1718. *A South Prospect of ye Flourishing City of New York...*, 1719-21, re-issued London, 1746; *A South East View of ye Great Town of Boston in New England in America*, *c.*1725, and other views; *...Plan of Boston in New England...*, *c.*1728 (engraved by Thomas Johnston). **Ref.** HOLMAN, Richard B. 'William Burgis' *in Boston prints and printmakers* (Colonial Society of Massachusetts, Boston, 1973).

Burgklehner, Matthias, *the younger* (1573-1642). Mineralogical map of Tyrol (12 sheets), 1608, re-issued 1611, and in 1629 by Andreas Spängler; map of Tyrol in the shape of an eagle, 1609, 1620.

Burgkmair, Hans, *the elder* (1473-1531). Copper engraver of Augsburg. N. Cusa's map of Germany, *c.*1500.

Burglechner [Burgklehner], Mathias, *the elder* (*d.*1603). Cartographer of Innsbruck.

Burgoci, Francisco de. *Geographia e descripcion de la parte septentrional del Polo Artico de la America*, Mexico, Juan Ruiz 1674.

Burgoyne, *General* John. British General active in the American War of Independence. *State of the Expedition from Canada...*, London 1780 (contained battle plans engraved by William Faden).

Burgstaller, Johann Christoph (1674-1758). Collaborated with others on maps of Slovakia.

Burhans, S.H. Maps of Illinois, 1860-1864 including: *Map of Cass County, Illinois*, Chicago 1860 (with J. van Vechten); *Map of Will County, Illinois*, Chicago, Edward Mendel 1862 (with Van Vechten); published W.L. Flower's *Map of Cook County*, 1862 (with Van Vechten); *Map of DeWitt County, Illinois*, Chicago 1864 (with L.M. Snyder); *Map of Logan County, Illinois*, Chicago 1864 (with Snyder).

Burkart, J. Der Bau der Erdrinde, 1838 (with J. Nöggerath).

Burkhardt, E.F. *Eisenbahn-Karte von Deutschland*, *c.*1845.

Burland Desbarats Lithographic Company. Lithographers of Montreal. *Johnston's New Topographical Map of the Whole Dominion of Canada*, 1874; *Map of the Country to be traversed by the Canadian Pacific Railway*, 1876; Alan Mcdougall's *Map shewing the Railway Systems in the Eastern Provinces of Canada*, Ottawa 1876; Charts of the Halifax Fishery Commission (4 sheets), 1877; *Chart shewing the relative positions of Bute Inlet, the Valdes Islands, Nodales Channel...* [British Colimbia], 1877; and others.

Burleigh, —. *Plan of the River Wear from Newbridge to Sunderland Barr* (4 sheets), 1737.

Burlingame, E.H. *...City of Williamsport*, 1872.

Burman [Burmen], Gerhard von. Skåne, Stockholm 1756.

Burmania, G. (*fl.*1736). Dutch military engineer and cartographer, worked in Groningen.

Burmeister, Herman K. (1807-1892). Argentine geographer.

Burmester, Johan (1664-1713). Swedish surveyor of Gotland.

Burn, George. *The Channel Pilot*, 1786, Sayer & Bennett 1791, Laurie & Whittle 1803 (with John Stephenson); *Coasting Pilot for Great Britain and Ireland* (27 sheets), 1791-1803; *Coast from Arundel Haven*, 1794 (revision of J. Avery, 1721); *Chart from the Nore to Orfordness*, *c.*1800; *New and Correct Chart of the Thames Mouth*, 1809.

Burn, Robert. *Map of Rome*, 1878.

Burnaby, *Captain* —. Surveyor. Ordnance Plan of the Parish of St Longfomacur (14 sheets), 1856.

Burncker, J.M. *Carte d'une partie de la Syrie*, 1782.

Burnes [Burns], *Lieutenant* Alexander. *A New Map of Cutch*, 1829; *Central Asia*, John Arrowsmith 1834.

Burnet, Thomas (1635-1715). Telluris theoria, London 1681; maps of Western Hemisphere, Europe &c.

Burnett, Gregory (*fl.c.*1820-1844). Surveyor. ...*Sutherland* (6 sheets), 1833 (with William Scott).

Burnett, J.C. *Plan of Moreton Bay*, 1846.

Burnett, *Lieutenant* R., *RN*. Charts of Tasmania, 1860; *Approaches to Hobart Town*, *c.*1868.

Burney, James (1750-1821). Survey of the south and east coast of Tasmania, [n.d.]; *Chart of the Coast of China*, 1811; *Chronological History of the Discoveries in the South Sea or Pacific Ocean* (5 volumes), 1803-1817 (with maps engraved by J.Russell); *Chronological History of the North Eastern Voyages of Discovery*, 1819.

Burnford, Thomas. Engraver. *A Map of the French Conquests*, *c.*1674; *A Mapp of the Seventeen Provinces*, 1674; maps for Richard Blome's *Cosmography*, 1682.

Burns, Alexander. See **Burnes**, Alexander.

Burns, Arthur. Geodaesia Improved, 1771.

Burns, John L. Publisher of Salt Lake City. *Map of Salt Lake City and suburbs*, 1871

Burow, A. *Plan der Ferstung Rendsburg*, *c.*1862; *Plan von Fridenica*, *c.*1872.

Burr, David H. (1803-1875). Topographer to the United States Post Office and Geographer to the House of Representatives of the United States. Also publisher of 189 Broadway, New York. Atlas of New York, 1829; *A Series of Maps for a General Atlas*, New York 1831; *A New Universal Atlas*, New York, Wm. Hall & Co. 1835; *Map of the Northern Parts of Ohio, Indiana and Illinois with Michigan...*, 1836, 1843; *Topographical Map of the City and Country of New-York and the Adjacent Country*, New York, J.H. Colton & Company 1836; *Map of the United States of North America with Parts of the adjacent countries*, 1839; American Atlas, 1839 (maps engraved and printed by John Arrowsmith, London, and were also available separately); United States, 1842; Ohio, J.H. Colton 1845; World, Boston, Haven 1850; *Map showing the Extent of Surveys in the Territory of Utah*, 1856; *Map of a Survey of the Indian Reservation on Spanish Fork*, 1856; and many others. **Ref.** COHEN, P.E. & AUGUSTYN, R.T. *Manhattan in maps 1527-1995* (1997) pp.120-121; RISTOW, W.W. *American maps and map makers* (1985) pp.103-108.

Burr, Henry A. Topographer to the Post Office Department, Washington DC. Draughtsman for [John] *Disturnell's New Map of the United States and Canada*, 1854.

Burr & Ballisat. J. Burr and G. Ballisat, surveyors of New South Wales. *Plan of the Allotments of Ground Granted from the Crown in New South Wales*, 1814.

Burrell, Peter (1723-1775). Surveyor General of Land Revenue, Fellow of the Royal Society. Directed Thomas Richardson's survey of the Royal Manour [*sic*]of Richmond (15 sheets), 1771; Maidstone Hill, East Greenwich, 1772 MS.

Burriel, *Father* Andres Marcos. *Mapa de la America Septent...*, 1757; maps for Fr Miguel Venegas' *Noticias de la California*, Madrid 1757.

Burritt, Elijah H. *Atlas, Designed to Illustrate the Geography of the Heavens*, New York 1835 (star charts engraved by F.J. Huntington).

Burrough, *Sir* James. *Ichnographia Eccles. Monasteri St. Edmundi*, *c.*1745.

Burrough, Stephen. See **Borough**, Stephen.

Burrough, William. See **Borough**, William.

Burroughs, H.N. *A New Map of Texas Oregon and California*, Philadelphia, S. Augustus Mitchell 1846; another version published Philadelphia, Thomas, Cowperthwait & Co. 1849; reworked as *Map of the State of California, the Territories of Oregon & Utah and the chief part of New Mexico*, published in: *A New Universal Atlas*, Philadelphia, Thomas, Cowperthwait & Co. 1850.

Burroughs [Burrows], Samuel (*fl.*1677-1707). Bookseller 'at the Crown in Cornhill' (1697); 'Bible and Three Leggs in the Poultry' (1702); 'Little-Britain' (1703-1705), London. One of the publishers of the 3rd edition of P. Gordon's *Geography Anatomiz'd: or, The Geographical Grammar*, 1702 (first published 1693); one of the sellers of Robert Morden's *The New Description and State of England*, 1704 edition (first published 1701).

Burrow, Edward J. [& Co.]. *The Royal Automobile Club County Road Map and Gazetteer*, 1938.

Burrow, Reuben (1747-1792). Mathematician of Yorkshire. Worked at the Royal Observatory with the Astronomer Royal, Nevil Maskelyne. Surveys for gravitational measurements, 1774-1775; recruited by the Board of Ordnance for geodetic surveys in India.

Burrowes, Thomas H. *State-Book of Pennsylvania* (60 maps), Uriah Hunt 1847 (second edition).

Burrows, S.E. *Roccos*, 1864.

Bursian, Conrad. *Geographie von Griechenland*, *c.*1872.

Burslem, F.H. Surveyor and lithographer. *Map of the Settled Portion of the Colony of South Australia*, 1842; *Barossa Special Survey*, 1843; *Green Hills*, 1843; *Rapid Bay*, 1845.

Burstall, *Lieutenant* (later *Commander*) Edward. Hydrographer. Newhaven Harbour, 1839; River Crouch, 1847; Folkestone, 1848; Beachy Head, 1852; *North West Coast of Mull*, 1860.

Burston, John (*fl.*1638-1666). 'Thames School' chartmaker, apprenticed to Nicholas Comberford in 1628, Freeman of the Drapers' Company, 1638. 'Dwelling at the Signe of the Platt in Ratcliff Highway neare London'. Mediterranean, 1658, *c.*1666; East Atlantic, 1660 MS. **Ref.** CAMPBELL, Tony 'The Drapers' Company and its school of chartmakers' in: WALLIS & TYACKE (Eds.) *My head is a map* (1973) pp.81-106.

Burt, Henry M. *Map of the White Mountain Region*, 1874.

Burt, R.K. Printer of Holborn Hill, London and Crystal Palace Sydenham. W.G. Harris's *Map of the Province of South Australia*, 1862.

Burton, *Captain* —. *Africa. West Coast*, 1861 (with Beddingfield).

Burton, Charles Edward (1846-1882). Astronomer.

Burton, Edmund Merceron. *Egyptian Campaign*, 1882, 1887.

Burton, Giles (*fl.*1612-1656). Estate surveyor of Sussex, mapped estates in Kent, Sussex and Surrey including *...Parish of Northiam* [Sussex], 1636.

Burton, James. Surveyor. Intended Improvements to the Bedford Estate, 1806 (engraved by J. Tuppin).

Burton, M. Surveyor. Proposed Roads in Buckingham, 1853 MS.

Burton, Richard [R.B.]. See **Crouch**, Nathaniel.

Burton, *Sir* Richard Francis (1821-1890). Traveller, orientalist, translator. Various manuscript sketch maps and plans to illustrate his travel works, including sources of the Nile (adapted from a map by Speke), Nigeria, and routes between Zanzibar and the Great Lakes in Eastern Africa in 1857, 1858 & 1859 (with J.H. Speke); Pangani River, 1857.

Burty, —. 'Rue de la Veille Estrapade 3 Paris'. Engraver for Victor Levasseur, 1851.

Burucker, Johann Michael (1763-1808).

Engraver and geographer of Nuremberg. Behring Straits, 1781.

Burwell, Lewis. ...*County of Brant, Canada West*, 1859.

Burwood, R. Engraver. ...*County of Pembroke*, 1827; ...*Parish of St. Pancras*, 1850.

Busbie, I. Co-publisher, with M. Lownes, J Helme & I. Browne, of M. Drayton's *Poly-Olbion*, 1612, 1613.

Busby, B. Surveyor. *Orangeburgh District South Carolina*, 1820, for Robert Mills's *Survey Atlas of South Carolina*, 1825.

Busby, Richard (*fl.*1748-1759). Surveyed estates in various English counties, including *Ly Farm* [Wiltshire], 1750.

Busch, Andreas (*fl.*1656-1674). Instrument maker of Limburg. Armillary sphere containing an orrery, 1657; Gottorp Globe, 1664 (under direction of Adam Olearius).

Busch, Cornelius von. Geographische Carten von gantz Teutschland, 1648.

Busch, Edward. Engraver. *Atlas of the late Borough of Germantown*, 1787.

Busch, Georg (*fl.*16th century). German astronomer.

Busch, Georg Paul (*d.*1756). Engraver. *Strasse von Gibraltar*, 1720; *Das Herzogthum Schlesien*, 1742; *Bayern*, 1745.

Busch, J.G. *Karte des Deutschen Reiches*, 1872; *Plan von Leipzig*, *c.*1890.

Büsching, Anton Friedrich (1724-1793). Lutheran theologian and geographer, professor of philosophy at Göttingen 1754-1761. *Neue Erdbeschreibung*, Hamburg 1754 onwards; Dutch edition, 1761; English edition, 1762; editor of the journal *Magazin für die Historie und Geographie*, 1767-1788; and of the review journal *Wöchentliche Nachrichten von neuen Landcharten...*, 1772-1787, both containing many maps. **Ref.** BUCHHEIM, G. *The Map Collector* 59 (1992) pp. 24-27; KRETSCHMER, DÖRFLINGER, WAWRIK *Lexicon zur Geschichte der Kartographie* (1986) vol. C/1, pp.126-127 [in German].

Busemann, M. Richard Kiepert's *Grosser Hand-Atlas*, 1893-1896.

Bush, B.F. ...*Bay County* [Michigan], 1869; ...*Saginaw Valley*, 1870.

Bush, William. Civil Engineer. *Plan of the Proposed London Steam Docks at Deptford*, 1841.

Bushnan, John. Midshipman and assistant surveyor under Captain William Edward Parry during the voyage in search of a Northwest Passage. Chart of the north east coast of America, 1824.

Busius [Bos; Bossius], Albertus. Publisher and painter of Düsseldorf. Gerard Mercator's *Atlas...*, 1595, 1602 editions.

Busius, Jacob. See **Bos**, Jacob.

Busk, *Captain* H. Isle of Man, 1861.

Bussac, —. *Plan de la forêt de Compiègne*, 1772.

Bussard, William. *A map of Georgetown in the District of Columbia*, 1830 (engraved by W. Harrison).

Bussche, Georg Wilhelm von dem. *Topographische Aufnahme vom Hochstift Osnabrück*, 1764-1767 (with F.C. Benoit); *Osnabrugensis Episcopatus*, *c.*1775.

Bussching, D. Grand atlas d'Allemagne, 1789.

Bussell, Charles E. Atlas of the City of Eau Claire, Wisconsin, 1888.

Bussemacher [Bussmacherus, Buxemacher], Johannes [Janus] (*fl.*1580-1613). Cartographer, engraver and printer of Cologne, 'Off. S. Maximini Strasse'. Westphalia, 1590; maps for Matthias Quad's *Europae...*, 1594, 1596, expanded to *Geographisch Handtbuch*, 1600, and *Fasciculus geographicus*, 1608 (maps also used by Botero, 1611); *Regni Hispaniae...*, 1600; *Novi orbis pars borealis...*, Cologne, 1600; M. Quad's ...*Rheni... descriptio*, 1604.

Bussemer, Ph. *Touristen-Karte des Schwarzwaldes*, *c.*1895.

Bussius Belga, Jacobus. See **Bos**, Jacob.

Bussmacherus. See **Bussemacher**.

Bussy, J.H. de. *De Bussy's Atlas van Nederland voor Kantoor, School und Huis*, 1915.

Bustamente, Daniel Sanchez. *Mapa de la república de Bolivia*, Gotha, J. Perthes 1913.

Bustillo, José Maria de. *Carta Esférica... del Canal do Sn Jorge*, 1857.

Bustron, Florio. Cipro, *c.*1560 MS.

Büthner, F. (1622-1701). Professor of Mathematics at Danzig. Cometa scepticus, 1681 (engraved by J.A. Seugel).

Butler, B.F. Lithographer in San Francisco. San Francisco, 1852; North Portion of San Francisco County, 1853; G. Woodman's *Map of the Mining Sections of Idaho & Oregon...*, San Francisco, A. Gensoul 1864.

Butler, E.W. *Plan of the Parish of Gruyere*, 1875.

Butler, Frederick. *A Modern Atlas*, Deming & Francis 1825.

Butler, *Reverend Dr* George. Edited *Public Schools Atlas*, 1872-1893.

Butler, H. *Dramatic Almanac Map for 1853 showing the principal Theatrical Towns*, 1853.

Butler, J., *junior*. Plan of Paddington, 1824.

Butler, J.P. *Manila Bay*, 1830.

Butler, James. Maps of the New York area including: ...*Peekskill*, 1852; ...*Sing-Sing*, 1852; ...*Staten Island*, 1853.

Butler, Peter. ...*City and Suburbs of Cork*, 1774.

Butler, Philip and Richard. *Map of the County of Catherlogh* [Ireland] (9 sheets), 1789.

Butler, *Dr* Samuel (1774-1839). Bishop of Lichfield and Coventry. *Atlas of Ancient Geography*, 1822; *Atlas of Modern Geography*, 1825 and later editions.

Butler, Thomas. *Chart of the Western Part of the Pacific Ocean*, 1799.

Butler, & Co. Publishers of Philadelphia.
• **Butler**, E.H. & Co. Successor to Cowperthwait, Desilver & Butler. *Mitchell's School Atlas*, 1859 edition; *Mitchell's New Intermediate Geography*, Philadelphia, E.H. Butler, 1860 and many other editions; *Mitchell's New Primary Geography*, 1860 edition; *Mitchell's New School Atlas. Mitchell's Modern Atlas*, 1869 edition; succeeded by J.H. Butler & Co..
•• **Butler**, J.H. & Co. Successor to E.H. Butler (above). Republished works that had appeared earlier under the imprints of Thomas, Cowperthwait & Co., Cowperthwait & Co., Cowperthwait Desilver & Butler, and E.H. Butler & Co.. *Mitchell's New Primary Geography*, 1876 edition.

Butlin, J. English surveyor. *Plan of Northampton*, Jefferys 1746 (with John Noble).

Butterfield, G.W. Surveyor. *Plan of Union Village in Medway*, 1855.

Butterfield, Michael (*c.*1635-1724). English instrument and globe maker.

Butters, R. Publisher & bookseller, 79 Fleet Street, London; later 22 Fetter Lane, Fleet Street. *Political Magazine*, 1789-1790 (with maps engraved by John Lodge); maps in *An Atlas of England*, 1803, 1805 (pocket atlas with maps after John Cary, many of them with North at the bottom); William Green's *The Picture of England illustrated with correct colour'd maps of the several counties*, 1804.

Butterworth, E. One of the engravers for John Thomson's *Atlas of Scotland*, completed and published in 1832; Aberdeen, 1826.

Butterworth, J. Engraved for Joseph Priestley, 1769; Proposed Canals, 1770 and 1790; Whitecloth Hall, 1775.

Butts, A.G. Revised J.R. Butts's map of Georgia, 1870.

Butts, J.R. *Map of the State of Georgia*, Macon [Georgia] 1870 (revised by A.G. Butts).

Buvignier, Armand. *Carte Géologique des Environs de Paris*, 1850, 1865.

Buwama Aardenburg, Sjoerd Cornelis (1758-*c.*1806). Dutch military engineer, surveyor and cartographer. Officer in Corps der Genie (1782); made series of surveys in Ommeland 1789-1794.

Buxemacher, Johannes. See **Bussemacher**, Janus.

Buxton, H.G. *De Soto's Route*, 1851.

Buxton, J.T. West Australian Goldfields, 1896.

Buxton, J.W. Stationer and publisher of Queen St, Brisbane, Queensland (1868). *The new map of Queensland...*, 1863.

Buy de Mornas, Claude (*d.*1783). 'Professeur de géographie et d'histoire'; 'géographe de Monseigneur le Duc de Berry'; 'géographe du roi' (1764). 'Rue et vis à vis le Prieuré du Cherche Midi' (1761); 'rue St Jacques près St Yves' (1762). *Atlas Méthodique et Elémentaire de Géographie et d'Histoire* (4 volumes), Paris, L.C. Desnos 1761, 1783; *Atlas Historique et Géographique*, 1762 (re-used some of the plates from the *Atlas Méthodique*).

Buys, Cornelis, *the younger* (pre-1524-1546). Dutch draughtsman, worked in Alkmaar, 1541-1545.

Buys, Fransz. Anthon. Dutch cartographic draughtsman, worked in Zuid-Holland. Untitled map of Ooltgensplaat, *c.*1524.

Buyskes, *Admiral* Arnold Adriaan (1770-1838). *Situatie-plan stadt Cabo de Goede Hoop*, *c.*1790 MS.

Buyskes, P.J. *Sea of Marmara*, 1855, 1859; *Hydrographische Kaart der Zeegaten van Goeree en de Maas*, 1869.

Buytendijck, E. van (*fl.*17th century). Dutch land surveyor and cartographer, worked in Utrecht.

Buzin, J.J. von. Croatien und Slavonien, 1877.

Buzzo, Francesco. ...*Val Telina e Val Ghiavena con suoi confini*, *c.*1640.

Bye, Cornelis de. See **Bije**, Cornelis de.

Bye, Joseph P. Engraver, son of the printer Deodatus Bye of London. Apprenticed to Benjamin Baker in 1793, Freeman of the Stationers Company 1801. Noted at 3, Duke Street, Smithfield (1806); Camberwell Grove (1810). Joseph Enouy's *A new map of Scotland*, 1803; *Doncaster*, William Faden 1805; 3 charts in Dessiou's *Le Petit Neptune Français; or, French Coasting Pilot*, London, W. Faden 1805 (3rd edition); John Purdy's *Harbours and Islands in the British Channel*, 1806; *A Sketch of the Island of Ceylon...*, 1807; Azores, 1807; for Charles Smith, 1808; John Purdy's *The Continent and Islands of Asia...*, London, Laurie & Whittle 1809; for Dr James Playfair, 1814; and others.
• **Bye & Smith**. Partnership noted in 1805.

Byers, W.N. City plans. *Excelsior, Nebraska*, *c.*1857; *Omaha City*, 1857 (with Wiltse); *Saint Vrain, Colorado*, *c.*1860 (with Wiltse).

Byles, A.D. Publisher in Philadelphia. Kent County, Delaware, 1859.

Bylica, Marcia [Martin] (1433-1493). Polish astronomer and mathematician. Studied at Krakow (1451-1459) and at Bologna (1463-1464); Professor at Bratislava 1467, and at Buda 1471. As its first owner, Bylica's name is attached to Hans Dorn's 40-cm metal celestial globe of 1480. **Ref.** AMEISENOWA, Zofia *Globus Marcina Bylicy z Olkuszai mapy nieba na wschodzie i zachodzie* (Wroclaw 1959) [in Polish].

Byram, Ephraim Niles (1809-1869). American clockmaker. Built a celestial globe, before 1834; 15-foot diameter planetarium, before 1836. **Ref.** WARNER, D. *Rittenhouse* Vol.2 N°.1 (1987) p.32.

Byrd, Theophilus (*fl.*1653-1666). Estate surveyor in East Anglian counties.

Byrd, William [William Byrd II] (1674-1744). Plantation owner and amateur naturalist of Virginia. Appointed head of the Virginia Commission on the boundary between Virginia and North Carolina and worked on an unfinished manuscript study of the area entitled *History of the Dividing Line*; supervised the survey of the boundary, 1728 (with other copperplates, this was probably intended as an illustration in his *History*). **Ref.** PRITCHARD, M.B. & SITES, V.L. *William Byrd and his lost history* (1993); GROVE, P. & WALLIS, H. *The Map Collector* 56 (1991) pp.14-15.

Byres, John. Chief Surveyor, Tobago. *Plan of the Island of Dominica*, 1776; *Plan of the Island of St. Vincent*, 1776; *Plan of the Island of Tobago*, 1776; *Plan of the Island of Bequia*, 1776, all engraved by John Bayly and published by Samuel Hooper.

Byrne, A.T. *Mapa de la República de Honduras*, 1886.

Byrne, George. Engraver. John Rocque's ...*County and City of Kilkenny*, 1758; Arthur Richard Nevill's ...*Survey of the County of Wicklow*..., 1760; John Drury's *Valentia Island*, 1788.

Byrne, Joseph James. *Plan of the Metropolis of Ireland,* 1819; estate plans of Galway, 1852.

Byron, *Commodore* John (1723-1786). Circumnavigator. Grandfather of the poet. Governor of Newfoundland, 1769-1772; manuscript charts of a world voyage, 1764-1766, including *The Isle of Danger in the South Sea*, 1765 MS; Voyage to the South Seas, 1766.

Byron, Samuel A. *A New Plan of Dublin*, 1791; *Plan of the City of Kilkenny*, *c.*1795.

Byström, *Colonel* A. Herman (1853-1933). Cartographic editor at Generalstabens Litografiska Anstalt, Stockholm. Took over Sven Hedin's manuscript maps from Asia expeditions. Karte öfwer Vermlands, Generalstabens Litografiska Anstalt, 1897.

C. See **Châtelain**, Henri Abraham.

C., F.I. *Atlas de géographie physique, politique et historique... à l'usage des écoles chrétiennes des Frères* (36 maps), 1878 (6th edition).

C., G.E.B. See **Ekeberg**, Carl Gustav Eric.

C., H. See **Courtier**, Henri.

C., J. *A Plan of the Canals ...*, *c.*1625; New Kind of Fire, 1628.

C., L.S. *Grundriss der Vestung Phillipsburghe*, *c.*1734.

C., M. See **Capellaris**, Giovanni Antonio de.

C., N. The principal passages of Germany, 1636.

C., P.V. See **Call**, Pieter van.

C., R. See **Crawford**, Robert.

C., T. A Short discourse of Newfoundland, 1623.

C., T. *A Chart of the Bahama Islands with a description of the Gulf of Florida*, London 1739.

C., W. *Port Royal with the Intended New Fortifications* [Jamaica], 1786.

Caamaño, Jacinto. Sailed aboard the *Aranzazu* on an exploration of the north-west coast of America. He produced several manuscript charts which accompanied his account of the voyage, 1792. **Ref.** WAGNER, H.R. *Cartography of the north-west coast of America to the year 1800* (Amsterdam 1968) pp.233-235, 355, 358.

Caballol, Ignasi. *Plano geometrico de la ciudad de Mataro*, 1852.

Cabañas, *Don* Roduga. *Descripcion y Derotero de Las Costas Pueros Bay as Yslas y Bajos del Mar del Sur* [Description and Rutter of the Coasts, Ports, Bays, Islands and Banks of the South Sea], 1692.

Cabay, —. Engineer and surveyor on the Cassini *Carte de France*. Worked with others on 12 sheets covering parts of central France, 1772-1783.

Cabeo, Nicholas (*fl.*1629). Italian author and savant. Two untitled world maps in *Philosophica magnetica*, Ferrara 1629. **Ref.** SHIRLEY, R. *The mapping of the World* (1983) N°328 p. 351.

Cabeza, Joan. *Mapa topografich de la montanya de Montserrat y de ses vessants*, 1909.

Cabeza de Vaca, Alvar Nuñez (1490-1561). Spanish explorer. Relation, 1542.

Cabillet, E. *Plan de la ville de Bourdeaux*, *c.*1840.

Cabot, John and Sebastian. Italian explorers who settled in England.

• **Cabot** [Caboto], John [Giovanni] (1450-1498). Venetian pilot, sailed westward from Bristol in 1497 to search for a route to Asia; in 1498 explored east and west coasts of Greenland, Baffin Island and the coast of Newfoundland. **Ref.** FUSON, R.H. 'The John Cabot mystique', in: PALMER & REINHARTZ (Eds.) *Essays on the history of North American discovery and exploration* (Arlington, Texas 1988).

•• **Cabot** [Gabato; Caboto; Cabotus; Chiaboto; Gabotta], Sebastian (*c.*1476-1557). Cosmographer and cartographer, son of John, thought to have sailed with his father on the voyages to the New World. Worked for the English crown before moving to Spain *c.*1512 and becoming an offical pilot for the Spanish Casa de la Contratación, 1518-1548; voyage to the Paranà and Paraguay rivers, 1526-1530; returned to England 1547. Various manuscript maps; map of Gascony and Guienne, 1512; untitled planisphere (eight sheets), Antwerp 1544 (known from one copy with text found in 1843 and now in Bibliothèque Nationale, Paris; and another copy, without text, found in 1984 at Zentralbibliothek der Deutschen Klassik, Weimar). **Ref.** KARROW, R. *Mapmakers of the sixteenth century and their maps* (1993) pp.103-112; MEURER, P. *Fontes cartographici Orteliani...* (1991) p.122-123 [in German]; KELSEY, H. 'The planispheres of Sebastian Cabot and Sancho Gutiérrez', *Terrae Incognitae* XIX (1987), pp. 41-58; SHIRLEY, R. *The mapping of the World* (1983) N°81 pp.92-93; WILLIAMSON, J.A. *The voyages of John and Sebastian Cabot* (1937).

Cabral, Francisco Augusto Monteiro. *Carta hydrographica do Porto de Goa*, 1842; *Planta da Ilha fortificada de Angediva*, 1847.

Cabrera, Francisco. Engraver for M. Rivera Maestre's *Atlas guatemaltico*, 1832.

Cacaut, François. *Plan de la Ville de Nantes* (4 sheets), 1757-1759.

Cacciamali, Giovanni Battista. Carte geologica della grupa del Monte della Maddalena, 1899.

Cacciatore, Leonardo (1775-*c.*1827). Historian of Naples. *Nuovo Atlante Istorico* (2 volumes), 1825, 1827, 1831.

Cacciatore, Niccolò (1780-1841). Astronomer and director of the Observatory at Palermo, professor of mathematics and astronomy at the university of Palermo. Instructor at the Istituto Topografico di Palermo. Handled the division of Sicily into 23 administrative districts, 1811; defined the difference in longitude between Palermo and Naples, 1833. **Ref.** VALERIO, V. *Società uomini e istituzioni cartografiche nel mezzogiorno d'Italia* (1993) pp.465-466.

Cadamosto [Ca da Mosto], Alysius Alvise [Luigi da], (1432-1480). Venetian navigator. Voyages to Gambia and Cape Verde Islands, 1455-1456, under commission from Prince Henry of Portugal. Nuovo Portolano... Levante, MS chart.

Caddy, *Captain* John, *RE*. Ordnance draughtsman, produced several manuscript surveys including fortifications at Berwick-on-Tweed, 1789; harbour at Sunderland, 1791 MS; survey of the coast from Blyth to Hartley, 1797; Sunderland proposed docks, 1832.

Cade, John. Reprinted Saxton's 1583 wall map of England and Wales (cartouche erased), 1678 (with William Morgan).

Cadell, family and company. Booksellers and publishers in the Strand, London (1767-1780); 141 Strand (1781-1845).

• **Cadell**, Thomas I (1742-1802). From 1767 Thomas was in partnership (as Thomas Cadell) with William Strahan and then Andrew Strahan (William's son). Successor to Andrew Millar as one of the sellers of E. Gibson's edition of Camden's *Britannia...*, 1772 (with maps by R. Morden, first published 1695); many travel books, including editions of Captain James Cook's *Voyages*, 1773-1784 (with William Strahan); *Map of the European Settlements in South America*, 1783; *Plan of the Siege of Charlestown in South Carolina*, 1787.

•• **Cadell**, Thomas II. Son of Thomas I, took over the company on the retirement of his father in 1793, working under the guidance of William Davies (as Cadell & Davies) until the death of the latter in 1819. Thomas then traded under his own name until 1845. Daniel Lysons's *The Environs of London*, 1792-1796; Lysons's *Magna Britannia* [Volume VI], 1822.

•• **Cadell & Davies** [T. Cadell Jun. and W. Davies]. Partnership of Thomas Cadell II, and

William Davies. Engravers and publishers of the Strand, London. *Palaestine*, 1797; John Pinkerton's *Modern Geography*, 1802, 1806, 1811; *North America*, 1803; Aaron Arrowsmith's atlas containing maps of Africa, 1803; maps for J. Barrow's *An account of travels*, 1804; *Chart of Table Bay at the Cape of Good Hope*, 1805; *Pinkerton's Modern Atlas*, 1809, 1812 (also published by Longman, Hurst, Rees, Orme & Brown of Paternoster Row, maps by L. Hebert); Daniel and Samuel Lysons' *Magna Britannia* [Volume I], 1806, [Volume II], 1808-1810, [Volume III], 1814, [Volume IV], 1816, [Volume V], 1817.

Cadell, *Captain* Francil (1820-1879). Scottish explorer in Australia. Explored the Murray River and tributaries, 1848-1859.

Cadell, Robert (1788-1849). Publisher of Edinburgh.

Cadet, Felix. *Atlas Spécial de la géographie physique...de la France*, J. Delalain 1855 (with F. Bazin).

Cadet de Metz, Jean Marcel (1751-1835). French mineralogist. Various works on the history of geography.

Cadle, J.C. Railway maps of New England, 1831.

Cadman, John (*fl.c.*1740-1757). Surveyed estates in various English counties including *A Survey of the Estates of John Earle of Sandwich*, 1757.

Cadorini, M. Publisher of Franciscus Schottus's *Itinerario Overo Nova Descrittione de'Viaggi Principali d'Italia*, Padua 1654.

Cadot, P. Lithographic printer of Santiago, Chile. *Carta que demuestra...la cuestion de limites Chileno-Arjentina...bajo la direccion de la Oficina Hidrográfica, en Agosto de 1881*; Carlos M. Prieto's *Carta Jeográfica de la Rejion Limítofe de los Paises Bellijerantes del Pacifico*, 1884 (2nd edition).

Cadwalader, A. ...*White Pine Mining District*, 1869.

Cadwalader, John (*d.*1786). *Plan of Princeton, Dec.21, 1776*, MS; Princeton and Vicinity [New Jersey], MS.

Cady & Burgess. Publisher of 60 John Street, New York, successor to Paine & Burgess. [Roswell C.] *Smith's Atlas Designed to Accompany the Geography*, New York, Cady & Burgess 1850 (first published Paine & Burgess 1839); *A Concise and Practical System of Geography*, New York, Cady & Burgess 1860 (first published Paine & Burgess 1846).

Caesius. See **Blaeu** family.

Cafferty, E.M. *Map of Norwich*, 1872.

Cagnat, René (1852-1937). French surveyor and archaeologist, worked in Tunisia. *Atlas archéologique de la Tunisie* 1883-1886.

Cagniard de la Tour, *baron* Charles (1777-1859). French engineer geographer.

Cagno [Canius], Paolo [Paulus]. Italian cartographer from Genoa. Compiled an untitled map of Naples, 1582, known from one example now at Biblioteca Angelica, Rome, later printings 1595, 1615. **Ref.** MEURER, P. *Fontes cartographici Orteliani...* (1991) pp.123-124 [in German].

Cagnoli, Antonio (1743-1816). Mathematician and astronomer of Zante. Works on mathematical geography.

Cagnoni, —. Engraver for Mauro Fornari, 1786-1789.

Cahill, Daniel. Irish cartographer. Queen's County [Leix] (4 sheets), 1806 (engraved by J. Ford).

Cahill, J.T. Draughtsman and surveyor of 130 Pitt Street, Sydney. *Oakville, North Willoughby*, *c.*1885; *Quakers Hill Estate*, 1886; *Harris estate* [Ultimo], 1892; *Town of Riverstone*, *c.*1892.

Cahill & Kirchiner. Surveyors, 112 Pitt Street, Sydney. *North Shore* [Sydney], [n.d.]; *Zachell's Estate, Coogee*, 1884.

Cahoone, John. Surveyor of Newport. Chart of Long Island Sound, from Montauk Point to Frog's Point, New York, Edmund Blunt 1805, 1806, 1813 (with Nicoll Fosdick).

Caignart de Saulcy, Louis Félicien Joseph. *The Imperial Map of Palestine*, *c.*1864.

Caillat, Philippe. *Carte de la Tunisie ancienne*, 1876.

Caillaux, Alfred. *Carte minière de la France*, *c.*1880.

Caille, J. de. See **La Caille**, Jean de.

Caille, L.A. du. See **Du Caille**, Louis Antoine.

Caille, *abbé* Nicolas Louis de la. See **La Caille**.

Cailleux, A. Atlas, ou collection des cartes d'après M. Bonne, 1802-1803.

Cailliaud, Frédéric (1787-1869). French geologist, archaeologist, egyptologist and explorer. Travelled in Egypt 1815-1819, and 1822. *Voyage à l'oasis de Thèbes et dans les déserts situès à l'orient et à l'occident de la Thébaïde*, 1821; *Voyage à Méroé*, 1826-1827; *Carte générale de l'Égypte et de la Nubie*, 1827; *Cartes géographiques du voyage à Méroé*, 1827; *Carte géologique de la Loire Inférieure*, 1851. **Ref.** BROC, N. *Dictionnaire illustré des explorateurs Français du XIXe siècle: I Afrique* (1988) pp.63-64 [in French].

Caillié, René (1799-1838). French explorer in west Africa in 1820s. *Journal d'un voyage à Tombouctou et à Jenné...1824, 1825, 1826, 1827, 1828*, 1830. **Ref.** BROC, N. *Dictionnaire illustré des explorateurs Français du XIXe siècle: I Afrique* (1988) pp.64-65 [in French].

Cailloue, Jean [John]. French bookseller in the Strand, London. *Situation of Mantua*, 1702; *Theatre of the War in the Sevennes*, 1703.

Cailloué, Pierre. Publisher and bookseller of 'Cour du Palais', Rouen. *Recueil Des Voyages Qui Ont Servie A L'Etablissement Et Au Progrez De La Compagnie Des Indes Orientales...* (10 volumes), 1725 (third edition of Renneville's voyages, first published Amsterdam 1702-1706).

Caimor, Balthasar. *Descriptio et situs Terrae Sanctae alio nomine Palestina*, 1617 (broadsheet version of G. de Jode's 1593 map). **Ref.** *Imago Mundi* 48 (1996) p.234.

Caimox, C. See **Caymox**, Cornelius.

Cairncross, Thomas W. Civil engineer. *Map of Cape Town*, 1891.

Caiz, Juan las. Descripcion del seno Mexicano Islas de Barvolento, 1708.

Cajamarca, —. *Atlas géografico del Peru*, 1865.

Calameus, J. See **Chaumeau**, Jean.

Calapoda, Giorgio. See **Callapoda**, Giorgio.

Calcura, P. Carta geognostica del Ustica, 1842; Carta geologica di Limosa, 1851.

Caldas, F.J. de. Rio Grande de la Magdalena (for atlas to von Humboldt's voyage), 1805-1834.

Calder, James Erskine (1808-1882). Admiralty Surveyor. *Map of Maria Island Van Diemen's Land*, *c.*1840; *Tasmania. Derwent River*, 1851; *Tasmania. Entrecasteaux Channel*, 1851; *Map of the Seat of War in New Zealand*, 1863.

Calder, *Captain* Patrick D. *RE*. Surveys at North Yarmouth [Norfolk], 1813.

Calderinus [Calderini], Domitius [Domizio] (*c.*1444-1476). Edited Virgil and other Latin poets, also translated Pausanias's guide to Greece into Latin. Edited Ptolemy's *Geographia*, Rome edition 1478 (maps printed by Conrad Sweynheim and Arnold Buckinck).

Calderon, Augusto. *Mapa de los Caminos de Hiero de España y Portugal*, 1878.

Caldwall, James. Engraver of Angel Court, Windmill Street, London (1768-1784). Peter Bell's *Environs of London*, 1769.

Caldwell, Joseph A. Publisher of Condit, Ohio, specialising in illustrated U.S. county atlases for the area from central Ohio to central Pennsylvania, including: *Knox County*, 1871; *Atlas of Franklin County Ohio*, 1872; *Caldwell's Illustrated, Historical Centennial Atlas of Washington County, Pennsylvania*, Condit, 1876; *Caldwell's Illustrated, Historical Centennial Atlas of Beaver County, Pennsylvania*, 1876; *Oil Belt of Clarion County*, 1878; *Monroe County*, 1898.

Caldwell, Richard. London mapseller. *The Traveller's Guide Thro' London*, 1739.

Calef, John (1725-1812). Surgeon of Massachusetts. *The Siege of Penobscot* [1779, Maine], 1781 (position of troops and fleet).

Calheiros da Graça, Francisco. Charts of Brazil, 1890-1894.

Calhoun, J. Surveyor General. *Plan of public surveys of Kansas & Nebraska*, 1856.

Calkoen, Jean Frederik van Beeck [Beeck Calkoen] (1772-1811). Dutch astronomer. Théorie de construction pour les mappemondes, Utrecht 1810.

Call, Jan van (*fl.*1647-1656). Dutch surveyor and cartographer, worked in Gelderland (1647-1656).

Call, John. *Arcot Fort*, *Carangoly*, and other maps in *History of the Military Transactions in Indostan*, 1778.

Call, Pieter van. Plans of towns in Belgium and France including Ath, 1698; Ghent, 1709; Bethune, 1710; Aire, 1711; Douay, 1718.

Callahan, Denis. Chart of the head of navigation of the Potomac River, 1857; Survey of the North and North-West Lakes, 1858.

Callan, Bernard. American surveyor and civil engineer. *Plan of the City of Rochester, New York*, 1851; *Map of the City of Cleveland, Ohio*, 1852; worked with I. Slatter on *Map of the City of Boston, Masssts...*, New York, Matthew Dripps, and Boston, L.N.N. Ide 1852 (lithographed by F. Mayer).

Callan, James. *A New Map of Arkansas*, 1836.

Callander, James (1758-1820). Surveyor. *Chart of the Knysna*, 1806; *Chart of Infanta or Broad River* [South Africa], 1817.

Callander, John. See **Callender**, John.

Callapoda [Calapoda; Sideri], Giorgio (*fl.*1537-1565). Portolan maker of Crete. World, 1550 MS, engraved on copper, 1552; manuscript atlas (10 charts), 1563. **Ref.** RATTI, A. *Imago Mundi* 40 (1988) pp.77-85.

Callaway, Thomas W. *Wilkes County*, 1877; *Putnam County, Georgia*, 1878.

Calléja[s], Félix del Rey (*b.*1750). Spanish General. *Mapa de la Colonia del Nuevo Santander* [Texas], *c.*1795 MS; *Plano geografyco de la provyncya de Texas...*, [n.d.].

Callendar, George. See **Callender**, George.

Callender, family. Engravers of Boston.
• **Callender**, Joseph (1751-1821). Engraver of Boston, working mostly on bookplates and pictures. Osgood Carleton's *Plan of Boston*, 1796; Osgood Carleton's *Map of Massachusetts Proper...*, Boston, B. & J. Loring 1801 (with Samuel Hill).
• **Callender**, Benjamin. Father of Benjamin and brother of Joseph.
•• **Callender**, Benjamin *Jr.* (1773-1856). Merchant and engraver of Boston, son of Benjamin, above. James Whitelaw's *A Map of the State of Vermont*, 1793, 1795 (published in *The Natural and Civil History of Vermont*, by Samuel Williams); for *Malham's Naval Gazetteer*, 1796, 1797, 1804, 1807; maps and charts for *The Naval Gazetteer*, 1797; maps for Jedidiah Morse's *American Gazetteer*, 1797; *Map of the District of Maine*, Boston 1799 (with Samuel Hill).

Callender [Callendar], George, *Master, RN*. *Boston Harbour surveyed*, 1769, published 1775.

Callender [Callander], John. *Terra Australis Cognita* (3 volumes), Edinburgh 1766-1768 (a compilation of many voyages to the southern seas from Vespucci, 1501, to Byron, 1764).

Callet, Jean François (1744-1798). French hydrographer.

Callewaert Frères, C. *Atlas général*, 1858; *Atlas élémentaire*, 1861; *Atlas classique...*, Brussels 1870 (edited by L. Piré).

Callier, Camille (1804-1889). French geographer and engineer, travelled extensively in the Levant, 1830-1845. His numerous plans and geographical descriptions are held by the Société de géographie, Paris. **Ref.** BROC, N. *Dictionnaire illustré des explorateurs Français du XIXe siècle: II Asie* (1992) pp.73-75.

Callot, *Freiherr* Eduard von. *Frankreich*, 1859.

Callot, Jacques (1592-1635). Painter and engraver of Nancy, teacher of Antonio Francesco Lucini. Engraved Jerusalem for Bernardino Amico's ...*Terra Sancta*, 1620 edition; *Tabula Obsidionis Bredanae* (6 sheets), 1624-1625; *Obsidio Rupellae*, 1628; *The Siege of La Rochelle...* (16 sheets), *c.*1629.

Calmar, Yvez. *Plan of Brava on the East Coast of Africa*, 1755, English edition 1780.

Calmet, Augustin (1672-1757). Historian and theologian. *Dictionnaire historique... géographique et litteral de la Bible*, 1730.

Calmet de Beauvoisin, Marie Antoine. French military cartographer. *Plan de la bataille d'Austerlitz...* [1805] (4 plans on one sheet), c.1805 MS; *Bataille de La Corogne*, 1809; *Itinéraires détaillés de plusieurs parties de l'Espagne*, 1813; map of Spain and Portugal, 1821. **Ref.** VANNEREAU, M.-A. in: *A la découverte de la terre* (Bibliothèque Nationale 1979) p.58 [in French].

Caloiro, Placido. See **Oliva**.

Caloiro e Oliva, Giovanni Battista. See **Oliva**.

Calon, *frères* [Calon de Beaulieu; Calon de Felcourt]. French surveyors, worked on seven sheets (5, 22, 23, 111, 116, 117, 148 in north and east France) of the Cassini survey of France between 1756 and 1764.
• **Calon**, Etienne-Nicolas de (1726-1807).
• **Calon de Felcourt**, Jean-Pierre (*b.*1729). Sheet 148 (Genève-Gex) of the Cassini *Carte de France*, 1761 (engraved by Dupain-Triel, père).

Calver, *Staff Commander* Edward Killwick, *RN*. Admiralty Surveyor. Chart of the Entrance to the River Tyne, 1838-1840; North and South Shields, 1838-1840; River Blyth with Southwold Harbour, 1844; Coasts of England, 1847-1856; River Tyne, 1852; Wick, 1857; *The Downs*, 1865.

Calver, W.B. *Master, RN*. Surveys of Tobago, 1864-1865, published 1867.

Calvert, Albert Frederick (1872-1946). Traveller and mining engineer. Expedition to north-west Australia, 1891-1893; *Albert F. Calvert's map of Western Australia showing goldfields*, *c.*1896; Coolgardie, 1896.

Calvert, Alexander. River Swale, 1792.

Calvert, Cecil. See **Baltimore**, 2nd Baron.

Calvert, Charles. See **Baltimore**, 6th Baron.

Calvert Lithographing Co. Lithographers of Detroit. *Map Showing the Line of the Port Huron & Lake Michigan and the Chicago & Michigan Grand Trunk Railway Companies...*, 1866; *Map of the State of Michigan*, 1869; Edward Molitor's *Sectional Map of the States of Michigan and Wisconsin*, 1871; City of Detroit, 1871, 1879.

Calvet, D. Ezequiel. *Plano del Puerto y rada de Barcelona*, *c.*1851.

Calvi, Ulisse. Engraver. Plan of Alexandria, 1882.

Calvin, John. Theologian. Map showing the location of the Garden of Eden, first published in *Commentaire*, 1554, and in Bibles from 1560. **Ref.** DELANO SMITH, C. & E.M. INGRAM *Maps in Bibles 1500-1600* (1991); DELANO SMITH, C. *The Map Collector* 39 (1987) pp.8-10; DELANO SMITH, C. *The Map Collector* 31 (1985) p.51 (letters).

Calvus, M. Fabius. *Antiquae urbis Romae* (20 plans), 1532.

Calwagen, Erich. Danish geographer. *Geographisk Charta öfwer Medelpad*, 1769.

Cam, D. See **Cão**, Diogo.

Camacho, Josef. 'Primer Piloto...de la Real Armada'. ...*La Costa septentrional de la California*, 1779.

Camacho, Lucio. *República de Bolivia*, 1842-1859.

Cambel, *Lieutenant* John. Military engineer. Surveyed the Mississippi, Iberville, Amite, Pearl, Pascagoulas, Mobile and Alabama Rivers between 1768-1769; manuscript plan of the proposed town of Dartmouth [Mississippi], 1774. **Ref.** BOSSE, D. *Imago Mundi* 41 (1989), pp.9-18.

Cambrensis, Giraldus. See **Giraldus** Cambrensis.

William Camden (1551-1623). (By courtesy of Rodney Shirley)

CAMDEN's

BRITANNIA
ABRIDG'D;

WITH

IMPROVEMENTS, and CONTINUATIONS, to this preſent Time.

To which are added,

Exact LISTS of the Preſent Nobility of *England*, *Scotland*, and *Ireland* :

Alſo a Valuation of all

Eccleſiaſtical Preferments At the End of each County.

With many other Uſeful Additions.

The whole Carefully Perform'd, and Illuſtrated with above Sixty Maps Exactly Engraven.

VOL. I.

London, Printed by *J. B.* for *Joſeph Wild*, at the *Elephant* at *Charing-Croſs*. 1701.

Title page to Camden's Britannia, *London 1701.*

Cambridge, Richard Owen (1717-1802). Poet. *An account of the War in India...* (with 12 maps and plans by Thomas Jefferys), 1761, 1780.

Camden, William (1551-1623). Antiquarian, historian and schoolmaster. Produced a study of Britain which was published in various editions over many years. *Britannia* (8 volumes with one map), 1586; *Britannia*, 1607 (Latin text, 57 maps by William Kip and William Hole); *Britain*, 1610, 1637 (in English translation with maps by William Kip, William Hole, John Norden, and others after Christopher Saxton); *Britannia*, 1617, 1637 (abridged version with Latin text and 46 miniature maps by Pieter van den Keere); *Britannia*, 1695 edition (with maps by Robert Morden); *Camden's Britannia Abridg'd...*, 1701 (maps by J. Seller from his *Anglia Contracta*, *c.*1694);*Britannia*, 1789 (with maps by John Cary). **Ref.** SKELTON, R.A. *County atlases of the British Isles 1579-1703* (London 1970) *passim.*

Camerarius [Kammermeister], Elias (*c.*1530-1581). Mathematician of Wittenberg, later of Frankfurt an der Oder. *Marca Brandenburgensis & Pomerania* 1585 (used by Gerard Mercator 1585, and Abraham Ortelius 1588).

Camere, —. Carte géologique des environs de Nice, 1877.

Camerino, Leonardo Miguel. *Descripsion de la Costa de la Luciana*, 1784 MS.

Cameron, David Nelson (1820-1916). Geographer. Higher Geography, 1865; Intermediate Geography, 1866.

Cameron, George. Engraver of Edinburgh. *An Exact Map of Breadalbane in Perth Shire*, 1770; Charles Ross's *A map of the Shire of Lanark* (4 sheets), 1773; Charles Ross's Dumbartonshire, *c.*1777.

Cameron, George. Publisher and mapseller of Edinburgh. *A Map Of The Island of New York and Environs with the March of the British Troops*, 1776 (engraved by Todericks Wynd of Edinburgh). **N.B.** Compare with George **Cameron**, engraver (above).

Cameron, H.C. Wall map of *Ancient Italy*, 1865; *Ancieant Greece*, 1866; *Roman Empire*, 1867 (all with A. Guyot).

Cameron, *Captain* John, *RE*. Director General of Ordnance Survey 1875-1878. Dean Forest, 1855; *Environs of Aldershot*, 1856; directed the *Ordnance Survey of London* (16 sheets), 1857.

Cameron, R.M. *The Penny Atlas*, 1873; *The Pupil's Atlas*, 1873.

Cameron, Verney Lovett *RN* (1844-1894). African explorer, leader of Royal Geographical Society's expedition to aid David Livingstone in 1873. *General Map of Tropical South Africa*, 1876; Tanganyika, 1876.

Camers, Johann. Humanist of Vienna. Author of *Solini Enerrationes*, 1520 (included a world map by Peter Apian and Lorenz Fries based on an earlier map by Waldseemüller).

Camesina, Albert. Reproductions of early plans of Vienna, 1863-1876.

Cammaecker, Jacob. Dutch map copyist, worked for Rekenkamer van Holland, 1536. Untitled manuscript map of north-western Friesland, Griend, Terschelling en Ameland (presumed for Jacob van Deventer 1545).

Cammen, Edmond Paridant van der. *Carte géologique de l'Ile de Chypre*, 1874.

Cammermeyer, A. *Reisekart over det Nordlige Norge* (4 sheets), 1882; *...Sydlige Norge* (2 sheets), 1887.

Cammor, W. Georg. ...*Original Plan der Stadt Magdeburg*, *c.*1853.

Camocio [Camotij; Camocius; Camotto; Cometti; Comotio], Giovanni [Joan] Francesco [Francisco] (*fl.*1558-*c.*1575). Bookseller, publisher, print and mapseller 'ad Signum Pyramidis', Venice. His plates were re-issued by Donato Bertelli and others. Published Paolo Forlani's world map *Nauigationi dil mondo nouo*, 1560 and later (like the Gastaldi world map of 1546, on which it was based, the earlier versions show Asia and America joined); Regiones septentrionales, 1562; Africa, 1563; Dalmatia, 1563; France, 1563; Great Britain, 1563; Friuli, 1563; Holland, 1563; Crete, 1564; *Disegno del Porto di Malta...*, 1565; Ireland, 1565; Sicily, 1566; Friesland, 1566;

Greece, 1566; Cyprus, 1566; Corfu, 1568; Morea, 1569, 1570; Venice, 1571; Ireland, 1572; *Isole famose, porti, fortezze e terre marittime sottoposte alla Ser. Sig. di Venetia*, 1572 (with maps by Camocio, Bertelli, Forlani and others); Europa (12 sheets), Venice 1573 (dedicated to Antonio Almarano); Asia, 1575; America (9 sheets), 1576; Europa, 1579; and many others. **Ref.** GANADO, A. & AGIUS-VADALÀ, M. *A study in depth of 143 maps representing the Great Siege of Malta 1565* (1994) pp.267 et seq.; SHIRLEY, R. *The mapping of the World* (1983) N°106 pp.121-122; MEURER, P. *Fontes cartographici Orteliani...* (1991) p.125 [in German]; GALLO, R. *Imago Mundi* 7 (1950), pp.93-102.

Camoens, Luis de. World map, Madrid 1639. **Ref.** SHIRLEY, R. *The mapping of the World* (1983) N°47 p.371.

Camotto, G.F. See **Camocio**, Giovanni Francesco.

Camp, John. Bommerlerwaard, 1765.

Camp, John de la. United States Mexican Boundary, 1857; Region between Gettysburg and Appomattox, 1869 (with John E. Weyss and Thompson).

Camp, *Captain* Wilhelm. ...*Fürstenthum Ostfriesland u. Harrlingerlande*, Berlin 1804.

Campa, Luis S. *Plano Topografico del Distrito de Mexico*, 1857.

Campana, Pietro. Engraver. *Pianta di Roma* , 1748 (for Giovanni Battista Nolli); Napoli, 1775 (for Giovanni Carafa).

Campana von Splügenberg, Anton R. (1776-1841). Military cartographer, head of the military geographical institute of Milan. *Carta di Cabottagio del mare Adriatico* (27 sheets in 1 volume), Milan, 1822-1824.

Campanius, Thomas Holm. See **Holm**, Thomas Campanius.

Campbell, *Admiral* -. Brazil, 1807.

Campbell, *Major* Albert H. In charge of surveying wagon roads to the Pacific, 1856-1859, Chief Topographer for the Army of Northern Virginia, established an army mapping bureau in Richmond in 1862. Pacific wagon roads, Washington 1859; supervised manuscript surveys of the counties of Virginia, early 1860s; *Seat of the War*, 1862.

Campbell, Alexander F. *Proposed Railway from England to India*, 1851; Northern Boundary of the United States, 1878.

Campbell, *Lieutenant* (later *General*) *Sir* Archibald (1739-1791). Governor of Jamaica. *A Map of the Island of Guadeloupe*, 1760; *Sketch of the Coast Round the Island of Dominique...*, 1761 MS; ...*Northern Frontiers of Georgia*, 1780; *Survey of the South Coast of Jamaica*, 1785.

Campbell, D. Surveyor. *A Plan of the Surprise of Stoney Point* [New York]...*1779*, 1784.

Campbell, Dugal. Engineer to William, Duke of Cumberland. *Plan of the town and castle of Stirling*, 1746 MS; *Plan of the battle of Culloden*, 1746 MS, published by C. Morley & M. Overton, 1746.

Campbell, *Lieutenant* Dugald (*c*.1758-1810). Military engineer and surveyor. *Plan of Fort George and the Battery at New York*, 1782 MS; *Plan of Sandy Hook and the Shoals near it...*, *c*.1782; *Saint John River*, 1784-1785; *Sketch of the route from Fort Cumberland to Fredericton*, 1799; *A Map of the province of New Brunswick*, *c*.1805. **Ref.** LEMON, D.P. *The Map Collector* 61 (1992) pp.34-38; *Dictionary of Canadian biography* Vol.V (Toronto 1983).

Campbell, I.C. *Map of the County of Pembroke*, 1827.

Campbell, *Major* J.L. *Plan of Ghuznee* [Afghanistan], 1839 (lithographed by L.J. Hebert); *Idaho*, 1860.

Campbell, *Captain* James. *Directions for navigating up Delaware Bay*, William Faden 1776.

Campbell, John. Engineer of Pensacola. *Charleston and the British attack of June 1776*, 1776; *A sketch of St. Augustin Harbour...*, 1780; Survey on Red River, 1800 MS.

Campbell, *Reverend* John. British missionary in South Africa from 1818. *South Africa*, London, F. Westley 1822 (published in Campbell's *Travels in South Africa...*).

Campbell, *Lieutenant* Marius H. *Map of the Southern Appalachians*, 1895; *Coal Fields of the United States*, 1908.

Campbell, R. Engraver of Philadelphia. C. Malte Brun's *Universel System of Geography* Vol.3, Boston 1834.

Campbell, R.J. Surveyor of Sydney. *Wellington*, 1881.

Campbell, *Captain* Robert. Surveyor, 40th Company of St. John's Loyalists. *A Map of the Great River St. John & Waters... from the Bay of Fundy, up to St. Anns...*[New Brunswick], 1788 (engraved by S.I. Neele).

Campbell, *Lieutenant* Robert. Admiralty Surveyor. *A new and correct map of Scotland or North Britain*, 1790 published in Robert Sayer's *A General Atlas*, re-issued by Laurie & Whittle, 1794; *Island of St. Kilda*, 1799; *Plan of the African Islands*, 1811; *A Plan of the Island of Ascension*, 1825.

Campbell, Robert A. *Atlas of the State of Illinois*, 1870 (with Henry F. Walling); *Atlas of Missouri*, 1873, 1880.

Campbell, Thomas. *City of Dublin*, 1811.

Campbell, W. Plan of the Proposed Harbour of Wick, 1814 MS; *Caithness-Shire*, 1822.

Campbell, W.M. Country round Kandahar, 1879.

Campe, Friedrich. Publisher of Nuremberg. C.G.T. Reichard's *Neuer Hand-Atlas über Theile der Erde*, 1813 and later editions; Östliche Halbkugel, 1816; Australien, 1817; F.W. Streit's *Charte von der Preussischen Provinz Schlesien...oder den Mährischen Kreisen Troppau und Teschen...*, 1824; Bayern, 1838.

Campell, Ulrich (*c.*1510-*c.*1582). Swiss reformer and historian, born in Susch, worked in Zernez and Tschlin. First printed map of Rhaetia, 1573.

Campen, Gerhardt von. Publisher of Cologne. Published Michael von Aitzing's *Leo Belgicus*, 1583

Campen, H.A. van. *Kaart der Stad Leiden*, 1870.

Campen, Jacob van. Illustration of the terrestrial and celestial maps which were inlaid in the floor of the City Hall in Amsterdam between 1648 and 1650, published in *Afbeelding van't Stadt Huys van Amsterdam*, Amsterdam 1661 **Ref.** SHIRLEY, R. *The mapping of the World* (1983) N° 423 p.446.

Campen, Joannes van. Manuscript atlas of Ceylon, 1719.

Campen, W.J. van. *Nieuwe Kaart der Stad Leyden*, 1850.

Campenhausen, Balthasar, *the elder* (1746-1808). Military cartographer.

Camperio, *Captain* Manfredo (1826-1899). Italian geographer and explorer. *Carta economíca della Tripolitania e Cirenaica*, 1884.

Camphuysen, J. (1760-1840). Plan of Waterloo, *c.*1816.

Campi [Campus], Antonio [Antonius] (1525-1587). Artist and cartographer of Cremona and Milan. *Agri Cremonensis typus*, 1579 (after map compiled 1571, no copy extant, used by Abraham Ortelius 1579), separate Italian single-sheet printing entitled *Tutto il Cremonese et soi confini et sua diocese... M.D.LXXXIII*, 1583.

Campi, Luigi (1784-1834). Topographical draughtsman and lithographer from Mantua, worked in the Officio Topografico di Napoli. Went to Naples in 1814 with Ferdinando Visconti. Maps of the coast of the Adriatic, 1818; *Pianta della città di Noja*, 1820; worked on the Naples sheet of the Officio map of the Kingdom of Naples, 1821; map of the towns around Vesuvius, 1821; *c.*20 hydrographic charts for *Portulano delle coste della Penisola di Spagna*, 1824-1827; *Carta generale dell'Africa*, 1830 (taken from the work of A.H. Brué). **Ref.** VALERIO, V. *Società uomini e istituzioni cartografiche nel mezzogiorno d'Italia* (1993) pp.467-469.

Campiglio, P. *Carta della Circoscrizioni Militari del Regno d'Italia*, 1883.

Campion, C.W. *Map of the Collieries and Iron Works... in South Wales*, 1877, 1894.

Campion, Edward. Description of Ireland, 1571, printed 1633 (with Robert Stonyhurst).

Campion frères. 18th-century Parisian publishers.

Campo, Jossef del. *Puerta del Boston*, 1780s; Plano de la Bahia de Laware y entrada de Filadelphie, 1784; *Descripcion Hidrografica... desde la Punta y Rio de Sn. Juan hasta Filadelfia...*, *c.*1785 MS.

Campomanes, Pedro Rodriquez. *Carta Hidrographica...à Navign de Hannon Cartagines*, 1754; *Vista por Mar de la Antigua Ciudad de Cartago*, 1754 (both published in *Antigüedad Maritima de la Republica de Cartago*, 1756).

Campos, Francisco Carneiro de. *Planta da cidade do Rio de Janeiro*, 1858.

Campos, Rafael Torres (*fl.*19th century). Spanish geographer.

Camps, D.J. Manuscript plans in the Philippines, 1833.

Campus, Antonius. See **Campi**, Antonio.

Camus, Charles-Etienne (1699-1768). French mathematician, member of the Académie des Sciences, worked with César-François Cassini as a member of the association formed to promote work on the survey of France. *Route de Paris à Calais* (11 sheets), 1766.

Canacci, Giuseppe. Tuscan engraver. Some maps in *Atlas Historique*, Florence, Molini & Landi 1807; *Carta dell'Antica Grecia*, 1811; *Carta Topografica... dalla Città di Firenze*, *c.*1840.

Canali, Giovanni Battista (*fl.*1696-1702). Mapmaker of Padua. Siciliae veteris typus, 1696; Italia gallica sive Gallia cisalpina, 1699; *Latium Campania et Samnium*, 1699; *Mappamondo overo carta generale del globo terrestre*, Padua 1699, 1701; *Africa*, 1701; *l'Asia*, 1701; *l'Europa*, Stamperia del Seminario 1701 (engraved by A. Recutti); *l'America*, 1702. **Ref.** SHIRLEY, R. *The mapping of the World* (1983) N°·594, p.589; ARRIGONI & BERTARELLI *Le carte geografiche dell'Italia* (Milan 1939), 768, 2167, 2537.

Cancrin, J.A. Engraver. *Libya et Aethiopia*, 1830.

Candargy, C.A. *Umgebung der Stadt Mytilini*, *c.*1890.

Candish *Sir* Thomas. See **Cavendish**, Thomas.

Candler, Benjamin. *A Chart of the Windward Passage to Jamaica*, *c.*1730.

Cane, Raffaello, *Carta Topografica dell'Agro Romano*, 1878.

Caneiro, Nicolo de. See **Caverio**, Nicolo de.

Canepa, Albino de. Portolan maker of Genoa. Portolan of the Black Sea area west to the Atlantic, 1481; Mediterranean, 1489.

Caneria, J.A. Engraver of Amsterdam. Pierre Lapie's *Océanie*, Amsterdam and The Hague, F.J. Weygand 1827; Lapie's *Océanie ou Australasie et Polynésie...*, Amsterdam and The Hague, F.J. Weygand 1827 (both re-engravings of earlier maps).

Canfield, *Captain* A. American surveyor. Clinton River, Michigan, 1853.

Canfield, C.A. *Geneva*, 1850.

Canfield, Thomas H. *The Country West of Dakota to the Pacific Ocean*, 1870.

Caniani, Giuseppe. Engraver of Milan. Maps, including Spain, for *Atlante della Geografia Antica e Moderna*, *Nuovo atlante di geografia universale*, Barbiellini 1807; for Luigi Rossi, 1820.

Canin, Jan. Publisher of Dordrecht. Bible (5 maps), 1580.

Canina, Luigi. *Pianta Topografica di Roma Antica*, 1832; *La Campagna Romana* (6 sheets), 1845.

Canius, Paulus. See **Cagno**, Paolo.

Cañizares, José de. *Plano del Puerto de San Francisco*, *c.*1775 MS; *Mapa del Puerto de Sn. Francisco*, 1776 MS. **Ref.** HARLOW & NEWMAN *The Map Collector* 63 (1993), pp.38-39.

Cannabich, Johann Gottfried Friedrich (1777-1859). German geographer. Geographical manual, 1816; *Vollständiges Handbuch der neuesten Erdbeschreibung*,

Weimar 1819 and later (with A. Gaspari, G. Hassel, J. GutsMuths and F. Ukert); one of the editors of the periodical *Allgemeine Geographische und Statistischer Ephemeriden*, 1822-1831; European Russia, 1833.

Canning, Alfred Wernam (1860-1936). Explorer and surveyor for the lands departments of New South Wales and Western Australia. *Wiluna - Kimberley Stock route*, 1906-1907.

Cannon, *Lieutenant* James, *RN*. *Harbour of Trincomalee*, Admiralty Chart 1832 (with Hector Loring), published 1834.

Cannon, John. Surveyor for Van Diemen's Land Company. *Outline of Forest Farms... on the main land at Circular Head*, London, Royston & Brown 1844.

Cano [Elcano, Del Cano], Juan Sebastian del (*d.*1526). Spanish navigator and portolan maker; served under Magellan. East Indies, *c.*1523 MS chart, now lost; map of the region of Rio de la Plata, the Patagonian Coast and Strait of Magellan, 1523. **Ref.** FURLONG, G. *Imago Mundi* 23 (1969) pp.77-79.

Cano y Olmedilla, Juan de la. See **Cruz** Cano y Olmedilla, Juan de la.

Canocchi, Giovanni. Italian engraver. Toscana, 1769.

Canot, Pierre Charles (*c.*1710-1777). French engraver, active in England, elected Associate Engraver of the Royal Academy, 1777. Milton's *Plan of Portsmouth Dockyard*, 1754; *A Geometrical Plan and West Elevation of His Majesty's Dock Yard near Plymouth*, 1756.

Canstein, Philipp von. Relief maps for school atlases, early 19th-century.

Cant, John. *Gulph of Persia*, 1786, 1787; *Part of the Coast of Persia*, 1787.

Cant, Samuel (*fl.c.*1731-1759). Schoolmaster and surveyor. Parish of Leonard's Winchelsea, 1731; Town of Hastings, 1746.

Cantel, — *Le sieur*. *La principauté de Catalogne avec les comtez de Roussillon et de Cerdagne...*, Paris, Nolin 1694 (published in *Le théatre du monde*); Carte de l'Italie, 1701.

Cantelli da Vignola, Giacomo [Jacobo] (1643-1719). Cartographer of Modena. Alta Lombardia, 1680; China, 1682; Venezia, 1682; Marca Brandenburgico, 1687; *Mercurio Geografico*, 1688; *Il Regno di Navarra*, 1690; *Il principato di Catalogna*, 1690; Francia, 1691; Italia, 1695; *La Biscaia*, 1696; pair of globes for the Duke of Modena, [n.d.].

Cantemir, *Prince* Constantin Dmitri (1673-1723). Cartographer in Russian Service. *Principatus Moldaviae*, 1716, *c.*1737; *Constantinople* (2 sheets), 1720.

Canter, Johannes Jacob (*fl.*18th century). Dutch cartographer.

Cantino, Alberto. Italian diplomat in Portugal. Commissioned planisphere *Carta da nauigar per le Isole nuouam tr...in le parte de l'India: dono Albert Cantino Al S. Duca Hercole* (on vellum in 6 sheets), 1501 or 1502 MS. **Ref.** KRETSCHMER, DÖRFLINGER & WAWRIK *Lexicon zur Geschichte der Kartographie* (1986) vol. C/I, p.128 (numerous references cited) [in German].

Cantino, Francisco. *Ichnographia Ville Tiburtinae*, 1751.

Cantle, J.M. Draughtsman of 50 Castlereagh Street, Sydney. *Rosehill Heights, Warren Grove Estate*, 1892.

Cantler, Fridrich Gottlieb. See **Canzler**, Friedrich Gottlieb.

Cantoni, C. Departemento de Reno (2 sheets), 1810.

Cantova, Juan Antonio. Islas de los Dolores [Carolines], 1731.

Cantzler, Barnhard. Mathematician. *Comitatus Wertheimici*, 1603. **Ref.** OEHME, Ruthardt *Barnhard Cantzler und seine Carte der Grafschaft Erbach* (Karlsruhe 1991).

Canu, Philippe. Engraver for Claude Sicard's *Route des Hébreux*, 1727.

Canus, D. See **Cão**, Diogo.

Canzler [Canzier, Cantler], Friedrich Gottlieb (1764-1811). Professor at Göttingen, worked for Homann's Heirs, Nuremberg. *Karte vom Fünften Erdtheil oder Polynaesien-Inselwelt oder Australien od. Süd Indien*, Nuremberg 1795; *Karte von Mittel-America oder Westindien*, 1796.

Cão [Cam, Cano, Canus], Diogo [Diego] (*d.c.*1486). Portuguese navigator, discovered and explored the estuary of the Congo River, 1484.

Capassi, P. *Plan de la baye et port de Rio de Janeiro*, Dezauche 1785.

Capella, Martianus (*fl.*410-439). Wrote an influential geography entitled *De nuptiis Mercurii et Philologiae*.

Capellaris, Giovanni Antonio de [Jean Antoine] (*b.c.*1724). Engineer. *Nuova Carta topografica del territorio di Friul*, 1778, 1793; Istria, 1803; *Carta novissima della Dalmazia*, 1806.

Capellini, Giovanni. *Carta Geologica... del Golfo della Spezia*, 1863.

Capellis, —. *comte* de. Sierra Leone River, 1779, *c.*1802.

Capellman, R. *Plan der Städte Aachen und Burtscheid*, 1877.

Capello, *Lieutenant* —. *Atlantic Ocean. Pilot Chart*, 1868.

Caper, Johannes. Plan of Prague, 1562 (used by Braun & Hogenberg).

Capitaine, H. See **Au Capitaine**, Henri.

Capitaine [du Chesnoy] family.
• **Capitaine**, Pierre Charles (*d.*1778). Engineer and military cartographer. Worked as an engineer on the Cassini survey of France from 1755. Was appointed 'Directeur du Dépôt de la Carte à l'Observatoire' in 1762.
•• **Capitaine du Chesnoy**, Louis (*c.*1749-1797). Son and successor of Pierre Charles, 'Ingenieur, Géographe du Roi', worked at the Paris Observatory with Cassini IV. Followed his father as senior engineer on the *Carte de France*, being involved, both as surveyor and engraver (with his father-in-law P. Bourgoin), in the preparation of several sheets of the survey, 1766-1781. *Carte du Théâtre de la Guerre dans l'Amerique Septentrionale*, *c.*1779; *Carte de la France, réduite d'après les 180 feuilles de la grande carte de la France levée géométriquement...* (18 sheets), Paris 1790; *Carte de la France suivant sa nouvelle division en départments et districts*, Paris 1790; *Carte géométrique des routes de postes de la France...*, 1795; *Carte chorographique de la Belgique* (69 sheets), 1797; worked with Belleyme and Chanlaire on topographical maps of France including *Carte Géométrique des Routes de Postes de la République française*, 1798 (with P.G. Chanlaire). **Ref.** DRAPEYRON, L. 'Cassini de Thury, les Capitaine et l'enquête sur la grande carte...' in: *Revue de Géographie* vol.38-39 (1896) pp.241-255.
•• **Capitaine, Pierre Nicolas** (1773-1838). Son of Pierre Charles. Abridged edition of Cassini's *France*, 1789.

Capitaine du Chesnoy, Michel (1746-1804). Lieutenant (later Major) in the French army, went to America with Lafayette, for whom he prepared campaign maps between 1776 and 1781. Work, mostly coloured manuscript, includes *A New Plan of Boston Harbour...*, 1776; *Plan de Rhode Islande...*, 1778; *Carte du Théatre de la Guerre dans l'Amérique Septentrionale...1775, 76, 77, et 78* (copper engraving), published 1779; *Plan de Carillon ou Ticonderoga...*, *c.*1777; *Plan de la Retraite de Barrenhill en Pensilvanie...*, 1778; *Carte de l'affaire de Montmouth...*[New Jersey], 1778; *Carte de la campagne en Virginie du Major Generale...*, 1781; *Plan of the Harbour and City of Annapolis*, 1781. **Ref.** COHEN, P.E. 'Michel Capitaine du Chesnoy, the marquis de Lafayette's cartographer' in: *Antiques* (January 1998) pp.171-177; GUTHORN, P.J. *American maps and map makers of the Revolution* (1966) pp.9-12.

Caplen, A.E. Lithographer of Cape Town. Railway map of Oliphants River [South Africa], 1881; *100 Mile Road Map from Cape Town*, *c.*1912.

Caplin, Jean-François. Engraver for Dépôt de la marine. Côtes de Brésil, 1822; Côtes du Pérou, 1824; Golfe du Mexique, 1826; Rade de Rio, 1829; for Victor Charles Lottin's maps of New Zealand from the voyages of J.S.C. Dumont d'Urville, 1833.

Capocacia, Giuseppe. *Ferrara*, 1602.

Capodilista, Gabriele (*d.*1475). Map of Palestine from the Spring Voyage of 1458, *c.*1475 MS.

Cappelier, Bernard. *Diocese de Tournay*, 1694, 1709, 1726.

Capper, Benjamin Pitts. *A Topographical Dictionary of the United Kingdom* (46 maps), London, R. Phillips 1808, 1813, 1825, 1826, 1829.

Capper, John. Emigrants Guide to Australia, 1852; Map of the Gold Fields, *c.*1852.

Capper, Robert. Revised James Wyld's *Plan of Swansea*, 1888.

Cappi, Giovanni [Johann von] (1780-1847). Viennese publisher, sometime partner in the firm of Artaria & Co.

Capriolo [Capreolus, Cavriolo, Gavriolo], Elia (*d.*1519). Italian cartographer. *Brixial et Agri Geographia*, *c.*1502.

Carabelli, Ferdinando. Compiled and engraved *Atlante geografia storico*, 1876.

Carafa [Caraffa], Giovanni, *duca di Noia* (1715-1768). *Mappa topografia città di Napoli*, 1775 (revised and engraved by Nicolo Carletti).

Caraffa, Giovanni. Neapolitan globemaker. Terrestrial globe, 1575.

Caraman, Adolphe de Riquet, *comte de* (1800-1876). French traveller in Palestine, 1837. Constructed maps at scale of 1:500 000 of routes from Cairo to Jerusalem via Sinai, and from Jerusalem to Aqaba, 1837.

Caramuel, Juan. Bishop of Campagna. World map in *Architectura Civil...*, Vergeven 1678. **Ref.** SHIRLEY, R. *The mapping of the World* (1983) N° 480 p.495.

Carandini, Giuseppe. *Reggio di Lombardia*, 1817.

Carballo, *Fr* Paulo Monso. Ucayale and Huallago rivers, 1814 MS; Plan del curso los Rios Huallaga y Ucayali, 1830.

Carbó, Benito Chias. See **Chias** y Carbó, Benito.

Carbonel, Esteban. American Northwest, *c.*1632 MS.

Carbonel, Juan José. Reino de Valencia, 1812.

Carcereri, —. *Carte du Cordofan et du Pays des Noubas*, 1874.

Cardano, J. Engraver. Costas de Yucatan, 1808; *Carta Esferica...Las Costas del Seno Mexicano*, 1814.

Carden, *Lieutenant* Hans (*fl.*1773-1782). *Plan, three sections and two elevations of Fort San Fernando d'Omoa...* [Honduras], 1779 MS; Plan of Black River on the Mosquito Shore, [n.d.].

Cardenas, Manuel Joseph de. *Planta Cosmographica del Lago de San Bernardo*, 1691 MS.

Cardim, António Francisco, *SJ* (1596-1649). Portuguese Jesuit, died Macao. *Iapponiae Nova & Accurata Descriptio*, in *Fasciculus e Japponicis Floribus*, Rome 1646. **Ref.** SCHÜTTE, J.F. *Imago Mundi* 23 (1969), pp.46-47.

Cardinael, Sybrandt Hanszoon (1578-1647). Cartographer and mathematician of Amsterdam.

Cardon, Antoine J. (1739-1822). Belgian painter, engraver and map maker. Forests of Meerdael, 1769 (surveyed by Sr. Joris); *Carte topographique de la Forêt de Soigne*, 1769, published 1770; *Carte générale des chasses royales ...du château et parc royal du Marimont*, 1776 (surveyed by Jean, Baron de Ravizza).

Cardona, J. *Map of the Country round Cadiz*, 1811.

Cardona, Nicolas. Western Hemisphere, *c.*1632 MS.

Cardono, —. Engraver for Isidoro de Antillon y Marzo, 1804.

Carel, D.W. See **Hattinga**, David Willem Carel.

Carette, Ernest (1809-1890). *Carte de l'Algérie*, Paris 1846; *Kabilie*, 1847.

Carew, Benjamin. See **Hallowell**, Benjamin.

Carew, *Sir* George (1555-1629). Survey of Kerry & Desmond (6 maps), 1617 MS; Wars of Ireland (maps), 1632.

Carew, T. See **Carve**, Thomas.

Carey, family and associated companies. Publishers and booksellers of Philadelphia.
• **Carey**, Mathew (1760-1839). Native of Dublin, settled in Philadelphia in 1784 as a publisher, noted at 122 Market Street, Philadelphia. Used maps prepared by Samuel Lewis and Amos Doolitle amongst others. *A General Atlas for the Present War*, 1794; *The General Atlas for Carey's Edition of Guthrie's Geography Improved*, 1795, and editions to 1818; *Carey's American Atlas* (20 maps), Philadelphia 1795, re-issued 1796, (26 maps) 1809 (thought to be the earliest atlas of the United States, it re-used some of the maps from the *General Atlas*); *Carey's American Pocket Atlas*, 1796, and editions to 1814; *Carey's Minor Atlas*, 1802, 1810; S.S. Moore and T.W. Jones' *The traveller's directory: or, a pocket companion*, 1802, 1804 (lithographed by T.W. Jones); Scripture Atlas, 1817; United States of America, 1820; publisher of the 'second American edition' of C.V. Lavoisne's *Complete Genealogical, Historical, Chronological, and Geographical Atlas*, 1820, 1821, 1840 (the first edition of 1807 had no maps). **Ref.** RISTOW, W. in: WOLTER, J.A. & GRIM, R.E. *Images of the World* (1997) pp.301-303; RISTOW, W. 'Early American atlases and their publishers' in *Images of the World* (1997); TYNER, Judith *Mercator's World* Vol.1, N°6 (1996) pp.36-41; WHEAT & BRUN *Maps and charts published in America before 1800* (1978) many entries.
•• **Carey**, Henry Charles (1793-1879). Eldest son of Mathew Carey, above. Worked in publishing from 1817-1835. Became partner in father's firm of Carey, Lea & Carey, in 1817, taking control with Isaac Lea, 1822.
••• **Carey, Lea & Carey** (*fl.*1817-1822). Partnership between Mathew Carey, I. Lea, and H.C. Carey.
••• **Carey & Lea**. Publishers of Chestnut Street, Philadelphia. *A complete historical, chronological, and geographical American Atlas...*, Philadelphia 1822 (a translation of C.F. Weiland's *Atlas von Amerika*); *Family Cabinet Atlas*, 1832-1836.
•• **Carey**, Edward L. (*d.*1845). Younger son of Mathew, later worked with Abraham Hart (see Carey & Hart).
••• **Carey & Hart**. Publishing firm founded by Edward Carey and Abraham Hart in 1828, and after the death of Carey was continued by Hart until 1854. Bought the publishing rights to H.S. Tanner's *A New Universal Atlas*, 1842, 1843 (the *Atlas* was sold on to S. Augustus Mitchell in 1845); New Mexico, 1847.

Carey, A.D. Chinese Turkestan, 1887.

Carey, *Captain* (later *Lt.-Colonel*) Constantine P. *RE*. Surveyor for the Ordnance Survey. Surveys on the Isle of Wight, 1860; surveys in London Upper Holloway-Plaistow Marsh - Tulse Hill - Brentford (326 sheets), Southampton 1868-1876 (with Colonel Bayly and W. le Poer Trench).

Carey, James. Printer and publisher of Philadelphia. *Plan of the French Invasion of England & Ireland*, 1798.

Carey, Thomas. *Observations sur les cartes des Isles de Guernsey, Aureny*, 1746.

Carey, William. *Westminster & Southwark*, 1808.

Carignano, Giovanni Maura da. Genoese chartmaker. Mappamundi, *c.*1310. **Ref.** DUKEN, A. *Imago Mundi* 50 (1988) pp.86-95.

Carkas, John. *The Barony of Naas*, 1655; *The Barony of Ballibrit*, 1657.

Carl, F. Maehren und Oesterreichisch Schlesien, Vienna 1810.

Carl, Ferdinand. *Bier Productions Karte von Mittel Europa*, 1876.

Carl, Georgius. See **Carolus**, Joris.

Carl, Hans (1587-1665). Town maps of Nuremberg.

Carl, J. *Hopfenbau Karte von Mittel Europa*, 1875.

Carl, Rupert. Gallia Benedictina, 1738; Italia Benedictina, for Homann's Heirs 1742.

Carlberg, Johann Eberhard. First town architect of Stockholm 1727. Manuscript plan of Stockholm, 1750.

Carles, W.R. Korea, 1886; Central China, 1898.

Carless, *Lieutenant* (later *Captain*) Thomas Grere. Hydrographer, Indian Navy. *Chart of the Red Sea*, 1830-1833; *Survey of the North East Coast of Africa*, 1838-1848; *Chart of the Gulf of Aden*, 1854.

Carlet, Louis François, *marquis* de la Rozière. *Hesse Cassel*, 1759; *The Bishoprick of Paderborn*, 1760; *Carte d'une Partie du Théâtre de la guerre en Allemagne* (4 sheets), 1761.

Carleton, Henry Richard (1850-1901). Engineer of Sydney. *Plan of Port Kembla*, 1895.

Carleton, I.D. Land agent Port Huron. *Port Huron, Sarnia & Gratiot...Canada West*, *c.*1859 (lithographed by Sage & Son, Buffalo, NY); *Map of the Great Oil Regions of Michigan & Canada*, *c.*1880.

Carleton, Osgood (1742-1816). Surveyor, cartographer and teacher of mathematics in Boston, Massachusetts. Carleton's certificate of accuracy appeared on many charts issued by John and William Norman and on the 18 sheets of Matthew Clark's *A Complete Set of Charts of the Coast of America...*, 1790. *New General Chart of the West Indies* (4 sheets), 1789; John Norman's *American Pilot*, 1791, 1792, 1794; *United States* (6 sheets), 1791, 1797, 1806; Jedidiah Morse's *American Universal Geography*, 1793; *An accurate Map of The District of Maine...* (4 sheets), 1795, 1798, 1801, 1802; *An Accurate Map of The Commonwealth of Massachusetts* (4 sheets), *c.*1795, 1798; William Norman's *Pilot for the West Indies*, 1795; *Plan of Boston*, 1796, 1797, 1800, 1803, 1805 (engraved by Joseph Callender); *Map of Massachusetts Proper...* (4 sheets), 1798, 1801, 1802, revised Amos Lay 1822. **Ref.** DANFORTH, S. *Imago Mundi* 35 (1983) pp.38-40; GUTHORN, P. *The Map Collector* 12 (1980) pp.27-28; WHEAT & BRUN *Maps and charts published in America before 1800* (1978) many entries.

Carletti, Niccolo. Corrected and engraved Giovanni Carafa's *Mappa topografia della città di Napoli*, 1775.

Carli, *conte* G. Carta argonautica, *c.*1745.

Carli, Vincenzo Pazzini. See **Pazzini** Carli, Vicenzo.

Carling, J. Territorio Portuguez de Goa, 1893.

Carlo, — *sieur du*. 'Géographe du roi'. *Plan de La Rochelle*, 1628.

Carlsteen, Jonas. Swedish surveyor. Wassemanneland, 1688 MS; Fråndesta, 1689 MS.

Carlton, William H. Map showing the areas occupied by Indian tribes in the Pacific Northwest, 1857.

Carmichael, John (1803-1857). Engraver of King St, Sydney (1832-1847). Major Thomas L. Mitchell's *Map of the Colony of New South Wales*, 1834; map of the Town of Sydney, 1837; *Plan of the North Entrance to Moreton Bay*, 1846; *City of Sydney*, 1854; *Plan of the Town of Launceston Van Diemen's Land*, 1855.

Carmona, Manuel Salvador. Engraver. Frontispiece to Vincente Tofiño de San Miguel's *Atlas Marítimo de España*, 1789.

Carnall, R. von. *Geognostiche Karte von Oberschlesien*, 1844; *Geologische Karte von dem niederschlesischen Gebirge*, *c.*1862-1868.

Carnan, Thomas (1737-1788). Publisher of 65 St. Pauls Churchyard, London; in partnership with his half-brother Francis Newbery. Daniel Paterson's *A new and accurate description of all the direct and principal cross roads in England and Wales*, 1771-1786; John Gibson's *New and Accurate Maps of the Counties of England and Wales*, *c.*1779 edition; *Gibson's Atlas Minimus*, 1779.

Carnbée, *Baron* Pieter Melvill van. Het eiland Java (5 sheets), 1849 and later; Profil over het eiland Java, 1842; *Algemeene Atlas van Nederlandsch Indië*, 1853-1862 (with W.F. Versteeg). **Ref.** KOEMAN, C. *Atlantes Neerlandici* (1967) Vol.II pp. 276-278.

Carnegie, David Wynford (1871-1900). Explorer. *Map shewing route of exploring & prospecting expedition Coolgardie & Kimberley, Western Australia*, 1898.

Carneiro, António de Maris (*d.*1642). Portuguese cartographer. *Descripçam da Fortaleza de Sofala* (48 maps), 1639; 11 maps of the coast of Portugal and Spain in: *Regimento de Pilotas*, 1642. **Ref.** CORTESÃO & TEIXEIRA DA MOTA *Portugaliae monumenta cartographica* (1960) Vol.5.

Caröe, Frederic Carl Wilhelm (1811-1882). Danish military cartographer.

Carolus, Frans. Mapmaker of Amsterdam. Holland, Frederick de Wit, *c.*1690, republished by Pieter Schenk *c.*1705.

Carolus [Carl], Joris [Georgius] (1601-1625). Pilot and map maker from Enkhuizen. Compiled various MS maps including the Arctic Ocean on a polar projection, 1614; map of Iceland published as *Tabula Islandiae*, Blaeu 1630 and later; charts of northern waters published by Jansson, 1634.

Caroly [Caroli], Francesco de. *Carta degli Stati di S.M. il re di Sardegna*, 1779, English edition, 1794; *Carta Topografica dei contorni della Città di Torino*, 1785.

Caron, Edmond-Jules (1857-1917). French naval officer and hydrographer. *Mission de Timbuktu. Carte du Niger*, 1891; *Carte de la région de Tombouctou*, 1895; *Atlas du haut-Niger* (21 sheets), 1898. **Ref.** BROC, N. *Dictionnaire illustré des explorateurs Français du XIXe siècle: I Afrique* (1988) pp.66-67 [in French].

Caron, François (*d.*1674). French Huguenot, travelled and lived in Japan for 20 years, official of Vereenigde Oostindische Compagnie (VOC or Dutch East India Company). Caron's description of Japan of 1636 was used frequently, first by Isaac Commelin in *Begin ende Voortgangh, Van De Vereenighde Nederlantsche Geoctroyeerde Ooste-indische Compagnie*, Amsterdam 1645 (without map), then by Joost Hartgers as *Beschrijvinghe Van het Machtigh Coninckrijcke Japan...*, Amsterdam 1648 (without map). Caron's map, *Perfeckte Kaert vande Geleggentheydt des landts van Iapan*, was first included in an edition of his description published in The Hague, 1661, London 1663, and many other editions. **Ref.** LANDWEHR, J. *VOC A bibliography of publications relating to the Dutch East India Company 1602-1800* (1991) entry n° 250 p.103; *ibid.* n^os.^516-522 pp.291-522; *ibid.* n°s. 666-669 pp.416-418; PAULY, U. in: LUTZ, W. [Ed.] *Japan - A cartographic vision* p.27 and catalogue n^os.^114-115 p.202.

Carondelet, *Baron* — de. Fortificaciones Nueva Orleans, 1792; Mississippi and Ohio Rivers, 1794.

Carpeau, Pierre F. *Dunkerque*, 1765.

Carpelan, Wilhelm Maximilian (1787-1830). Surveyor of Sweden. Organised the engraving department of the 'Ingenjörskår', of which he was chief 1826-1830.

Carpenter, Benjamin D. *Map of the roads in Washington County, D.C.*, Washington, J.F. Gedney 1870.

Carpenter, Nathaniel. *Geography delineated forth in two bookes*, Oxford 1625.

Carpenter, Samuel. American township surveyor, worked in Michigan, *c.*1817.

Carpentier, —. French engraver. L. Brion's *Mappemonde philosophique et politique...*, 1800.

Carpinetti, João Silveiro. Mapmaker of Lisbon. *Mappas das provincias de Portugal*, *c.*1762.

Carpini [Pian del Carpine], Giovanni de Plano (*c.*1182-1252). Franciscan traveller, undertook a mission to the Great Mongol Empire, 1245-1247, an account of which was published by Hakluyt, 1598.

Carr, L. Lake Nyassa (9 sheets), [n.d.] MS.

Carr, *Captain* Richard. *A Description of al* [sic] *the postroads in England, from London to Edinburgh...drawn and perfected by R. Carr*, 1668 (one of the earliest general maps to show the postroads throughout England and Wales). **Ref.** SHIRLEY, R. *Printed maps of the British Isles 1650-1750* (1988) pp.40-41.

Carr, *Captain* Richard [Richardo] (*fl.*1663-1676). Dutch military engineer and cartographer of Amsterdam, worked in Zeeland 1676. Plano de Cavite, 1663; *Caerte vande Landen in Vlaenderen...*, 1676.

Carr, *Captain* Robert. Chart of the China Seas and Philippines, 1734, 1778.

Carranza, Domingo Gonzalez. Principal Pilot to the Flota of New Spain, 1718. Coasts, Harbours & Sea Ports, Spanish West Indies, 1740.

Carraro, Giuseppe (*d.*1886). Italian geographer. Wrote on physical and statistical geography.

Carrasco, Eduardo (1779-1865). Peruvian hydrographer; Director Marine Academy, Lima. Province de Lambeyegue, Wyld 1819.

Carrascosa, Gaetano (*b.*1834). Geographical engineer in the 'Officio Topografico di Napoli'. Survey of Torricella, 1856; survey of the city of Atri, 1859-1860. **Ref.** VALERIO, V. *Società uomini e istituzioni cartografiche nel mezzogiorno d'Italia* (1993) pp.470-471.

Carré, V. Engraver for Dépôt des cartes et plans de la Marine, 1843-1861.

Carreras y Candi, Francesc. *Geografia general de Catalunya*, Barcelona, Alberto Martin 1913-1918 (maps by Benito Chias y Carbó).

Carres, Louis (*d.*1833). Atlas of Jesuit Missionary Provinces.

Carrigain, Philip. *New Hampshire*, 1816 (engraved by William Harrison).

Carrington, Albert. American topographical engineer. Worked with H. Stansbury and J.W. Gunnison on surveys of Utah for J.J. Abert. *Map of a Reconnaissance between Fort Leavenworth on the Missouri River and the Great Salt Lake in the Territory of Utah*, 1850; *Map of the Great Salt Lake...*, 1850.

Carrington, Frederick A. *A Map of Scotland* (6 sheets), 1842 and later editions (engraved by John Dower).

Carrington, G.W. *A Map of Scotland*, 1846, 1851, *c.*1855, 1859.

Carrington, Henry Beebee (1824-1912). *Battles of the American Revolution, 1775-1781*, published 1876.

Carrington, Octavius. Provincial surveyor. New Plymouth, New Zealand, 1844, 1846; Plan of Pekapeka, 1860; *Province of Taranaki*, 1863.

Carrington, Robert Christopher. Admiralty hydrographer. Tonga, 1866; various charts of the coast of England, 1894.

Carroll, M. *This Harbor of Boston...*, *c.*1688 MS.

Carront, A. *Carte du Théatre de la Guerre ou l'on Voit les Endroits ou les Battailles Se Sont*, Liège [after 1758] (engraved by I. Lovinsosse).

Carruthears, —. *Ordnance Plan of the Parish of Hoddom*, 1859.

Carson, C.W. Engraver and publisher, 68 State Street, Albany, New York. Albany, 1845.

Carson, W. Publisher. Family Atlas, 1863 (with John Betts).

Cartaro [Cartarius, Cartero, Kartaro, Karterus], Mario (*fl.*1562-1611, *d.*1620). Cartographer and engraver of Rome from 1560, and later Naples. Worked for Ferdinando Bertelli and Antonio de Angelis. Cyprus, 1562; Crete, 1562; Palestine, 1563; *Malta di nouo...*, 1565; Raab, 1566; Famagusta, 1571; Tunis, 1574; Rome, 1575; Rome, 1576; untitled globes, 1577; Italy, 1577; Maastricht, 1579; Rome, 1579; Hierusalem, 1578; Lisbon, 1580; Venice, 1580; Egnazio Danti's map of Perugia, 1580; Messina, 1581; Palermo, 1581; Genoa, 1581; Milan, 1581; Pozzuoli, 1584; Sicily, 1585; Rome, 1693; Kingdom of Naples (20 sheets), 1611. **Ref.** GANADO, A. & AGIUS-VADALÀ, M. *A study in depth of 143 maps representing the Great Siege of Malta 1565* (1994) pp.86-91.

Carte, John (*fl.*1695-1700). 'Watch-Maker in Garden-Court, near the Fountain the Middle-Temple', London. World map on a north polar projection entitled *The Frontispiece of the Cosmographical Clock...*, *c.*1700. **Ref.** SHIRLEY, R. *The mapping of the World* (1983) N°614, p.607.

Carte, Samuel. Plan of Leicester, *c.*1722.

Cartee, Cornelius S. School Atlas, 1856.

Carter, *Lieutenant* —. *Abyssinia*, 1869.

Carter, E.M. *Sketch map of the County of Kent*, A. & C. Black 1907.

Carter, E.O. *Islas de Lobos de Fuera* [Peru], 1852 MS; four manuscript surveys of Lobos de Tierra [Peru], 1852; *Plan of Crescent City, California*, 1855.

Carter, George. Editor of various school atlases, 1895-1912.

Carter, Henry John. Geological Atlas of Western India, 1857.

Carter, James G. *Essex County*, *c.*1845.

Carter, S.B. *Map of LaSalle County and Part of Marshall County, Illinois*, 1859 (with Th. Neubarth).

Carter, S.B. *Marseilles*, 1868.

Carter, William T. American geological surveyor. *Soil Map: Virginia - Maryland - West Virginia, Leesburg sheet*, 1903; *Soil Map: Virginia - Maryland - West Virginia ....Alexandria Counties sheet*, 1915.

Carter & Wilkinson. Publisher of Providence. *A Map of the State of Rhode Island*, 1795.

Carteret, *Lieutenant* (later *Rear Admiral*) Philip (*d.*1796). Served under Byron and Wallis aboard *Swallow*, 1766. His charts of Queen Charlotte's Islands were used by Beautemps-Beaupré, 1807.

Cartero, Mario. See **Cartaro**, Mario.

Cartier, Jacques (1491-1557). French navigator and explorer from St. Malo. Made 3 voyages to the coast of Canada and the St. Lawrence in search of a route to Asia, 1534-1542; Nova Francia, Hochelaga [Montréal], 1535, in Giovanni Battista Ramusio's *Navigationi e viaggi*, 1556 and later. **Ref.** BOISSONHAULT, R. *Jacques Cartier, Explorer and Navigator* (Ottawa 1987) [French or English]; LANGELIER, G. 'Jacques Cartier's contribution to cartography', *The Archivist* Vol.II, N°·1 (January 1984) pp.13-17.

Cartilia, Carmelo (*fl.*1720). Italian globe and instrument maker.

Cartwright, E. Lithographer. Leeds, 1832; Plan of Evesham, Ordnance Survey 1832.

Cartwright, R. Map of Bridport for Robert Dawson's *Plans of the Cities and Boroughs of England and Wales*, 1832; Bedford Level (6 sheets), 1842.

Cartwright, Samuel. Publisher of London. Gerard Mercator's *Historia Mundi*, English edition 1635, 1637, 1639 (with Michael Sparke).

Cartwright, Samuel A. Yellow fever map of Natchez, Mississippi, 1826

Cartwright, Thomas (*fl.*1818-1828). Surveyor, produced various manuscript maps of Essex and Sussex.

Carvalho, I.J.M. Imperio do Brasil, Rio de Janeiro 1856.

Carvalho da Costa, Antonio Nuñes de (1650-1715). Portuguese cartographer. Chorographica portugueza, 1706-1712.

Carve [Carew; 0'Corrain], Thomas (*c.*1590-1672). Irish traveller. Itinerarium, 1639.

Carver, *Captain* Jonathan (1732-1780). Traveller, surveyor, cartographer. *Travels Through the Interior Parts of North America in the Years 1766, 1767 and 1768*, London 1778; *A New Map of the Province of Quebec* (for Thomas Jefferys's *American Atlas*), 1776; Laconia North America, 1777 MS; *A New Map of North America*, 1778; New Universal Traveller, 1779. **Ref.** BOSSE, D. 'The maps of Robert Rogers and Jonathan Carver' *The American magazine and historical chronicle* Vol.2, N°·1 (1986), pp. 45-61; GUTHORN, P.J. *British maps of the American Revolution* (1972); PARKER, J. *Jonathan Carver's map of his travels* (c.1964).

Carver, Marmaduke. Rector of Harthill in Yorkshire. ...*Paradisi Terrestris*..., in *A Discourse of the Terrestrial Paradise*, 1666.

Carver, S. *Boulogne*, 1846.

Carver, Samuel. See **Bowles** & Carver.

Carwitham, John (*fl.*1723-1742). Engraver, worked for John Bowles 1739. William Gordon's *Bedfordshire*, 1736; *A Plan of the City of New York*, and *A Plan of the Harbour of New York* (2 plans on one sheet), *c.*1740; A Draught of the Bay of Biscay, 1742. **Ref.** LOWENTHAL & WALLIS *The Map Collector* 63 (1993) pp.10-11; COHEN, P.E. &

AUGUSTYN, R.T. *Manhattan in maps 1527-1995* (1997) pp.57-59.

Cary, family. Producers of maps, charts and globes. **Ref.** SMITH, D. in: *Dictionary of national biography: Missing persons* (Oxford 1993) p.118; SMITH, D. *The Map Collector* 43 (1988) pp.40-47.

• **Cary**, George, *the elder* (*d.*1830). Brother and partner of John.

• **Cary**, John, *the elder* (1755-1835). English cartographer, engraver, globe maker and publisher, 188 Strand (1783-1787); Near Norfolk Street, 181 Strand (1791-1820); 86 St James's Street (1820-1832). Worked with his brother George. Apprenticed to William Palmer 1770; surveyor of roads to the General Post Office, 1794. Best known for his English county atlases. *Cary's New and Correct English Atlas* (in 12 parts), 1787-1789, amended edition, 1793 and later; engraved the plates for Robert Gough's edition of Camden's *Britannia*, 1789, 1806; *Cary's Traveller's Companion...*, 1790 and later; *New Itinerary*, 1798 and later editions; terrestrial and celestial globes 1799 and later (with brother William, below); *The New British Atlas*, 1805 (with John Stockdale); *New Universal Atlas*, 1808 and later; *Viceroyalty of New Granada and Government of Caracas*, 1809; *Cary's New English Atlas*, 1809 and later (first published as a part-work from 1801); England and Wales for the General Post Office, 1832. **Ref.** BURDEN, E. *The Map Collector* 57 (1991) pp.32-37; FORDHAM, H.G. *John Cary...* (Cambridge 1925).

•• **Cary**, George, *the younger* (*d.*1859). Son of John Cary the elder.

•• **Cary**, John, *the younger* (1791-1852). Son of John the elder, whose business he continued. The plates passed to G.F. Cruchley *c.*1844, and from him to Gall & Inglis, *c.*1876.

•• **Cary**, G. & J. Publishing company consisting of John Cary the elder, John Cary the younger and George Cary. *Catalogue of maps, atlases, globes and other works published by G. and J. Cary*, 1800; *Cary's Traveller's Companion...*, 1822 edition (first published 1790).

• **Cary**, Francis (1756-1836). Engraver. Brother of John *the elder* above.

• **Cary**, William (1759-1825). Cartographer and globe maker, brother of John *the elder*, above, with whom he collaborated. Celestial Globe, 1799; maps of Africa, 1805, 1811 and 1828.

Casabona, —. *Pianta Topografica de Genova*, 1877.

Casalis, Godefroi (*d.*1856). Italian geographer. Dictionnaire géographique et statistique des États Sardes, [n.d.].

Casanes, F. de. See **Cesanis**, Francesco de.

Casati, Paolo. Uso del compasso di proportione, Bologna 1664; Hydrostaticae dessertationes, Parma 1695.

Case, John. Estate surveyor. *A Survey of an Estate belonging to Carew Mildmay* [Essex], 1775; Terling Hall in Essex, for Robert Cope, 1775

Case, N. Publisher. Bible Atlas, 1832.

Case, O.D. & Co. Publishers of Hartford, Connecticut,USA. *Case's 12-inch Library Globe*, 1877; Bible Atlas, 1877.

Case, Zophar. 'Clerk in the Auditor's Office' (Illinois). *A Map of the Military Bounty Lands in Illinois*, after 1830 (drawn by James Eddy for Pendleton's Lithography); *Clinton County, Ohio*, *c.*1860.

Casebourne, Thomas. Surveyor. *London and Liverpool Mail Road*, 1826 (with T. Telford and J. Easton); maps in Thomas Telford's *Atlas*, 1838.

Casembrot, A. Dutch engraver at Messina. Maps and views of Messina, some published by the Allard family, *c.*1660s.

Casement, [*Sir*] Roger. Consul at Lourenço Marques, Mozambique. Manuscript maps of parts of Calabar province [Nigeria], 1894; amended Colonel Antonio Juan de Aranjo's 1887 plan of Lourenço Marques, 1896.

Cashee, Nathaniel. Plans of Manila and Tondo, 1819.

Casilini, O. de. Topografia dello Stato d'Ascoli, 1680.

'Casket Mappamundi', engraved by Paulus Ageminius. Cordiform world map engraved on lid of a casket in the Trivulziana collection at Milan. A copy of Sylvanus's world map of 1511.

Caslon, William, father & son.
• **Caslon**, William *the elder* (1692-1766). Engraver bookbinder and printer from Cradley Heath, Worcestershire, England. Father of William, below. Began work as an ornamental engraver of guns in The Minories, London. Caslon started a type foundry in Helmet Row, later Ironmonger Row, and then Chiswell Street. The typeface he devised still carries his name. Known to have engraved up to 7 of the maps in Thomas Gardner's *Pocket Guide to the English Traveller*, 1719 (only one map is signed by Caslon).
•• **Caslon**, William *the younger* (1720-1778). Typefounder, son of William, above. Joined his father in partnership, to form William Caslon & Son. He took over the firm in 1766.

Casparie, W.R. *Atlas van het koningrÿk der Nederlanden*, Groningen 1868; another issue *c*.1872.

Cass, Lewis (1782-1866). Governor of Michigan Territory. *Map of North West Territories of the United States*, *c*.1821; Map of the surveyed part of Michigan, 1822.

Cassas, Louis François. Atlas of Istria and Dalmatia, *c*.1802.

Cassell, John (1817-1865). Temperance missionary and entrepreneur, son of a Manchester publican. Moved to London in 1836, and founded the publishing company which bore his name in 1848. **Ref.** SMITH, D. 'Cassell and Company, 1848-c.1890' in: *IMCoS Journal* 70 (Autumn 1997) pp.7-17.
• **Cassell & Company** (*fl*.1848-*c*.1890). London publishing firm founded by John Cassell. First cartographic publication *The London Conductor*, 1851 (a guide to the Great Exhibition, containing a plan of London by Wyld); also published part works and a magazine of art. The company collapsed, and in 1855 it was taken over by printers Petter & Galpin.
•• **Cassell, Petter & Galpin**. Publishers, '...La Belle Sauvage Yard, Ludgate Hill, EC...', [London]. Using cheap lithographic production techniques they updated and reproduced maps first published in the *Dispatch*, using them to compile a long list of atlases. Cassell's flourished and eventually opened offices in New York, Paris and Melbourne, but was eventually liquidated at the beginning of the 20th century. *Cassell's Home and School Geography Atlas*, 1855; *Cassell's British Atlas* (issued as a part work), 1867-1868 (a re-publication of *The Dispatch Atlas*, 1863, itself originally published in the *Weekly Dispatch* as a part work, 1856-1862, plates later used yet again by Bacon & Co.); *Cassell's Complete Atlas*, *c*.1863; Cassells's County Maps for Road and Rail, published separately from 1863; *Cassell's General Atlas*, *c*.1868; *Cassell's Popular Atlas*, 1890; took a share in the Atlas Publishing Co. which published *Universal Atlas*, 1891-1893 (a translation of *Andree's Allgemeiner Handatlas*); *Cassell's Gazetteer of Great Britain and Ireland*, 1893-1898 (with W. & A.K. Johnston); *Times Atlas*, 1895-1896; *Cassell's Atlas of the World*, 1909 (with maps by Bartholomew); and many others.

Cassella, Andrea. America settentrionale, Torino 1849.

Cassianus Da Silva, D.F. Engraver of Antonio Bulifon's Napoli (21 sheets), 1692.

Cassine, Jean-Baptiste. Monk and cartographer. Giovanni Montecalerio's *Chorographica descriptio provinciarum...S. Francisci Capucinorum*, Milan 1712 edition.

Cassini, family. One of the foremost names in French cartography. Through several generations, members of this family held the office of director of the Paris Observatory, and worked on the well known survey and mapping of France in the 18th century. **Ref.** PELLETIER, Monique *La Carte de Cassini* (Paris 1990) [in French]; KONVITZ, J. *Cartography in France 1660-1848* (Chicago 1987) pp.1-31.
• **Cassini** [Maraldi], Angela. Sister of Jean-Dominique (below), married Francesco Maraldi. Mother of Giacomo Filippo Maraldi and grandmother of Giovanni Domenico Maraldi. See **Maraldi** family.
• **Cassini**, Jean-Dominique [Giovanni Domenico] [**Cassini I**] (1625-1712). Astronomer and geodesist from Imperia, became professor of astronomy at Bologna. He settled in France 1671 when he was appointed director of the Paris Observatory, and became a French citizen in 1673. Supervised the Académie des Sciences' measurement by triangulation of a meridian for France. He died in Paris. Contributed to *Neptune françois*;

Jean-Dominique Cassini [Cassini I] by Cossin. (By courtesy of the Bibliothéque Nationale, Paris)

Hydrographia Galliae, 1685, London edition issued by Robert Morden 1699. **Ref.** *Dictionary of Scientific Biography* Vol. 3 (1971) pp.100-104.

•• **Cassini de Thury**, Jacques [Giovanni] [**Cassini II**] (1677-1756). Member of the Royal Society and l'Académie Royale des Sciences. Born in Paris, son of Jean Dominique, whom he succeeded as director of the Paris Observatory. He died at Thury. *Planisphère terrestre*, Paris, Jean Baptiste Nolin 1696 (a reduced copy of the Great Planisphere on the floor of the Paris Observatory); joined his father and his cousin, Giacomo Maraldi, on the survey along the Paris Meridian in 1700; continued triangulation of France with his son César François [Cassini III], working on the 18-sheet survey of France completed in 1744; *L'Europe suivant les Nouvelles Observations...*, 1747. **Ref.** SHIRLEY, R.W. 'The Maraldi-Cassini triangulated map of France 1744' in: *IMCoS Journal* N°34 (Autumn 1988) pp.4-7.

••• **Cassini de Thury**, César-François [**Cassini III**] (1714-1784). Born at Thury, son of Jacques, studied astronomy at the Paris Observatory under Giacomo Filippo Maraldi

César-François Cassini de Thury [Cassini III] (1714-1784) (By courtesy of the Observatoire de Paris)

(his father's cousin). Member of the Royal Society, the Berlin Academy and l'Academie des Sciences. Based at the 'Quai de l'Horloge dans le même maison que le feu M. Deslisle' (1745). Succeeded his father as director of the Paris Observatory and of the Triangulation of France. Worked with Giovanni Domenico Maraldi (a nephew of G.F. Maraldi) on establishing the Cherbourg-Nantes-Bayonne meridian, from 1737. In 1747 Cassini was asked by Louis XV to undertake a more detailed survey of France; he proposed 180 sheets over 18 years and worked on it until his death in Paris in 1784. Responsibility for the completion of this great map fell to his son Jean-Dominique [Cassini IV]. *La méridienne de l'Observatoire, Paris*, Paris 1744; *Nouvelle Carte Qui Comprend les principaux Triangles qui servent... à la Description Géometrique de la France...* (18 sheets), 1744 (with Maraldi, shows the triangulation network); *Carte qui comprend touts les lieux de la France qui ont été déterminés par les opérations géométriques*, 1746-1747; *La France divisée...*, 1756; *Carte de France...*, 1776; *La description géometrique de la France*, 1783; *Carte Générale et Topographique de la France...* (2 volumes), 1789.

•••• **Cassini**, Jean [Jacques]-Dominique, *comte* de [**Cassini IV**] (1748-1845). Born Paris, died Thury. Astronomer and surveyor, son of César François with whom he collaborated. Astronomer at l'Academie des Sciences 1770, became director of the Paris Observatory on the death of his father in 1784. Worked on the great Cassini map until ownership and control was assumed by the revolutionary government in 1793. Assisted in the post-revolutionary division of France into départements, 1790. *Atlas National de France* [Carte de Cassini; *Carte de France*; Carte de l'Académie] (180 sheets), 1749-1815; *Atlas topographique, minéralogique et statistique de la France*, 1818.

Cassini, Giovanni Maria (1745-*c*.1824). Painter and engraver, Calcografia camerale, Rome. L'America, 1788; *Emisfero Terrestre meridionale delineato...*, 1789; gores for terrestrial and celestial globes, 1790-1792; map of Latium [Italy] (4 sheets), 1791; edited and engraved 182 maps for *Nuovo Atlante Geografico Universale...* (3 volumes), Rome 1792, 1797, 1801; *Carta generale dell' Italia* (15 sheets), 1793; *La Nuova Olanda*, 1798; *La Terra Ferma e la Guiana co suoi Dipartimenti*, Rome 1798; engraved 80 tables in *Nuova raccolta delle vedute antiche e moderne di Roma*, 1799; *Stato ecclesiastico* (7 sheets), 1805; *Stato ecclesiastico* (15 sheets), 1824.

Cassiodorus (*fl.* AD 500). Instructor of geography for monastic students.

Casson, Russel. *A Plan of the Towns of Manchester and Salford* (2 sheets), 1741 (with J. Berry).

Castaldo [Castaldi]. See **Gastaldi**.

Castañera, Ignacio. *Mapa Indicador de la División Militar de España y Portugal* (4 sheets), 1889.

Castanso, Miguel. See **Costansó**, Miguel.

Castell, Prosper von. District of Kastell, 1792.

Castell, William. A Short discourse on the coasts and continent of America, 1644.

Castelli, D. Carte ... della Romagna, 1621.

Castello, Pablo. Plano de San Augustin de la Florida, 1763.

Castelnau, Francis de la Porte, *comte* de (1812-1880). French traveller in the Americas. *Géographie des parties centrales de l'Amérique du Sud...*, Paris 1854.

Castera, Ignacio de. Plano geometrico... ciudad de Mexico, 1776-*c*.1785.

Castiglioni [Castiglione; Castilioneus], Bonaventura (1487-1555). Inquisitor of Milan (1553-1555). Manuscript map of the Alps near Milan (now lost), probably intended for binding into *De Gallorum insubrum antiquae sedes*, Milan 1541 (printed by Giovanni Antonio Castiglioni). **Ref.** KARROW, R. *Mapmakers of the sixteenth century and their maps* (1993) p.113; MEURER, P. *Fontes cartographici Orteliani...* (1991) p.126 [in German].

Castillo, Domingo del. Pilot. Manuscript map of the gulf of California, 1541 (suggests that California might be a peninsula rather than an island), copied by Francisco Antonio Lorenzana, and published in *Historia de Nueva Espana* [Mexico], 1770.

Castillo, Rafael de. Filipinas, 1890.

Castle, H.J. Lithographer, York Street, Toronto. *Map of the Great Lakes*, 1836.

Castleman, Henry. (*fl.*1776-1798). Ordnance draughtsman. *Plan of St Nicholas Island* [Plymouth], *c.*1780 MS; Plan and sections of Mount Wise battery [Plymouth], *c.*1780 MS; and other manuscript surveys of Plymouth harbour, 1780-1781.

Castman, Adolf. Swedish surveyor. Lilla Söd, Östergötland, 1696 MS.

Castner, Gaspard, *SJ* (1665-1709). German Jesuit, worked in China. Relatio sepulturas S. Francisco Xavieris, 1700 (with 3 maps).

Caston, William. **N.B.** A misreading of the signature of William **Caslon** *q.v.*

Castorius (*fl.*4th century AD). Roman grammarian, geographical writer and surveyor. At one time was erroneously credited with authorship of *Tabula Peutingeriana* (Peutinger Table). **Ref.** DILKE, O.A.W. *Greek and Roman maps* (1985) p.113.

Castos, Dominicus. See **Custos**, Dominicos.

Castries, Henry de (1852-1927). French surveyor, worked in Algeria and Morocco. Map of western Draa, 1880; Figuig, 1882; map of southern Oran [Algeria] (15 sheets), 1886; route from Casablanca to Fez, *c.*1887.

Castro, João de (1500-1548). Born Lisbon, died Goa. Portuguese admiral and explorer, Viceroy of the Indies. Compiled three celebrated coastal pilots: Lisbon to Goa, 1538; Goa to Diu, 1538-1539; Goa to Suez, 1541-1542. **Ref.** *Charts of the Rutters of India of Dom João de Castro* [facsimile with an introduction by Luis de Albuquerque] (Lisbon, Edicoes Inapa 1988).

Castro, Joseph de. Carte von Inner Oesterreich (6 sheets), Vienna, Tranquillo Mollo 1812.

Castro, Juan José. *Map of the Railways of South America...*, 1893 (engraved by P. Ludwig).

Castro y Andrade, *Don* Tomas de. Plan of Manila, 1733 MS; *Balabac*, 1753, published Alexander Dalrymple 1774; Plans Cavite Manila, 1762; *Ypoloté*, 1774.

Catalan. See **Este**.

Catalgne, M. de. 'Ingénieur du roi'. Plan de la ville Marie, 1720.

Cataneo, Aniello (*b.*1743). Engraver in the workshop at Herculaneum, the Stamperia Reale, the Officio Topografico di Napoli, and the Officina del Galiani. Engraved for Rizzi Zannoni, 1781 and later; part of a 9 sheet map of Calabria for Eliseo della Concezione, 1783; maps for Giovanni Vivenzio's *Istoria et teoria d'tremuoti*, 1783; *Atlante Marittimo*, Rizzi Zannoni 1784 (with Giuseppe Guerra); Bay of Naples, 1785. **Ref.** VALERIO, V. *Società uomini e istituzioni cartografiche nel mezzogiorno d'Italia* (1993) pp.472-475.

Catarelo, J. de. *La Chia Caballar en España*, *c.*1868.

Catat, Louis (1859-1933). French navigator, doctor and traveller. Carried out many topographical surveys during his explorations of Madagascar 1889-1890. *Voyage à Madagascar (1889-1890)*, 1895.

Catesby, Mark (*c.*1679-1749). English artist, naturalist and etcher. *A Map of Carolina, Florida and the Bahama Islands*, published in his *Natural History of Carolina, Florida, and the Bahama Islands* (2 volumes), London, 1731-1743.

Cathcart, —. *Map of Laporte Co., Indiana*, 1862.

Catherwood, Frederick. *Splendid Panorama of Jerusalem* (a cyclorama in New York City, later destroyed by fire), [n.d.]; *Plan of Jerusalem*, 1833; *New Map of Palestine*, 1862.

Catlin, George (1796-1872). American explorer, geographer and ethnologist. *Niagara River*, 1827, published 1831; Indian localities in 1833, *c.*1841; Indian frontier in 1840; best known for *Catlin's North American Indian Portfolio*, 1844.

Cato, William W. Surveyor. *Plan of Durban*, 1877.

Caton, I.L. (*fl.* 18th century). Engraver.

Cattin, J.B. Celestial globe, 1728.

Cattlin, F. Town plans of Canada: *Montréal*, 1830; *Québec*, 1830; Bytown, 1832; *Goderich*, 1835; *Guelf*, *c.*1835.

Catuffe, Jan [Jean]. Bookseller of Amsterdam 'in de Kalverstraet op de hoek'. *La figure de la terre*, 1738; French editions of Peter Kolb's *Description du Cap de Bonne-Esperance*, 1741.

Catullo, —. *Esquisse d'une Carte géologique d'Italie*, 1846.

Cau, Bonifacius. *Nieuwe Paskaart van Oostende tot den Hoek van Schouwen*, 1799.

Cauche, François. *Isle Dauphine...* in Sanson's *Atlas*, *c.*1651.

Cauchy, François Thomas (1795-1842). Belgian engineer and geologist. *Carte administrative et industrielle comprenant les mines, minières, carrières, usines &c de la Belgique* (9 sheets), 1841.

Caucia, M. (*fl.* 18th century). German cartographer.

Caucigh, Michael, *SJ* (18th century). Mathematician and globemaker.

Caulfield, *Major* William (*fl.*1733-1767). Survey of the Roads of Scotland, 1746-1754 MS.

Caulin, Antonio. Spanish geographer. Orénoque, 1759; Historia corografia... Nueva Andalucia, 1779.

Caumartin, François. 'Graveur en taille-douce' [copper]. Worked for Alexis-Hubert Jaillot in 1672.

Caundish, Richard. See **Cavendish**, Richard.

Caus, Isaac de. Hortus Pembrochianus, c.1645.

Caus, Solomon de. Engineer. Italy, Sardinia and Corsica (from the Richelieu Atlas), 1624 MS.

Causton, Arthur (*fl.*1837-1843). Surveyor of Gloucester, produced manuscript maps for parts of various English counties including City of Gloucester, 1843.

Causton, Isaac (*fl.*1719-1725). Surveyor, produced manuscript maps of Suffolk. *A Mapp of Briset Hall*, 1725.

Cauvanilles, A.J. Mapa del Reyno de Valencia, 1795.

Cauvin, Thomas (1762-1846). French geographer and archaeologist. *Geographie ancienne du diocese du Mans*, 1845.

Cavada y Mendez de Vigo, Augustin de la. Manila, Philippines, 1876.

Cavalier, Jean. Engineer and cartographer of Agde, appointed 'géographe du roi', 1621. *Carte ...de Languedoc*, *c.*1643, 1671.

Cavalier de La Salle, Robert. See **La Salle**, René Robert Cavelier, *Sieur* de.

Cavalieri, Luigi (1811-1880). Architect and topographical engineer in the Officio Topografico di Napoli. Began work on field surveys along the border between the Kingdom of Naples and the Papal States in 1851. Map of Brindisi and Naples, 1863 (with others); conducted surveys in Sicily, 1865; map of Messina, 1866; *Siracusa e dintorni*, 1868. **Ref.** VALERIO, V. *Società uomini e istituzioni cartografiche nel mezzogiorno d'Italia* (1993) pp.475-476.

Cavalleri[s], Giovanni Battista de (1530-1597). Italian engraver working in Rome. *Descriptio Ducatus Polocencis*, 1580; *Terra et Castello di Dieppa*, 1589.

Cavalleri[s], Piero de. Engraver. *Vero disegno del Lago di Geneva*, 1589.

Cavallini, Giovanni Battista (*fl.*1630-1669). Portolan maker of Livorno, often worked with the Oliva family. The Portolan Atlas of Ferdinando de Medici, Grand Duke of Tuscany (13 maps), 1635 MS; Portolan atlas of 6 charts, 1636 (with Jaume Oliva); chart of the western Mediterranean and eastern Atlantic, 1637; portolan chart of the Mediterranean, 1656.

Cavallini, Pietro (*fl.*1654-1688). Mallorcan portolan chartmaker working in Livorno. 4 charts and 6 atlases known. Portolan atlas, 1669 MS (British Library).

Cavassi, J. Archivolti. *L'Isle et Royaume de Sicile* (2 sheets), 1714.

Cavazza, Giovanni Battista, SJ. Engraver. *Nova totius terrarum orbis geographica ac hydrographica tabula* [after Claes Jansz. Visscher, 1639], for Riccioli's *Geographicae crucis...*, Bologna 1643. **Ref.** SHIRLEY, R. *The mapping of the World* (1983) N°357 p.382.

Cave, Edward (1691-1754). Publisher and editor from Newton, near Rugby, England. Apprenticed to Freeman Collins. Later became editor of *Gentleman's Magazine*. which published at least 50 maps between 1739 and 1754. *An Exact Map of the Crim...* (woodcut), 1739; *An Accurate Map of the West Indies*, 1740; *A New Map or Chart of the Western or Atlantic Ocean...*, 1740; *Italy* (woodcut), 1742; *A New Chart of the Coast of New England*, 1746 (engraved Thomas Jefferys); *The Geography of the Great Solar Eclipse of 14th July, MDCCVIII*, 1748; *South America Drawn from the best Maps...*, Jefferys 1749; *A Map of the British American Plantations...*, 1754. **Ref.** REITAN, E.A. *Imago Mundi* 37 (1985) pp.54-62.

Cave, John. Plan of *The Porcelain Manufactory of Worcester*, 1752.

Cavelier, Guillaume. Bookseller and publisher of 'rue St. Jacques à la Fleur de Lys d'Or', Paris. *Les délices de l'Italie*, 1707; Cornelis de Bruyn's *Voyage au Levant*, 1714; publisher of Jean-Baptiste Labat's *Nouvelle Relation de l'Afrique Occidentale...* (5 volumes), 1728.

Cavelier de La Salle. See **La Salle**.

Cavendish, *Captain* Alfred Edward John. *Korea*, 1894.

Cavendish, father and son.
• **Cavendish** [Caundish; Candys], *Sir* Richard (*fl.*1513-1549). Hydrographer and fortifications engineer from Suffolk, father of Thomas, below. Essex and Suffolk coasts, 1525 MS; Thames Estuary, 1533 MS; Dover Harbour, 1541.
•• **Cavendish** [Candish], Thomas (1560-1592). Son of Richard, brother-in-law of Robert Dudley. Circumnavigation of the World aboard *Delight*, 1586-1588; manuscript map of the Pacific shore of America, *c.*1588; map of the South Atlantic, *c.*1588; Voyage of the *Delight* to the Straits of Magellan, 1591 MS. **Ref.** QUINN, D.B. (Ed.) *The last voyage of Thomas Cavendish 1591-1592* (1975).

Caverio [Caneiro; Canerio; Caveri], Nicolo di. Mapmaker of Genoa. Produced an influential manuscript world chart which was one of the first maps to incorporate a scale of latitude. Many of its geographical details and place names were adopted by Waldseemüller. Planisphere, c.1505 MS (Bibliothèque Nationale, Paris). **Ref.** MOLLAT DU JOURDIN M., and LA RONCIERE, M. de. *Les portulans: Cartes marines du XIIIe au XVIIe siècle* (Fribourg 1984) [in French]; also published in English as *Sea charts of the early explorers 13th to 17th century* (London 1984) pp.216-217.

Cavis, C.H.V. *Map of the White Mountains*, 1872.

Cavriolo, Elia. See **Capriolo**, Elia.

Cavrois, N. *Carte Routière du Pas-de-Calais*, 1867.

Cawston, George. Nyasaland, Edward Stanford 1891; *Map of the Matabili, Mashona... countries*, 1888; *Africa south of the Equator*, 1889.

Cawthorne, James. Publisher in London, Catherine Street, Strand. Jamaica, 1807.

Cay, Robert. Completed John Horseley's *Northumberland* (2 sheets), 1753.

Caylus, — de. *Plan of the Town of Fort Grenada, Plan of the Town and Citadel of Fort Royal*, [Martinique], in Thomas Jefferys' *North America and the West Indies*, 1760.

Caymox, firm. Publishing company of Nuremberg, consisting of Cornelis and Balthasar Caymox.
• **Caymox** [Caimox; Chaÿmox], Cornelis. Merchant of Nuremberg. *Nürnberg*, in Braun and Hogenberg's *Civitates orbis terrarum*, 1575. **Ref.** MEURER, P. 'Der Nürnberger Verlag Caymox und die Kartographie' in: *Quaerendo: a quarterly journal...* 23,1 (1993) pp.24-43 [in German, with English summary].
• **Caymox**, Balthasar (1561-1635).

Edward Cave (1691-1754). (Taken from Map Collectors' Circle*)*

Christophorus Cellarius (1638-1707).

Caziaic, *Lieutenant* L.V. Artillery Map of the Siege of Yorktown, [Virginia], 1781-1782.

Cazzul, Matteo and Carlo. *Il Piemonte con il Monferrato*, 1615.

Ceard, Nicolas (1745-1821). Swiss engineer of French descent, road superintendent and councillor in Bresse, Bugey, Versoix and Simplon. Route du Simplon, 1820.

Cecil, *Sir* William, *First Baron* Burghley (1520-1598). Secretary of State and Lord Treasurer to Elizabeth I. Regarded as a key figure in English cartography during the Renaissance, both as a map collector and patron of mapmakers including Christopher Saxton and John Rudd. Part of his collection of maps is preserved at Hatfield House, Hertfordshire. Burghley himself drew many sketch maps. Liddesdale [Scottish borders], 1552; Ireland, *c.*1558; Rough sketch of the River Lea, from Bow Bridge to Hackney..., *c.*1560; and others. **Ref.** HARLEY, J.B. *The Map Collector* 3 (1978) pp.12-19.

Cecill, Thomas (*fl.*1627-1635). Engraver of London. *Nova Terrae-Mariae tabula*, in *A*

Relation of Maryland [Lord Baltimore's map], 1635.

Cécille, *Captain* Jean-Baptiste Thomas (1787-1873). French naval officer, served in the Indian Ocean, West Indies, etc. from 1815, New Zealand in 1838 (with Fournier and Durand-Dubraye), China from 1842. His voyages contributed greatly to knowledge of the geography and economy of China. *Campagne dans les mers de l'Inde et de la Chine à bord de la frégate l'Erigone... 1841... 1844* (4 volumes), 1847-1850.

Cederborg, L. *Postkarta öfver Sverige*, *c.*1870.

Cederbourg, Eric. *Plan af Gotheborgs Stad*, *c.*1739.

Cederstolpe, *Lieutenant* Theodor von. Prussian infantry officer. *Historischer Plan der Bundesfestung Luxembourg*, Luxembourg, M. Kronner 1845.

Celi, Francis Mathew. Maps for Thomas Jefferys' *Topography of North America*, 1768.

Cella, C. See **Zell,** C.

Cella, Philipp. German cartographer and globemaker. *Der gestirnte Himmel*, 1824; *Pneumatisch-portativer Erdglobus* [inflatable globe], 1831.

Cellarius, Andreas (*fl.*1656-1702). Dutch mathematician and geographer, rector of Hoorn. *Descriptio Poloniae et Lithuaniae*, 1659, 1660; *Atlas Coelestis*, 1660, 1661, re-issued 1708 by Pieter Schenck.

Cellarius [Keller], Christophorus [Christoph] (1638-1707). Geographer, professor at Halle. *Geographia antiqua*, 1686 and editions to 1812; *Notitia orbis antiqui*, 1701.

Cellarius Ferimontanus, Daniel. German geographer and physician from Wildberg, settled at Vlissingen, Netherlands. Compiled text for Gerard de Jode's *Speculum orbis terrarum*, 1578.

Celle, Benedetto. Publisher of Genoa. Issued Francesco Maria Levanto's *Specchio del mare*, 1664.

Cellier, *veuve* A. Publisher of Paris. Henri Justel's *Recueil de divers voyages*, 1684.

Cellius. See **Zell.**

Celtis [Celtes; Meissel; Pickel], Konrad [Conrad] (1459-1508). Geographer, professor at Vienna. He collected and used globes and maps, amongst them a road map of the Roman Empire, which he left to Konrad Peutinger (it was published as *Tabula Peutingeriana*, 1598 and many later editions). His poetic description of Germany with 12 woodcut illustrations (of which 4 were pictorial maps) was published as *...Quator Libri Amorum...*, Nuremberg, Hieronymus Holzel 1502. **Ref.** DREYER-EIMBCKE, O. *The Map Collector* 74 (1996) pp.18-21.

Centeno y Garcia, José. *Philippines*, 1876.

Center, *Lieutenant* A.T., US Army. Survey of Hawe Bay, Michigan, 1836.

Central School Supply House. Publishers and globe makers of Chicago. 12-inch terrestrial globe, 1900.

Centurion, Manuel. *Plano General dela Provincia de Guaiana*, 1770 MS.

Cenus, N. See **Zeno,** Nicolo.

Cerda, Ildefons (1815-1876). Catalan engraver. Plano de los alrededores de la ciudad de Barcelona, 1855; Plano de los alrededores de la ciudad de Barcelona y proyecto de su reforma y ensache, 1859.

Cernotti, Leonardo. Translated Giovanni Antonio Magini's Latin edition of Ptolemy's *Geographia* into Italian for Simon Galignani's heirs, Venice 1597-1598 edition.

Cerri, Carlo. *Carta stradale e postale dell' Italia*, 1849; *Italia carta generale*, 1859.

Cerro, Giacomo. *Nova ac verissima totius Terrae Sanctae descriptio*, 1569.

Cerruti, Agostino. *Corso del Po per la Lombardia...* (6 sheets), Rome, D. de Rossi 1703 (engraved by A. Barbey).

Cerruti, G.E. Carta dello Stretto di Galero, Torino 1873.

Konrad Celtis 1459-1508. (By courtesy of László Gróf)

Certes, Jean. Publisher in Lyons for Pierre Du Val, 1676, 1688, 1712.

Cervino, Pedro Antonio. Mapa geográfica de la provincia de Chiquitos, 1777; *Plano topográfico de la ciudad de Buenos Ayres*, 1814, republished Paris, Darmet 1817.

Cesani [Alysius Cesanis Ydrontius], Luigi. World, 1574 MS; manuscript atlas (4 maps), 1581.

Cesanis [Cexano], Alvixe [Aloyse; Alvise; Luigi]. Copy of a chart of the Mediterranean, 1489 MS (British Library).

Cesanis [Casanes, Cexano], Francesco de. Venetian chartmaker. Chart of the Mediterranean, 1421 MS (Museo Civico, Venice).

Cesanis, Jean. Mapmaker of Dieppe. World, 1570.

Cesario, Agostino. Portolan, 1570.

Cesaris, Giovanni Agusto (1749-1832). Italian astronomer and topographer.

Cespedes, Andreas Garcia [Carcia] de (*fl.*16th century). Spanish geographer and chronicler of the Indies. Geographia general e historia, 1598; *Regimiento de navegacion*, 1606.

Ceulen. See **Keulen**.

Cexano, Alvise. See **Cesanis**.

Chabannes, — *Amiral vicomte* de. *Atlantic Ocean Pilot Chart*, 1868.

Chabbert, —. *Plan de la Ville d'Orleans*, *c.*1867.

Chabert de Cogolin, Joseph Bernard, *marquis* de (1723-1805). French Admiral and hydrographer. Voyage en l'Amérique septentrionale 1750-1751, Paris 1753; *Carte réduite des côtes de l'Acadie, de l'Isle Royale et de la partie méridionale de l'Isle de Terre-Neuve* [voyage of 1750-1751], Bellin 1756 and later (published in *Hydrographie Françoise*); *Carte des côtes de la Grèce...*, 1797, published Depôt de la Marine *c.*1800 (in *Neptune de la Mediterranée*).

Chace, John, *Junior*. Civil engineer and surveyor of New York. Survey for *Clark's Map of Fairfield County Connecticut...*, Philadelphia, Richard Clark 1856, 1858; *Fulton County, New York*, 1856; *Rockingham County, New Hampshire*, 1857; *Suffolk County, New York*, 1858.

Chadov, Aleksandr. Collaborator in *Atlas Russicus*, 1742.

Chadwick, James. Surveyor of Liverpool, produced manuscript surveys of estates and towns in Kent and Lancashire, 1725.

Chadwick, Joseph. Surveyor. North America, 1764-1772. *Plan of part of Penobscot River...*, 1771 MS.

Chaffat, Antoine du (*d.*1740). French military cartographer and engraver in Bavarian service. Philippsburg, 1734; plan of Alt-Breisach, 1735; plan of Memmingen, 1737; Crimea, 1738; *Provinciarum Turcio Tartaricum*, *c.*1738.

Chaffers, E.M. Surveyor. *Chart of Port Nicholson, New Zealand*, 1840; *Tory Channel*, 1841.

Chaffrion [Chafrion], José [Giuseppe] (1653-1698). Mathematician and cartographer of Valencia. *Mappa Geografica... delle Provincie del Tortona*, *c.*1665; *Carta de la Rivera de Genova* (8 sheets), 1685; *Topographía da Liguria*, 1685; La Liguria, 1697.

Chaigneau, William. *...Map of the Post Towns & Barracks in Ireland...*, 1730.

Chaillé-Long [Long], Charles Rollinson Whittingham (1842-1917). Francophile and native of Maryland. More correctly surnamed Long, he adopted 'Chaillé-Long' in acknowledgement of Huguenot ancestry. Veteran of the Union army, appointed Lieutenant-Colonel [*bey*] in the Egyptian Army, serving for a while as Aide to Charles Gordon [of Khartoum]. Explored Nile and Lake Victoria Nyanza regions, 1874-1875, discovered Lake Kyoga on the White Nile; reconnaissance of the Juba River [Somalia], 1876; later sent as Consul General to Korea. **Ref.** ICENOGLE, D. *The Map Collector* 46 (1989) p.31; ICENOGLE, D. 'The Romantic Adventurer as Explorer? The Case of Charles Chaillé-Long' in: *Terrae Incognitae* Volume XX (1988) pp.21-42.

Chailley-Bert, Joseph (1854-1928). French traveller and geographer, worked in Indochina and Netherlands East Indies 1880s-1922. *La colonisation de l'Indo-Chine*, 1892; *Les hollandais à Java*, 1898.

Chaillou, François. One of the surveyors for Cassini's *Carte de France*, working with others on surveys in and around Paris, and parts of the Loire valley (sheets 1, 7, 8, 48 & 49), 1754-1758; Brie, 1762.

Chain & Hardy. Publishers of Denver, Colorado. Map of Arizona and New Mexico, 1881.

Chaix, A. Atlas des chemins de fer, Paris *c.*1860.

Chaix, Emilio. *Carta volcanologia e topografica dell'Etna*, 1892.

Chaix, Napoléon [et Cie]. '...rue Bergère, 20, a Paris...'. *Plan de Londres pour le guide*, Paris 1851.

Chaix, *Professor* Paul (1808-1901). French geographer and historian, worked in Geneva. Atlas élémentaire, 1842; *Carte du Duché de Savoie*, 1832; Vallée Beaufort, 1856; William Gell's Environs of Rome, 1861.

Chalk, Meggy & Chalk. Publishers of the later edition of Chapman and André's map of Essex (25 sheets), 1833 (first published in 24 sheets 1777)

Challamel, A. [*ainé*]. Publisher of Paris. *Atlas des colonies françaises*, 1866; P. Pelet's *Nouvel atlas des colonies françaises...*, 1891.

Challoner, *Sir* Thomas. De republica anglorum, 1579.

Chalmandrier, Nicolas. Engraver and publisher of Paris. Engraved 6 sheets of the Cassini *Carte de France*, from 1756; *Plan de la ville de Madrid*, 1761; engraved most of the maps in Jean Chappé d'Auteroche's *Carte Générale la France et l'Empire, la Pologne et la Russie*, 1768 (volume III of Chappé d'Auteroche and Krasheninnikov's *Voyage en Sibérie*); *Carte Generale du Canal Royal de la Province de Languedoc...* (3 sheets), 1771; *Carte du Comté d'Ablois...*, 1772; *Plan du Varsovie*, 1772; *Carte du Canal royale... du Languedoc* (16 sheets), Toulouse 1774; maps for Rigobert Bonne 1762, 1779; *Plan de Gibraltar, attaqué par Terre et par Mer Par l'Armée Espagnole et Française...*, Paris, Esnauts et Rapilly 1782.

Chalmers, Cathcart. British surveyor. Revised Sir R. Schomburgk's *Map of British Guiana*, 1875 (original map dated 1841-1844).

Chalmers, *Reverend* P. Dunfermline Coalfields, 1841.

Chalonnois, Jean Jubrien. See **Jubrien,** J.

Chamberlain brothers. Globe and orrery makers and sellers of Boston, Massachusetts. Ref. WARNER, D. Rittenhouse Vol.2 N°2 (1987) p.53.
• **Chamberlain,** Nathan B. (1809-1878). Moved from western Massachusetts to Boston in 1831. Made many school globes and orreries, at least one with his brother Daniel. 13-inch globe, *c.*1840s.
• **Chamberlain,** Daniel (*fl.*1840s). Worked with his brother Nathan, on at least one orrery.

Chamberlain, E.L. Surveyor. Atlas map of St. Charles County, Missouri, 1875.

Chamberlain, John. Portsea Island, 1757; St Ouens Bay, Jersey, 1758 MS; Milford Haven, 1765.

Chamberlain, Thomas C. Geological atlas of Wisconsin, 1876.

Chamberlaine, Thomas. Ireland, 1776.

Chamberlaine, William *junior*. *An Exact and Correct Plan of Cordwainers Ward*, London 1768 (engraved by James Kirk).

Chamberlin, Thomas Crowder (1843-1928). Geologist. Survey of Wisconsin, 1877.

Chambers, B. Engraver of Washington D.C. *Plan of Lovejoys Narrows... Kennebeck River...*, 1826; S.L. Dashiell's *Map of the Northern Part of... Maine and... adjacent British Provinces...*, 1830; New Map of the State of Alabama, 1830.

Chambers, Ephraim (1680-1740). Apprenticed to John Senex.

Chambers, Francis (*fl.*1836-1859). Surveyor. *Plan of the Ward of Farringdon Within* [London], 1851 (with Richard Tress); *Plan of the Parish of St. Michael le Querne* [London], 1853 (with Richard Tress).

Chambers, Frederick C. Surveyor. *Town of Greenburgh* [New York], 1857; *Town of Cortlandt, Westchester County, New York*, 1858.

Chambers, Julius. *The New Europe* ['Acco' peace map, N°1], New York, American Code Co. 1915 (with F. Cunliffe Owen); *The New Europe* ['Acco' peace map, N°2], New York, American Code Co. 1915.

Chambers, R. Map of Canada, copied from Samuel Augustus Mitchell's map, 1839.

Chambers, *Captain* William (*fl.*1758-1816). Royal Navy surveyor. Many manuscript plans of bays, etc., in Canada, including a plan of Lake Champlain, 1782. **Ref.** *Atlas of Lake Champlain 1779-1780* (facsimile with introductory essay by J. Kevin Graffagnino) (Vermont 1984); GUTHORN, P. *British maps of the American Revolution* p.19.

Chambers, William and Robert (*fl.*1800-1883). Publishers of Edinburgh. *Atlas of Ancient and Modern Geography*, 1845 and later editions; School Atlas, 1845; Atlas for the People, 1846; Atlas to Chambers' *Encyclopaedia*, 1869; Handy Atlas, 1877; Atlas to Chambers' Gazetteer, 1895.

Chambers, Zachary. *Spring Gardens* [Westminster], 1742 MS; *Bailiwick of St. James*, 1767 MS.

Chambertin, G.L. Surveyor. *Map of the Cities of Marietta and Harmer, Ohio*, 1869.

Chambon, Gobert-Denis (*d.*1781). Engraver of Paris. J. Janvier's *Carte du Gouvernement Militaire de l'Isle de France*, Paris, J.B. Nolin 1746; engraver for S.G. Longchamps & Jean Janvier's Amérique, 1754; Longchamps' *Carte des Possessions Françoises... dans le Canada et... Louisiane*, 1756; engraved lettering on some sheets of the Cassini *Carte de France*, published from 1756.

Chambray, *Marquis* de. Atlas of the campaign of 1812 (5 sheets), Paris 1823.

Chamereau, Nicolas. Géographie pratique, Amsterdam 1715; Moyen de trouver la longitude, Amsterdam 1715.

Chamisso, Adelbert von (1781-1838). German writer and natural scientist, circumnavigated the World under the command of Otto von Kotzebue. Reise um die Welt, 1815-1818, Leipzig 1836.

Chamot, Jean. Plan de Paris divisé en 12 mairies, Paris 1809.

Chamouin, Jean-Baptiste-Marie (*b.*1768). Editor and engraver in Paris, 'Rue de la Harpe N^os.^20 & 242'; 'N^o.^35' (1809). *Carte de la France*, [n.d.]; engraved the maps for Pascal François Joseph Gosselin's *Recherches sur la Géographie Systématique et Positive des Anciens*, 1798-1813; Plan de Paris, 1803; Tyrol, 1808; engraver for Alexandre Emile Lapie, 1808, 1812; engraved Pierre Lapie's *Carte des systèmes de Ptolemée*, *Géographie du Moyen Age*, and several other maps of Europe, Asia and Africa, 1809, used by Malte-Brun in *Précis de la géographie universelle*, 1810 (printed by Prudhomme).

Champ, —. *Quatres chartes pour la campagne d'Italie*, 1801.

Champain, *Lieutenant* I.U. Surveyor. *Plan of Lucknow*, 1858.

Champante, William. Partner (with Benjamin Whitrow) in the firm of Champante & Whitrow.
• **Champante & Whitrow**. Publishers of Jewry Street, Aldgate, London (1785-1816). Jigsaw puzzle map, *c.*1770.

Champeaux, —. Plan of Vavao for J.S.C. Dumont d'Urville, 1847 (with Duranty).

Champias, —. Worked with Chaillou and Pouillard on the survey for sheet 7 [Fontainebleau-Etampes] of the Cassini *Carte de France*, 1750-1755

Champion, —. *Plan of the Bristol Channel, from The Key to the Hot Wells*, 1767.

Champion, J.N. Engineer, geographer and engraver of Leipzig. *Plan von der Festung Luxemburg*, Leipzig, Johann Conrad Hinrichs 1802; Nouvelle carte de l' Allemagne (30 sheets), Leipzig 1806; Carte routière de l'Europe, 1811; Mappemonde, 1816; reduced Aaron Arrowsmith's map of America for A.R. Frémin in 1828.

Championnet, Jean Etienne. Royaume de Naples, 1779.

Champlain, Samuel, *sieur* de (1567-1635). Explorer and Governor of Québec. *Carte Geographique de la Nouvelle Franse...* [St Lawrence valley], 1612 and other maps in: *...Les Voyages...ou journal trés fidele des observations...de la Nouvelle France*, Paris, Jean Berjon 1613; untitled map of Canada, *c.*1616 (engraved by Melchior Tavernier, amended and issued by Pierre Duval 1653); *Carte de la Nouvelle France*, 1632. **Ref.** BURDEN, P. *The mapping of North America* (1996) pp.197-200; *ibid.* pp.206-220; *ibid.* pp.228-231; *ibid.* pp.294-297; *ibid.* 395-396; HEIDENREICH & DAHL 'The two states of Champlain's Carte Géographique', *Canadian Cartographer* 16, N^o.^1 (1979) pp.1-16; HEIDENREICH, C.E. *Explorations and mapping of Samuel de Champlain, 1603-1632* (Cartographica Monograph N^o.^17, 1976); MORISON, S.E. *Samuel de Champlain: Father of New France* (Boston 1972).

Chanas, Jean Baptiste Adolphe (1810-1865). Campagne de 1815, 1858.

Chancel, Charles François Joseph de. Prepared sheet 70 [Périgeux] of the Cassini *Carte de France*, 1773-1779 (engraved by Capitaine and Bourgoin).

Chancellor, Richard (*d.c.*1555). Captain of the *Edward Bonaventure*. Sailed initially with Sir Hugh Willoughby's expedition in search of a north-east passage through the Arctic to China, 1753-1754. Separated from the other two vessels Chancellor reached the White Sea and Moscow, to which he made a second voyage in 1555. He died when his ship was wrecked off the Scottish coast on the return voyage. An account of the voyages was written by Clement Adams, 1555, and later used by R. Hakluyt, 1589, and Samuel Purchas, 1625.

Chancourtois, A.E. (1820-1886). French engineer and geologist.

Chancy [Chanzy], *General* Antoine Eugene Alfred (1823-1883). Campagne de l'armée de Loire 1870-1871, 1871.

Chandia, Antonio de. Philippines, 1727.

Chandler, *Lieutenant* —. Salt Lake City to Cajon Pass, 1859.

Chandler, John. Surveyor. *A Map of ye Moheagan Sachims Hereditary Country...* [Connecticut], 1705 MS published 1769.

Chandler, John. Hydrographer, pilot and master *RN*. Drew charts for David Steel, including: British Channel, 1782; *A chart of the north east coast of Great Britain*, 1782; North Sea, 1784, 1795; prepared a major part of *Steel's Seaman's Guide and New Coaster's Companion*, 1792.

Chandler, R.W. United States, 1829.

Chandler, Richard (*d.*1744). Publisher and bookseller of London, in partnership with Caesar Ward 'at the Ship without Temple Bar, and at their Shops in Coney Street, York, and at Scarborough Spaw'. In 1738 they bought the copyright, plates and stock of Robert Morden's *Magna Britannia*, selling it complete and as individual maps. After the suicide of Chandler, the shop at Temple Bar was taken over by William Sandby. *Magna Britannia Antiqua & Nova...* (6 volumes), 1739. **Ref.** HODSON, D. *County atlases of the British Isles published after 1703* Vol.I pp.27-31.

Chandler, W. Maps of Brazil, 1864-1869.

Chandler, William. Surveyor in North America. *An Exact Plan of the Sea Coast...from Paucatuck River... to Slocums Harbour*, 1741 (with James Helme).

Chandless, William (1829-1896). Royal Geographical Society Gold Medallist. British explorer and surveyor of Amazon and tributaries. Maps were engraved by Edward Weller to accompany his published notes. *Map of the R*[iver] *Purüs...*, 1865 MS, (engraved on two sheets by Edward Weller), 1866; *Map of the River Aquiry...*, 1866; *Map of the River Juruá...*, 1869.

Chang Jen-chun. Kuang tung yu ti chuan-tu [Complete Atlas of Kwantung], 1897.

Chang Heng (*c.*AD 70-140). Chinese astronomer. Star maps.

Changuion, D.J. Bookseller and publisher of Amsterdam. G.T.F. Raynal's *Atlas portatif...*, 1773 (with Harrevelt); M. Robertson's *Histoire de l'Amérique*, 1779.

Changuion, François. Bookseller of Amsterdam. Map of Algiers, 1755.

Chanier, L. *Carte de la Montagne de l'Epine*, 1869; *Ville des Sables d'Olonne*, 1875.

Chanlaire, Pierre Grégoire (1758-1817). Geographer and publisher of Paris, produced atlases and separate maps. Rue Serpente N°·15 ('Bureau de l'Atlas National' 1793); then '26 rue de la Harpe' (from 1791); 'rue Geoffroy-Langevin N°·328' (1797); 'rue Geoffroy-Langevin N°·7' (from 1807). Collaborated with geographer Pierre Dumez and engraver François d'Houdan on *Atlas national de la France*, 1790-1811 (after the execution of Dumez in 1794, his widow and Chanlaire continued to update and publish the *Atlas national*); reduced edition appeared as *Atlas national portatif de la France*, 1791; *Carte de l'ancien royaume de Pologne...*, 1795 (with E. Mentelle); *Carte Géométrique des Routes de Postes de la France*, 1796 (with

L. Capitaine); Carte chorographique de la Belgique (69 sheets), Paris 1797 (with Louis Capitaine de Chesnoy); *l'Atlas universel de géographie*, 1797 (with Edme Mentelle); *l'Atlas des commençans*, 1798 (with Edme Mentelle); *Département du Loiret*, 1800; Cappaduce, 1800; Arménie, 1801; Description topographique et statistique de la France, 1801-1811; Europe (48 sheets), 1801; Nouvel atlas de la France, 1802; *Atlas de la République française composé de 102 feuilles*, 1802 (with P.C. Herbin de Halle); *l'Atlas élémentaire portatif*, 1807 (with Edme Mentelle); *Description topographique et statistique de la France*, Paris, P.G. Chanlaire & la Veuve Dumez 1810 (with J. Peuchet); Isles Britanniques, 1818. **Ref.** PACHA, B. in: PACHA, B. & MIRAN, L. *Cartes et plans imprimés de 1564 à 1815* (Bibliothèque Nationale de France, Paris 1996) p.44 [in French].

Channeau, F. Engraver for Jaillot's *Atlas Nouveau*, *c.*1681.

Chanoine, Jules (1835-1915). French explorer attached to China expedition 1859, and in Japan from 1866 where he adapted an earlier map of east Asia for the Japanese government. *Les délimitations de frontières et les traités avec la Chine*, 1889. **Ref.** BROC, C. *Dictionnaire illustré des explorateurs Français du XIXe siècle: II Asie* (1992) pp.81-83.

Chantovoine, —. *Carte de la Virginie*, 1781 MS.

Chantreau, Pierre Nicolas. *Carte de l'Irlande Pour le Voyage dans les trois Royaumes*, Paris, Briand 1792 (engraved by Boutrois); *Carte de Russie*, 1794.

Chantry, John (*fl.*1660-1669). *The Course of the Great River Wolga* (2 sheets), 1662; engraver for London edition of Johann Albrecht von Mandelslo's Voyage, 1682.

Chantry, Thomas. Surveyor. *A New and Accurate Plan of the City of Bath*, 1793, 1805.

Chanvalon, Jean Baptiste. *Isle de Martinique*, 1763.

Chanzy, A.E.A. See **Chancy**, Antoine Eugene Alfred.

Chapaux, Albert. *Carte générale de l'Etat Indépendant du Congo* (4 sheets), 1894.

Chapeaurouge, Carlos de. *Mapa de la Provincia de Santa Fé*, 1872; Buenos Aires, 1883; Republica Argentina, 1897; *Mapa general de la República Argentina. Construido especialmente para ilustrar el canso national de 1914*, Buenos Aires L.J. Rosso y cia. 1914.

Chapellier, —. Engraver for Dépôt des cartes et plans de la marine, 1874.

Chapelot & Cie.. Publishers of Paris. *Carte de la frontière nord-est de la France...*, 1914 (engraved and printed by Erhard brothers).

Chapin, A.M. *City of Manchester, Hillsborough County, New Hampshire*, 1850.

Chapin, William (1802-1888). Engraver of New York. Lay's map of the United States, 1832, 1839.

Chapman, *Captain* Charles. Azores, 1782.

Chapman, C.F. Plan of the Township of Papineau, Toronto 1878; Township of Proudfoot, 1878.

Chapman, George T. *Map of the northern part of the province of Auckland*, *c.*1865; *Waikato*, 1866; *Bay of Plenty*, *c.*1873.

Chapman, Isaac A. *Plan of the position taken by Genl. Burgoyne on the 10th of Octr. 1777* [at Saratoga], Philadelphia, M. Thomas 1818 (engraved by G. Fairman).

Chapman, James (*fl.*1784-1834). Surveyor, produced various manuscript maps of parts of Scotland including *Island of Lewis*, 1807-1809, 1821.

Chapman, James. *A Map of the Gold Fields of South Eastern Africa*, 1876.

Chapman, John. Cartographer and publisher, Royal Arcade, Pall Mall. Engraved *Staffordshire* for William Yates, 1775; *A Map of the County of Essex* (24 sheets), surveyed 1772-1776, published 1777, 1785, (25 sheets) 1833 (with Peter André); *Nottinghamshire* (4 sheets), 1776, 1785, 1792; *Plan of Newmarket*, 1787.

Chapman, Leander (*fl.*19th century). Surveyor General of Michigan. *Sketch of the Public Surveys in Michigan*, 1854.

Chapman, R.M. Lands of Saco Water Power Company, 1848.

Chapman, Silas. Publisher of Milwaukee. *Wisconsin*, 1853; Milwaukee, 1856; *New Sectional map of Minnesota*, 1856; *Township map the North West*, 1858; *Chapman's Rail Road Map of Ohio, Indiana, Michigan, Illinois, Missouri, Minnesota & Wisconsin*, 1859 (lithography by Chas. Schober); *Complete map of Colorado*, 1861; *Chapman's New Township Map of the Northwest...*, 1862; *Sectional map of the surveyed part of Dakota*, 1869.

Chapman, William (1749-1833). Engineer. Canal from Newcastle to Maryport, 1795; plan of part of the rivers of Derwent and Hertford, 1795; Newcastle to Haydon Bridge, 1796.

Chapman, William, *RN*. Admiralty hydrographer, Master of *Illustrious*. Surveys of the coast of Portugal, including *...entrance of the River Tagus*, 1806 published in Thomas Hurd's Channel Atlas, London, Ballintyne & Byworth 1811; survey off the banks of the Thames and the East Coast, 1809; survey of Plymouth Sound (amendment of Murdoch Mackenzie's survey of 1779), published by Hurd as *A Survey of Plymouth...*, *c.*1810; *...coast of France from Pte de Conq to Pte Trevignan...*, 1810, published in Hurd's Channel Atlas, London, Ballintyne & Byworth 1811.

Chapman & Hall. Publishers of 186 Strand, London (1846-1850), then 193 Piccadilly (1850-1880). John Gorton's *A Topographical Dictionary of Great Britain and Ireland*, 1833 (maps drawn and engraved by Sidney Hall, often republished in the later atlases of Sidney Hall); *A New British Atlas*, 1833; took over from Baldwin and Cradock as the publisher of maps for the Society for the Diffusion of Useful Knowledge [SDUK], 1842-1844; Sidney Hall's *A Travelling County Atlas*, 1842 and later; Sidney Hall's *A New County Atlas*, 1847; Penny Maps, 1852; Lowry's Universal Atlas, 1853; Sidney Hall's *The English Counties*, 1860. **Ref.** CAIN, M.T. 'The maps of the Society for the Diffusion of Useful Knowledge: A publishing history' in: *Imago Mundi* 46 (1994) pp.159-163.

Chappé d'Auteroche, *abbé* Jean (1722-1769). French astronomer and geographer, born in the Auvergne, died California. Member of the Académie Royale des Sciences. Worked with Rittman on sheet 161 [Landau-Wissenbourg] of the Cassini *Carte de France*, 1755-1761; undertook a journey to Siberia in 1761, information from which was recounted in the 4 -volume work he prepared with Stepan Petrovitch Krasheninnikov - *Voyage en Sibérie...en 1761* (volume II has 7 maps, Volume III, 30 maps), Paris, Debure 1768 (maps engraved by J.B. Tilliard, Croisey and N. Chalmandrier); *Plan de la Ville de Mexico*, 1769 (engraved by de la Gardette) published in his *Voyage en Californie...*, Paris 1772.

Chappell, G.G. Plan of Bewdley, 1832; Calne, *c.*1832.

Chapperon, —. Worked with others on surveys for sheet 17 [Albi] of the Cassini *Carte de France*, 1769-1774 (engraved by Aldring and Bourgoin).

Chapuy, Jean Baptiste (1760-1813). Engraver.

Charbonnier, Pierre-François (*c.*1703-post 1751). 'Imprimeur en taille-douce, rue du Foin', Paris (1725), 'rue St Jacques au Chariot d'Or' (1738, 1750). Plan de Paris, 1744.

Chardin, Jean (1643-1713). French traveller in Persia and India 1665-1677. *Journal du Voyage du Chevr. Chardin en Perse & Aux Indes Orientales...* (with 8 maps), Westminster, 'Moyse Pitt' 1686 (English version *Travels of Sir John Chardin into Persia and the East Indies*, 1686); *Voyages du Chevalier Chardin...* (4 volumes, with 15 maps) 1735 (an expanded edition of the former).

Chardin, Charles. Printer of Paris for Vivien de Saint Martin. Archipels de l'Océanie, 1885.

Charland, Louis (1772-1813). Land survey maps including: *Plan de la Ville et Cité de Montréal...*, 1801; *A New Topographical Map of the Province of Lower Canada... from... surveys... by... Samuel Holland... 1587*, 1803 (with William Vondenvelden).

Charle, Jean Baptiste Louis. French geographer. 'rue du Condé 14', Paris. *Carte minéralogique des Pyrénées*, 1831; *Nouvel Atlas national de la France*, 1833; *Nouvel Atlas de Géographie Universelle* (15 sheets), Paris 1840; *Atlas élémentaire*, 1845; Carte du royaume de Siam, 1854; *Carte des deux Amériques*, 1865, 1868.

Charles, François. Engraver. For Philippe Marie Guillaume Vandermaelen, 1829-1846; *Carte de la Belgique dediée à l'armée 1832*, Brussels, L. Rottigni.

Charles, Henry. Publisher of 57 Mary Street, Dublin. *A Complete System of Modern Geography*, 1816.

Charles, Henry. Publisher and printer of Philadelphia. United States, 1819.

Charles & Dandelle. France par divisions militaires (22 sheets), 1822.

Charlesworth, L.C. *Map of Eagle Lake*, 1897.

Charlevoix, Pierre François Xavier de *SJ* (1682-1761). French explorer, in Canada 1705-1720. *Histoire de l'Isle Espagnole ou de S. Domingue...*, Paris, François Didot 1730-1731; *Histoire et Description Générale de Japon*, 1736 (with maps by Bellin); *Histoire et Description Generale de la Nouvelle France*, Paris, Didot 1744; *L'Amérique suivant... Nouvelles Observations...*, Paris, Le Rouge 1746, Paris, Crepy 1774; *Histoire de Paraguay*, 1751, 1756, 1757.

Charlton, John. Ilchester Meadow, Somerset, 1792 MS.

Charlton, L. Plan of the Town of Whitby, 1788.

Charmetant, F. *Carte partielle des Missions de l'Afrique Equatoriale*, 1879.

Charmont, —. *Carte itinéraire du Royaume de France*, *c.*1835.

Charnières, Charles F.P. de (1740-1780). French sailor and astronomer. Théorie et pratique des longitudes en mer, 1772.

Charpentier, Le. See **Le Charpentier**.

Charpentier, Charles. *Plan de la ville de Lunéville*, 1875.

Charpentier, F. *Nouvelle carte de la France*, 1869.

Charpentier, H. Île de la Réunion, 1867.

Charpentier, J. *Carte officielle des Chemins de Fer Belges*, 1875.

Charpentier, Jean de. Carte Géognostique des Pyrénées, *c.*1823.

Charpentier, Johann Friedrich Wilhelm von (1728-1805). Geologist and director of mines Freiberg. *Mineralogische Geographie der Chursächsischen Lande*, Leipzig 1778.

Charpentier, W.H. 'Artists Repository 50 High Street Portsmouth' (1850); '46 High Street' (1897). *Map of Portsmouth*, *c.*1850; *Charpentier's New map of Portsmouth...*, *c.*1897.

Chartier, A.T. *Nouvelle Carte...de l'Europe*, 1867; *Carte de la Gaule*, 1867; *Carte Générale de la Chaine des Alpes françaises*, 1868; European Railways, 1873; maps for J. Migeon, 1874.

Chase, G.E. *Plan of the new city of Pensacola* [Florida], 1836.

Chase, J.G. *Rail Road Map showing the Street Rail Road Routes in and leading from Boston*, 1865; *York County, Maine*, 1856; *Plan of Cambridge, Massachusetts*, 1874.

Chase, John Centlivres (1806-1876). Cape Colony public servant. *Map of the Eastern Frontier of the Colony of the Cape of Good Hope*, London, John Arrowsmith 1836, 1838; *Cape of Good Hope*, 1843; *Emigrants Guide to the Cape of Good Hope*, 1845.

Chase, William. Bookseller in Cockey Lane, Norwich. *A New Mapp of the City of Norwich*, 1728; *...Map of Norfolk* (with Thomas Goddard), 1731, 1740, 1745.

Chassant, —. Engraver. Mappemonde, 1823; Martinique, 1826; one of the engravers for maps of the expeditions of J.S.C. Dumont d'Urville, 1833-1842; for Dépôt de la marine, 1848, 1860.

Chasse, Barthelemy (*fl.*early 1700s). Designer and engraver for Laurent Brémond, 1730.

Chassegros de Léry. See **Chaussegros** de Léry.

Chassereau, Pierre [Peter]. Member of French Huguenot family which settled in Soho, London. Surveyor and cartographer, worked at 'the Sign of the Golden Head, Berwick St.', London. Worked with John Rocque on survey of York, *A plan of the City of York*; ...*City and Harbor of Havana*, T. Bowles 1739, also Homann's Heirs 1739-1740; *A New and correct Map of the Islands of Malta, Gozo & Cuming...*, 1740; *A New and Correct Plan of the Harbour of Carthagena in America...*, London, Tho. & John Bowles 1740; *An actual Survey of the Parish of St. Leonard in Shoreditch, Middsx*, 1745; Plan of Ipswich, 1745; *Plan de la Ville...de York*, 1750.

Chassignet, Daniel. Engraver. Celestial globe, Rome, 1616.

Chassun, *sieur* de. See **Classen**.

Chasteler-Courcelles, Jean Gabriel Joseph, *marquis* du (1763-1825). Austrian military cartographer.

Chastenet-Puiségur, Jacques, *comte* de. *Le Pilote de l'Île de Saint Dominigue*, 1787; *A General Chart of the Atlantic or Western Ocean*, 1804.

Chastillon [Châtillon], Claude de (1559-1616). 'Topographe du roi' (Henri IV), 1591; 'ingénier du roi' (for Champagne), 1595. Surveyed and drew maps of his native province of Champagne and its borders, 1599; surveys in Guyenne and Poitou, 1606; other drawings and surveys of sites throughout France, 1589-1616, culminating in the posthumous publication of *Topographie françoise*, J. Boisseau 1641, 1648, L. Boissevin 1655. **Ref.** BUISSERET, D. *Imago Mundi* 36 (1984) pp.75-78; PASTOREAU, M. *Imago Mundi* 32 (1980) p.57.

Chatel, P. *Nouvelle Carte de S. Stevensweert*, 1633.

Châtelain, family. Best known for *Atlas Historique*, parts of which appeared under various Châtelain imprints from 1705-1739. **Ref.** KOEMAN, C. *Atlantes Neerlandici* Vol.II (1969) pp.33-38.
• **Châtelain**, Zacharie (*d.*1723). Bookseller and publisher of Amsterdam.
• • **Châtelain**, Henri Abraham (1684-1743). Son of Zacharie. *Atlas historique...* (7 volumes), Amsterdam, 1705-1720 and later (with text by N. Gueudeville); *Mappemonde...*, Amsterdam 1723; *Plan de l'Histoire universelle...*, *c.*1730. **Ref.** SHIRLEY, R. *Printed maps of the British Isles 1650-1750* (1988) pp.41-43.
• • **Châtelain frères** (*fl.c.*1707-1715). Partnership of Henri Abraham and Zacharie junior. Credited as publishers of volume 4 of *Atlas historique...*; *La science des personnes de la Cour, de l'Epee et de la Robe...*, 1710 (with a map of England and Wales).
• • [**l'Honoré**] & **Châtelain**. Publishers of volumes 2, 3, 5, 6 & 7 of *Atlas historique...*, 1705-1735; F. Froger's *Relation d'un Voyage de la Mer du Sud, Detroit de Magellan, Bresil, Cayenne et les Isles Antilles*, 1715.
• • **Châtelain**, Zacharie (1690-1754). Brother of Henri Abraham above. A.A. Bruzen la Martinière's *Introduction à l'histoire de l'Asie, de l'Afrique et de l'Amérique*, 1735; Baron de Pufendorff's *Introduction à l'Histoire... De L'Univers*, 1743.

Châtelaine, Anatole. *Atlas chronologique des chemins de fers de France*, Paris 1855; *Voies de communication... dans le monde entier*, 1862, 1868.

Chatfield & Brown. Surveyors of 47 Castlereagh Street, Sydney. *Dickson's Estate*, 1892.

Châtillon, C. See **Chastillon**, Claude de.

Chatillon [Chastillon], Hugues de. 'Ingénieur et géographe du roi', 1616.

Chatillon, Jean de. 'Ingénieur et géographe du roi', worked also for Melchior Tavernier. *Carte de la coste de la Rochelle à Brouiage et de l'Isle d'Olèron*, 1627 (with Tavernier), published de Beaurain, 1751.

Chatillon, *Captain-Lieutenant* Z.G. de (*fl.*1770s). Dutch military engineer and surveyor, worked in Limburg.

Chatilov, B. Map of the Russian-Chinese border according to the Treaty of 1689, [n.d.] MS; Map of Ienissei and Angara, [n.d.] MS (both with P. Skobel'tsyn).

Chatov, *General* —. Black Sea and Caspian Sea, 1819; Map of Walachia and Bulgaria (4 sheets), 1854.

Chatwind, Philip. See **Chetwind**, Philip.

Chauchard, *Captain* C.A. French military cartographer. *Carte générale de l'Empire d'Allemagne* (10 sheets), Paris, Dezauche 1784 (reduced single sheet version published by Dezauche the same year); *Carte de la partie septentrionale de l'Italie*, S. Dezauche 1791; *A General Map of the Empire of Germany, Holland, Switzerland, the grisons, Sicily, Corsica and Sardinia* (23 sheets), London, John Stockdale 1800.

Chaucheprat, —. *Plan de la Baie de la Concepcion* [Chile], 1827; *Plan du Mouillage de Chorrilos*, 1828; *Plan du Port de l'Ancon*, 1828.

Chaudière, Guillaume. Publisher of 'Ruë S.Iaques, à l'enseigne du Temps & de l'Homme Sauvage', Paris. André Thevet's *La Cosmographie Universelle*, 1575, 1581.

Chaulmier, —. Engraver. *Plan de Gibraltar*, Paris 1782.

Chaumeau [Calameus], Jean [Johannes], *sieur* de Lassay (*fl.c.*1560s). Jurist of Bourges. Wrote *Histoire de Berry*, Lyon, Antoine Gryphius 1566 (with woodcut map *La carte de Berry*, later copied by Ortelius, Belleforest and De Jode). **Ref.** KARROW, R. *Mapmakers of the sixteenth century and their maps* (1993) pp.114-115; MEURER, P. *Fontes cartographici Orteliani...* (1991) p.127 [in German].

Chaumette Des Fosses, Amadée. Corso de los Rios Huallaga y Ucayali, 1830; Pampa del Sacramento, 1836.

Chaumier, C.J. Draughtsman and engraver of Paris. Corrected Jaillot's *Carte du théatre de la guerre...France vers le Brabant, le Païs Liègeois, le Luxembourg...*, Paris, Basset 1792; amended Jaillot's *Carte de la guerre actuelle dans le Départment du Nord le comté de Flandre et de Hainaut...*, Paris, Basset 1792; *Carte Routière de la France...*, Paris, Basset 1793; *Carte routière de la République française*, Paris, Basset 1794; *Carte routière du Thèâtre de la guerre Dans la partie Septentrionale de l'Italie...*, Paris, Basset 1800 (with H. Jaillot); *Carte Routière du Théatre de la guerre... Flandre, ...Brabant*, 1806; *Tableau général et itinéraire du Royaume de France*, 1817.

Chauncy, *Sir* Henry (1632-1719). Antiquarian and historian of Hertfordshire. *The Historical Antiquities of Hertfordshire...*, 1700 (with a large map by Herman Moll and plans of St. Albans and Hitchin).

Chauncy, H.S. Surveyor of 191 Pitt Street, Sydney. *Plan of the Auriferous region of Mount Alexander*, 1852; *Gladstone* [Leura], 1885.

Chauncy, Philip. Local maps of Victoria, Australia, 1856-1859.

Chauncy, *Captain* R. Publisher of Charles Sholl's Military Topographical Map of Eastern Virginia, 1854.

Chaurand, Enrico de. *Carta... della Ethiopia* (8 sheets), Florence 1894.

Chaussegros [Chassegros] de Léry, *père et fils*. **Ref.** HEIDENREICH, C.E. *The Map Collector* 27 (1984) pp.22-24; *Dictionary of Canadian biography* Vol.3 pp.116-119.

• **Chaussegros de Léry**, Gaspard-Joseph *père* (1682-1756). Chief engineer of the French army in New France. Drew many maps of the country for the Dépôt des cartes, which were later published by others such as Bellin. *Plan du perron ou terrasse du nouveau Palais de Kebec*, 1719; Québec, 1722; *Plan de la Ville de Montréal en Canada...*, 1724; *Plan de la Ville de Montréal dans la Nouvelle France*, 1729; *Carte...des Lacs du Canada*, Bellin 1744; Lake Champlain, 1748; *Plan, profil et élévation des Forges du Roy...*, 1749.

•• **Chaussegros de Léry**, Gaspard-Joseph *fils* (1721-1779). Joined his father in the above work in the 1730s, becoming assistant engineer in 1739. Untitled chart of the east shore of Lake Superior, 1735; *Plan du Fort du Détroit...*, 1749.

Chauveau, François (1613-1676). French engraver. Sanson d'Abbeville's *Carnutes Partie Merid.le de l'Evesché de Chartres...*, Paris, Mariette 1660; worked on Alexis-Hubert Jaillot's *Atlas nouveau*, 1681.

Chauvigné-Jaillot. See **Jaillot**.

Chavanne, *Doctor* Josef (1846-1902). Austrian geographer, Vienna. *Physikalische Wandkarte von Afrika*, 1878; *Central Asien*, 1880; *Amerika und Westindien*, 1882; *Physikalisch-statistischer Hand-Atlas von Oesterreich-Ungarn* 1882-1887; *Carte de l'Afrique Equatoriale*, Brussels 1883; *Congo*, 1885; *Mapa de los Ferro-Carriles...de la República Argentina*, 1889. **Ref.** DAMMERER, F.P. *Leben und Werk der österreichischen Kartographen Josef Chavanne und Franz Ritter von Le Monnier* (1995) [in German].

Chaves, — (*fl.*18th century). Islas Malvinas, MS.

Chaves [Chiaves; Chavez] family.
• **Chaves**, Alonso de (1492-*c.*1590). Spanish pilot and cosmographer to Charles V of Spain. *Quatri partitu en cosmografia practica...*, 1530; *Carta moderna general*, 1536 (now lost). **Ref.** KELLEY, J.E. *Imago Mundi* 42 (1990) pp.26-49.
•• **Chaves**, Gerónimo [Hieronymo] de (1523-1574). Son of Alonso. Successor to Sebastian Cabot as 'Piloto mayor de la Casa de la Contratación' (1552). Translated Sacrobosco's *Tractus de sphaera*, 1545 (with map); manuscript maps of Andalusia and Florida used later by Abraham Ortelius in 1579 and 1584 respectively; published *Chronografia o repertorio de los tiempos*, Seville, Alonso Escrivano 1548-1588. **Ref.** KARROW, R. *Mapmakers of the sixteenth century and their maps* (1993) pp.116-118; MEURER, P. *Fontes cartographici Orteliani...* (1991) p.127 [in German].

Chavignaud, L. *Plan de la ville de Lille*, 1873.

Chawner, Thomas (1774-1851). Surveyor and architect, prepared many maps and plans, mostly in the London area, many of them with James Pennethorne. St. Marylebone, 1794 (with Thomas Leverton); plan of the town of Newark, 1804; Ranelagh Estates, Chelsea, 1807 MS; Duke of Queensberry's House [Richmond], 1829 (with H. Rhodes); grounds west of Kensington Palace, 1838; *New Oxford Street*, 1840; St. Marylebone, 1841; plan for Victoria Park [Hackney], 1841; Windsor Castle, 1842; and many others.

Chaÿmox, Corneille. see **Caymox**, Cornelis.

Chazelles, Jean Matthieu de (1657-1710). French hydrographer and astronomer in Marseilles. Côtes de Provence, *c.*1689; various charts in *Neptune François*, 1693 and later; *Carte de France*, 1693. **Ref.** DEBARBAT, S. & DUMONT S. 'Ces Messieurs de l'Académie sur les côtes de Gascogne à la fin du XVIIe siècle' in: *Actes du 118e Congrès nationale des sciences historiques et Scientifiques Pau 1993* (Paris, Comité des Travaux historiques et scientifiques 1995) pp.19-30 [in French].

Chedeville, A. Plan des chutes de Niagara, 1851.

Chédufau, *Doctor* — (*fl.*1830-1840). Physician in Egyptian army under Ahmed Pasha, travelled in Arabia and surveyed Hijaz. Géographie de l'Arabie, 1843.

Cheevers [Chevers], John. Engraver. Otaheite, 1769; *Chart of Part of the Coast of New South Wales*, 1770 (for Captain James Cook's *Voyages*, 1773); General William Roy's Scotland, 1774, *c.*1793; *The Carnatic*, 1778; *Rapids in the River Ohio*, 1778; *A New Map of the Western Parts of Virginia, Pennsylvania, Maryland and North Carolina* (4 sheets), 1778; Henry Pelham's Clare (12 sheets), 1787 (with T. Harmar).

Cheffins, family [C.F.Cheffins & Son]. Surveyors and publishers of London.
• **Cheffins**, Charles Frederick (1807-1861). Republic of Texas, London 1841; *London and its Environs*, 1854; map of London in: *Post Office London Directory*, Kelly & Co. 1856 and many later editions; 'Map of the London Postal Districts' in *Post Office London Suburban Directory*, 1860, 1863, 1865, 1868, 1872.
•• **Cheffins**, Charles Richard. Surveyors, publishers, lithographers, 11 Southampton Buildings, Holborn; 23, Abingdon Street, Westminster, (*c.*1862). English & Scotch Railways, 1841-1856; *Hertfordshire...*, *c.*1857; for Kelly's Directories, 1860; *The Visitors' Railway Station Map of London and its Environs*, *c.*1862.

Chelius, C. *Übersichts-Karte des Odenwalds*, 1898.

Chelsham, Robert (*fl.*1659-1672). Surveys of parts of Berkshire, Surrey and Wiltshire, including the parish of Bray, 1672.

Chelyuskin, Semyon. Pilot and member of the Great Northern Expedition 1735-1742. Maps of various rivers and capes, including Mys Severovostochny (later named in his honour).

Chemitte, François. *Plan...de la ville de Candie*, 1692.

Ch'en Lun Chiung (1684-1747). Chinese hydrographer.

Chenevière, Jean Jacques. Geographical game, *c.*1720.

Cheney, brothers. American engravers, two of the seven sons of George Cheney of Connecticut. E. Bissell's *Map of Manchester* [Connecticut], 1849.
• **Cheney**, John (1801-1885). Engraver.
• **Cheney**, Seth Well (1810-1856). Crayon artist and engraver.

Cheney, Flavius [Cheney Globe Co.]. Terrestrial and celestial globes from 1888-*c.*1900. 12-inch terrestrial globe, 1896. **Ref.** WARNER, D. *Rittenhouse* Vol.2 N°2 (1987) pp.53-54.

Cheng Ho (*fl.*1405-1433). Naval commander of the Ming dynasty. Nautical chart showing the east coast of Africa, 1621. **Ref.** *Imago Mundi* 24 (1970) pp.25-26.

Cheng Shum Kung. Chinese cartographer. Map of Japan, 1557.

Cherbuliez, A. [& Compagnie] (*fl.*19th century). Publishers in Paris.

Chéreau, Jacques *le jeune* (1688-1776). Engraver, 'rue St Jacques au Coq'. *Bourges Ville Capitale du Berri Archeveché et primat d'Aquitaine*, *c.*1739; Bethlehem cité de David, [n.d.]; Jérusalem ville capitale, [n.d.]; *Le Mississipi ou la Louisiane...*, 1740; *Barcelone...*, 1750.

Chéreau, Jacques-Simon (*d.*1808). 'Rue St Jacques', Paris. Son-in-law of engraver André Basset. *Vue de la ville de Richelieu en Poitou...*, 1788 (from the work published by Basset, 1763).

Cherette, F. 'Ingénieur royale'. Plans of châteaux and gardens (3 sheets), Paris 1720 MS.

Chermont, Jean Gabriel, *comte* de. Said to have worked in association with Louis Charles Desnos, 1790.

Chermside, *Lieutenant* Herbert Charles *RE*. Assistant Military Attaché at Constantinople, 1878-1879. *Survey of defensive position near Bulair...* [Turkey], 1876 (with G.A. Cockburn); Rough Sketch of Hadji Oghlou Basardjik... [Bulgaria], 1877; *Sketch map shewing approximate position of Russian's and Turk's near Constantinople...*, 1878; *Sketch of Varna & Environs*, 1878.

Cheron, L. Designed title cartouche for Herman Moll's *World*, 1719.

Chéron, Louis. *Plan de Golfo Dulce* [Costa Rica], 1850.

Cherubini, Claudio. Italian relief cartographer. *Carta in rilievo delle Alpi occidentali e dell'Appennino ligure*, 1882.

Chesbrough, E.S. City engineer, Boston, Massachusetts. *Boston Harbour*, *c.*1852; Boston Water Works, 1852.

Chéseaux, Jean Pierre Loys de. See **Loys de Chéseaux**, Jean-Pierre.

Chesebrough, Ephraim. *A New and Correct Chart of the East End of Long Island Sound...*, *c.*1814 (engraved by Amos Doolittle).

Chesham, F. Engraver. *Plan of the intended... Canal... from Stroud to the River Severn*, *c.*1775.

Chesneau, Nicolas. Thought to be one of the editors (with M. Sonnius) of F. de Belleforest's *La Cosmographie universelle*, Paris 1575.

Chesney, *Colonel* Charles Cornwallis. *The Rivers Euphrates and Tigris* (12 sheets), 1850; *Outline Maps for Military Students*, 1868.

Chesney, *Lieutenant* G.T. *Plan of the Siege Works of Delhi*, London, James Wyld 1857.

Chesnoy. See **Capitaine du Chesnoy**.

Chesser, William. Nepean Bay, South Australia, London, John Arrowsmith 1839.

Chester, *Captain* James, Third Artillery. Commander of the surveying party for *Yorktown, Virginia and the ground occupied in the siege of 1781*, 1880.

Chesteven, Robert. A Relation of the present state of the Summer Islands, 1629 MS.

Chetwind [Chatwind, Chetwynd], Philip. On the death of Henry Seile in 1662, Chetwind acquired the plates for Peter Heylin's *Cosmography* (4 maps), the Chetwind imprint appears from 1666.

Chevalier [Chevallier], François (*d*.1738). French mathematician and cartographer. In 1707 he presented to the Académie des Sciences a method by which priests could survey their own parishes from the church tower, each contributing a part to a larger diocesan map (the method was used by d'Anville at Lisieux, 1723). Chevalier was also Master of Mathematics to Louis XV (1717). Map of the diocese of meaux, 1698.

Chevalier, Marcel. *Alt Urgell, Andorra*, 1913; *Garrotxa* [Catalunya], 1921.

Chevalier, Michel (1806-1879). French statesman, professor of political economy, cartographer, advocate of free trade. Voies de communication aux Etats Unis, 1841.

Chevalier, Stanislas (1852-1930). French Jesuit missionary in China; geographer, astronomer and meteorologist. Made a study of the course of the Yang-Tse. *Atlas du Haut Yang-Tse-Kiang*, *c*.1899; *Atlas de la Lune*, [before 1924]. **Ref.** BROC, N. *Dictionnaire illustré des explorateurs Français du XIXe siècle: II Asie* (1992) pp.94-95 [in French].

Chevallier, François. See **Chevalier**, François.

Chevallier, Henri. *Atlas de Géographie Ancienne*, *c*.1867.

Chevard, Vincent (1748-1826). French mapmaker. *Carte des états provinciaux Projetés pour le Païs Chartrain*, Chartres, F. Le Tellier 1788.

Chevers. See **Cheevers**.

Chevillot, —. *Plan de la ville et fort Louis* [Louisiana], 1713-1714.

Chevreau, H. *Carte topographique du Département du Rhône*, 1868.

Chewett [Chewitt] family. Family of mapmakers of Toronto, active in the late 18th and early 19th centuries, produced many maps of Ontario. **Ref.** WINEARLS, J. *Mapping Upper Canada 1780-1867* (1991); *Dictionary of Canadian biography*.

• **Chewett**, William (1753-1849). Senior Surveyor, Deputy Surveyor-General and Acting Surveyor-General, Upper Canada. Produced many township plans and maps of Upper Canada and its regions. *Plan of Two Townships...*, 1791; *Plan of part of the Town of Cornwall*, 1792; *Plan of the Province of Upper Canada*, 1792; *Plan of Upper Canada showing Simcoe's journeys...*, 1794; *Sketch of the Town of York* [Toronto], 1799; *Plan of Maryville*, 1802; *A Map of the Located Districts in the Province of Upper Canada*, William Faden 1813, 1819, 1821, 1825, 1835, 1838; and many others.

•• **Chewett**, James Grant (1793-1862). Son of William, (above), surveyor and architect, Deputy Surveyor-General, Upper Canada 1832-1841. Did many township surveys and produced many important maps. *Plan of the central part of the Province of Upper Canada Shewing the seat of War in the Years 1812, 1813 & 1814*, 1819; *Map of the Proposed Canal through the District of Niagara and Gore* [Welland Canal], for the Commissioners of Internal Navigation, 1823; *A Map of Part of the Province of Canada showing the proposed route for a Canal* [Rideau Canal], New York, Samuel Maverick 1825; *A Map of the Province of Upper Canada* (3 sheets), 1825, published for the Canada Company by C. Smith & Son 1826 and many later editions; plan of the town of York [Toronto], 1828; *Plan Shewing the Survey of the Military Reserve at Niagara*, 1831; military plan of Penetangushine, 1835; *Townships in the Newcastle District...* [Lake Ontario], 1835; and many others.

••• **Chewett**, William Cameron (1828-1897). Publisher and printer, eldest son of James, (above). Initially in partnership with Thomas Maclear as Thomas Maclear & Co., took over the company in 1857, renaming it W.C. Chewett and Co. (see below).

•••• **Chewett**, W.C. [& Co.]. Printers of Toronto, became Copp, Clark and Co. in 1869. Published many maps including railway, canal, and county maps, and new editions of Maclear maps. *Map of British North*

America, *c.*1862; *Map of the County of Huron*, 1862; *Map of the County of Renfrew*, 1862; *Plan of the North Shore of Lake Superior*, 1863; *Map of Canada West...*, 1863; Edward Hodder's *Chart of Lake Ontario...*, 1863; *Plan of Survey for Georgian Bay Canal...*, 1863; *Chart of Lake Huron*, 1864; *Emigration Map of Canada Western Section*, 1865; *Pocket Map of Upper Canada*, 1865; colour lithograph *Geological Map of Canada West*, 1865; *City of Toronto*, 1866; *That Part of Canada now invaded by Fenians* [Niagara Peninsula], 1866; *Map of the Town of Galt*, 1867; and many others.

Cheyne, Andrew. *Master*, Merchant Navy. Sailing Directions from New South Wales to China & Japan, 1862.

Cheyne, N.B. Travelling Directory through Scotland, 1792; Roads of Great Britain, 1797.

Chézaud, —. *Profil en long du Canal... de Suez*, *c.*1859.

Chia Tan (730-805). Chinese cartographer. Large map of China 33 ft by 30 ft.

Chianale, Saverio. Engraver of Tuscany. Greece and Canada for *Atlante delle geografia antica e moderna*, Barbiellini 1807.

Chiandi, Giovambattista (*b.*1785, *fl.* to 1821). Geodesist, surveyor and draughtsman. One of the team sent from Naples by the 'Deposito della guerra di Milano' to survey the Adriatic Sea for Napoleon. Triangulation of Corfu, 1812-1813, published by F. Visconti, *c.*1825; sailed with William Henry Smyth to survey the Adriatic, *c.*1819. **Ref.** VALERIO, V. *Società uomini e istituzioni cartografiche nel mezzogiorno d'Italia* (1993) pp.476-478.

Chiang Kuo Chang. Yangtze River Gorges, 1899.

Chiappeti, Vincenzo. Idriografia teorica practica, Parma 1715 MS.

Chiara de Arimino, Matheo de. Supposedly a portolan maker of Rimini credited with a manuscript atlas containing four nautical charts: Europe and the Mediterranean; Asia from Eastern Mediterranean to China; Africa and Arabia; and North and South America, 1519 MS. **N.B.** This atlas is now believed to be a fake. **Ref.** WOODWARD, D. *The Map Collector* 67 (1994) pp.2-10.

Chias y Carbó, Benito. Spanish engineer and cartographer. *Atlas geográfico ibero-americano*, Barcelona *c.*1900 (edited by Alberto Martin); *Navarra*, 1902 (for Alberto Martin); other maps for Alberto Martin include *Alava*, *Guipúzcoa* and *Vizcaya*, all published Barcelona 1903; *Mapa de la provincia de Barcelona*, 1903; maps for Ceferino Rocafort y Casimiro Dalmau's *Espana regional* 1903; maps for Carreras y Candi's *Geografia general de Catalunya*, 1913-1918.

Chiavacci, Bastiano dell'. Immagine del mondo, 1579.

Chiaves, Hieronymo. See **Chaves**, Gerónimo.

Chicago Standard Publishing Co. *The Standard atlas and gazetteer of the World*, 1888.

Chichagov, Pyotr. Russian topographer-engineer, worked in Siberia 1717-1738. Maps of various districts, rivers, forts and towns, as well as astronomical observations for 1302 points in Tobol'sk province.

Chichagov, Semyon. Russian topographer. Maps of Novgorod and Tver provinces, 1732-1739.

Chichagov, *Captain* Vasiliy. Russian navy. Maps of Northern Hemisphere and area between Lapland and Spitsbergen, *c.*1765.

Chicot, —. Engravers of Paris. *Plan du Fort St. Philip* [Minorca], 1781.

Chidley, I.J. Publisher, 123 Aldersgate St, London. Arrowsmith's *New General Atlas*, 1820.

Chiesa, Andrea. Cartographer. Tiber, 1744 (with B. Gambarini); *Carta topografica di tutta la Pianova del Bolognese*, 1762.

Chiesa, Innocento (18th century). Italian engraver.

Chieze, Jacques de. *La Principauté d'Orange et Comtat de Venaissin*, Blaeu 1627.

Chiffelle, Thomas P. City surveyor of Baltimore. *Map of the City of Baltimore and of Part of Baltimore County...*, 1852. Chiffelle was succeeded by his former pupil, Simon Martenet.

Chikaraishi Chūichi. Japanese cartographer. Worked with Matsumara Kiyoyuki and Kawai Tsunehisa on a sight-seeing map of Kamakura, 1685.

Chikhachev, P. *Asie mineure*, 1853; *Essai d'une Carte générale de l'Altai* (2 sheets), 1854; *Carte géologique de l'Asie mineure*, *c.*1877.

Chilcott, J. *A Map of the country Eleven miles round Bristol*, *c.*1842.

Child, G. Engraver. For Thomas Salmon's *Universal Traveller* (2 volumes), 1752-1753.

Child, J. American surveyor. Sectional View of Copper Falls Mine, 1849.

Child, Nathaniel. North half of the Township of Woodstock, 1772.

Child [Childe], Timothy. Publisher of London, 'at the Sign of the Unicorn at the West end of St. Pauls Churchyard', later 'White Hart West end of St. Pauls Churchyard' (from 1701). Worked with Awnsham and John Churchill. *A New Body of Geography* (maps by Herman Moll), 1695 (with Abel Swall); published G. Hornius's *A full and exact Description of the Earth, or Ancient Geography...*, 1700; *A System of Geography* (maps by Moll), 1701, 1709.

Childs, C.G. Engraver for Samuel Cumings's *The Western Pilot*, 1829.

Childs, O.W. Surveyor. *Canal from the Pacific to Atlantic Oceans*, 1850-1851, 1852; *Part of the City of New York*, 1862.

Chill, D.H. *Map of the Indus River* (4 sheets), 1835 (with Edward Winston).

Chillas, D. Lithographer of Philadelphia. Huron Harbour, 1854.

Chilmead, J. English globemaker. Hues' Use of Globes, 1638.

Chimmo, *Commander* William, *RN* (*d.*1891). English hydrographer aboard HMS *Gannet*. survey of the south west Pacific, 1852-1856; surveys on the island of Trinidad, 1866-1868; Labrador, 1867; Borneo, 1871; Sulu, 1872.

Ch'in Hun-Keung. Navigational map, 1744.

Ch'ing Chün. Ming dynasty Buddhist monk and cartographer. Map of the World. **Ref.** *Imago Mundi* 24 (1970) p.23; KUEI-SHENG CHANG *Imago Mundi* 31 (1979) p.15.

Chino, Eusebio Francisco. See **Kino**, Eusebio Francisco.

Chinshōtei –. Japanese mapmaker. Map of China in the Ching dynasty, 1835 (with Seitaien).

Chipet, —. Plan of Yokohama, 1865.

Chipperton, *Captain* J. Bay of Guayaquil, H. More and T. King 1725-1728.

Chiquet, E.L.M. World, 1729.

Chiquet, Jacques (*c.*1673-1721). French cartographer, engraver and 'marchand d'estampes, rue St Jacques', Paris. Antoine Menard's *Le Nouveau et Curieux Atlas géographiques et historique*, 1711; *Nouveau atlas français*, 1719; *Atlas Géographique et Historique*, 1719; *Géographie Sacrée Ancienne et Moderne*, Paris 1719.

Chirikov, *Lieutenant* Aleksei Il'ich. Lieutenant under Vitus Bering in first Kamchatka Expedition 1725-1730, also on 1741 Aleutian Islands voyage; supervised and produced numerous maps and journals. Observations astronomiques, 1741; *Nouvelle carte des Découvertes des vaisseaux russiens aux côtes de l'Amérique Septentrionale*, St Petersburg 1758. **Ref.** LONGENBAUGH, L. *The Map Collector* 29 (1984) pp.29-32.

Chisholm, G.G. Atlases for Longman, 1859-1895.

Chisun'en Baikei. Japanese mapmaker and publisher. Plan of Sumpu and vicinity (woodblock), 1868.

Chiswell, Richard (1639-1711). Publisher and mapseller at 'Two Angels and Crown

Little Britain', later 'Rose and Crown St. Pauls Churchyard'. Sold Samuel Hobson's *A Ground Plot of Londonderry*, *c.*1693; John Speed's *Theatre of the Empire of Great Britaine* [with Prospect], 1676 (with Thomas Basset); *A New Map of the Cityes of London, Westminster...*, 1707; maps for the Royal Society.

Chittenden, G.B. South West Colorado, *c.*1876.

Chittenden, *First Lieutenant* Hiram Martin (1858-1917). American military engineer, became assistant to the officer in charge of road construction in Yellowstone National Park. Map of Utah, Colorado, Wyoming and Southeast Idaho (2 sheets), 1889.

Chitty, M. *North West Quarter of Van Dieman's Land...*,Van Diemens Land Company, 1833 (engraved by J. Cross).

Chitty, Simon Casie. *The Ceylon Gazetteer*, 1884.

Chkatov, —. Engraver of St Petersburg, for Pyadishchev 1823-1826.

Chlebowski (18th century). Prussian general and cartographer.

Cho Sekisui. See **Nagakubo** Sekisui.

Choate, Columbus D. *Map of Fairfax County, Virginia*, 1911.

Chocarne, —. Engraver for Pilote français. Golfo Porto Vecchio, 1828.

Chodźko, Josef (1800-1881). Polish geographer and historian 'à la Librarie Polonaise 20 rue de Seine', Paris. Carte géographique de la Pologne, 1830; Table de la Pologne, 1830; Atlas des sept partages, 183l-1846; Poland, 1863.

Choffard, Pierre Philippe (1730-1809). Artist and engraver of Paris. 'Rue des Francs-Bourgeois, place St Michel, entre une porte-cochère & un patissier'; 'r. des Cordeliers, la 1re porte cochère a droite en entrant par la rue de la Com-Fr chez un Sellier'; 'quai Voltaire n°.21 (where he died). Plan of London, 1756; cartouches for the Robert de Vaugondy's *Atlas Universel*, 1757; frontispiece for *Entretiennes géographiques*, 1761; West Indies, 1762; title-page for Jacques-Nicolas Bellin's *Petit atlas maritime*, 1764; Boullanger's World, 1781.

Choiseul-Gouffier, Marie Gabriel Auguste Florent, *comte* de. *A New Chart of the Mediterranean Sea*, 1776, English edition 1794; *Voyage Pittoresque a la Grèce* (volume 1), Paris 1782, (volume 2), 1809, (volume 3), 1822 (engraved by J. Perrier).

Chōjiya Genjirō. Japanese publisher at Kyoto. Shokoku Junran Dainihon Saiken Dōchū Zu [travel map of Japan], 1843 (by Okada Shuntōsai and Okada Mohei).

Chollet, Louis. Essequibo Colony [Guyana], *c.*1791; Pomeroon Coast, 1794.

Cholnoky, Jenö (1870-1950). Hungarian geographer, professor of geography at Budapest and Kolozsvár [now Cluj, Romania]; travelled in Asia, and surveyed the Hoang Ho [Yellow] river. **Ref.** STEGENA, L. *Térképtörténet* [Map history] Budapest 1981 [in Hungarian].

Chopin, M. Russie, 1838.

Chopy, Antoine (1674-1760). French theologian, poet, painter and engraver, born Narbonne, worked in Geneva. *Carte du lac de Genève*, 1730.

Chōsan'An Imboshi. Japanese mapmaker. Guide plan of the *haiku* poets' residences in Kyoto, 1856.

Chou Jen-Chi. Atlas of Chekiang, 1755.

Choudiakoff, —. See **Chudjakov**, Yefim.

Chovin, I.A. Engraver. *Carte de la Partie Septentrionale et Occidentale de l'Amérique*, 1764; *Carte de la Partie Septentrionale et Orientale de l'Asie*, 1764; Samuel Engel's Extraits raisonnes des voyages, 1779.

Christensson, Marten (*fl.*17th century). Swedish cartographer.

Christiaensz., Leenhart (*fl.*17th century). Amsterdam engraver.

Christian, Robert. *Master* HMS *Shrewsbury*. Plan of the Harbour of St. Johns Newfoundland, 1762 MS; Plan of eastern Canada including New Brunswick, Nova Scotia, and the St. Lawrence, [n.d.] used by G. Nicolls, 1814.

Christiani, F. Maps of Denmark, 1859-1878.

Christie, Charles. Water Bay on Sumbawa [Netherlands East Indies], published Alexander Dalrymple 1780.

Christie, *Major* Charles. New York, New Jersey, 1759 MS.

Christiersson, Thomas Hedraeus (1606-1672). Swedish mining surveyor, worked in Uppland, 1635 and Falun, 1637. Maps of copper mines, 1645-1654.

Christman, F. Maps for Ludwig Wilhelm Ewald, 1861.

Christoforo [Cristofano], Girolamo da (*fl.*16th century). Florentine cartographer, worked in Lyon.

Christoph [Christoffel, Christopheri], Johan [Joannes] (*fl.*1598-1619). Painter, publisher and cartographer of Cologne. Metellus' *Germaniae Superior*, 1598 (from which *Ducatus Breme & Ferdae* was used by Joan Blaeu in the 1662 *Atlas major*); German edition of José de Acosta's *Geographische und historische... Landschafft America*, 1598 (with maps by Johannes Matalius Metellus, derived from the maps of Wytfliet); *Eynkommen, Reichthumb und Schatz*, (22 maps), 1599 (a re-issue of Lambert Andreas' *Theatrum Oder Schawspiegal*, 1596, itself derived from Botero's *Delle relationi universali* which had no maps); Insularium, Cologne 1601.

Christopher, *Lieutenant* —. Indian Navy hydrographer. Survey of the East Coast of Africa, 1838-1848.

Christopher, Joseph Steer. Writer. *Emigrant's Guide to the Cape of Good Hope*, 1845.

Christopher, R. Publisher and mapseller. With John Cary, Christopher was one of the co-publishers of Andrew Armstrong's *County Palatine of Durham* (4 sheets), 1791 edition (first published by Thomas Jefferys, 1768).

Christy, William. Surveyor. Directory Map of Huntingdon County, Pennsylvania, 1856.

Chromer, Martin. *Mitnächtischer Völckeren Historien in welcher viler Nationen....*, 1562 (incorporated maps taken from S. Munster's *Cosmographia*).

Chrysanthos, *Presbyter, Patriarch of Jerusalem*. Greek geographer. *Index Geographicus tum veteris tum novae*, Padua 1700; reduced version in *Notara introductio geographiam et sphaeram graecae*, Padua 1700. **Ref.** SHIRLEY, R. *The mapping of the World* (1983) N°s. 198, 199, pp.592-593, 595.

Chrysologue, Noel André (1728-1808). French Capucin. *Hemisphère inférieur de la Mappemonde...*, 1774; *Abrégé de l'astronomie*, 1778; *Planisphères célestes*, 1778; *Théorie de la Surface de la Terre*, *c.*1813.

Chrysoloras, Manuel (*c.*1355-1415). Greek scholar, born Constantinople. Took manuscript of Ptolemy's *Geographia* to Florence in 1395 and passed on the task of translation to Jacopo d'Angelus.

Chrystin, Jean-Baptiste. *Les Délices des Pais-Bas*, François Foppens 1711, 1713 (included Harrewijn's 1697 plan of Luxembourg).

Chrzánowski, Wojciech (1793-1861). Cartographer of Warsaw and Paris. *Karta dawnéj Polski...*, 1859.

Chu Ssu-pên. Chinese geographer of the Mongol dynasty. *Yü T'u* [The Terrestrial Map; sometimes known in the West as the Mongol Atlas] 1320, (a large map of China and surrounding countries, later enlarged and republished by Lo Hung-hsien as *Kuang Yü T'u*, *c.*1555, (used by M. Martini as the basis for his *Novus Atlas Sinensis*). **Ref.** KUEI-SHENG CHANG *Imago Mundi* 31 (1979) p.15; *Imago Mundi* 24 (1970) pp.22-23.

Chu Yü. Atlas of China, 1895.

Ch'üan Chin [Kwon Keun]. With Yi Hwei, produced a world map entitled *Hun-yi-chi-ang-li li-tai kuo-tu chih-t'u*, 1402 (based on an earlier map by Ch'ing Chün).

Chudeau, René (1864-1921). French geologist and explorer in Saharan Africa 1904-1913. Produced several geological maps of the sahara desert.

Chudjakov [Choudiakoff], Yefim. *Plan der... Stadt St. Petersburg*, 1790.

Chukei, Inō Tadataka. See **Inō** Tadataka.

Chun Hsu Chao. North and South Hemispheres, 1807.

Church, Ambrose F. Surveyor. Township maps of Canadian counties: *Colchester, Nova Scotia*, 1864; *King's County*, 1864; *Cumberland*, 1870; *Hampshire*, 1871.

Church, B.S. *Reconnaissance in advance of Camp Mansfield by 12th Regiment Engr.*, *c.*1860.

Church, George Earl. *A Railway Map of the Argentine Republic*, 1898.

Church, William (*fl.*1794-*c.*1829). Surveyor of Abingdon, Berkshire. Produced estate and inclosure maps in various English counties.

Church, *Commander* W.H. Admiralty Surveyor. Ireland, 1849, 1855.

Churchill, Awnsham and John. Booksellers and publishers at the Black Swan, Paternoster Row, London. With A. Swale, jointly sold Edmund Gibson's edition of William Camden's *Britannia*, 1695; *A New set of Maps...*, 1708; Herman Moll's *Compleat Geographer*, 1709; *Atlas Manuale*, 1709.
• **Churchill**, Awnsham (*fl.*1681-*d.c.*1728). In partnership with his brother John; Member of Parliament for Dorsetshire 1705-1710; shop and stock were passed to William Taylor in 1719, but Churchill retained copyright in *Britannia* until his death. *Britannia*, (with revisions by Gibson), was published by Awnsham Churchill, 1722 and sold by Taylor.
• **Churchill**, John (*fl.*1690-1714). Brother and partner of Awnsham from 1690.

Churchill, Henry A. *The Fortress & Field Defences of Kars*, 1855; *Turco-Persian Frontier*, 1855.

Churchman, John (1753-1805). Surveyor and geometrician of Maryland, 'Surveyor-General for Part of North America'; Member of the Imperial Academy of Sciences at St Petersburg; worked on the development of contour, or isoline, mapping. *Map of the Peninsula between Delaware & Chesapeak Bays*, *c.*1770; *Magnetic Atlas or Variation Chart*, New York 1790; *Essay on the Improvement of Geography*, 1804. **Ref.** RAVENHILL, W. *The Map Collector* 34 (1986) pp.23-25.

Churruca, *Don* Cosmo de. Las Caribes, 1793; Ysola de Trinidad, 1793 MS; Puerto Rico, 1794; Islas Antillas, 1802 (with Joaquin Francisco Fidalgo); Islas Caribes, 1804; West Indies, Laurie & Whittle 1813.

Churruca, Evaristo de. *Port San Juan*, published British Admiralty 1874.

Churton, William (*d.*1767). Surveyor of Edenton, North Carolina. Appointed Commissioner with Daniel Weldon for South Carolina (boundary line). Supplied material for Joshua Fry and Peter Jefferson, 1751, John Abraham Collet, 1770, and Henry Mouzon, 1775.

Ciantar, Giovannantonio. *Malta Illustrata* (2 volumes, 3 maps), 1772-1780 (revised edition of the work by G.F. Abela).

Cid, Francisco. *Planta hydrographica do Porto da Furna*, 1894.

Ciera, Michele Antonio (*d.*1815). Astronomer of Coimbra. Maps of Portugal rectified, 1804 for Dépôt de la Marine.

Cierto, Andres. *Descripcion de las Costas de Tierra Firme, en la America Septentrional...*, 1755 MS.

Cieza, Peter de. *A New Map of the Kingdom of Peru*, engraved by John Senex, in *Travels in Peru*, London 1709.

Cigerlus. See **Ziegler**, Jacob.

Cigni, Domenico. Engraver in Rome. *Provincia Quitensis Societatis Jesu in America*, *c.*1751 (drawn by G. Cigni).

Cigni, Giulio Cesare. Draughtsman in Rome, for *Provincia Quitensis Societatis Jesu in America*, *c.*1751 (engraved by D. Cigni).

Cimerlini [Cimerlino, Cimerlinus, Cunerlinus], Giovanni [Joannes] Paolo. Engraver from Verona, worked in Venice. *Cosmographia universalis ab Orontio olim descripta*, 1566 (cordiform map after Oronce Fine). **Ref.** SHIRLEY, R. *The mapping of the World* (1983) N°116, p.135.

Cingolani, Giovanni Battista. Cartographer of Rome. *Topografia geometria dell' Agro Romano*, 1692; Campagna di Roma (6 sheets), 1704.

Cintra, C. Provincia do Espirito Santo, Rio de Janeiro, 1878 (with C. Riviere).

Cipri, G. *Chemin de fer Indo-Européen*, 1881.

Cipriani, Giovanni Battista. *Pianta topografica della città di Roma e dintorni*, 1851.

Cirkoff, *Lieutenant* Axence. Corps of Austrian Imperial Topographers. *Carte de la principauté de Serbie*, 1848.

Cirni, Anton Francesco. *La Rocca con il Forte delle Gerbe*, 1560.

Ciroldis, J. See **Giroldi**, Giacomo.

Ciseri, —. Large plan of Jerusalem (after Adrichom), Rome 1609.

Cisneros, Carlos B. Peruvian geographer. Bolivia, Lima 1897.

Citters, E. van. *Kaart van de Spoorwegen... in Nederland*, 1888.

City Press. See **Collingridge**, W. H. & L.

Ciurana, A. *Barcelonès*, 1915.

Civelli, Giuseppe. *Gran carta d'Italia* (28 sheets), 1853.

Claesson [Felterus], Kjetil (*d*.1690). Swedish surveyor and cartographer. Manuscript survey *Væstergötlandh, Dalslandh samt Hallan och Bahuslahn*, *c*.1650, later used for Karta över Ålvsborgs læn, 1654 MS; Skærborgs och Övreholma, 1655 MS; Special Landhkort... öfwer Daahls Landh, 1657 MS; Elfsborgs lahn, 1654 MS; Væstergötlandh, 1689 MS.

Claesz. [Claeszoon], Baert. Dutch surveyor and cartographer, worked in Noord-Holland. Untitled map of Heerhugowaard, 1631.

Claesz. [Claeszoon], Cornelis. Dutch painter, printer and mapmaker (also called 'Cornelis den Schilder aen 't kerckhoff'), lived in The Hague, worked for the Rekenkamer van Holland. Untitled map of Wieringen, 1540-1541.

Claesz. [Claeszoon, Nicolay, Nicolas], Cornelis [Corneille] (1560-1609). Dutch printer, publisher and bookseller, 'op't Water in't Schryf-boeck - a l'enseigne du livre a escrire', Amsterdam. Hessel Gerritsz's *Zeevaert*, 1588; Lucas Janszoon Waghenaer's *Spieghel der Zeevaert*, 1588; Pieter Plancius's world map *Nova et Exacta Terrarum Orbis Tabula Geographica Ac Hydrographica* (18 sheets), 1592 (co-published with Vrients); *Nova Francia*, Amsterdam, 1594; worked for Jan Huyghen van Linschoten, 1595-1596; world map showing the voyage of Olivier van Noort, 1602; America, 1605; Willem Lodewijcksz's *Nieuwe caerte van de eylanden van Java, Sumatra, Borneo tot Malacca...*, 1598; Langenes' *Caert-Thresoor*, 1598-1609; and others; joint publisher of Mercator-Hondius Atlas, 1606-1608 and *Atlas Minor*, 1607. **Ref.** SHIRLEY, R. *The mapping of the World* (1983) N°183 pp.199-202; *Ibid.* N°387 p.408.

Claeyssens [Clayssius], Pierre [Pieter] (*fl*.c.1600). Dutch map copyist, worked in Flanders.

Clairaut, Alexis Claude (1713-1765). French cosmographer, one of the leaders of an expedition to Lapland commissioned with the task of establishing the true shape of the earth, 1736. Their conclusion (that the earth is an oblate spheroid) was published in Pierre Bouguer's *La Figure de la Terre*, 1749. Théorie de la lune, 1752.

Clamorgan, Jean de (1480-1562). Chartmaker of Dieppe. Manuscript charts of the Atlantic.

Clampe, Richard. *A Description of the Siege of Newarke...*, *c*.1646.

Clampitt, Nathan A. County surveyor. Washington County, 1841 MS.

Clancy, Jonathon. *Master*, HMS *Centurion*. Draught of Lisbon Harbour, 1763 MS; *The Faro and Mole of Messina* [Sicily], 1763 MS; many marine surveys in the Mediterranean.

Claparède, Arthur de (1852-1911). Swiss geographer, historian and diplomat, worked in Vienna, Bern and Geneva. President Geographical Society of Geneva. Japan, 1889; A travers le monde, 1894.

Clapiès, Jean de (1670-1740). Astronomer, cartographer and mathematician from Montpellier.

Clapp, *Lieutenant* Edward S. *Harbours and Anchorages in the Sandwich Islands*, 1878.

Clapp, O.W. Publisher of St. Louis. *Map of the Chicago-New York Electric Airline*, 1907.

Clapp, William W., *junior*. *Plan of Boston corrected*, 1864.

Clapperton, *Lieutenant* (later *Captain*) Hugh (1788-1827). African explorer. Narrative of Travels in North and Central Africa, 1826 (includes a translation of an Arabic manuscript on the geography of the interior of Africa).

Claret de Fleurieu [d'Eveau de Fleurieu], Charles Pierre, *comte* de (1738-1810). French mariner, mathematician and circumnavigator. 'Directeur général des ports et arsenaux de la marine', 1770, 'ministre de la marine', 1790. Voyages, 1768-1769; Découvertes au sud est de la Nouvelle Guinée, 1768-1769, *c.*1790; Azores, 1772; Atlantic, 1777; West Africa, 1797; *Neptune du Cattegat et de la Baltique* (65 maps, plans and views), 1786-1790, published 1809 (only 100 copies printed, edited by Philippe Buache and prepared by Charles François Beautemps-Beaupré). **Ref.** TOOLEY, R.V. *The Map Collector* I (1977) p.24.

Clarici, Giovanni Battista. Ducato di Urbino, 1572.

Clarici, Paolo Bartolommeo (1664-1725). Painter and cartographer of Padua. *Diocesi Padovana*, 1720; *Il Polesine di Rovigo*, 1721.

Claringbull, W. *Chart of Plymouth Sound and map of the towns of Plymouth, Devonport and Stonehouse...* (completed by J. Claringbull), 1841.

Clark. See also **Clarke**.

Clark, *Captain* —. *Plan of the city of Cadis*, 1761.

Clark, Benjamin A. Surveyor. *Map of the Town of Deerfield, Massachusetts*, 1855; *Map of Hamilton, Madison County, New York*, [n.d.].

Clark, F.A. White Pine Mining District, 1870-1880.

Clark, F.C. *New Sectional map of Missouri*, *c.*1860.

Clark, G.T. American surveyor. *City of Flint, Michigan*, 1873

Clark, H. London printer. Richard Blome's *Present state of His Majesties Isles and Territories in America*, 1687.

Clark, J. Engraver of map of Hertfordshire in N. Salmon's *The History of Hertfordshire*, 1728.

Clark, J. Prepared maps for William Bell's *A New General Atlas*, Edinburgh, Brown & Nelson 1834, 1837.

Clark, J. Surveyor. Wenona, Doriphan County, Kansas, *c.*1860.

Clark [Clarke, Clerk], James [I]. 'Engraver & Printseller in Gray's Inn' [London]. For John Seller's New Jersey, *c.*1664; English Pilot, 1671; for Wenceslaus Hollar's *London*, *c.*1675; star charts for Edmond Halley, 1678; Thornton's *New Mapp of the World*, 1685; for Captain Greenvile Collins's *Great Britain's Coasting Pilot*, 1693.

Clark, James (*d.*1719). Engraver of Edinburgh. Charts for John Adair's *The Description of the Sea Coast and Islands of Scotland*, from 1693, published 1703; John Adair's *Nova Scotiae Tabula*, published in George Buchanan's *Rerum Scoticarum Historia*, 1727.

Clark, John (*d.*1746). Bookseller at 'the Golden Ball, in St. Paul's Churchyard', London. One of the partners (with C. Hitch and W.H. Toms) in Thomas Badeslade and William Toms' *Chorographia Britanniae*, *c.*1745. Succeeded by his widow, Anne.
• **Clark**, Anne. Widow of John Clark, whom she succeeded in the Hitch and Toms partnership. Her share passed to William Johnston, *c.*1748. **Ref.** HODSON, D. *County atlases of the British Isles published after 1703* Vol.I pp.160-162.

Clark, John, *junior*. (*d.*1819). *A Map of the Raritan River...with a plan of the roads*, [n.d.].

Clark, John. Land surveyor. Roxburghshire in John Thomson's *Atlas of Scotland*, 1832.

Clark, Matthew (1714-1798). Boston trader and auctioneer. Undertook the publication of a set of charts of the American coasts at the advanced age of 76. *A Complete set of Charts of the Coast of America from Cape Breton to...the Gulf of Mexico* (18 charts), 1790 (charts engraved by John Norman and Joseph H. Seymour amongst others, and certified by Osgood Carleton, who also sold them). **Ref.** GUTHORN, P.J. *United States coastal charts 1783-1861* (1984) p.7; WHEAT & BRUN *Maps and charts published in America before 1800* (1978) many entries.

Clark, *Captain* R.J. Treffey. *A Chart of the River Magdalena from the source to the mouth*... [Colombia], 1852.

Clark, Richard. Publisher of Philadelphia; also worked in partnership with the Tackaburys. Published many plans of Connecticut towns prepared by E.M. Woodford and Lawrence Fagan, 1851-1854. e.g. Fagan's *Map of the Town of Newtown* [Fairfield County, Connecticut], 1854; *Clark's Map of Fairfield County Connecticut*, Philadelphia 1858 (from a survey by J. Chace, Junior, printed by Wagner & McGuigan); *Clark & Tackabury's new Topographical Map of the State of Connecticut*, 1859 (with R.M. & G.N. Tackabury, engraved by Friend & Aub); *Clark's Map of Litchfield County*, 1859 (surveyed by G.M. Hopkins).

Clark, *Reverend* Samuel (1810-1875). Worked with publishers Edward Stanford and August Petermann on thematic maps, wall maps and atlases for educational use. *Darton and Clark's School Room Map of Ireland*, 1841; British Empire, 1848-1851; *Map showing the distribution of the population of the British Isles*, also a map of occupations in the British Isles, in *Maps Illustrative...*, 1851; *Bible Atlas*, 1868. **Ref.** ROBINSON, A.H. *Early thematic mapping in the history of cartography* (1982) pp.124-126; *ibid.* pp.142-143.

Clark, Thomas. *Map of the district Five miles round the Town of Bridgewater*, 1853 (with Robert Down).

Clark, W.C. *Pocket map of Sydenham*, 1875.

Clark, *Captain* William (1770-1838). American explorer and governor of Missouri Territory. With Meriwether Lewis undertook an influential exploration of the Western United States from the Mississippi to the Pacific. He worked on manuscript maps, including one of the region west of the Great Lakes, 1806-1810 (copied by Samuel Lewis, and engraved by Samuel Harrison for publication in Nicholas Biddle's *History of the expedition under the command of Captains Lewis and Clark*, Philadelphia, Bradford & Inskeep, 1814). **Ref.** HECKROTTE, W. *The Map Collector* 39 (1987) pp.16-20; MOULTON, G.E. 'Lewis & Clark', in WILLINGHAM, W.F. & INGRAM, L.W. (eds.) *Enlightenment science in the Pacific Northwest: the Lewis Clark expedition* (1984) pp.19-25; WOOD, W.R. 'John Evans & William Clark: two early western explorers' maps re-examined' in *We proceeded on* Vol.9 N°1 (1983) pp.10-16; MOULTON, G.E. 'Another look at Clark's map of 1805' *ibid.* pp.19-22; MOULTON, G.E. (ed.) *Atlas of the Lewis & Clark expedition* (1983); WOOD, R.W. & MOULTON, G.E. 'Prince Maximilian and new maps of the Missouri and Yellowstone Rivers by William Clark' in *The Western Historical Quarterly* Vol. XII N°4 (October 1981) pp.373-386.

Clark, William Bullock. American geologist involved in the foundation and running of the Maryland Geological Survey from 1896. *Map of Maryland showing the Geological Formations*, 1907.

Clarke, *Captain* Alexander Ross *RE* (1828-1914). Joined the Ordnance Survey in 1850; posted to Canada 1851-1854; contributed greatly to the technical development and methodology of the Ordnance Survey; published theoretical works on geodesy; retired 1881. Colony of Victoria, 1855; worked on the findings of the Ordnance Survey's triangulation of Great Britain, published as *Account of the Observations and Calculations of the Principal Triangulation...*, 1858. **Ref.** SECORD, J.A. in: *Dictionary of national biography: Missing persons* (Oxford 1993) p.136; PAPWORTH, K.M. 'Clarke and the Principal Triangulation' in: SEYMOUR, W.A. *A History of the Ordnance Survey* (Folkestone 1980) pp.139-146.

Clarke, Benjamin. British Gazetteer (43 maps), 1852 (with Henry George Collins).

Clarke, Charles Russell. *Map of the Lake States* in: *Clarke's New School Geography*, San Francisco, H.H. Bancroft & Company 1866.

Clarke, H.G. [& Co.]. Publishers of London. *Clarke's plan of London as it is to day* [showing the Great Exhibition site in Hyde Park], *c.*1851; *The Environs of London...1851* [map for the Great Exhibition], 1851; *London in all its glory*, 1851.

Clarke, H. Wadsworth. *Map showing the Location of Salt Wells at Syracuse*, 1868; *The Highlands of Syracuse, New York*, 1872.

Clarke, J. *Map of the Vales of Clwyd & Langollen* [Wales], *c.*1862.

Clarke, Jacob Richard. Printseller and publisher of George Street North, Sydney. North East Australia, 1860; North East Australian Gold Fields, 1860; *Map of the colonies of New South Wales and Queensland*, 1860.

Clarke, James [II] (*fl.c.*1772-1789). Various manuscript maps of parts of north-west England including *A Survey of the Lakes* [English Lake District] (11 copper engraved maps), London 1787.

Clarke, James [III] (*fl.*1812-1815). English cartographer, produced inclosure and road maps of parts of Hampshire, Isle of Wight and northern France. *A Military, Marine and Topographical Survey of the Isle of Wight*, 1812.

Clarke, Samuel. Geographical Description of the Known World, 1646.

Clarke, Samuel. *Geological map of the United Kingdom of Great Britain and Ireland* (21 sheets), 1860; *Railway and Steam Navigation map of the United Kingdom* (21 sheets), 1860.

Clarke, Stephen Reynolds. *The New Yorkshire Gazetteer*, 1828.

Clarke, *Captain* T.J. Naval operation before Canton, 1840 MS.

Clarke, Thomas. Draughtsman working in the style of the Thames School. *A Mapp...of Sommer Islands* [Bermuda], 1678 (taken from a survey by Richard Norwood).

Clarke, *Captain* Thomas. Bonny River [African Pilot], *c.*1785.

Clarke, Thomas (*fl.* 1797-1800). Engraver in New York and Philadelphia, including work for the *American Universal Magazine*. *Plan of the Attacks of Landau in 1713* and a plan of the fortress of Philipburg for Thomas Dobson's *Encyclopaedia*, 1798; George Gilpins's *Plan of the Town of Alexandria in the District of Colombia*, published I.V. Thomas 1798; *Sketch of the Harbour and Environs of Toulon* for John Gifford's *History of France*, 1798.

Clarke, W.B. Drew town plans for the Society for the Diffusion of Useful Knowledge, including *Dublin*, Baldwin & Cradock 1836.

Clarke, William Branwhite (1798-1878). Geologist. *Geological sketch map of New South Wales*, 1880; *Geological map of New South Wales* (2 sheets), 1893.

Clarkson, Matthew. Publisher and Mayor of Philadelphia. ...*Plan of the Improved Part of the City* [Philadelphia], 1762; publisher of Henry Dawkins' *A General Chart of all the Coast of...Louisiana*, 1763.

Clas, George. See **Glas**, George.

Clasen, Herman (*fl.*1637-1657). Dutch surveyor and cartographer, worked in Groningen, succeeded Egbert Haubois as *stadsbouwmeester* of Groningen 1650.

Clason Map Co.. Publishers of Denver. *Clason's map of the european war zone*, 1914; *Clason's war map of the western front*, 1918.

Classen [Chassun; Classun], — de. *Carte du diocèse d'Aire*, 1635; *Carte du Siege Presidial Dax*, 1638. **N.B.** His work is said to have been used by Melchior **Tavernier**, 1634.

Claudianus [Klaudian; Kulhanek], Nicolaus [Mikulaš] (*d.*1522). Physician and printer of Jungbunzlau, Bohemia. Woodcut map of Bohemia, 1518 (believed to be the first separate map of Bohemia).

Claudius, Hendrik. German apothecary working for the Dutch East India Company, sent to study the flora of the Cape of Good Hope. Manuscript map of the Cape published as *Carte des Pays et des Peuples du Cap de Bonne Esperance Nouvellements Decouverts* *par les Hollandois*, Paris, Seneuze 1686 (in G. Tachard's *Voyage de Siam des Pères Jésuits*). **Ref.** STONE, J.C. (Ed.) *Norwich's maps of Africa* (1997) N°.161, p.183.

Claus, Jacob. *Afteykeninge van de Stadt Philadelphia*, 1684 (Dutch edition of Thomas Holme's plan).

Clausen [Claussøn]. See **Clavus**, Claudius.

Clausen, J.C.G. Leonhard Euler's map of France, Basle 1753.

Clauser, Jacob. Wood-engraver. Worked on Sebastian Münster's *Cosmographia universalis*, 1550 and later editions.

Clausner, Jakob Joseph (1744-1797). Swiss archive secretary, surveyor and engraver, born in Zug, worked in Zug, Rheinau and Strasbourg. Suisse, 1799.

Clausolles, P. *Atlas historique et géographique de la France*, 1846.

Clauson, C. Mappa topografica della città di Napoli, 1775.

Clavero, Gregorio. Manuscript plans of Manila, 1792-1800.

Clavigero [Clavijero], Francisco Xavier (1731-1787). Teacher of rhetoric and philosophy from Veracruz, Mexico, exiled to Italy. Best known for his study of pre-conquest Mexico: *Storia Antica del Messico...* (4 volumes, 2 maps, 1 view), Cesena, Gregorio Biasini 1780.

Clavus [Clausen; Clavius; Cymbricus], Claudius (*b.*1388). Danish geographer. Map of the Nordic lands, 1427 (used in early editions of Ptolemy, and as sources for later maps by J. Ziegler, 1532, N. Zeno, 1558, and others).

Clay, A.M. *Bergen County, New Jersey*, 1874.

Clay, H. *An introductory atlas of international relations* (47 maps), London, Headley Bros. 1916 (with A. Greenwood, and with maps by H.S. Hattin).

Clay, Cossack & Co. Lithographers of Buffalo. C. Collins' *Map of the Lake Shore and Michigan Southern Railway...*, 1875.

Claye, Jacques de Vaulx de. Brazil, *c.*1579 (sea charts).

Clayssius, Pieter. See **Claeyssens**, Pierre.

Clayton, Henry (*fl.*1805-1846). English surveyor. Various manuscript maps of Essex and London including *West Horndon, Childerditch and Ingrave*, 1805.

Clayton, P. Surveyor. Woodson, Illinois, *c.*1857; *Weston, Platte County, Missouri*, 1858.

Cleaton & Gittins. Book and printsellers of Shrewsbury. Basil Wood's *New Map of Shropshire* (4 sheets), *c.*1710.

Cleef, Gebroeders van. Publishers. Van den Bosch's *Nederlandsche Bezittingen in Azie, Amerika en Afrika*, 1818.

Cleer, Thomas (*fl.*1685-1706). English estate surveyor. Produced various manuscript maps of parts of England including West Lodge Enfield Chase, 1702 MS; Pansole in Boarstall [Buckinghamshire], 1705.

Clegg, Ernest. Produced many pictorial maps of English counties, also *Kent: Battle of Britain*, 1945.

Clegg, John. *Elements of Geography...* (with maps), Liverpool, J. McCreery 1795, 1796.

Clegg, W.E. *Oldham and District*, *c.*1898.

Cleghorn, *Doctor* A. Mountain regions of Punjab, 1864.

Cleghorn, J. Engraver of a plan of Oxford which was later attributed to James Bingley; some wood engraving for vignettes in Moule's *The English Counties Delineated*, 1837.

Cleland, James (1770-1840). Statistician. Plan of Glasgow, *c.*1822.

Clemens, M. *Map of the State of Ohio*, 1861 (with T.W. Baker).

Clement, T. *Plan de la ville d'Anvers*, 1814.

Clement, Timothy (1706-1766). Surveyor. *...Plan of Hudsons River from Albany...to Lake George*, 1756.

Clements, James E. *James E. Clements' Map of Washington City and Surrounding Country*, 1891.

Clements, *Captain* John. Plan of the Harbour of Chusan [China], 1756, published Dalrymple *c.*1792; plan of Ningpo, Tchekiang, 1756, published Dalrymple *c.*1792; Natunas Islands [East Indies], 1773.

Clements, S. *Chart of the Entrance to the River Thames*, 1791.

Clerc, le. See **Le Clerc**, Jean.

Clerc, *Capitaine* —. *Plan en relief de La Spezzia*, 1809-1811.

Clerc, Pierre Antoine (1774-1843). French surveyor.

Clerck [Klerck], Nicolaes de (*fl.* early-17th century). Engraver and cartographer of Delft. Worked for the Hondius family and J. Londerseel.

Clericus, J. See **Le Clerc**, Jean.

Clerk, Ja. See **Clark**, James [I].

Clerk, Thomas. Engraver and lithographer of Edinburgh, 265 High Street (1813-1823), 22 St. James's Square (1831-1835). John Thomson's *New General Atlas*, 1814 and later; John Wood's *Town Atlas of Scotland*, 1828.

Clerke, *Captain* Charles (1741-1779). Served with Captain James Cook's voyages from 1768, including the exploration of the North West Coast of America, 1778-1779. Clerke took command following Cook's death in 1779.

Clerke, Thomas. *Lordshippe or Mannor of Holkeham* [Norfolk], 1590 MS; manuscript maps of Kent estates for All Souls College, Oxford, 1589 MS. **Ref.** EDEN, P. 'Three Elizabethan estate surveyors: Peter Kempe, Thomas Clerke, and Thomas Langdon' in TYACKE, S. (ed.) *English map-making 1500-1650* (London 1983) pp.68-84.

Clermont, —. *Carte de l'Isle de Corse*, 1769; *Carte de la Province de Normandie* (2 sheets), 1780.

Clérot, Victor. *Carte routière de l'Auvergne et Lyonnais*, *c.*1866; *Cosmorama Universel*, *c.*1866; *Nouvelle Carte routière de la Bretagne*, 1867; *Nouvelle carte de tous les Chemins de fer de l'Europe*, 1869; *Environs de Paris*, *c.*1871; *Nouvelle Carte de l'Amérique*, 1877; *Nouvelle Carte de l'Afrique*, 1878.

Clerq, J.H.W. le. See **Le Clerq**.

Clerville, Louis Nicolas de (*d.*1677). French cartographer. *Carte générale de Montagnes de la Haute Auvergne*, 1642; *Carte de la Haute Auvergne*, 1670.

Clesmeur. See **Du Clesmeur**.

Cletcher, D. Engraver and mapmaker. Maestricht, 1633.

Clevelly, A. River Dart, 1772.

Cleynhens, Bernard (*fl.*1700-1742). Dutch publisher. *Accuraat geographisch Kaart-boekje...*, c.1700.

Cleynhens, V. *Een Accuraet Kaert-Boekje...* [pocket atlas of western Europe], c.1750.

Cliffe, J. Waterford, 1725.

Clifford, J. *Map of Fourteen Miles round Tunbridge Wells, Kent*, *c.*1820.

Clifford, *The Honourable* Robert Edward (1767-1817). *A Sketch of the Russian Campaign in 1812*, 1813. **Ref.** RAVENHILL, W. 'The Honourable Robert Edward Clifford, 1767-1817: a cartographer's response to Napoleon' in: *The Geographical Journal* 160 (July 1994) pp.159-172; RAVENHILL, W. 'The Honourable Robert Edward Clifford: a Devon map collector and military cartographer' in: *Report and Transactions of the Devonshire Association for the Advancement of Sciences, Literature and the Arts* (Torquay) 125 (December 1993) pp.113-125.

Clifton, Leonard Worsley (*b.*1830, *fl.* 1849-1891). Postmaster, Banbury, Western Australia, later Collector of Customs. *Plan of the intended town of Australind*, 1870.

Clint, Raphael (1797?-1849). Surveyor, engraver and lithographer of Sydney, New South Wales, located at Bridge St (1837); George St (1842); Old Bank of Australia,

George St (1847). *A Chart of the North Island of New Zealand...*, [n.d.]; Southwest side of the Firth of Thames, New Zealand, 1839; *The Brighton estate at Lady Darling's Point*, 1841; *Camperdown Terrace*, 1841; Plan of the City of Wellington, Port Nicholson, New Zealand, 1841; *A map of the county of Cumberland...*, *c.*1848.

Clinton, family.
• **Clinton**, *Admiral* George (1686-1761). Surveyor in North America, Colonial Royal Governor of New York. New Jersey, 1781.
•• **Clinton**, *Sir* Henry (1730-1795). Son of George Clinton. Commander of British troops in North Carolina during the Revolution. Manuscript plans of battles and seige operations in New Jersey during the American War of Independence, including *British camp on west bank of the Schuylkill River at Philadelphia*.
••• **Clinton**, *Sir* William Henry (1769-1846). Son of Sir Henry. *A Plan of the attack of Fort Sullivan near Charles Town...1776*, 1791. **Ref.** GUTHORN, P. *British maps of the American Revolution* (1972) p.17.

Cloard, Charles. Manuscript map of East & West Florida, 1739.

Clobu ciarich, [Clobucarius; Globuciarich; Klobučarić], Johann (*c.*1550-1605). Monk, cartographer and engraver. Large scale map of Austria (109 sheets), 1601-1605 MS.

Clod [Clodiensis]. See **Ré**, Sebastiano di.

Clodi[an]us, Marcus [Marcellus]. Publisher of Rome. *Britaniae Insulae*, 1589 (after George Lily); Milan, 1589; Parma, [n.d.].

Cloete, A.J. South East Africa... Cape... Natal, John Arrowsmith 1850.

Cloete, Lawrence. Government surveyor, active in the area around Pietermaritzburg before 1849. His maps and plans were later used by others.

Clogher, William. *Map of Chicago and Vicinity*, 1849.

Clooster, Geleyn Pietersz. (*fl.*1627-1652). Dutch surveyor and cartographer of Oude Niedorp, worked in Noord-Holland. *Caerte ofte afbeeldinghe vande Braeckges ende Sijgerrijdt inden banne van Winckel ghelegen...*, 1636-1637 MS.

Clooster, Johannes (*fl.*1650-1670). Dutch cartographer and surveyor, worked in Noord-Holland. *Afbeeldinghe... vande wateringhe in het cavel Gheestmerambacht*, 1670 MS.

Cloppenburgh, Evertszoon [Everardus]. Publisher in Amsterdam, 'op't Water in den Vergulden Bijbel'. *Germania*, 1630; Jan Huyghen van Linschoten's *Itinerarium*, 1644; *Nova Hispaniae Descriptio*, [n.d.].

Cloppenburgh, Jan [Johannes] Evertsz. Publisher in Amsterdam, 'op't Water in den Vergulden Bijbel'. Jan Huyghen van Linschoten's *Le Grand Routier de Mer...*, 1619; *Gerardi Mercatoris Atlas sive Cosmographicae Meditationes...*, 1630, 1632, 1636.

Clopton, J.W. *Map of the City of Helena, Arkansas*, 1859.

Closche, G. Engraver of Charles d'Argentré's map of *Rennes*, 1616

Close, *Sir* Charles. Director General of the Ordnance Survey, 1911-1922. **Ref.** SEYMOUR, W.A. (Ed.) *A history of the Ordnance Survey* (1980).

Cloué, —. Map of Port Mahe, Seychelles, 1846 (with Bellot and Leblanc).

Clouét, G.Ch. (1817-1889). French admiral and hydrographer.

Clouet, *abbé* Jean-Baptiste-Louis (*b.*1730). Académie royale des sciences de Rouen. *Géographie moderne*, Paris, Mondhare 1767; *Des Possessions françoises aujourd'hui sous la domination angloises...*, *c.*1770; *Carte d'Amérique* (4 sheets), 1776, 1793; *Carte d'Afrique*, 1780; Mappe monde, 1785.

Clough, Charles Thomas (1852-1916). Yorkshire geologist. Joined the Geological Survey in 1875, moving to Edinburgh in 1884. Responsible for many geological maps of Scotland prepared throughout his career. **Ref.** HARRIS, A.L. in: *Dictionary of national biography: Missing persons* (Oxford 1993) pp.138-139.

Clouzier, François. Publisher 'au Palais à l'Image Nostre-Dame, proche l'Hostel de Mr le premier President, à la Seconde Boutique', Paris. Souchu de Rennefort's *Relation du premier Voyage de la Compagnie des Indes Orientales en l'isle de Madagascar*, 1668.

Clouzier [Clousier], Gervais (*fl.*1656-1679). Published Jean Baptiste Tavernier's *Les Six Voyages...*, 1676; *Recüeil de plusieurs*

Correction:

Clobucciarich, [Clobucarius; Globuciarich; Klobučarić], Johann (*c.*1550-1605). Monk, cartographer and engraver. Large scale map of Austria (109 sheets), 1601-1605 MS.

Geography, settled in Leiden, later works published posthumously by Elzevier. *Germania antiqua* (11 maps), Leiden 1616 (engraved by N. Geelkercken); *Alpinae seu Foederatae Rhaetiae... nova descriptio*, Basle 1618 and later editions [titles vary] (with F. Sprecher von Bernegg); *Sicilia Antiqua*, 1619 and later editions; *Italia Antiqua* (2 volumes, 15 maps), Elzevier 1624, 1629, Hondius 1629, Bunon 1659, 1674, 1724 and many other editions (engraved N. Geelkercken); *Introductio in universam geographiam*, 1624 and many later editions. **Ref.** KRETSCHMER, DÖRFLINGER & WAWRIK *Lexicon zur Geschichte der Kartographie* (1986) vol.C/1, p.144 [in German]; BROMBERG, S.A. *The Map Collector* 11 (1980) pp.20-25.

Cnobbart, Joannis. Publisher of Antwerp for Famiano Strada's *De bello Belgico*, 1635, 1636 (with Leo Belgicus maps as titles).
• **Cnobbart**, —. Widow and heir of Joannis e. Published later editions of *De bello co*, 1640-1649.

bert, Michiel. Publisher of Antwerp. *che Merkurius of Bondige Beschrijving eheel Frankrijk*, 1666.

f, Matthaeus Ferdinand (1715-1771). *cipataus Brandenburg...*, 1740; *tatus Oettingensis*, 1744.

e, Marmaduke. Surveyor. Lexington, berry & Richland Districts, South lina, 1820; *Atlas of the State of South ina*, 1825 (for Robert Mills).

, Johann Wilhelm Fortunat (1822-). Swiss forestry superintendent and grapher, worked in Geneva, Chur and , and for the Swiss Topographical au, 1844-1851. Worked with G.H. ır. Manuscript surveys for the Grisons 1844-1851 (used for the topographical of Switzerland).

, Francis R. *Chart of the Harbour of Cruz*, 1843.

, Henry (*fl.*1841-1855). Surveyor of ln's Inn Fields. Estate and tithe maps of of Essex and Kent, including Park & nds of Lee Priory, 1841 MS.

é, Jacque André. Ganges, Eugène Henri 1726.

Cobbett, William (1762-1835). London publisher and author, 11, Bolt Court, Fleet Street. *Plan of the blockade of Cadiz*, 1797; *A Geographical Dictionary of England and Wales* (52 maps), 1832, 1854 (the second edition used larger maps).

Cobo, Juan, *SJ*. Doctrina Christiana en lengua China, Manila 1593; also woodcut map of China and America (first Western printed map in Chinese).

Philipp Clüver (1580-1623). (By courtesy of University of Michigan Library, Ann Arbor, MI)

Coburn, W.A. *Plan of the Isthmus of Tehuantepec*, 1851.

Coccejo [Coccejus], Johannes Heinrich (1603-1669). *Ierusalem...*, Nuremberg, Samuel Faber 1720.

Cocchi, I. Carta geologica dell'Italia, 1867; Carta geologica della parte orientale dell'isola d'Elba, 1872.

Cocchini, Giuseppe. Printer 'all 'insegna della stella', Florence. Sir Robert Dudley's *Dell' Arcano del mare*, 1661.

Cochado, António Vicente. Portuguese pilot. *Descripcao do Rios Para Curupa e Amazonas*, *c.*1623.

Cochelet, Adrien-Louis (1788-1858). French geographer.

Cocheris, Hippolyte. *Tabula Dioecesis Versaliensis*, 1874.

Cochet, —. *Carte Archéologique du Département de la Seine-Inférieure...*, 1859.

Cochin, family of Troyes.
• **Cochin**, Jacques (*fl.*1533-1590). Artist and wood engraver of Troyes. Father of Noël I (below).
•• **Cochin**, Noël I (*c.*1560-*c.*1638). Painter and engraver of Troyes, son of Jacques, father of Nicolas and Noël II. Incorrectly credited with Sébastian de Ponthault de Beaulieu's *Les glorieuses conquestes de Louis le Grand.*
••• **Cochin**, Nicolas (1610-1686). Engraver and artist of Troyes, son of Noël I (above). Worked with his brother Noël II on some of the town plans, battle plans and views for Sébastian de Ponthault de Beaulieu's narrative of the military campaigns of Louis XIV entitled *Les glorieuses conquestes de Louis le Grand* [known as the *Grand Beaulieu*], from 1643.
••• **Cochin**, Noël II [Cochin de Venise] (1622-1695). Engraver, son of Noël I, brother of Nicolas, with whom he worked. *Vue générale de paris*, 1669

Cochin, family of Paris, descended from the Cochins of Troyes. Engravers of prints who also worked on maps. **Ref.** PORTALIS, R. & BERALDI, H. *Les graveurs du dix-huitième siècle* (c.1880) [in French].
• **Cochin**, Charles. Painter of Paris, godson of Charles Duchesne.
•• **Cochin**, Charles Nicolas I [*père*] (1688-1754). Engraver 'galeries du Louvre', Paris. Cartographic work seems to be limited to ornaments for the French edition of George Anson's *Voyage de l'amiral Anson*, 1751.
•• **Cochin** [Hortemels], Louise Magdeleine (1686-1767). Engraver of Paris, worked with her two sisters until her marriage to Charles Nicolas Cochin (above) in 1713. After the death of her husband, worked with her son Charles Nicolas II (below).
••• **Cochin**, Charles Nicolas II [*fils*] (1715-1790). Prolific artist and engraver of Paris, 'secretaire de l'Académie de peintre-sculpture, galeries du Louvre'. *Mappemonde*, 1752; engraved some of the maps for Robert de Vaugondy's *Atlas Universel*, 1758-1776; frontispiece for Bonne - Desmarest *Atlas encyclopédique*, 1787.

Cochran, *Lieutenant* Alexander, *RN*. *China Province of Se Chuen*, 1892.

Cochran, A.B. *Atlas of Schuylkill* [Pennsylvania], 1875 (for Frederick W. Beers).

Cochrane, A.S. *Map of part of the Athabasca River*, 1885.

Cochrane, Joseph. Publisher. M.P. Goodwin's *Map of Guernsey*, 1832; *Plan de la ville de Boulogne sur Mer*, 1844.

Cock [Coquus], Hendrik [Henrique] (*c.*1550/1555-after 1598). Dutch cartographer from Gorinchem, lived in Rome, later Spain from 1574. *Hispaniae nova delineatio... MDLXXXI*, Abraham Ortelius 1581 (one example only known, at Darmstadt); later printing 1583. **Ref.** MEURER, P. *Fontes cartographici Orteliani...* (1991) p.129 [in German].

Cock [Coc; Coquis; Coccius], Hieronymus [Girolamo; Jeroen] (1507/1510-1570). Publisher, engraver and printseller of Antwerp. *Nova Celeberrimi Ducatus Geldriae...* (6 sheets), *c.*1545, 1601; map of Piedmont, 1552; *Nova Descriptio Hispaniae* (4 sheets), 1553 (copied from the work of Vincenzo Paletino); *Nova et Exactissima Insulae Siciliae Descriptio* (2 sheets), 1553; *Totius Illius Regionis... Turcicam vocant...*, 1555; *...Sabaudiae Ducati Descriptio*, 1556;

Antverpia..., 1557; Map of Netherlands (3 sheets), 1557; *Elaboratissima Mediolanensis Ducatus...*, 1562; Gutierrez's *Americae sive Quartae Orbis Partis...*, 1562; Christiaan s'Grooten's *Nova Totius Germaniae* (9 sheets), *c.*1565; *Malta*, 1565; published Peter Laicksteen's *Nova Descriptio Amplissimae Sanctae Terrae*, Christian Sgrooten 1570.

Cockburn, William. Barony of Balquhidder, 1756 MS.

Cockerill, family.
• **Cockerill**, Thomas, *the elder* (*fl.*1674-1702). Served part of apprenticeship with John Overton, became a publisher at 'the Atlas in Cornhill' (with Robert Morden 1677); 'Three Bears in Poultry over against the Stock Market' (1678); 'Corner of Warwick Lane Paternoster Row' (1697); 'Three Legs and Bible against Grocers Hall in the Poultry' (1699). *Geography Rectified*, 1680 (with R. Morden); involved with others in the publication of Patrick Gordon's *Geography Anatomiz'd...*, 1693, 1702; *The New Description and State of England*, 1701 (contained the so-called 'miniature Morden' maps; published with R. Morden and R. Smith).
•• **Cockerill**, Thomas. Nephew of Thomas *the elder*, served as his apprentice 1688-1695.

Cockrem, Edward. *Plan of the Town of Torquay...*, 1840.

Cockrem, Philip. North American publisher, 'Princess Street near Smith Street', New-York. Chart of an area of North America, covering the mouth of the Mississippi in the west, to Barbados in the east, 1730.

Cockson, Thomas. Engraved Giovanni Baptista Boazio's *An Exact Map of the Town of Cales* [Cadiz], 1596.

Cocquart [Coquart], A. *Carte generale des Monts Pyrenées*, *c.*1730.

Codazzi, Agostino [Giovanni Battista Agostino] (1793-1859). Geographer and engineer of Italian origin, worked extensively in South America. His name is retained in the title of the official mapping agency. *Atlas Fisico y Politico de la Republica de Venezuela*, Caracas 1840 (first survey of Venezuela); *Colombie*, 1865; surveys used in the preparation of *Atlas de los Estados Unidos de Colombia, Antigua Nueva Granada*, Paris, Renous and Maulde 1865 (engraved by Thierry frères of Paris); *Atlas de Colombia*, 1889; *Mapa de la República de Colombia*, 1890.

Codde, *Captain* Pieter. Navigator and chart-maker of Enkhuizen. *Afbeeldinghe vande... Seehaven ende stadt van Duynkercken*, 1631 (used by Blaeu, 1634, and later by Joannes Janssonius).

Codine, J. Mer des Indes, 1867.

Codogno, O. Itinerario delle poste per tutto il mondo, 1620.

Coeck, Gerard [Geeraerd; Gerardus] (*b.*1608). Engraver of Amsterdam. Joannes Janssonius's Holland, 1647; Jan Adriaensz. Leeghwater's *De Haerlemmermeer en Omstreken*, 1641; Pieter Goos's *Lichtende Colom*, 1650 and later; Nicolaas Ten Have's *Transisalania provincia; vulgo Over-IJssel*, Gerard Valk 1650; for Joan Blaeu's *Atlas maior*, 1662.

Coeler [Köller], George [Joris] (*fl.*1640). Engraver and publisher of Amsterdam, worked for Hondius.

Coelho, Gomes. *Plano Hydrographico da Costa de Loanda*, 1883.

Coelho, João. *Mapa general dos Estados Unidos do Brazil* (2 sheets), 1891.

Coello de Portugal y Quesada, *Lieutenant-Colonel* Francisco (1822-1898). Spanish military engineer, explorer and geographer. *Atlas de España y sus posesiones de Ultramar*, 1847-1868; Canarias, 1849; Filipinas, 1849-1852; Isla de Cuba, 1853; *Mapa geologico de España y Portugal*, 1879. **Ref.** KRETSCHMER, DÖRFLINGER & WAWRIK *Lexicon zur Geschichte der Kartographie* (1986) vol. C/I, p.144 (numerous references cited).

Coenders van Helpen [Conders ab Helpen], family. Dutch cartographers, worked in Groningen. *Prov. Groningae et Omlandiae tabula* (4 sheets), *c.*1677/1678 (engraved by Cornelis Appeus).
• **Coenders van Helpen**, Wilhelm [Willem] (*b.*1629).
• **Coenders van Helpen**, Frederik (*b.*1631).

Coenraetsz. [Coenraetszoon; Coenraeds; Koenss.], Jacob (*fl.c.*1549-*c.*1582). Dutch surveyor and cartographer, lived and worked in Zuid-Holland, in particular Delfland and Schieland. Produced numerous manuscript maps.

Coenraetsz., Odulphus. Dutch surveyor and cartographer, worked in Zuid-Holland 1564-1579.

Coentgen, Franz. *Plan der Stadt und Festung Mainz*, 1795; *Plan der... Attaque auf... das Fort Kehl*, *c.*1797.

Coentgen, Heinrich Hugo. See **Cöntgen**.

Coffey, James. County of Limerick, G. McKern, 1825.

Coffin, J.A. City of New York, 1847.

Coffin, R. Engraver for Benjamin Donn's *Bristol*, 1769; *Plan of the City of Bath*, 1773.

Cogan, *Lieutenant* Robert, Indian Navy. West coast of India, 1828; Trigonometrical Survey of Bombay Harbour, published James Horsburgh 1833.

Cogeur, --. *Carte topographique de la Forêt de Soigne*, 1770.

Cogger, Ambrose. Estate plans in Sussex and Kent, l627-1628 MS.

Coghlan, David. *City of Scranton, Pennsylvania*, 1870.

Coghlan, Francis. Publisher of 'West Strand', London. *A visit to London or Stranger's guide*, *c.* 1831, *c.*1833; *The Stranger's Guide In London*, 1835; *Guide to the Rhine*, 1841; *Map of Routes to the principal Cities...in Europe*, 1841; B.R. Davies's *London and its Environs*, 1843 edition (with others, first published 1840); *Coghlan's pocket picture of London & its environs*, London, H. Hughes 1847, 1848 (compiled by A.C. Lee).

Coghlan, John. *Buenos Ayres Gran Ferro-Carril del Sud*, 1860.

Coghlan, William C. *Business Map of Pittsburgh*, 1868; *The City of Louisville, Kentucky*, 1873.

Cogniard, Frédéric (1856-1904). French explorer and surveyor in Indochina. Surveyed frontier between French Indochina and China, 1889; later surveyed Mekong and Annam regions, 1890-1891.

Cohen, *Doctor* E. *Karte der Umgegend von Heidelberg*, 1874-1877; *South African Republic Transvaal*, 1877.

Cohrs, Edvard. Swedish publisher. *Atlas över Sverige*, 1890.

Coignard, Jean-Baptiste. Printer of Paris, 'imprimeur du roi', and from 1751 'secrétaire du roi', at which time his stepson, Antoine Boudet, bought half his business. *La Carte d'Orleanois, de la Beauce, de la Sologne, du Berry, de la Marche, du Bourbonnois, du Nivernois, de la Bourgogne, et de la Bresse*, Paris 1791 (engraved by Roussel, published in Audiffret's *Le Géographie ancienne et moderne*).

Coignet, Michel (1549-1623). Mathematician and cosmographer of Antwerp. Navigation Guide, 1580; wrote text for miniature editions of Ortelius published by Jan Keerbergen, *Epitome theatri orbis terrarum Abrahami Ortelij*, Antwerp, 1601-1612 (engraved by Ambrose and Ferdinand Arsenius); *Tabula geographica indicans iter novum inter Medioanum et Antverpiam*, 1621 (printed by A. Verhoeven).

Coijet, Carl Fredrik. Swedish cartographer. Plan of St Petersburg, 1722 MS.

Coimbra, Heitor de (*fl.* early 16th century). Portuguese cartographer.

Coindé, *Madame* --. Additional maps for Emmanuel Auguste Dieudonné Le Sage's *Atlas*, 1813.

Col, Juan. *Provincia de Corrientes*, 1891.

Cola, Giovanni [Ioanne]. Italian author. *Viazo da Venezia al Sancto Iherusalem*, 1500 and later (with woodcuts based on the work of Bernhard von Breydenbach).

Colao, Bruno (*b.c.*1818). Topographical draughtsman in the Officio Topografico di Napoli, head of the artistic division of the Istituto Topografico Militare. Worked on several sheets of the Officio's map of the Kingdom of Naples, 1836; *Pianta di Napoli*, 1844, published 1845; engraved *Carta geologica dell'isola d'Ischia*, 1849; *Bonificamento del bacino inferiore del Volturno*, 1856 (by G. Savarese); route map of Naples (4 sheets), 1858-1859; school atlas, 1859 (with Federico Gatti); map of Gaeto, 1861 (for Francesco Saverio Anfora); *Carta della Regione perturbata dai Fenomeni Vesuviani*, *c.*1862. **Ref.** VALERIO, V. *Società uomini e istituzioni cartografiche nel mezzogiorno d'Italia* (1993) pp.479-480.

Colart, L.S. Histoires de France et d'Angleterre, 1841.

Colbeck, Charles. Editor *Public Schools Historical Atlas*, 1885.

Colbeck, Joseph (*fl.*1719-1741). Surveyor, produced estate and inclosure maps of various parts of England including Great Kimble [Buckinghamshire], 1741 MS.

Colberg. See **Kolberg**.

Colbert, Edmond, Surveyor. Land allotment in Victoria, 1854-1859.

Colbert, Edouard Charles Victurnien, *comte de* Maulevrier (1758-1820). Manuscript maps of the war in Virginia and New York, 1781-1783.

Colbert, Jean Baptiste [Ioannes Baptista] (1619-1683). Statesman, finance minister of Louis XIV and promoter of cartography in France, dedicatee of French text edition of Joan Blaeu's *Atlas maior*, 1662. Ordered a survey of the condition of France in 1663, requiring maps of the provinces.

Colbourne, Thomas. *A Map of... Navigation... on the River Shannon*, 1808; *A Map of the Bogs in... County Clare*, 1811-*c.*1814.

Colbran, J. *Map of Tunbridge Wells*, 1850.

Colbran, Thomas (*fl.*1768-1796). Estate surveyor. Parish of Easthurst, 1768; Beacon Estate, 1768; Lidham Hill, 1774.

Colbrant, Jules. *Plan de la ville de St. Omer*, 1865.

Colburn, Henry. Publisher of Conduit Street, London. H. Lichtenstein's *A Map of the European Territory of the Cape of Good Hope...*, 1815 (engraved by M. Thomson, published in Lichtenstein's *Travels in Southern Africa*); George Thompson's *A Map of Southern Africa...*, 1827 (published in Thompson's *Travels and Adventures in Southern Africa*); *Sonora and Gulf of California*, 1829.

Colby, C.J. Union County, Iowa, 1876.

Colby, Charles G. (1830-1866). Editor of [S.E.] *Morse's General Atlas of the World*, 1856; *The Diamond Atlas*, New York, Samuel Gaston 1857; *The World in Miniature...The Western Hemisphere*, New Orleans 1857.

Colby, George N. *Rockland,... Maine*, 1873; *City of Lancaster, Pennsylvania*, 1874; Aroostook County, Maine, 1877 (with F.B. Roe); *Atlas of Hancock County, Maine*, S.F. Colby & Co. 1881; Colby & Co's Washington County, Maine, 1881; Somerset County, 1883; Maine, 1884.

Colby, James M. *Map of the City of Big River Rapids, Michigan*, Detroit, The Calvert Lithographing Co. [before 1880].

Colby, *Captain* (later *Major General*) Thomas Frederick (1784-1852). Superintendent of the Ordnance Survey 1820-1846. Surveyor for the Ordnance Survey of England and Wales, 1799-1811 (with Lieutenant-Colonel William Mudge); Ireland Six-Inch Survey, 1820-1847.

Colden, Cadwallader (1688-1776). *A Map of the Countrey of the Five* [Indian] *Nations belonging to the Province of New York*, 1724, 1735.

Coldewey, Ehrenreich Gerhard (*d.*1773). *Tabula Frisiae Orientalis*, 1730 (used in Homann's *Atlas Germaniae*, 1753, 1762).

Cole, Benjamin. Engraver and mapseller. 'New Wadham College, Oxon' (1695); 'At the Sun and Key near Snow Hill, Conduit Street', London; 'At London Horse Yard St. Pauls Churchyard'; 'At the corner of Kings Head Court near Fetter Lane', London. Maps for Edward Wells's *A New Sett of Maps*,

Jean Baptiste Colbert (1619-1683). (By courtesy of Valerie Scott)

1700 and later editions; Twenty miles round Oxford, 1706; *Twenty miles round Cambridge*, 1710; Ward maps for William Maitland's *Survey of London*, 1754; *A New Map of the Countries Ten miles round... London*, 1756; Jamaica, 1760.

Cole, Charles Nelson (1723-1804). Lawyer. Registrar of the Bedford Level Corporation. Prepared a large scale map of the *Bedford Level*, 1789 (engraved S.J. Neele).

Cole, G. Drew the maps which were engraved by J. Roper to accompany E. Brayley and J. Britton's partwork *The Beauties of England and Wales* (18 volumes), *c.*1804-1810, published as *The British Atlas*, Vernor, Hood & Sharpe 1810; maps also re-issued to illustrate Revd. J. Nightingale's *English Topography*, Baldwin, Cradock & Joy, 1816 and other later issues.

Cole, Humphrey [Humfray] (*c.*1530-1591). Engraver, goldsmith and the first English instrument maker to be known by name. *Canaan... at this daye... called the holie land*, London, Richard Jugge, 1572 (published in the second edition of the 'Bishop's Bible', credited as the first map known to have been engraved by an Englishman); Armillary Sphere, 1582. **Ref.** ACKERMANN, S. (Ed.) *Humphrey Cole: Mint, measurement and maps in Elizabethan England*, British Museum occasional paper nº 126 (1998); '...the first great English line-engraver of maps' in *Imago Mundi* 27 (1975) p. 8.

Cole, James. Engraver and mapseller at the 'Crown in Great Kirby Street, Hatton Garden', London. Gibraltar, 1725; *The Description of Romney Marsh*, 1737; *An Actual Survey of the City of Bath*, 1743; Pierre Chassereau's St. Leonards, 1745.

Cole, William (*c.*1735-1824). Estate surveyor and schoolmaster of Colchester, Essex. Produced various manuscript maps of parts of East Anglia and Sussex.

Cole, William. Publisher, 10 Newgate Street, London. *Geographical Companion through England*, 1824; took over from Hodgson as publisher of Ebden's county maps, 1825-1826. **Ref.** SMITH, D. *Imago Mundi* 43 (1991) pp.48-58.

Colebrook, Pawlett. *Plan of the River of Firth of Forth...*, 1782 MS.

Colebrook, R.H. Course of the Ganges, 1806.

Coleman, G.D. Surveyor. Town and Environs of Singapore, 1836.

Coleman, *Reverend* Lyman. Editor of Wellington Williams's *Atlas of Biblical Geography* (7 maps), Philadelphia 1860, 1864; *A Map of Palestine*, *c.*1868; *A Landscape Map of Egypt*, 1874.

Colepepyr, Robert. *A Map of Rye Harbour*, *c.*1698.

Coles, Edward. Map of the village of Peoria, 1820.

Coles, John. Boston publisher. *An Accurate Map of the Four New England States*, 1785.

Coles, Joseph. Engraver. *A True plan of the city of Excester*, 1709.

Coles, Thomas. *A Chart of the Coast of North Carolina between Cape Hatteras and Cape Fear*, 1806 (with Jonathan Price); and other coastal surveys.

Colfax & Holt. *Plan of the City of New-London*, New London 1847.

Colijn [Colin], Michiel. Dutch publisher, Amsterdam. *Novus Orbis, sive descriptio Indiae Occidentalis...*, 1622 (expanded Dutch edition of A. Herrera y Tordesilla's description of the New World, first published Madrid, 1601); *Polemographia Auraico-Belgica*, 1622; de Jonghe's *Hydrographica, planéque Nova Indiae Occidentalis*, 1664. **Ref.** BURDEN. P. *The mapping of North America* (1996) pp.241-248.

Colin, —. Philippines, 1659.

Colin, A. Publisher of Paris. L. Gallois' *Carte murale du théâtre de la guerre*, *c.*1914.

Colin, Elie (1852-1923). French cartographer, worked in Madagascar, established observatory at Tananarive, 1894; surveyed Madagascar during the 1890s. *Carte de l'Imerina-sud au 1:100 000*, 1909 (with Pierre Roblet). **Ref.** BROC, N. *Dictionnaire illustré des explorateurs Français du XIXe siècle: I Afrique* (1988) p.84 [in French].

Colin, Emanuel. *Description des Indes Occidentales*, Amsterdam 1622 (French edition of A. Herrera y Tordesilla's description of the New World, first published Madrid, 1601). **N.B.** See also Michiel *Colijn* (above).

Colines [Colinaeus], Simon de (1480-1546). Engraver and publisher in Paris.

Colinet, Charles. *Nouvelle Carte Topographique de la Forêt de Fontainebleau*, 1870.

Colins. See **Collins**, Charles Hubbard.

Colius, Jacobus. See **Cool**, Jacob.

Coll, Luis. *Croquis de un proyecto de carreteras* [Catalonia], 1903.

Collard [Collaert], Adrian (1560-1618). Engraver for Gerard de Jode; *Americae Retectio*, Antwerp, Filips Galle 1592. **Ref.** STEVENS & TOOLEY *The Map Collector* 2 (1978) pp.22-24.

Collard, Richard (*fl.*1822-1826). Surveyor and publisher of Broadstairs. *Map of the Town and Royal Harbour of Ramsgate*, 1822 (with G. Hurst); *Broadstairs*, 1824.

Collard, Thomas W. Surveyor. *A Plan of the City of Canterbury...*, 1843, 1860 and later; reduced plan of Canterbury, 1860.

Collard, W. Engraver of Newcastle-upon-Tyne. Roads in the vale of Derwent, 1833; *Plan of the Coal and Iron Mine Districts in the Forest* [Forest of Dean] (15 sheets), 1835; *Country round Newcastle*, 1850.

Collegno, H. de. *Esquisse d'une Carte géologique d'Italie*, Paris 1846.

Collen, George William. *Britannia Saxonica*, 1833.

Colles, father and daughter.
• **Colles**, Christopher (1739-1816). Engineer, inventor and publisher, born in Ireland and emigrated to the United States. *A Survey of the Roads of the United States of America* (86 plates), 1789 (engraved by C. Tiebout, said to be the first American roadbook); *The Geographical Ledger and Systematized Atlas...*, New York 1794; Part of the State of New Jersey, 1808. **Ref.** RISTOW, W.W. (Ed.) Introduction to facsimile edition of *A Survey of the Roads of the United States* (Harvard University Press 1961).
•• **Colles**, Eliza (1776-1799). American engraver; daughter of Christopher, above. Engraved maps in Christopher Colles's *Geographical Ledger*, 1794. **Ref.** RISTOW, W.W. *The Map Collector* 10 (1980) pp.14-17.

Collet, George Henry Victor. See **Collot**, Georges Henri Victor.

Collet, J. *Les cartes topographiques. La carte de France*, 1887.

Collet [Collett], *Captain* John Abraham. Swiss-born engineer, surveyor, governor of Fort Johnston, North Carolina. *Plan of Johnston Fort*, 1767 MS; *Map of the Back Country of North Carolina*, 1768 MS; *A Compleat map of North Carolina*, London, S. Hooper 1770 (engraved by J. Bayly); *A General Map of the Southern British Colonies in America*, 1776 (Collet & others).

Colleton, *Captain* James Roupill. Battle of Salamanca, 1814.

Colley, *Colonel* G. Pomeroy. *Plan of the pensioners village...*[King Williams Town], 1855; *Map of part of Cape Province around East London...*, *c.*1875; Military Sketch Transkei, 1875; *Map of the South African Republic* [Transvaal], 1877.

Collie, J. *Plan and elevations, sections, details and views of the Cathedral of Glasgow*, 1835.

Collie, John Norman. *Sketch Map of the Canadian Rocky Mountians*, 1897-1902.

Collier, *Sir* George, *RN*. *Plan of Princes Island* [Gulf of Guinea], 1819 MS.

Collier, George Ralph. *Survey of the channels into the Texel*, 1799-*c.*1803.

Collier, P.F. & Co. Publishers of New York. *The new encyclopedic atlas and gazetteer of the World*, 1914, 1917 (edition revised by Francis J. Reynolds).

Collier, Richard. River Clyde, 1776; *A Plan of Glasgow*, 1776.

Christopher Colles (1739-1816). (By courtesy of the Library of Congress, Washington DC)

Collier, William. ...*Plan of the Town and Castle of Windsor...*, 1742.

Collier, William Francis (1807-1890). Worked with Leonard Schmitz. *The Crown Atlas of Historical Geography*, 1871; *International Atlas...*, 1873; *Library Atlas...*, 1875.

Collignon [Colignon], François [Francesco] (1610-*c*.1671). Engraver, born in Nancy, worked in Paris and Rome (where he died). *Plan de la ville de Théonville*, 1644 (for Sébastien de Pontault de Beaulieu); *Vero desegno...di Valenza*, 1656; ...*Villes de l'Europe* (93 plates), 1672; ...*Città e Fortezza di Brisach*, 1677; *Pianta delle Fortificationi...di Malta*, *c*.1685; Fifteen Plans of Fortifications in Candia, 1699; *Isola di Candia*, 1699; P. Mariette's perspective view of Tours, Paris, 1700.

Collignon, Romain Charles Edouard (1831-1897). Chemins de fer russes, 1857, 1862; Atlas, 1868.

Collimitius [Tanstetter], Georgius [Georg] (1482-1535). Bavarian cartographer and mathematician. Mapped Turkish theatre of war, 1522; corrected Cuspinianus's map of Austria, 1526; edited Lázár and Ziegler's *Tabula Hungariae*, engraved and printed by Apian and published by Cuspinianus, 1528. **Ref.** KARROW, R.W. *Mapmakers of the sixteenth century and their maps* (1993) p.140.

Collin, E. *père et fils*. French engravers, 'Rue St Thomas du Louvre N°·22, Paris' (1808); 'Quai des Augustins N°·23' (1812). **N.B.** Sources usually do not separate the individuals or the work. Thomas Lopez's *Carte des royaumes d'Espagne et de Portugal oú l'on a marqué les Routes de Postes...*, Paris, E. Collin 1808; *Carte Itinéraire de la France...*, E. Collin 1812 (by De Simencourt); *Plan du siège de Badajoz*, E. Collin 1812; *Bataille de la Gevora*, 1812; Cattegat, 1813; Mer Baltique, 1815, Rade du Brest, 1822; *Carte des routes royales de la France...*, 1824, amended and republished as *Carte des routes nationales de la France...*, 1848; Afrique Occidentale, 1828; Mimizan, 1829.
• **Collin**, Charles Etienne (*fl.*1798-1829). Engraver for Dépôt de la marine. Golfe de Suez, 1798; seven maps for C.P. Claret de Fleurieu's *Voyage Autour du Monde, Pendant les Années 1790, 1791 et 1792 par Etienne Marchand...* (15 maps), 1800; charts and maps of New Zealand for C.F. Beautemps-Beaupré, 1807.
•• **Collin**, Etienne [*fils*] (1790-1852). Engraver for Firmin Didot's *Atlas des mémoires du maréchal Suchet...sur ses campagnes en Espagne*, 1828; Chart of the island of Martinique, 1831; Léon Emanuel Simon Joseph de Laborde's *L'Arabie Petrée*, 1834; for Dortel de Dessin's maps of the expedition of Abel Aubert du Petit-Thouars, 1838; also charts for Beautemps-Beaupré.

Collin, L. Louisiane et pays voisins, 1802.

Collin, Richard. Engraver. *Status Belgicus*, *c.*1630; Fernandez De Medrano's world map, 1688.

Collin, T.I. *La Carte topographique de Spa et de toutes les fontaines minerales*, 1788.

Collingridge, James (*fl.*1749-1778). Estate and inclosure maps of various parts of England including Buckinghamshire, 1773.

Collingridge [W.H. & L. Collingridge], William Hill and Leonard. Printers and publishers of London, at 117-120 Aldersgate Street (1871-1880); 154 Aldersgate Street (1881); 'City Press, 148 & 149 Aldersgate Street, London E.C.' (from 1882). Map for *City of London Directory* (scale 14 inch: mile), 1871 and editions to 1884; *Collingridge's City of London Directory Map*, 1885 and editions to 1909 (scale *c.*25 inches to the mile); *N. Thierry's Map of the City of London...*, 1893.

Collingwood, *Lieutenant* W. Enceinte of Bagdad, 1856 (with Felix Jones); *Trigonometrical Survey of a part of Mesopotamia*, *c.*1868; *Plan of the supposed ruins of Babylon*, *c.*1874.

Collins brothers. Map publishers of 22 Paternoster Row, London.
• **Collins**, William Paprill.
• **Collins**, Henry Edward.
Perhaps sons of Henry George (below) whose residual stock they bought in 1858. Stock and plates were sold again in 1859 due to bankruptcy. *Collins' New Map of the Environs of London, 30 miles from the Metropolis*, *c.*1858.

Collins, B.C. Printer of W. Tunnicliff's A Topographical Survey of the Counties of Hants, Wilts, Dorset, Somerset and Cornwall, commonly called the Western Circuit, 1791.

Collins, Benjamin. Printer of Salisbury, 'on the New Canal'. Naish's plan of Salisbury, [n.d.]. **N.B.** Compare with B.C. **Collins**, above.

Collins, Charles. Engineer. *Map of the Lake Shore & Michigan Southern Railway*, 1875 (printed by lithographers Clay, Cossack & Co.).

Collins [Colins], Charles Hubbard. Map of Chignectou Basin, 1754 MS (plans of English and French forts in Nova Scotia).

Collins, Christopher. *Master* HMS *Cumberland*. Charts of the coast of Great Britain, 1781-1802; Survey Varne and Ridge, 1793.

Collins, David. Early Australian settler and explorer. Map of Botany Bay in *Account of the English Colony of New South Wales*, London 1798, 1802.

Collins, F. Howard. *Twelve Charts of the Tidal Streams of the North Sea*, 1894; ...*Tidal Streams on the West Coast of Scotland*, 1894.

Collins, Freeman. Possibly a brother of Greenvile Collins. Publisher and printer of Greenvile Collins's *Great Britain's Coasting Pilot*, 1693; also printed Edmund Gibson's edition of William Camden's *Britannia*, 1695 (with maps by Robert Morden).

Collins, *Captain* Greenvile (*fl.*1669-1694). Officer in the Royal Navy, sailed with John Narborough's expedition to Patagonia and Chile 1669. Created Commander, and granted the title 'Hydrographer to the King' by Charles II in 1679. Between 1681 and 1688, surveyed the coasts of Britain, producing 120 manuscript charts (only 48 were finally engraved) which formed the basis of his atlas. *Great Britain's Coasting Pilot; Being a New and exact Survey of the Sea Coasts of England, Scotland, and chief Harbours of Ireland...*, 1693, re-issued in at least 14 editions to 1792 with revisions and additions. **Ref.** MOORE, J.N. *The Map Collector* 30 (1985) pp.30-32; VERNER, C. 'Captain Collins' Coasting Pilot' in *Map Collectors' Circle* 58 (1969); ROBINSON, A.W. *Marine cartography in Britain* (1962) pp.34-46 plus Appendix C.

Collins, Henry George (*fl.*1850-1858). Publisher of maps, atlases and globes, 22 Paternoster Row, London. Firm dissolved in 1859. *England*, 1840 (a series of coloured maps); editions of Henry Teesdale's *The New British Atlas*, *c.*1848 (a reworking of Rowe's *The English Atlas* of 1816); *Travelling Atlas of England & Wales...*, 1850; map of Cambridgeshire, *c.*1850; *Collins' London and Environs...*, 1851 (map for 1851 Great Exhibition); *London in 1851*, 1851; *Plan of London*, 1851; British Gazetteer, 1852 (with Benjamin Clarke); *Illustrated Atlas of London with 7,000 References*, 1854; *Collins' London*, 1855; Indestructible Atlas, *c.*1858; One Shilling Atlas, *c.*1858; Junior Classical Atlas, 1859; *Collins' New Map of the County of Kent*, 1859; *Collins' Standard Map of London*, Edward Stanford *c.*1859. **Ref.** HYDE, R. *Printed maps of Victorian London* pp.2, 43-44.

Collins, James. Engraver. Signed two maps in Greenvile Collins's *Great Britain's Coasting Pilot*, 1693.

Collins, James. Engraved maps for Henry George Collins.

Collins, John. Deputy Surveyor General, Quebec. Produced numerous manuscript surveys of parts of the province of Quebec, many including proposed settlements, boundaries and allocation of land parcels. As Deputy Surveyor General he also verified the maps of others. Several surveys of areas in the Bay of Chaleur, marking out fifty acre plots for the curing of fish, 1765 MSS; the bay and harbour of Gaspé, 1765 MS; *A Survey of Lake Champlain...*, 1765 MS; *A Plan of Lake Champlain including Crown Point and St John's on which are affixed such of the Grants made by the Crown of France as are at present known*, 1766 MS; *A Plan of the... Line between... Quebec and New-York*, 1772 MS; Lake Huron, 1785; Kingston [Quebec], 1783 MS; *Gulf of St Lawrence*, 1790; map of the northern shore of Lake Ontario, 1792 MS.

Collins, R. (*d.*1756). Engraver. J. Kirby's map of Suffolk (4 sheets), before 1737; Thomas Warren's Plan of St. Edmunds Bury, 1747.

Collins, William [Sons & Co.]. Publishers in Glasgow, later also London and Edinburgh. Many 19th-century works were also published in New York through G.P. Putnam's Sons. In the 20th century William Collins has also published educational atlases in association with Longman. It is now HarperCollins Cartographic, part of the HarperCollins publishing group, which incorporates Times Books, Bartholomew, and Nicholson. William Collins and the succeeding companies have published a vast number of world, school, specialist and regional atlases, as well as wall and sheet maps. Recent output includes a large number of national atlases for countries all over the World. Earlier atlases include: *Collins' Imperial Outline Atlas*, 1860; *My First Atlas* (12 maps by J. Bartholomew), *c.*1870; *Academic Atlas* (32 maps), Glasgow, W. Collins, Sons & Co. 1870 (compiled by J. Bartholomew); *Collins' Imperial Atlas of Blank Projections*, 1870; *The Student's Atlas*, Glasgow 1870, Glasgow, London & Edinburgh *c.*1885, 1888 (maps by J. Bartholomew, also published by G.P.

Putnam's Sons in New York); *The Advanced Atlas* (32 maps), Glasgow, W. Collins, Sons & Co. *c.*1870 (prepared by J. Bartholomew); L. Schmitz's *International Atlas*, Glasgow, W. Collins, Sons & Co. *c.*1870, 1873, New York, G.P. Putnam's Sons 1879, London & Glasgow, W. Collins, Sons, & Co. *c.*1880, New York, G.P. Putnam's Sons *c.*1881 (maps by J. Bartholomew); *Collins' Imperial Atlas of Australia & New Zealand*, 1871; *Crown Atlas of Classical Geography*, 1873; *Collins' Crown Atlas of the Australian Colonies*, 1875; *Collins' Imperial Atlas of the British Empire*, 1875; *Collins' Junior Atlas of Physical Geography*, 1875; *The Handy School Atlas* (32 maps), *c.*1885; *Collins' Large Shilling Atlas* (24 maps), 1893; *The Graphic Atlas* (114 maps), 1896; *Collins' Complete Atlas* (80 maps), 1897; Clear School Atlas (34 maps), London & Glasgow, William Collins, Sons & Company *c.*1895, second edition published as *Collins' Clear School Atlas* (47 maps and an index), *c.*1920, 1921; *The Indian Popular Atlas of Modern Geography*, *c.*1900, *c.*1905; *Collins' Indian Clear School Atlas* (43 maps), London & Glasgow *c.*1920; *Collins' Graphic Atlas*, *c.*1930, 1940, 1952; and many others.

Collins & Clark. Publishers of Philadelphia. *Map of the City of Bridgeport, Conn.*, 1850; *Plan of the City of New London...*, 1850; *Map of New Haven, Connecticut*, 1851.

Collinson, *Lieutenant* (later *Admiral*) *Sir* Richard *RN* (1811-1883). Hydrographer. In 1850, in command of the *Enterprise*, joined the voyage to find the lost expedition of Captain John Franklin to the Bering Straits. Survey China, 1840-1845; *The Ordnance Map of Hong Kong* (4 sheets), 1846; track of HMS Enterprise, 1850-1854 MS; Chart of the Polar Sea, 1854; *Journal of HMS Enterprise*, 1889.

Collinson, Richard. *Plan of Eccles and District*, 1887.

Collinson, William S. ...*Passage from Tanjong Bon... to the Straits of Durian*, 1823; *Survey of the Straits of Durian*, 1823; *Straits of Singapore*, 1835.

Collis, William. Lithographer of 19 Smith St., Collingwood, Victoria, Australia (1859). *Victoria, census districts*, *c.*1858; *County of Bourke, Victoria*, 1865, 1866; *County of Hampden*, 1867.

Collomb, Edouard. *Carte géologique des environs de Paris*, 1865; *Carte géologique de l'Espagne et du Portugal*, 1868.

Collombat, Jacques. Publisher of Paris, 'rue St. Jacques au Pelican...' *Carte Générale Du Monde*, Paris 1695; *Nouvelle Carte Générale Du Monde*, Paris 1695 (a slightly larger map, engraved by Berey). **Ref.** SHIRLEY, R. *The mapping of the World* (1983) N°s. 565 and 566, pp.563 and 565.

Collon, —. *Plan communal de la ville de Liège*, 1820.

Collon, Jacques. *Carte topographique de l'Île de Corse*, 1827.

Colloredo, Fabius Antonio, *marquis* (1672-1742). Contributed to the compilation and engraving of *Parvus Atlas Ungariae*, 1689 (with Gábor Hevenesi).

Collot [Collet], Georges [George] Henri [Henry] Victor (1752-1805). Military cartographer. Plan topographique du Détroit, *c.*1796; Voyage dans l'Amérique septentrionale, 1826; Map of the Country of the Illinois, 1826.

Collyer, J. *A New System of Geography*, London, 1765 (with D. Fenning).

Collyer, Joseph [*the younger*] (1748-1827). Engraver for John Taylor's Ireland, 1793; for William Faden, 1797.

Colmenas, Juan Alvarez de. Délices d'Espagne et de Portugal, Leiden 1707.

Colnett, *Captain* James, *RN*. Chart of California, *c.*1791 MS; survey of the Canadian coast to Alaska, 1791; Japan, 1791; *Plan of the Anchoring Place at the Island Quibo*, 1794; Voyage to the South Atlantic, 1798; *Plan of the Island Cocos*, Aaron Arrowsmith 1798; *Chart of the Galapagos*, 1798; *Plan of the Islands Felix and Ambrose*, 1798.

Colom, family.

• **Colom**, Jacob Aertsz., *the elder* (1600-1673). Printer, publisher, mapmaker, globe-maker, and bookseller of Amsterdam. In competition with Willem Blaeu. Founded Navigation School in Amsterdam. *De Vyerighe Colom*, 1632-1662; French edition

L'Ardante ou flamboyante colomne de la mer, 1633-1662; English edition *Fierie Sea Colomne*, 1633-1655; *Upright Fierie Colomne*, 1648 and later;*Atlas of Werelts-Water-Deel...* (23 maps), 1663 (re-used some maps from Vyerighe Colom); *Atlas Maritimo*, 1669; and many similar titles and editions. **Ref.** KOEMAN, C. *Atlantes Neerlandici* Vol.IV (1970) pp.143-152.

•• **Colom**, Arnold [Aernout; Arent] (1624-1668). Son of Jacob Aertsz. Chart publisher, engraver and bookseller, 'Op de Texelsche Kay inde Lichtende Colom', Amsterdam. *Ora maritima orbis universi*, 1654, re-issued as *Zee-Atlas ofte Water-Wereldt*, 1658; *Lighting Colom of the Midland-Sea*, 1660, 1661, J & C Lootsman edition 1678; *Europa* (single chart), *c.*1665. **Ref.** KOEMAN, C. *Atlantes Neerlandici* Vol.IV (1967) pp.116-118.

Colomb, Louis de (1823-1902). French explorer; member of a scientific expedition to the Sahara 1854; undertook a comparative study of routes across the desert. *Exploration des ksours et du Sahara de la province d'Oran*, 1858. **Ref.** BROC, N. *Dictionnaire illustré des explorateurs Français du XIXe siècle: I Afrique* (1988) p.85 [in French].

Colombo, Barthélemi [Bartolomeo]. *Les provinces de Veronese, du Vicentin,...*, 1736.

Colombo family. See **Columbus** family.

Colon family. See **Columbus** family.

Colquhoun, Archibald Ross. Indo China, South China, 1882; *South Western China*, 1889.

Colson, John. Mathematician, teacher of navigation. Joint publisher with John Seller and others of *Atlas Maritimus*, 1675 and later editions.

Colston, Raleigh Edward (1825-1896). Joined General Charles Pomeroy Stone's American mission to Egypt. Surveys of the Nubian desert, 1873-1874; surveys of Sudan, 1874-1876. **Ref.** ICENOGLE, David *The Map Collector* 46 (1989) pp.27-28.

Coltellini, Marco. Publisher in Leghorn, N.W. Italy. *Gazettiere Americano*, 1763 (maps engraved by V. Vanni and G. Pazzi).

Coltman, Nathaniel (*fl.*1752-*c.*1807). Surveyor, employed by Post Office; publisher in Greenwalk, Blackfriars Road. South Wales, 1797; *Laurie and Whittle's Welsh Atlas...*, 1805; *Kent* (2 maps), Laurie & Whittle 1807; *A new map of Hampshire including the Isle of Wight...*, Laurie & Whittle 1809; Coach and Crossroads of England and Wales, 1809; Inland Navigation, 1808; Plan of London, 1831; York, 1833; *A New map of the County of Kent*, 1856.

Colton firm. Publishing and mapmaking family and firm of New York, founded 1831. The Coltons' atlases were later re-published by Ivison *et al*, and by Sheldon & Company. **Ref.** RISTOW, W.W. in: WOLTER, J.A. and GRIM, R.E. *Images of the World* (Washington 1997) pp.323-326; HECKROTTE, W. *The Map Collector* 62 (1993) pp.10-11; RISTOW, Walter W. *American Maps and Mapmakers* (1985) ch.19.

• **Colton**, Joseph Hutchins (1800-1893). Publisher and founder of the Colton firm in 1831, 9 Wall Street, New York (1833); 86 Cedar Street, New York (1849, 1854); 174 William Street (1854); and 172 William Street (1854-1865). David Burr's *Topographical Map of the City and Country of New-York and the Adjacent Country*, New York, J.H. Colton & Company 1836; *The Environs of London*, 1836; J.C. Smith's *Guide through Ohio, Michigan, Indiana, Illinois, Missouri, Wisconsin & Iowa...*, 1839 (engraved by S. Stiles, Sherman & Smith); South America, 1845; *Map of California, Oregon, Texas and the territories adjoining...*, 1849; *Colton's Railroad & Township Map Western States...*, New York 1853 (printed by D. McLellan); *General map showing the countries explored & surveyed by the United States & Mexican boundary commission*, 1854; *Map to illustrate Capt. Bonneville's Adventures among the Rocky Mountains*, 1854; *Connecticut with portions of New York & Rhode Island*, 1854, 1855, 1857, 1859; *Colton's Atlas of the World* (2 volumes), 1855; *Colton's General Atlas*, 1857 and editions to 1888 (compiled by G. Woolworth Colton); *Colton's Illustrated Cabinet Atlas*, 1859; *J.H. Colton's School Atlas...*, 1860; *J.H. Colton's American School Quarto Geography with an atlas...drawn by G. Woolworth Colton*, 1860 (1865 edition published by Ivison, Phinney, Blakeman & Co.); *Colton's Octavo Atlas of the World*, 1862; *New Illustrated (Steel Plate) Family*

Atlas, New York, Johnson & Browning 1861 and later editions (with A.J. Johnson); *Colton's new topographical map of the states of Virginia, Maryland and Delaware...*, 1862; *Colton's Condensed Octavo Atlas of the Union...*, 1864; *Quarto Atlas of the World*, 1865; *Colton's Octavo Atlas of the World*, 1865; and other school and general atlases.

•• **Colton**, George Woolworth [& Co] (1827-1901). Publisher, geographer, mapmaker and engraver of 172 William Street, New York. Son of Joseph, whose business he took over (together with brother Charles) *c.*1865. Worked also with George W. Fitch. Missouri, 1851; *Colton's Atlas of America*, 1855; worked with G.W. Fitch on maps for *Colton & Fitch's Introductory School Geography*, New York, J.H. Colton 1856, and other editions including, Ivison & Phinney 1859, Ivison, Phinney & Co. 1863, Sheldon & Company 1876; worked with G.W. Fitch on maps for *Colton & Fitch's Modern School Geography*, New York, J.H. Colton & Company 1856 and other editions published by Ivison & Phinney 1857, 1859, Ivison, Phinney & Co. 1860, 1863, Ivison, Phinney, Blakeman & Co. 1867, 1868, 1871 etc. (1867 and later editions were revised by Charles Carroll Morgan); *Atlas of the World*, New York 1857 (with R.S. Fisher); Series of railway maps, 1858; *G. Woolworth Colton's map of the country 500 miles around the city of Washington*, 1861; *Colton's New Brunswick, Nova Scotia, Newfoundland and Prince Edward Island*, 1871; *G. Woolworth Colton's County & Township new Railroad Map of the Ohio, Indiana & Michigan*, New York, G.W. & C.B. Colton 1865 (published simultaneously in Chicago by Andrews & Bigelow); maps for *Colton & Fitch's Primer of Geography*, 1866; *Map Showing the Route of the Central Pacific Railroad*, 1867; *Colton's General Atlas*, 1869, 1871, 1874, 1879; *Colton's Wyoming, Colorado and Utah*, 1876; Central America, 1889; Complete Ward Atlas of New York City, *c.*1892; and others.

•• **Colton**, Charles B. (1832-1916). Son of Joseph, took over control of the company with brother George *c.*1865.

••• **Colton**, G.W. & C.B. & Co. Publishers of 172 William Street, New York (1868). *Colton's General Atlas*, 1868 edition; *Colton's Map of Lake Superior...*, 1868; *Map showing the route of the Michigan Airline Railroad...*, 1869; *Colton's Guide Map of Michigan*, Rufus Blanchard 1870; *Map of the Land Grant of the Grand Rapids and Indiana Railroad Company*, 1870; Louisiana, 1870; *Complete Ward Atlas of New York City*, 1896. **N.B.** See also **Johnson** & Browning (who describe themselves as 'successors to J.H. Colton & Company'); **Ivison, Phinney** & Co.; **Ivison, Phinney, Blakeman** & Co.; **Sheldon** & Co.; and **Thayer** & Colton.

Columbine, *Lieutenant* (later *Captain*) Edward H., *RN* (*c.*1760-1811). Hydrographer and artist. *The Road and Harbour of St. John in the Island of Antigua*, 1789, published Faden 1793; *A Survey of the Island of Trinidad*, 1803, published 1816; *Chart of the Virgin Islands*, 1808.

Columbine, G.A. The Mole of Naples, *c.*1790.

Columbus [Colombo; Colon], family.

• **Columbus**, Christoforo [Christopher] (1451-1506). Navigator from Genoa, settled in Lisbon with his brother Bartolomeo. Best known for his transatlantic voyages and discovery of the West Indies and coast of Central and South America. Possibly produced a chart of the Atlantic with the coasts of Europe and Africa, *c.*1492 (now in Bibliothèque Nationale, Paris); also possiblya sketch map of northern coast of Hispaniola, December 1492 or January 1493. **Ref.** BURDEN, P. *The mapping of North America* (1996) pp.xiv-xvi; *The Map Collector* 58 (1992) pp. 2-5, 20-21, 24-28; HENIGE, D. *In search of Columbus: the sources for the first voyage* (1991); DREYER-EIMBCKE, O. *Kolumbus. Entdeckungen und irrtumer in der Deutschen cartographie* (Frankfurt am Main 1991) [in German]; KELLEY J.E. *Imago Mundi* 42 (1990) pp.26-49; RANDLES, W.G.L. *Ibid.* pp.50-64; REVELLI, P. *Cristoforo Colombo e la Scuola Cartografica Genovese* (Genoa 1937) [in Italian].

•• **Columbus**, Fernan[do] [Hernando] (1488-1539). Son of Christopher, priest of Cordobá. Known to have produced notes of a survey of Spain; possibly associated with an anonymous map of the World, printed in Venice, 1527. **Ref.** SHIRLEY, R. *The mapping of the World* (1983) N°58 p.64.

• **Columbus**, Bartolomeo [Bartholomew] (1460-1514). Cartographer, brother of Christopher. African coasts, 1506 MS (attributed).

Colvill, H. Engraver. Windsor Castle, 1847.

Christopher Columbus (1451-1506). (By courtesy of the National Maritime Museum, Greenwich)

Colvin, Verplank. *Adirondack Survey*, 1872.

Colyer, Walter. *Edwards County, Illinois*, Albion 1907 (with Ben Mayne).

Colyns, Alexander. ...*Oenipontis Prospectus*... [Plan Innsbruck], 1598 (used by Braun and Hogenberg).

Combatti, Bernardo. Publisher. *Mappa Generale delle Lagune di Venezia*, 1815, *c.*1850.

Combe, Querenet de la. *Carte des Environs de York en Virginie... la position des armées*, 1781.

Combe, T. *Leicestershire and the surrounding country*, 1834.

Comberford, father and son. Chart or plattmakers of the 'Thames School'. **Ref.** SMITH, T.R. in: THROWER, N.J.W. (Ed.) *The compleat plattmaker* (1978) pp.45-100; CAMPBELL, T. in: WALLIS & TYACKE (Eds.) *My head is a map* (1973) pp.81-106.

• **Comberford**, Nicholas (*c.*1600-1673). Apprenticed to the London Drapers Company school of portolan chart makers. Until 1628 worked for his former master John Daniel at Irongate. Later moved to Ratcliff [East London], 'Dwelling at the signe of the Platt neare the west end of the School House in Ratcliffe'. Untitled portolan of the Mediterranean and Black Seas, 1626; completed about 27 manuscript charts at Ratcliffe between 1641 and 1670, including untitled portolan charts of the British Isles and North France, 1641; Biscay to Channel, 1641, 1666, 1668; North Atlantic, 1646, 1650, 1657, 1659; Mediterranean, 1647, 1657, 1663, 1664, 1666; *The south part of Virginia, now the north part of Carolina*, 1657; English Channel, North Europe, 1665; and many more, all manuscript. **Ref.** SMITH, T.R. *Imago Mundi* 24 (1970) p.95.

•• **Comberford**, Thomas (*fl.*1656-1657). Son of Nicholas and freeman (by patrimony) of the Drapers Company. Although listed in the company records as a plattmaker with three apprentices, no work is yet known on which his name appears.

Comberousse, J. See **Decomberousse**, Joséphine.

Combette, A. Publisher of Paris, 'Rue de Parcheminerie 15'. Victor Levasseur's *Atlas National...de la France*, 1845 and later editions.

Combis D'Augustine, *Lieutenant* M. de. French hydrographer. *Plan de la Cremaillière*, and *Plan du havre de St. Antoine*, 1786 published in *Neptune de l'Amerique Septentrionale*, volume VII of *Neptune François*, Paris, Dépôt Général de la Marine 1800-1824.

Combitis, Nicolaus de (15th century). Venetian artist. Mediterranean Sea.

Combletee, D. Rivière et forts de Choueguen, 1756.

Comenius [Comenio, Komensky], Johann [Jan] Amos (1592-1670). Czech theologian, philosopher, educationalist, writer and geographer. During peripatetic early life lived in Hungary, Poland, England, Prussia. Best known work *Orbis pictus*, Sárospatak 1685 (published in German, Hungarian and Slovak). **Ref.** SPERLING, W. in: KÖHL, P.H. & MEURER, P.H. (Eds.) *Florilegium Cartographicum* (Leipzig Bad Neustadt a.d.S. 1993) pp.145-150 [in German].

Cometti, G.F. See **Camocio**, Giovanni Francesco.

Cominotti, C. *Pianta della Città di Cagliari*, 1832 (with E. Marchesi); *Pianta della Città di Sassari*, 1832.

Commelin [Commelijn], Isaac [Isak] (1598-1676). Compiled a work describing voyages made for the Dutch East India Company. *Begin ende voortgangh*... (2 volumes), Amsterdam, Jan Jansz. 1644, 1645, 1646, French edition by René Augustin de Renneville, 1702-1706, 1725, English edition, London, W. Freeman 1703; Histoire de la vie & actes mémorables de Frédéric Henri de Nassau, Amsterdam 1656; Beschrijvinghe van Amsterdam, 1693. **Ref.** LANDWEHR, J. *VOC A bibliography of publications relating to the Dutch East India Company 1602-1800* (Utrecht 1991) *passim*..

Comminelli, Ugo. Worked with Pietro del Massaio on Ptolemaic codices, *c.*1450s.

Comotti, Giovanni Francesco. See **Camocio**, Giovanni Franceso.

Compton, Thomas (*fl.*1801-*c.*1810). Surveyor, produced various manuscript maps of parts of England and Ireland, including *The Island of Lundy*, 1804.

Comrie, A. Hull & Selby Railway, 1834.

Comstock, *Captain* Cyrus Ballou (1831-1910). US Army Corps of Engineers. New York cartographer. Siege of Vicksburg, 1863; Mouth of the Detroit River, 1874.

Comstock, Theodore B. Geologist in Texas. Maps of Mineral Region, 1880.

Comyn, Tomas de. Islas Filipinas, 1821.

Conaway, —. See **Conway**, Michael.

Concepción, Juan de la. Yslas Philipinas, 1744-[1788].

Concezione, *father* Elisio della. Map of Calabria (9 sheets), 1783.

Concha Miera, Francisco de la. Plan Oviedo, 1777.

Conclin, George. New River Guide, Cincinnati 1849, 1850, 1855.

Condamine, Charles Marie de la (1701-1774). Mathematician, geographer and astronomer of Paris, undertook voyage to Peru, 1743-1744. Observations in the region used as a definitive source for many 18th-century maps of Peru including: *Carte du Cours du Maragnon ou de la Grande Rivière des Amazones*, 1743-1744 published Paris, Prévost 1756; for Robert de Vaugondy's maps of Peru in *Atlas Portatif*, 1748; for D'Anville's *Carte de la Province de Quito au Perou*, Prévost 1756; and others.

Conde, Pedro Garcia. Nouvelle Espagne, 1807 (from Miguel Costansó's surveys of 1769-1770); Plano de Mexico, 1811; *Carta geográfica general de la Republica Mexicana*, 1845, 1848.

Conder, *Lieutenant* (later *Major*) Claude Regnier (1848-1910). Archaeologist and cartographer. *Map of Western Palestine* (26 sheets), 1872-1877, published Southampton, Ordnance Survey 1880 (with Horatio Henry Kitchener), reduced edition 1881; *Palestine*, 1890. **Ref.** HOPKINS, I.W.J. *Imago Mundi* 22 (1968) pp.30-32.

Conder, Thomas. Engraver and cartographer. Maritime survey of Ireland, 1775; Moore's Voyages, 1778; *York Town and Gloucester as besieged by the allied army* [Virginia], 1781; James Stobie's Perth and Clackmannan, 1783; engraved maps for George Augustus Walpoole's *The New British Traveller*, London, A. Hogg 1784 (later republished as *The New Complete British Traveller*, 1794); *A plan of the town and township of Liverpool...*, 1785 (from a survey by Chas Eyres); Virginia, 1788; Elihu Barker's Kentucky, 1794; map of the English canals and navigable rivers, 1795; engraved *A Plan of the Rapids of the Ohio*, (from the work of T. Hutchins) for the third English edition of Imlay's *A Topographical Description of Western Territory...*, 1797; Wells's Geography Old and New Testament, 1801; *Map of Wales Divided into counties...*, 1801 (based on William Owen map of 1788).

Conders ab Helpen. See **Coenders** van Helpen.

Condet [Kondet]. Family of engravers, the relationships between them are unknown.
- **Condet**, Claas (1685-1725). Worked for J.B.B. d'Anville.
- **Condet**, Johannes (1711-1781). *Mappemonde à l'usage du roi par Guillaume Delisle... nouv. corrigée*, J. & C. Mortier *c.*1757.
- **Condet**, Harmen[us] (1725-1757).
- **Condet**, Gerard[us] (1731-1764). Worked for J.B.B. d'Anville, Cornelis Mortier, Henry Popple and Guillaume Delisle.

Condie, Thomas. Bookseller of Philadelphia. Plan of the Jail at Philadelphia, 1797-1798.

Condina [Condima], Sebastiano. Chartmaker of Messina. Manuscript chart of the Mediterranean, *c.*1615.

Condit, —. *Marian County, Indiana*, 1855.

Cone, J. Engraver for Fielding Lucas, 1823; for Samuel Cumings, 1829.

Coney, W. Intelligible panoramic map of the seat of war [North Italy], 1859.

Coneys, W.T. *Fond du Lac County, Wisconsin*, 1862.

Conincx [Coninx], Arnold [Aernout]. Printer of Antwerp. French edition of Peter Apian's *Cosmographicus liber*, 1584; Cornelis de Jode's *Speculum*, 1593; Abraham Ortelius's *Epitome*, 1595.

Conklin, E.W. *Map of the Village of Jamaica*, 1868.

Conklin, George W. *Handy Manual... & Atlas of the World*, Chicago, Laird & Lee 1888.

Connor, J. *City and suburbs of Cork*, 1774.

Conrad, Frederik Willem (1769-1808). Surveyor of Delft. Rhynstroom, 1793; Land van Heusden en Altena, 1798 (with J. Engelman).

Conrad, George. *Der Fränckische Kräyss neu delinirt*, 1638-1641.

Conrad, J.F.W. *Kaart van het Eiland Zuid-Beveland*, 1858; *Waterstaatskaart van Nederland*, 1865.

Conrad, John [& Co.]. Publisher of Philadelphia. Aaron Arrowsmith and Samuel Lewis' *A New and Elegant General Atlas*, 1804.

Conrad, Timothy Anthony (1803-1877). Alabama, 1835 (said to be the first geological map of Alabama).

Conradi, F.L. *A Plan of Fort St. George* [Madras], 1755 MS.

Conradi, Karl Theodor. *Accurate Geographische Carte über... Bretten*, 1779.

Consag, Fernandó. See **Koncság**, Ferdinando.

Consitt and Goodwill. *A New plan of the Town of Kingston upon Hull*, 1817.

Consoni, *Lieutenant* Joseph. Plan der Haupt- und Residenzstadt München, 1806.

Constable, A. [& Co.]. *Hand Atlas of India*, 1873-1893.

Constable, Archibald [& Co.]. Publishers of Edinburgh and London. For William Mackenzie, 1811; for Aaron Arrowsmith, 1817; Map of Edinburgh District, 1822.

Constable, *Lieutenant* (later *Commander*) C.G., *RN*. Admiralty surveyor. Charts of the Persian Gulf surveyed, 1857-1860.

Constanlino, —. *Plano... da Cidade de Lisboa*, 1812.

Constans, —. *Carte de la France...*, 1873.

Constantin de Ramière. See **Ramière**.

Constantinus of Antioch. See **Cosmas** Indicopleustes.

Contarini [Contareni], Giovanni Matteo (1452-1507). Italian cartographer. Untitled world map, 1506 (engraved on copper by Francesco Rosselli) thought to be the oldest known printed map to show America, it was rediscovered in 1922, and is now in the British Library. **Ref.** BURDEN, P. *The mapping of North America* (1996) pp.xii-xxiii; SHIRLEY, R. *The mapping of the World* (1983) N°24 pp.23-24.

Contarini [Contareni], Giovanni Pietro (1549-1603). Painter and cartographer of Venice. Candia, 1564; Europe (16 sheets), Venice 1564.

Conte, —. *Carte du Département de la Savoie*, 1866.

Conte Hoctomanno. See **Freducci**, Conte di Ottomano.

Cöntgen [Coentgen], father and son.
• **Cöntgen**, Henrich Hugo (1727-1792). Engraver in Mainz to the Court and the University. Military maps and plans for Laurids de Therbu (after 1763); Samuel Holland's *The Provinces of New York and New Jersey*, Heinrich Ludwig Broenner 1777.
•• **Cöntgen**, Georg Joseph (1752-1799). Engraver, son of Henrich Hugo, above; moved to Frankfurt am Main, 1777. *Les plans de la Guerre de Sept Ans...* (42 plans), 1789.

Conti, Filippo. Engraver of three maps (of Tuscany, Europe, and the State of the Church) for Bartolomeo Borghi's *Atlante geografico*, Siena, Pazzini Carli, 1788-1800.

Conti, Niccoló dei (*c*.1395-1469). Venetian merchant and traveller of Chioggia. His information was used by Fra Mauro.

Conty, Henri A. de. *Carte pratique et officielle des Excursions de Normandie*, *c.*1867.

Converse, M.S. City plan of Elmira, New York, 1876.

Conway [Conaway], *Captain* Michael, *RN*. Deputy Master of Trinity House. Manuscript charts of the English coast including: Girdler Bank to Whitstable [approaches to the Thames], *c.*1780.

Conway, *Sir* William Martin. *The Karakoram Himalayas*, 1894.

Conynenberg [Conijnenberg], Jacob. Nephew of Jacob Jacobsz., assisted Caspar Jacobsz. in the continuation of the family business after the death of Jacob Jacobsz., his name appearing on the imprint of *Nieuwe Water-Wereldt Ofte Zee-Atlas*, 1705 edition. **N.B.** See also **Jacobsz.** family.

Conzatti, Jo.-Baptiste [Giovanni Battista]. Printer and publisher of Padua. Associated with Lasor A Varea's [Savonarola] *Universus Terrarum Orbis...* (2 volumes), 1713.

Coode, *Sir* John (1816-1892). Harbour engineer. *Douglas Bay* [Isle of Man], 1873; *Melbourne Harbour Trust*, 1879; *Victorian Harbours*-Belfast, 1879; *Lakes Entrance*, 1879; *Warrnambool*, 1880; *Portland*, 1887 etc. **Ref.** (Obit.) *Proceedings of the Institution of Civil Engineers* CXII (1892-1893).

Cook, Alexander. *A New Royal Authentic and Complete System of Universal Geography* (2 volumes), London, 1787-1810 (with Thomas Bankes & Edward W. Blake).

Cook, Ezra A. Lithographer. *Ashland, Saunders County, Nebraska*, 1870; *Kansas*, 1870.

Cook, Franklin. Drew and published *Map of Minneapolis, Minnesota*, 1876.

Cook, Frederick. Fire insurance maps of the city of Chicago, 1865.

Cook, George H. (1818-1889). American geologist.

Cook, J. Engraver. *Plan of the River Thames*, 1847.

Cook, *Captain* James, *RN* [James Cook N°.1] (1728-1779). Explorer, circumnavigator and hydrographer. Began the surveying part of his career with brief training from Samuel Holland in 1758, going on to survey the Bay of Gaspé in the same year. Cook went on to command several notable voyages. First Voyage, 1768-1771; Second Voyage, 1772-1775; Third Voyage, 1776-1779. ...*Draught of the Bay and Harbour of Gaspee in the Gulf of St.Laurence taken in 1758*...(2 sheets), London, Mount & Page 1759; Newfoundland surveys, 1758-1765, published in atlas form as *The Newfoundland Pilot*, T. Jefferys 1769; *A New Chart of the River St. Laurence* (12 sheets), 1760. **Ref.** DAVID, A. *The Map Collector* 61 (1992) p.56; DAVID, A. (ed.) *The Charts and Coastal Views of Captain Cook's Voyages* Vol.1 (Hakluyt Society, London 1988) Vol.2 (1992) Vol.3 (1998); DAVID, A. *The Map Collector* 52 (1990) pp.2-6; SPENCER, J.R.H. 'A New Zealand Draught Chart by James Cook, RN' in *Cartography* Vol.13 N°4 (1984) pp.316-322; SKELTON, R.A. & TOOLEY, R.V. 'The marine surveys of James Cook in North America 1758-1768' in: TOOLEY, R.V. (Ed.) *The mapping of America* (1980) pp.173-206; BARTON, P.L. *The Map Collector* 11 (1980) pp.29-33; BEDDIE, M.K. (ed.) *Bibliography of Captain James Cook* second edition (Sydney 1970).

Cook, *Lieutenant* James *RN* [James Cook N°.2]. *A Plan of the Road and Harbour of Fowey or Foy*, London, Sayer & Bennett 1779. **Ref.** BLACK & SKELTON *The Map Collector* 34 (1986) p.10.

Cook, James [James Cook Nos.3 & 4] (*fl.*1762-1775). Master of the surveying ships *Mars* 1762-1763, and *Alarm* 1763-1766, later provincial surveyor in South Carolina. *A Draught of Spirito Sancto...* 1765; *A Draught of West Florida...*, 1765; *Directions for Three Charts...* 1766; *A Draught of the Harbour of Halifax...*, 1766; *A Draught of Port Royal Harbour in South Carolina*, 1766; *A Map of South Carolina*, 1771; *An Exact Map of the Boundary Line between the Provinces of North & South Carolina...*, 1772 MS; *A Plan of the Province Line from the Cherokee Line to Salisbury Road...*, 1772 MS; *South & North Carolina...*, 1772; *A Map of the Province of South Carolina*, 1773. **Ref.** REA, Robert *The Map Collector* 51 (1990) pp.34-37; BLACK & SKELTON *The Map Collector* 34 (1986) pp.10-15; REA, R. 'Master James Cook and Gulf Coast cartography' in *Florida Historical Quarterly* 63 (1985).

Cook, James B. *Memphis, Tennessee and Its Environs*, 1860.

Cook, Robert Jewell [& Co.]. Lithographers of 29 Charing Cross, Westminster. From *c.*1872 published as Robert J. Cook & Hammond (see below). England and Wales, 1866; *London Railway Travelling Made Easy...*, 1870 and editions to 1879; Railway maps of the United States, 1872.

Signature of James Cook RN (1728-1779). (By courtesy of the Public Record Office, Kew)

• **Cook**, Robert J. & Hammond. Lithographers of Broadway, Westminster (1872, 1892); 2 & 3 Tothill Street, Westminster (1903). Founded by R.J. Cook, above, agents for the Ordnance Survey. *Hertfordshire*, 1888; South America, 1890; two maps of London and its water companies in Cripp's *The Position of the London Water Companies*, 1892.

Cook, Thomas (1808-1892). English tourist agent. *Circular Tours*, 1873; *Cook's Literary and Historical Map of London*, 1899.

Cooke, father and son.
• **Cooke**, John (1731-1810). Father of Charles Cooke, below. Engraver, draughtsman, and publisher of London; Clare Court, Drury Lane (1790); 50 Howland Street (1799); 11 Platt Place (1805). Engraver for David Steel, 1790; Bryan Edwards's *History of the British West Indies*, 1794; *Plan of the City of Philadelphia*, 1796; Thomas Smith's *Universal Atlas*, John Harris 1802, 1804; Aaron Arrowsmith, 1807.
•• **Cooke**, Charles (1750-1816). Publisher of N°·17 Paternoster Row, London, son of, and successor to, John Cooke, above. Published George Alexander Cooke's *The Modern British Traveller* (47 parts, 25 volumes), 1802-1810; *Topography of Great Britain*, 1810.

Cooke, —. *Royal Map of Dublin &c.*, 1831.

Cooke, *Captain* (later *Major*) Anthony Charles, *RE*. Director Topographical Department of the War Office, New Zealand. *Plan of Sevastopol*, 1857; Peking, 1859; *Plan of the neighbourhood of Verona*, 1859; Waikato River, New Zealand, 1864.

Cooke, Charles. See **Cooke**, father and son (above).

Cooke, D.B. [& Company]. *D.B. Cooke & Co's Great Western Railway Guide...* [North America], 1856 (printed by Middleton Wallace & Co.).

Cooke, *Captain* Edward. Navigator. *A Map of the World* and *Map of the River Amazon*, in *A Voyage to the South Sea, and round the World, perform'd in the years 1708, 1709, 1710 and 1711...* (2 volumes), London, B. Lintot and R. Gosling 1712.

Cooke, Edward. Lithographer of Bedford Street, Bedford Row, London. W.P. Spalding's *Cambridge*, 1875 and later.

Cooke, Francis. Principles of Geometry, Astronomy and Geography, 1591.

Cooke, G. One of the engravers of the decorative cartouche for Aaron Arrowsmith's *Map of America*, 1804 (with W. Cooke).

Cooke, George Alexander (*fl.*1802-1816). No relation to Charles Cooke, above, his publisher. *Universal Geography*, 1802; *The modern British Traveller, or tourist's pocket directory...* [also known as *Cooke's Topography of Great Britain*] (47 parts, 25 volumes published in two versions, one with coloured maps), Charles Cooke *c.*1806, 1810, Sherwood, Neely & Jones 1822, 1824, 1830 (maps used in Gray's *New Book of Roads*, Sherwood, Jones & Co. 1824).

Cooke, Henry Thomas [& Son]. Publishers of Warwick. *Cooke's Map of Warwickshire...*, 1861 (first published as Harris's New Map of Warwickshire, 1842); map of Warwickshire in *An historical and descriptive guide to Warwick Castle, Kenilworth Castle..., and other places of interest...*, 1870; *The County of Warwick*, 1875 1876, 1877 (from a plate by John Bartholomew, 1872).

Cooke, John. See **Cooke**, father and son (above).

A picture of James Cook which was engraved for Hervey's Naval History, *published March, 1780.*

Cooke, John. Engraver of Hendon. *Isle of St. Domingo or Hispaniola...*, London, William Faden 1796 (taken from the survey by Don Joseph Solano, 1776). **N.B.** Compare with **Cooke**, father & son, (above).

Cooke, John. Engraver and geographer of Plymouth. *Borough of Plymouth*, 1820; *Plymouth Breakwater*, 1823; *Chart of Plymouth Sound*, 1824; *The Environs of Plymouth Devonport and Stonehouse*, 1830; *Plan... of Plymouth, Devonport and Stonehouse*, 1834.

Cooke, Oliver D. [& Co.; & Sons]. Publishers of Hartford, Connecticut. William C. Woodbridge's *School Atlas to accompany Woodbridge's Rudiments of Geography*, 1821, 1823; William C. Woodbridge's *Modern Atlas on a New Plan*, 1823; Emma Hart Willard's *Atlas*, *c.*1826; *Woodbridge's school atlas*, 1831, 1833, 1835.

Cooke, *Lieutenant-Colonel* Philip St. George (1809-1895). Santa Fe to Pacific, 1847 MS.

Cooke, S.R. Engraver. John Bumpus's *New Plan of London and Its Environs*, 1827, 1828, 1829, 1830.

Cooke, S.W. Engraver. Laurie & Whittle's *New Plan of London, Westminster and Southwark*, 1801, and editions to 1815.

Cooke, Thomas. *Map of the City & Liberty of Westminster*, 1847.

Cooke, W. One of the engravers of the decorative cartouche for Aaron Arrowsmith's *Map of America*, 1804 (with G. Cooke).

Cooke, William D. Relief globe for the blind, 1853; *Map of the Routes to the Virginia Springs*, 1858; *New Map of the State of North Carolina*, *c.*1860.

Cookingham, E.R. Sanilac County, Michigan, 1894.

Cookson, Arthur Augustus. Surveyor. *Chart of the Islands Westward of Socotra* [South Yemen], 1850.

Cookson, Cecil H. *Sketch Map of the Province of British Columbia*, 1899.

Cool, [Colins; Colius; Cools; Ortelianus], Jacob [Jacques; Jacobus] (1563-1628). Publisher and antiquary of Antwerp and London, nephew of Abraham Ortelius. Prepared the *Tabula Peutingeriana* for publication by his uncle, 1598.

Cooley, William Desborough (1795-1883). Edited Thomas Maynarde's manuscript account: *Sir Francis Drake his Voyage, 1595*, London, Hakluyt Society 1849; *Africa from the Equator to the Southern Tropic*, 1853; *Map of part of Africa south of the Equator...*, *c.*1864.

Cooling, *Lieutenant* William John, *RN*. Admiralty Surveyor. Mediterranean, 1830; *The Gulf of Arta*, 1834.

Cools, M. Engraver of Antwerp. Worked for Josua van den Ende; maps for Blaeu's atlases.

Cooper, Adam. Mapmaker of Preston, Lancashire. Plan of Liverpool Castle, 1704.

Cooper, *Captain* Allen. ... *the Track of Ship Atlas thro' the islands to the east of Banka*, 1785 published A. Dalrymple 1789.

Cooper, H. Engraver of 28 Chancery Lane, London. Abraham Rees's *Cyclopaedia*, 1806-1810; maps for Benjamin Pitts Capper's *A Topographical Dictionary of the United Kingdom...*, 1808 and later editions; for James Playfair, 1809; for William Miller's Red Sea, 1809; for George Atkinson's Ceylon, 1813; *Map of London and its Environs...*, 1813; for J. Philips, 1823.

Cooper, J. Publisher of 36 Fetter Lane, Fleet Street, London. *British Isles*, 1835 and later; *England, Wales and the Southern Part of Scotland*, 1849.

Cooper, J.F. ...*Western & Atlantic Rail Road of Georgia*, U.S. Topographical Bureau 1837.

Cooper, *Mrs* Mary (*fl.*1740-1761). Took over as bookseller and publisher at the Globe in Paternoster Row after the death of her husband Thomas (not listed). *A Plan of the Town of Dunkirk*, 1743; *A Survey of Westminster Bridge*, 1748; T. Kitchin and T. Jefferys's *The Small English Atlas*, 1749; John Cowley's *New Set of Pocket Maps of all the Counties of England*, 1745; *A Map of*

Nicolaus Copernicus (1473-1543). (Taken from Map Collectors' Circle*)*

London, Westminster and Southwark with ye New Buildings to ye Year 1758, 1758.

Cooper, Richard (father & son). Engravers of Edinburgh.

• **Cooper**, Richard, *the elder* (*d.*1764). Engraver and publisher of London and Edinburgh. Engraved John Adair's maps of the three Lothian counties, 1730-1745; for Alexander Bruce, *A Plan of Loch Sunart*, 1733; for John Jaffray, 1739; William Moir, 1739; William Edgar's *A New and Correct Map of the Shire of Peebles or Tweeddale*, 1741; *A map of the King's Roads...*, 1742; Alexander Bryce's *A map of the north coast of Britain...*, 1744; Plan of Leith, 1759; and others.

•• **Cooper**, Richard, *the younger* (*c.*1740-1814). Son of Richard Cooper the elder. Painter and engraver of Edinburgh.

Cooper, Richard (*fl.*1737-1764). Surveyor and engraver of Edinburgh. *A Map of the Forest of Mamlorne*, 1737. **N.B.** Compare with Richard **Cooper**, father & son (above).

Cooper, Robert (*fl.*1830-1846). Surveyor,

engraver and publisher of York. Various manuscript maps of Yorkshire; maps for R. Dawson, 1832; *A Map of the County...of York*, 1832.

Cooper, T. Publisher of Edward Wells's *A New Sett of Maps*, 1700.

Cooper, T.W. Assistant Surveyor. Plans of Land Grants in Victoria, 1857-1859.

Cooper, Thomas. Map of Middle States of America, in *Some Information Respecting America*, 1794.

Coordes, G. (1839-1890). German geographer. *Klimatologische Schulwandkarte von Europa* [school wall chart], Berlin, Chun 1888.

Coote, John. Bookseller of 16 Paternoster Row, London. *New Geographical Dictionary* (text, plates and 44 maps), 1759-1760; sold P. Russell & Owen Price's *England Displayed*, 1769 (with several others).

Copeland, *Captain* Richard. Admiralty Surveyor, Commander of HMS *Mastiff*. Worked on the British Naval Aegean Survey from 1825-1836 (covering Cape Malea to the Dardanelles), producing many charts. He was succeeded by Thomas Graves aboard HMS *Beacon*. **Ref.** FISHER, Susanna *The Map Collector* 54 (1991) pp.18-23.

Copernicus [Kopernicus], Nicolaus (1473-1543). Astronomer and mathematician of Cracow. His work *De Revolutionibus*, 1530, published 1543, proved that the planets revolve around the sun. Maps of Prussia, 1529; Lithuania, now lost.

Copland, Robert. Translator of the Rutter of the Sea, 1550, 1555.

Copley, Charles (*fl.*1846-1877). Engraver and publisher of Brooklyn, worked with the U.S. Coast Survey. Charts of the coasts of America, including West Indies, Gulf of Mexico, and the North Atlantic. Charts of *Tierra del Fuego*, 1849; pair of 16-inch globes, 1852 (later sold by Franklin Globes and Gilman Joslin up to 1890s); *Coast of the United States*, 1855, 1857. **Ref.** WARNER, D. *Rittenhouse* Vol.2 N°2 (1987) pp.54-55.

Copmayer. See **Koppmayer**, Jakob.

Coppenius, Aegidus. See **Diesth**, Aegidus.

Coppens, J.B. *Plan de la Bataille de Waterloo*, *c.*1820.

Coppens, P. Atlas de la Belgique, 1852.

Coppo [Coppus; Copus; Kopić], Pietro [Petar; Petrus] (1469/1470-1565/1566). Venetian cartographer, lived and worked mostly in Izola, Istria, where he died. *De toto orbe...* (22 manuscript maps), 1520, he also prepared a summary of it entitled *De summa totius orbis*, c.1520 for which he prepared 15 woodcut maps, 1524-1526 (mostly based on the earlier manuscript maps); *Portolano*, Venice, Augustino di Bindoni 1528; *Istria*, 1528 (later copied by Ferdinando Bertelli , 1569, Giovanni Francesco Camocio, 1569, and Abraham Ortelius, 1573, and others. **Ref.** KARROW, R. *Mapmakers of the sixteenth century and their maps* (1993) pp.119-124; MEURER, P. *Fontes cartographici Orteliani...* (1991) p.130 [in German]; LAGO, L. & ROSSIT, C. *Pietro Coppo. Le 'Tabulae' 1524-1526* (Trieste 1986); SHIRLEY, R. *The mapping of the World* (1983) N°52 p.57; SHIRLEY, R. *Early printed maps of the British Isles 1477-1650* (1980) N°21 p.13; DEGRASSI, A. 'Di Pietro Coppo e delle sue opere' in DEGRASSI, A. *Scritti vari di antichità* (1962); ALMAGIÀ, R. *Imago Mundi* VII (1950).

Coquand, H. *Carte géologique du Département de la Charente*, 1859.

Coquart, Antoine (1668-1707). Draughtsman and engraver. Camps Ortnaw, 1690; for Nicolas de Fer, 1696; engraved the first five of the six historical plans of Paris for N. Delamare's *Traité de la Police*, 1705; plans of Turin, 1705; Gibraltar, 1706; Anvers, 1708; Barcelona, 1711; for Blottière, 1730; Roussel's *Carte Generale des Monts Pyrenées*, 1730 (with J.B. Delahaye); *Plan de Luxembourg* for Bernard Jaillot, 1741.

Coquus, Henrique. See **Cock**, Hendrik.

Cora, Guido (*b.*1851). Geographer of Turin. *Carta speciale della Baia d'Assab*, 1883; *Carta della Colonia Eritrea*, 1891; *Carta oltimetrica...dell'Italia*, 1891.

Coran, T. South Carolina, Charleston 1802.

Corbara, Carlo (1823-1901). Mathematician and topographical engineer in the Officio Topografico di Napoli. Worked on several sheets of the Officio's map of the Kingdom of Naples, 1851-1860; contributed to the topographical survey of the southern provinces of Sicily, 1866-1870 including: *Pianta della città di Catania*, 1867. **Ref.** VALERIO, V. *Società uomini e istituzioni cartografiche nel mezzogiorno d'Italia* (1993) pp.481-483.

Corbelletti, Francesco. Publisher of Rome. Famiano Strada's *De Bello Belgico*, 1632 and 1647; Rome, 1658; João José de Santa Theresa's *Istoria delle guerre del regno del Brasilia*, 1698.

Corbet, C. Plan of Culloden, 1746.

Corbet, W. *New Plan of the City of Dublin*, 1803.

Corbett, Daniel. Engraver. County Cork, 1750.

Corbett, John A. Surveyor in Australia. *Country Lands on the Reedy Creek*, 1856; *Agricultural Lots at Wangaratta*, 1857; *Agricultural Lots in the Parish of Oxley, Victoria*, 1857.

Corbett, V.P. Publisher of Washington. *Seat of War showing battles of July 18th, 21st and Oct. 21st. 1861*, 1861; *Sketch of the Seat of War in Alexandria & Fairfax Counties* [Virginia], 1861; *Map of the seat of war... 1861* [Washington], 1861 (printed by A. Hoen & Co.); and others.

Corbigny, Charles-Paul. See **Brossard de Corbigny**, Charles-Paul.

Corbridge, James (*fl.*1720-1737). Surveyor. *Newcastle upon Tyne*, 1723; *...This Mapp of the City of Norwich...*, 1727; *An Actual Survey of the County of Norfolk*, 1730 and 1755 (2 sheets); *Suffolk*, 1765.

Corby, Robert (*fl.*1797-1845). Surveyor, of Norfolk, produced various manuscript maps of parts of East Anglia and Hertfordshire. Surveys of inclosure awards, Suffolk, 1803-1817.

Cordeiro, Luciano (1844-1900). Portuguese geographer. Founder of Societa de Geografia.

Cordier, Henri (1849-1925). Orientalist, geographer and scholar from New Orleans, studied in England and France and was sent to Shanghai by his American employers in 1869. Became a professor of the history and geography of the Far East in 1881. *Atlas Sino-Coréen*, 1896; also extensive bibliographies for the region, notably *Bibliotheca sinica* (6 vols.), 1904-1922; *Bibliotheca japonica* 1912; *Bibliotheca indosinica*, 1912-1913.

Cordier, L. Engraver. Chorographica Regni Poloniae, 1729.

Cordier, Louis [Ludovicus] (*d.*1711). Engraver for Nicolas Sanson d'Abbeville the elder, 1661-1675; maps for A.H. Jaillot, *c.*1690-1707, published in *Atlas françois*, from 1696, and later by Dezauche in *Atlas Géographique et Universel*, 1789.

Cordier, Louis Etienne Joseph (1775-1849). French military engineer, responsible for many canal and route improvements in northern France. 'Inspecteur divisionnaire des ponts et chaussées', 1827-1849. Published several works on canals, railways and routeways in France, England and America including an atlas, *c.*1829.

Cordier, Nicolas. Two hydrographers of this name are mentioned in reference, probably father and son. Both are credited with maps of the coast of France from Bordeaux to Calais.

• **Cordier**, Nicolas (1660-1728). French geographer and hydrographer at Havre. Published various works on navigation and shipbuilding. Map of the coast of France from Bordeaux to Calais, [n.d.].

•• **Cordier**, Nicolas (1682-1766). French geographer, teacher of hydrography at Dieppe. Map of the coast of France from Bordeaux to Calais, [n.d.].

Cordier, Robert (*d.c.*1673). French engraver. Engraved several maps for Sanson d'Abbeville 1648-1667 (published in *Cartes Generales de la Geographie Ancienne et Nouvelle...*, Paris 1676); engraved for Hubert Jaillot, 1669; for Pierre Du Val, 1670.

Cordier, Robert. Engraver. Cercle d'Autriche, 1782.

Cordova, Jacob de. *Map of the State of Texas*, 1849, published as *Colton's New Map of the State of Texas*, 1865.

Coreal, Francisco. *Voyages de François Coreal aux Indes Occidentales... depuis 1666 jusqu'en 1697* (3 volumes), Amsterdam, J. Frederic Bernard 1722.

Corenwinder, N. *Plan de la Ville et port de Dunkerque*, 1841.

Corfield, William. Surveyor, produced estate and inclosure maps in Dorset, Shropshire and Wiltshire including *Plan of the Manor of Staunton St. Bernard* [Wiltshire], 1784.

Corne, Louis (*fl.*1730s). Engraver for Jacques Ayrouard *Recueil de plusieurs plans des ports... Mediterranée*, 1732-1746.

Corneille, Jean-Baptiste (1649-1695). Painter and engraver of Paris. Reputedly worked on the figures of the constellations on the Coronelli Louis XIV celestial globe.

Cornelis, Lambert (1546-1601). Engraver. Zaltbommel for Jan Huyghen van Linschoten, 1596; for Mercator-Hondius Atlas, 1606.

Corneliszoon, Cornelis (1493-1544). Painter and engraver of Leiden. Friesland, 1531; Zuydt Holland, 1537.

Cornelius, Flaminius. *Creta Insula*, 1755.

Corneliussen, O.A. Norwegian geologist. *Geologisk Kart over det Nordlige Norge*, 1879.

Cornell, Sarah Sophia. *The World in Hemispheres*, 1855; *Companion Atlas to Cornell's High School Geography*, 1859, later editions to 1879; *Physical Geography*, 1870; *Primary Geography*, *c.*1871.

Cornell, Silas (1789-1864). Map of the City of Rochester, New York 1838 and later; *District School Globe* (5-inch), before 1848; Cornell's *Improved Terrestrial Globe* (5-inch and 9-inch globes on patented mount), before 1848.

Cornell, Thomas. Caricature Map of England, Ireland & France, 1785.

Cornell, Thomas C. *Map of...the Town of Yonkers*, [N.Y.], 1892.

Cörner, Elek [Pater Alexius] (1714-1747). Slovakian geographer, graduated at Rome, taught cartography in Nyitra [now Nitra, Slovakia] and Pest. Important work Compendium ... (textbook on surveying), published Nyitra 1747.

Cornet, Charles Joseph. French explorer on Ouadaï, Kanem and around Lake Chad. Produced maps of the areas he covered, particularly Ouadaï, *c.*1906-1907. **Ref.** BROC, N. *Dictionnaire illustré des explorateurs Français du XIXe siècle: I Afrique* (1988) p.89-90 [in French].

Cornet, F.L. (*d.*1887). Belgian geologist.

Cornet, J. *Geognostische Übersicht des... Kongo-Gebietes*, 1894.

Cornetus, Petrus. Chartmaker of Rotterdam. Portolan, 1618.

Cornish-Bowden, Athelstan Hall. Surveyor-General for the Cape of Good Hope, 1903.

Cornput, Jan van den. See **Corputius**, Johannes.

Cornuau, Pierre. French cartographer, worked with others on several sheets of the Cassini *Carte de France*, most of them in Alsace and Lorraine (sheets 110, 141, 142, 161, 162, 164, 165, 175), 1757-1764; later around the Limousin (sheets 14 & 34), 1767-1780; worked for Louis Capitaine du Chesnoy, 1781.

Cornwall, *Captain* Henry. Aden, Alexander Dalrymple 1703; *Voyages to India*, 1720.

Cornwall, James. School atlases, 1852-1862.

Corny, *Carte des environs d'Alger*, 1886; *Plan de la ville d'Alger*, 1888.

Coromina i Faralt, Josep [José] (1756-1834). Catalonian engraver. Map of Catalonia, *c.*1800 (one sheet only preserved); *Plan de los canales proyectados de riego y navegacion de Urgel*, 1816.

Coronel, Hernando de los Rios. Mapa de Luzon e Hermosa, 1597.

Coronelli, *Father* Vincenzo (1650-1718). Venetian map and globe maker. Minorite friar at San Nicolò della Lattuga, Venice

Vincenzo Coronelli (1650-1718). (By courtesy of the British Library, London)

(1665-1671), later at Santa Maria Gloriosa dei Frari (1671-1674); Doctor of Theology at Collegium Santo Bonaventura, Rome 1673; founder of the Accademia Cosmografica degli Argonauti, 1684 at Venice; became Cosmographer to the Republic of Venice 1685. From 1681-1683 he worked in Paris on a pair of 3.85 metre diameter globes for Louis XIV; set up a globe-making workshop in the convent in Venice, 1686. Various globes, 1683-1716; *Description géographique et historique de la Morée...*, 1686; *Memorie Istoriogeografiche Della Morea...*, 1687; *Conquiste della Serenissima Republica di Venezia...*, *c.*1688; *Isola di Rodi geografica-storica, antica e moderna...*, Venice 1688; pair of 110-cm printed globes, 1688 (the celestial globe was engraved by J.B. Nolin in Paris); *Mediterraneo*, 1688; *Atlante veneto* (65 charts, maps, and plates), 1690-1696; *Arme delli Regni di Spagna*, 1691; *Corso geografico universale...*, 1692-1694; *Epitome cosmografica...*, 1693; *Libro dei globi*, 1695; *Historia del Regno di Negroponte...*, 1695; *Isolario dell' Atlante Veneto* (88 charts, maps & views), 1696-1697; *Teatro delle città e porti pricipali dell'Europa in Pianta, in profilo ed in elevatione* (3 volumes), 1697; *Libro dei Globi* (with globe gores), 1697; *Specchio del mare*, 1698; *La Francia divisa in XII prefetture...*, Venice *c.*1706; *Belgio confederato*, Venice 1706; *Lombardia*, 1706; *Isole britanniche*, 1707; *Europa vivente...*, 1715; and many others. **Ref.** DEKKER, E. and KROGT, P. van der *Globes from the western World* (1993) pp.59-67; BONASERA, F. *I globi di V. Coronelli, sec. 17.-18. conservati nella Biblioteca di Jesi...* (Jesi, La Truganina 1987) [in Italian]; RHODES, D.E. *Imago Mundi* 39 (1986) pp.77-78; BONASERA, F. *L'opera geografica e cartografica di Vincenzo Coronelli* (Jesi, La Truganina 1984) [in Italian]; DILKE, O.A.W. & M. *The Map Collector* 29 (1984) pp.10-12; PELLETIER, M. *Imago Mundi* 34 (1982) pp.72-89; WALLIS, H. *The Map Collector* 13 (1980) pp.22-26 (globes); WALLIS, H. introduction to facsimile ed. of *Libro dei globi* (Amsterdam 1969) (contains long biographical essay with numerous references cited).

Corputius [Corput Broedanas], Johannes van (1542-1611). Student of Mercator, from whom he learnt mathematics and engraving, later becoming a military engineer. ...*Topographia Duisburgi*, 1566, (later used by Braun and Hogenberg, *q.v.*); Steenwyck, 1581. **Ref.** POSTEMA, J. *Johannes van den Corput 1542-1611: kaartmaker, vestingbouwer, krijgsman.* (Kampen 1993) [in Dutch].

Correa, Manuel. *An exact and true description of the coasts, Ports...on the Coast of Luzon...*, published A. Dalrymple 1759.

Corréard, Joseph (1792-1870). *Théâtre de la guerre en Orient*, 1853; Atlas de la Mer Noire, 1854.

Corry, *Captain* Armar L., *RN*. Observations used by Arrowsmith on map of Syria, 1823.

Corsali, Andreas. Italian navigator. Polo antartico (celestial chart), 1516, 1550, 1553.

Corsulensis, Vicentius. See **Paletino**, Vicenzo.

Cortambert, family. French geographers.

• **Cortambert**, Pierre François Eugène (1805-1881). French geographer and teacher of Toulouse. *Géographie Universelle...*, 1826; *Elements de géographie ancienne...*, 1834; *Tableau de la géographie universelle*, 1835; *Leçons de géographie* (with atlas of 37 maps), 1839; *Océanie...*, Paris 1857; Cochinchine française, 1860; Atlas général de la France, 1868; Géographie élémentaire, 1874; revised and augmented Malte Brun's *Géographie universelle*, 1875; and many more works.

•• **Cortambert**, Richard (1836-1884). Geographer attached to the geography section of the Bibliothèque Nationale. *Géographie de l'Italie*, 1859; *Géographie commerciale*, 1868.

Cortazar, Daniel de. *Mapa geologico y topográfico de la provincia de Almeria*, 1873-1874.

Corte, Carlo da. Chartmaker of Genoa. Chart, 1592.

Corte, Ludovico. *Disegno di Pavia*, 1654.

Cortelyou, Jacques. Surveyor general of the New Netherlands. Said to have produced manuscript maps of the New Amsterdam area, 1657, 1658, 1660, 1661, and possibly 1664. **Ref.** COHEN, P.E. & AUGUSTYN, R.T. *Manhattan in maps 1527-1995* (1997) pp.38-43

Corte Real [Cortereal], brothers. Portuguese landowners from the island of Terceira, Azores. **Ref.** MORISON, S.E. *The great explorers* (Oxford University Press 1978) pp.79-83.

• **Corte Real**, Vasco Annes. Eldest of the brothers, was refused permission by the king to go in search of Miguel and Gaspar.

• **Corte Real**, Miguel (*d.*1502). Sailed (in a separate ship) with his younger brother Gaspar's second expedition to Greenland and Newfoundland, in 1501, from which Gaspar's ship never returned. Was himself lost when he set sail for Newfoundland in 1502 on a fruitless search for his lost brother.

• **Corte Real** [Cortereal], Gaspar (*b.c.*1450-*d.*1501). Discovered 'Terra Verde', 1500, but disappeared during a second voyage from Greenland to Newfoundland and beyond in 1501. His findings first appeared on a manuscript portolan associated with Alberto Cantino, 1502.

Cortés, Hernándo (1485-1547). Spanish conquistador from Estremadura, conqueror of Mexico. *La preclara narratione*, Venice 1524, published as *Praeclara Ferdinandi Cortessi...*, Nuremberg, Friedrich Peypus 1524 (contains an untitled woodcut map of Mexico City and Gulf of Mexico marking Florida for the first time on a printed map).

Cortes, José. *Mapa... de la America Sept...*, 1799 MS.

Cortés, Phelipe Feringan. *...Isla de Santa Rosa*, 1775 MS; Pensacola, 1755 MS.

Cortés, Tomás. Plan of Manila, 1823.

Cortés de Albacar, Martin. Spanish cosmographer, geographer and mathematician from Aragon, described the magnetic variation of the compass. *Breve compendio de la Sphera y de la arte de naveguar*, Cadiz 1551, English translation *The art of navigation*, 1561 and editions to 1630.

Cortese, Emilio. *Carta geologica dell' intorni di Parghelia*, 1891; *Carta geologica della Calabria*, 1894-1895.

Cortier, Maurice (1879-1914). French explorer in Africa, including Sénégal, Tchad, the Sahara and Asia Minor. Killed on the French front during the First World War. Maps of his travels published in various issues of *La Géographie*, 1906-1914. **Ref.** BROC, N. *Dictionnaire illustré des explorateurs Français du XIXe siècle: l Afrique* (1988) p.90 [in French].

Cortiro, Nicola dal. Venetian cartographer. Friuli (24 sheets), 1535 MS.

Cortona, Giovanni Antonio di (*d. c.*1560). Painter of Udine. *Udine*, *c.*1540 MS.

Corvinus, F. *Grundriss der Stadt Posen*, 1856.

Corvinus, Johann August (1682-1738). Engraver in Augsburg. S. Kleiner's view of Mainz, Augsburg 1720.

Corzula, Vincenzo. See **Paletino**, Vicenzo.

Cosa, Juan de la (*d.*1510). Spanish navigator, born at Santoña, Cantabria, died at Cartagena de Indias, Colombia. Pilot to Columbus, worked at Santa Maria near Cadiz from 1496 as chartmaker, took part in Alonso de Ojeda's expedition to West Indies 1496, undertook his own voyages to the Caribbean 1504, 1507. Manuscript chart of the New World, 1500 (believed to be the earliest map to show part of North America, including English voyages to Labrador. Now in Museo Naval, Madrid). **Ref.** KELLEY, J.E. *Imago Mundi* 42 (1990) pp.26-49.

Cosel, E. von. *Topographische Karte der Provinz Brandenburg*, 1861.

Cosentino, Nicola (1814-1881). Topographical draughtsman, artist, and map colourist, worked under Bruno Colao. Coloured V. Di Carlo's map of Caserta, 1857; part of the city of Naples, 1858; 2 sheets of a map of the provinces of Napoli, 1870. **Ref.** VALERIO, V. *Società uomini e istituzioni cartografiche nel mezzogiorno d'Italia* (1993) pp.483-484.

Cosin, Samuel (*fl.*1633-1639). Estate surveyor of Essex.

Cosmas Indicopleustes [Constantinus of Antioch] (*fl.*540-550). Merchant of Alexandria, 'sailor to India'. Advocated the theory of a flat earth in his *Christian topography*, based on biblical references and illustrated with maps and diagrams.

Cosmo de Churruca. See **Churruca**, Cosmo de.

Cossin, Jehan. French marine cartographer. *Carte géografique universelle*, Dieppe 1570.

Cossins, John (*c.*1697-1743). Surveyor. *A new map of ye City of York*, *c.*1720s MS; *New plann of the Town of Leeds*, *c.*1720s

MS; manuscript plan of Beverley, *c.*1727-1729; *New map of the Town of Whitby*, *c.*1728-1729 MS; *A new and exact plan of the City of York*, 1748.

Costa, Agostino (*b.*1754). Engraver of Tuscany. The two hemispheres and other maps in B. Borghi's *Atlante Geografico*, Siena, Pazzini Carli 1788-1800; Sicily, 1789; Lombardy, 1794; maps (including some re-used from the Carli Atlas) in B. Borghi's *Atlante Generale*, Florence, 1816-1819.

Costa, Antonio Carvalho da. See **Carvalho da Costa**, Antonio Nuñes de.

Costa, Bento da. Chief pilot and cartographer to P. Teixeira's Amazon expedition, 1638-1639.

Costa, Euzebio da. Geographer. Planisphere, Lisbon 1720 MS.

Costa, Giovanni Francesco (*d.*1773). Venetian engineer and engraver.

Costa, J. da. See **De Costa**, J.

Costa Azevedo, José da. Hydrographer. *Trabalhos Hydrographicos* [charts of the Amazon], 1863-1868.

Costa e Miranda, José da. See **Miranda**, José da Costa.

Costansó [Costanzo], Miguel (*fl.*1769-1811). Spanish chartmaker, accompanied Gaspar de Portolá to California, 1769-1770. Manuscript plan of the port of San Blas [Mexico], 1768; chart of the California coast, 1770 MS; map of the Spanish expedition to Monterey, 1770 MS; *Carta Reducida del Oceano Asiático Ó Mar del Súr...*, Madrid 1771; *Carta ó mapa geográfico de una gran parte del reino de Neuva España*, 1777 (maps published in *An Historical Journal of the Expeditions, by Sea and Land, to the North of California in 1768, 1769, and 1770*, A. Dalrymple 1790.

Costaz, *baron* Louis (1767-1842). French mathematician and geographer.

Coste, Pascal (1787-1879). French engineer and architect of Marseilles. Worked also on irrigation canals in Egypt. Carried out archaeological surveys in Persia with Eugène Flandin, 1840-1841. Basse Egypte, 1818-1827, published 1830 (with Eugène Flandin); *Voyage en Perse* 1851-1854.

Costello, Thomas H. Inventor of Chicago. Registered patents on a variety of globe mounts. Established Weber Costello Co. (incorporating C.F. Weber & Co.), 1907, becoming major globe producers in the USA.

Costo, Gieronimo (*fl.* late 16th century). Genoese chartmaker working in Barcelona.

Costo, Giovanni. Mapmaker of Genoa. Portolan world, 1602 MS.

Cotes, J. *fl.c.*1791-1819). Surveyor and schoolmaster of Derbyshire. The Surveyor's Guide, 1806.

Cotes [Coates], Thomas. Printer of *Historia Mundi*, Sparke and Cartwright 1635 (W. Saltonstall's English translation of Mercator's Atlas).

Cotilla, Juan. *Plano de la Isla de San Carlos... Misisipi*, 1769 MS.

Cotsford, Edward. *Itchapour District*, 1778.

Cotta, *Verlag*. Publishers of Stuttgart. J.G.F. Bohnenberger's *Charte von Wirtemburg*, Tübingen, 1797; C. Eckhardt's *Fortsetzung der topographischen Karte von Schwaben* (4 sheets), 1820-1823 (second part); A. von Coulon's *Karte von Deutschland* (27 sheets), 1823.

Cotta, Bernard von (1804-1879). German geologist. *Kohlen-Karte...im Koenigreich Sachsen*, 1830-1846 (with K.F. Naumann); *Geognostische Karte von Thüringen*, 1836; Umgegend von Dresden, 1868.

Cotta, Johann Georg. Founded J.G. Cotta'sche Buchhandlung. *Das Osmanische Reich in Europa*, 1828; *Die Bayerische Pfalz*, 1867.

Cottam, Arthur. *Charts of the Constellations*, 1889.

Cottart, P. Engraver for Claude de Chastillon, *Topographie Françoise*, 1641.

Cotte, Jean van de. Belgian geographer. Carte topographique de la Palestine, 1847.

Cotterell, H.F. *A New Map of the Country round Bath and Bristol*, *c.*1840.

Cotterell, J.H. *Plan of the City and Borough of Bath*, 1852.

Cotterell, Thomas Sturge. *Historic map of Bath*, 1898.

Cottes, Augustin (1871-1913). French surveyor working in Africa and Indochina, particularly Tonkin and Annam regions. Member of the Service géographique in Vietnam from 1903, also worked on establishing and mapping the frontier between the French Congo and Cameroun. Provided survey information for Auguste Pavies's map *Carte de l'Indochine* (4 sheets), 1899; survey of the Mekong river (28 sheets), *c.*1903; other work published in: *Annales de géographie*, 1905. **Ref.** BROC, N. *Dictionnaire illustré des explorateurs Français du XIXe siècle: I Afrique* (1988) pp.93-94; *ibid.* Vol.II (Asie) pp.109-110 [in French].

Cotton, George H. *Plat of St. John, Harrison County, Iowa*, 1857.

Cotton, *Sir* Robert Bruce (1571-1631). Antiquary, collector of books, maps and manuscripts. His collection was transferred to the nation in 1700, and is now in the British Library.

Couagne, Jean-Baptiste de. See **Decouagne**, Jean-Baptiste.

Couceiro, L. *Africa oriental portugueza*, 1892.

Couch, P.J. *Grand National Exhibition: Couch's Stranger's Map of London. Drawn and lithographed by P.J. Couch*, London, W.A. Couch 1851.

Couch, William A. Publisher of 9 Tudor Street, Blackfriars, London. *Grand National Exhibition: Couch's Stranger's Map of London. Drawn and lithographed by P.J. Couch*, 1851 (with W. and T. Piper).

Couché, father and son. Engravers of Paris.
• **Couché**, Jacques (*b.*1750). Engraver of Paris, work may have included maps.

•• **Couché**, François-Louis, [Couché fils] (1782-1849). Prolific engraver of Paris, specialising in military subjects. Engraved maps for the Dépôt de la guerre; for Pierre Didot, 1806; for Aristide Michel Perrot, 1845.

Couchman, *Sir* Harold John (1882-1956). Surveyor in the Royal Engineers, joined Survey of India, 1906, served in France and Belgium, 1914-1919; Surveyor General of India from 1933. Surveys of south-west India, 1908-1909; Burma, 1910-1911; southern and central India, 1912-1914. **Ref.** HOWARTH, R.J. in: *Dictionary of national biography: Missing persons* (Oxford 1993) pp.152-153.

Couchman, Thomas. Revised T. Richardson, W. King and A. & W. Driver's Map of Hampshire (New Forest sheet, for 1849 edition), 1849; *Plan Pental Island* [New South Wales], 1870.

Coucke, J. (1783-1853). Painter and engraver. Town plans and views.

Coulier, Philippe Jean. Atlas général, Paris 1850.

Coullomb, François. Naval architect of Toulon. Designer of ship construction plates, 1683, later used by Laurent Brémond in *Ports et Rades de la Mer Mediterranée*, 1730 edition.

Coulon, —. *Plan de la Ville de Nantes*, 1795.

Coulon, A. von (1779-1855). *Post Karte von Baiern*, 1810; *Süd Deutschland*, 1820; *Karte von Deutschland* (27 sheets), Cotta 1823.

Coulon, Louis. *Le fidèle conducteur pour le voyage d'Angleterre*, Troyes 1654 (road maps).

Coulubrier, —. French engraver. Louis Brion de la Tour's *Carte de la Partie de La Virginie...*, *c.*1787.

Counradi, M. *Grundriss der Stadt Braunschweig*, *c.*1735.

Counsell, E.A. Surveyor-General for Tasmania. Hobart, 1897.

Couplet, Philippe, *SJ*. *Tabula chronologica monarchiae sinicae*, 1686.

Coupvent-Desbois, Aimé Auguste Elie (*fl.*1830s). Maps for Jules Sébastien César Dumont d'Urville, 1837-1840; *Plan du Canal Bowen...*, 1839.

Courbé [Courbet], Augustin (*fl.*1635-1649). Publisher 'dans la petite salle du Palais, à la Palme', Paris. Published for Pierre d'Avity, 1635; Isaac La Peyrère's *Relation du Groenland* (1 map), 1647.

Courbet, A. *Plan du Mouillage d'Alexandrette*, 1854-1856.

Couret de Villeneuve, Louis-Pierre (1749-1806). French printer, publisher, bookseller and 'Auteur de Journal du Département [Loiret], rue de Colombier, N°.74', Orleans. *Nouveau plan d'Orléans augmenté de ses Faubourgs...*, 1778 (engraved by Lattré); *Département du Loiret...*, 1790.

Court, *Captain* Charles, *RN*. Admiralty Surveyor. *Chart of the Red Sea*, 1804-1806; *A Survey of...Palmiras*, 1817; Moscos Islands, 1826; *Survey of Mootapilly Bay*, 1830.

Court, Stephen, *Second Master, RN*. Chart of Land discovered by HMS *Investigator*, 1851 MS.

Courtalon, *l'abbé* J.C. *Atlas élémentaire de l'Empire d'Allemagne*, Paris 1774.

Courtier, Henri. Drayning of the Fens, 1629.

Courtier, Henry. *The Revised Map of Canaan*, 1881; *Biblical Atlas*, 1890.

Courtier, L. *Plan de Reims*, 1899.

Courtin, Pierre. Nouveau traité de la sphère, Paris 1670.

Courtry, Alexandre. *Congo Français*, 1897-1898.

Cousin, David. Architect. Ground plan for improving meadows, Edinburgh 1864.

Cousin [Cusinus], Hugues [Hugo] (*fl.c.*1532-1566). Family originated at Nozeroy in Franche-Comté. In the service of Charles V he travelled to Hungary (1532), Provence, Tunis, Dalmatia (1539), and the Spanish Court (1542). Map of Burgundy, later published by Ortelius as *Burgundiae Comitatus*, 1589. **Ref.** MEURER, P. *Fontes cartographici Orteliani...* (1991) p.131 [in German].

Cousin, Paul. *Topographical Plan of the City of Quebec*, 1875.

Coussin [Cousin], Honoré. Engraver for Jacques Ayrouard, 1732-1746; *Carte... de Provence*, 1758; *Nouvelle Plan de la Ville d'Aix*, 1765.

Coustolle, *Lieutenant* -- (*fl.c.*1891-1901). French naval officer, hydrographer, surveyed the course of the Yangtze, from 1891 (it was completed by others in 1895).

Coutans, *Dom* Guillaume (*b.*1724). Mapmaker, 'bénédictin de la Congrégation de Saint-Maur'. *Atlas topographique des environs de Paris* (16 sheets), 1775, revised 1800 (Charles Picquet); *Description historique et topographique de la grande route de Paris à Reims* (22 maps), 1775.

Coutant, --. Engraver for Hyacinthe-Yves-Philippe Florentin de Bougainville, 1837.

Coutaut, E. *Carte routière et vinicole du Département de la Gironde*, 1877.

Coutgon, --. Engraver. Course of the Rhine, Neuwied to Mainz, *c.*1780.

Couto, Mateus do. Portuguese cartographer. Codex of Portugal (55 maps), 1693 (with João Roiz Mouro).

Couts, *Lieutenant* Cave Johnson. Colorado and Gila Rivers, California, 1849 MS.

Couture, Peter (*fl.* 1773-1778). Ordnance draughtsman. *A Plan of the Royal Citadel and Town of Plymouth*, 1776 MS.

Couzens, M.K. American civil engineer and draughtsman. *The Key to East Virginia...* New York, W. Schans 1861; *The Key of the South*, 1861.

Covarrubias, Francisco Diaz. *Carta hydrografica del Valle de Mexico*, 1862.

Covel, Thomas. Surveyor. *Ickworth Survey Boocke*, 1665.

Covens, family. Map publishers of Amsterdam, associated with the Mortier family.
• **Covens**, Johannes [Jean, Jan] (1697-1774). Brother-in-law of Cornelis Mortier, with whom he founded a company in 1721.
•• **Covens**, Johannes II (1722-1794).
••• **Covens**, Cornelis (1764-1825).
• **Covens & Mortier** [J. Covens & C. Mortier] (*fl.*1721-1774). Map printers and publishers of Amsterdam. Founded in 1721 by brothers-in-law Johannes Covens and Cornelis Mortier, based on the business of Cornelis's father Pieter Mortier *q.v.* (1661-1711). Business continued by heirs until 1866. Their prolific output over the years included wall maps of many countries and continents, a large number of town plans, pocket atlases, and re-issues of general atlases by Sanson, Jaillot, Visscher, Delisle and De Wit. Atlases included: F. de Wit's *Atlas Maior* (190 sheets in 2 volumes), 1725 and later; *Atlas Antiquus...* (89 maps), *c.*1725; Delisles's *Atlas Nouveau...*, 1730 and later; Van der Aa's *Nouvel Atlas*, *c.*1730; *Les forces de l'Europe, Asie, Afrique et Amerique*, 1734; *Atlas Portatif et Nouveau du Voyageur* [Netherlands], 1740; *Nieuwe Atlas* (3 volumes), *c.*1741 (with maps by many different cartographers); E.H. Fricx's *Atlas des Pays Bas*, 1745 (maps first published 1703-1712); *Atlas Nouvel Militaire*, *c.*1747; *Nouveau théatre de la guerre au Pays Bas...*, *c.*1747; and many others. **Ref.** KROGT, Peter van der (Ed.) *Stock catalogues of maps and atlases by Covens & Mortier* (facsimile with introduction) (Utrecht 1992); KOEMAN, C. *Atlantes Neerlandici* Vol.II (1969) pp.45-86; see also **Mortier**, family.
•• **Covens & Mortier & Covens Jr.** (*fl.*1774-1783).
••• **C. Mortier & J. Covens & Zn**, (*fl.*1783-*c.*1788).
•••• **J. Covens & Zoon** (*fl.c.*1788-1794).
••••• **Mortier Covens & Zoon** (*fl.*from 1794).

Coverdale, Miles (1488-1569). English ecclesiastic, translator of the Bible into English, 1535 (with woodcut map by an unknown maker entitled *The description of the londe of promes called Palestina / Canaan or the holy londe.*) **Ref.** INGRAM, E.M. *The Map Collector* 64 (1993) pp.26-31; DELANO SMITH, C. *The Map Collector* 39 (1987) pp.5-6.

Covert, *Sir* Walter [Waltar] (1543-1631). Survey of the coast of Sussex, 1587 (with Sir Thomas Palmer, charts by N. Reynolds), published as *A Survey of the Coast of Sussex in 1587*, Lewes, W.E. Baxter 1870 (engraved by Maclure, Macdonald and Macgregor).

Cowan, Frank. *South Western Pennsylvania*, 1874.

Cowan, John. Chart of the River Shannon, 1771.

Cowell's Anastatic Press. Printer of London, known for its use of the short-lived anastatic process. Peter Brutt's *Plan ...for Harwich Docks*, 1852; R.A.S. Adair's *Military Sketch-Map of England and Wales*, 1855. **Ref.** WALLIS & ROBINSON *Cartographical innovations* (1982) N°.7.012 p.288.

Cowie, George. Publisher of 31 Poultry, London (1813-1836). *Environs of London*, 1828; *Plan of the City of London*, 1828. **N.B.** Compare with **Cowie & Strange**.

Cowie & Strange. Publishers of 64 Paternoster Row, London. *Plan of the City of London divided into wards*, 1828 (published in Thomas Allen's *A new history of London, Westminster and the Borough of Southwark*); James Bingley's *Environs of London*, 1828.

Cowl, Richard (*d.*1789). Schoolmaster of Plymouth. *A Plan of the Town and Citadel of Plymouth* (2 sheets), 1778 published Faden 1780; *...Plan of the Borough of Plymouth*, 1779 (engraved by G. Terry). **Ref.** STUART, E. *Lost landscapes of Plymouth* (1991) pp.135-136, 139, 186.

Cowles, *Captain* Calvin D. *Atlas to accompany the official records of the Union and Confederate armies*, Washington, United States War Department 1862-1895.

Cowley, H. *Plan... Annapolis Royal*, 1752 MS.

Cowley, John. Engineer and 'Geographer to His Majesty'. *A Hydrographical Description or Chart of the Sea Coasts of Great Britain*, 1734; North Britain, 1734; Coasts of Florida and Carolina, Samuel Vaughan 1740; Mediterranean, 1744; collaborated with Robert Dodsley on the preparation of *The Geography of England*, published by Robert Dodsley 1744; *New Set of Pocket Maps of all the Counties of England*, London, R. Dodsley 1745.

Cowley, R. *Charles Town South Carolina*, London, Fielding & Walker 1780 (engraved by J. Lodge).

Cowley, *Captain* William Ambrose. Circumnavigator, prepared charts and kept a journal of world voyage 1683-1686, used by William Hack in *A collection of Original Voyages*, 1699 (includes a world map by Herman Moll showing Cowley's voyage); discovered and charted the Galapagos Islands, 1684, used by W. Hack, 1685 and later. **Ref.** WORAM, J.M. *The Map Collector* 44 (1988) pp.17-20.

Cowper, H. Swainson. *A Map of Parts of Gharian... Tripoli*, 1897.

Cowper, Thomas. Almanack for 1637... with principal highways in England and Wales, 1637.

Cowperthwait, [& Co.]. The Cowperthwait name is attached to a confusing group of Philadelphia publishing companies. They often published the same works at different dates, and sometimes concurrently. Succeeded by the Butler family of Philadelphia.
• **Cowperthwait**, J. [& Co.]. Publisher of Philadelphia. *Warren's Physical Geography*, 1845.
• **Cowperthwait**, Huling. Partner in Thomas, Cowperthwait & Co. Published H.N. Burrough's *A New Map of Texas Oregon and California...*, 1849; *A New Universal Atlas*, 1851.
•• **Cowperthwait & Co.** Publisher of Philadelphia. *Mitchell's Primary Geography*, 1849 edition; *The Common-School Geography*, 1865 edition; *A New Primary Geography*, 1868; *A Brief Course in Geography*, 1874; *Warren's Common School Geography*, 1877.
•• **Thomas, Cowperthwait & Co.**. Publishers of 253 Market Street, Philadelphia (1849). Some stock later passed to Charles Desilver. *A New Map of Texas, Oregon and California*, 1849; Samuel Augustus Mitchell's *A New Universal Atlas*, 1850, 1853; Mitchell's *New Atlas of North and South America*, 1851; School Atlas, 1856 & 1858.
••• **Cowperthwait, Desilver & Butler**. Republished works earlier published by Thomas, Cowperthwait & Co. *Mitchell's A New Universal Atlas*, edition of 1854; *Mitchell's School Atlas*, edition of 1854. **N.B.** See also **Desilver** and Butler.
•••• **Cowperthwait**, H. [& Co.]. *Mitchell's School Atlas*, editions of 1855, 1856, 1857, 1858 (first published Thomas, Cowperthwait & Co.); D. M. Warren's *The Common-School Geography*, 1857, 1858, 1864; *A Primary Geography*, 1858, 1867.

Cox, *Professor* E.T. *Map showing the Coalfields of South Western Indiana*, 1874.

Cox, F.G. *War map of Europe*, San Francisco, F.G. Cox 1914.

Cox, George. Publisher in London, 18 King Street, Covent Garden. Issued an Atlas for the Society for Diffusion of Useful Knowledge, 1852-1856; *The Environs of London*, 1853.

Cox, Gustavo. *Mapa de los Distrito Minerales de San Antonio*, 1865.

Cox, *Commander* H.L. Admiralty Surveyor. England, 1853-1860; Salcombe River, 1863.

Cox, J. Engraver. *Crosby's Complete Pocket Gazetteer of England and Wales*, 1818.

Cox, John Henry. Supplied information from Spanish manuscript charts of San Francisco and the Inlet of Bucareli, Alaska to A. Dalrymple, who used them as a basis for charts published in 1789.

Cox, Leonard M. American civil engineer. *The Island of Guam*, Washington 1904.

Cox, Thomas (*c.*1660-1734). Bookseller and topographer. 'Corner of St. Swithin's Alley, Cornhill' (1724); 'at the Lamb, under the Royal Exchange, Cornhill' (1730), London. Seller of William Camden's *Britannia*, as *Magna Britannia et Hibernia*, 1724-1731. **Ref.** HODSON, D. *County atlases of the British Isles published after 1704* Vol.I (1984) pp.9-25, 31.

Cox, W.J. *Schuylkill County, Pennsylvania*, 1864 (with W. Scott).

Cox, Zachariah. Manuscript map of part of Mississippi Territory, *c.*1815.

Cox & Sharland. 'Booksellers, Stationers & Printers 150 High Street, Southampton'. *Map of the New Forest*, *c.*1870.

Coxe, Daniel, father & son.
• **Coxe**, *Doctor* Daniel (1640-1730). Physician to Charles II and Queen Anne; Governor of West Jersey, 1687-1692. *A Map of Carolana* [sic] *and of the River Meschacebe...*, published in Coxe's *A Description of the English Province of Carolana...*, 1722, 1726, 1741 (published by his son).
•• **Coxe**, Daniel II (1673-1739). Published his father's description and map of Carolina.

Coxe, *Reverend* William (1747-1828). Historian of Gloucester; translator of Stepan Petrovich Krasheninnikov's *Opisanie Zemli Kamtschatki* (St Petersburg 1766), as *Russian Discoveries between Asia and America*, 1780; Travels in Poland, 1784; *Map of Switzerland*, 1789.

Coxen, J.U. *City of Nagpoor*, 1865-1870.

Coypel [Coipel], Charles-Antoine (1694-1752). *L'Italie... par le Sr. d'Anville*, 1743 (engraved by Aveline).

Cozens, William H. *Map of the City of Pilot Knob, Iron County, Missouri*, 1858.

Cozuela, Vasquez de. *Royaumes de Portugal et d'Algarve*, 1762.

Cozzaglio, A. de. Carta geologica dei d'intorni di Esine, 1893.

Craalinge, Jan. See **Cralingen**, Jan.

Craan, W.B. Waterloo, 1816; Plan de la Ville de Bruxelles, 1836.

Craandijk, J. *Atlas van Wandelkaarten...door Nederland* (22 maps), Haarlem, H.D. Tjeenk Willink *c.*1878.

Crabb, Thomas. Publisher of '15 St John Street, Blackfriars Road', London. Set of geographical cards, *c.*1820.

Cracalescu, D.M. *Romania*, 1888, 1892.

Cracklow, Charles Thomas. *Plan of the proposed Docks...on the Surrey side of the River Thames*, 1796, 1803.

Cracroft, John. Greenland, Iceland... Nova Zembla, 1740.

Cradock, companies. **Ref.** SMITH, D.,'John Heywood and others' in: *IMCoS Journal* 69 (1997) pp.23-30; see also **Baldwin**.
• **Cradock & Joy** (*fl.c.*1809-1815). Publisher of 3 Ave Maria Lane (*c.*1809), 32 Paternoster Row (1810-1815). Later joined by Robert Baldwin to form Baldwin, Cradock & Joy (1816-*c.*1828), then Baldwin & Cradock (1829-1841). Publishers of John Lambert's Travels through Canada and United States, 1813 (with Map of the British Settlement and plan of Quebec).
•• **Cradock**, T.M. [& Co.]. (*fl.c.*1840-1842). Publishing partnership at 48 Paternoster Row, joined by Charles Cradock in 1840.
••• **Cradock**, Charles [& Co.]. (*fl.c.*1842-1857). Publishers, 48 Paternoster Row, London. F.P. Becker's *The Visitor's Map of London*, 1851.

Cradock, James. Plans of the Parish of Clarendon, Jamaica, 1759 MS.

Crafts, Charles P. Map of Douglas County, Illinois, 1904.

Crafts, N. Henry. *Plan of Boston*, 1864; *Plan of Boston and Roxbury*, 1867.

Cragie, —. Admiralty Surveyor, HMS *Pelorus*. Chart of Pelorus Sound, 1839 MS.

Craig, Francis S. Surveyor for W.P. Spalding's *Plan of Cambridge and its Environs*, 1875, 1881, 1885.

Craig, George A. Spencer Town, Massachusetts, 1884.

Craig, James (*d.*1795). Architect of Edinburgh. *Plan Bridge over North Loch* [Edinburgh], 1763; *Plan of the new streets and squares intended for the City of Edinburgh*, 1768, 1773 (engraved by P. Begbie).

Craig, John (1796-1854). Mineral surveyor of Lanarkshire, emigrated to the USA in 1850. Geological map of Lanarkshire, 1839.

Craig, John (*fl.*1804-1812). Surveyor. Plans of Edinburgh and Leith, 1804; Ross and Cromarty, 1826 (for John Thomson's *Atlas of Scotland*).

Crajenschot, Theodore. Publisher of Amsterdam. *Atlas der 17 Provintien in 19 kaartjes*, *c.*1780.

Cralingen [Craalinge], Jan. Publisher of Amsterdam. *Leo Belgicus*, 1636; plan of London, 1666.

Cram, D.H. *A Map of Part of Alabama and West Florida*, 1875.

Cram, George Franklin (1842-1928). Teacher, soldier, and founder of the Cram firm of engravers and publishers in Chicago. After service in the Union Army, Cram became a clerk in his uncle Rufus Blanchard's firm, and in 1867 became a partner.
• **Cram**, George F. [& Co.]. Established his own company, George F. Cram & Co., in 1869. Recorded at many addresses in Chicago including: 134 Lake Street (*c.*1867); 148 Lake Street (1870); Cram Map Depot, 55 W. Lake Street (*c.*1871); 66 Lake St., Chicago (1879); 415-417 Dearborn Street, Chicago (1889); proprietor of the Western Map Depot, 262-264 Wabash Avenue, Chicago (1879, 1880, 1883). The business was sold to the National Map Company in 1920. *Guide map of Chicago*, 1868; *New railroad and township map of Iowa*, 1868; *New railroad and township map of Missouri and Kansas*, 1869; *New sectional map of the state of Iowa*, 1869; *New sectional map of the state of Missouri*, 1869; *Cram's township and railroad map of the north-western states*, 1870; *New sectional map of the state of Illinois*, 1870; *Cram's Rail Road, County and Township Map, of Ohio, Michigan, Indiana, and Kentucky*, 1870 (engraved by G. Ismall); *Sectional map of Nebraska*, 1871; Wisconsin, 1872; Colorado, Indiana, Iowa, Kansas, Missouri, 1873; *Cram's New Sectional Map of Michigan*, 1874; *The New Commercial Atlas of the United States*, 1875; *Railroad map of Michigan*, 1878; *Cram's Railroad and Township Map of Utah*, 1879; *The Standard Atlas of the United States*, 1880; *Cram's Unrivaled Family Atlas of the World*, 1883 and many other editions; *The People's Illustrated and descriptive Family Atlas*, 1886; *The Imperial Office Directory and Reference Atlas*, 1889; *Cram's Big American Railway System Atlas of the World*, 1891; *Cram's Big War Atlas*, 1898; *Cram's bankers' & brokers' railroad atlas*, 1899; *Cram's Travelers' Atlas of the United States*, Chicago 1912; *The European war...the whole situation made plain*, 1914; *The ideal reference atlas of the World*, Chicago & New York, 1914, 1916; *Atlas of the war in Europe*, New York 1915; *Historical atlas of Europe*, 1916; *Cram's modern reference atlas*, 1917; *United States at war*, Chicago 1917; *Cram's globular world map*, Chicago & New York, G.F. Cram 1917; *Cram's commercial and library map of the United States...*, 1917; *History Atlas: Story of the Great War*, 1919; and many others. **Ref.** DANZER, G.A. 'George Cram and the American perception of space' in: CONZEN, M.P. (Ed.) *Chicago mapmakers* (1984) pp.32-45

Cram, *Captain* T.J. Captain in the U.S. Topographical Engineers. *Report on the survey of the boundary between the State of Michigan and the Territory of Wiskonsin* (with 5 maps), 1840 (with J.D. Webster, engraved by W.J. Stone), a later edition with a sixth map, 1842; *York River Harbor, Maine*, 1854.

Cramer, Charles. *Map exhibiting that portion of the State of Pennsylvania traversed by the surveys for a continuous rail road from Harrisburg to Pittsburg...*, Philadelphia, J.T. Bowen 1840; *Map of the rail road surveys between Hillsborough & Chillicothe* [Ohio], 1851.

Cramer, *Captain* John, *RN*. Captain of HMS *Rattlesnake*. Kersteman's *Saldanha Bay*, 1803.

Cramer, John Anthony. *Italiae antiquae*, 1825; *Graecia Antiqua et Nova*, 1827; *Asia, vulgo Minor dicta*, 1832; *Italy*, 1836.

Cramer, Louis H. Saratoga & Ballston, 1876 (with F.W. Beers).

Cramer, Zadok (1773-1813). Born in New Jersey, moved to Pittsburgh in 1800, where he established a printshop, bindery, library and bookshop. He was joined in partnership by John Spear in 1808, and William Eichman in 1811. He was succeeded by his widow and his partners. *Ohio Navigator* (no maps), Pittsburgh 1801, third edition published as *The Ohio and Mississippi Navigator* (no maps), 1802 republished as *The Navigator* (with charts), 1806 and later (believed to be the earliest navigation guide for inland waters of North America); *Louisiana and Mississippi Almanack*, 1811, 1813. **Ref.** RISTOW, W.W. *American maps and mapmakers* (1985) pp.236-237.

Cramm, Gustav. Plan of Hamburg, 1873-1876; Karte von Kephalonia, 1874?.

Cramoisy, Sébastien and Gabriel. Publisher of Paris. Atlases and travel literature including Nicolas Tassin's *Plans des villes de France*, 1634-1636; Philip Briet's *Parallela Geographiae*, 1648; François Du Creux's *Historiae Canadiensis*, including *Tabula Nova Franciae*, 1664.

Cramp, K.R. *Australasian school atlas...* (58 maps), Melbourne, H. Milford 1915 (with J.G. Bartholomew).

Crampel, Paul (1864-1891). French explorer and surveyor in Tchad and Congo regions, studied Oubangui, Chad and Chari basins. Map of l'Oubangui, 1890; work published in *Le Tour du Monde*, 1890, 1892. **Ref.** BROC, N. *Dictionnaire illustré des explorateurs Français du XIXe siècle: l Afrique* (1988) p.95-97 [in French].

Crampton, *Professor* R.C. *New Map of the City of Jacksonville, Illinois*, 1868.

Cranach, Lucas *the elder* (1472-1553). Painter, etcher and designer of woodcuts. Went to Wittenburg in 1505, where he became friendly with Martin Luther. Produced a nine-block map of the Promised Land, *c.*1508, a reduced version appeared in Lutheran Bibles as a map of the Exodus (woodcut), Zurich, 1525 and later.

Crane, *Captain* Gilbert. Surveyor. Charts in John Seller's *English Pilot*, 1671 and later.

Crane, John. *Butler County, Ohio*, 1855.

Crane, Walter. Victorian artist, designer and socialist. *Map of the World showing the extent of the British Empire*, 1886.

Cranston, George. Road surveyor. Roxburgh, for John Thomson's *Atlas of Scotland*, 1832.

Cranz [Crantz], David. *Historia om Gronland* (with 3 maps), 1769.

Craskell, Thomas. Engineer, worked with the surveyor James Simpson on a series of maps of Jamaica for D. Fournier of London. *...the Island of Jamaica*, 1763; *Map of the County of Cornwall* [Jamaica], 1763; *Map of the County of Middlesex* [Jamaica], 1763; *...Map of the County of Surry* [Jamaica], 1763.

Crassus, Valerius. *Carta delle tre Valli di Piemonte*, 1640.

Crates of Mallos [or of Pergamum] (*b.c.*215 BC). Literary and linguistic critic. Constructed large terrestrial globe to illustrate theory of Homer's account of the Aethiopians in Odyssey, in which four land masses were separated by wide belts of water.

Crato. See **Krafft**, Johann.

Crause [Crausenstein], Hans Heinrich. Engineer and military cartographer. *Pomerania citerior*, *c.*1664.

Craven, A. *State Geological Survey of California*, 1874.

Craven, T.A.M. (1813-1864). Surveyor US Navy. *Harbour of Bridgeport, Connecticut*, *c.*1835 (with other surveyors of the U.S. Coastal Survey).

Craven & Co. Publishers of Nottingham. *Craven & Co's. Directory Map of Hertfordshire*, *c.*1854.

Crawford, D.C. Ionia County, Michigan, 1891.

Crawford, Fraser Smith (*c.*1829-1890). Government photolithographer, South Australia. *Map showing the area of the Northern Territory and South Australia...*, *c.*1874.

Crawford, *Lieutenant* John, *RN*. Hydrographer. *Plan of Macassar*, 1814; *Plan of the Straits of Bally* [Bali], 1816; *China Sea*, 1821; *Survey of the Coast of Ava* [Burma], 1849.

Crawford, Joseph. *Map of Lee County, Illinois*, Chicago, Edward Mendel 1863 (with Jason C. Ayres).

Crawford, Joseph T. *Galveston Bay* [Texas] (2 sheets), London, Quarter Master General's Office 1837 (drawn by L.J. Hebert).

Crawford, J.W. *Map of Richmond, Union County, Ohio*, 1875; *Road map of Union County*, *c.*1875.

Crawford, Robert. Draughtsman. Farm plans East Grinstead, *c.*1840 MS (attributed).

Crawford, William [& Son.]. Surveyors of Edinburgh, produced many maps of parts of Scotland. *Plan of the Town... of Dundee*, 1793.
• **Crawford**, William (*fl.*1774-1828). *Dundee*, 1777 (engraved by D. Lizars); *Map of Dumfries-Shire* (4 sheets), *c.*1804 (engraved by Kirkwood & Sons); *Northern & Southern part of Dumfriesshire*, 1828.
•• **Crawford**, David (*fl.*1802-1829). Son of William, above. Manuscript estate plans.
•• **Crawford**, William (*fl.*1793-*c.*1845). Son of William, above. Dalswinton, Dumfries-shire, 1817.

Crawley, Richard. Estate Surveyor. Bedford, 1634.

Crawter, Henry (*fl.*1796-1849). Surveyor of Cheshunt, Hertfordshire. Produced manuscript surveys of various parts of England including Cheshunt inclosure award [Hertfordshire], 1804.

Crealock, H. Hope. *Plan of the action of Palechiao* [China], 1861.

Crease, *Major* Anthony R.V. *Plan of Kertch* [Crimea], 1856; *Plan of Yeni-Kale*, 1856.

Creassy, James. River Kyme Eau and New Sleaford, 1773.

Credner, Heinrich. *Geognostische Karte des Thüringer Waldes*, 1855; *Geognostische Karte der Umgegend von Hannover*, 1865, 1874.

Credner, Hermann (1841-1913). Head of the regional geological institute in Saxony and Thuringia. *Geologische Specialkarte des Königreichs Sachsen*, 1875.

Creech, William. Publisher of Edinburgh. J. Laurie's *Plan of Edinburgh and the country adjacent...*, 1786, 1811.

Creed, William. Engraver. *Europe* for Thomas Dobson's *Encyclopaedia*, 1792.

Creffeldt, Martinus [Martin?] Carolus [Karel?]. *Egentlicke Beschrijvinghe des Iselstrooms, sampt Ittelicke[n] omliggende[n] Steden unde Vlecken*, before 1579 (known only from one example in a private collection). **Ref.** MEURER, P. *Fontes cartographici Orteliani...* (1991) p.131 [in German].

Creighton, R. Draughtsman. English counties for Christopher and John Greenwood's *Atlas of the Counties of England*, from 1818; maps for Samuel Lewis's *A Topographical Dictionary...*, from 1831, including *A Plan of London and its Environs*, 1831 (engraved by J. & C. Walker); and *Map of North & South Wales...*, 1833; County Atlas, 1842; *A Parliamentary map of Kent* (4 sheets), 1845; England and Wales (9 sheets), *c.*1855.

Creitz, Cornelis. See **Cruys**, Cornelis.

Cremer. See **Mercator**, Gerard.

Crenne, Verdun de la. See **Verdun** de la Crenne.

Crépy [Crespy] family. Geographers, engravers, publishers and map sellers of Paris and Marseille. Noted at 'rue St. Jacques devant la rue du Plâtre' and 'rue St. Jacques à St. Pierre près la rue de la Parcheminerie'. In general maps are attributed simply to 'Crépy' rather than to individuals, these include: *Plan de Luxembourg Ville Forte Capitale du Duché*, *c.*1730; L'Empire anglois en Amérique, 1733 (after Henry Popple); Asie, 1735; *Le Royaume d'Espagne divisé En deux grands Etats d'Aragon et de Castille..., 1735; Les Provinces de Berry, Bourbonnois et Nivernois*, 1737, 1780; *Carte du duché de Brabant...*, 1745; *Carthagène appellé par les Espagnols Cartagene le Inda...*, 1745; *Blois*, c.1753 (birds-eye view); for J.B. Nolin's l'Afrique, 1755; re-issued George-Louis Le Rouge's *Atlas Nouveau*, 1767-1773; *La Bretagne*, 1767; *Le Duché de Bourgogne et partie de la Bresse*, 1767. **N.B.** Family relationships are uncertain.
• **Crépy** [Crespy], Jean (*c.*1650-1739). Engraver of Paris, believed to be the father of Etienne-Louis, Louis and possibly Jean-Baptiste. Apart from his association with the work listed above, Jean is credited with the engraving of N. Jaugeon's *Sciences Du Jeu Du Monde Ou La Carte Generale Contenante Les Mondes Coeleste Terrestre Et Civile...* (6 sheets), 1688; issued *Nouvelle Methode de Geographie ou Voiage du Monde... par un jeu*, *c.*1690. **Ref.** SHIRLEY, R. *The mapping of the World* (1983) N$^{os.}$538 and 543 pp.535, 540-541 and 545.
•• **Crépy**, Etienne-Louis (*c.*1706-post 1759). Son of Jean. Publisher and engraver, in partnership with brother Louis (1750) selling 'tout ce qui concerne la géographie' at 'rue St Jacques à St Pierre près a rue de la Parcheminerie'. **N.B.** Possibly the same person as Louis.

•• **Crépy**, Louis (*fl.*1720-1760). Son of Jean, noted as 'marchand imager' at the above address.
••• **Crépy**, Pierre-Louis (*fl.*1751-1771). Traded from 'rue St Jacques', Paris. Possibly son of Louis.
••? **Crépy**, Jean-Baptiste (*fl.*1753-1796). Map and printseller of 'Rue St Jacques à St Pierre, près la rue de la Parcheminerie' (1776). Possibly son of Louis, or brother of Louis and Etienne-Louis, father of Jean-Baptiste 'marchand d'estampes' of Marseille (below). *Possessions Angloises en Amérique*, 1772; *L'Amerique Suivant Le R.P. Charlevoix Jte, Mr. de la Condamine et Plusieurs autres Nouv^le Observations*, 1774; M. Moithey's *L'Europe Divisée en tous ses Royaumes...*, 1785.
•••• **Crépy**, Jean-Baptiste, *fils* (*b.*1753 or 1754). Son of Jean-Baptiste of Paris, 'marchand d'estampes' of Marseille.

Crépy, Paul (1835-1899). Founder, Société géologique de Lille.

Crescentio, Bartholomeo. *Nautica Mediterraneana*, Rome 1596.

Crespi, Juan, *SJ* (1721-1782). *Carta de Puerto... de San Francisco*, 1772 MS.

Crespy, Jean. See **Crépy** family.

Cresques, father and son. Cartographers of Mallorca. Worked together on the so-called 'Catalan Atlas', 1375-1377; also said to have prepared world maps for the Kings of France and Aragon, as well as Prince Henry of Portugal.
Ref. *Imago Mundi* XXII (1968) pp.107-109.
• **Cresques**, Abraham (*d.*1387). Father of Jafuda.
•• **Cresques** [Riba; Ribes], Jafuda [Jacme] (*c.*1360-after 1420). Moved to Barcelona *c.*1391 following his conversion to Christianity and the adoption of the name Jacme Riba. Credited with a portolan chart of the Mediterranean, 1416; he also possibly worked in Portugal, 1420-1427.

Cressy, J. Publisher of Philadelphia. John Marshall's *Life of George Washington*, 1834 (maps engraved by J. Yeager).

Crest, Jacques Bartholomé Micheli du. See **Micheli du Crest**, Jacques-Barthélemy.

Creuse --. Worked for Dépôt de la guerre on map of France, 1832.

Creutzig [Cruciger], Johann [Johannes] (*fl.*1553-1583). Printer of Neisse (Silesia) 'Zur Neiss, bey Johann Creutzig'. Printed Martin Helwig's woodcut map of Silesia, 1561.

Creux[ius], F. du. See **Du Creux**, François.

Creuzbauer, W. *Karte von dem Grossherzogtum Baden*, *c.*1860.

Creuzbaur, Robert. Communication maps of United States including *A Map to illustrate...the Gulf of Mexico and the Mississippi Valley to California and the Pacific Ocean*, 1849; draughtsman for J. de Cordova's map of Texas, 1854.

Crevaux, *Doctor* Jules Nicolas (1847-1882). *Fleuves de l'Amérique du Sud*, 1877-1879.

Crèvecoeur, Michel Guillaume St. Jean de [pseud. J. Hector St. John] (1735-1813). *Map of the Island of Nantucket*, and *...Martha's Vinyard*, both published in *Letters from an American Farmer, describing... the British Colonies in North America*, London 1782.

Crevel, Jean. One of the publishers of Nicolai's world map in *L'Art de Naviguer* (woodcut edition), Rouen 1573.

Crevier, Jean Baptiste Louis (1693-1765). *Atlas de géographie ancienne*, [*n.d.*].

Crew, J. Engraver and printer. Plan of Cape Town in Cape of Good Hope Almanac and Annual Register for 1852, 1851; South Africa, 1857.

Crew [Crewe], Randolph (1631-1657). *Comitatus Palatinus Cestriae*, in Daniel King's *Vale-Royall of England*, 1656 (after William Smith 1585).

Cribb, Robert. Publisher and printseller of 288 High Holborn, London 'near Great Turnstile'. *...Settlement at Sydney Cove, Port Jackson*, 1789.

Crichton, *Captain* C. David. Plan of Patta, 1751; *Plan of Kisseen Bay* [Arabia], *c.*1744, published Alexander Dalrymple 1775.

Crichton & Bell. Engravers. District of Kantyre, 1793.

Crickenburg, Johannes van. Painter of Ghent. Boundaries of France, *c.*1506.

Criddle, *Lieutenant* J.S. *A Survey of the Island of Thwart-the-Way*, 1820.

Crighton, George. Ooloogan Bay, Palawan [Philippines], 1836 MS.

Criginger [Crigingerus; Crispingius; Grisviger; Kriegengerus; Krieger; Kriegner; Kringerius; Krüger; Kruginger], Johannes (1521-1571). Theologian, teacher and mapmaker of Joachimsthal, Bohemia. *Bohemia regni... descriptio*, 1568, used by Ortelius 1570 and Gerard de Jode 1578; *Chorographia nova Misniae et Thuringiae*, 1568. **Ref.** KARROW, R. *Mapmakers of the sixteenth century and their maps* (1993) pp. 125-128; MEURER P. *Fontes cartographici Orteliani...* (1991) p.132 [in German].

Cring, Henry. *Atlas of Holmes County, Ohio*, 1875.

Crippen, John (*fl.*1723-1731). Estate Surveyor of Canterbury. Crablice and Pairs Farms, Essex, 1731 MS.

Crisp, Edward. Surveyor and publisher 'at the Carolina Coffee House in Birchin Lane', London. *Plan of Charles Town*, 1704; *A Compleat Description of the Province of Carolina in 3 Parts*, 1711 (including an inset map of South Carolina by Thomas Nairn, both drawings later used by Moll, 1720).

Crisp, May. *Plan of Reception Bay* [Tristan d'Acunha], 1814.

Crispi, Giambattista. *Gallipolis*, 1591, used by Braun & Hogenberg, *q.v.*.

Crispingius, Johannes. See **Criginger**, Johannes.

Crispinianus, Johannes. See **Cuspinianus**, Johannes.

Cristofano, Girolamo da. See **Christoforo**, Girolamo da.

Crivellari, F. *Topografia della regia Città di Vicenza*, 1821.

Crivelli, Taddeo (*c.*1425-*c.*1480). Painter of miniatures in Ferrara 1452-1471. Credited with maps for 1477 edition of Ptolemy; also untitled Ptolemaic world map, *c.*1480. **Ref.** SHIRLEY, R. *The mapping of the World* (1983) N°5 pp.5-6; CAMPBELL, T. *Earliest printed maps* (1987) pp.27-30, 126, 213-214.

Croc, H. de. See **Croock**, Hubert de.

Croce, Enrico. *Carta d'Italia*, 1875; *Carta geografica dello Stato di S. Paulo*, 1893.

Crocker, family. Surveyors of Frome in Somerset. **Ref.** BENDALL, S. *Dictionary of land surveyors and local map-makers of Great Britain and Ireland 1530-1850* 2nd edition (1997) p.122.

• **Crocker**, Abraham (*c.*1742-1821). Prepared manuscript maps of parts of Dorset and Somerset. Elements of Land Surveying, 1806, 1817.

• **Crocker & Son** (*fl.*1801-1802).

• **Crocker & Sons** (*fl.c.*1808-*c.*1815).

•• **Crocker**, John (*c.*1773-1831). Surveyor, worked with father Abraham. Also worked with his brother Philip and Samuel Mudge as a civilian surveyor on the Ordnance Survey from 1805-*c.*1817.

•• **Crocker**, Philip (*c.*1780-1829). Conducted a survey of the Winchester-Old Sarum road for Sir Richard Colt Hoare, 1801; worked with brother John and Samuel Mudge as a civilian surveyor for the Ordnance Survey from 1805, in West Sussex and East Hampshire, 1805-1806; Southern Wiltshire and parts of Hampshire, 1807-1811; prepared illustrations and maps for Hoare's *Ancient History of North and South Wiltshire; The Hundred of Mere*, 1822.

•• **Crocker**, Edmund (*c.*1784-1829). Worked with Samuel Mudge for the Ordnance Survey, c.1807-1818; with brother Philip in southern Wiltshire and parts of Hampshire, 1807-1811.

Crocker, H.S. [& Co.] of San Francisco. *County of Sacramento*, 1871; map of the Alaska and Yukon goldfields, Dewey & Co. 1897.

Crocker, William P. Stansteed-Montreal Rail Road, 1845; *Plan and Profile of the Alton & Terrehaute Rail Road*, 1850.

Croes, J.J.R. (1834-1906). West Side New York, 1873.

Crofutt, George A. *Crofutt's Trans Continental Tourists Guide*, New York 1871; *Salt Lake City, prepared expressly for Crofutt's Salt Lake City Directory*, 1885 (map drawn by A.M. Froiseth).

Croghan, William. American surveyor, 1800-1804.

Croiset, —. Engraver for Jean Baptiste Léonard Durand's *Atlas du voyage au Sénégal*, 1802.

Croisey [Crossey], Jean-Baptiste. Draughtsman and engraver. Jacques-Nicolas Bellin, 1763; worked with other engravers on five sheets of Cassini's *Carte de France* (51, 98, 101, 134, 162), 1763-1771; some maps for J. Chappé d'Auteroche's *Voyage en Sibérie...*, 1768 (others were by Tilliard and Chalmandrier); Dépôt de la Marine, 1768-1798; Louis Brion de la Tour, 1770; worked for Jean de Beaurain, 1774; *The Province of New Jersey*, 1777; Louis Charles Desnos, 1790-1792; Joseph-Antoine Bruny d'Entrecasteaux, 1807.

Croissant, J. Engraver of Antwerp. Map for Nicolas de Nicolay's *...voyages, faicts en la Turquie*.

Croix, F. de la. Engraver. North America, 1687.

Croix, A. Pherotee de la. See **De La Croix**, A.P.

Croix, L.A.N., *abbé* de la. See **Delacroix**, *abbé* Louis Antoine Nicolle.

Cromack, Irwin C. *Map of the City of Boston*, 1896.

Cromberger, Jacobus. Printer in Seville. Peter Martyr's *Opera*, 1511. **NB.** In 1544, when Cabot's planisphere was sent to Germany for engraving, Lazaro Noremberger and Gabriel Micel were noted as partners in the Cromberger printing house of Seville.

Crome, August Friedrich Wilhelm (1753-1833). German teacher of geography and history. *Neue Carte von Europa...*, 1782; *Produktenkarte von Europa* [map of the goods produced around Europe], 1782; Netherlands, 1785. **Ref.** HARMS, H. *Cartographica Helvetica* 3 (1991) pp.33-38 [in German].

Cronenberg, F. *Kaart van het Gouvernement Sumatras West Kust*, 1852; *Kaart van Boni*, 1860.

Cronenberg, Johann. *Charte over Eilandet St Croix udi America*, *c.*1750 (with Johann von Jaegersberg). **Ref.** HOPKINS, D. *Imago Mundi* 41 (1989) pp.44-58.

Cronk, Henry H. *New Map of Tunbridge Wells*, for Brackett, 1863.

Cronk, William. Surveyor of Kent. Chart of the coast around Sandwich [Kent], 1775; *A Map... of the parish of St. John the Baptist in the Isle of Thanet* [Kent], 1776.

Cronstedt, F.A.U. *Charta öfver Hellsingland*, 1796; *Charta öfver Herjeaedalen*, 1797.

Croock [Croc], Hubert de (1490-1554). Painter, engraver and publisher of Bruges.

Croock [de Croc], Willem Hendrickszoon. Surveyor and engraver of Amsterdam. Amstelland, 1525; Noord Holland, 1529.

Crophius, Martin Gottfried (*c.*1717-1765). Engraver in Augsburg, for Georg Matthäus Seutter's *Atlas minor*, 1744; also Tobias Conrad Lotter's *Novus Atlas*, 1770.

Cropper, C.W. Map of gold mining leases of Western Australia, 1890.

Cropsey, I.E. Plan for Panama Canal, 1852.

Crosby, Benjamin (1768-1815). Bookseller of London, successor to 'Mr Stalker of Stationers' Hall Court', in 1814 he was himself succeeded by Simpkin & Marshall (his two assistants), and by Baldwin, Cradock & Joy. *A Plan of London, Westminster and Southwark*, 1803 (engraved by Barlow, used in Tegg & Castleman's *New Picture of London*); *Crosby's Complete Pocket Gazetteer of England and Wales...*, 1807, Baldwin, Cradock & joy, Simpkin & Marshall, J. Bumpus 1818, J. Bumpus 1822 (the maps were later used by S. Tymms). **Ref.** BURGESS, T. and BURDEN, E. *The Map Collector* 38 (1987) p.52; SMITH, D. *The Map Collector* 29 (1984) pp.34-35.

Crosby, J. Publisher. *A Plan of London, Westminster and Southwark*, 1803.

Crosley [Crossley], William (*d*.1797). Surveyor and engineer, worked with Joseph Lindley as Lindley & Crosley. *Map of the County of Surrey* (2 sheets, with Joseph Lindley), surveyed 1789-1790, published 1793, 1814 (editions to 1874); plan of the proposed Rochdale Canal, 1794.

Cross, Edward (1798-1887). Surveyor. Arkansas, 1837.

Cross, Joseph. Engraver and publisher, '18 Holborn Hill opposite Furnival's Inn', London. William Edgeworth's Roscommon, 1817; Lambeth, 1824; *Chart of Van Diemen's Land*, 1826-1842; John Oxley's *Chart of Part of New South Wales*, 1827; *A Map of the County of Hertford*, 1827; *New Plan of London*, 1828, editions to 1854; *Map of South West Part of Australia*, 1833; *London Guide*, 1837, editions to 1851; Robert Dixon's *Map of the Colony of New South Wales*, 1837 (engraved by J. & C. Walker); *Map of the United States*, 1850; and others.
• **Cross & Son** (*fl.* from 1844).

Cross, Thomas, father and son.
• **Cross**, Thomas (*fl*.1632-1682). Engraver. Fens, 1642; maps for Thomas Fuller's *Pisgah-Sight of Palestine*, 1650; Faroe Is., 1676; *An Exact Map of England*, 1682.
•• **Cross**, Thomas. Son of Thomas (above), also an engraver, but not known to be associated with maps.

Crossey. See **Croisey**.

Crossley, William. See **Crosley**, William.

Crosthwaite, Peter (1735-1800). *Maps of the Lake District Surveyed & Planed* [sic] *by P. Crosthwaite. Admiral at Keswick Regatta, who keeps the Museum at Keswick & is Guide, Pilot, Geographer & Hydrographer to the Nobility & Gentry who make the Tour of the Lakes* (7 maps), Keswick 1783, 1788, 1794, 1800 and later (engraved by S. Neele, H. Ashby and J. Ellis).

Croswell, William (1760-1834). Mercator Map of the Starry Heavens, *c*.1810 (earliest known American star map).

Crouch, Nathaniel [pseudonym RB, Robert Burton]. Publisher and mapseller at the Royal Exchange. Map of New England in *English Empire in America*, 1685.

Crousaz, *Colonel* Abraham de (1629-1710). Swiss architect and councillor in Lausanne. Grundriss Pässe des Bern Gebiets, 1670.

Crow, Frederick. (*fl*.1838-1839). Estate surveyor of Daventry. Estate maps of Buckinghamshire and Northamptonshire.

Crow, James (1711-1786). Surveyor of London, prepared various manuscript maps of parts of England and Ireland, many with Thomas Marsh.

Crowder, John (1742-1816). Estate surveyor in Nottinghamshire and Yorkshire. Maps including Pontefract Park, 1781 MS; Middleton Crown Estate, 1787 MS; Manor of Cowick, 1800 MS.

Crowe, Benjamin. *A New & Exact Map of Part of the Coast of Norway*, Mount and Page 1740.

Crozat, François (1858-1893). French physician and explorer in the Sudan from 1889. Assisted with a boundary survey of the Ivory Coast, 1892.

Crozet, Claude (1790-1864). French soldier and engineer, worked on the road systems of Virginia. *A map of the internal improvements of Virginia*, Philadelphia 1848.

Crozet, *Lieutenant* Julien. Assumed command of a French expedition to Tasmania, New Zealand and Guam after Marion-Dufresne was killed by the Maories. *Nouveau Voyage à la Mer du Sud, Commencé sous les ordres de M. Marion* [-Dufresne]... (5 charts on 2 sheets), Paris, Barrois l'ainé 1783.

Crozier, *Commander* (later *Captain*) F.R.M. Admiralty Surveyor. Antarctic exploration 1839-1843; Northwest Passage, 1845 (lost with John Franklin's third Arctic expedition).

Cruchley, George Frederick (1787-1880). Mapseller, engraver and globemaker, 349 Oxford St. (1823-1825); 38 Ludgate St. (1825-1833); 81 Fleet St. (1833-1875). Trained by Aaron Arrowsmith and re-issued some of his maps. Bought John Cary's plates 1844. *Cruchley's environs of London* *extending thirty miles from the Metropolis*, 1824 and later; *Cruchley's New Plan of London and Its Environs*, 1828, 1834, *c.*1851; Lancashire, 1836; *Australia, New Zealand, East India Islands &c.*, 1842 and editions to 1856; General Atlas, 1843; *A New Topographical map of the environs of Canterbury*, 1850; ...*Map of the Country 14 miles round London*, 1852 (re-publication of Cary); Railway and Telegraph County Atlas, 1855-1858; Cary's County Atlas, 1862; guide books of London, 1867; *Cruchley's County Map of Bedford Showing all the Railways...*, 1875; *Cruchley's County map of Berkshire...*, Gall & Inglis 1877; and others. **Ref.** SMITH, David *The Map Collector* 49 (1989) pp.16-22.

Cruciger, Johannes. See **Creutzig**, Johann.

Cruger, *Captain* S. *A New Chart of False Bay*, 1808 (appeared as an inset on a map of the Cape of Good Hope published by W. Heather in *Marine Atlas or Seamen's Complete Pilot*).

Cruijs, Cornelis. See **Cruys**, Cornelis.

Cruise, Richard J. *Geological Survey of Ireland*, Leitrim 1876-1877.

Cruls, Luiz (1848-1908). Central Brasil, 1894.

Cruquius, brothers. Mapmakers of Delft.
• **Cruquius**, Nicolas Samuelszoon (1678-1754). Hydrographer, engineer and Supervisor of Sluices and Dams. *'t Hooge Heemraedschap van Delfland* (25 leaves forming a wall map), 1712 (with brother Jacob); Isobath map *Caarte...van de Rivier Merwede* (3 sheets showing contours of the bed of the river), Leiden 1730 (the earliest Western printed map with contour lines). **Ref.** WALLIS, H. & ROBINSON, A.H. *Cartographical innovations* p.225.
• **Cruquius**, Jacob. Surveyor, worked with brother Nicolas, above.

Cruttenden [Cruttendon], R. Publisher. Neal's *History of New England* (with map), 1720.

Cruttwell, Clement (1743-1808). Publisher and surveyor of Bath. Atlas to Cruttwell's Gazetteer, 1799-1808.

Cruttwell, *Lieutenant* James, *RN*. *Map of the Roads of Portugal*, 1832.

Crutwell, J.A. *General Atlas*, London 1834.

Cruys [Cruis; Crys; Cruijs; Creitz], *Admiral* Cornelis (1657-1727). Dutch admiral in Russian service at invitation of Tsar Peter I (Peter the Great). *Nauw-keurige afbeelding vande Rivier Don...*, Amsterdam, Hendrick Doncker 1705. **Ref.** KOEMAN, C. *Atlantes Neerlandici* Vol.IV (1967) pp.188-189.

Cruz, Alonso de Santa. See **Santa Cruz**.

Cruz, *Lieutenant* Manuel de la. Plano Puerto de Braga, 1838 MS.

Cruz, Manuel. *Mapa parcial de la República Argentina, Chile y Bolivia*, 1896.

Cruz Bagay, Nicolas de la. See **Bagay**, Nicolás de la Cruz.

Cruz Cano y Olmedilla, Juan de la (1734-1790). Spanish cartographer and engraver. *Plano de Charlestown*, 1750; *Mapa Maritimo del Golfo de Mexico* (2 sheets), 1755 (with Tomás López); Florida and Louisiana, 1755; Straits of Magellan, for Casimiro Ortego, 1769. *Mapa geográfico de América Meridional...* (8 sheets), 1775, re-issued London, William Faden 1800; worked for Vincente Tofino de San Miguel, 1789. **Ref.** SMITH, T.R. *Imago Mundi* 20 (1966) pp.49-78; DONOSO, R. 'El Mapa de la América Meridional de la Cruz Cano y Olmedilla', in: *Revista Chileña de Historia y Geografia* N°31 (1963) pp.121-175 [in Spanish].

Cryfftz, N. See **Cusanus**, Nicolas.

Csaplovics, Johann von (1780-1847). Ethnographer. Karte Ungern, 1829.

Cseh, Ferenc. See **Böhm**, Ferenc.

Cséti [Chrismár], Ottó (1836-1906). Hungarian engineer and cartographer. Graduated from Vienna, 1862, taught at the academy at Selmecbánya 1872-1902. Constructed a number of surveying and cartographic instruments. **Ref.** VAJDA, P. *Nagy magyar feltalálók* [Great Hungarian inventors] (Budapest 1958) [in Hungarian].

Csibi, Lörinc (1849-1903). Hungarian forestry engineer, born Ditró [now Ditráu, Romania], died Selmecbánya [now Banska Stiavnica, Slovakia]. Forestry engineer. Received a prize for forestry maps at the Paris Exposition, 1900.

Csikszépvizi Deák, Lajos. Hungarian cartographer. Produced diocesan maps of Transylvania, 1841.

Csintalan, József. Hungarian professor, produced the first Hungarian language map of Pest-Buda, 1828.

Cuaranta, B. Spanish lithographer. Maps for Doroteo Bachiller, 1852; Miguel Grilo, 1865, 1876.

Cubitt, Thomas (*fl.*1776-1795). Architect and surveyor, employed in the drawing office of the Tower of London. Various maps including Scilly Isles, 1791; *An accurate Survey... of the Island of Jersey* (4 sheets), 1795 (with William Gardner and Thomas Gream).

Cubitt, *Sir* William (1785-1861). Civil engineer. Various plans of parts of Britain and Ireland associated with engineering works, including plans of Belfast Harbour, 1835.

Cucinello, Domenico. Lithographer of Naples, 1860.

Cuco, P. Engraver for Tomás López, 1778-1795.

Cudlip, *Lieutenant* (later *Commander*) F.A., *RN*. Admiralty surveyor. Singapore, Durian & Rio, 1831, 1840; Heligoland, 1855; Plan of Cowes, Isle of Wight, 1856-1861.

Cuipers, P. *Republicas de America Central*, 1860.

Culbertson, Alexander. Yellowstone and Platte Rivers, 1857 MS.

Culemann, Friedrich. Grundriss der Stadt Braunschweig, 1789.

Culenburgh, Jacob. Rhynberck, published Claes Jansz. Visscher, 1633.

Cullen, *Doctor* Edward. *Rio Savana or Chaparti, with the Route thence to the Atlantic*, 1851 (lithographed by Harrison & Sons).

Cullingworth, J. Publisher and Stationer, Natal Star Office, Durban. *Colony of Port Natal*, 1862 (with Joseph F. Masser).

Cullingworth, William. Plan of Stonely, Huntingdonshire, 1769.

Culver, John P. Civil engineer and surveyor. *Covington, ... Nebraska*, 1857 (with Henry D. Betts); *St. Helena, ... Nebraska*, 1858.

Cumberland, *Captain* Charles Edward. *Plan of the attacks... before Sebastopol, 1854-1855*, 1858.

Cumberlege, Stephen Austin. '...at the Kings Arms, Paternoster Row...', London. *The Island of Minorca*, 1781.

Cumings, Samuel. *The Western Pilot* [Ohio and Mississippi Rivers], 1822, and editions to 1854.

Cumming, G.W. *A Survey of the Islands and Harbour of Banda*, 1818.

Cumming, John. Publisher. 16 Lower Ormond Quay, Dublin. *New Traveller's Guide through Ireland*, 1815; *A New General Atlas*, 1817 (for John Thomson of Edinburgh and London); *The Traveller's New Guide through Ireland...*, 1819 (also published by others in London); *Atlas of Scotland*, 1832 (also for John Thomson).

Cumming, William. Plan of Inverness, 1802; River Ness and Caledonian Canal, 1802; Plan of the proposed turnpike road through the County of Nairn, 1819.

Cummings, Benjamin F. Salmon River, 1856 MS.

Cummings, J.A.J. *Columbia and Montour Counties, Pennsylvania*, 1860.

Cummings, J.B. *Map of...Portland, Maine*, 1854.

Cummings, Jacob Abbot (1773-1820). American geographer. *A School Atlas accompanying ancient and modern Geography*, Boston, Cummings & Hilliard *c.*1815 and other editions; *First Lessons in Geography and Astronomy*, Boston, Cummings & Hilliard 1822 and later.

Cummings & Hilliard. Publisher of N°·1 Cornhill, Boston. Jacob A. Cummings' *A School Atlas...*, *c.*1815 and other editions;

Maps accompanying Worcester's Epitome of Modern Geography, *c.*1820 (another atlas by Worcester was published by Hilliard, Gray & Co.).

Cundall, H.J. *Prince Edward Island*, 1851, 1861.

Cundee, James (*fl.*1812-1814). Publisher, Albion Press. Plan of London, 1808; one of the sellers of G. Cole and J. Roper's *The British Atlas*, 1810; county maps, 1812-1814, used in *Robins's Atlas of England and Wales*, 1819, and James Dugdale's *New British Traveller*, J. Robins & Co. *c.*1819.
• **Cundee**, J. & J. (*fl.*1814-1815). Imprint appears on some maps dated 1814-1815 for J. Dugdale and J. Robins & Co.

Cundell, F. Cork, showing fortifications, 1587 MS.

Cunerlinus, J. See **Cimerlini**, Giovanni Paulus.

Cunerth, O. *Handkarte der Province Westpreussen*, 1891?

Cuningham, W. See **Cunningham**, William.

Cuninghame, William. Plan of Port Royal Ratan, 1743.

Cunningham, Allan (1791-1839). Botanist and explorer in Australia, 1823-1829. His information was used on later maps including the SDUK map of New South Wales, Baldwin & Cradock, 1835.

Cunningham, George Godfrey. Associated with two works, capacity unknown. Gazetteer of the World, 1850-1856; James Bell's *System of Geography*, 1860 (first published Glasgow, 1830).

Cunningham, P.A. *City of Auburn, New York*, 1871; *City of Decatur...Illinois*, 1872; *Howard County, Indiana*, 1875.

Cunningham, R.S. [& Co.]. Atlas of Shenectady, New York, 1892.

Cunningham, Robert James Hay (1815-1842). Geological surveyor. Lothians, 1838; *Geognostical Map of Sutherland*, 1841; *Geognostical Map of the Stewarty of Kirkcudbright*, 1843.

Cunningham [Cuningham], William (1531-*c.*1586). Mathematician and surveyor. *The Cosmographical glasse...* (with woodcut plan of Norwich), 1559 (said to be the oldest printed British town plan, later used by Braun and Hogenberg).

Cuomy, V. Geologische Karte der Insel Capri, 1889.

Cupet, *Captain* Pierre Paul (1859-1907). Surveyed frontier between Siam and Indochina 1888-1890. *La Grande Carte de Indochine* (4 sheets), 1893. **Ref.** BROC, N. *Dictionnaire illustré des explorateurs Français du XIXe siècle: II Asie* (1992) p.114.

Curado, Paiva. *Oceano Atlantico Norte... Plano da Bahia do Turrafal*, 1890.

Cureau, Adolphe (1864-1913). French colonial administrator in Gabon, Bangui and Congo regions. Made boundary survey of German and French claims in Cameroun region 1900-1902. *Carte des pays Nzakara et Zandé entre le M'Bomou, le Bomou et le Soueh*, 1899. **Ref.** BROC, N. *Dictionnaire illustré des explorateurs Français du XIXe siècle: I Afrique* (1988) p.99.

Curioni, G. Carta geologica delle provincie Lombarda, 1876.

Currer, William. Surveyor of Crown Lands for the Duchy of Lancaster. Worked with Christopher Saxton on the survey of Scholes Wood in the West Riding of Yorkshire, 1609-1610; later worked with Robert Saxton on a survey of Tanshelf, near Pontefract.

Currey, Edward. *General Plan... Witwatersrand Gold Fields*, 1892.

Currie, Mark John, *RN* (1795-1874). *Map of the country to the South of Lake George...* [New York], 1825.

Currier, Nathaniel (1813-1888). Lithographer, '148 Nassau, corner of Spruce Street', New York. Ann Arbor, Michigan, 1836; Northern Part of Illinois, 1836; surveyed part of Wisconsin, 1838; produced Chauncey Barnard's surveys of Connecticut towns: *Thompsonville*, 1849; *Tariffville*, 1850.

Curry, William. Publisher of 9 Upper Sackville Street, Dublin. James Fraser's *Guide through Ireland*, 1837; road maps for tourists in Ireland, 1844.

Curtice & Stateler. *Map of St. Paul, West St. Paul and Minneapolis, Minnesota*, 1876.

Curtis, Cockburn. Norwich Navigation, 1846.

Curtis, G.D. *Map of Trail Creek Mining Camp*, 1897.

Curtis, H.K. (*fl*.1834-1839). American surveyor.

Curtis, L.A. Surveyor of 30 Sydney Arcade, Sydney. *Sunny Bank Estate near Auburn Park* [Bass Hill], 1884.

Curtis, R. *Ground Plan of the British Cemetery... near Sebastopol*, 1856.

Curtis, *Lieutenant* Roger, *RN*. Chart of part of Labrador, 1773.

Curtis, Samuel R. Map of Des Moines River, 1849.

Curtis, W.J. Engineer of 25 Throgmorton Street. *A New Map of the Island of Trinidad*, 1845; Trinidad railway, 1848; *The London Railway*, 1861.

Curtiss & Sons. *Curtiss & Sons' Map of Portsmouth*, *c*.1881.

Curtius, Ernst (1814-1896). *Sieben Karten... von Athen*, 1868; *Atlas von Athen*, Berlin, Dietrich Riemer 1878 (with Johann August Kaupert); *Karten von Attika* (25 sheets), 1881-1894 (with Kaupert); *Karten von Attika* (12 sheets), 1881 (with Kaupert).

Curtius, Ludovicus. *Antiquissimae ac celeberrimae regiaq. civitatis Papiae icon*, 1600.

Curtz, Johannes Laurentius. Administrative map of Tambach [Franconia], 1735.

Curtze, H.L. Pyrmont, 1817.

Cusanus [Cusa; Cryffts; Gryftz; Khryffts; Koebs; Krebes; Krifts; Kusa], *cardinal* Nicolas (1401-1464). German polymath and cartographer, studied at Deventer and Heidelberg. Attended Synod of Basle, 1432; Papal Nuncio, later cardinal (1448). Met János Dominis who provided Hungarian cartographic information. Manuscript map of Germany and central Europe, *c*.1450, later published in many editions including *Quod picta est parva Germania tota tabella...*, 1478 and later issues, also used in the Nuremberg Chronicle, *c*.1491; used as a source for many later maps.
Ref. KARROW, R. *Mapmakers of the sixteenth century and their maps* (1993) pp.129-137; MEURER, P. *Fontes cartographici Orteliani...* (1991) pp.133-135 [in German]; CAMPBELL, T. *Earliest printed maps* (1987) pp.35-52.

Cushee, family. Surveyors and globemakers at 'The Globe and Sun, between St Dunstan's and Chancery Lane, Fleet Street', London.
• **Cushee**, Richard (*c*.1696-1748). Surveyor, globemaker, instrument maker, and member of the Merchant Taylors of London. Worked with Joseph Harris, John Senex (1722-1729) and Thomas Wright (1731-1732). Produced globes, also estate and inclosure maps of parts of England including *Manor of Ruckholt, Essex*, 1728 MS; *The Description and Use of the Globes and the Orrery*, 1731 (with Thomas Wright); pocket globe, 1731; Berden, 1732.
•• **Cushee**, E. [Elizabeth?]. (*fl*.1729-1772). Thought to be the daughter of Richard, with whom she worked as globemaker and publisher, married Richard's apprentice, Nathaniel Hill. Succeeded at the Globe and Sun by Nathaniel's apprentice Thomas Bateman. Surrey, 1729; *A New Globe of the Earth*, *c*.1750.

Cushing, S.B. City of Providence, 1849 (with H.F. Walling).

Cusi, Giuseppe. *Topografia della Provincia di Sondrio*, 1825.

Cusinus, Hugo. See **Cousin**, Hugues.

Cuspinianus [Grispinianus; Spiesshaimer; Spiesshamer], Johannes (1473-1529). Diplomat, physician, historian and geographer. Served Maximilian I from 1506 in Poland and Hungary. *Austria... cum omnibus eiusdem marchionatibus, ducibus, archducibus*, *c*.1506; *Descriptio urbis Viennensis Danubique*, (both lost); published *Tabula Hungariae*, 1528 (surveyed by Lázár with the assistance of Ziegler, edited by Collimitius and printed by Apian). **Ref.** KARROW, R. *Mapmakers of the sixteenth century and their maps* (1993) pp.138-141; MEURER, P. *Fontes cartographici Orteliani...* (1991) pp.135-136 [in German].

Cust, Robert Needham. *A Language map of Africa*, 1884; *Missionary Map of Africa*, 1891.

Custance, William (*fl.c.*1790-1833). Surveyor and publisher from Chesterton, Cambridgeshire. Various manuscript maps of parts of England including *A New Plan of the University and Town of Cambridge to the Present Year 1798*, 1798; Estate plans, 1808.

Custos [Custis; Custodis], brothers.
• **Custos**, David. Engraver. Hungary and Austria, *c.*1618; Prussia, 1627; prepared reduced versions of Mercator maps for Nicolaus Helvicus' *Theatrum Historiae Universalis Catho.-Protest*, Frankfurt am Main, Schönwetter 1644.
• **Custos**, Raphael (1591-1651). Engraver and publisher, brother of David above. Honter's Swabia, 1619; *Bavariae, olim Vindeliciae Ileriae Amnis...descriptio*, 1619.

Custos, Dominicus (1560-1615). Born in Antwerp, settled at Augsburg 1584. Maps of Transylvania, *c.*1596; Kanisza, 1601.

Custos, Jakob. *Posthauss zu Augspurg*, 1628.

Cutfield, *Commander* W., *RN*. Admiralty surveyor. Africa, 1821-1826 (with Commander Alexander Thomas Emeric Vidal).

Cuthbertson, Walter R. *Map of part of British New Guinea*, 1887.

Cutler, Manasseh (1742-1823). Geologist and botanist in Massachusetts. *Map of the Federal Territory from the Western Boundary of Pennsylvania to the Scioto River...*, 1787 (thought to be the first map of Ohio). **Ref.** WHEAT & BRUN *Maps and charts published in America before 1800* (1978) pp.145-146.

Cutler, Nathaniel. *Atlas Maritimus et Commercialis*, 1728 (with Edmond Halley); *A General Coasting Pilot*, London, James & John Knapton 1728.

Cutts, *Lieutenant* R. *Mare Island Straits*, Washington 1851; *San Diego Entrance and Approaches*, Washington 1851 (the first US government survey of San Diego harbor).

Cuvier, Georges L.C.F.D. *baron* de (1769-1832). French zoologist and mapmaker. *Carte géognostique des Environs de Paris*, 1810 (with Alexandre Brongniart); *Carte géologique des Environs de Paris*, 1811, 1821 (with Alexandre Brongniart).

Cuyk, P. (1720-1787). Engraver of 's-Gravenhage. Texel, 1780.

Cuylenburg, John van. Island of Singapore, 1898.

Cuynat, C.S. (1774-1853). French geographer. Topographie de Barcelone, 1841; Topographie des Asturias, 1849.

Cylov, Nikolai Ivanovich (1801-1879). Military cartographer of St Petersburg.

Cymbricus, C. See **Clavus**, Claudius.

Cynthius, M. See **Zündt**, Matthias.

Cysat, father and son.
• **Cysat**, Renwart (1545-1614). Chemist and historian of Lucerne, father of Johann Baptist. *Der...Japponischen Insel*, Fribourg, Abraham Gemperlin 1586. **Ref.** FREI, Walter *Der Luzerner Stadschreiber Renward Cysat 1545-1614* (Lucerne 1963) [in German].
••**Cysat**, *Professor* Johann Baptist *SJ* (1587-1657). Swiss theologian, mathematician and astronomer in Lucerne, Ingolstadt, Madrid, Innsbruck and Eichstätt. *Tabula Cosmographica*, *c.*1618. **Ref.** DÜRST, A. in: SCHARFE, W. (Ed.) *Gerhard Mercator und seine Zeit. 7. Kartographiehistorisches Colloquium Duisburg 1994.* Vorträge und Berichte Duisburg (1996) pp.103-120 [in German].

Cysat Johann Leopold (1601-1663). Historian of Lucerne. Lake Lucerne, 1645, 1661.

Czajkowski, Franciszek, *SJ* (1742-1821). Historian and cartographer of Warsaw. Palatinate of Sandomierz, 1786 MS.

Czaki, Ferensz [Francis] Florian (*d.*1772). Cartographer to Stanisław II Poniatowski, last King of independent Poland. Worked on an atlas of Poland with Charles Perthées and Jan Bakalowicz. Vistula, 1760; Spisz [Hungary], 1762.

Czeel. See **Zell**, Heinrich.

Czermak, E. *Karte von Schlesien*, 1869.

Czerny, Joseph (*fl.*early-19th century). Cartographer and globemaker of Vienna.

Czerwenka, J. *Plan der Stadt Mittweida*, 1880.

Czetter, Sámuel (1765-*after* 1819). Hungarian engraver, born Felpéc, on the invitation of Tsar went to work at the Kremlin *c.*1818, died in Russia. Engraved a number of maps for Korabinsky, 1817.

Czetz, János (1822-1904). Hungarian military cartographer, born Gidófalva [now Ghidfalau, Transylvania, Romania], died Buenos Aires. Took part in the War of Independence, 1848-1849, after which he had to emigrate, settling in Argentina in 1860. Took part in the mapping of South America, became chairman of the Argentinian Cartographic Institute in 1875. Founded the Argentine military engineering academy. **Ref.** RAUM, F. *Magyar térképészek Délamerikában* [Hungarian cartographers in South America] in: *Geodézia és Kartográfia* [Geodesy and cartography] (1983/3).

Czjzek, Johann. Geologist. *Geognostische Karte der Umgebungen von Krems*, 1849; *Geologische Karte der Umgebungen Wiens*, 1860.

Czoernig von Czernhausen, Karl Joseph, *Freiherr von* (1804-1889). Austrian geographer. *Ethnograph. Karte der öesterreichischen Monarchie* (4 sheets), 1855; *Die ethnographischen Verhältnisse des Österreichischen Küstenlandes*, Trieste 1885.

Czszkowski, —. Exploration géologique de la Région de l'île d'Elba, 1880.

ATLAS FRANÇOIS,
CONTENANT LES CARTES GEOGRAPHIQUES
dans lesquelles sont tres exactement remarquez
LES EMPIRES, MONARCHIES,
ROYAUMES et ESTATS
DE L'EUROPE, DE L'ASIE, DE L'AFRIQUE
ET DE L'AMÉRIQUE :
AVEC LES TABLES ET CARTES PARTICULIERES,
DE FRANCE, DE FLANDRE, D'ALLEMAGNE
D'ESPAGNE ET D'ITALIE.
DEDIÉ AU ROY,
Par son tres-humble, tres-obeissant, tres-fidele Sujet et Serviteur
HUBERT IAILLOT, Geographe ordinaire de sa Majesté

D., S. Plan de Pensacole, 1719.

D.B. See **Bertelli**, Donato.

D., C. See **Doedtsz.**, Cornelis.

D., J. See **Davies**, John.

D.V.T.Y. See **Avity**, Pierre de Montmartin.

D'Abbeville, S. See **Sanson** family.

Dabenzara, J. See **Debenzara**, Jehu.

Dabey, J. See **Deventer**, Jacob van.

D'Ablaincourt, P. See **Perrot** d'Ablancourt.

Dablon, Claude, *SJ* (1618/1619-1697). Missionary in Canada. *Relation... en Nouvelle France 1671 et 1672*, Paris 1673; *Lac Superieur*, 1672 (with Claude Allouez).

Daboll, Nathan. *Present Seat of War...New York*, 1777.

Da Cadamosto. See **Cadamosto**, Alysius Alvise.

Da Carignano, L. See **Carignano**, Giovanni Maura da.

Da Corte, C. & D. See **Corte**, Carlo da.

Da Costa, A. Carvalho [Carvalão]. See **Carvalho** da Costa, Antonio Nuñes de.

Da Costa, E. See **Costa**, Euzebio da.

D'Acosta, José. See **Acosta**, José de.

Da Costa e Miranda, J. See **Miranda**, José da Costa.

D'Acuna. See **Acuna**.

Daddow, S. Harries. *Anthracite coal fields of Pennsylvania*, *c.*1868.

Dade, Thomas (*fl.*1690-1704). Land surveyor. Rattesden [Suffolk], 1704.

Daetszoon, C. See **Doedtsz.**, Cornelis.

Dagelet, *Mme* Le Pauté de. See **Le Pauté** Dagelet.

Dagen, Hendrik Rochusz. van (*b.*1634). Engraver of Amsterdam. Worked for Joan Blaeu.

Dagg, George A. *Devia Hibernia*, 1893.

Daggett, A. Engraver and publisher of New Haven, Connecticut. *Map of Connecticut*, *c.*1827, later editions included: Hartford, Willis Thrall 1831; E. Huntingdon 1836, 1837; Brown & Parsons 1847-1854; Brown & Gross 1858.
• **Daggett & Ely**. Engravers of New Haven, Connecticut. The name Daggett & Ely appears on the later edition of above Connecticut map, published in Hartford by Willis Thrall, 1831; William Lester's *Map of*

New London and Windham Counties, 1833.
•• **Daggett, Hinman & Co.** Engravers. J.G. Percival's *A Geological Map of Connecticut*, 1842.

D'Agnolo, J. See **Angelo**, Jacopo d'.

Dagustine, M. de Combis. See **Combis** D'Augustine, M. de.

Dahl, Peter M. Plat Book Hennepin & Ramsey Counties, Minnesota, 1898.

Dahlberg[h] [Dalberck], *count* Erik Jonsson (1625-1703). Military engineer and cartographer of Stockholm. Funen, 1657; Bogense, 1658; Denmark, 1660; *Suecia antiqua...*, 1667-1716; Atlas of Sweden, 1698 (presented to Charles XII); ...*Tabula chorographica Regni Poloniae*, *c.*1729.

Dahll, Tellef (*b.*1825). Norwegian geologist. *Geologisk Kart over det Söndenfjeldske*, 1866; *Geologisk Kart over det Nordige Norge*, 1879.

Dahlman, Carl Edward (1828-1900). Swedish cartographer. *Karta öfver...Frövi-Fala Jernbanan*, 1867; *Karta öfver Stockholm*, 1870; ...*Sverige*, 1876; ...*Svenska arktiska...* [map of Swedish expeditions to the Arctic], 1880.

Dahse, Paul. *Karte der Goldküste*, 1881.

Daikyoji Kahe. Japanese publisher of the early Edo period. *Edo koezu* [Plan of Edo], [n.d.] revised in 1654, now lost; Map of Japan, *c.*1660.

D'Ailly, P. See **Ailly**, Pierre d'.

Daines, Isaac C. Surveyor, 20 St. Swithin's Lane, London. *County Courts Districts*, *c.*1846.

Daintree, Richard (1831-l878). Geologist and photographer, specialised in the mines of Australia. *Map of Queensland shewing mineral areas*, 1872, *c.*1882.

Dajot, —. *Carte routière de Seine et Marne*, 1872.

D'Albe, L.A.G.B. See **Bacler** D'Albe, Louis Albert Ghislain.

D'Albeca, A.L. See **Albéca**, Alexandre d'.

Dalberck, E.J. See **Dahlberg**, Erik Jonsson.

D'Albertis, L.M. See **Albertis**, L.M. d'.

Dal Buono, F. See **Buono**, Florino dal.

Dalby, Isaac (1744-1824). Professor of Mathematics at Woolwich Military Academy; worked with William Roy on the triangulation of England from 1787. He was appointed to the Drawing Room of the Board of Ordnance after Roy's death in 1790 and worked with William Mudge in the early years of the Ordnance Survey.

Dale, *Lieutenant* —. *Map of the River Jordan and the Dead Sea*, 1849.

Dale, *Corporal* J.H., Royal Engineers. Many manuscript maps and plans of parts of Bermuda, 1888-1904 [Public Record Office, Kew].

Daleković, Feliks. *Plan Srbsko-Turskoj Granjze* [Political map Serbia], 1878.

Dalen, Cornelis van (1602-1665). Engraver of Amsterdam. Lewes Roberts's *The Marchants Map of Commerce*, 1638; Vegetius's *De re militari*, 1645.

Dalgleish, William. Worked on the Ordnance Survey of Ireland; pioneer of the electrotype process for making reproduction intaglio printing plates for large scale maps, 1837-1840. **Ref.** ANDREWS, J.H. *A paper landscape. The Ordnance Survey in nineteenth-century Ireland* (Oxford 1975).

Dalier, —. French engineer and surveyor, one of the more prolific contributors to the Cassini *Carte de France*. Worked with Pasumot and others as a surveyor or checker on 24 sheets of the survey, 1754-*c.*1768.

Dall, Nicholas Thomas (*d.*1777). Danish landscape painter who settled in London *c.*1760. Designed cartouches for Thomas Jefferys's map of Yorkshire (20 sheets), 1771-1772.

Dall, P. *Master, RN*. Leith Harbour, 1831.

Dall, William Healey (1845-1927). Geographer, naturalist and explorer of Boston. *Alaska*, 1869, 1884; *Map showing... the Indian tribes of Washington Territory*, 1876; Pacific coast pilot, 1879, 1883.

Dall'Acqua, Giovanni Domenico. Draughtsman. *Iconographia Vicenza*, 1711.

Dallas, W.L. *Charts of the Bay of Bengal*, 1886; *Meteorological Atlas of the Indian Seas*, 1908.

Dalla Vedova, Giuseppe (*b*.1834). Geographer of Padua.

Dallaway, James (1763-1834). Historian and topographer. *A history of the western division of the county of Sussex...*, London, T. Bensley 1815, 1819 (with maps by Dallaway and others). **Ref.** KINGSLEY, D. *Printed maps of Sussex* (1982) pp.384-385.

Dallinger, J. Engraver. W.S. Millard & Joseph Manning's *Plan of the City of Norwich*, 1830.

Dallinger, Robert (*fl*.1768-1806). Estate surveyor and schoolmaster in Essex. Rivenhall, 1773 MS; Wickham Bishops, 1775 MS.

Dalli Sonetti, B. See **Bartolommeo** dalli Sonetti.

Dallington, *Sir* Robert. Survey of the State of Tuscany, 1596 MS, printed 1605.

Dallman, *Captain* Eduard (*d*.1896). German seaman and explorer. *Grönland*, 1873-1874; *Kara-See*, 1878; *Neu-Guinea*, 1884-1885.

Dall'Olmo, R. See **Olmo**.

Dally, A. *Carte des Régions militaires de la France*, *c*.1874.

Dally, N. *Amérique centrale et les Antilles*, Brussels 1840.

D'Almada, A. See **Almada**, Andreas d'.

Dalmas, J.B. *Carte géologique du Département de l'Ardèche*, 1859.

Dalmatino, Natale. See **Bonifacio**, Natale.

D'Almeida. See **Almeida**.

Dalmont, Isidore. Letter engraver. For Carl Müller's *Geographi Graeci Minores*, 1855; *Carte des Phares des Côtes de France*, 1855; Paris historique, *c*.1865; also for J. Migeon, *c*.1874.

Dalorto [Dell'Orto], Angelo [Angellino]. Catalan or Italian portolan maker. La carta nautica, 1325; World, Genoa 1330. **N.B.** Angelino **Dulcert**, who produced a Mappemonde in 1339, is often considered to be the same man.

Dal Pian, Giovanni Maria (1764-1800). Engraver, active in Venice, and in Vienna from 1797. Engraved maps for Santini's *Atlas Universel*, 1776.

Dal Re, Marc Antonio. See **Re**.

Dalrymple, Alexander (1737-1808). Prolific writer and hydrographer in the service of the East India Company, 1752-1771; became first Hydrographer to East India Co., 1779; appointed director of the newly-founded Hydrographical Office, 1795. Published many of his own works as well as those of others, established at Soho Square, London (1771), later High Street, Marylebone (1806). Many manuscript surveys in East Indies, 1761-1764; *An Account of the Discoveries Made in the South Pacifick Ocean previous to 1764*, London, A. Dalrymple 1767; *An Essay on the Most Comodious Methods of Marine Surveying*, 1771; *A collection of charts and memoirs*, 1772; *A collection of plans of ports in the East Indies*, London 1774-1775, 1782, 1787; *A collection of Voyages chiefly in the Southern Atlantick Ocean*, London 1775; *Collection of Nautical Papers concerning the Bay of Bengal*, 1785; Plan of Port San Francisco in New-Albion, 1790 (taken from a manuscript plan by Cañizares *c*.1775); Formosa, 1792; Town of Cheltenham, 1806; and many other maps, charts, essays and memoirs. **Ref.** DAVID, Andrew C.F. 'Alexander Dalrymple and the emergence of the Admiralty Chart' in: HOWSE, D. (Ed.) *Five hundred years of nautical science 1400-1900* (Greenwich 1981) pp.153-164; COOK, A. *Imago Mundi* 33 (1981) pp.46-64; EDWARDS, H.S. *The Map Collector* 4 (1978) pp.19-29.

Dalrymple, *Captain* John (*d*.1766). Scottish officer. Organised publication of revised edition of Joshua Fry and Peter Jefferson's ...*Virginia*, 1755.

Alexander Dalrymple (1737-1808). (Taken from Map Collectors' Circle*)*

D'Altischofen, P. See **Pfyffer** von Altischofen.

Dalton, I. *Plan of the Level of Ancholme,...Lincolnshire*, 1791 (with E. Johnson).

Dalton, J. *The Coti River,...Borneo*, 1828-*c.*1837.

Dalton, William Hugh (pseud.). Supposedly revised and improved Alexander Hogg's part-work *The New British Traveller*, 1784 as *The New and Complete English Traveller*, *c.*1794. **Ref.** HODSON, D. *County atlases of the British Isles published after 1703* Vol.III (1997) pp.118-120.

Dal Trozzo, A. Publisher. Poland, Warsaw 1810.

Daly, D.D. Survey of River Linghy, 1875; Part of the Malay Peninsula, 1876; Northern Perak, 1882; British North Borneo, 1888; Sketch Map of Native States adjoining Malacca, 1875.

Dalyell, *Sir* Robert. *Vilayet of the Danube...*, 1869.

Dalzel, *Captain* Archibald. Cato or Regio Creek, 1785; Coast of Africa, 1797, 1799; Plan of Loanda St. Paul, 1797; Atlantic, 1802.

Da Massaio, P. See **Massaio.**

Dameld, J. See **Daniel**, John.

Damell, Giovanni. See **Daniel**, John.

d'Amendale. See **Damme**, Jan van.

Damerum, William. American surveyor. *Map of the Southern part of the State of New York...*, New York, Wm. Damerum 1815 (engraved by P. Maverick), the plate was altered and re-used by Roderick Nevers of Hartford, 1851, 1852.

Damian, Josef. *Tiefenkarte des Caldonazzo- und des Levico-Sees*, 1892.

Damianus. See **Schoonebeck**, P.D.

Damien de Templeux. See **Templeux.**

Damilano, J. Plan of Alexandria, 1887.

Damme, Alfred (*d.*1892). German cartographer.

Damme, Jan van, *sieur* d'Armendale [Aumale]. 'Géographe du Roy de France', 1634. ...*Comté de Charolois*, H. Hondius, 1631, used also by J. Blaeu, 1635, M. Tavernier and others.

Dammert, I.C. *Plan von der Neustadt Pirmont...*, 1790.

Damoreau, —. See **Smith & Damoreau.**

Da Mosto. See **Cadamosto.**

Damour, Amedée. Geological maps of Cyprus, 1854 (surveys used by J. Wyld, 1878).

Dampier, *Captain* William (1652-1715). Hydrographer and circumnavigator. *A New Voyage around the World*, 1697-1709; *A View of the General & Coasting Trade-Winds in the Atlantick & Indian Oceans*, 1699; [Winds of the]...*great South Ocean*, 1699; *A Draught of Sharks Bay*, 1701.

Dampierre, *Captain* —. Naval officer. A new chart of Saldanha Bay, inset on William Heather's ...*chart of the Cape of Good Hope*, 1796. **Ref.** WORAM, J.M. *The Map Collector* 44 (1988) pp.19-20.

Damrong, *Prince* Rajanubhad. Siam, Calcutta 1897.

Dana, James Dwight (1813-1895). American scientist and geologist at Yale University. *Manual of geology*, 1862 (used by the U.S. Geological Survey).

d'Anania, G.L. See **Anania**, Giovanni Lorenzo d'.

Dance, father & son.
• **Dance**, George, *the elder* (1700-1768). Architect and Surveyor to Corporation of the City of London.
•• **Dance**, George, *the younger* (1741-1825). Architect and Clerk of Works to the City of London, 1768-1815. Plan of the Excise Office, 1768; Plan of Newgate, 1780; *The Thames between London Bridge and Blackwall Reach showing proposed Docks at Rotherhithe and the Isle of Dogs*, 1796; Port of London Improvements, Luke Hansard 1799-1803; A Survey of the River Thames, 1803.

Dance, Nathaniel. Plan of Fleet Market, London 1737.

Danckerts [Danckaerts; Danckers; Dankerts; Danquerts] family. Publishers and engravers of Amsterdam. **Ref.** KRETSCHMER, DÖRFLINGER & WAWRIK *Lexicon zur Geschichte der Kartographie* (1986) p.150 [in German]; SHIRLEY, R. *The mapping of the World* (1983) several entries; KOEMAN, C. *Atlantes Neerlandici* Vol.II (1969) pp.88-97; KEUNING, J. *Imago Mundi* 12 (1955) pp.136-139.
• **Danckerts**, Cornelis (1561-1631). Printseller, born in Amsterdam, active in Antwerp.
•• **Danckerts**, Cornelis, *the elder* (1603-1656). Engraver and publisher in Amsterdam and Nieuwendijk. Son of Cornelis (above) whom he succeeded in 1631; not to be confused with his first cousin, Cornelis Danckerts de Rij. Engraved for John Speed, 1627; Charte universelle de toute le monde, 1628 (with Melchior Tavernier); Augustine

Captain William Dampier (1652-1715). Copperplate engraving by Charles Sherwin. (Taken from Map Collectors' Circle*)*

Ryther's *the Cittie of London*, *c.*1633; *Nouvelle et Exacte Carte du Duche de Brabant*, Sanson 1635; Pieter Bertius's Asia, 1640; Europe, 1643; Danube, 1647; Panorama of London, 1647; Seven United Provinces, 1651; Holland, 1656.
••• **Danckerts** [Danckert], Dancker (1634-1666). Son of Cornelis the elder. Publisher in 'de Danckbaerheyt', Amsterdam. Noord Holland, 1657; *Daniae Regni Typus*, 1657; *Orbis Terrarum...*, 1658; Germany, 1661; *Plan des principales cités de l'Univers*, [n.d.].
Ref. LOWENHARDT, W. *The Map Collector* 25 (1983) pp.10-12.
••• **Danckerts**, Justus [Justinus], *the elder* (1635-1701). Son of Cornelis the elder. Engraver, mapseller and publisher in the Kalverstraat, Amsterdam. Hungary, 1662; assembled and issued many atlases, *c.*1670-1700 (with his sons); *Accuratissima Regnorum Seuciae, Daniae, at Norvegiae, tabula...*, Amsterdam *c.*1688-1700; *Nova orbis terrarum tabula* [large wall map], *c.*1680; *Asiae Tabula...*, *c.*1680; *Geographische-en Historische Beschryving Der Tien Spaansche Provintien*, c.1690 (pocket atlas), French edition published 1694; *D'Forteresse der tien spaansche Provintien of Belgii Regii...* (26 plans on one sheet), *c.*1690 (re-published by Cornelis, 1696, and by Reinier & Josua Ottens, *c.*1727 and later); also made globes.
•••• **Danckerts**, Justus, *the younger* (*d.*1692). Son of Justus the elder. Engraver and publisher in Amsterdam. *Nova totius terrarum orbis tabula*, *c.*1685.
•••• **Danckerts**, Theodore, *the younger* (1663-c.1727). Son of Justus the elder, worked on many maps published in Danckerts Atlases. *Regni Sicilae et Insulae Maltae et Gozoe*, *c.*1695.
•••• **Danckerts**, Cornelis, *the younger* (1664-1717). Son of Justus the elder. Engraver and publisher in Amsterdam. Fortresse van Naples, 1680; *Nova totius terrarum orbis tabula*, *c.*1695, engraved Jacob Falkema; *De Werelt Caart*, *c.*1700. Business continued by his widow until her death in 1719.

Danckerts de Rij, Cornelis (1596-1662). Surveyor of Amsterdam, first cousin of Cornelis Danckerts the elder. Plan of Grol, Claes Jansz. Visscher 1627; *Kaertboeck ...stadt Amsterdam*, 1642-1643.

Danckwerth, brothers.
• **Danckwerth**, Caspar (*d.*1672). *Newe Landesbeschreibung der zwey Hertzogthumer Schleswich und Holstein*, 1652 (with Johannes Mejer of Husum).
• **Danckwerth**, Joachim. Collaborated with brother Caspar above.

Danckwerth, F. See **Dankworth**, F.

Dancy, *Lieutenant* Francis L. Surveyor General. Portsmouth, Ohio, to Linville, North Carolina, 1836; State of Florida, 1860.

Dandalle, —. *Carte de France* (22 sheets), Paris, Charles Picquet 1822.

Dandeleux, H. Engraver. Plan de Dantzig et ses environs, 1813; for Laurent Gouvion de Saint-Cyr, 1821; for Auguste-Henri Dufour, 1822; for Adrien Hubert Brué, 1822; for A. Lorrain, 1836.

Dandeleux, Pierre-Léonard (*b.*1783). French engraver. *Carte générale physique et routière des 22 cantons de la confédération Suisse*, Paris, A.H. Brué 1822, 1835 (published in Brué's *Atlas universel*).

Danes [Daynes], Richard (*fl.*1628-1665). Estate surveyor, produced manuscript maps of various parts of England including Epping, 1634 MS.

Danet, publishers of Paris.
• **Danet**, Guillaume (*c.*1670-1732). Publisher and paper merchant, 'sur le Pont N. Dame a la Sphere Royale', Paris. Son-in-law and part successor (with Jacques-François Bénard) of Nicolas de Fer 1720, whose maps he re-issued. *Les frontières de France et d'Espagne...*, Paris 1709; India, 1721; Santo Domingo, 1723; Paris, 1724; Gibraltar, 1727; London, 1727; Roussillon, 1730; *L'Amérique Méridionale et Septentrionale...*, 1731, 1740; *L'Afrique*, 1732; *Plan de Luxembourg, Ville Forte Capitale...*, Paris 1734, 1742.
• **Danet**, *veuve* [Marguerite de Fer] (*fl.*1732-1746). Took over on the death of her husband and continued in business until her own death in 1746. Business sold by her daughter Marie Bertrand to sister Marie-Anne and Nicolas-Jean-Baptiste Desbois.

Danet, Thomas. *The Description of the Low Countries into English*, 1591 (translation of the work by L. Guicciardini).

Dangeau, Louis de Courcillon, *abbé* de (1643-1723). Cartographer of Paris. *Planisfere*, 1697.

Danger, Henry. Surveyor of Sydney. *Map of the River Hunter*, 1828.

Dangerfield, Frederick [& Co.]. Lithographer and printer of 22 Bedford Street, Covent Garden, London. Output included many official maps associated with the colonies. *Geological Map of British Guiana*, 1873; *Plan showing the Parochial Division of the City of London*, 1880; *Meteorological Atlas of the British Isles*, Meteorological Council 1883; *Map of part of Benin...*, 1883; *Town of Katanu*, 1884; *Map showing the boundaries of the South African Republic...*, 1884; *South Coast of Newfoundland*, 1885; *Newfoundland. Fisheries Commission 1884-1885*, 1885; ...*Venezuela*..., 1886; South America, 1888 (copy of 1748 d'Anville map); Guyana, 1888; *Sketch map of the Territory in dispute between Venezuela and British Guiana...*, 1888; and many others.

d'Anghiera [d'Angleria]. See **Martyr**, P.

Daniel, H.A. (1812-1871). German geographer. Manuals of geography, 1845-1849.

Daniel [Dameld, Damell, Daniell], John [Giovanni] (*c*.1565-1649). Chartmaker 'in St. Katherines near unto the iron gate by the Tower of London'; Freeman of the Drapers' Company, 1590; compass maker and manuscript chartmaker of the 'Thames School'. *Chart of Spitsbergen*, 1612 (made for the Muscovy Company but spirited to Holland and published by Hessel Gerritsz.); Chart of the southern Atlantic, 1614; West of America, 1637 MSS. **Ref.** CAMPBELL, T. in: WALLIS & TYACKE (eds.) *My head is a map* (1973) pp.81-106.

Daniel, R.W.T. Mississippi, 1857.

Daniel, Richard. *Compendium of useful hands*, 1664; *A Map of ye English Empire in ye continent of America*, 1679.

Danielis, C. Engraver at the Romanian Ministry of War. *Harta principatelar unite Romane...*, 1863.

Daniell, C.B. Publisher of Beaumont, Texas. Map of Jefferson County, 1897.

Daniell, John. See **Daniel**, John.

Danielow [Danielov], Ivan. *Empire de Russie* (12 sheets), Vienna 1812; *L'Empire Ottoman* (9 sheets), 1813.

Danielson, J. Karta ofver Mellersta, 1880.

Danit, —. Printer of Amsterdam. Willem Barentsz.'s *Carta Trium Navigationum...* [Polar Regions], in Jan Huyghen van Linschoten's *Itinerario*, 1599.

Danizy, A. See **Danysi**, Auguste Henri.

Dankerts. See **Danckerts** family.

Dankworth [Dankwerth], F. Engraver of Philadelphia. Worked for Henry Schenck Tanner, 1833, 1844; World, 1838; *Mitchell's Reference and Distance Map of the United States*, 1838 (with J.H. Young, E. Yeager and E.F. Woodward); also for Samuel Augustus Mitchell, 1834-1849; *Map of the State of Maine...*, 1844 (with J.H. Young); *Mouth of the Chester River* [USA], 1849 (with O.A. Lawson, W. Smith and S.T. Pettit); *Salem Harbour Massachusetts From a Trigonometrical Survey...*, 1855 (with H.M. Knight, J. Knight, S.E. Stull and J. Young); engravings used in the preparation of *Preliminary Chart of Charleston Harbor and its approaches*, U.S. Coast Survey Chart 1855.

D'Anna, Alessandro (1746-1810). Painter and draughtsman in Naples (1779-1810), worked in G.A. Rizzi-Zannoni's Officina Topografica.

Danov, H.V. (1828-1911). Bulgarian cartographer. European Turkey (4 sheets), printed Vienna 1863, re-issued 1870.

Danquerts. See **Danckerts** family.

Danti [Dante], Egnazio [Carlo Pellegrino; Ignazio] (1536-1586). Astronomer, mapmaker, instrument maker and monk of Perugia. Adopted the name Egnazio when he joined the Dominican Order in 1555. Professor of mathematics at the University of Bologna from 1576; cosmographer to Pope Gregory XIII, 1580; Bishop of Alatri, 1583. Designed and began painting 57 wall maps of the countries of the world in the Guardaroba Nuova at the Palazzo Vecchio, Florence, 1563-1574 (he

completed 30, the remaining 27 were finished by S. Bonsignori); wrote several theoretical works including *Trattato della Sfera*, 1567, published 1571; Globes, 1567; surveys of Perugia, Bologna, Umbria and other areas, from 1577; map of Perugia, 1580 (engraved by Cartaro, later used by Ortelius); ...*Perugia*..., 1580 (on wall of Palazzo del Governato, Bologna); designed a sequence of maps of Italy for the Galleria del Belvedere of the Vatican, completed under his supervision *c*.1581; Orvieto, 1583 (later used by Ortelius).
Ref. GAMBI, L. *The gallery of maps in the Vatican* (New York 1997) also in Italian as *La Galleria delle Carte geografiche in Vaticano* (Italy 1997); GANADO, A. & AGIUS-VADALÀ, M. *A study in depth of 143 maps representing the Great Siege of Malta 1565* (1994) pp.478-482; MEURER, P. *Fontes cartographici Orteliani...* (1991) pp.136-137 [in German] .

Danti, Pietro Vincenzo di Bartolomeo Rinaldi (c.1430-1513). Mathematician and globe-maker of Perugia.

d'Anville, J.B.B. See **Anville**, J.B.B. d'.

Danysi [Danizy], Auguste Henri (1698-1777). French astronomer and cartographer.

Danz, C.F. Schmalkalden, 1848 (with Caspar Friedrich Fuchs).

Daoust, Claude. Printer of Lyons. World map for Dyamatier, 1506.

Dapper, Olfert (1636-1689). Doctor and geographer of Amsterdam. Translated works of Montanus into German. Mesopotamia, 1661; Africa, 1668; China, 1670; Persia, 1672; Asia, 1672; America, 1673; African Islands, 1676; Syria & Palestine, 1677; Morea, 1688; *Carte de l'Archepel*, 1690; *Description exacte des isles de l'Archipel*, Amsterdam 1703.

D'Après de Mannevillette, J.B.N.D. See **Après** de Mannevillette, Jean Baptiste Nicolas Denis d'.

Darbamont, --. Toul & surrounding country [France], *c*.1700.

Darbishire, Bernhard Vernon (*fl*.1895-1920). Mapmaker in his own right, also in partnership with William Stanford. *Map of Mashona Land and Manika* [Zimbabwe], Royal Geographical Society 1894 (With E.G. Ravenstein); Asia Minor, 1895; explorations of Sir William MacGregor, 1896; *War Atlas* (12 maps), Oxford University Press 1915; plus numerous maps of all parts of the World.
• **Darbishire and Stanford** (*fl.c.*1902-1904). Partnership of B.V. Darbishire and William Stanford at the Oxford Geographical Institute. *Map of the World shewing British Possessions...*, 1903; *The Autograph Handmaps*, from 1903 (outline maps with relief).

Darby, Henry William. *London Guide*, 1842; *The Environs of Cheltenham*, 1843; *Plan of the Town of Cheltenham*, 1843.

Darby, J.G. Engraver of Buffalo. O.G. Steele's *Steele's Map of the States of Ohio, Indiana and Illinois...*, 1835.

Darby, John, father and son. Publishers and printers in Bartholomew Close, London. John Seller's *English Pilot*, 1671-1677; *Atlas Maritimus*, 1675; Senex's New General Atlas, 1721; Nathaniel Cutler and Edmond Halley's *Atlas Maritimus et Commercialis*, 1728.

Darby, M. Engraver. John Grundy's *Plan of the East Fen*, 1774.

Darby, William (1775-1854). Surveyor and geographer of Hanover, Pennsylvania. *Louisiana*, Philadelphia 1816; Plan of Pittsburgh, 1817; *Chart of Mobile Bay*, 1818; *A Map of the United States Including Louisiana*, 1818 (engraved by Jas. D. Stout), published in Darby's *The Emigrant's Guide to the Western and Southwestern States and territories*, New York, Kirk & Mercein 1818; *A Tour from the city of New York to Detroit...*, New York, Kirk & Mercein 1819 (maps engraved by W. Hooker); map of Florida, 1821; *Universal Gazetteer*, 1823, 1827; *Map of the Southern Part of the Province of Texas*, [n.d.].

Darbyshire, B.V. See **Darbishire**, Bernhard Vernon.

Darbyshire, George C. *Plan showing the Railway from Geelong to Ballarat* [Victoria, Australia], 1857.

Darbyshire, John. *Township of Gooramadda* [Victoria], 1859.

D'Arcy, *Lieutenant* Robert, Royal Engineers. *Plan of Mahon Harbour the Fortress of St Philips...* [Minorca], 1782 MS; manuscript maps and charts of the Bahama Islands and Barbados, 1785-1800; *Plan of Fort George in the Island of Minorca...*, 1799 MS; *Plan of the Entrance of the Harbour of Mahon* [Minorca], 1801 MS; Flushing, 1809 (all at the Public Record Office, Kew).

Dardano, A. *La Regione Veneta e le Alpi...*, Novara 1917.

Dare, Nathaniel B. *Cleveland*, 1868.

Daret [Darret], Pierre (*c.*1604-1678). 'Graveur ordinaire du Roy', at 'rue St Denis au Chateau de Pontoise' (1634); 'rue Quincampois' (1635); 'rue St Jacques...à la Perle' (1639); 'sur le Quai des Gesvres près du Palais, à la Patience' (1652); Paris. Worked for Louis Boissevin.

D'Argentré, B. See **Argentré**, Bertrand d'.

Darine, —. Engraver. Oakley's plan of Metz, 1754.

Darinel. See **Boileau** de Bouillon, Giles.

D'Aristotile. See **Aristotile**, Nicolo d'.

Da Riv, N. Engraver. Carta topografica dei colli, 1836.

Darke, Sydney W. *Salt Lake City*, 1887.

Darling, Charles. Engraver for the Ordnance Survey, 1848.

Darling, Eugene F. Map of Ogden [Utah], 1889.

Darling, Orlando. Surveyor. *Fort Scott, Bourbon County*, 1860.

Darly, Pierre [Petrus]. French engraver for Jean Moreau. *Peregrinatio Israelitarum...*, 1621, 1622.

D'Armendale. See **Damme**, Jan van.

Darmentier, L.J. Engraver. Maps for Nicolas de Fer.

Darmet, J.M. Engraver in Paris. *Plano topográfico de la Ciudad de Buenos Ayres... Levantado en el Año 1814...*, 1817; Napoleonic battle plans, 1820-1824; l'Amérique Méridionale, 1825; *Plan de la Ville de Lyon*, 1830; *Carte de la Colonie Française d'Alger*, 1836; *Carte géographique, statistique, historique et politique de la Colombie*, 1840.

Darmstadt, Johann Adolph. Engraver. Plan von Ratzeburg, 1799.

Darnius, Oleguer de Taverner d'Ardena *conde* de (1676?-1727). Cartographer from the village of Darnius in the Catalan Pyrenees. Produced many military and administrative maps of Catalonia, most of them manuscript. He died in Barcelona. *Nuevo mapa del Principado de Cataluña y sus confines*, 1726 (map showing the varying condition of roads in Catalonia, engraved by Marcos Lomelin).

D'Arnoullet, B. See **Arnoullet**, Balthazar d'.

Darondeau, Benoit-Henri (1805-1869). French hydrographer. Prepared sailing instructions and charts for the Dépôt de la Marine including a chart of the north coast of Sicily, 1858.

Darré, C. *Plan de la Ville d'Elbeuf*, 1866.

Darre, Nils Stockfleth (1765-1809). Scandinavian geographer. Worked on surveys from 1799, becoming Director of the Topographical Survey from 1806.

Darret, P. See **Daret**, Pierre.

Darsy, Eugène. Atlas de géographie physique, politique et historique, Paris 1890.

Dartois, Jean. *Mappemonde Historique*, 1750.

Darton, family and company. Engravers and publishers, 55 Gracechurch Street, London (from 1787); 'Map, Print, and Chart Warehouse, 40 Holborn Hill', London (to 1810); then 58 Holborn Hill. The succession of business and variety of imprints is confusing.

• **Darton**, William *Senior* (1755-1819). Engraver in Tottenham, London for Dilly & Robinson, 1785. In 1787, set up as printseller and publisher with Josiah Harvey. *A Chart of* *the World...*, 1785; *The British Colonies in North America from the best authorities*, 1785 (published in The Atlas to *Guthrie's system of geography*).

• **Darton & Harvey** (*fl.*1787-1814). Partnership of William Darton Senior and Josiah Harvey at 55 Gracechurch Street, London. *A new Pocket Plan of London Westminster and Southwark*, Darton & Harvey 1792, 1800; *Maps of the several counties and shires in England...*, Darton & Harvey 1804.

•• **Darton**, William *Junior* (*fl.*1804-1835). Son of William, above. Set up on his own as publisher at Holborn Hill House in 1804, using a confusing variety of imprints. *Dix's Juvenile Atlas*, William Darton 1811; *The Union Atlas*, William Darton 1811 and later; Thomas Dix's *A complete atlas of the English counties...*, William Darton 1822 (republished as *The Counties of England*, William Darton & Son 1835); *Darton's New Miniature Atlas...*, William Darton 1822, 1825.

•• **Darton**, W. & T. (*fl.*1807-1810). William Junior and his son Thomas. *The Strangers Guide through...London and Westminster*, W. & T. Darton 1807.

•• **Darton, Harvey & Darton** (1814). *London Westminster & Southwark...*, Darton, Harvey & Darton *c.*1814.

•• **Darton**, William & Son (*fl.c.*1830-1837). William Junior and his son John. *Darton & Son's New Map of...London...*, William Darton & Son *c.*1835; *The Counties of England*, William Darton & Son 1835 (taken from Thomas Dix's *A complete atlas of the English Counties...*, William Darton 1822).

••• **Darton**, Thomas Gates. Son of William Junior, above, with whom he worked from 1807-1810, then on his own as printseller and publisher, 25 Great Surrey St., near Blackfriars Bridge, London. Robert Rowe's plan of London, 1811 edition.

••• **Darton**, John Maw. Son of William Junior, above. Joined the family firm in 1830.

•••• **Darton & Clark** (*fl.c.*1841-1845). Partnership of John Darton and Sam Clark. *Darby's London Guide*, Darton & Clark 1841; *Twelve Miles round London*, Darton & Clark 1843; *Reuben Ramble's travels through the counties of England*, Darton & Clark 1845.

•••• **Darton & Co.** (*fl.c.*1850-1858). Republished *Reuben Ramble's travels...*, 1850; *Collins' railway & pedestrian atlas of England...*, Darton and Co. 1858.

•••• **Darton**, John & Co. *Hertford...*, John Darton & Co. *c.*1860.

•••• **Darton & Hodge**. Publishers of 58 Holborn Hill. *The Stranger's Guide. A New Map of London*, Darton & Hodge *c.*1862, republished as *Darton & Hodge's New Map of London*, Darton & Hodge 1864.

Darton, L. Engraver. Vignette view of York on John Tuke's map of Yorkshire, 1787.

Darton, N.H. Dakota artesian basin, [n.d.].

Das, N.C. Ancient India, *c.*1899.

Das, Sarat Chandra. Travelled in Tibet from 1879; *Journey to Lhasa and Central Tibet*, 1904.

Dasaques, Louis de. Plan of Courtrai, in J. Blaeu's *Tooneel der Steden van der Vereenighde Nederlanden* ['Town books'], 1649.

Dasauville, W. See **Dassauville**, William.

Daser, P. Plan of Tellichery, published by Alexander Dalrymple 1780.

Da Serra, Maximiano José. Survey of the River Minho and environs [Portugal], 1798.

Dash, T.A. London Parks, 1862-1863; *Brompton Cemetery*, 1884.

Dashiell, S.L. North America, *c.*1800; *Map of the Northern Part of the state of Maine*, Washington DC, 1830.

Da Silva, F.M.P. Surveys of the coast of Estremadura, 1842-1843 (with C.M. Batalha and C.F.B. De Vasconcellos).

Dasori, Giovanni Battista. Engraver. Pontine Marshes [Italy], 1795.

Dassauville [Dasauville], William. Engraver of Edinburgh. *Scotland*, 1805; maps for John Thomson's *Atlas of Scotland*, 1815-1832; Western Isles (3 sheets), 1822-1823.

Dassie, F. Routier des Indes, Paris 1677; Description générale des costes de l'Amérique, Rouen 1677; Le pilote expert, 1683.

Dasypodius [Hasenfratz], Conrad [Konrad] (*c.*1530-1600). Swiss cartographer in Strasbourg. Globes, 1574 (in the 'Strasbourg clock').

Datan, E. *Stadsplan over Borås*, 1728.

Dati, brothers. Credited with *La Sfera*, *c.*1400, consisting of manuscript charts, drawings and views.
• **Dati**, Leonardo (1360-1425). Florentine cosmographer and poet.
• **Dati**, Gregorio di Anastasio [Gorio di Stagio] (1363-1436).

Dau, Friedrich von (1796-1867). German cartographer.

Dau, W. German economic cartographer. Produced maps showing regional economy and employment patterns. *Kartogramme der Erwerbsdichte*, Petermann 1906.

D'Aubant, *Major* Abraham (*fl.*1754-9). Ordnance draughtsman. *Plan of the town, citadel, dock and country round Plymouth*, 1756; Jersey, 1737; Goat Island, *c.*1778; *Plan of the Town and environs of Newport, Rhode Island*, 1779; plan and sections of the Cobb, Lyme Regis, Dorset, 1783; Fort Brown, 1799.

Daubenton, P.P., *SJ*. Islas Carolinas, 1757.

Daubeny, Charles. Geological Map of Sicily, 1825.

D'Aubert, B. See **Aubert**, Benoni.

Daublebsky, Moritz, *Freiherr* von (1884-1917). Austrian general and cartographer.

Daubrée, Auguste (1814-1896). French geologist. *Carte géologique du Département du Bas-Rhin*, 1851; Eaux souterraines, 1887; Régions invisibles du globe, 1888.

Daudet. Mapseller of Grande rue Mercière, Lyons.
• **Daudet**, —. For Nicolas Baillieul le jeune from 1747. Bailleul's *L'Italie divisée selon des differents Etats qui la partagent*, 1747; Gaspard Bailleul's *L'Afrique Divisée...*, 1752.
• **Daudet**, *veuve* —. Widow and successor of above. Sold work for Louis Joubert, 1767.

Daudet, *chevalier* Louis Pierre (*fl.*1722-1744). Graveur et Géographe du Roy. *Carte du cordon sanitaire établi lors de la peste de 1721, entre le Rhône et la Dordogne...*, 1721; *Plan de la Ville de Reims*, 1722; Nouveau guide des chemins de France, 1724-1768; *Route de Paris à Reims*, [n.d.]; *Carte des différentes routes de Paris à Compiègne, de Compiègne à Soissons, et de Paris à Soissons*, 1728; Antoine Chopy's *Carte du Lac de Genève*, 1730.

Daugustine, M. de Combis. See **Combis** D'Augustine, M. de.

Daulphinois, N. See **Nicolay**, Nicolas de.

Daumas, *General* Eugène (1803-1871). French cavalry officer. Director of the Bureaux arabes in Algeria (1841-1847). Conducted a series of detailed interviews with the Arab population which resulted in several descriptive works on the Sahara. His information formed the basis of Captain Gaboriaud's *Carte du Sahara Algérien*, 1845 (engraved by J. Schwaerzlé).

Daumas, *Reverend* F. Voyage to Basutoland, 1836; Missionary stations in South Africa, 1840; Sketch of Basutoland, 1868 MS.

Daumont, Jean-Francois (*d.c.*1775). Publisher and mapseller, 'rue de la Ferronnerie à l'aigle d'or' (1754) and 'rue St Martin près St Julien' (1758), Paris. Successor to Jean-Baptiste Nolin. Nolin's Amérique 1754, 1759; *La Terre Sainte Divisée en ses Douze Tribus...*, 1756; *Carte Generale du Royaume de Bohème du Duché de Silesie...*, 1757; *Le Duché de Silesie*, 1757 (a republication of Tobias Mayer's map of 1749); *Le Cercle de Basse Saxe Dressée sur les Cartes des Héritiers d'Homann à Nuremberg*, 1757 (engraved by Desbrulins); *Nouvelle carte du gouvernement general d'Orleanois...*, 1761.

Daurkin, Nikolay. Cossack of Chukchi origin. Map of the coast from the Kolyma to the Anadyr Rivers [E. Siberia], 1765.

Daussy, Pierre (1792-1860). French engineer and astronomer. Directeur d'hydrographie, Dépôt de la Marine. Côte occidentale d'Afrique, 1833; Océan Atlantique Méridionale et Septentrionale, 1834-1866; *Cartes des Côtes Méridionales d'Afrique...*, Paris 1838; *Carte des Mers Australes... en 1842*, 1865 (engraved by Jacobs).

D'Auteroche, Chappé. See **Chappé** d'Auteroche, Jean.

Dauthendey [Dauthendeij], Caspar. Architect and mathematician. *Ducatus Brunsuicensis*, Blaeu 1640.

Dauty & Rovet. Publishers of Paris. J.B.L. Charle's *Nouvel Atlas national de la France*, 1833.

D'Auvergne, Philip, *prince* de Bouillon (1754-1816). Plan of Fairhaven, 1744. **NB.** Compare with **Auvergne**, Philip de.

D'Aux, *chevalier* —. Plan of Boston, 1693.

Daux, A. (*d.*1881). French archaeologist. Plan of Carthage, *c.*1884.

Davenet, *Captain* —. *Renseignements sur les communications de Sébastopol, Balaclava et Beïdar à Simférol* [Crimea], 1855 MS.

Davenport. See **Dewitt** & Davenport.

Davenport, J. *Road Book from London to Naples*, 1835.

Davenport, Samuel. Engraver. *Quebec*, 1835.

Daventerius [Daventria]. See **Deventer**, Jacob van.

Da Verrazano. See **Verrazano**.

Davet, P. Engraver. *L'Isle St. Honoré de Lérin*, 1635.

Davey. Surveyor. Estates at Weston, 1774.

Davey, J.M. (*fl.c.*1830-1842). Land surveyor of Canterbury, produced estate and tithe maps of parts of Kent including Staple & Ash, *c.*1830.

Davey, N.T. Surveyor. Cachar, 1870, 1877.

Davey, Peter. Bookseller, involved in the publication of Thomas Hutchinson's *Geographia Magnae Britanniae*, 1756 edition; Kitchin's *Geographia Scotiae*, 1756 (with others); Thomas Salmon's *Geographical Grammar*, 1758.

David, Johan (1796-1846). Cartographer and lithographer of Vienna. Engraved 6 sheets (on copper) of *Karte des Erz-Herzogthums Oesterreich ob- und unter- der Enns... reduzirt und gezeichnet im Jahr 1813...* (29 maps), Vienna, General Quartiermeister Stab 1813; Tirol, *c.*1823; for Joseph Meyer, *c.*1835.

David, Johann. Engraver of Vienna. Engraved maps of Moravia and Bohemia for K.K. Militaerischen Geographischen Institute, c.1844-1847 (often worked with J. Kiss). *Special Karte der Markgrafschaft Maehrenmit den Antheilen des Herzogthums Schlesien...*, 1844; *General-Karte der Markgrafschaft Mähren mit den Antheilen des Herzogthums Schlesien...* [Czech Republic], 1846 (with J. Kiss); engraved at least 3 sheets of *Special-Karte des Koenigreiches Boehmen* (39 maps), 1847-1860 (with J. Kiss). **N.B.** Compare with Johan **David**, above.

David, M. Cartographer. Hannover, *c.*1803.

David, Martin Alois (1757-1836). Mathematician and astronomer, director of the Prague Observatory. *Charte vom Koenigreiche Boheim...*, 1800, 1834.

David, Pierre. Publisher, 'sur le Pont Neuf devant la Samaritaine', Paris. *Voyage aux Indes Orientales*, 1646.

David, T. *Topographische Karte... Wien*, 1823 and later.

David, *Sir* Tannet William Edgeworth (1858-1934). Welsh geologist, born Cardiff. Geological map of Vegetable Creek, New South Wales, Sydney 1885; Tingha & Elsmore Districts, 1895.

Davidos, C.F. Mapmaker. *Spanien und Portugal...nach dem Atlas Thomas Lopez und andern sichern Hülfsmitteln* (9 sheets), Vienna, T. Mollo 1820 (title etc. in both German and French), 1835 edition published by F. Mollo.

Davidson, A.F. Lower Platte River, 1845 MS.

Davidson, Allan Arthur (*c.*1878-1930). Mining engineer and metallurgist. *Map of explorations in Central Australia*, 1902.

Davidson, C. Wright. Atlas of Minneapolis, 1887.

Davidson, George (1825-1911). Astronomer and geographer, served under A.D. Bache and F.M. Thorn as assistant in the Coast Survey [United States]. Later produced works on the history of the discovery of the north-west coast. Worked with others on: *San Diego Entrance and Approaches*, 1853; *Cape Flattery and Neé-ah Harbor, Washington*, 1853; Canal de Haro & Strait of Rosario, 1854 (with J. Alden); *Reconnaissance of the West Coast of the United States... From San Francisco to Umpquah River*, 1854 (with J. Alden); *Western Coast of the United States (Northern Sheet)...*, 1855 (with J. Alden); Coast Pilot of Alaska, 1869; *Wilmington and San Pedro Harbors*, 1888-1889; *Santa Barbara and Approaches*, 1888.

Davidson, James (*d.*1801). Publisher, and partner in the firm of Mount & Davidson, which succeeded Mount and Page as publishers of *The English Pilot*, and Greenvile Collins's *Coasting Pilot*, 1792.
• **Davidson, Smith & Venner**. This imprint appears on some work previously published by Mount & Page, *c.*1800.

Davidson, *Captain* John Wynn (1823-1881). Fort Tejon to Utah, 1859 MS.

Davidson, R. *Plan of...Leamington Spa*, 1876.

Davidson, Walker Rennie. Surveyor-General for New South Wales. Map of New South Wales copied by J. Noone for publication in Bailliere's *New South Wales gazetteer*, Sydney 1866, 1870.

Davidson, William. Engineer. *Watts River Scheme, Melbourne*, 1885.

Davidsz. [Davidszoon], Dirck. Sea charts for Pieter Goos, 1666; and Jacob Aertsz. Colom, 1669.

Davie, —. Publisher of Chicago. *Atlas*, 1863.

Davies, Arthur (*fl.*1782-*c.*1822). Surveyor of Oswestry, produced manuscript maps of Wales and the Marches. *A Plan of the Ellesmere Canal*, 1796 (with Richard Jebb, engraved by Neele).

Davies, B. *A Plan of the River Loddon*, 1769.

Davies, Benjamin. Bookseller at N°·68 Market Street, Philadelphia. Published *Some Account of the City of Philadelphia*, 1794 containing A.P. Folie's map *...plan of the city and suburbs of Philadelphia...*, 1794 (engraved by R. Scot & S. Allardice); *A map of Philadelphia...*, 1799.

Davies, Benjamin Rees. London engraver and publisher at 30 Goodge Street (1811); 10 Compton Street, Brunswick Square (1813); 34 Compton Street (1823); 16 George Street, Euston Square (1840-1848). Some Davies maps were later republished by Edward Stanford, Letts, Son & Co. and others. Cadiz, 1811; John Arrowsmith's Bible Atlas, 1835; *London and its environs*, 1840 and editions to 1854; *London*, 1843 and later; *Post Office directory of the six home counties...*, W. Kelly and Co. 1845; *London*, 1847 and later; *Map of Central Europe*, 1848; *Davies's Map of the British Metropolis...*, 1848; *Adams's Guide Map to the Environs of London 25 miles round*, *c.*1850; *London 1851*, 1851; *Track of the West India Hurricane of 1851*, 1851; *Davies's New Map of the British Metropolis*, 1852 and later editions; *Collins' London*, 1855; engraver for *Map of Sebastopol and Surrounding Country...*, London, Ackermann & Co. 1857; maps for *Weekly Dispatch Atlas*, 1858-1860 (with Edward Weller); for Frederic Kelly (Directory publisher), 1860; *Davies's New Map of London*, 1863; *Davies' New Map of London and its Environs*, 1869 and many later editions; and many others.

Davies, Charles Lennox. Surveyor in Ontario. Produced numerous plans of building lots in Ontario settlements including London, Napier, Port Stanley, Stratford, Strathroy and Watford, 1855-1867; *Plan of part of the Town of Shakespeare in the County of Perth*, Toronto 1855.

Davies, Griffith (*fl.*1732-1765). Essex surveyor. Harwich, 1745 MS.

Davies, Henry. *Environs of Cheltenham*, 1834.

Davies, Henry (*fl.*1837-1845). Surveyor of Kimpton, Hertfordshire. Produced manuscript maps of Bedfordshire and Hertfordshire including *Plan of the Town of Luton*, 1842.

Davies, J. *Map of the Country round Bath*, 1865.

Davies, John (*c.*1627-1693). Translator of Johann Albrecht von Mandelslo's Voyages, with maps of India, Livonia and the Volga, 1662.

Davies, Richard. See **Davis**, Richard.

Davies, Richard. *New Map of Birmingham*, 1858.

Davies, Thomas (*c.*1712-1785). Actor and bookseller of 8 Russell Street, Covent Garden, London. Sold William Camden's *Britannia*, 1772 edition.

Davies, *Lieutenant-General* Thomas (*c.*1737-1812). Draught of the River St. Lawrence, *c.*1760.

Davies, Thomas. Edinburgh publisher. *New Atlas of the World*, 1859 (with Thomas Nelson).

Davies, William. Map of Virginia, 1807 (with William Prentis, engraved on 9 plates by Frederick Bossler).

Davies, William (*d.*1819). Assistant to Thomas Cadell II, in Cadell & Davies of the Strand. **N.B.** See **Cadell.**

Davies, William Augustus. Aston Manor [Warwickshire], 1895.

Davies, William Henry (*fl.*1840-1867). Land surveyor of Abingdon, Berkshire. Produced manuscript maps of Berkshire and Oxfordshire.

Davies & Bryer [Davies, Bryer & Co.]. Engravers, known as Davies, Bryer & Co. from *c.*1868. Admiralty charts: *Harbours and Anchorages on the West Coast of Newfoundland*, 1866; *West Indies, Island of Trinidad*, 1866; *West Indies, Tobago Island*, 1867; *East Coast of Newfoundland*, 1868; *Black Sea Odessa Bay*, 1868; *West Coast of Africa*, 1869; Coast of the Gulf of St Malo, 1869; *North Atlantic Ocean Eastern Portion*, 1870; *South Atlantic Ocean*, 1871; *Mediterranean. The Dardanelles*, 1871; *Baltic Gulf of Bothnia*, 1872; *Mediterranean Sicily*, 1872; Suez, 1873; Ports and anchorages on the East coast of Spain, 1891; Genoa, *c.*1914; and others. **N.B.** Compare with **Davies & Company** and **Davies & Powell**.

Davies & Company. Engravers of '19 Craven Street, Strand'. Engraved many charts for the Admiralty. *Passamaquoddy Bay and St. Croix River* [New Brunswick], 1848; *East Coast of Newfoundland*, 1873; *Cape Verde Islands*, 1874; *Walfisch Bay Settlement* [West Africa], 1878; *Mediterranean Adriatic Sea Gulf of Cattaro to Corfu*, Admiralty 1878; *Island of Saint Helena*, 1879; *Queen Charlotte Islands* [Canada], 1880; *Map of the Island of Mauritius*, 1881; *Africa East Coast...* (several sheets), 1881-1891; *Newfoundland*, 1883; *Hudson Bay and Strait*, 1884; *River Cameroon*, 1896; *Mediterranean. Malta Valetta Harbour*, 1909; *Calabar River* [Nigeria], 1913; and many others. **N.B.** Compare with **Davies & Bryer** and **Davies & Powell**.

Davies & Eldridge. Booksellers, Exeter. [James] *Wallis's New Pocket Edition of the English Counties or Travellers Companion...*, 1810.

Davies & Powell (*fl.*1862-1865). Engravers for the Admiralty. All the maps were engraved from drawings by E.J. Powell. *Her Majesty's Dockyard at Bermuda*, 1862; *Ras Bulaou to Alexandria* [Egypt], 1864; *West Indies. Leeward Islands*, 1864; *Esquimalt & Victoria harbours* [Canada], 1864; *South East Coast of Newfoundland*, 1864; *South West coast of Newfoundland*, 1865; *Vancouver Island*, 1865; and others. **N.B.** Compare with **Davies & Company** and **Davies & Bryer** (above).

Da Vignola, C. See **Cantelli** da Vignola, Giacomo.

Da Vinchio, P.M. See **Vinchio**.

da Vinci, Leonardo. See **Vinci**, Leonardo da.

Davis, —. See **McCarthy** & Davis.

Davis, B. Printer of Philadelphia. Scott's *Atlas of the United States*, 1796.

Davis, Charles (*d.*1755). London bookseller of Fleet St; Paternoster Row; 'Holborn opposite Gray's Inn Gate'. Succeeded by nephew Lockyer Davis *q.v.*. Sold William Camden's *Britannia*, 1753 edition.

Davis, *Admiral* Charles Henry (1807-1877). American hydrographer from Boston, died Washington. American Nautical Almanac, 1849-1862; *Report on Interoceanic Canals and Railroads*, Washington, 1867.

Davis, Cornelius Butter (*fl.*1838-1855). Surveyor of East Woodhay, Hampshire, England. Prepared manuscript surveys of parts of Hampshire and Berkshire.

Davis, F.A. [& Co.]. Historical atlases of American counties. Berks County, Pennsylvania; Rockland County, New York, 1876.

Davis, George Horace (*b.*1881). Perspective maps for the *Illustrated London News* including *The tremendous range of RAF bombers pictorially shown*, 1939; panoramic map showing the D-Day invasion of Normandy in June 1944, published in *Illustrated London News*, 17th June 1944. **Ref.** COOK, K.S. *The Map Collector* 51 (1990) pp.39-40.

Davis, J. Publisher of 36 Essex St, Strand, London. Ancient Egypt, 1814.

Davis, Jefferson (1808-1889). President, Confederate States of America. As Secretary of War he commissioned *Explorations and Surveys for a Railroad Route from the Mississippi River to the Pacific Ocean...*, Washington 1859 (surveys were actually made by many others).

Davis [Davys], John (*c.*1550-1605). Voyages to discover a Northwest Passage, 1585-1587 (with Adrian Gilbert). Assisted with Emery Molyneux's globe; *The World's Hydrographical Description*, 1595.

Davis, John. Printer of Philadelphia. *...chart of Delaware Bay...*, 1756 (with Joshua Fisher); Nicholas Scull's *...Improved part of the province Pennsylvania...*, 1759.

Davis, *Reverend* John (1821-1903). Patented a 'Lunatellus Globe' (sun, moon and 12-inch terrestrial globe), 1884; 12-inch globes incorporating maps of W. & A.K. Johnston of Edinburgh, from 1884. **Ref.** WARNER, D. *Rittenhouse* Vol.2 N°2 pp.58-59.

Davis, John. *Bird's Eye View Map of Eastbourne*, 1897.

Davis, *Sir* John Francis. Island of Chusan, 1853 MS.

Davis, John Lambe (*fl.*1760-1778). Estate surveyor, produced manuscript maps of various parts of England including White End Farm Chesham [Buckinghamshire], 1760; plans of Chesham, 1762; estate maps Hertfordshire, 1770-1776.

Davis, Lockyer (1719-1791). Bookseller and publisher, 'Holborn, opposite Gray's Inn Gate', succeeded his uncle Charles Davis [*q.v.*] in 1755. Sold William Camden's *Britannia*, 1772 edition.

Davis, O.F. *Kearny City, Nebraska*, *c.*1857; *Dakota County, Nebraska*, *c.*1860.

Davis, R.A. *Basingstoke*, 1851.

Davis [Davies], Richard. Publisher at 'The 3 Ink-Bottles in Castle-Ally at the Royal Exchange', London. Co-publisher, with Henry Overton, of *A New And Exact Map of Great Britain...*, 1708; Theatrum Historiae, 1709; John Seller's *Middlesex* (2 sheets), 1710; Herman Moll's *...Map of Africa*, 1710 edition (with D. Midwinter).

Davis [Davies], Richard (1750-1814). Farmer, agricultural writer, land surveyor and inclosure commissioner from Lewknor, Oxfordshire. Topographer to George III. *A New Map of the County of Oxfordshire* (16 sheets), 1797 (engraved by J. Cary); *A Plan of the County of Oxford* (in a book on the agriculture of Oxford), 1794; also many estate plans. **Ref.** GRÓF, L. *The Map Collector* 71 (1995) pp.32-35.

Davis, Thomas W. Plans of districts of Boston, 1868-1873.

Davis, U.J. *Gloucestershire*, *c.*1858.

Davis, W.T. Engraver for Robert Wilkinson, 1797.

Davis, William (1771-1807). Surveyor, mathematician and bookseller of London. *A Complete Treatise on Land Surveying*, 1798, 1804, 1813 editions; surveys in the south-eastern counties of England.

Davis, William J. Geography of Kentucky, 1881.

Davis, William Morris (1850-1934). Geographer and geologist of Cambridge, Massachusetts. Worked for the Northern Pacific Railroad Company on the Northern Transcontinental Survey, 1881-1884 (with others); New England States, 1895.

Davison, Robert. *Oristano Bay* [Sardinia], 1803.

D'Avity, P. See **Avity**, Pierre de Montmartin.

Davodeau, Antoine. Map of the Garden of Eden in a French Bible, Geneva 1560 (with F. Jaquy and J. Bourgeois).

Davy, —. Atlas of Mitchell County, Kansas, 1884 (with Gillen).

Davy, J. Chart of Tor Bay, 1761.

Davy, John. *Map of the Island of Ceylon*, 1821.

Davy, Marie. World Communications Map, 1868.

Davydov, G.I. Dvukratnoe pateshtsvie v Amerika, St Petersburg 1802-1804; German translation, published Dorpat [Tartu, Estonia] 1816.

Davys, J. See **Davis**, John.

Daw, Edmund. Publisher of 114 Fetter Lane, London. *Parish of St. Pancras*, 1849; *Parish of St. Mary Abbotts*, 1852, 1863.

Daw, M.E. [& Co.]. *The British Isles*, 1875.

Dawes, Henry. Photo-lithographer of Cape Town. Cape Division, 1890.

Dawes, W. Governor Colony of Sierra Leone. Sierra Leone, 1794, in Wilkinson's Africa, 1800.

Dawes, William (1762-1836). Scientist, officer of marines and surveyor. Port Jackson [Sydney], John Stockdale 1789; Territory of New South Wales, John Stockdale *c.*1791

Dawes, *Reverend* William Rutter (1799-1868). Astronomer. Star maps for Society for Diffusion of Useful Knowledge *Atlas*, 1844 and later.

Dawkins, Henry (*fl.*1753-80). Engraver of Philadelphia, moved to New York 1774. The following maps are attributed to him as engraver: *...plan of the improved part of the city...* [Philadelphia], 1762; *A General Chart of the Coast of the Province of Louisiana*, Matthew Clarkson 1763; *A topographical plan of...Indian Country..* [Fort Pitt to Wyandotte Town, Ohio], 1765; *...boundary lines between the province of Maryland and the three lower counties on Delaware...*, 1768, and *...boundary between the provinces of Maryland and Pensylvania*, 1768 (both with James Smither, showing parts of the Mason and Dixon line survey); Nicholas Scull's *...map of the province of Pennsylvania...*, 1770; *...chart of Delaware Bay and river...*, 1776; *Map of the peninsula between Delaware & Chesopeake Bays...*, 1778. **Ref.** WHEAT & BRUN *Maps and charts published in America before 1800* (1978) many entries.

Dawkins, W. Boyd. *A Geological map of the neighbourhood of Dover*, 1895.

Dawson, —. Scottish Schoolmaster. Plan of Stonehaven, 1818.

Dawson, Benjamin [& Son]. Publisher of Montreal. *Canada, New Brunswick, Nova Scotia &c.*, *c.*1860 (taken from a Bartholomew map of 1856), later republished by Dawson Frères.

Dawson, C.B. Surveyor of 88 Pitt Street, Sydney. *Oatleys, Georges River*, 1844; *Lymerston, St Peters* [Tempe], 1885; *Quakers Hill Estate*, 1886.
• **Dawson & Ewing**. 88 Pitt Street, Sydney. *Mount View Estate, Coogee Heights*, 1886.
•• **Dawson & Dawson**. 88 Pitt Street, Sydney. Tempe, 1892; *Bundarra Estate*, 1892.

Dawson, E.B. American engraver. Assisted and engraved for Henry Schenck Tanner, 1829-1836, including *The Travellers' Pocket Map of Michigan*, Philadelphia 1832; maps in *A New Universal Atlas*, Philadelphia, Tanner 1836.

Dawson, Edward J. *Yukon Gold Fields*, 1897.

Dawson, F. *A planisphere of the stars...*, *c.*1830.

Robert Kearsley Dawson (1798-1861). (By courtesy of the Dawson family)

Dawson, G. Herbert. Surveyor General of British Columbia, 1913.

Dawson, George Mercier. Geologist. Director of the Geological Survey of Canada. Queen Charlotte Is., Gotha 1881; *Map shewing the Distribution of the Indian Tribes of British Columbia*, 1883; *Relief map of Canada and the United States*, 1900; *Geological Map of the Dominion of Canada*, 1901; other maps of Canada.

Dawson, John. Printer of John Speed's *Theatre...*, editions of 1627 and 1631.

Dawson, *Doctor* John William (1820-1899). Geologist, worked on the Canadian Geological Survey with Professor James Robb, J.B. Jukes, and others. *Geological Map of Canada*, Montreal, Dawson Brothers 1865 (printed by Stanford's in London).

Dawson, Joseph (*fl.*1754-1787). Surveyor of Essex, produced manuscript estate maps. Good Easter, 1754; Danbury, 1756-1758.

Dawson, *Lieutenant* (later *Commander*) Llewellyn Styles, *RN*. Hydrographer, contributed to many Admiralty charts from 1863. Fiji, 1876; Corinth Bay, 1888 (with E.H. Day).

Dawson, Robert, father and son.
• **Dawson**, Robert (1771-1860). Royal Military Surveyor and draughtsman, appointed draughtsman to the Board of Ordnance in 1791. He produced carefully executed maps and plans, and was particularly noted for his influence on the depiction of relief on maps. **Ref.** HODSON, Y. *The Map Collector* 54 (1991) pp.28-30.
•• **Dawson**, *Lieutenant* (later *Lieutenant-Colonel*) Robert Kearsley, *RE* (1798-1861). Eldest son of Robert, above. Head of the Survey Department of the Commons Enclosure and Copyhold Commission. Superintended the preparation of 19th-century tithe maps, county maps, city plans and boundary surveys. He also devised a table of conventional signs for use on them. *Plans of the Cities and Boroughs of England and Wales*, Hansard 1832. **Ref.** WALLIS & ROBINSON (Eds.) *Cartographical innovations* (Tring 1987) pp.131-132; BEECH, G. *The Map Collector* 33 (1985) pp.23-25; KAIN, R.J.P. *Cartographical Journal* 12 (1975) pp.81-88; CLOSE, C. *The early years of the Ordnance Survey* (1969).

Dawson, Simon James (1820-1902). Exploratory surveys of the route between Lake Superior and the Red River, 1857 (used and published by various companies 1857-1863).

Dawson, William. Publishers at N°·7 Paternoster Row, London. Rev. John Prior and Joseph Whyman's *A Map of the County of Leicester*, 1779.

Dawson Brothers [**Frères**]. Publishers of Montreal. Geological Survey of Canada [*Report...* with *Atlas of Maps and Sections*], 1865, 1879 (with maps by J.W. Dawson, J. Robb and J.B. Jukes, published in French and English); New Map of Canada, New Brunswick, Nova Scotia, 1867 (engraved and printed by J. Bartholomew of Edinburgh, first published 1856). **NB.** Compare with **Benjamin Dawson & Son**.

Dawson & Stephen. Surveyors of 139 Pitt Street, Sydney. *Graham's Garden, Marrickville*, 1880.

Dax, Paul (1503-1561). Cartographer of Innsbruck. Detailed maps of parts of Tyrol, 1530-1550s; Relief models of the area of Kufstein, up to 1540; Acherthal, 1545 (lost).

Day, family and companies. Lithographers of London. '59 Great Queen Street, Lincoln's Inn Fields' (from *c.*1823); '17 Gate Street, Lincoln's Inn Fields' (from *c.*1829); 'Savoy Steam Press', Savoy Street, Strand (from *c.*1869). Early users of the chromolithographic process to produce printed block colour.
• **Day**, William *the elder* (1797-1845). Initially worked in his own right, later in partnership with the Belgian painter Louis Haghe as Day & Haghe. Plan of Hull, *c.*1830; maps for *Journal of the Royal Geographical Society*, from 1833; Ebenezer Pocock's Patent Paper Globe, *c.*1835.
• **Day & Haghe**. Partnership of William Day and Louis Haghe at '17 Gate Street, Lincoln's Inn Fields', London. Lithographers to the King (1831), lithographers to the Queen. *Plan and Elevation of the Union Bridges - Ottawa River near the falls of Chaudiere*, 1831; *Wyld's Sketch of the Country around Montreal shewing the Villages and Military Positions*, James Wyld 1837; *East Sussex Polling Districts*, 1837; W. Allen's *A Part of the Rivers Niger & Chadda*, 1841; *Reach, Burwell and Wicken Lodes* [Cambridgeshire], 1841; *Bird's-eye View of the Falls of Niagara & Adjacent Country, Coloured Geologically*, 1845; and many others.
•• **Day**, William. Eldest son of William Day the elder.
•• **Day**, John B. (1823-1905). Son of William Day the elder. Traded in his own name and as Day & Son at 139 Fleet Street (1855); 'Savoy Steam Press' 3 Savoy Street, Strand (1869-1879). *King's School Atlas*, 1869; *Parochial Plan of the City of London...* 1870.
•• **Day & Son** (*fl.*1850-*c.*1868). Gate Street, Lincoln's Inn Fields, London. Lithographers to the Queen. *Western & Central Parts of the Valley of the St Lawrence*, 1850 (and other maps of the same date for J.J. Bigsby's *The Shoe and Canoe*, London, Chapman and Hall 1850); Divisional maps for the 1851 census, 1852; *Sketch of the Island of Mauritius*, 1852; charts of the east coast of Canada, 1853-1854; *Plan of Cape Town*, 1854; *Plano Topografico de la ciudad y puerto de Valparaiso*, 1854; *Map of Canada West...*, 1855; *Metropolis Water Supply...*, 1856; maps for the *Weekly Dispatch*, 1856-1863 (later published as *The Dispatch Atlas*); Benjamin Rees Davies's *Map of the County of Sussex Shewing the Turnpike Roads*, 1860 edition; *Geological map of Trinidad*, 1860 (by G.P. Wall and J.G. Sawkins); *Map of*

Victoria and Esquimalt Districts [Canada], 1861; B.R. Davies's *Leeds*, 1863 (for the *Dispatch Atlas*); Charles Noyce's *City of London*, 1864; *Map of the railways proposed by the Bills of the session of 1864 in the metropolitan district*, 1864; *Map showing the Distribution of Cholera in London... in 1866*, 1867; *A Map of British Honduras...*, 1867.
••• **Vincent Brooks, Day & Son.** *Map of London and its environs*, 1868; *The County of Sussex*, *c.*1872.
•• **Day**, Joseph. Son of William Day *the elder*.

Daye, John (fl.1577). Printer of London. William Cuningham's *The Cosmographical glasse*, 1559.

Day, Benjamin (*fl.* 1880s). Printer of Hoboken, New Jersey, creator of the 'Ben Day' process for transferring line, stipple, tint and pattern to lithographic plates. **Ref.** WALLIS & ROBINSON *Cartographical innovations* (Tring 1978) pp.289.

Day, Charles (*fl.*1838-1852). Land surveyor. Various maps in midland counties, including Keyston, Huntingdonshire, 1842.

Day, *Lieutenant* E.H., *RN*. Admiralty surveys. Corinth Bay, 1888 (with L.S. Dawson).

Day, *Lieutenant* G.F. Admiralty surveys. *Sketch of the River Paraguay*, 1853, published 1856.

Day, John B. See **Day**, family and companies.

Day, John Ogden. Printer of Albany. Improved and printed D.H. Vance's map of the western part of the State of New York, 1825.

Day, William (*fl.*1767-1782). Surveyor and publisher of Blagdon, Engalnd. Produced county and local maps of Somerset, including *County of Somerset* (9 sheets), Blagdon, W. Day 1782, W. Faden 1800, 1808 (with Charles Harcourt Masters, engraved by P. Begbie).

Day, William. See **Day**, family and companies.

Dayes, Edward (1763-1804). Engraver. Town plans of Durham, Hereford and Worcester for Cole & Roper's *The British Atlas*, 1810.

Dayles, Matthew (1703/4-1764). Farmer and estate surveyor of Buttsbury, Essex. Mayland, 1739; Prittlewell, 1746 MSS.

Dayman, Charles Orchard. Revised John William Lubbock's star maps for Society for Diffusion Useful Knowledge *Atlas*, 1860 and later editions.

Dayman, *Lieutenant* Joseph, *RN*. Admiralty surveyor. *Great Sandy Island Strait*, 1847; charts of Cape Province, Natal and Mozambique, 1848-1878; *Port Natal*, 1854; maps of the coast of parts of Cape Province, 1855; *Algoa Bay*, 1855; surveys in the Mediterranean, 1859; Africa, South coast, Admiralty charts 1867, 1870.

Daynes, John. Plan of the Manor Bloomsbury, 1664-1665 MS.

Daynes, Richard. See **Danes**.

Dayot [Ayot; D'Ayot], *Captain* Jean Marie. Officer of the French hydrographic service, promoted to Captain in the Cochin China navy. Prepared charts of the coasts, bays and harbours of Cochin China of which 11 were published by the Dépôt de la Marine in 1807: *...côtes de la Cochinchine et du Camboge..*; *...rivière de Saigon*; *ville de Saigon*; *...cap St Jacques jusq'àu faux cap Varelle*; *...baye de Camraigne jusq'à la baye de Tamquam*; *...port du Camraigne*; *baie de Nhiatrang...*; *...isle Shala jusq'au cap Varella...*; *...ports de Xuan-Dai, Vung-Cam et Van-Chao...*; *Port de Quinhn*; *port de Touran....* Dayot's surveys were used for the following Admiralty charts: *Hone Cohe and Caca be Harbours*, *c.*1835; *...gulf of Siam to Nhatrang*, 1840; *Nhatrang to Touron*, 1840; *River Dong Nai from Cape St James to the city of Saigon*, 1840. **Ref.** PALSKY, G. *Imago Mundi* 41 (1989) pp.59-69 [in French].

Dayton, A.W. Frederick County, Virginia, 1855 (with J.M. Lathrop).

De. N.B. Depending on usage, names with the prefix **De** may be found at **La**, or under terminal name.

Deacon, Charles William [later & Co.]. Publisher, Charing Cross Chambers, London. *Yorkshire*, 1893.

Deacon, Thomas. Publisher, 15 Furnival's Inn, Holborn, London. *Environs of London...*, 1831 (pocket atlas based on Cruchley's *Environs of London...*, 1824).

De Aefferden, F. See **Aefferden**, Francisco de.

Deakins, *Colonel* Francis. Plat of Carrollsburg [City of Washington], 1770 MS; survey of Western Maryland from Fairfax Stone to the Pennsylvania border, 1787.

Dealy, *Midshipman* William J. Survey of Magdalena Bay, 1818; Hamburg Creek, 1818.

De Amezaga, *Captain* Carlo. Italian Navy. Charts published in Viaggio di circumnavigazione, Rome 1886.

Dean, —. *National Atlas*, 1852 (with Gilmour); Seat of war in Italy, *c.*1859.

Dean, Charles Youle. *Kimberley District, Western Australia*, 1886.

Dean, James (*fl.*1810-1829). Surveyor, produced estate and industrial maps of Devon, London and Somerset, including the Manor of Doddiscombe Sleigh, Devon, 1814.

Dean, R. & W. [& Co]. Printers, publishers, and mapsellers, 33 Market Street Lane, Manchester (1811), later 80 Market Street. James Pigot's Plans of Manchester and Salford, from 1804.

Dean, Thomas. See **Dean & Munday**.

Dean & Munday. Publishers of 35 and 40 Threadneedle St., London (1837-1847). Edward Smith's Coast of Syria, 1840.

Deane, G.A. *Agricultural Map of Hertfordshire*, *c.*1883.

De Angelis, Antonio. See **Angelis**, Antonio de.

Dean-Pitt, *Lieutenant* S.T. Admiralty surveys. Corinth Bay, 1888 (with L.S. Dawson and others).

Dearborn, Benjamin (*b.*1755). Parts of Boston, 1814; plan of Boston and Roxbury Mill Dam [Massachusetts], 1834.

Dearborn, William L. Civil engineer. *A map illustrative of the route of the proposed Railroad from St. Louis to the Bay of San Francisco*, 1850.

Dearsley, William (*fl.*1819-1829). Estate surveyor, produced manuscript estate maps of Essex and Suffolk.

Deas, James. *Firth of Clyde*, 1898 (for Andrew Reid).

Dease, Peter W. Discoveries in the Arctic, *c.*1836-1839, published on an Admiralty Chart of Arctic America, 1845.

Deasy, *Captain* H.H.P. Tibet, 1896; W. China & Tibet, 1897.

Debar, J.H.D. *Map of the Central Oil Regions of West Virginia*, *c.*1864.

Debarsy, —. *Plan de la Ville de Namur*, *c.*1840 (with Leroy).

Debbieg, *Lieutenant* (later *General*) Hugh (1731-1810). Engineer. Assisted William Roy's survey of Scotland, 1749-1751; *A Plan of the City of Barcelona in Catalonia...*, 1767 MS; *Plan of the Town and Harbour of Cartagena...*, 1767 MS; *Plan of the Harbour of ferrol in Galicia...*, 1767 MS; *A Plan of Québec and environs... during the siege... in 1759* MS published *c.*1780 (with J.F.W. Des Barres and Samuel Holland); *Plan of the Harbour, Town and Fort of St Johns in Newfoundland*, 1762; *Plan of the Harbour of St John's...*, 1769 MS; *A Plan of the Ground called South Head* [St John's], 1772, 1775, 1810.

De Beins, J. See **Beins**, Jean de.

Debenzara [Dabenzara], Jehu (fl.1480-1505). Mapmaker of Alexandria. Charts of the Mediterranean, 1497, 1505 MSS.

De Berniere, Henry. Manuscript maps and sketches of the area around Boston on the eve of the Revolution, 1775 (with Captain Brown).

Debes, Ernst (1840-1923). Publisher and mapmaker of Leipzig. *Special-Karte von Süd-Schleswig...*, Gotha, Perthes 1864 (with Welker and H. Habenicht, for A. Petermann); *Das*

Grossherzogthum Hessen, 1875; *Kleiner Schul Atlas*, 1897; *Schul Atlas*, 1893; *Neuer Handatlas über alle Teile der Erde*, 1895 and later editions (with H. Wagner); *Neuer Handatlas...*, Leipzig, H. Wagner & E. Debes 1914.

Debes, Lucas Jacobson (1623-1676). Danish cartographer and naturalist, pastor of Thorshavn, Faeroes. Map of the Faeroes and map of the Maelstrom, both from *Faerooe et Foeroa reserata*, 1673.

De Blosseville, Jules. See **Blosseville**.

Debombourg, Georges. *Atlas Historique du Département... du Rhône*, 1862.

De Bosis, Francesco. Carta geologica della provincia di Ancona, 1867.

De Bouge, Jean-Baptiste. See **Bouge**, Jean Baptiste de.

De Brahm, Ferdinand Joseph Sebastian. Major of Engineers in the Continental Forces in North America from 1778. *The New York Campaign of 1776*, 1777; *Plan of the Town of Purisburg...*, 1779 MS.

De Brahm, [John] William Gerard (1717-1799). Military engineer from Coblenz, went to Georgia with a group of German emigrants in 1751. Appointed a surveyor for Georgia in 1754, and Surveyor-General of Southern District of North America, 1764. Designed fortifications for Charles Town and Fort Loudoun. *A Map of the inhabited Part of Georgia...*, *c.*1755 MS; *The Plan and profile of fort Loudoun...*, 1757 MS; *Charles Town*, c.1757 MS; *A Map of South Carolina And a Part of Georgia Containing the Whole of the Sea Coast* (4 sheets), 1757 (engraved by Thomas Jeffreys) and other editions; surveys of Florida, 1765-1771; charts of the coast and Gulf Stream from 1765 published in *The Atlantic Pilot*, London 1772; *Chart of the Atlantic Ocean with its Currents and Eddies...*, 1775 MS; contributed to *American Military Pocket Atlas*, Sayer & Bennet 1776; Georgia, 1780. **Ref.** CUMMING, W.P. *The Southeast in early maps* (3rd edition revised by L. De Vorsey, 1998) pp.29-30, 279-281, 314-321; DE VORSEY, L. *The Map Collector* 19 (1982) pp.27-30; DE VORSEY, L. *Imago Mundi* 28 (1976) pp.105-120.

Debray, —. *Carta General del Imperio Mexicano...* (2 sheets), Mexico 1865 (with Decaen); *Republica Mexicana*, 1867; Antonio García y Cubas's *Atlas pintoresco estados unidos mexicanos*, Mexico 1885.

De Bree, A. and J. See **Bree**, A. and J. de.

Debrett, John (*d.*1822). Publisher 'opposite Burlington House, Piccadilly', London. Succeeded John Almon 1781, retired 1814. For James Imray, 1797; Egypt, 1799; Chart Indian Ocean, 1800; John Thomas Serres *Little Sea Torch*, 1801.

Debrie, G.F.L. Engraver, for the last Mercator edition of Ptolemy's *Geographia*, Amsterdam, R. & J. Wetstein and G. Smith 1730.

De Bruls, Michael [Michaelson] Godhart (*fl.*1759-1763). 'Publisher and [copper] Engraver... at the lower end of New Street, next door to Col. Thodey, in New York.' *A Plan of the Landing Encampment and Attack against Fort Niagara, on Lake Ontario*, 1762; *Plan of Niagara with the adjacent country...* (2 plates), 1762; *A plan of the streets, etc. of the City* [New York], *c.*1763 (no surviving copy is known).

De Bruyn, Henry (*fl.c.*1800-1825). Estate surveyor in several English and Welsh counties, works included Fambridge, 1807 MS.

De Bruyne, *Kommandant* A. Commander of the ship *Willem Barents* on its voyage in the Arctic in 1878. Two resulting maps were: A. Braakenzick's *Kaart der Noordelijke Isjzee met reisroute...*, Amsterdam 1879 (given away to readers of *Het Nieuws van den Dag*); Swedish and Dutch Arctic Expeditions, 1878, published Edward Stanford 1879.

De Bry. See **Bry** family.

Debure, —, *l'aîné*. Bookseller of Paris at 'Quai des Augustins à l'image S. Paul'; 'hôtel Ferrand, rue Serpente N°·6' (1788). Sold Robert de Vaugondy's *Atlas universel*, 1798; also Jean-Guillaume Barbié du Bocage's *Recueil des cartes géographiques*, 1788.

DeCaen, Litoga [Lithography] de. *Carta General del Imperio Mexicano*, Mexico 1865.
• **Decaen y Debray**. Lithographers. **NB.** See also **Debray**.

Decan, —. Printer of 'rue Richer 19', Paris. C. Arnaud's Californie, 1849.

Décard, —. Publisher '...à Bordeaux...Rue des argentiers à la Sphere'. *Carte reduite de l'Isle de Saint Domingue...*, 1754.

De Cespedes, A.G. See **Cespedes**, Andreas Garcia de.

DeCew, family. Surveyors in Upper Canada, worked alone and together. Edmund and John DeCew: *Plan of that part of the town of Cayuga lying upon the west side of the Grand River...*, 1842; Port Colborne, 1860; Canboro, 1860.
• **DeCew**, Edmund (1812-1892). Provincial land surveyor. *Town of Cayuga...*, 1851; surveys of Cayuga, 1853-1867; Welland, 1853; Seneca, 1854; Caistorville, 1854; *Plan of the Town of Azoff in the Township of Cayuga...*, 1855.
• **DeCew**, John (1835-1884). Surveyor and draughtsman. Port Colborne, 1863.

Dechen, *Doctor* Ernest Heinrich von (1800-1889). Geologist and maker of geological maps. Rheinlaender, 1825; Skye, 1829; Deutschland, 1838, 1869, 1880; large geological maps of western and middle Europe, Berlin, S. Schropp 1838-1839; *Geologische Karte der Rheinprovinz und der Provinz Westfalen...* (35 sheets), 1856, 1872, 1876; Wiesbaden, *c.*1876.

Déchy, Mór (1851-1917). University professor, led expeditions to the high mountains of Europe and Asia, which he surveyed. Published first surveys of the Caucasian region. **Ref.** 'Déchy Mór' in: *Földrajzi Közlemények* [Geographical Bulletin] 1917/1 [in Hungarian].

Decimator, Heinrich (1544-1615). German theologian. *Libellus de stellis fixis*, 1587.

Deckar, *Captain* —, *RN*. *Island of Barbuda*, Admiralty 1814.

Decken, *Baron* C. von der. Tanganyika [later Deutsch Ost-Afrika], 1862.

Decker, Carl. Engraver. *De Stadet Bantam*, *c.*1700, for Pieter van der Aa's *Galérie agréable du monde*, 1729; *De Stadt Coylang*, *c.*1730.

Decker, *Major* Carl D. von (1784-1844). Prussian military cartographer. *Umgebung von Berlin* (10 sheets), 1816-1819; *Batailles ...de la Guerre de Sept Ans*, 1839.

Decker, Coenraet (1651-1709). Engraver and draughtsman of Amsterdam, worked for a while in Hamburg with Heinrich Heuss, also worked for Joannes Janssonius's heirs. *...Kennemerlandt ende West Vrieslandt*, 1650; Delft, 1676; Prague, 1679; *Luxenburg*, Hamburg, Heinrich Heuss 1684.

Decker, J.A. Ballon géographique [inflatable globe], 1889.

Decker, Paul (1677-1713). Architect, draughtsman and engraver. Attack on Gibralter, 1703.

Decluy. See **Le Gendre**-Decluy.

Decomberousse [Comberousse], Joséphine. *Plan de la Ville de Lyon*, 1813, 1816.

Decombes, Victor. *Atlas militaire*, 1876-1880.

Decorbie, —. Engraver of Madrid. *Plano de Madrid*, D.F. Coello and D.P. Madoz 1848 (with Leclercq).

De Costa, J. Publisher of Red Cross Street, Southwark, London. *A Plan of the Town and Harbour of Boston...*, London 1775 (engraved by C. Hall, the earliest battle plan of the American War of Independence).

Decouagne, Jean-Baptiste. Surveyor. *Carte du Gouvernement de Québec...*, 1709; *Carte du Gouvernement des Trois Rivières qui comprent en descendant le Fleuve St. Laurent...*, 1709; *Carte du Canada...*, 1711; *Plan de la Ville de Montréal en Canada...*, 1749; Cours de la Rivière d'Orange, 1750.

Decremps, —. *Plan de Londres*, 1789.

Dedenev, Mikhail. Russian army officer. Plans of Kristinestad [Kristiinankaupunki], Jakobstad [Pietarsaari] and Brahestad [Raahe], 1743 (all in Grand Duchy of Finland); road map St Petersburg-Moscow, 1766; Tvertsa River, 1766.

Dedenroth, Heinrich von. *Karte der Lombardei und Venedig*, *c.*1859.

John Dee (1527-1608)

Dedier, —. Poolo Condor, Alexander Dalrymple 1807.

Dedovich, —. 'Ingenieurs Oberstleutnant'. Oesterreich (14 sheets), 1803 MS.

Dee, *Doctor* John (1527-1608). Astronomer, astrologer, alchemist, mathematician, mystic, friend of navigators and geographers; strong advocate for Northwest Passage. *Anglorum ad Cathaivm per Sythicum Oceanus Secundi Labores*, 1580 MS; *Navigationes ad Cathayam per Septentrionalia Scythia & Tartariae littora delineatio Hydrographica*, 1580 MS (prepared for the proposed Pet-Jackman voyage to find the Northeast Passage); *Hemisphaerii Borealis Geographia atque Hydrographica descriptivo...*, 1583 MS (made for Humphrey Gilbert's quest for the Northwest Passage). **Ref.** de SMET, Antoine in: WALLIS & TYACKE (eds.) *My head is a map* (London 1973) pp.107-113 [in French]; TAYLOR, E.G.R. *Tudor geography 1485-1583* (London 1930).

Deeley, William (*fl.c.*1684-1701). Survey of Edgbaston, Birmingham, England, 1701.

Deelstra, F. Schoolmaster of Groningen, worked with J.D. Bakker. *Schoolatlas*, 1885; *Atlasje voor School en Huis...*, 1900; *Schoolatlas de Geheele Aarde*, 1901 and many later editions. **Ref.** KOEMAN, C. *Atlantes Neerlandici* Vol.VI (1985) pp.70-72.

Deering, Charles. Plan of Nottingham, 1751.

Defehrt, —. See **Desehrt**.

De Fer, Antoine & Nicolas. See **Fer**.

De Fer de la Nouerre, —. Carte élémentaire de la navigation du royaume de France, Paris 1787.

Deferes, L. Lithographer of Ghent. Amèrique Septentrionale, 1844.

De Ferrari, P.E. Cantina geologica del Monte Amiata, 1890.

De Foreest, Phillip. Map of New York engraved on a powder horn, 1762.

Deforrest, J.H. Jr. Gold Regions [North America], 1859.

Defremery, Charles (1822-1883). Fragments de géographie et d'histoire des arabes et persanes, 1849.

Deftedar-Bey, — (*d.*1831). Upper Nubia, for expeditions of Ishmael Pasha.

Degaulle, J.B. (1732-1810). French hydrographer and mathematician. Usage d'un nouveau calendrier perpetuel, 1768; Usage d'un compas azimuthal, 1779; maps in *Neptune François*, 1788.

Degen, Joseph Vincenz. Publisher in Vienna. Plan de la ville de Vienne (4 sheets), *c.*1803.

Degenhardt, O. *Der Oberschlesisch-polnische Berg-District*, 1871.

Degler, Johann. Frontispiece for each of the seven parts of Heinrich Scherer's *Atlas Novus*, *c.*1700.

degli Uberti, L. See **Uberti**.

Degni, Demetrio. Plan and view of Luxembourg in: *Raggvaglio historico...*, Modena 1698 (copied by Giàcomo Bertan, 1699).

De Goeje, M.J. See **Goeje**, M.J. de.

De Gomme, *Sir* Bernard. See **Gomme**, *Sir* Bernard de.

Degotardi, Johann Nepomuk (1823-1882). Photolithographer of Sydney. 20 York St (1855); 313 George St (1859-1859); Robinhood Lane off George St (1862); George St (1873-1875). Photopanorama, 1860; New South Wales, 1862; Map of Bathurst, 1862; Plan of Garrorigang, 1859; Plan of properties of the late Francis King, 1863.

De Graaff. See **Graaf**.

De Groot, —. Washoe Mines, San Francisco 1860; California mine surveys, 1863 MS; *Map of Nevada Territory*, 1863.

De Gruchy & Leigh [De Gruchy & Co]. Engravers and lithographers of Melbourne, and Sydney. 47 Elizabeth St. 7 Flinders Lane West, Melbourne (1859); George St., Sydney. Isometrical plan of Melbourne and suburbs, 1866; De Gruchy & Co's new squatting map of Riverina District, 1877. **NB.** Compare with H.G. **Gruchy** & Co., and Stephen Thomas **Leigh** & Co.

De Guignes, Christian Louis Joseph, *SJ* (1750-1849). *Planisphère célèste chinois*, Paris 1785.

Deharme, L.F. Engraver for John Rocque's *Berkshire*, 1761; *Plan de la Ville et Fauxbourgs de Paris*, 1763.

De Haven, *Lieutenant* Edwin H. Officer of the U.S. Coastal Survey. Worked with others on the Charles Wilkes survey of the Pacific coast, 1841.

De Haven, *Lieutenant* Edwin Jesse (1816-1865). American naval officer. Wind chart of the Atlantic, 1851; American Arctic expedition, 1851; Wellington Channel, 1852.

De Hooghe, R. See **Hooghe**, Romein de.

Deichmann, Louis (*fl.*1881-1899). Geographer of Kassel, Germany. Relief maps of Black Forest, *c.*1893; Waldeck, *c.*1893.

Deidier, *l'abbé* — (1696-1746). Published a study of the fortifications of cities. *Le parfait Ingenieur françois, ou La Fortification offensive et defensive...*, Amsterdam 1734, Paris, Charles-Antoine Jombert 1736 and later editions.

De Jode. See **Jode**.

De Jonghe, C. See **Jonge**, Clement de.

Dekinder, K. Engraver for Heinrich Kiepert, 1860.

Dekker, Errico. Draughtsman. Plan of the waterfront of Naples, 1822.

De Krafft, F.C. Surveyor of Washington, DC. Plan of the City of Washington, 1828.

De La Barre, Le Febvre. See **Barre**, J.F. Le Febvre de la.

De la Bèche, H.T. See **Bèche**, Henry Thomas de la.

De la Bella. See **Macaneo**, Domenico.

Delaborde, *comte* Alexandre. See **Laborde**, *comte* Alexandre.

De la Caille. See **La Caille**.

De la Cases. See **Las Cases**.

Delachaux, Enrique. *Provincia de Catamarca*, 1893; Provincia de Buenos Aires, 1893; Sud-Amérique, 1895; Central Argentina, 1899.

De la Cosa, J. See **Cosa**, Juan de la.

Delacre, Louis. Charleroi, 1875.

De la Croix, A. Pherotee. 'Géographe du roi'. *Nouvelle méthode de... la Géographie Universelle*, Lyon 1690; *Géographie Universelle* (4 volumes), 1693; *Allgemeene Wereld-Beschryving*, Amsterdam, François Halma 1705 (after Guillaume Sanson, 1690).

De la Croix, F. See **Croix**, F. de la.

Delacroix, *abbé* Louis Antoine Nicolle. French geographer. Work used by Guillaume Delisle; Jean Charles Maclot; Louis Brion de la Tour. *Relation Universelle de L'Afrique Ancienne et Moderne*, Lyons 1688.

De la Croyère, D. See **Delisle**, family.

De Lacy, Walter Washington (1819-1892). Military engineer and surveyor of Helena, Montana. Territory of Montana, 1865, 1870, 1879; Idaho Territory, 1866 MS.

De Laet. See **Laet**.

Delafage, N. *Plan du passage...de Ségre...*, 1646.

De La Farellière, —. Milford Haven, 1666.

De la Feuille. See **Feuille**.

Delafield & Haven. American surveyors and engineers. *Eddytha Farm* [Madison County, Illinois], *c.*1856.

Delafosse, *sieur* —. Geographer. L'Amérique, Lyon 1771.

Delafosse [de la Fosse], D. French geographer and engraver. *Carte du Gouvernement Générale du Maine et du Perche...*, Paris, Basset 1773 (edited by Daumont, script engraved by E. Voysard).

Delafosse, Jean-Baptiste (1721-1806). French geographer and mapmaker. *Nouvelle carte du gouvernement general d'Orleanois...*, Paris, Daumont 1761; *Cartes des provinces-unies Connües sous le nom de Hollande*, Paris, Mondhare 1779; *Carte de France divisée en ses provinces...*, 1786 (copied from Bourgoin's *Carte itineraire...*, of 1773); *Introduction à la géographie carte des diverses positions de la sphere, des Systhèmes de l'Univers, des Planettes des Eclypses &c.*, Paris, Jean *c.*1791.

De la Fosse [Delafosse], — (1721-1775). French engraver, for Jean-Baptiste Nolin and Louis Charles Desnos, also engraved cartouches for J.B.B. d'Anville, 1751-1755.

De la Garde, J. Terrestrial globe, 1552.

Delagardette, C.M. *Ruines de Paestum*, 1840.

De la Gardette, Pierre-Claude. See **Gardette**.

Delagrave, Charles (1842-1934). Publisher for Société Géographique. '15 rue Soufflot' and '58 rue des Ecoles', Paris. Atlas de géographie moderne, physique et politique, 1870; Adrien Hubert Brué's *Atlas universel de géographie...*, 1875 (first published 1822); Eugène Darsy's Atlas, 1890; *France: les départements, les corps d'armée*, 1899; *France Carte Michelin. France en 47 feuilles...*, 1911-1914.

Delagrive, *abbé* Jean (1689-1757). Priest and philosopher, abandoned the church to learn geography, drawing, engraving and triangulation. Plan of Paris, *after* 1731 (copied by Homann Heirs 1739); manuscript atlas of the River Seine and its tributaries, *c.*1738; *Plan de Paris*, 1741; plan of Versailles, 1746; reduced plan of Berlin, 1757.

Delaguette, *veuve* —. Printer, 'rue St Jacques à l'Olivier Duschesne', Paris. Isles de Jersey et Guernesey, 1757.

Delahaye [De La Haye]. Family of engravers, 'graveurs en géographie', worked for the principal French geographers of the day including D'Anville, Delisle, Jaillot, Robert de Vaugondy and others. **Ref.** PEDLEY, M. *Bel et Utile: The work of the Robert de Vaugondy family of mapmakers* (1992) pp.54-60; PEDLEY, M.S. *Imago Mundi* 36 (1984) pp.52-56.

• **Delahaye**, Jean-Baptiste (*fl.*1708-1723). 'Rue des Orties' (1714); 'rue Fromenteau' (1717, 1723). Engraver, worked for Guillaume and Adrien Sanson, and for Pierre Moullart-Sanson. Hemisphaerium superiori globi, 1708.

• **Delahaye**, Marie Denise Gervais [de la Rivière]. Engraver, wife of Jean-Baptiste.

•• **Delahaye**, Jean-Baptiste-Henri [*l'ainé*] (*d.c.*1755). Son of Jean Baptiste, above, elder brother of Guillaume Nicolas. Engraver for Guillaume Delisle, 1724; Bernard Jaillot, 1721-1725; J.B.B. d'Anville from 1727; engraver for la Blottière and Roussel's *Carte generale des Monts Pyrenées*, *c.*1730 (drawn by A. Coquart); *Imperium Caroli Magni*, 1752; and *Le Tyrol*, 1753 (both for the Robert de Vaugondy *Atlas Universel*).

•• **Delahaye**, Guillaume-Nicolas (1727-1802). 'Rue de l'arbre sec' (1767); 'place du chevalier du guet' (1781). *Amerique Septentrionale*, D'Anville 1746; D'Anville's *L'Amérique méridionale*, 1748; D'Anville's *Canada, Louisiane et Terres Angloises...*, 1755; engraved for Robert de Vaugondy, 1750-1756; Dépôt de la Marine, 1752-1785; Philippe Buache, 1753; engraved Robert de Vaugondy's maps for Charles de Brosses' *Histoire des Navigations aux Terres Australes*, 1756; D'Anville's *Hémisphère Occidental ou du Nouveau Monde*, 1761; *Nouvelle Topographie...*, 1784; maps for Jean-Guillaume Barbié du Bocage, 1786-1788; for Dezauche, 1790; and others.

• **Delahaye**, Matthieu. Engraver, apprenticed to his brother Jean-Baptiste.

De la Haye, Mark W. Surveyor-General, US Public Surveys Kansas and Nebraska, 1863.

Delahaye [de la Haye], Gabriel. *The Map of the Island of Minorca*, 1708 MS; map of Alicante.

De la Hire, Philippe. See **La** Hire.

Delaistre, H. *Carte de la Navigation...de l'Empire Français*, 1812.

Delalain, J. Publisher of 'rue des Ecoles', Paris. François Bazin and Félix Cadet's *Atlas Spécial de la Géographie Physique, Politique et Historique de la France*, *c.*1862.

Delalain, Nicolas-Auguste (*d.*1807). Editor and publisher of Paris. Le géographe parisien, 1769; *Atlas moderne...*, 1771 (with Jean Lattré).

Delamarche, father & son.

• **Delamarche** [Lamarche], Charles François (1740-1817). Geographer, publisher, globe-maker, of Paris. 'Rue du Foin St. Jacques au College de Mtre. Gervais', Paris (1778-1804); 'Rue du Jardinet N°·13 vis à vis celle de l'Eperon' (or Quartier St. André des Arts) (1805-1807). Bought the stock of Didier Robert de Vaugondy from Jean Fortin, and the stock of Lattré (which included work by Bonne, Desnos & Janvier). Etats unis de l'Amérique, 1785; Globes, 1785-1800; up-dated and re-published the Vaugondy *Nouvel Atlas* and *Nouvel Atlas Portatif*, 1784 and later; *Petit Atlas Moderne*, *c.*1790; Les usages de la sphère et des globes, 1791; *Amerique ou Indes Occidentales*, 1792 and later; *Le Royaume de France...*, 1793; Tableaux géographiques, 1794; Institutions géographiques, 1795; Atlas d'étude, 1797; Atlas élémentaire, 1800 and later; World and Continents (each 4 sheets), 1805; *Mappe-Monde...*, 1812; and others.

•• **Delamarche**, Félix [*fils*]. 'Ingénieur mécanicien pour les Globes et Sphères, Rue du Jardinet N°·13', Paris. Son of Charles François, above; worked with Charles Dien 1819-1820. *Atlas de géographie ancienne et moderne*, 1823 and later.

Delamarche, François Alexandre. *Carte de France après la Guerre de 1870-1871*, 1871.

Delamare, —. Engraver. *Plan...de la Ville de Lyon*, *c.*1874.

Delamare, F. Engraver of 'rue St. André-des-Ares' (1854); also noted at 'Rue de la Harpe 26', Paris. For Eugène Andriveau-Goujon, 1850; *Carte de la République de Costa Rica*, 1851; *Première carte du théatre de la guerre en orient (Turquie d'Europe)*, Paris 1854; *Géographie des parties centrales de l'Amerique du Sud*, Paris, P. Bertrand 1854 (with L. Boussard); for George Woolworth Colton, 1855; *Carte du Canton de Genève*, 1856; *Voyage en Turquie et en Perse... les années 1846, 1847, et 1848 par Xavier Hommaire de Hell...*, Paris, Pierre Bertrand 1859 (maps by Barthelemy Semino); *Carte de la République de Paraguay*, Paris, A. Bry 1862; *Atlas Géographique de la République du Pérou*, Paris, Auguste Durande 1865 (from an 1861 atlas by Mariano Felipe Paz Soldan).

Delamare, Nicolas (1639-1723). Produced a chronological series of maps of Paris known as *Traité de la Police* (7 maps), 1705 and later (drawn and engraved by A. Cocquart).

Delamarre, Casimir. *Carte linguistique, ethnographique... de l'Europe orientale*, 1868.

Delambre, Jean Baptiste Joseph (1749-1822). Parisian astronomer and geodesist, determined meridian of France. Base du système métrique, 1810; Histoire de l'astronomie, 1817.

Delamonce, —. *Nouvelle Mappe Monde*, *c.*1750.

Delamotte, Guérin. Engraver. Carte chorographique des environs de Lisbonne, 1821.

Delamotte, William. Draughtsman. Wren's plan of the City of London, 1800.

De la Pérouse. See **La** Pérouse, Jean François Galoup.

De la Peyrère. See **La** Peyrère.

De la Pointe, François. See **La** Pointe, François de.

Delaporte, *Lieutenant* Louis. French naval officer. Charts Indo-Chine, 1873 (for Dépôt de la marine).

Delarageaz, Louis-Henri (1807-1891). Swiss surveyor in Lausanne. His work was used in a topographical map of Switzerland, and in a topographical map of the canton of Vaud, Lausanne 1865.

Delaram, Francis. (*fl.*1615-1624). Engraver. Map of Middle East, 1615 (in G. Sandy's *Relation of a journey*).

Delaroche, Charles F. *Plan de la Ville d'Alger*, *c.*1850; *Algérie*, 1851; *Carte du Sahara Algérien*, 1853.

Delarochette [de la Rochette], Louis Stanislas d'Arcy (1731-1802). Cartographer and engraver, associated with William Faden. North America, Thomas Bowles, 1762 and later; 14 maps for John Blair's *Chronology and History of the World*, 1768 edition; Caribee Is., 1768; *A plan of the Town, Citadel, Dock and Sound of Plymouth*, Faden 1780; *A Map of North America and the West Indies*, London, Robt. Wilkinson 1781; chart of the Channel Islands and adjacent coast of France, Faden 1781; *The Dutch colony of the Cape of Good Hope*, Faden 1782, 1795; Map of Europe, W. Faden 1786; *Italy with the Addition of the Southern Parts of Germany...*, London, W. Faden 1800; Lower Egypt, 1802; *A Chart of the Indian Ocean...with Part of the Pacific Ocean* [includes Australia], Faden 1803 (engraved by J. Hatchett).

Delarue, François. Publisher in Paris for Alexandre Vuillemin, 1864.

De la Rue, Philippe. Cartographer of Paris. Maps of the Holy Land and Middle East for Pierre Mariette, 1651 (engraved by J. Sommer).

De Lasaux, P. See **Lasaux**, P.D.

Delaserre, Léon. *Carte géologique et zoologique du Languedoc*, 1899.

De la Tour. See **Brion** de la Tour.

De la Tour, Le Blond. See **Le** Blond de la Tour.

De la Tour, *veuve* — (*fl.*1764). Printer, 'Rue de la Harpe', Paris.

Delaune, —. Engraver for Louis Vivien de St. Martin, 1883-1900; one of the engravers for G. Bagge's *Russie Orientale et Caucasie*, Hachette & Co. 1900.

Delaval, J.B. Publisher and mapseller, 'Rue Geoffroy-Langevin N°.7', Paris. Départements de Drome, Ardèche, Loire, Gard, 1792, 1818; Asie, 1819, 1823.

De Lavaux, Alexander. Engineer. *Colonie of Provintie van Suriname* (2 sheets), 1700; *Plan of the City & Castle of Chester...*, 1745; *A Plan of the Castle of Chester*, 1745.

Delavigne, N. Chantilly, 1725.

Delaville, *abbé* —. *Carte de la Pensilvania*, 1756.

Delaw, —. *La Peur du Tunnel*, 1908 (caricature of Britain as an old woman from *La Sourire*). **Ref.** *The Map Collector* 46 (1989) p.48.

Delbalat, Jean Baptiste James. Plan des bouches de Tigre, 1848; *Plan du port de Gènes*, 1853, published 1857.

Delbos, Joseph. *Carte géologique du Haut Rhein*, 1865.

Del Cano [Delcano], J.S. See **Cano**, Juan Sebastian del.

Del Canto, Manuel. *Puerto de Rio Janeyro y Ciudad de S. Sebastian*, *c*.1790.

Del Castillo, *Lieutenant* F. *Plano de las ruinas de Cempoala*, 1891.

Delcros [Delcroz], François Joseph (1777-1865). Engineer and geodesist, born in Florence, worked in France and Switzerland.

Delebecque, André (*b*.1861). Atlas des Lacs Français, 1895.

De Leeuw, Arent Martenez. See **Leeuw**, Arent Martenez de.

Delerue, Louis. *Plan de la Ville de St. Omer*, 1847.

De l'Escluse, C. See **Clusius**, Carolus.

Delesquellen, —. *Fernando de Noronha*, 1735 MS, published Alexander Dalrymple 1781.

Delesse, Achille E.O.J. (1817-1881). French geologist and mineralogist, Director 'Lithologie des fonds des Mers, Rue de Géologie' 1860-1878. *Carte hydrologique de Paris*, 1858; hydrographical charts of French départements, 1862-1873; *Carte lithologique des mers de l'Amérique du Nord*, 1870; and others.

Delestre, —. Engraver of one sheet for Jean François de Galaup de La Pérouse *Atlas...*, 1797.

De Leth. See **Leth**.

Deletre, —. Engraver for Auguste Henri Dufour, 1857.

Delfino, Frederico (1477-1547). Professor of Astronomy at Venice. De fluxo et refluxo aquae maris, 1599.

Delfos, Abraham (1731-1820). Engraver of Leiden. Maps for van Mieris's Beschryving van Leyden, 1762.

Delgado, J.F.N. Carta geografica de Portugal, 1876, 1899.

Delight, E. Boxley Farm, Maidstone, Kent, England *c*.1760 MS.

Delisle [De L'Isle; L'Isle de] family. **Ref.** PEDLEY, M. *The Map Collector* 48 (1989) pp.14-18; LAGARDE, L. *Imago Mundi* 41 (1989) pp.19-43 [in French]; SCHWARTZ, S.I. & TALIAFERRO, H. *The Map Collector* 26 (1984) pp.2-6; PELLETIER, M. *The Map Collector* 24 (1983) pp.12-13.

• **Delisle**, Claude (1644-1720). French geographer and historian, pupil of Nicolas Sanson. Three of his twelve children, Guillaume, Joseph Nicolas and Louis, became eminent geographers in their turn. Prepared maps for Pierre Le Moyne d'Iberville's voyage and a historical atlas, 1684.

•• **Delisle**, Guillaume [Insulanus] (1675-1726). Pupil of Jean Dominique Cassini. Member of the French Académie Royale des Sciences (1702), 'Premier géographe du roi' (1718). One of the key figures in the development of French cartography, he strongly believed in the importance of accuracy. During his lifetime his one hundred or more maps were constantly updated to reflect widening knowledge of the World. Addresses recorded as 'Rue des Canettes près de St. Sulphice' (1700-1707); 'Quai de l'Horloge à la Couronne de Diamans' (1707-1708); 'Quai

Guillaume Delisle (1675-1726). Copperplate drawn by Cless after a painting. (Taken from Map Collectors' Circle*)*

de l'Horloge à l'Aigle d'or' (1708 onwards); 'Quai de l'Horloge'; and 'Quai de l'Horloge au Palais'. Compiled globe, 1699; map of the World and the 4 continents, 1700; *Atlas de Géographie*, 1700-1712 and many later editions; Mississippi, 1701 MS; *Carte de Moscovie*, Paris 1706; *Siciliae Antiquae...*, 1714; *Hemisphere meridional pour voir plus distinctement les terres Australes*, 1714; *Carte de la Louisiane et du Cours du Mississippi*, 1718 (the first printed map to name Texas); *Carte d'Amerique*, 1722; duo-hemispheric world map, 1724; posthumous *Atlas Nouveau*, 1730 and later editions, Italian edition *Atlante Novissimo*, 1740-1750. Guillaume was succeeded by his widow Marie Angélique, then by son-in-law Philippe Buache. Stock was later sold to Jean-Claude Dezauche, who republished Delisle's atlases.
•• **Delisle**, Marie Angélique [*Veuve*] (*d.*1745). Daughter of Pierre Duval, wife and successor to Guillaume. She continued to sell Guillaume's maps until her own death, when the bulk of the stock passed to her son-in-law Philippe Buache. Brought to publication Guillaume's *Carte de l'Afrique Françoise...*, 1727.
•• **Delisle de la Croyère**, Louis (1687-1741). Brother of Guillaume and Joseph-Nicolas. Astronomer. Joined his brother Joseph Nicolas in Russia in 1726 and worked with him until his death in 1741 on an expedition with Chirikov.
•• **Delisle**, Joseph-Nicolas (1688-1768). Younger brother of Guillaume. Studied with Cassini as an astronomer and cartographer. Took service in Russia, where he founded the Academy of Sciences of St. Petersburg. He returned to Paris in 1747 with an important map collection. Worked with Kirilov on the first atlas of Russia, *Atlas Russicus*, 1745; N. Pacific, 1731 MS, printed 1750; Georgie, 1766. **Ref.** HECKROTTE & DAHL *The Map Collector* 64 (1993) pp.18-23.

Delisle de Sales, Jean-Baptist-Claude Izouard (1741-1816). Compiled *Histoire des Hommes...* (with *c.*15 maps of ancient Europe), 1779-1782.

Delitsch, *Professor* Otto (1821-1882). Engraver and geographer of Leipzig, also worked with Jan Felkl on German language globes. *Europa*, 1860; *Die beiden Hemisphären* (wall chart), Leipzig 1862; *Mitteleuropa, orographische-hypsometrisch und hydrographische* (wall chart), Leipzig 1862; *Neuer Atlas der ganzen Erde...*, 1863 (possibly C.G.D. Stein's *Neuer Atlas*, which dates back to 1819); his Professorial thesis, which covered population mapping, was entitled *Kartographische Darstellung der Bevölkerungs-Dichtigkeit von Westdeutschland...* (with 3 multi-coloured maps), Leipzig 1866; *Großer Hand- Atlas des Himmels under der Erde*, 1871 (with H. Kiepert, C.F. Weiland, C. & A. Graef, C. Bruhns); *Schul-Wandkarte*, 1874.

Delius, C. *Plan von Berlin*, *c.*1850; *Neuester Plan von Berlin und Charlottenburg...*, 1869.

Delius, Christoph [Kristóf] (1728-1779). German cartographer, studied at Wittenbergben and Selmecbánya [now Banska Stiavnica, Slovakia]. Professor at the mining academy in Oravica [now Oravita, Romania], and later professor of the academy of Selmecbánya, 1770-1772, died in Florence. Mining atlas, 1771.

Delius, Hermann. Lithographer for Heinrich C. Kiepert. Engraver for Theodor von Liechtenstern's *Schulatlas zur Erd- und Staatenkunde für höhere Schulen...*, Berlin, G. Reimer 1838, 1846; *Plan von Berlin 1841*; Bromberger Kreis, 1843; *Karte der Herzogthümer Schleswig Holstein & Lauenburg...*, Berlin, Nicolai c.1848; *Karte des Isthmus von Nicaragua und Panama...*, Duncker 1851; plans of Berlin, 1859, 1867; Schweiz, 1866.
• **Lith. Anstalt von H. Delius.** An imprint of Hermann Delius.

Delius, Thomas. Plans of Berlin, 1874, 1875; Grünewald, 1875; Potsdam, 1876; Brandenburg, *c.*1890.

Delkeskamp, Friedrich Wilhelm (1794-1872). German panorama painter and copper engraver from Bielefeld. Worked in Frankfurt-am-Main. *Panorama des Rheins*, editions from 1825 to 1868 (various language versions e.g. English in 1835); *Malerisches Relief des klassischen Bodens der Schweiz*, 1830; *Malerisches Rheinpanorama von Mainz bis Köln*, 1837 (in 1842 the map was extended to cover Speyer); panorama of the Mosel, 1839; town plan of Frankfurt, 1859-1864; and others.

Della Bella. See **Macaneo**.

Della Bella, Stefano. See **Bella**, Stefano della.

Della Gatta, Giovanni Francesco. Publisher in Rome. *Palaestinae sive Terre Sancte descriptio*, 1557 (reduced version of Wolfgang Wissenburg's *Descriptio Palestinae Nova*, *c.*1537).

Delleker, George. Engraver of Philadelphia, partner of James Young (1822-*c.*1823). Engraved for D. Macpherson, 1806; Mathew Carey, 1817; Carey & Lea, 1822, 1826; Fielding Lucas, 1823; as Young & Delleker engraved the maps for Anthony Finley's *A New General Atlas*, 1824.

Dellinger, A. Engraver for Heinrich C. Kiepert, 1869.

Dell'Orto, A. See **Dalorto**, Angelo.

Delmas, Jacques. French geographer. L'Aude, 1867; Provence, 1878.

Delmenhorst, —. Münster & Osnabrück, 1730; *L'Evêché de Münster*, 1757; *Ducatus Bremae*, 1767.

Deloche, Jean Edouard Maximin (1817-1900). French historian. Géographie historique de la Gaule, 1864.

Deloncle, François. *Carte politique de l'Indo-Chine*, 1889.

De Lorenzo, —. Carta geologica d'intorni di Lagonegro, 1898.

Delpech, Ernest. *Plan de la ville de Bordeaux*, 1869, 1876.

Delphinas, O.F. See **Fine**, Oronce.

Delpino, —. Spanish cartographer. Pyrenees, 1768 (unfinished).

Delsenbach, Johann Adam (1687-1765). Artist and engraver of Nuremberg. *Nurnbergischer Prospecten*, 1716.

Delsol, Théodore (*b.*1819). Copper engraver of Paris. One of the engravers for *Carte topographique de la France*, Depôt de la Guerre 1832-1847; For Gérard Lemaitre, *c.*1840; E.H. Michaelis's *Topographische Karte des... Kantons Aargau...* (4 sheets), 1845-1848; A. Stryieński's *Carte Topographique du Canton de Fribourg...* (4 sheets), 1855.

Deluc, *Professor* Jean-André (1727-1817). Swiss physicist and geologist, born in Geneva, worked also in London, Berlin, Hanover and Brunswick.

Delure, Jean Baptiste (d.1736). Map, instrument and globemaker of 'Quai de l'Horloge', Paris. 12-cm globe, Paris 1707; *Gibraltar*, 1727.

Delves-Broughton, *Captain* W.E. Royal Engineers. One of the surveyors for the British Boundary Commission in Canada. *A Map of a portion of the Country in dispute with the United States including the Plain that separates the Highlands claimed by that Government from the Highlands of the Treaty of 1783...* [Quebec / Maine], 1842, 1848 (with James D. Featherstonhaugh, engraved by James Wyld); and other maps of the area.

Demanne, — (*fl.*1770s-1780s). Mapseller, 'Rue de l'Ortie, vis-à-vis le Logement du Premier géographe du roi', Paris. For J.B.B. d'Anville; J.-D. Barbié du Bocage, 1786.

De Mannevillette, A. See **Après** de Mannevillette, J.B.N.D. d'.

De Marchi, Francesco (1504-1576). Military and civil engineer and architect of Bologna, moved to Rome in 1535. He wrote an important treatise on fortifications, which included many maps and plans. Plan of Rome, 1551 (with Leonardo Bufalini); plan of Parma, 1552 (published without the intended text in *Architettura militare di Francesco Marchi*, Venice, Francesco Franceschi 1577); *Ritratto dallo istesso disegno mandato da Malta*, 1567 MS (published in 1597 without text, and with text in *Della architettura militare del Capitano Francesco de' Marchi*, Brescia, Dall'Oglio 1599, 1600, Bozzoli fratelli 1603, Pietro Maria Marchetti 1609). **Ref.** GANADO, A. & AGIUS-VADALÀ, M. *A study in depth of 143 maps representing the Great Siege of Malta 1565* (1994) pp.381-387; *ibid.* pp.452-454

Demarest, P. Plan de Marseille, 1808.

Demarque-Geoffroy, —. Plan de la Forêt de Compiègne, 1874.

Demasso, M. Engraver of *L'Afrique selon les Auteurs les plus Modernes*, mid-18th century (this map shows similarities to R.P. de Murillo Velarde's miniature map of 1752).

De Mayne, Anthony. Admiralty surveyor. Surveys of the West Indies, 1811-1828; Chesapeake Bay, 1820; *The Island of Jamaica*, 1821, 1823; *A General Chart of the West Indies and Gulf of Mexico*, 1824, 1832; *A Chart of the Gulf of Florida...*, 1824; Cape Palmas, 1831.

Dembsher, Giuseppe. Geographer of Venice. Edited first foreign edition of Stieler's Hand-Atlas as *Atlante Universale del Globo*, 1829.

De Mechel, Christophe. See **Mechel**.

Demers, Jean Baptiste. *Plan Figuratif de la Seigneurie de Ste. Croix...*, 1831.

Demersay, L.M. Alfred. French geographer and traveller. Carte physique Paraguay (16 sheets), 1860-1864.

Demezynski [Demezynshi], —. *Pianta della Città di Milano*, 1844.

Demidov, [Demidoff] Anatoly Nikolayevich. *Carte du Voyage dans la Russie Méridionale exécuté en 1837...*, Paris, Ernst Bourdin 1841 (drawn and engraved by P. Tardieu, printed by Chardon & Son).

Demiege, —. Maconnais (4 sheets), 1776.

Deming & Francis. Publishers of Wethersfield [USA]. Frederick Butler's *A Modern Atlas*, 1825.

Demler, George. Engineer. *Plan of Fort Niagara* (inset on Michael de Bruls *Plan of Niagara*, 1762).

De Mole, G.E. *Rivoli Bay*, 1858; *Lacépede Bay* [South Australia], 1859 (both with P.A. Nation).

Demongenet, François. See **Mongenet**.

Demonville, —. 'Imprimeur-libraire de l'Académie Française, Rue S. Severin, aux Armes de Dombes', Paris. Co-publisher, with Malassis of the expanded edition of J.B.N.D. d'Après de Mannevillette's *Le Neptune Oriental...*, 1775.

De Morphey, —. Archipel des Seychelles, 1756 MS.

De Mosquera, T.C. See **Mosquera**.

Demotier, —. Lithographer. *Plan de la Ville et du Port de Calais*, 1845.

Dempster, Henry (1807-1877). *Sketch of Table Bay* (woodcut), 1841.

De Munk, F. See **Monachus**, Franciscus.

De Musis, Julius. See **Musi**, Giulio de.

Denaix [Denais], Maxime Auguste (*d.*1844). 'Chef de Bataillon au Corps Royal d'Etat Major'. *Atlas physique, politique et historique de l'Europe*, Paris 1829; Carte de l'Analyse géographique, 1831; Atlas de la France, 1836-1837.

Denarowski, Karol [Carl]. Polish cartographer. Sanitäts Karte der Bukowina, 1880.

Dencède, A. France kilométrique, 1893.

Denecourt, C.F. *Carte topographique de la Forêt de Fontainebleau*, *c.*1844, 1870, 1878.

Dengelsted, —. Maps for Pieter Schenk, 1752-1758.

Denham, *Lieutenant* (later *Captain*) Henry Mangles, *RN*. Admiralty surveyor. Fiji, 1822; Bristol Channel, 1827-1835; *A Chart of the North East Coast of Devonshire*, 1833; West coast of England, 1842; Niger & Guinea coast, 1845-1846; River Dee, 1847; South West Pacific, 1852-1860.

Denham, W.W. Sketch of proposed Railway between Maharanji and Port Albert, New Zealand, 1870.

Denhardt, C. Äquatoriales Ost-Afrika, Justus Perthes 1881; Untere Tana Gebiet, Berlin 1884.

Denis, —. Umgebungen von Metz, 1872.

Denis, Ferdinand (1736-1805). *Gegend von Mannheim*, 1780, 1814; Mannheim u. Umgebung, 1782.

Denis, Louis (1725-1794). Geographer and engraver of the 'Rue de Bievre', Paris. Geographer to Duc de Berry, associate of L.C. Desnos (1760). Plan de Paris, 1758 (with J.J. Pasquier); *Mappemonde*, 1764; Atlas géographique, 1764; *Environs du Caen*, 1765; Guide des Voyageurs, 1768; Atlas de Normandie, 1770; *Guide royale... des grandes routes de Paris au villes de royaume*, 1774; *Carte du Théâtre de la guerre en Amérique*, 1779; *Carte du theatre de la guerre dans l'Inde...*, Paris, Basset 1781 (engraved by Bellanger); *Carte de l'Isle Minorque*, 1781; *La Mer Mediterranée*, Paris, Basset 1782; corrected J.B. Nolin's *Mappemonde*, 1785 edition; *Carte de la Provence...*, Paris, Basset 1785 (engraved by Croisey); Amérique, 1788; updated several maps by Nolin, including those of Africa, America, and Asia, all republished in Paris by Basset, 1793.

Denison, Charles. *Pocket atlas of Climatic maps of the United States*, 1885.

Denison, J. United States cartographer. Maps for Jedediah Morse's *The American Universal Geography*, including *...New Hampshire & Vermont* (engraved by Amos Doolittle); *Massachusetts from the best authorities* (engraved by Samuel Hill); *The State of New York* (engraved by Amos Doolittle); *...Maryland & Delaware* (engraved by Amos Doolittle); *Map of North & South Carolina* (engraved by Amos Doolittle), all drawn *c.*1778, engraved later, and published in Boston by Thomas & Andrews 1796.

Denman, James F. Vintners of London. Map showing the wine growing districts of Paarl, South Africa, *c.*1890s.

Dennis, A. (1779-1817). Monmouth County, 1781 MS (with I. Williams, reduced from John Hills).

Dennis, W. Luke. Transport map of Birmingham, 1886.

Dennis, William. Charts of Great Britain, 1781-1802.

Denores, Jason. Breve tratatto del mondo, Venezia, A. Muschio 1571.

De Nobilibus, P. See **Nobili**, Pietro de.

De Nova, P. Straits of Malacca, Robert Sayer 1779.

Denovan & Co. Publishers of Edinburgh. *Edinburgh and Leith Directory*, 1804-1805, containing *...This Plan of the City of Edinburgh...* (engraved by D. Lizars [senior]).

Dent, *Lieutenant* Albert, *RN*. Admiralty surveyor. *Sound of Harris*, 1859; *Hebrides*, 1865.

Dent, *Captain* Frederick Tracy (1821-1892). American soldier. Assisted with James H. Simpson's surveys of New Mexico, 1849.

Dent, R. *Plan of Islington Parish*, 1805-1806, 1822.

Denton, Guillermo. See **Denton**, William.

Denton, John Bailey (1814-1893). Various maps in different English counties including *Parish of Hitchin*, 1844.

Denton, Thomas (*fl.*1809-1841). Surveyor, produced estate and inclosure maps in various English counties including plan of Kirtling, 1814; plan of East Bedfont, 1816 MS (with William Denton); *Plan of the parish at the time of the inclosure...* [Egham], 1816 MS.

Denton, William (*fl.*1800-*c.*1819). Estate and inclosure plans in the English home counties.

Denton, William [Guillermo] (1823-1863). American geologist. *Distritos Minerales de San Antonio*, 1865.

D'Entrecasteaux, J.A.B. See **Bruni** D'Entrecasteaux, Antoine Raymond Joseph.

Dentu, —. Printer and bookseller, 'Palais du Tribunat, galéries de bois N°·240', Paris. Recueil de Cartes, 1802.

Denys, Jehan. Norman pilot of Honfleur. Newfoundland, *c.*1506; Canada, *c.*1506; St Lawrence, *c.*1508.

Denys, Nicolas. *Description géographique et historique des costes de l'Amérique septentrionale*, 1672.

Denys, William. Hydrographic surveyor. *A new chart of Mounts Bay in Cornwall*,

London, Sayer & Bennett 1780 (with John Thomas), Sayer, Stephenson & Burn 1786; *A nautic survey of Mounts Bay...*, Laurie & Whittle 1794 and later (with John Thomas).

Denza, Francesco. *Carta delle Correnti Marine*, *c.*1875.

Denzler, Johann Heinrich (1814-1876). Swiss surveyor and topographer, director of the land register in Zurich, Bern and Solothurn. Collaborated on G.H. Dufour's *Topographische Karte der Schweiz* (copper engraving, 25 sheets), 1845-1864 (known as the 'Dufour map').

De Palmeus, G. See **Palmeus**.

De Panger, Roger. 'Ingénieur royal'. Plan particulière de l'embouchure du fleuve Saint Louis, 1724.

Depéret, Charles. *Carte géologique de la France*, 1889.

De Pierre, —. Austrian pilgrim. Map of the City of Jerusalem and part of the Holy Land, 1728. **Ref.** RUBIN, R. *The Map Collector* 55 (1991) pp.30-31.

Depósito Hydrográfico, Madrid. Spanish nautical charting organisation, founded 1797 by José de Espinosa y Tello.

Deposito Topografico de Manila. Publisher of Y. de Aragon y Abollado's plans of the Philippines, 1819-1821

Dépôt de la Guerre, Paris. Founded by the Marquis de Louvois 1688. Published Roussel and La Blottière's *Pyrenées*, 1730; Atlas des campagnes de Napoléon, 1844; Atlas de l'expédition de Chine, 1861-1862.

Dépôt des Cartes et Plans de la Marine. French state hydrographic service in Paris. Founded in 1720 to provide the French navy with accurate charts. Jacques Nicolas Bellin was the first chief hydrographer.

Depping, George Bernard (1784-1853). A writer on geography. Born in Munster, later moved to France. *L'Angleterre ou Description Historique et Topographique du Royaume Uni de la Grande-Bretagne* (6 volumes), Paris, Etienne Ledoux 1823, 1824, 1828, 1835, Brussels, L.J. Brohez, 1828 (maps drawn by Aristide Michel Perrot, engraved by Madame Migneret).

De Puy, William Harrison (1821-1901). People's Atlas, Boston 1885.

Dequevauviller, —. Engraver, for La Pérouse, 1796; maps in the Catalonia volume of Alexandre de Laborde's *Voyage pittoresque et historique de l'Espagne*, Paris, Pierre Didot l'ainé 1806.

De Ram, J. See **Ram**, Johannes de.

Derbishire & Desbarats. Publishers of Toronto. Thomas Devine's *Map of the North West Part of Canada*, 1857.

Derby, *Lieutenant* George Horatio (1823-1861). United States topographical engineer. Colorado River, 1850.

Derendinger, Johann Stephan (*fl.*mid-18th century). Swiss cartographer and surveyor in Solothurn.

Derfelden van Hinderstein, *Baron* Gijsbert Franco von (1783-1857). Dutch geographer. *Algemeene Kaart van Nederlandsch Oostindië* (8 sheets), 1842 (with G.A. Tindal); Sumatra, *c.*1872.

Derksen, J. See **Diricks**, J.

Dermott, James R. *Map of the Public Appropriations of the City of Washington*, *c.*1796 (known as the 'Tin Case Map').

Derocco, D.J. 'Professeur de géographie à Belgrade'. *Carte de l'extension ethnique de la nation serbo-croato-slovène...*, Belgrade 1915 (with St. Stanoïévitch); *Railways of Yugoslav Countries*, Paris 1919.

De Rogier, Johan (1600-1684). Swedish cartographer.

Derond, J. (*fl.*1770). Engraver and cartographer in Amsterdam.

De Roos, *Commander the Hon.* John Frederick Fitzgerald. Admiralty charts. *Port Maceio*, 1834; *River Parahyba*, 1834.

Derosaire, Michel. Lithographer. Allahabad, 1862; Bareilly, 1886; and others.

Derozier, —. See **Desrosiers**, —.

Derrécagaix, Victor (1833-1915). French geographer, assigned to the topographical service in Algeria in 1862; from 1888-1896, director of the Service géographique de l'armée; organised surveys of Algeria and Tunisia. *Le sud de la province d'Oran*, 1877. **Ref.** BROC, N. *Dictionnaire illustré des explorateurs Français du XIXe siècle: I Afrique* (1988) pp.109-110 [in French].

Derrien, —. *Carte...du Mont-Cenis*, 1821.

Derrien, *Captain* I. Galilée, 1870; Oran, 1874; Algérie, 1876; Haut Sénégal, 1882.

Desaguiliers, Henri. *Table des 17 Provinces des Pais Bas*, *c.*1710.

Desaint, J. Ch. Associated with Cassini's *Carte de France*, 1783.

De Sanctis, Gabriello. See **Sanctis**, Gabriello de.

Desandrouins, Jean-Nicolas. *Carte des environs de Williamsburg Armée de Rochambeau*, 1782; Entrée de la Baye de Chesapeack, 1782; Partie du Nord de l'Isle de New-York, 1781.

De S'Angelo. See **Angelo**, M. de S.

Desaulces de Freycinet, L.-C. See **Freycinet**, brothers.

Desbarats, G.E. [& Co.] Lithographers of Montreal. *Map of Canada and Part of the United States...*, Montreal 1873; *Dominion of Canada Telegraph & Signal Service* [Saskatchewan, Manitoba, Alberta], 1883. See also **Burland**-Desbarats and **Derbishire** & Desbarats.

Des Barres [Barres; Desbarres], *Captain* Joseph Frederick Wallet (*c.*1729-1827). Hydrographer, cartographer, military engineer and landscape artist from Basle, Switzerland. Emigrated to England and trained at the Royal Military Academy, Woolwich. Became a colonial administrator in Halifax, Nova Scotia, where he died. Lieutenant Royal American Regiment, 1756; engineer under General Wolfe at Quebec, 1759; Governor of Prince Edward Island. *A Plan of Quebec and Environs... Siege of ...1759*, *c.*1759 MS, published *c.*1780 (with Samuel Holland and Hugh Debbieg); surveyed coasts of Nova Scotia, 1764-1774; Rhode Island Harbour, 1776; *The Atlantic Neptune* (about 800 charts & views in various states), 1775-1781. **Ref.** DEBARD, J.M. *The family origins of Joseph Frederick Wallet DesBarres: A riddle finally solved* in: *Nova Scotia Historical Review* (Autumn 1994) and *Actes du Colloque Montbeliard sans frontières* (1994) pp.141-171; EVANS, G.N.D. *Uncommon Obdurate: the Several Public Careers of J.F.W. Des Barres* (Toronto 1989); DE VORSEY, L. *The Map Collector* 19 (1982) p.31.

Desbleyns, —. *Carte ecclésiastique...du Royaume de France*, 1624.

Desbois, Nicolas-Jean-Baptiste (*fl.*1731-1749). Publisher, 'Rue St. Jacques à la Sphère Royale', Paris. Took over veuve Danet's business (1746) was himself succeeded by his wife Marie-Anne, below. Re-issued *Amérique...*, *c.*1746.
• **Desbois**, Marie-Anne (*fl.*1749-1752). Publisher. Daughter of Guillaume Danet, succeeded her husband Nicolas-Jean-Baptiste (above).

Desbordes, Charles. Plan of Turin, 1705; plan of Menin, published by 'weduwe' [widow] Visscher *c.*1706; *Zelon* [Ceylon], 1706 (curiously reversed copy of de Bry's map of 1605).

Desbrosiers, —. Engraver. French provinces, 1707-1715. **N.B.** Compare with **Desrosiers**.

Desbruslins, father & son. French engravers. Both worked as letter engravers on several sheets of the Cassini *Carte de France*, published from 1756; *Le Cercle de Basse Saxe...*, Paris, J.-F. Daumont 1757.
• **Desbruslins**, F., *père*. Engraver for Bernard Jaillot, 1719; Lakeman, 1727; Philippe Buache, 1741-1757; Pierre François Xavier de Charlevoix, 1744; Pedro Gendron, 1754; père Claude Buffier, 1754.
•• **Desbruslins**, —, *fils*. Engraver. Hesse Cassel, 1760-1761; maps for l'abbé Courtalon's *Atlas élémentaire de l'Empire d'Allemagne*, Paris 1774.

Desbuissons, —. Engraver, 'Rue des Noyers N^o.^8', Paris. For Aristide Michel Perrot, 1827.

Desbuissons, Louis Edouard (*b.*1827). French cartographer and geographer, 'géographe du Ministère des Affaires Etrangères' (1865-1900). Worked with E. Cortambert.

Produced maps of Africa and of the French railway system. F. Andriveau' s *Carte Physique et Routière De l'Italie...*, Paris, Andriveau-Goujon 1859; Opérations en Mexique, 1862; Basse Egypte, 1869; *Carte de la Basse Egypte et du Canal Maritime de Suez...*, Paris, E. Andriveau-Goujon 1882; Guinée, 1886; and others.

Descalier. See **Descelliers**, Pierre.

Descalvi, *Doctor* N. Plan del Rio Bermeju [Argentina], 1831.

Descelliers [Desceliers; Descalier], Pierre (1487-1553). Cartographer and leading figure in the 'Dieppe School of cartographers'. Painted miniatures, and made charts and atlases from 1546. Produced several world maps including *Mappemonde*, 1546 (with emphasis on Jacques Cartier's exploration along the St. Lawrence and in New France).

Deschmann, K. (*d.*1889). Austrian geographer and naturalist.

Descubes, A. *Map of the Island of Mauritius*, 1880.

Desehrt, —. Engraver, publisher and map seller of Copenhagen. Sjælland, 1768; C. Niebuhr's *Arabien*, 1774; Jersey, *c.*1780.

Deseine, François Jacques (*d.*1715). Parisian geographer and bookseller working in Rome.

Desenne. See **Menard & Desenne**.

Desessards, C. *Carte géologique du département de Deux Sèvres*, 1870.

Desfontaines, B. French engraver, credited with work on the Cassini *Carte de France*, 1744-1760.

Desfontaines, J.B. California, 1849.

Desfossés, Romain Joseph. *Baie de l'Ouest*, 1832; *Plan des Îsles du Salut*, 1834.

Desgodins, *abbé* C.H. Brother of French explorer père Auguste Desgodins, whose work he used as a basis for his publications. Frontière du Thibet oriental, 1885.

Desgranges, —. 'Géographe du roy'. Palatinat du Rhin, 1688; *La Carte des Royaumes d'Angleterre...*, 1689; Greece, 1702.

Desgranges, —. *Plan de la Ville du Mans*, 1862 (with A. Imbault).

Deshayes [Deshaies; Deshaise], Jean *sieur de* (*d.*1706). Sept îsles, 1686; Rivière de S. Laurens, *c.*1695; Parish map of Canada, *c.*1700; *Carte Marine du l'Embouchure de la Rivière de S. Laurens*, *c.*1702; Coste du Canada, 1704; *De la Grande Rivière de Canada appellée par les Européens de St. Laurens*, N. de Fer 1715.

Des Hayes, Louis, *baron* de Courmenin (1592-1632). Voyage de Levant, 1624; Voyage en Danemarc, 1664; Itinéraire à Constantinople, 1665.

Desilver, Charles. Publisher of N°·251 Market Street, Philadelphia (1856); 716 Chestnut Street (1856-1857); 714 Chestnut Street, Philadelphia (1858). *Mitchell's A New Universal Atlas*, edition of 1856 (later edition of a work publishd by S.A. Mitchell in 1846); *A new universal atlas*, Philadelphia 1857; James S. Richie's *Wisconsin and its resources...*, 1857; J.L. Hazzard's *Map of Lake Superior with its Railroad & Steamboat Connection...*, 1857. **N.B.** See also **Cowperthwait**, Desilver & Butler.

Desimoni, Cornelio (1813-1899). Italian geographer.

Desing, P. Anselm (*fl.*1720-1740). Benedictine monk of Ensdorff. *Katholic Kinder Atlas*, 1723; Haupt Karte Engelland, Schottland, Irland, one of several maps in Desing's *Historica Auxilia*, 1733-1734, 1741, 1746-1747.

Desjardins, —. Insets on Thomas Jefferys's chart of the St. Lawrence, published Sayer & Bennett 1777 and later.

Desjardins, *Professor* Constant. Hauteurs des montagnes, 1830; Atlas von Europa, Vienna 1836-1838; Länge der Ströme, *c.*1855 (thematic charts).

Desjardins, Ernest Emile A. (1823-1886). Geographer of Paris. *Atlas géographique de l'Italie*, 1852; *Table de Peutinger*, 1869; Atlas universel, 1877.

Desjeans, Jean Bernard Louis, *sieur* de Pointis (1645-1707). *Rencontre de l'Escadre de France et del Armee Angloise et la route qui fit Le Baron de Pontis dans sa retraite* (French and English fleets between Cartagena and Jamaica), 1697.

Desliens, Nicolas. Chartmaker of Dieppe. Manuscript world maps, 1541, 1566, 1567.

Deslongchamps, —. One of the surveyors for the Cassini *Carte de France*, worked with others on seven sheets, all in the south of France, 1769-1780. **N.B.** Compare with **Longchamps**.

Desmadryl, frères. Draughtsmen for A. Schneider's *Îles Ioniennes*, 1823.

Desmadryl, A. (*fl.c.*1830). Lithographer. **N.B.** Possibly one of the *frères*, above.

Desmadryl, N. *Plan de la Ville du Mans*, *c.*1855. **N.B.** Possibly one of the *frères*, above.

Desmarest, Nicolas (1725-1815). French physician and geologist, 'Hôtel de Thou, rue des Poitevins', Paris. Dictionnaire de géographie physique; *Atlas encyclopédique...*, Paris 1787 (with Rigobert Bonne and Bory de St. Vincent); Puy de Dôme, 1823.

Desmarests, Jean. Designed a pack of cards entitled *Le Jeu de geographie*, Paris 1644 and later printings to 1698. **Ref.** SHIRLEY, R. *The mapping of the World* (1983) N°361 p.384-385.

Desmarets, Jean Baptiste François, *marquis* de Maillebois. Commander of troops in Corsica at the time of an uprising in 1739. Events were illustrated in *Carte de L'Isle de Corse...*, 1741; *Cartes géographique et topographique de l'Italie*, 1775.

Desmaretz, A. Lithographer of Paris. *Carte des environs de Constantinople...*, Paris, Picquet 1829.

Desmaretz, *Captain* John Peter. Engineer who became Chief Ordnance Draughtsman. Survey of Westminster, Chelsea and Kensington, 1717; River Medway between Rochester and Sharpness, 1724; *The harbour and vicinity* [Harwich], 1732; Landguard Fort, 1736; Sheerness, 1738; Portsmouth Harbour, 1750; proposals for Shoreham harbour, 1753; plan of Ramsgate harbour, 1755 (with Sir Piercy Brett); Plan of Dunkirk Harbour, 1763-1765; and many other forts and harbours. **Ref.** ROBINSON, A.H.W. *Marine cartography in Britain* (Leicester University Press 1962) pp.92-93.

De Smet, Pierre-Jean, *SJ* (1801-1873). Map of the Rockies, 1844-1847.

De Smidt, Abraham (1829-1908). Cape Colony Surveyor-General 1872. New Map of Cape Colony, 1895.

Desmond, Hugo A. Engineer. Railway from Mejillones to La Paz, 1880.

Desmond, John T. Atlas of Haverhill & Bradford, Massachusetts, 1892.

Desmoulins, Léandre Eugène. *Côtes de Syrie*, 1861-1862.

Desnos, Louis-Charles (1725-1805). Mapmaker, globemaker and publisher of Paris, worked with most of the eminent cartographers of his day. 'Ingénieur géographe pour les globes et sphères'; 'ingenieur géographe du roi de Danemark'. In 1749 married Marie-Charlotte Loye, young widow of Nicolas Hardy *q.v.* of 'rue Saint-Julien-le-Pauvre', from where he began to trade as a globemaker (1753); later 'Rue St. Jacques au Globe' (from 1757). Map seller for some sheets of the Cassini *Carte de France*, and for work by Dumez and Chanlaire. By the time of his death in Paris in 1805, the plates for works by Buy de Mornas, Brion de la Tour, Delamare (*Traité de la Police*), Robert de Vaugondy, Fricx and others were under his control. Globes, made in association with Jacques Desnos or Jean-Baptiste Nolin, *c.*1750-1782; George Louis Le Rouge's *Atlas nouveau portatif*, 1756; re-issued Guillaume Danet's Asie, 1760; *Nouvel atlas de la généralité de Paris*, 1760; C. Buy de Mornas's *Atlas Méthodique et Élémentaire de Géographie et d'Histoire*, 1761, 1783; amended and re-engraved Nicolas de Fer's *Les Routes des postes du royaume de France*, 1761; Atlas chorographique de la France, 1763; *L'indicateur fidèle ou Guide des voyageurs...*, 1764 and later editions (with Michel); *Petit Atlas Elementaire...*, Paris *c.*1765; Nouvelle histoire générale, 1766; *Nouvel itinéraire général*, 1766; Isles Britanniques (4 sheets), 1766; *Nouvel atlas* *d'Angleterre*, 1766; *Atlas général méthodique*, 1768; *Atlas de France*, 1775; *La France dans toute son etendue,...*, 1789; *Atlas national et générale de la France*, Paris 1790 (shows new départements); *Carte de France divisée en vingt trois divisions militaires...*, 1792; Amérique, 1795; *Mappe Monde Celeste, Terrestre et Historique*, 1796. **Ref.** PACHA, B. in: PACHA, B. & MIRAN, L. *Cartes et plans imprimés de 1564 à 1815* (Bibliothèque Nationale de France 1996) pp.44-46 [in French].

Desobry, P. Publisher of New York. *The Fireman's Guide*, 1834 (plan of the city compiled by U. Wenman of the New York Fire Department). **Ref.** COHEN, P.E. & AUGUSTYN, R.T. *Manhattan in maps 1527-1995* (1997) pp.116-117.

Desor, [Peter Johann?] Edouard (1811-1882). Born near Frankfurt-am-Main, worked in Switzerland as geologist and paleontologist. Orographie des Alpes, 1862.

De Soto, Hernando (*d.*1542). Said to be the first European to explore the interior of North America, in particular the Mississippi and Appalachians, 1539-1542. His travels first appeared on an anonymous manuscript map, *c.*1544.

Desparées, *Sieur*. See **Erault**.

Desprez, Germain. Guide chemins Angleterre, 1571 MS.

Despriez, —. *Plan de Caen*, 1840.

Despuig i Dameto, *Cardinal* Antoni [Antonio] (1745-1813). Mapa de la Ysla de Mallorca (8 sheets), Madrid 1785.

Desray, —. Publisher, 'Rue Haute-feuille N°.4, pres celle St. André des Arts', Paris. Eustache Hérisson's *Atlas portatif*, 1807; for Jean Denis Barbié du Bocage, 1811; Adrien Hubert Brué's *Grand Atlas Universel*, 1815, 1821.

• **Desray**, *Veuve*. Widow of, and successor to, Desray, above. A.H. Brué's *Atlas géographique, historique, politique et administratif de la France...*, 1828.

Des Roches [D.R.], Jean Baptiste Hamont *sieur* (*fl.*1645-*d.c.*1693). French military engineer, married Reine-Michèle, niece of Sébastien de Pontault de Beaulieu. Succeeded de Beaulieu, 1674. Completed the engraving of the so-called 'Grand Beaulieu' (using the monogram D.R.), but died before its publication in 1694. **N.B.** See also **Beaulieu**. **Ref.** PASTOREAU, M. *Imago Mundi* 32 (1980) pp.59-61 [in French].

Desrosiers, —. Engraver for Blaeu's *Atlas Novus*, *c.*1635.

Desrosiers [des Rosiers; Derozier], —. Engraver for Guillaume Delisle, 1705-1715, including: *Theatrum Historicum...*, 1705; *Carte de Tartarie...*, 1706; *Carte des Comtez de Hainault de Namur et de Cambresis...*, 1706; *Carte de l'Egypte de la Nubie de l'Abyssinie &c.*, 1707; *Carte de la Prevosté et Vicomté de Paris...*, 1711.

Dessalines d'Orbigny, Alcide (1802-1857). Naturalist, geologist and traveller. Voyage en Amérique Méridionale, 1835-1847; *Partie République Argentine*, 1835; *Carte de l'Amérique Méridionale*, 1838; République de Bolivie (2 sheets), 1839.

Dessans, —. Entrée de Charles Town, 1777.

Dessel, Camille van. *Carte archéologique de la Belgique*, 1877.

Dessel, F. van. *Nieuwe Caert van Frieslant &c.*, *c.*1720.

Dessingy, —. Côte de la Guyane, 1765 (for Jacques-Nicolas Bellin).

Dessiou, Joseph Foss, *Master, RN* (*fl.*1790-1842). Chart compiler and tidal expert of 3 (later 10) Albion Row, Commercial Road, London. *...new and correct survey of Dartmouth Harbour...*, 1770 MS; River Dart, 1782; *Le Petit Neptune Français; or, French Coasting Pilot*, William Faden 1805 (3rd edition); charts used in the preparation of Heather's *New Mediterranean Pilot*, 1802; *A Survey of Tor-Bay*, P. Steel 1804; Dartmouth, W. Faden 1804; *General Chart of the Mediterranean Sea*, 1805; *New Chart of the West Indies*, 1810 (in Jefferys's West India Atlas, Laurie & Whittle 1810 and later); *Directions for navigating in and throughout the North Sea... written to accompany a new chart of the North Sea and Kattegat...*(6 sheets by Dessiou), W. Faden 1816; Brazil, W. Faden 1818 (compiled from the surveys of others); *Chart of Burry River & Rosilly Bay*, 1821; *Chart of Burry Roads and Loughor Estuary*, 1824, and others.

Dessoutt, Simon. *Sketch of the River Coa* [Portugal], 1812 MS.

Destar, John. Navigation London to Lowestoft (5 sheets), [advertised 1778].

De Stefano, C. Carta geologica del terreno lignifero di Garfagnana, 1888.

Destenave, Georges (1854-1928). French colonial officer, undertook a geographical study of Lake Tchad. Carte du macina, [n.d.]; *Reconnaissance de la région du Tchad* in: *La géographie*, 1903. **Ref.** BROC, N. *Dictionnaire illustré des explorateurs Français du XIXe siècle: I Afrique* (1988) p.111 [in French].

Desterbecq, François (*fl.*1807-1849). Artist and lithographer. Director of 'Etablissement géographique', The Hague. *Atlas van het Koningrijk der Nederlanden...* (14 maps), 1841.

De St Hilaire Mallet, —. Engraver and engineer. *Nouveau Plan de la Ville de Luxembourg...*, The Hague, I.M. Husson 1743.

Destouches, *Ensign* —. French navy. *Côte occidentale d'Afrique...*, 1887.

Destrée, Jacques [or Maillard]. Plan d'Ypres, 1564.

Detaille, —. *Topographical Map of St. Lucia*, 1847.

De Thury, Cassini. See **Cassini**, family.

Detterline, T. *Battle of Gettysburg*, 1863.

Dettmers, A. Printer for Heinrich C. Kiepert, 1857.

Detz, L.B.J.S. City of Bangkok, Singapore 1871.

Deur, family.
• **Deur** [Deuer], Abraham Jansz. (*fl.*1666-1714). Engraver of Amsterdam for Frederick de Wit, the Visschers and others. Pieter Goos's Vlaemsche, Soute en Caribische Eylanden, 1668 (the revised issue); Canada, 1670; Gelderland, 1684.
•• **Deur**, Jacob (*fl.*1709-1714). Engraver of Amsterdam, son of Abraham, collaborated with brother Jan.
•• **Deur**, Jan. Engraver of Amsterdam, collaborated with brother Jacob. Italy, 1706; globes, 1720.

Deur, Johannes (*c.*1667-1734). Map, chart and globemaker of Amsterdam. Received a charter for a *Ressort* (clockwork spring) which could be placed inside a celestial or terrestrial globe to make it revolve. Made pocket globes of which 2 copies are known today. Pair of 6-cm globes, Amsterdam *c.*1720. **N.B.** Compare with Jan **Deur**, above, possibly one and the same.

Deutecum family. See **Doetichum** family.

Deutsch, H. *Plan of Ballarat Municipalities*, *c.*1865.

Deutsch, Hans Rudolf Manuel [Emanuel]. Wood engraver. Würzburg, 1548; Nördlingen, 1549, both used by Sebastian Münster (bearing HRMD monogram) in the *Cosmographia universalis*.

Deuvez, Arnold. Artist of Paris. Painted Jean-Baptiste Nolin's globe, 1693.

Devanne, —. *Plan de la ville de Bordeaux*, 1867.

De Vasconcellos, C.F.B. Surveys of the coast of Estremadura, 1842-1843 (with C.M. Batalha and F.M.P. Da Silva).

De Vaugondy, R. See **Robert** family.

Devaux, Pierre. Portolan of the Atlantic, Le Havre 1613.

Devaux, V. Lithographic printer of Port Louis, Mauritius. Isle Maurice, 1848; *Plan du Port Louis et de ses environs...*, 1850.

Devecis del Vecchio, Ferenc (1850-1924). Hungarian engineer of Italian descent (ancestors came from Florence *c.*1440 and settled in Hungary), specialised in town plans. New survey of Budapest, 1896.

De Veer. See **Veer**, Gerrit de.

Devel, Pieter. Engraver of Brussels. *Bouchain Ville Forte du Comte de Hainaut...*, Brussels,

Eugène Henri Fricx 1711; worked on Fricx's *Carte des Pays Bas*, *c.*1712.

Develet-i Aliye [Devlet-i Aliyenin], Hududu (*fl.*18th century). Turkish cartographer in London. Map of Turkey in Turkish.

Deventer [Dabey; Daventerius; Daventria], Jacob van [Jacob Roelofs; Roelofszoon] (*c.*1500-1575). Geographer to Charles V (1536-1547) and also Philip II (1555-1575), worked in Mechelen. *Brabantiae Belgarum...* (copper engraving), 1536, 1556, 1566, 1567, 6-sheet woodcut edition published as *Ducatus Brabantiae*, 1558; *Hollandt*, 1537 MS, copper engraving printed 1542, 1553, 1556, and other editions including a 9-sheet edition 1558; Gelderland (9 sheets), 1543, and other editions; *Dordrecht...*, 1545 MS; Frisia (9 sheets), 1545 and other editions; *Zelandia*, 1546 and other editions; plan of Basle, Sebastian Münster 1555; surveyed and compiled 300 manuscript plans of Dutch and Belgian towns, 1558-1572 of which 220 survive, several used in Braun and Hogenberg, *q.v.*; and others. **Ref.** KARROW, R. *Mapmakers of the sixteenth century and their maps* (1993) pp.142-158; MEURER, P. *Fontes cartographici Orteliani...* (1991) pp.137-140 [in German]; SCHILDER, G. *The Map Collector* 17 (1981) pp.4-5; HOFF, B. van't. *De kaarten van de Nederlandsche provinciën in de zestiende eeuw door Jacob van Deventer* ('s-Gravenhage, M. Nijhoff, 1941) [in Dutch].

De Vere, *Lieutenant* (later *Major*) Francis Horatio *RE*. Military engineer associated with plans of fortifications in Gibralter, 1851; Crimea, 1854-1855; Corfu, 1863.

Deverell, F.H. *Mapa de las Valls de Andorra*, 1890.

Deverell, W. Gold Lakes Otago, New Zealand, Wellington 1888.

Devereux, —. Engraver. North America, *c.*1849.

Devert, B.A.H. Plan de Paris, *c.*1830.

Deville, E. (1849-1924). French Canadian surveyor. North West Territories, 1881; Province of Manitoba, Ottawa 1884.

Devilliers, P. Engraver for Alexandre A. Vuillemin, 1873.

Devin, —. Louisiana, 1719-1720 MS; Baye de St. Louis, 1720.

Devine, Thomas. Draughtsman, Deputy Surveyor-General for Ontario. *Map of the North West part of Canada*, Toronto, Derbishire & Desbarats 1857; *Crown Land Survey of Canada, Topographical Plan of the North Shore of Lake Huron*, Toronto, 1858, 1860; *Government Map of Canada, from Red River to the Gulf of St Lawrence*, Quebec & Toronto, Thompson & Co 1859 (engraved Rae Smith); *Government Map of part of the Huron and Ottawa Territory, Upper Canada*, 1861 (engraved Rae Smith); Canada, 1873; North America, 1878.

Devi Prasada, Munshi. Hindi Atlas for Schools, 1882.

Devos, *Reverend* —. Ortos Country & parts of Mongolia, Calcutta 1876.

De Vou, Johannes. Plan of Rotterdam, 1694.

Devoux, Esprit. *Carte de Provence*, 1758; Plan de la Ville d'Aix, *c.*1765.

De Vries. See **Vries**.

Dew, father and son. Estate surveyors of Brentwood, Essex.
• **Dew**, John [I] (*fl.c.*1790-1809). Estate surveyor of Brentwood [Essex]. South Hanningfield, 1799 MS.
•• **Dew**, John [II]. Recorded as working with his father in 1809.

Dewald, G.A. Stephan (*fl.*1845-1876). *Westliche (Östliche) Halbkugel*, 1860; Deutsches Reich, 1876.

De Wale, P. See **Wale**.

Dewalque, Gustave (*b.*1826). Belgian geologist. Carte géologique de la Belgique, 1879.

Dewar, J. Engraver. Henry Teesdale's World, 1844.

Dewar, *Captain* James. Soundings on the Chagos, 1763, published Alexander Dalrymple.

Dewarat, Peter (*fl.*1759-*d.*1800). Surveyor of Mannheim in the service of the Kurpfalz.

Forestry maps, 1768; *Kriegstheater der teutschen und französischen Grenzlanden zwischen dem Rhein und der Mosel* (6 sheets), 1793-1799; *Specialkarte des Rheinlaufes van Straßburg bis Bingen, nebst des beiden angrenzenden Gegenden zu beiden Ufern bis an das Gebirge* (2 sheets), 1795; Course Neckar, 1798.

De Ward, John (*fl.*1618-*d.*1625. Estate surveyor in Kent and Sussex. Work included survey of Franefield, 1618 MS.

Dewasme [De Wasme], —. Belgian lithographer. *Operations du Siège de la Citadelle d'Anvers*, before 1832.
• **Dewasme-Pletinckx**. 'Etablissement Royal de Lithographie'. *Plan du Quartier Léopold* [Brussels], *c.*1835 (drawn by M. Suys); Malta, Brussels 1835.

Dewe, I. & R. Pictorial plan of Oxford, *c.*1850.

Dewe, John. Post Office Inspector. *Map of Canada West Shewing the Post Offices and Mail Routes*, Toronto, 1860 (printed by Maclear & Co.).

Dewe, Thomas. One of the publishers of Michael Drayton's *Poly-Olbion*, 1622.

Dewerthen, H. Draughtsman for *Map of Utah and Colorado*, St Louis 1869.

Dewey, William. *Des Moines River Improvement*, 1849.

Dewey & Co. (*fl.*1897). Publishers of San Francisco. Published maps of the Alaskan N.W. Territory goldfields through the Mining and Scientific Press.

Dewhurst & Nichols. *Map of the borough of Hertford*, 1835; *Plan of Bedford*, 1836.

Dewick, Joseph Dalton. Printer, partner in Dewick & Clarke.
• **Dewick & Clarke**. Printers for J. Wilson's *A Plan of the Cities of London, Westminster and the Borough of Southwark ...*, 1791.

De Winter. See **Winter**, Antony de.

Dewing, *Lt. Col.* Edward John *RE*. Associated with numerous plans of fortifications on Gibralter, 1897-1900.

Dewing, Francis (*fl.*1716, *d.* after 1745). Engraver and printer from London, established in Providence, Rhode Island, then Boston, later returned to England. Cyprian Southack's *A New Chart of the English Empire of North America*, Boston 1717 (copper engraving, first chart of eastern North America printed in North America), amended editions 1717, 1721, 1746; Southack's *The Harbour and Islands of Canso... Nova Scotia*, 1720; John Bonner's *The Town of Boston in New England*, Boston 1722, 1725 (in some later states, Dewing's name was erased); *A New Plan of ye Great Town of Boston in New England... in the Year 1743*, 1743. **Ref.** WHEAT & BRUN *Maps and charts published in America before 1800* (1978) several entries.

De Wit, Frederick. See **Wit**, Frederick de.

DeWitt, Simeon (1756-1834). Pioneer American cartographer and geographer; assistant to Robert Erskine, whom he succeeded as Geographer to the Continental Army, 1780; Surveyor General of the State of New York, 1784. Winter cantonment of the American army, 1778; Part of Rockland Modern County, 1780; *De Witt's State-Map of New York*, c.1793 (engraved C. Tiebout); *A Plan of the City of Albany... MDCCXCIV*, 1794; *A Map of the State of New York*, 1802 (engraved by Gideon Fairman). **Ref.** ALLEN, D.Y. *Long Island maps and their makers - Five centuries of cartographic history* (1997) pp.56-62; GUTHORN, P.J. *American maps and map makers of the Revolution* (1966) pp.13-16.

Dewitt & Davenport. Booksellers & publishers, Tribune Buildings, Nassau St., New York. J.T. Lawson's Mines of Upper California, 1849.

Dexmier de Saint Simon, Etienne Jules Adolphe, *viscomte* d'Archiac. *Carte géologique du Département de l'Aisne*, 1845; *Carte géologique des environs de Paris*, 1865.

Dezauche, father and son.
• **Dezauche**, Jean Claude (*fl.*1770-1824). French engraver, geographer, editor and map seller of 'rue Saint-Severin en face de la rue de la Harpe', Paris (1778); 'rue des Noyers N°·33 près celle des Anglois' (1781, 1789); 'N°·40 rue des Noyers' (1816-1824). Produced maps and plans for the Dépôt de la

Marine. Acquired the work of many other geographers and cartographers (much of which he re-edited and republished) including that of Guillaume Delisle and Philippe Buache, 1780; Jean Nicolas Buache de la Neuville, *c.*1780; Renou de Chauvigné, 1781; d'Après de Mannevillette, 1792. *Carte générale des méridiens et de l'equator magnétique sur l'année 1778*, [n.d.]; *Atlas Géographique et Universel*, 1780 and many later editions (re-using the maps of Delisle, Buache and Jaillot); M. Chauchard's *Carte Générale de l'Empire d'Allemagne...* (10 sheets), 1784; Topographie de la Zélande, 1790; *Carte itinéraire de la France* (4 sheets), 1796; *Atlas géographique des quatre parties du monde*, 1800; and many others. **Ref.** PACHA, B. in: PACHA, B. & MIRAN, L. *Cartes et plans imprimés de 1564 à 1815* (Bibliothèque Nationale de France 1996) pp.46-48 [in French].

•• **Dezauche**, J.-A. Son of Jean Claude. Succeeded his father at the Dépôt de la Marine, 1824-1849; Histoire Sainte, 1783.

Dezendorf, J.F. Surveyor. *Map of the Cities of Norfolk & Portsmouth* (4 sheets), 1876.

Dezharden, K. Map showing the extent of the Serbian language, Belgrade 1853.

Dezhnev, Semen Ivanovich (*c.*1605-1673). Russian explorer. Siberian Sea, 1648.

Dezoteux, —. Staff officer, French army. Prepared a chart for marquis de Chastellux's Journal, 1786, English edition 1787.

Dheulland [D'Heulland; Dheulan], Guillaume (1700-1770). Publisher, engraver and draughtsman of Paris. *Nouveau Plan de la ville d'Angers* (4 sheets), 1736; engraved for P.F.X. de Charlevoix, 1736; Bellin's *Carte de l'Empire du Japon*, Paris, J.M. Gandouin 1736; for Jacques Nicolas Bellin, 1744; César François Cassini de Thury's *Nouvelle carte Qui Comprend les principaux Triangles qui servent à la description Géométrique de la France...*, 1744; *Brabant et partie de la Holland*, Paris 1747; *Plan Topographique de la Ville... de Gibraltar*, 1750; for Antoine François Prévost, 1752; *Plan de la ville de Paris*, 1756; *Theatre de la Guerre... en Allemagne*, Paris 1758; and others.

D'Houdan, —. See **Houdan**, François d'.

Diamanti, E. *Nouveau Plan d'Alexandrie*, 1877.

Dias, Antonio. Portuguese cartographer of the mid-18th century.

Dias, Manuel *the younger* [Yang ma-no] *SJ* (1574-1659). Portuguese Jesuit. 23-inch Chinese terrestrial globe, 1623 MS (with Nicolo Longobardi and Chinese scholars). **Ref.** WALLIS, H. *The Map Collector* 41 (1987) p.16; HODSON, Y. *The Map Collector* 28 (1984) p.14.

Dias, *Lieutenant* V.P. Rio Amazones, 1865.

Diaz Romero, F. See **Romero**, Francisco

Diaz y Magne, Juan. *Carta... los Dardanelles, Mer de Marmara...*, 1784.

Dibbets, G.J. *Topographische Kaart van de Omstreken van Arnhem*, 1821.

Dibdin, Thomas. *A Correct Chart of the River and Bay of Nagore*, 1779, 1785.

Dicaearchus [Dikaiarkhos] (350-290 BC). Philosopher and historian of Messina, pupil of Aristotle. In his *Periodos ges* [circuit of the earth] established a parallel of latitude including Gibraltar, Messina, Rhodes, the Taurus Mountains and the Himalayas; map of Alexander the Great's campaigns, 320 BC.

Dicey family of printers and mapsellers. **Ref.** HODSON, D. *County atlases of the British Isles* Vol.I (1984) pp.66-70, pp.149-150.

• **Dicey**, William (*d.*1756). Printer, publisher and mapseller of Northampton and then London, 'at the printing office in Bow Church Yard'. Founder of W. & C. Dicey, *c.*1720.

•• **Dicey**, W. & C. Partnership of William and Cluer Dicey. Credited with R.W. Seale's Europe, *c.*1745; *A Correct Map of Great Britain and Ireland*, 1746; Scotland, 1746.

•• **Dicey**, Cluer [C. Dicey & Co.]. Son and partner of William, above. He joined his father in business in Aldermary Church Yard, London, *c.*1736. Bought the Speed plates *c.*1754, non-Speed Overton county maps, and possibly the plates for Saxton's maps, which were reprinted and sold cheaply as loose sheets; in partnership with Richard Marshall, 1764. *A New & accurate Plan of the Cities of London & Westminster & Borough of Southwark...*, 1765; edition of John Speed's *Theatre*, *c.*1770.

••• **Dicey**, Thomas. Son of, and successor to, Cluer Dicey, above.

Dickens, *Lieutenant-Colonel* C.H. Canals from River Soane, South Behar, 1861.

Dickens, *Captain* Thomas Mark. Manuscript plans of fortifications. *Environs of Scylla Castle* [Italy], 1810; *Plan of the Fortifications and harbours of the City Valetta* [Malta], 1811; *Environs of Cadiz*, 1811.

Dickert, Lorenz. *Relief Karte von Central Europa*, 1882.

Dickert, Theodore (1801-1883). German cartographer. Numerous relief maps. Giant moon globe [6m diameter], 1849.

Dickinson [Dickenson], family.
• **Dickinson**, John (1699-1778). Land surveyor of Yorkshire, produced various surveys in northern England, including survey of Sutton super Darwent, East Riding of Yorkshire, 1725 MS; County of York, 1740; West Coatham in Redcar and Lazenby in Wilton, 1751.
•• **Dickinson**, Joseph (1745-1823). Land surveyor, nephew of John (above). Produced maps of estates and inclosures throughout England, including a plan of the Level of Axholme, Lincolnshire, 1767-1768, published 1791; worked in Cumberland, Northumberland and Durham (170 plans of the estates of Greenwich Hospital), 1805, 1824.
••• **Dickinson**, Samuel (1770-1830). Surveyor, son of Joseph (above), with whom he worked. Produced various maps in Lincolnshire and Yorkshire including the course of Rivers Baine & Waring, 1792 (with Robert Stickney); East Halton, 1802; Swallow, 1806; inclosure map of Croxton, 1810.

Dickinson, —. Chart of the Tagus estuary from Cascaes to Lisbon, 1699.

Dickinson, Bispham. Printer and map seller with T. Millward at Inigo Jones Head, London. Frederick de Wit's Edinburgh, *c.*1690; Groundplot of Hamburg, *c.*1706; *The Traveller's Guide containing a Map of... Roads of England and Wales...*, 1737.

Dickinson, Rudolphus. Elements of Geography, 1813.

Dickinson, S.N. *Plan of the City of Boston*, 1844.

Dickinson, *Lieutenant* Sebastian. Royal Engineers. *A Plan of the Island of Heligoland*, 1808 MS; *Plan of the Lower Town at Heligoland*, 1808 MS; Estremadura, before 1811; Plan of Porto Praya, 1812; Cape Verd Islands, 1813.

Dickinson, Thomas (*fl.*1766-1767). Surveyor. Lands in Goudhurst, Kent, 1767 MS.

Dickinson, *Captain* Thomas. Surveyor. Island of Bombay, 1843.

Dickison, A. Draughtsman for Doctor Robert Bell's *Geological Survey of Canada*, 1905-1906 (with C.O. Senécal).

Dickmann, Aegidius. Danzig (7 sheets), 1617.

Dickson, George K. Atlas of Madison County, Illinois, 1892 (with H. Riniker & R. Hagnauer).

Dickson, J. Bookseller of Edinburgh. John Ainslie's Selkirkshire, 1772.

Dickson, John. Seaman's Guide round Great Britain, Liverpool 1700.

Dickson, Samuel (*fl.*1839-1859). Estate surveyor, produced manuscript maps of estates in Essex and Suffolk including Chelmsford, 1846; Hadleigh, 1852; Paslow Wood, 1859.

Dickson, *Lieutenant* W.B., *RN*. Gulf of Bengal, 1861; South Andaman Islands, 1863.

Dicquemare [Diquemare], *l'abbé* Jacques-François (1733-1789). Index géographique, 1769; revised plates for Après de Mannevillette's *Neptune oriental*, 1775; British Channel & Bay of Biscay, 1778.

Dicuil (*b.c.*780). Irish monastic scholar, grammarian, explorer and geographer, said to have visited Iceland [Thule]. De mensura orbis terrae, finished in 825 AD.

Diderich, H. Region du Mayumbe, Congo, 1899.

Diderot, Denis (1713-1784). French author and philosopher. Major work was *Encyclopédie*, 1770-1779 (Gilles Robert de Vaugondy contributed articles on geography and globemaking, as well as maps).

Didier, G. Ost-Afrika, 1861.

Didier & Tebbett. Publishers of 75 St. James St., London. *A New Plan of the Cities of London, Westminster &c....*, 1807.

Didiet, —. Côte d'Or, 1867.

Didot, family. Printers, publishers and booksellers of Paris. **Ref.** MELLOT, J.-D. & QUEVAL, E. *Répertoire d'imprimeurs/libraires XVIe - XVIIIe siècle* pp.231-232 [in French].

• **Didot**, François (*d.c.*1758). Publisher of 'Quay des Augustins, près le Pont S.Michel, à la Bible d'Or', Paris. Pierre François Xavier de Charlevoix' *Histoire de l'Isle Espagnole ou de S. Domingue...* (2 volumes), 1730-1731; Charlevoix' *Histoire et Description Generale de la Nouvelle France* (3 volumes), 1744; Antoine François Prévost's *Histoire Generale des Voyages...* (17 volumes), 1746-1761 (volumes 15-16 were published by his widow, 1760-1761, volume 17 by Arkstée et Merkus of Amsterdam, 1761); Charlevoix' *Histoire du Paraguay*, 1756.

• **Didot**, *Veuve* Marguerite [Ravenel] (*d.*1766). Widow of François, above. 'Quai des Augustins, À la Bible d'or' (1757-1762). Published volumes 15-16 of Prévost's *Histoire Generale des Voyages...* (17 volumes, 1746-1761), 1760-1761. **Ref.** MICKWITZ, MIEKKAVAARA & RANTANEN *The A.E. Nordenskiold Collection* Vol.3 pp.41-43; *ibid.* pp.132-142.

•• **Didot**, François-Ambroise [Didot *l'aîné*] (1730-1804). Printer and bookseller at the 'Quai des Augustins, À la Bible d'or', Paris (1753-1789), eldest son of François Didot. In 1789, he passed the printing side of his business to his elder son Pierre, and the type-foundry to his younger son Firmin Didot.

••• **Didot**, Pierre [Didot *l'aîné*] (1761-1853). Elder son of François-Ambroise. Started as bookseller in 1785, developed his role as a printer when he inherited the printing side of his father's business in 1789. 'Rue Dauphine N°.116'; 'Rue-Pavée-Saint-André-des-Arts'; 'Aux galeries du Louvre'; 'Rue des Orties'; 'Au Louvre, galerie N°.3 des Orties'; 'Rue du Pont-de-Lodi N°.6'. 'Imprimeur du Roi'. Also worked with his brother-in-law Louis-Alexandre Jombert (1787), and his brother, Firmin Didot. Alexandre de Laborde's *Voyage pittoresque et historique de l'Espagne*, 1806; *Mappemonde historique*, 1808.

•••• **Didot**, Jules (1794-1871). Printer and bookseller, son of Pierre Didot. Succeeded his father in 1822. 'Rue du Pont-de-Lodi N°.6', 'Boulevard d'Enfer N°.4'. Europe politique, 1823.

••• **Didot**, Firmin (1764-1836). Younger son of François-Ambroise, inherited his father's type-foundry in 1789. Publisher engraver, typographer, bookseller, writer and poet of Paris, 'imprimeur du Roi'. Recorded at several addresses including: 'Rue Dauphine N°. 116' (1790); 'Rue Thionville N°. 116' (1797-1802), 'N°. 1850' (1799), 'N°. 10' (1806-1810); 'Rue Jacob N°. 24, Faubourg St Germain près la rue des Saint-Pères' (1811-1813). Ceded the business to his three sons *c.*1827. *Relation de l'Enlevement du Navire le Bounty* (with 3 maps), 1790; *Atlas des mémoires du maréchal Suchet...en Espagne*, 1810; *Cartes Typogeographiques...*, 1823; also numerous travel and illustrated books.

•••• **Firmin-Didot** *frères*. Sons of Firmin Didot, with whom they worked, and to whom they succeeded *c.*1827. Publishers, 'Rue Jacob 56', Paris. France, 1840.

•••• **Firmin-Didot**, Ambroise (1790-1876). Son of Firmin Didot, printer and classical scholar, after travels in the near east he traded with his brothers as Firmin-Didot frères.

•••• **Firmin-Didot**, Hyacinthe. Son of Firmin Didot, traded with his brothers as Firmin-Didot frères.

•••• **Firmin-Didot**, Frédéric. Son of Firmin Didot, traded with his brothers as Firmin-Didot frères.

•• **Didot**, Pierre-François [*le jeune*] (1731-1795). Son of François, brother of François-Ambroise. Printer and bookseller, 'imprimeur des États de Languedoc' (1779-1789). 'Quai des Augustins'; 'Rue du Hurepoix'; 'Quai des Augustins N°.22'.

••• **Didot**, Pierre-Nicolas [*le jeune*]. Son of Pierre-François, succeeded to his father's printing business in 1795. Recueil de cartes géographiques, 1799.

Didrichsen, Ferdinand (1814-1887). Danish geographer and botanist.

Diebel, Elias. Plan of Lübeck, *c.*1550 (used by Braun & Hogenberg *q.v.*).

Diebison, C. Lithographer. *Plan von Breslau*, 1858; *Neuester Plan von Breslau*, 1867.

Diebitch, Marshall. Sketch of ground near Varna [Bulgaria], 1847 MS.

Diebitsch-Narten, Karl Friedrich Wilhelm, *Freiherr* von (1738-1822). Military cartographer of Wittenberg, in Russian service.

Diedrich, Rudolf. *Koningrijk der Nederlanden*, 1850.

Diefenbach, C. *Schulwandkarte... Wiesbaden*, *c.*1872.

Diéguez, Juan. Bolinao [Philippines], 1800.

Diehl, Johann Philip. Panorames de Francfort, 1834; *Schul-Atlas für den Unterricht in... Erdkunde...*, 8th edition 1865.

Dielhelm, Johann Hermann (*d.*1784). Map of the Rhine, 1739, 1744, 1736; Neckar, Main, Lahn und Mosel, 1740, 1781; Elbe, 1741; hydrographical dictionary of the Rhine entitled *Antiquarii*, 1768 (with maps of German rivers); Donau, 1785.

Dien, father and son.
• **Dien**, Charles, *l'ainé*. Engraver, publisher, and globemaker of Paris, 'Rue du Foin St Jacques, au College de Mtre Gervais' and 'Rue de Jardinet N°.13'. Father of Charles Dien, le jeune, below. Script engraver for Barbié de Bocage's *L'Ecosse avec ses Isles...*, 1800; engraved for P.L. Griwtonn, 1801; French edition of Thomas Jefferys's map of St. Lucia, 1802 (first published 1775); *Carte générale Des Etats Prussiens ou Théâtre de la guerre 1806*, 1806; G.T. Raynal's Atlas du globe terrestre, 1820.
•• **Dien**, Charles, *le jeune* (1809-1870). French cosmographer, son of above. Atlas des phénomènes célèstes, 1843; Atlas célèste, 1855; *Uranographie*, *c.*1870.

Diener, Carl. *Mittel Syrien*, 1885.

Dienz, —. *Saar & Mosel Weinbau-Karte*, 1868.

Dieppe School (*fl.*1535-1660). Group of chartmakers at Dieppe, charts known as 'cartes dieppois'. **N.B.** For individual chartmakers see **Brouscon**, Guillaume; **Clamorgan**, Jean de; **Descelliers**, Pierre; **Desliens**, Nicolas; **Guérard**, Jean; **Le Testu**, Guillaume; **Le Vasseur**, Guillaume; **Parmentier**, Jean; **Rotz**, Jean; **Vallard**, Nicolas; **Vaulx**, Jacques de.

Diercke, father & son. Worked separately and together on school atlases. The Diercke firm was absorbed by Westermann Verlag in 1937.
• **Diercke**, Carl [Karl] (1842-1913). Compiled school maps. *Karte des Harzes*, 1887; *Schul-Atlas über alle Teile der Erde*, 1883 and later editions (with E. Gaebler and Paul Diercke); *Diercke Schul-Atlas für höhere Lehranstatten*, Braunschweig, G. Westermann 1899 (with E. Gaebler)
•• **Diercke**, Paul (1874-1937). Compiler of wall maps and school atlases. *Atlas für Mittelschulen*, 1912 and after (with his father, above). **Ref.** KRETSCHMER, DÖRFLINGER and WAWRIK *Lexicon zur Geschichte der Kartographie* (1986) vol.C/1, pp.174-175 (numerous works cited).

Dieskau, *General* —. French soldier. Carte du Lac St. Sacrement, 1755.

Diesner, L. *Schulwandkarte... Hessen-Nassau*, 1877.

Diespecker, L. Environs de Cognac, 1864.

Diest, Gillis Coppens. See **Diesth**, Aegidius Coppenius.

Diest, Wilhelm von (1851-1932). *Itinerar-Aufnahmen... Kleinasien*, 1895; Anatolische Eisenbahn-Linien, 1898.

Diestel, —. *Plan von Travemünde*, 1891.

Diesth [Diest], Aegidius [Gillis] Coppenius [Coppens]. Printer of Antwerp for Peter Apian, 1534-1544; Johannes Honter, 1552; Peter Apian's *Cosmographia*, 1564; plan of Antwerp, 1565; publisher of the first edition of Abraham Ortelius's *Theatrum orbis terrarum*, 1570, 1571, 1573, 1574.

Diet, Leo. *Carte de l'Egypte*, 1884.

Dietell, Christoph (*fl.*1723-1759). Engraver of Vienna. Maps to accompany the descriptions of Joseph Stöcklein's travels, 1728-1832.

Dieterici, *Professor Doctor* Carl Friedrich Wilhelm (1790-1859). Director of the Statistical Bureau in Berlin, 1844-1859. *Karte vom Jordanthal*, *c.*1855; *Bevölkerung der Erde*, Petermann 1859.

Dietlein, Woldemar. Weltgeschichte (8 maps), 1874.

Dietrich, *Lieutenant —. Plan von Moskau*, 1812 (with Goetz); *Plan der Stadt... Posen*, *c.*1863; *Plan... Cilli*, *c.*1873.

Dietrich, A. *Karte des Kohlenrevieres... von Lugau*, 1859.

Dietrich, G. Engraver for Joseph Meyer's *Hand-Atlas*, 1866, 1867; for L. Ravenstein and Joseph Meyer's *Special-Karte von Deutschland, der Schweiz...*, 1868.

Dietrich, Wilhelm. *Karte der Bayer. Eisenbahnen...*, 1896.

Dietz, Adrian. *Karte von Schweinfurt*, 1892.

Dietz, C.W. *Residenz Stad Kiobenhavn*, 1769.

Dietz, Christoph Moritz (*b.*1729). German surveyor. *Plan von dem Elb-Strohn...*, Dresden 1766.

Dietz, Franz Maximilian (*d.c.*1750). *Post-Karte von Teutschland*, Berlin, 1795, and many subsequent editions; *Deutschland, Niederlande &c.*, 1838; *Post- und Reise-Karte von Deutschland*, 1833, 1855.

Dieu, —. Designed frontispiece for Hubert Jaillot's *Atlas François*, 1695.

Diewald, Johann Nepomuk (1774-1830). Cartographer active in Nuremberg. *Karten österr. Territorien*, Homann's Heirs 1806-1808; *Das Herzogthum Holstein*, 1815; Niederlande, 1817; *Weltcharte*, 1829; *Atlas des Inn-Kreises* (10 sheets), 1831; *General-Charte des Traun-Kreises*, 1832; *Chorographische Karte des Mühlkreises...*, 1840.

Diez, F.M. *Postroutenkarten von Deutschland*, Frankfurt 1795.

Diezelski, E. The Yukon River, Washington DC 1871.

Digges, family. English mathematicians.
• **Digges**, Leonard (*b.c.*1510/1530-*d.c.*1574). Mathematician. *A Boke named Teactonicon* (first important English book on surveying), 1556; *A geometrical Treatise named Pantometria*, 1571 and later editions (includes descriptions of the theodolite, invented by Digges, and of a simple method of triangulation).
•• **Digges**, Thomas (*c.*1543-1595). Mathematician, hydrographer and geographical theorist, son of Leonard (above), father of Dudley (below); published his father's *Pantometria*. Winchelsea Harbour, *c.*1577; Platte of Dover Haven, 1581, 1595 MSS.
••• **Digges**, *Sir* Dudley (1583-1639). Son of Thomas. Circumference of the Earth, 1612.

Dignoscyo, L. de. *Plan de la ville de Lyon*, 1847, 1874.

Diguja, D.J. Cumana, 1898.

Dijck, H. See **Dyck**, Heyman van den.

Dijk, P. van. Java, 1872-1873.

Dikaiarkhos. See **Dicaearchus**.

Diko, W. *Schul-Atlas*, *c.*1868.

Dilcher [Dilger], Heinrich (1824-1885). Lithographer in Amsterdam.

Dilich [-Schäffer], Wilhelm Schaeffer (*c.*1571-1655). Geographer, map-colourist and mining surveyor. *Hessische Chronica* (120 views), 1605; *Landtafeln hessischer Ämter zwischen Rhein und Weser* (27 maps), 1607-1617; map of Landgrafschaft Hessen-Kassel, 1607-1624.

Dillon, C.H. Master of HMS *Vernon*. Hydrographic surveys of *Beirout Bay*, 1842; Bay of Acre, 1843; Tripoli, 1843; *Montevideo Bay*, 1850; and others.

Dillon, J.P. *Master, RN*. Assisted J. Parsons with surveys for Admiralty charts including: *West Indies. Part of the Grenadines from Carriacou to Battowia*, 1861; *West Indies. Tobago Island*, 1864-1865 published 1867 (engraved by E. Stanford); *West Indies - Tobago, Scarborough, Rockly Bay*, 1864-1865 published 1867 (engraved by Davied & Bryer).

Dillon, Theobald. *Plan des Mouvements des Armées sur le Rhin*, 1796.

Dillon, William, *Second Master*, HMS *Wellesley*. S. Juan de Nicaragua, 1833 MS; Mia-Tao, 1840.

Dilly, family.
• **Dilly**, Charles (1739-1807). Publisher and bookseller, in partnership with brother Edward at 22 Poultry, London; on his own from 1779, collaborated with G.G.J. & J. Robinson 1783-1801, succeeded by J. Mawman *c.*1801. *North and South America*, 1769; United States, 1783; Atlas to *Guthrie's System of Geography*, 1785 and later editions; *Map of the County of Somerset*, 1792.
• **Dilly**, Edward. Brother and partner of Charles, above.

Dilworth, Thomas (*d.*1780). English globe seller and schoolteacher.

Dimes, W. Piercy. Ecclesiastical Divisions of England & Wales, 1874.

Dimmock, C.H. American civil engineer, assisted Captain J. Macomb with his exploration and surveys of New Mexico and Utah, 1860, published New York, Egloffstein 1864.

Dinale, Niccolo. Dialogo sopra la sphera del mondo, Venice 1579.

Dinant, Michelange de (*fl.*1709-1712). Monk and engraver. Giovanni Montecalerio's *Chorographia descriptio provinciarum...S. Francisci Capucinorum*, Milan 1712 edition.

Dinesen, Jørgen. Danish cartographer. Øresund, 1688.

Dingelsted, Christopher Adolph. *Tabula geographica Halberstadtiensis*, *c.*1710.

Dingelstedt, V. von. Ethnographical maps. *Map of Armenia and Kurdistan*, 1898; *Ethnolographical Map of Eastern Turkey... and Western Turkey*, London, Royal Geographical Society 1910-1916.

Dinges, J. Three dimensional relief map of the Bavarian Alps, 1901.

Dingle, Edwin John. Map editor. *The new map of China*, 1916 (with H.J. Fruin); *The new atlas and commercial gazetteer of China* (28 maps), Shanghai, North China Daily News & Herald 1917.

Dinglinger, Georg Friedrich. *Plan der Stadt Hannover*, 1748.

Diniz, M. *Africa Oriental Portugueza*, 1896; *Carta de Moçambique*, 1903.

Dinomé, *abbé* S.E. Achille (1787-1871). French geographer, translated many geographical works. Edited *Annales des voyages*.

Dinsmore, —. *Railroads & Canals of the United States & Canada*, 1856.

Dinsmore, A.F. *Territory of Utah*, 1884, 1889 (for the General Land Office, printed by Julius Bien).

Diodati, Fabio. Viaggi in Alemagna, 1635 MS.

Diogo, Mestre (*fl.*16th century). Portuguese cartographer.

Dion, Philippe. *Carte vinicole de l'Algérie*, 1891.

Dionysius Periegetes [Alexandrinas] (*fl.*2nd century AD). Worked in Alexandria. Wrote a description of the world in Greek hexameter verse entitled *Periegesis...*, translated in 6th century by Priscian of Caesarea; printed as *Cosmographia de situ orbis*, Venice 1478.

Dionysius. Greek Orthodox monk at St. Catherine's monastery in Sinai. *Athanasius Sacri Montis Sinai in Arabi Archiepiscopus...*, Naxos, Michel Forro 1736.

Diot, *Madame* —. Map colourist of Paris. Coloured the maps for Malte-Brun's *Précis de la géographie universelle*, 1810 (maps by J.B. Poirson and P. Lapie, printed by Prudhomme).

Di Pietro, Giovanni. Engraver of Rome. Map of Switzerland for Barbiellini's *Atlante...*, 1807.

Di Pietro, Marco (1757-1837). Engraver of figures and lettering active in Rome until 1805. Moved to Naples in 1806, where he worked in the topographical office until his death. Engraved the five continents for Barbiellini's *Atlante...*,1807; maps in *Atlante Universale*, Gabinetto Letterario 1813; plan of Palermo for G. Lossieux, 1818.

Diquemare, J.F. See **Dicquemare**, *l'abbé* Jacques-François.

Dirat, —. One of the surveyors for sheet 101 [La Rochelle] of the Cassini *Carte de France*, 1765-1769.

Dircks, Henry. *Plan of the Castle and Citadel of Raglan*, 1862.

Dirckx, Jan. Portolan maker 'in't Vergulde Compas tot Edam'. Manuscript chart of the Atlantic, 1599.

Diricks [van Campen; Derksen], J. Plan of Hamburg, 1611.

Diricksen, Jan. Engraver. Copenhagen, 1611.

Diriks, C. *Norske kyst fra Arnedal til Christiansanda*, 1856.

Dirom, Alexander. Mineralogical survey of Dumfriesshire, 1802, used on W. Crawford's 4-sheet map of the county, *c.*1804, re-issued Faden 1812.

Diron, —. Baye de la Mobile[Alabama], 1725.

Diruf, Arthur. *Stromkarte der Elbe*, 1885.

Dirwalt [Dirwald; Dirwaldt], Joseph. Schwaben (4 sheets), 1809; Baiern (12 sheets), 1813; *Allgemeiner Hand Atlas*, Vienna *c.*1818; Süd-America (2 sheets), 1818; *Charte von Nord America...*, Vienna, T. Mollo, 1819.

Dispatch. See **Weekly Dispatch**.

Diston, John. Trinity House pilot and chart-maker. Collection of six charts of the east coast of England from the Thames to the Humber, *c.*1760, published by Sayer in several editions, and Laurie & Whittle.

Disturnell, John (1801-1877). Publisher and guide book compiler of 102 Broadway, New York (1847). Founder member of the American Geographical Society of New York. *A guide between New York, Philadelphia, [etc.]*, 1837; 30 miles round New York, 1839; *Map of the State of New York...*, 1845; *Mapa de los Estados Unidos de Méjico...*, 1846 and 24 further editions; *New Map of California, Oregon, Washington, Utah and New Mexico*, 1853 (surveyed and engraved by G. Schroeter); *Disturnell's New Map of the United States and Canada*, 1855; *A Trip Through the Lakes of North America; Embracing a Full Description of the St. Lawrence River...*, 1857; J.G. Bruff's *Army map of the seat of war in Virginia*, 1861; J.G. Bruff's *New map of the seat of war in Virginia and Maryland*, 1863; *The Great Lakes or Inland Seas of America...*, New York, Charles Scribner 1865; *Map of the Region of the Great Lakes and Upper Mississippi River*, 1868 (printed by J. & R. McLellan); *Township Map of Lake Superior region and Upper Mississippi*, 1871; *Sectional Map of the Copper, Iron and Silver Region of Lake Superior*, 1874. **Ref.** RISTOW, W. 'John Disturnell's Map of the United States' in *A la carte* (Library of Congress, Washington 1972).

Dittenberger, Friedrich. *Hand- und Schul-Atlas der neuesten Erdbeschreibung*, Karlsruhe, Braun 1819.

Dittlinger, Albrecht Anton (1704-1780). Surveyor of Bern.

Dittmar, Heinrich (1792-1866). *Historischer Atlas*, 1856.

Dittrich, —. *Plan der Stadt Magdeburg* (8 sheets), 1882.

Di Tucci, P. Carta dei termi Caziali, 1879.

Ditzinger, Ludwig (*fl.*1600). Goldsmith and engraver of Tübingen.

Divinus, Eustachius de (*fl.*17th century). Italian astronomer.

Dix, J.C. & W. Lithographers. Map illustrating Ancient History, 1847.

Dix, Thomas (1769/70-1813). Schoolmaster in Northamptonshire, then surveyor at North Walsham, Norfolk. *Dix's Juvenile Atlas*, William Darton 1811; *New Map of the County* [Suffolk], 1816; *A New Map of the County Palatine of Chester...*, London 1818, 1830; Bedfordshire, 1818; County of York, 1820, 1835; *A Complete Atlas of the English Counties* (completed and published posthumously by William Darton) 1822 and later editions.

Dix, W. See **Dix**, J.C.

Dixon, *Captain* —. Master of the *Skelton*, from Whitby, Yorkshire. Associated with Thomas Scott's *Chart of Van Diemen's Land*, 1824 (possibly as the carrier of Scott's surveys, rather than as contributing surveyor).

Dixon —. See **Griffing, Dixon & Co.**; also **Stannard & Dixon.**

Dixon, Arthur. Wangaratta, Owen's River, 1855.

Dixon, C.G. British Guiana, 1894.

Dixon, *Captain* George (*d.c.*1800). Sailed with James Cook on his Third Voyage. *A Voyage round the World... 1785...[-]...1788...*, published Geo. Goulding 1789; Chart of the North-West Coast of America, 1788.

Dixon, Jeremiah (*d.*1779). English astronomer, mathematician and surveyor. Best known for the famous map of the boundary between Maryland and Pennsylvania (with Charles Mason). Later continued as a surveyor in his native County Durham. *A Plan of the Boundary Lines between the Province of Maryland and the Three Lower Counties on Delaware with Part of the Parallel of Latitude which is the Boundary between the Provinces of Maryland and Pennsylvania* [Mason-Dixon line], 1763-1768. **Ref.** HOWSE, D. in: *Dictionary of national biography: Missing persons* (Oxford 1993) pp.185-186. See also Henry **Dawkins**, engraver.

Dixon, John (1740-1811). Engraver. John Rocque's Dublin (4 sheets), 1760.

Dixon, John. Civil engineer. *Parts of Durham and Yorkshire, showing the Stockton and Darlington, Wear Valley and Redcar, and proposed Middlesborough and Guisborough, railways...*, 1851.

Dixon, *Lieutenant* Joseph, US Topographic Engineers. Fort Dallas to Gt. Salt Lake, 1859.

Dixon, Matthew. Plan of Dock at Annapolis Royal, 1756.

Dixon, Richard (*fl.*1832-1845). Surveyor of Godalming, Surrey, produced manuscript maps of southern England including a tithe map of Warlingham and Chelsham, 1844.

Dixon, Robert (1800-1858). Surveyor and explorer of Sydney, New South Wales,; Surveyor General's Dept (*c.*1829-1841). Aracati River, 1831; *Map of the Colony of New South Wales*, 1837; Moreton Bay, 1840; Map of Gippsland, *c.*1845.

Dixon, Thomas. *Borough of Bradford*, 1856.

Dixson & Kasson. Overland mail routes, San Francisco 1859.

Dixwell, James. Printer of St. Martin's Lane, London. John Rocque's map of Berkshire, 1761.

Djurberg, Daniel (1744-1834). Member of the Cosmographical Society of Uppsala. *Karta över Polynesien...*, Stockholm 1780, 1789 (first Swedish map to show James Cook's discoveries; introduced name *Ulimaroa* for Australia).

Dlholuczky, János. Hungarian military surveyor and cartographer. Mapped Máramaros in Transylvania, 1774.

Doane, Thomas (1821-1897). American engineer. Stanstead & Montreal Railroad, 1845.

Dobbie, W.H. *Plan of the Bay of Kossier*, 1799, published Alexander Dalrymple 1802.

Dobbs, Arthur (1689-1765). Ulster landowner, member of the Irish House of Commons, Surveyor General of Ireland (from 1733) and Governor of North Carolina, 1754. Early advocate of the search for a Northwest Passage. He later accused Captain Middleton, who had undertaken exploratory voyages in Hudson's Bay, of falsifying his charts and logs to rule out a possible route. **Ref.** BARR, W. & WILLIAMS, G. (Eds.) *Voyages to Hudson's Bay in search of a Northwest Passage 1741-1747 Vol.I The voyage of Christopher Middleton 1741-1742* (Hakluyt Society, London 1994); CLARKE, D. *Arthur Dobbs Esquire 1689-1765 Surveyor General of Ireland prospector and Governor of North Carolina* (London 1958).

Dobeli, Samuel (1858-1919). Swiss teacher and designer of relief maps, worked in Strengelbach and Aarau.

Dobert, Wilhelm. Wandkarte Sachsen, 1876.

Doblado, J. Published José Asencio's Madrid, 1800.

Dobler, Alfred. *Eisenbahn und Postkarte Deutschland*, 1864; *Karte des Rheins*, 1873.

Dobner von Dobenau, Johann (1816-1889). Viennese general and cartographer.

Dobrée, *Captain* Nicholas. Agent to the Admiralty. Alderney, 1746; Guernsey, 1746; Chart of the Channel Islands, 1746, updated chart published Laurie & Whittle 1794.

Dobson, E. Proposed Terminus of the Lyttleton Railway, New Zealand, 1864.

Dobson, James. Chart of the globe, 1794 (based on Edmond Halley).

Dobson, Joseph. Pilot. *Chart of the entrance of the River Tees*, 1762.

Dobson, Thomas. Publisher of Philadelphia. Fisher's *A chart of the Delaware Bay and River*, 1786 edition; Encyclopaedia (with maps), 1792; Map of Paradise in Bible, 1799; C. Hauducoeur's *A map of the head of Chesapeake Bay and Susquehanna River...*, 1799 (engraved by Allardice).

Dobson & Rhoades. *Atlas of the Town of Calumet*, 1875.

Dodd, family.
• **Dodd**, Ralph (1756-1822). Civil engineer and surveyor, father of George (below). Produced civil engineering maps of sites throughout England. Newcastle-Carlisle Canal, 1795; Plans of the Port of London Improvements, published by Luke Hansard for the House of Commons 1798-1800; Hartlepool, 1802; *Plan of the intended South London Union Docks*, 1806.
•• **Dodd**, George (1783-1827). Civil engineer, son of Ralph (above). Strand Bridge [London], 1808.

Dodd, J.J. Lithographer. *Dame Venodotia*, [map of North Wales in the shape of a woman], 1840 (engraved by Hugh Hughes).

Dodd, James Solus. Travellers' Directory through Ireland, 1801.

Dodd, Moses. *Ten miles round Reading*, 1846.

Dodd, Mead & Co. Universal Atlas, 1892.

Dodens, R. Di sphaera cosmographica isagoge, 1584.

Doderlein, P. Carta geologica della provincia di Reggio Emilia, 1870.

Dodge, *Major General* Grenville Mellon (1831-1916). American civil engineer and politician. Chief engineer of the Union Pacific Railroad. *Map of the Military District, Kansas and the Territories*, 1866; *Map of the Union Pacific Rail Road...*, 1869.

Dodge, Robert Perly (1817-1887). Plats of Squares West Washington, 1833.

Dodsley, family. Publishers of London.
• **Dodsley**, Robert (1703-1764). Publisher, writer, poet, dramatist from Mansfield, Nottinghamshire. Founded the Annual Register 1758 (with Edmund Burke). 'At Tully's Head in Pall Mall', London from 1735. John Cowley's *The Geography of England*, London 1743 (used some maps prepared for Dodsley's *Publick Register*); Cowley's *A New Sett of Pocket Mapps of... England & Wales*, 1745. **Ref.** HODSON, D. *County Atlases of the British Isles* Vol.II (1989) pp.3-13.
• **Dodsley**, R. & J. (*fl.*1755-1764). Partnership of Robert and James. Published first edition of Lewis Evans' *A General Map of the Middle British Colonies in America*, 1755; *A New and Accurate Description of the Present Great Roads...of England & Wales*, 1756; *A New and Correct Plan of London*, 1761; *London and its Environs Described* (6 volumes), 1761; *England Illustrated* (2 volumes), 1763 and later editions including a partwork. **Ref.** HODSON, D. *County Atlases of the British Isles* Vol.II (1989) pp.165-173.
• **Dodsley**, James (1724-1797). Publisher and bookseller, partner with brother Robert. Continued business alone from 1764. *Kitchen's English Atlas*, 1765, 1770 (unrevised maps from *England Illustrated*, but without text); republished Robert's *The Geography of England* as *The Geography and History of England*, 1765 (without county maps).

Dodson, James, *FRS* (*d.*1757). Teacher of mathematics. Revised Edmond Halley's chart of compass variations as *A Correct Chart of the Terraqueous Globe*, 1756 and editions to 1794 (with William Montaine).

Dodson, *Colonel* William. *Mapp of the Great Levell of Fenns*, 1665.

Dodsworth, —. *Siege of Delhi...1857*, 1876.

Dodt, H. Prussia, 1862.

Dodwell, —. Andaman and Nicobar Islands, Alexander Dalrymple 1778.

Doedtsz. [Doedes; Doetszoon; Doetz.; Daetszoon], Cornelis. Chartmaker of Edam 'inde vier heems Kinderen'. Baltic, 1589, re-issued by Claes Jansz. Visscher 1610; Asia, 1598; Europe, 1600; portolan of the coasts of the Atlantic, 1600; Ganges, 1600; said to have worked with Adriaen Block on an untitled manuscript map of the area of New England from Chesapeake Bay to Penobscot, 1614. **Ref.** COHEN, P.E. & AUGUSTYN, R.T. *Manhattan in maps 1527-1995* (1997) pp.22-23.

Doelter, C. Geologische Karte des Monzoni Gebirges, 1875; Karte des Vulcan Ferra, 1878.

Doergens, —. *Imperial map of Palestine*, 1864.

Doering [Döring], Ferdinand von (1820-1889). Map editor. *Admin. Statistischer Atlas v. Preussischen Staate*, 1827-1828 (and later editions); maps of Prussia, 1827, 1836; *Special-Karte vom Sundewitt...*, Berlin, J.H. Neumann 1864.

Doerrbecker [Dörrbecker], P.A. *Authentic Map of Beira Harbour*, 1891.

Doesborgh [Doesburgh], Thomas (*fl.*1667-1714). Engraver. *Kaap Goed Hoop*, Huych Allard 1667; maps for Carel Allard's *Orbis habitabilis oppida* (town views and plans), 1698; map of Utrecht province, *c.*1700; Cruquius' *Delflandt*, 1712.

Doetichum [Deutecum; Doet; Doetecum; Doetechum; Doetecomius; Dotecum; Duetechum]. Family of engravers and map publishers working in Deventer and Haarlem (from 1587). **Ref.** KRETSCHMER, DÖRFLINGER & WAWRIK *Lexicon zur Geschichte der Kartographie* (1986) pp.161-162 [in German]; SHIRLEY, R. *The mapping of the World* (1983) N° 124 pp.146-147, N° 177 pp.194-195, N° 187 p.207.

• **Doetichum**, Johan [Jan] van, *the elder* (*d.*1600). Engraved for Christopher Plantin and Officina Plantiniana, 1559 onwards; De Jode's *Nova Totius Terrarum Orbis Descriptio*, 1571 (with Lucas); Gerard de Jode's Speculum, 1578; for Lucas Jansz. Waghenaer, 1584-1592; Abraham Ortelius and Jan Huyghen van Linschoten. *Peruvianae pars....fretum Magalanicum*, *c.*1590; Map of the 17 provinces, 1594.

•• **Doetichum**, Baptista van (*d.*1611). Son of Johan (above), with whom he worked. Lucas Jansz. Waghenaer's *Spieghel der Zeevaert*, 1584 and other editions (with Johan); Belgium, 1588; Doedtszoon's Baltic, 1589; for Pieter Plancius, e.g. *Orbis terrarum typus*, 1590; worked with his father on engravings for Jan Huyghen van Linschoten, 1595; Bible maps, 1609; Hannonia, published later by Joannes Janssonius 1633.

•• **Doetichum**, Johan [Jan] van, *the younger* (*fl.*1592-1630). *Nova Francia*, 1594; *Leo Belgicus*, 1598; Africa, 1610; Claes Jansz. Visscher's Hungary, 1624.

• **Doetichum**, Lucas van (*fl.*1558-*c.*1579). Worked with his brother, Johan the elder. Giacomo Gastaldi's Africa, 1560; engravings for Gerard de Jode, including *Nova Totius Terrarum Orbis Descriptio*, 1571 (with Johan); America, 1578; Bohemia, *c.*1590.

Doetsch, Johann Anton von. *Imperium Russicum*, 1785.

Doetszoon [Doetz], C. See **Doedtsz.**, Cornelis.

D'Ogerolles, J. See **Ogerolles**, Jean de.

Doggett, John *Jr. Railroads in Michigan with steamboat Routes on the Great Lakes*, New York 1848 (for *Doggett's Railroad Guide & Gazetteer*); United States, 1851.

Doglioni, G.N. *Anfiteatro di Europa* (33 miniature maps in text), Venice 1623.

Doharty, John. See **Dougharty**, John.

Doi Empei. Japanese cartographer. Map of the eight provinces of the Kantō area, Izumiya Hanbei 1849.

Doidge, W. & H. Surveyors, worked together on *Plan of the Antient City of Canterbury*, 1752.

• **Doidge**, William (*fl.*1730-1752). Also worked in Devon and Cornwall.

• **Doidge**, Henry (*fl.*1752).

Doidge & Co. Publishers of 169 & 170 Union Street, Plymouth. *Doidge's New Large Scale Cycling and Touring Road Map of Plymouth and District*, 1890.

Doin, Ochikochi. See **Ochikochi** Doin.

Doino, Catharino. *Vicenza*, 1611.

Dokumujishorō. Japanese publisher. Hyōhyō Sanjin's map of Gokinai Shōran, 1841.

Dolben, Willlam Digby Mackworth, *RN*. Admiralty chart of the River Volta, 1862.

Dolce, Lodovico (1508-1568). Venetian translator of classical texts. Untitled world map in *Le transformationi di M. Lodovico Dolce*, Venice 1553 **Ref.** SHIRLEY, R. *The mapping of the World* (1983) N°95 p.108.

Dolcetta, Giuseppe. Engraver for Vincenzo Maria Coronelli's *Atlante veneto*, 1690-1696.

Dolendo, Bartholomeus (1571-1629). Engraver of Leiden. Maps for Wahrhaftige Beschreybung aller Züge und Victorien, 1612.

Doležal, Anton. Cartographer specialising in economic and population maps. *Graphische Darstellung der Waaren-Durchfuhr durch das oesterreichische Zollgebeit 1845 und 1863*, Vienna, 1865; *Statistische Karte des Weinbaues in Nieder-Oesterreich*, 1866; industrial maps of parts of the Austrian Empire, 1867; *Oesterreichische-Ungarische Monarchie*, 1870; *Galicya*, *c.*1875; contributed to J.R. Lorenz's *Atlas der Urproduction Oesterreichs*, Vienna, R. v. Waldheim *c.*1876.

Doležal, Franz [František]. *Plan von Brünn* [Brno], 1858.

Dolfinado [Dolfinatto], N. del. See **Nicolay**, Nicolas de.

Dolier, F. See **Dollier** de Casson, François.

Dolivar [Dolivart], Juan. *La Carta des Royaumes d'Angleterre...*, 1689.

Doljatov, I. See **Dolyatov**, Ivan.

Doll, A. *Canarien Inseln*, Vienna 1805.

Doll, G. Valdivia, 1850; Atacama, 1854.

Doll, J. Zuid-America, Amsterdam 1783.

Doll, M. *Umgebung von Karlsruhe*, 1895.

Dolland, —. Miniature globe, London 1839.

Dollfus, Auguste (1810-1869). French traveller and geologist. Voyage géologique dans Guatemala, 1868.

Dollier de Casson [Dolier], François. *Carte du Lac Ontario*, 1670.

Dolling, James. *Pocket Map of London*, *c.*1862, (from *c.*1872 published as Dolling's *Penny Waistcoat Pocket Map of London*); Paddington, 1863.

Dolter, Cornelio (*b.*1850). Austrian geologist. Reiseskizze aus Sardinien, 1878; Vulkane Kapverden, 1882.

Dolyatov [Doljatov], Ivan. Map of north-east Asia and north-west America [Russian], 1772 MS.

Domann, Bruno. Draughtsman. Maps for *Afrika*, J. Perthes 1891, 1892, 1898 (with H. Habenicht); maps for Stieler's *Hand-Atlas*, Justus Perthes 1891 and later; *Atlas portatil*, 1894 (for Justus Perthes).

Domenech [Domenche; Domeneth], Arnald. Catalan portolan maker working in Naples, pupil of Pietro Roselli. Mediterranean, 1486.

Domenico, Machaneus. See **Macaneo**, Domenico.

Domergue, Eugène (*b.*1849). French geographical publisher. Géographie pittoresque, 1875.

Domeyko, Ignacy (1802-1889). Chilean geologist and cartographer.

Domingues, Francisco Atanacio, *SJ*. Mapa de la provincia de Moqui, 1777 MS; New Mexico, 1777 MS.

Dominguez, Francisco. Mar del Zur, 1584 MS.

Dominus, N.G. See **Germanus**, Nicolaus.

Doms, P.J. Engraver. J.F. Keyser's *Nouvelle Carte Générale de la Belgique*, Etablissement Géographique de Bruxelles *c.*1840; one of the engravers for Van der Maelen's *Carte Topographique de la Belgique* (248 sheets), *c.*1854.

Domvill, Silas. Harwich, 1667.

Domville, William. Bookseller, Royal Exchange, London. One of the sellers of P. Russell & Owen Price's *England Displayed*, 1769 (originally by Thomas Read).

Don, J. *Noul plan al capitalei Bucuresti...*, Bucharest, E. Graeve 1897.

Donald, Thomas (*fl.*.1750-*c.*1797). Surveyor, produced county and local maps including Bedfordshire, 1765; Buckinghamshire, 1770; Yorkshire, 1771-1772 (all with John Ainslie); *The County of Cumberland Surveyed* (6 sheets), 1770-1771, published 1774; Environs Keswick, 1789; Norfolk, 1790-1794, published 1797 (with Thomas Milne).

Donaldson, Thomas. *The Public Domain...* [America], 1884 (history of settlement with 15 maps).

Donat, *Le Sieur* —. Chart of Tristan da Cunha, 1767 published A. Dalrymple *c.*1792.

Donck, Adriaen van der (1620-1699). Lawyer and Dutch colonial administrator. *Nova Belgica sive Nieuw Nederlandt* (map of New England) in *Beschryvinghe van Nieuw Nederlant*, 1656. **N.B.** Doubt is cast upon this attribution in BURDEN, P. *The mapping of North America* (1996) p.409.

Doncker, Hendrik, father and son. **Ref.** KOEMAN, C. *Atlantes Neerlandici* (1967-1971) Vol.IV pp.152-189.
• **Doncker**, Hendrik [Hendrick I] (1626-1699). Bookseller, instrument maker, hydrographer and publisher 'Inde Nieuwbrugh Steeg in't Stuurmans gereetschap'; 'a l'enseigne des utensils de Pilots', Amsterdam. British Isles, 1658; *De Zee atlas ofte Water-Waereld*, 1659 and many other editions; *Nieuw Groot Stuurmans Straets-boeck...*, 1664, republished by Hendrik II in 1704 and 1708; *La Atlas del Mundo...*, *c.*1665 and later; Grand et nouveau miroir ou flambeau de la mer, 1667-1684; *De Nieuwe Groote Vermeerderde Zee-Atlas*, 1675; *Nieuwe Nederlandtsch Caertboeck...*, 1685 (from the maps of Abraham Goos); *Nieuwe Wereld* (4 sheets), 1687 (with Jacob Robijn, engraved by Jacob Harrewijn); Portulan pour la Mer Baltique, 1694; and many others. **Ref.** KOEMAN, C. *Atlantes Neerlandici* Vol.IV (1970) pp.152-189 (with individual entries and collations).
•• **Doncker**, Hendrik [Hendrick II] (1664-1739). Son of Hendrik, above. Published Cornelis Cruys's *Nieuw Pas-Kaart Boek...*, *c.*1705; and others.

Dondey-Dupré *Père et Fils*. 'Imprimeurs-Libraires, Rue Saint-Louis, N°·46, Au Marais', Paris. *Atlas portatif et complet de la Royaume de France...*, 1823.

Donegan, P.H. Surveyor of the city of Washington. Assisted A. Bastert and J. Enthoffer with their *Map of the city of Washington* (10 sheets), 1872.

Donelly, J.P. Mining District of Lake Superior, 1872.

Donelson, John. *Plan of the Boundary between Virginia & the Cherokee Indian*, 1771.

Donia, Francesco. Engraver, for Giovanni Giacomo de Rossi's *Mercurio geografico*, 1688-1690.

Donis. See **Germanus**, Nicolaus.

Donn [Donne], Benjamin (1729-1798). Surveyor and mathematician of Bideford, Devon. *A map of Plymouth*, 1759; plans of Bristol, from 1759; *A Map of the County of Devon* (12 sheets), 1765; *Eleven miles round Bristol*, 1769; Gloucestershire, 1769; *Plan of the City of Salisbury*, 1781; *Western Circuit of England containing the counties of Cornwal, Devon, Dorset, Somerset, Wilts & Hants*, 1784; *Eleven miles round Bath*, 1790; *City of Bath*, 1810. **Ref.** RAVENHILL, W.L.D. (Introduction to facsimile) *Benjamin Donn: A map of the County of Devon 1765* (Devon and Cornwall Record Society and the University of Exeter 1965).

Donne, Samuel (*fl.*1754-1786). Surveyor of Melbury Osmond, Dorset. Produced various maps of Dorset and Somerset, including the manor of Leigh upon Mendip, Dorset, 1764; Manor of Little Winsor, Dorset, 1775 MS.

Johann Doppelmayr (1677-1750). (By courtesy of Deutsches Museum München)

Donnell, Henry Clay. *The Presidential Elections of the United States* (23 maps - one for each of the elections), 1877.

Donnelley, Reuben H. Publisher of Chicago. Sectional atlas of Chicago, 1891; Atlas of St. Paul, Minnesota, 1892.

Donnelly, *Captain —, RN*. Chart of Rio de la Plata, A. Arrowsmith 1807.

Donnelly, Dominick (*fl*.1678-1719). Surveyed estates in Ireland and Essex, including Epping, 1719 MS.

Donnelly, John Fretcheville Dykes. *Plan of the Siege Works... for the Attack of Sebastopol...*, 1855 MS (with C.E.S. Baynes); *Plan of Sevastopol*, 1857; *Plan of the attacks on Sevastopol*, 1858.

Donnelly, Ross. *Faro, Shetland & Orkney Isles...*, 1797; Skye to Aberdeen, 1797.

Donnelly, Thomas. Surveyor. Part of the town and Borough of Charlemont, Co. Armagh, 1815.

Donnet, Alexis [Alejo]. 'Ingénieur-géographe'. France (25 sheets), 1818 (reduction of Cassini's *Carte de France*); *Mapa civil y militar de España y Portugal...*, Paris 1823; Arrondissement de Corbeil, 1834; *Atlas de la France*, 1844 (with A.R. Frémin); *Environs de Paris*, 1867.

Donnet, Hendrik (fl.1766). Worked in Amsterdam. Dantzig, 1785.

Donnet, Johannes (1754-1779). Engraver in Gdansk.

Donnet, Samuel (1699-1734). Artist, publisher and engraver of Amsterdam.

Donnot, Didier. 'Géographe du roi', 1622.

Donnus. See **Germanus**, Nicolaus.

Donosa, *Don* Juan. *Carte hydrographique de la province du Chaco* [Colombia], 1827.

Donoso, —. 'Capitan de Ing[eniero]s'. *Costa de Chile*, 1855.

Donovan, W. *Lands of Jackson, Lansing & Saginaw Rail Road Co.*, 1873.

Donus. See **Germanus**, Nicolaus.

Donzel, Anton. French script engraver. João José de Santa Teresa's *Istoria delle guerre del regno del Brasile*, Rome 1700; for Pietro Petrini, 1700.

Doolittle, A.J. California, 1863 (with Leander Ransom).

Doolittle, Amos (1754-1832). Engraver and mapmaker of New Haven, Connecticut. Assisted in engraving Abel Buell's wall map of the territories of the United States..., 1784; *A Map of the United States* for *Bailey's Pocket Almanac*, 1785; Joseph Southern Purcell's *A Map of the States of Virginia North Carolina South Carolina and Georgia... East and West Florida...*, 1788; William Blodget's *A topographical map of the State of Vermont*, 1789; maps for Jedidiah Morse in *Geography Made Easy*, 1790 and *The American Universal Geography*, 1793; *Connecticut*, 1794; maps for Ezra Stiles's *A History of Three of the Judges*, 1794; compiled some of the maps for Mathew Carey's *The General Atlas*, 1795; Mathew Carey's *American Atlas*, 1795; *A map of the district of Maine...*, for Sullivan's *History of Maine*, 1795; *Province of Maine*, for *Carey's American Pocket Atlas*, 1796; James Whitelaw's *A correct map of the State of Vermont*, 1796, 1810, 1821, 1838, 1851; *A Correct Map of Connecticut*, 1797; *A chart of St George's Bank...*, 1797; *Plan of New Haven*, 1812, 1817, 1824; *An Improved Map of the United States by Shelton and Kensett*, *c*.1813 (with Thomas Kensett); *Map of New London and its vicinity*, 1813; *A Map of the United States and British Provinces of Upper and Lower Canada...by Shelton & Kensett*, 1816 (engraved by A. Doolittle and J. Kensett).
Ref. WHEAT & BRUN *Maps and charts published in America before 1800* (1978), many entries.

Doolittle, Isaac. With William A. Barnet, set up what is believed to be the first lithographic printing shop in America at 23, Lumber Street, New York City in 1821. *Barton on the Catskills*, 1822 (a geological map to accompany an article by D.W. Barton).

Doolittle & Munson. Draughtsmen and engravers. United States, Cincinnati 1847.

Doornick, Marcus Willemsz. Publisher of Amsterdam. *Platte Grondt der Stadt London*, 1666 (with a description of the Great Fire); Thames, River and Estuary (2 sheets), 1667.

Dopff, Daniël Wolff von (1655-1718). *Plan der Turckischen Attacquen voor Wien*, 1683 MS; *Endwurff der endsetzten Verstung Gran...* [Hungary], 1685 MS; *Plan de l'Attaque de Neuheusel...1685* [Slovakia; assault by German troops on the Turks], 1685 MS.

Doppelmayr [Doppelmayer; Doppelmaier], Johann Gabriel (1677-1750). Astronomer, mathematician and cartographer of Nuremberg. Celestial charts from 1704 onwards issued by Johann Baptist Homann; issued first German translation of Nicolas Bion's *Traité de la construction et des principaux usages des instrumens de mathématique*, as *Mathematische Werck-Schule*, 1712 (first published by Bion, 1709); *Basis Geographicae Recentioris Astronomica...*, Homann *c.*1724; globes, 1728-1740; Hemisphaerium coeli, 1730; completed *Atlas novus coelestis*, Homann's Heirs 1742; Tabula selenographica, *c.*1750 (after Johannes Hevelius). **Ref.** SHIRLEY, R. *The Map Collector* 46 (1989) pp.34-38; KRETSCHMER, DÖRFLINGER & WAWRIK *Lexicon zur Geschichte der Kartographie* (1986) vol.C/1, pp.177-178 [in German].

Dopter, —. Printer of Paris. France, 1837.

Doran, father and son. Irish portolan makers.
• **Doran**, Edmond (*fl.*1586). Irish portolan maker.
•• **Doran**, Hercules. Portolan maker, son of Edmond above. Mediterranean, 1586.

Doran, Thomas. *Map of Grundy County, Illinois*, Chicago 1863; published W.H. McChesney's *Map of Henderson County, Illinios*, Philadelphia 1868.

D'Orbigny, A. See **Dessalines** d'Orbigny, Alcide.

Dore, Edward (*fl.*1759-1769). Surveyor. Plan of the Borough of Devizes (2 sheets), *c.*1759; Estate of Baseldon, Berkshire, 1769 MS.

Dore, *Captain* Hastings. Manuscript maps of the routes taken by the East India Company's troops in Sumatra, 1805.

Dorfinger, József András. Hungarian cartographer and royal topographer. Gazetteer of Pest, 1827 (with maps); Dunaharaszti, 1828.

Doring, F. van. See **Doering**, Ferdinand von.

Dormier, —. Engraver. Plans of Madras and Pondicherry for *Monuments de l'Hindoustan*, 1821.

Dorn, *Captain* —. Santo Tomas de Guatemala, 1850.

Dorn, Hans (1430-1509). Austrian Dominican monk and globemaker, librarian at the court of the Hungarian King. 40-cm metal celestial globe [sometimes known as Martin Bylica's globe], 1480; armillary sphere, [*n.d.*].

Dorn, Johann Martin (*fl.*18th century). Engraver of Nuremberg. Map of Cyprus in J.P. Reinhard's *Vollständige Geschichte des Königreichs Cypern*, 1766-1768.

Dorn, Sebastian (*d.c.*1778). Engraver for Homann's Heirs, 1746-1768, e.g. Tobias Maier's *Indes orientales*, 1748.

Dorn & Merfeld. Lithographers of Leipzig. *Der Nord-Ostsee Kanal Unter Benutzung der Generalstabskarte...*, Kiel, E. Homann 1886.

Dornseiffen, *Doctor* I. *Atlas van Nederland...*, Amsterdam 1861 and later; *Nederland en Zijne Overzeesche Bezittingen* (8 maps), Amsterdam 1863; Atlas van Nederlandsch Oost- en West-Indië, 1884, 1894, 1909.

Dorph, E. Tasermint, Greenland, 1874.

Dorr, *Captain* F.W. *Battlefield of Chattanooga*, 1864.

Dörrbecker, P.A. See **Doerrbecker**, P.A.

Dorret, James (*fl.*1744-1761). Land surveyor to the Duke of Argyll, producing work in Argyllshire and Perthshire. *Scotland*, 1750; *A Correct Map of Scotland*, 1751, also used by Bowen & Kitchin, 1787, and Thomas Bowles, 1794.

Dorrington, George. Engraver. *Chart of the Baltic Sea*, 1854; River Danube, 1854.

Dort, Franciscus à. *Beschreibung und Abcontrafeyung das Closter S. Laurenty, gelegen in Escuriali*, Hamburg, Philipp von Ohr 1597.

Dortel de Tessan, A. Marine surveyor with Du Petit-Thouars aboard the *Vénus* from 1836. *Plan de la Baie des Iles a la Nouvelle-Zélande*, 1845; *Plan de la Rade de Cherbourg*, 1853.

Dorward, *Lieutenant-Colonel* A.R.F. Trigonometrical Survey of the Country round Kingston Jamaica, 1879; *Up Park Camp* [Kingston], 1897, 1905, 1908.

Dosio, Giovanni Antonio (1560-1605). Roma, Bartolommeo Faleti 1561 (engraved by Sebastiano de Ré).

Dossaiga [Dossaija], Jaime. Spanish portolan maker. Portolan atlas of Europe and North Africa, 1596.

Dosseray, J. *Plan de Bruxelles*, 1876.

Dosseville, —. Printer 'en taille-douce' [copper-plate] of Paris, 'au conservatoire de musique, rue de Faubourg-Poissoniere'. Guillaume Coutans's Environs de Paris, 1800.

Dossow, R. von. *Fredericia* [Denmark] *und Umgegend*, *c.*1864.

D'Ostening, H. von. *Geographische Carten von gantz Teutschlandt*, 1650.

Doswell, George (1810-1861). Land surveyor and civil engineer of 35 Bernard Street, Southampton. Produced various maps of Hampshire and Surrey including *Plan of the town of Southampton*, 1842 (designed by J.D. Doswell). **N.B.** Compare with J. Doswell and J.D. **Doswell** (below).

Doswell, John (*fl.*1782-1802). 'John Doswell and Son Land, Timber and Building Surveyors. Romsey Hants.' *A plan of the new roads from Southampton...*, 1802. **N.B.** Compare with G. **Doswell** (above) and J.D. **Doswell**, (below).

Doswell, John Doswell (1782-1856). Estate agent and civil engineer of Southampton. *Plan of the town of Southampton including the race course, and principal roads*, 1822.**N.B.** Compare with J. **Doswell** and G. Doswell, (above).

Dotecum. See **Doetichum** family.

Dou [Douw], Jan Jansz. [Janszoon] (1615-1682). Dutch cartographer, mathematician and notary of Leiden, son of Jan Pietersz. *Alckmaer en Bergen*, Joan Blaeu 1640; *Rijnland*, 1647 (with S. van Broeckhuysen); *Kennemerland ende West-Vrieslandt* (12 sheets), 1647 (with Balthasar Florrisz. van Berckenrode), reprinted 1687, 1746.

Dou [Douw], Jan Pietersz. [Pieterszoon]. Surveyor and cartographer at Leiden; author of a Dutch book on surveying, 1625.

Doublebsky, Robert (1839-1910). Czech astronomer and military geographer.

Doubleday, N. [inc.]. Russell J. Walrath's *Complete war map of western Europe*, New York 1917.

Doublet, —. One of the engravers for *Atlas géographique, et historique des départements de la France...*, Paris, Libraire Générale 1869.

Doudan, —. See **Houdan**, François de.

Doué [Doue], Martin [Martino; Martinus] (1572-1638). Artist of Lille. *Gallo-Flandria*, Willem Blaeu 1635.

Dougal, W.H. American engraver. Lake Michigan, 1852; *Boundary between the United States & Mexico*, 1855; New Mexico, 1859; *Territory and Military Department of Utah...*, 1860 (compiled by the Bureau of Topographical Engineers).

Dougal, W.M. American engraver. *Potomac river from Washington to Chesapeake Bay*, Washington, J.C. Entwistle 1887.

Dougharty [Doharty], John.
• **Dougharty**, John, *the elder* (1677-1755). Estate surveyor and mathematician of Worcester, England. Produced maps of parts of Hertfordshire, Staffordshire and Worcestershire, including an estate map of Lindridge, 1721 MS; plan of Worcester, 1741-1742.
•• **Dougharty**, John *the younger* (1709-*d.c.*1773). Surveyor of Worcester, son of John, above. Produced maps of estates in England, including ones at Droitwich, 1731-1733; Ruislip, 1750; Kidderminster, 1753. **Ref.** SMITH, Brian, *Transactions of the Worcestershire Archaeological Society* 3rd series, vol.15 (1996).

Doughty, Charles Montagu. (1843-1926). Explorer and orientalist. *Travels in Arabia Deserta*, 1888; Western Arabia, 1884.

Doughty, W.E. *Map of Lapeer Co., Michigan*, Samuel Geil 1862 (with I.M. Gross).

Douglas, *Captain* —, *RN*. *A new chart of Table Bay*, 1796 (inset on J.W. Norie's ...*chart of the Cape of Good Hope*).

Douglas, —. *Chart of the Hawaii Archipelago*, 1840 (2nd edition).

Douglas, Bloomfield, *RN*. Admiralty charts of the waters around South Australia, 1856-1860, including: Port Elliott, 1856; Sea Mouth of the River Murray, 1857; Victor Harbour, 1857.

Douglas, C.E. Plan of Waiho Country, Wellington, New Zealand, 1893.

Douglas, D.B. Map of the Niagara River, 1814; Siege and Defence of Fort Erie, 1816.

Douglas, *Lieutenant* H.H., *RN*. Admiralty surveyor. Corinth Bay, 1888.

Douglas, J. Sketch map of the north-west district of Otago, New Zealand, 1863; Canterbury-Otago Boundary, 1863.

Douglas, James. Worked with the engineer Thomas Telford. *General Plan by Messrs. Telford & Douglass for the further improvement of the Port of London...*, 1800 (plate ix of *Third Report from the Select Committee Upon the Improvements of the Port of London*) .

Douglas, N. Engraver of Edinburgh. Worked on John Wood's *Town Atlas of Scotland. Plan of St. Andrews*, 1820; *Plan of Brechin*, 1822.

Douglas, *Doctor* Robert. *Roxburgh & Selkirk*, 1798 (engraved by John Ainslie).

Douglas, William (1692-1752). *British Dominions of New England...*, 1753.

Douglas, William (*fl*.1760-*d*.*c*.1790). Produced various maps of parts of Scotland, including *Glasgow*, 1777 (engraved J. Urie); *A plan of the environs of Greenock*, 1778 MS; *A plan of Hamilton Haugh and part of the town*, 1780.

Douglass, C.C. Worked with S.W. Higgins and Bela Hubbard on surveys of Michigan counties. *Map of Calhoun county*; *Map of Jackson county*; *Map of Lenawee county*; *Map of Washtenaw County*; all published Washington DC *c*.1850.

Douglass, L.F. Surveyor. Jersey City, 1841-1876.

Doumerc, J. *Province d'Oran*, 1889.

Doupe, J.N. Saskatchewan, Ottawa 1885.

Dourado, Fernão Vaz (*c*.1520-*c*.1580). Portuguese navigator, map and portolan maker, born and worked mostly in Goa. Mainly known through six surviving illuminated manuscript atlases dated 1568, 1570, 1571, 1575, *c*.1576, and 1580. Those of 1575 and *c*.1576 were made in Portugal, the others in Goa. **Ref.** PFLEDERER, R. 'The Portuguese century' in: *Mercator's World* (July/August 1997) pp.24-28.

Doutre, J.M. Publisher. *Cours du Rhône de Lyon à la mer Méditérranée*, 1839 (with O. Martinon, drawn and engraved by L.G. Sampierdaréna).

Douw, Jan Jansz. See **Dou**.

Douw, Jan Pietersz. See **Dou**.

Douw, Jan Rokuszoon. See-caart van den Hollandse Wal, *c*.1760.

Dove, Heinrich Wilhelm (1803-1879). Physicist and meteorologist. *Nördlichen Hemisphäre*, 1855; *Nordpolar-Länder*, 1868; *Atlantic Ocean*, 1868.

Dove, Henry Percy (*fl*.1885-1923). Surveyor of 19 Wentworth Court, Sydney, New South Wales. A new and complete wharf, street and building plan directory...[Sydney], 1880; plans of Sydney, 1880; Burwood, 1880.

Dovers, *Captain* William, *RN*. *Admiralty Bay, Bequia*, 1811.

Dowcett, Ralph (*fl*.1659-1677). Surveyor, produced maps of estates in Essex and Sussex including Danbury, 1667 MS; Little Baddow, 1677 MS.

Dowdall, *Captain* John. *The measure of the Castell of Eneskillin*, 1592 MS, reproduced by the Ordnance Survey 1860.

Dowde, J. Lake of Killarney, 1800.

Dower, father and sons. **N.B.** There is some confusion between John Dower, father, and John Dower, son. Works attributed to John Dower after the father's death in 1847, are assumed to be the work of his son.

• **Dower**, John Crane (*fl.*1820-1847). Engraver, draughtsman and publisher of 6 Cumming Place, Pentonville Hill, London. Engraved for Sharp, 1825-1828 (with Greenwood and Fowler); W. Swire and W.F. Hutchings's Cheshire, 1830; for C. and J. Greenwood, 1830; Henry Teesdale, 1831-1856; New General Atlas, 1831; *Van Diemen's Land*, 1831; for Thomas Moule, 1834-1847; *A New General Atlas of the World*, 1835; *The Stranger's Companion through London and its Suburbs*, Allan Bell & Comp^y. 1835.

•• **Dower**, John James (*b.*1825). Map engraver of Pentonville, also 108 Fleet Street (1870), son of John Dower, above, to whose business he succeeded. Engraved for Orr & Co., 1850-1860; *Penny Map of London, or Exhibition Guide*, 1851; *London and Twenty Miles Round*, 1853; *Environs of London*, 1855; *...the toll-gates and principal bars within six miles of Charing Cross* [London] (in *Illustrated London News*, 1857); maps in *Weekly Dispatch Atlas*, 1858-1863; *The Seat of War in Italy* (first published in the *New York Times*), July 7th 1859; *London. Guide to the International Exhibition 1862* (in *Illustrated London News* 1862); *...Environs of London*, 1862; Telegraph Map, 1865; *Watering Places of Kent*, 1865; *Cassell's Pocket Map of London*, 1866; revised G.W. Bacon's Thames, 1869; *Map of Quebec, New Brunswick, Nova Scotia &c...*, 1869; *Bacon's Modern London...*, 1870; *Bacon's Gem Map of London and Suburbs*, 1876.

•• **Dower**, Frederick James (*b.*1829). Another son of John Crane Dower, above. Map engraver who served his apprenticeship with Alexander Findlay.

Dowie, David. *A Map of the Forest of Mamlorne*, 1735-1737.

Dowland, James (*b.c.*1751-1823). Buckinghamshire estate surveyor, produced various manuscript maps of Midland counties.

Dowling, P.H. Maps of townships for L.H. Everts and D.J. Stewart's *Combination Atlas of Washington County*, 1874.

Down, Robert. *Five miles round Bridgewater*, 1853 (with Thomas Clark).

Downer, William. *Topographical Map of the County of Berbice, British Columbia* (3 sheets), John Arrowsmith 1844.

Downes, Charles John. Engraver and printer of 112 Fetter Lane, London. For Hasted's *County of Kent*, 1783; Alexander Taylor's Kildare, 1783; *Twenty Miles round London*, 1784; *A Plan of the Town of Nassau*, 1788; Robert Whitworth's Canal Ashby de la Zouch-Coventry, 1792.

Downes, William (*fl.*1817-1845). Surveyor, produced manuscript maps of estates throughout England, including Layer Marney, 1817 MS.

Downie, J. Cartographer. *Plan of the City and Harbour of Aberdeen*, published in Walter Thom's *The History of Aberdeen*, 1811.

Downie, James. Pt. Romania to Anambas Is., published Alexander Dalrymple 1805; Achen Road, &c., published Dalrymple 1806.

Downie, James. Sketch of the River Thames, New Zealand, 1820.

Downie, Murdo, *Master RN*. Surveys of the coast of Scotland. *A chart of the Tay from Barr to Dundee*, 1789; *A New Pilot for the East Coast of Scotland*, 1792; collaborated with Thomas Telford on a road map of Scotland, 1803.

Downing, Alexander. *Plan of Bury St. Edmunds*, 1740.

Downsborough, J. *Bexhill-on-Sea*, 1888.

Dowthwaite, A. Tottenham [River Lea, London], 1834.

Doyen, —. Lithographer of Turin. *Map of Malta*, 1893.

Doyle, D. Gaza Land, 1891.

Doyle, William. *The Harbours of Rineshark and Waterford*, 1737-8; Waterford, 1738.

Doyley, John, father and son.
• **Doyley**, John, *the elder* (*fl.*1794-*c.*1848). Land surveyor, 10 Greville St., Holborn, London. Produced manuscript maps of areas in various parts of England and Wales, including farms in Felsted and Gt. Waltham [Essex], 1796; Fyfield, 1798 MSS; Leyton, Wanstead and Woodford (90 maps), 1811-1812 MSS; Aldersbrook, 1818.
•• **Doyley**, John, *the younger* (*fl.*1825-1835). Surveyor of Grays Inn, London. Produced estate and railway maps in Essex and Middlesex.

D'Oyley, William. *Map of Epping Forest* [England], 1874.

Doyne, W.T. Plan of the East Coast Province of Canterbury, New Zealand, 1864; Levels of Waionakakariri, 1864; Portion of the Plains of Canterbury, 1865.

Dozy, *Doctor* G.J. *Historische Atlas der Algemeene Geschiedenis*, Zutphen, W.J. Thieme 1876; *Schoolatlas der geheele aarde* (26 maps), Arnhem, J. Voltelen 1877, 1879; *Kleine schoolatlas der geheele aarde*, Arnhem, J. Voltelen 1883 and other editions; *Historische Atlas, ten gebruike bij het onderwijs in Algemeene en Vaderlandsche Geschiedenis*, Zutphen, W.J. Thieme & Cie. *c.*1893 and other editions.

D'Perac, E. See **Du Perac**, Stefano.

Draak, F. Engraver for Pieter van der Aa, *c.*1715.

Drach [Draeck], F. See **Drake,** *Sir* Francis.

Draeck [Dragk], Pieter. 16th-century Dutch cartographer.

Drage, Theodore Swaine. North-west parts of Hudson Bay, 1746-*c.*1768.

Dragon, Demetrio. Engraver of Rome, worked on G.M. Cassini's *Nuovo Atlante Geografico Universale...*, Rome 1792, 1797, 1801.

Drake, C.F. Tyrwhitt. *Route map of the Negeb*, *c.*1871.

Drake, Daniel. *Drake's Statistical View*, 1815 (with a plan of Cincinnati engraved by J.H. Seymour).

Drake [Drach; Draeck], *Sir* Francis (*c.*1545-1596). Admiral, circumnavigator 1577-1580, voyage to the West Indies 1585-1586. Although Drake created his own manuscript maps and views relating to his voyages, none have survived. Existing maps are work inspired by his voyages. **Ref.** WARD, B. *The Map Collector* 59 (1992) pp.14-15; SEELER, O. *The Map Collector* 54 (1991) pp.32-34; WALLIS, H. *The Map Collector* 49 (1989) pp.34-41; WARD, B. *The Map Collector* 45 (1988) pp.2-8; pp.32-34; STUART, E. *The Map Collector* 42 (1988) pp.2-5; THROWER, N.J.W. *Sir Francis Drake and the famous voyage 1577-1588* (1984); *Sir Francis Drake* (British Library exhibition catalogue, 1977).
• **Drake**, *Sir* Francis, *Bart.* (1588-1637). Nephew of above. The World Encompassed, London, Nicholas Bourne 1628.

Drake, Francis. Plan of the City of York, 1736; *Roman roads in the County of York*, 1745.

Drapentier, J. *Town of Hitchin*, 1700.

Draper, Henry (1837-1882). American geographer and meteorologist.

Draper, John. Engraver. Moore & Jones' Travellers' Directory, 1802.

Draper, W.A. Surveyor. *Complete Map of Colorado*, 1861.

Drasche, Richard von. Geological map of Luzon, 1878.

Drayton, F. Surveyor. *Plan of the City of Philadelphia*, Carey & Lea 1824.

Drayton, John. Engraver. *South Carolina*, Carey & Lea 1822; *Alabama*, &c. for Cabinet Atlas, 1832.

Drayton, Michael (1563-1631). Poet and historian, wrote a collection of songs describing Britain, published as *Poly-Olbion, or a Chorographicall Description of tracts, rivers, mountaines, forests and other parts of this renowned Isle of Great Britaine*, 1612 (with decorative maps by William Hole), 2nd part 1622 (allegorical maps of rivers of England and Wales). **Ref.** COPE, G. *The Map Collector* 17

Sir Francis Drake (c. 1545-1596). A portrait engraved for Hervey's Naval History.

Sir Francis Drake (c. 1545-1596). The well known portrait from an original engraving by Jodocus Hondius. (By courtesy of the British Library)

(1981) pp.16-20; HEBEL, J.W. (ed.) *The works of Michael Drayton* 5 Volumes (Oxford, Blackwell 1941/1961); NEWDIGATE, B.H. *Michael Drayton and his circle* (Oxford, Blackwell 1961).

Drayton, *Lieutenant* T.F. One of the U.S. army surveyors under the command of Captain W.G. Williams of the Topographical Engineers. *Survey for a Ship Canal around the Falls of Niagara...*, 1835 (with J.G. Reed); *Survey for a Ship Canal To connect the lakes Erie & Ontario*, 1835.

Drebbe [Drebber; Drebbell], Cornelis Jacobsz. [Jacobszoon] (1572-1634). Mathematician, engraver and publisher. Plan Alcmaer, 1597; globe, presented to James I, 1621.

Drebes, F. Engraver. *General Map of the Northwestern Part of the Dominion of Canada*, Jacob Smith 1898.

Drechsel, Wolf. See **Trexel**, Wolf.

Drège, Carl Friedrich (1791-1867). Apothecary and botanist. *Map of Cape Town*, 1826.

Michael Drayton's tomb in Westminster Abbey. (By courtesy of Westminster Abbey)

Dreissig, M. Lithographer of Hamburg. Fr. Aeuckens's *Plan der Insel Heligoland*, Heligoland, P.F.L. Uterhark 1886.

Drenckan, G.A. (*fl.*1750). Cartographer of Kiel, Germany.

Drentwett, Abraham, *the younger* (1647-1729). Goldsmith and draughtsman in Augsburg. Maps for Johann Baptist Homann, e.g. Terra Sancta, *c.*1724; specialised in cartouches and vignettes for George Matthäus Seutter, 1725-1741.

Dresel. See **Kuchel & Dresel**.

Dreselly, A. *Schliersee*, 1896.

Dreuer, Domenicus. Lower Elbe, *c.*1569 MS.

Drew, Frederick. *Geology of the country between Folkestone & Rye*, 1864.

Drew, H. Bookseller and stationer of Jacksonville, Florida. *Drew's New Map of the State of Florida*, 1867, 1870, 1874, 1880, 1884.

Drew, J.W. Atlas of Clarke County, Ohio, 1870 (with D.J. Lake & Co.).

Drew, John. Publisher, N°·31 Fetter Lane, Fleet St., London. *Plan of London*, 1799; *Chronological charts of Ancient History & Geography*, 1835.

Drew, *Lieutenant* (later *Captain*) John Clarke, *RN*. Admiralty surveyor. *Plan of Petaldi Bay* [Greece], 1865 MS.

Michael Drayton (1563-1631). (By courtesy of The British Library)

Drewry, W.S. Slocan Mining Camp, British Columbia, 1897; Topographical survey Canada, 1900.

Drewyer, G. See **Drouillard**, George.

Drexel, —. Forest County, Pennsylvania, 1868.

Dreyer, F.A. Plan of Bremerhaven, *c.*1898.

Dreyskorn, G.R. Engraver for Friedrich Wilhelm Streit, 1817; Schlieben, 1828-1830.

Driat, —. *Plan de Lyon*, *c.*1874.

Driesemann, C. *Plan der Stadt Halle*, 1876, 1892.

Drigalski, Guido de. *Mapa de los ferro-carriles vapores... de Argentina*, 1896.

Drikerhoff, —. Environs of Mannheim, 1814.

Dring, Thomas (*fl.*1655-1695). Publisher. Adam Olearius's Voyage, London edition 1662.

Drinkwater, *Captain* —, *RN*. Labouan Island & Bruni River, Admiralty 1845.

Drinkwater, John. Publisher of Liverpool. Isle of Man, 1826 (survey by Benjamin Smythe, engraved by J. & A. Walker); Regents Canal, 1830.

Drioux, Claude-Joseph (1820-1898). *Atlas de l'histoire et de la géographie*, 1867; *Atlas universel et classique*, 1867 and other editions; *Nouvel atlas de géographie moderne*, 1876; *Atlas universel*, 1890.

Dripps, Matthew. Irish emigrant cartographer and publisher of 103 Fulton Street, New York. Marcus Smith's *Map of the City of Hartford*, 1850 (printed by Mayer & Korff); *Map of the City of New-York Extending Northward to Fiftieth Street*, 1851 (drawn by John F. Harrison); *Map of That Part of the City and Country of New-York North of 50th St.*, 1851 (drawn by H.A. Jones); I. Slatter and B. Callan's Map of the City of Boston, Massts..., 1852 (also published by L.N.N. Ide of Boston); Atlas of New Utrecht, 1887. **Ref.** COHEN, P.E. & AUGUSTYN, R.T. *Manhattan in maps 1527-1995* (1997) pp.124-127.

Driver, family. Surveyors of London over several generations, produced a large number of estate and local maps.

• **Driver**, Samuel [I] (1720-1779). Surveyor of London, father of Abraham. Produced manuscript maps of areas throughout England, including Dane-End Farm, Weston [Hertfordshire], 1768.

•• **Driver**, Abraham Purshouse (1755-*c.*1821). Surveyor of 3 New Bridge Street, London (1820), father of Edward, Samuel II and George. Produced various manuscript maps of areas throughout England, including the New Forest, 1789 (with William Driver, Thomas Richardson & William King); Romford, 1797 MS; East and West Ham, 1799 MS; Estates in Burnham [Buckinghamshire], 1800-1805; Stapleford Abbots, 1803 MS (with William Driver); Crown lands in Eltham, Lee, Mottingham and Bexley [Kent], 1806 (with William Driver); Manor of Lambeth, 1812; Leigh Swatch [Essex], 1814 (with Edward, William and Samuel Driver); *Frogmore and Shaw Farm in New and Old Windsor...*, 1819 MS (with Edward Driver); Higham House estate [Walthamstow], 1820 MS (with Edward Driver); Leyton, 1820 MS (with Edward Driver); and many others.

••• **Driver**, Edward (1783-1852). Estate surveyor of London, worked all over England on maps and plans. Collaborated with his brother George, 1824-1846. Leigh Swatch [Essex], 1814 (with Abraham, William and Samuel Driver). Hornchurch, 1814 MS; Leyton 1820 MS (with Abraham Driver); Scilly Is., 1833; and many others.

••• **Driver**, Samuel [II] (1785-1857). Surveyor of London, produced surveys of estates in the Home Counties, son of Abraham. Manor Lambeth, 1812; Leigh Swatch [Essex], 1814 (with Abraham, William, and Edward Driver).

••• **Driver**, George Neale (1794-1855). Estate surveyor, New Bridge St., London, later Richmond Terrace, produced manuscript maps of areas throughout England. Collaborated with Edward Driver 1824-1846. Stapleford Abbots, 1824 MS; Scilly Is., 1833.

•••• **Driver**, Robert Collier (1816-1898). Surveyor of London, nephew of Edward and George Driver. Plan waste lands Epping Forest, 1877.

•• **Driver**, William (1758-1819). Estate surveyor of London, worked throughout England, brother of Abraham. Plan of the New Forest [Hampshire], 1789 (with Abraham Driver, Thomas Richardson and

William King); Stapleford Abbots, 1803 MS (with Abraham Driver); Leigh Swatch [Essex], 1814 (with Abraham, Edward and Samuel Driver).

Drivet, F. Relief map of the environs of Paris, 1875; *Afrique, Amérique, Asie, Europe*, 1878.

Dröber, Wolfgang (*fl.*1903). Wrote theoretical works on early and aboriginal mapmakers.

Droeshout [Drowshot], Michiel (*b.*1570). Allegory on London and the Gunpowder Plot, 1605; Plan Breitstein, 1631.

Drogenham, Gerard. See **Droogenham**, Gerrit.

Dromeslawer, Robert. Plan of Hartlepool Harbour, *c.*1585 MS.

Dron, Henri. *La France Militaire*, 1867; *L'Europe des Points Noirs*, *c.*1869.

Droogenham [Drogenham], Gerrit [Gerard]. Land surveyor and engraver of Amsterdam. Terrestrial and celestial globe gores for Abraham van Keulen's 2-inch pocket globe, *c.*1697; plan of Amsterdam, N. Visscher 1700; *Nova tabula Imperii Russici*, 1704; for Henri Du Sauzet, 1734.

Drouaillet, —. Lithographer, Corner Washington & Kearney, San Francisco, for M. Milleson 1864.

Drouet, Santiago. Letter engraver for V. Tofiño de San Miguel, 1786-1787, including *Plano Geométrico de la Bahia de Algeciras y Gibralter*, 1786.

Drouillard [Drewyer], George. Yellowstone and Big Horn [USA], 1808 MSS.

Drouville, *Colonel* G. Contrées entre l'Indus et l'Euphrate, 1827.

Drowshot, M. See **Droeshout**, Michiel.

Droysen, *Professor* Gustav *the younger* (1838-1908). Historian. *Allgemeiner historischer Handatlas* (96 maps), Bielefeld and Leipzig, Velhagen & Klasing 1886 (under the direction of R. Andree); and other works for Velhagen & Klasing.

Droysen, Wilhelm. Engraver. H. Kiepart's *Cartes des nouvelles frontières entre la Serbie, la Roumainie, la Bulgarie, la Roumélie Orientale et les provinces immédiates de la Turquie selon les decisions du Congrès de Berlin, Juillet 1878* (4 sheets), Berlin, Dietrich Reimer 1881 (printed by Wilhelm Greve); H. Kiepart's *Ubersicht der Administrativ-Eintheilung und der Ortsbevölkerung der Neuen Nördlichen Provinzen des Griechischen Königreiches...*, Berlin, D. Reimer 1883; *Goldminen Gebiete... West-Australien*, 1896.

Dru, Léon (*b.*1837). French geographer and engineer. Mission géologique au Caucasie, 1882.

Drude, *Doctor* Oscar (1852-1933). Botanist and geographer in Dresden. *Atlas der Pflanzenverbreitung* in *Physikalischer Atlas*, 1886-1892.

Drugeon, —. *Petit atlas commercial de la France*, 1824; *Carte routière de la France*, 1868.

Drummer, Edmund. See **Dummer**, Edmund.

Drummond, Alexander. English Consul at Aleppo. Compiled a map of Cyprus, published in *Travels through different cities of Germany and several parts of Asia*, 1754.

Drummond, M.G. Short Treatise on Geography, 1708, 1740.

Drummond, Willis. *United States*, 1871.

Drury, *Commander* Byron, *RN*. Admiralty surveyor. Lowang Channel, 1840 (with B. Woodcombe); Survey of New Zealand, 1850-1856; Ahuriri Road & Port Napier, 1857.

Drury, Luke (*d.*1845). *A Geography for Schools*, Providence, Rhode Island, 1822.

Drury, *Captain* William O'Brien, *RN*. Charts of Ireland, Bantry Bay &c. 1788; Valentia Island, 1788; Surveys of the Harbours of Rutland... Blacksod... Corke & Waterford, 1789; Cork, 1790.

Drusiani, S. Linie telegrafiche d'Italia, 1869.

Druten, J. van. Guinea, Utrecht 1895.

Dry, Camille. Engraver. *Plan Général des Installation de la Compagnie à Colón* [Panama Canal Company], 1882.

Dryander [Eichmann], Johannes (1500-1560). Mathematician, physician and astronomer, professor at Marburg. De globulo terrestri, 1537; De usu instrumenti, 1538; *Hesse*, used by Sebastian Münster, 1540, also Abraham Ortelius, 1579; Cosmographiae introductio, 1543. **Ref.** MEURER, P. *Fontes cartographici Orteliani...* (1991) p.140 [in German].

Drygalski, Erich von (1865-1949). *Karte der Karajat-Eisströme* [Greenland], 1897.

Drysdale, Robert (*fl.*1797-1843). Surveyor of Dunfermline, produced manuscript maps of areas in Scotland and England. Corrected Fife for John Thomson, 1832.

Duams-Vorzet, Er. Engraver of the lettering on G. Bagge's *Russie Orientale et Caucasie*, Hachette & Co. 1900.

Du Bac, —. *Vicomté de Turenne*, *c.*1670.

Dubail, *Lieutenant* Augustin von Edmond (*b.*1851). *Europe centrale*, 1875; *Atlas classique de géographie universelle*, 1877; *Atlas de l'Europe militaire*, 1880.

Dubar, —.*Carte de la Belgique*, 1832.

Dubercelle, —. Engraver. Besançon, 1735; Ville et Abbaye de Tournes, *c.*1753.

Duberger, John Baptiste. Plan of part of the Province of Lower Canada, 1795 MS (with Samuel Gale); Country west of Lake Ontario 1800 MS; Plan of Quebec, 1800 MS.

Du Bocage, Georges-Boissaye. See **Boissaye** du Bocage, Georges.

Dubocage, J.D. Barbié. See **Barbié** du Bocage, Jean-Denis.

Dubois, —. One of the surveyors for the Cassini *Carte de France*, worked with others on fourteen sheets mostly in central and southern France, 1759-1787.

Dubois, —. Engraver for Alexandre Emile Lapie, 1816.

Du Bois, A. *La Géographie Moderne*, 1729.

Dubois, A. *Carte routière et hydrographique... Seine et Oise*, 1876.

Dubois, François. 'Imprimeur du roy'. Toul [Lorraine], 1608.

Dubois, J. *Cours du Weser* [Germany], 1758.

Du Bois, *Lieutenant* J.V. Camp Floyd to Fort Union, 1860 MS.

Dubois, Jacques. Geographer. Carte topographique de la Capitanerie d'Halatte, 1723 MS.

Dubois, Jean. *Carte de Mont Blanc*, 1825.

Dubois, M. *La Guerre en Orient*, Paris, A. Challamel 1915 (with J.G. Kergomard, engraved by R. & A. Hausermann, printed by Monrocq).

DuBois, N. Draughtsman for *National Map of the Territory of the United States from the Mississippi River to the Pacific Ocean*, Washington 1867 (printed by J.L. Gedney).

Dubois, T. Plan of the Battle of Culloden, 1746 MS.

Dubois de Montpéreux [Montperreux], Frédéric (1798-1850). Swiss archaeologist and geologist. *Voyage en Caucase chez les Tcherkesses, & les Abkhases en Colchide, en Géorgie, en Arménie et en Crimée* (21 maps and plans), Neuchatel 1843.

Dubon, —. *Atlas*, 1878.

Dubosc, Claude. Engraver. Béthune, 1731; Arras, 1736; Blenheim, *c.*1736; Plan de Gibraltar, 1750.

Du Bouchet, Michel. *Carte du Pais d'Auvergne*, 1645.

Dubouzet, *Admiral* —. *Océan Pacifique*, 1863.

d'Ubraye, L.A. See **Durand**-Dubraye, Louis Alexandre Jean François Léopold de.

Dubré, Jean. See **Du Pré**, Jean.

Dubrena, V. Table graphique de la navigation

intérieure de l'Empire Français, 1811; *Carte hydrographique de la France*, 1828; *Carte de la Navigation de France*, 1838.

Dubreuil, Pierre Justin Charles. *Plan de la baie Coquimbo, Chile; Iquique*, 1826-1836.

Dubrovin, Mark. Russian seaman and cartographer. Bokhara, 1731 (for Ivan Kirilov).

Dubuisson, —. Script engraver, worked with Pierre François Tardieu, *c.*1765, 1779, 1780.

Dubuisson, François Robert André (1763-1836). French chemist and mineralogist, conducted geological studies of Loire Inférieure. Essai d'une méthode géologique, Nantes 1819.

Dubure, —. Publisher of Paris. J. Chappé d'Auteroche's *Voyage en Sibérie...*, 1768.

Duby, *Père —. Mission de Sénégambie*, 1877.

Du Caille, Louis Antoine. French geographer and draughtsman. *Etrennes géographiques* (26 maps), Paris 1760, 1768.

Ducarla-Bonifas, Marcellin (1738-1816). French engineer and natural scientist. Worked in Geneva, Paris, Clermont, Castres, Lavaur and Villeneuve sur Tarn. Presented ideas concerning the use of contours to the Académie Royale des Sciences in 1771 (published as *Expression des nivellemens, ou méthode nouvelle pour marquer rigoureusement sur les cartes terrestres & marines les hauteurs & les configurations du terrain*, Paris 1782); Histoire naturelle du monde, 1782.

Du Carlo. See **Carlo**.

Du Chaffat, Antoine. See **Chaffat**, Antoine du.

Duchapt de Breuil, —. Worked with Duffau-Davray, Cabay and Dubois on sheet 15 [Aurillac] of the Cassini *Carte de France*, 1766-1775.

Duchatel, —. Publisher of 'rue de la Tacherie', Paris. *Carte de Catalogne et de Roussillon*, *c.*1700.

Duchaxel, Marie Edme Félix. *Mer Mediterranée. Côte d'Afrique*, 1857.

Duchemin, Noël-Laurent. Tableaux des villes de France, 1777; *Royaume de France*, 1815.

Duchêne, André (1584-1640). French geographer and historian; 'géographe du roi', 1618.

Duchesne, —. Bookseller, Rue St. Jacques, Paris. Histoire Jersey et Guernsey, 1757.

Du Chesnoy. See **Capitaine** du Chesnoy.

Duchetti, family. Declared successors to Lafreri's business when his assets were divided in 1580. Claudio and Francesco were the sons of Lafreri's sister, and her husband Etienne Duchet. **Ref.** GANADO, A. & AGIUS-VADALÀ, M. *A study in depth of 143 maps representing the Great Siege of Malta of 1565* (1994) pp.26-29

• **Duchetti** [Duchet; Ducheti; Ducheto; Duchettj; Duchettum], Claudio (*fl.*1554-1585). Publisher of Venice and Rome, nephew and part successor to Lafreri. World, 1570; Europe, 1571; Africa, 1579; *Palermo*, 1580 (engraved by Natale Bonifacio); plan of Ostia, 1581; plan of Genoa, 1581; map of Italy, 1582; *Bologna*, 1582; re-issued Nicolò Nelli's *Il Porto Dell'Isola di Malta*, *c.*1582; plan of the city of Rome, 1582; plan of Ancona, 1585; Naples, 1585; plan of Augsburg, 1585; and others.

•• **Duchetti**, haeredes Claudii. Much of Duchetti's stock was dispersed after his death, but he had made arrangements for the continuation of the name. Plan of Pozzuoli, 1586 (engraved by Brambilla).

• **Duchetti**, Francesco. Brother of Claudio, also a part successor to Lafreri, but not apparently involved in the business.

•• **Duchetti**, Stefano (*d.*1583). Son of Francesco and successor to Lafreri through his father's share. Worked in Claudio's workshop from *c.*1581.

Duchier, E. Engraver for Alexandre A. Vuillemin, 1873.

Ducker, Edward. Estate surveyor of Essex. Fryerning, 1770 MS; East Tilbury, 1773 MS.

Ducker, F. See **Duecker** von Hasslau zu Urstein und Winkel, Franz.

Duclaux, A. Map colourist of Paris. F. Bianconi's *Carte Ethnographique de l'Ancienne Voyévodie Serbe de Hongrie ou Banat de Temesvar*, *c.*1915.

Sir Robert Dudley (1573-1649). (By courtesy of Rodney Shirley)

Du Clesmeur, —, *chevalier*. Commander of vessel *Castries* in Marc-Joseph Marion-Dufresne's expedition to New Zealand, 1772. *Terres de Diemen*, 1772; *Mouth of River Dwina*, 1785.

Duclout, Jorge. *Mapa... la Pampa central*, 1887; *Mapa de la Republica Argentina*, 1888.

Ducourtioux et Huillard. Engravers and printers of Paris. *Limites entre le Brésil et la Guyane Anglaise* (90 maps), 1903.

Ducrest, Charles Louis, *marquis* de (1747-1824). Machines hydrauliques, 1777; Courants d'eau, 1800; Nouveau système de navigation, 1809.

Du Creux [Creuxius], François, *SJ* (1596-1666). *Historie Canadiensis*, 1664 (with a map of north-east Canada). **Ref.** Burden, P. *The mapping of North America* (1996) pp.449-450.

Ducros de Saint Germain, A.M.P. *Département du Puy-de-Dôme*, 1881.

Dudley, J. *Fort William & Mary... Piscataqua River* [New Hampshire], 1699 MS; *The South Battery in the Town of Boston... Massachusetts*, 1705; *The North Battery...*, 1705; *Marblehead Fort... Massachusetts Bay...*, 1705.

Dudley, *Sir* Robert [self-styled Duke of Northumberland, Earl of Warwick] (1573-1649). Born in Richmond, the illegitimate son of the Earl of Leicester. Engineer, explorer and geographer, active in Italy, died in Florence. Compiled chart atlas *Dell'arcano del mare*, 1646-1647 (printed by Francesco Onofri; some proof plates of the first edition were amended in manuscript by L. Holstein); revised edition, Jacopo Bagnoni and Lucini 1661. Both editions include charts engraved by Antonio Francesco Lucini. First nautical atlas by an Englishman, and the first nautical atlas with charts on Mercator projection. **Ref.** WARD, B. *The Map Collector* 59 (1992) pp.14-15; DILKE, O.A.W. and M. *The Map Collector* 19 (1982) pp.10-14; LEADER, J.T. *Life of Sir Robert Dudley* facsimile edition (Amsterdam 1977); SCHUTTE, J.F. *Imago Mundi* 23 (1969) pp.28-58.

Duecker [Dücker; Dükher] **von Hasslau zu Urstein und Winkel**, Franz (1609-1671). Born Innsbruck. *Saltzburgische Chronica*, 1666.

Duems, W. *Reisekarte von Mittel-Europa*, 1889.

Duentzfeld [Düntzfeld], Johann Friedrich. *Nieder-Isenburgische Länder*, 1772.

Duerer, Albrecht. See **Dürer**, Albrecht.

Duerkop [Dürkop], A. Das Saalgebiet, *c*.1898.

Duerrich [Dürrich], Ferdinand von. *Atlas der Schlachten*, 1792-1815, 1857.

Duetechum. See **Doetichum** family.

Dufart, P. Atlas de la géographie ancien historique et comparée, 1839.

Du Fayen, J. See **Fayen**, Antoine Jean du.

Duff, Patrick. *Elgin and Nairn Shires*, 1842.

Duffau-Davray, —. Worked with Duchapt de Breuil, Cabay and Dubois on sheet 15 [Aurillac] of the Cassini *Carte de France*, 1766-1775.

Duffee, William. Printer for a map of North America by Henry Schenck Tanner, 1822.

Duffield, W.W. Superintendent of the United States Coast and Geodetic Survey, 1890, 1897.

Duffin, *Captain* Robert. Pescadore Is., published Alexander Dalrymple 1792.

Du Fief, Jean-Baptiste-Antoine-Joseph. *Carte de l'État Indépendant du Congo* (4 sheets), 1893, 1895, 1900, 1905; Atlas de Belgique, 1894; Atlas général, 1895.

Duflot de Mofras, Eugène (1810-1884). *Explorations du territoire de l'Orégon, des Californies et de la Mer Vermeille...*, Paris 1844 (contains the first plan of Sitka, Alaska).

Du Foss, Nicolas. Publisher, 'rue St-Jean de Beauvais'; 'rue Sainct Jacques au Vase d'Or, Paris'. Pierre d'Avity's *Les estats... du monde*, 1626.

Dufour, Alphonse. *Nouvelle Carte routiere de France*, 1878.

Dufour, Auguste-Henri (1798-1865). Cartographer and publisher of Paris. Maps for J.A. Letronne, 1827; *Carte Physique, Politique & Comparée de la Turquie d'Europe...* (4 sheets), Paris, P.J. Lameau 1827 (engraved by R. Wahl and Hacq); maps in C.A. Walckenaer's *Histoire générale des voyages*, 1831; *Atlas classique et universel de géographie*, 1834; *Atlas nacional de España*, 1837, 1838, 1845; Atlas de géographie numismatique, 1838; *La terre, atlas historique...*, 1840; worked with Thunot Duvotenay, 1852; *Atlas universel, physique, historique et politique*, 1856-1858; *Carte de la Sicile*, 1859 (with M. Amari); A. Germond de Lavigne's *Itinaire descriptif...artistique de l'Espagne et du Portugal*, L. Hachette et Cie. 1859; Amérique du nord, 1860; *Atlas de France et ses 86 départements et colonies*, 1860 (with T. Duvotenay); Atlas historique, 1864; also road maps of France, Spain and central Europe.

Dufour, *General* Guillaume-Henri (1787-1875). Swiss army officer, engineer and councillor, worked in France, Thun and Geneva. Became cantonal surveyor at Geneva 1833. *Carte topographique de Canton de Geneve*, 1842, 1862; topographical survey of Switzerland, published as *Topographische Karte der Schweiz* (25 sheets), *c.*1845-1865 (generally known as the 'Dufourkarte' series, noted for its fine hachure and shadow technique); Eisenbahn-Karte Schweiz (4 sheets), 1874. **Ref.** DURAND, Roger *Guillaume-Henri Dufour dans son temps 1787-1875* in *Actes du colloque 1987* (Geneva 1991) [in French]; LANGENDORF, Jean-Jacques *Guillaume-Henri Dufour. General - Kartograph - Humanist. ine Bibliographie* (Zurich 1987) [in German]; KRETSCHMER, DÖRFLINGER & WAWRIK *Lexicon zur Geschichte der Kartographie* (1986) vol. C/1, p.181.

Dufour de Montland, —. Surveyor and survey checker on the Cassini *Carte de France*. Worked with others on seven sheets, mostly in southern France, 1775-1779.

Dufrénoy, —. Printer for L. Vivien de Saint Martin's *Australie*, Hachette 1889; *Russie Orientale et Caucasie*, Hachette & Co.1900 (drawn by G. Bagge, engraved by E. Delaune, E. Duams-Vorzet and D. Aïtoff).

Dufrénoy, Pierre Armand (1792-1857). Geologist, studied geological mapping in England in the 1820s. Worked with J. Beaumont on a geological survey of France, 1825-1836 (formed the basis of *La Carte Géologique de la France* (6 sheets), 1841); *Übersichts-Karte von Deutschland*, 1838; *Carte géologique des environs de Paris*, 1865.

Dufresne, A. *Possessions et colonies françaises*, for Victor Levasseur, 1879.

Dufresne, L.N. *Coast Telegraph Chart of the Gulf and Lower St. Lawrence...*, 1876, 1879.

Du Fresne, M. See **Marion**-Dufresne.

Du Fresnoy. See **Lenglet**-Dufresnoy, Nicolas.

Dugan, Frances L.S. *Map of the State of Kentucky*, 1818.

Dugay, *Captain* A. Plan de l'entrée du fleuve du Mississippi, 1803.

Dugdale, James. *Kent*, 1815 (engraved by Samuel John Neele); *The New British Traveller*, James Robins 1819.

Dugdale, Thomas. *Curiosities of Great Britain* (3 volumes), 1835 and later editions.

Dugdale, *Sir* William (1605-1686). Garter King of Arms, antiquary and geographer. Warwick (4 sheets), 1656; *History of Imbanking and Drayning of Divers Fenns and Marshes*, 1662 (map by J. Aerlebout); Romney Marsh, 1662; Rye Harbour, 1662.

Du Guay-Trouin, René. French seaman. Expédition de Rio de Janeiro, 1712.

Du Halde, Jean-Baptiste, *SJ* (1674-1743). Geographer of Paris. *Description géographique...de l'empire de la Chine et de la Tartarie Chinoise*, 1735.

Duhamel, A. *Carte géologique de la Haute-Marne*, 1857-1860.

Duka von Kadar, Friedrich Peter, *Freiherr* von (1756-1822). Military cartographer of Vienna.

Düker, F. See **Duecker** von Hasslau zu Urstein und Winkel, Franz.

'Duke's Plan'. Anonymous manuscript map of New York, now in the British Library. *A Description of the Towne of Mannados or*

New Amsterdam as it was in... 1661, 1664 MS. Information for the map was possibly taken from an earlier plan by Jacques Courtelyou. The map was lithographed by Geo. Hayward of 171 Pearl Street, New York 1859. **Ref.** COHEN, P. and AUGUSTYN, R.T. *Manhattan in maps 1527-1995* (New York 1997) pp.42-43.

Dukes, Thomas. *The Planisphere*, 1820.

Du Lac. See **Perrin du Lac**, François.

Dulague, Vincent François Jacques Nicolas (1729-1805). French hydrographer, from Dieppe, died Rouen. Leçons de navigation, 1768; Principe de navigation, 1787.

Dulany, J.C. Publisher of Baltimore. Published the League of American Wheelmen's *Road book of Maryland... Pennsylvania, New Jersey, District of Columbia, Virginia, West Virginia, North Carolina...*, 1897 (surveys by C. Watters).

Dulaure, Jacques Antoine (1755-1835). *Nouvelle topographie*, 1784; Historical plans of Paris, 1823, 1829.

Dulcert [Dulceti], Angellino [Angelino]. Portolan chartmaker of Majorca or Genoa, possibly another name for Dalorto *q.v.*. Portolan chart of Europe and North Africa, 1339.

Duler, J. Plan des Isles du Salut, 1764 MS.

Duller, Eduard. *Rhenish Atlas*, 1845.

Dülp, Johann Georg (1684-1722). At the request of J.A. Riediger, Dülp made two maps of the Duchy of Kulmbach-Bayreuth, published Augsburg, M. Seutter [n.d.].

Dumaresq, John. Fortifications and distances in the St Helier district, Jersey, 1786. **N.B.** Compare with Jean **Dumeresq**, below.

Dumaresq, Philip, *seigneur* de Saumarez (*c.*1650-1690). Jersey, in Philip Falle's *Account of the Isle Jersey*, 1694; also used in William Camden's *Britannia*, 1789 (Richard Gough's edition with maps by John Cary).

Dumas, Emilien. *Carte géologique du Département du Gard*, 1844-1875.

Dumas, *comte* Guillaume Mathieu (1753-1837). *Précis des evènements militaires ou essais historiques sur les Campagnes de 1799 à 1814*, 1816-1826 (with maps by the Tardieu family).

Dumas de Fores, Isaac. *La Clef de la géographie générale*, 1645 (with Jean Boisseau).

Dumas-Vorzet, E. Letter engraver for L. Vivien de Saint Martin's *Australie*, Hachette 1889.

Dumbleton, John. Plan of the Parish of St. Neots, Huntingdonshire, 1770.

Du Mée, L. Baay van Vigos, & Cadix, used by Anna Beeck 1702.

Duménil, —. Printer for Jules Sébastien César Dumont d'Urville, 1833-1834.

Dumeresq, Jean (1749-1819). Surveyor. **N.B.** Compare with John **Dumaresq**, above.

Dumez, Pierre (1757-1794). French engineer of 'Bureau de l'Atlas National, rue de la Harpe, N°·26', Paris, (1793). Instigator and director of the committee for the *Atlas National de la France*, on which he worked with Pierre Grégoire Chanlaire, and engraver François d'Houdan, from 1790. After his execution *c.*1794, the project was continued and completed by his widow, Chanlaire, and d'Houdan. *Atlas National Portatif de la France*, 1791 and later; *Atlas National de la France*, 1793 and later editions.

• **Dumez**, *veuve* —. Widow of above, publisher at 'Rue de la Harpe N°·26', Paris. After the execution of her husband she became co-publisher (with Pierre Grégoire Chanlaire) of *Atlas national de la France*, 1794; *Atlas national portatif de la France*, 1794 (with Pierre Grégoire Chanlaire).

Duminy, François Renier. Dutch colonial official at Cape of Good Hope. Coast of Africa, 1781-1785 published 1787; Plan of Beschermer's Harbour, Alexander Dalrymple 1793.

Dummer [Drummer], Edmund (*fl.*1677-1713). Naval hydrographer. Plymouth Sound, 1682; Portsmouth and Southampton Water, 1698 MS; Ports of the South Coast of England (17 plans), 1698 MSS (with Thomas Wiltshaw).

Dümmler, F. Publisher of Berlin. C. Ritter's *Entwurf zu einer Karte vom ganzen Gebirgssysteme des Himálaja*, 1832.

Dumont, André H. (1809-1857). *Carte géologique de la Belgique* (9 sheets), 1856; *Carte géologique de l'Europe*, 1875.

Dumont de Montigny, —. Carte de la Province de la Louisiane, 1740.

Dumont d'Urville, Jules Sébastian César (1790-1842). French aristocrat, navigator and geographer, second in command to Louis Isidore Duperrey on the *Coquille* 1824, later led his own expeditions to the Pacific aboard the *Astrolabe*. Great contribution to the mapping of Australia, New Zealand and Pacific. *Voyage de la Corvette Astrolabe executé pendant les années 1826, 1827, 1828, 1829*, Paris, J. Tastu 1833; *Voyage pittoresque autour du monde*, Paris, L. Tenré 1834-1834; *Voyage au pole sud et dans l'Océanie... 1837, 1838, 1838, 1840*, Paris 1847 (includes *Atlas Hydrographique* with maps prepared by C.A. Vincendon-Dumoulin). **Ref.** HOOKER, B. *The Map Collector* 43 (1988) pp.22 & 24.

Dumortier, Philippe. Engraver for Laurent Gouvion de Saint Cyr, 1821; engraver for plate 2 in the Atlas volume of *Mémoires du Maréchal Suchet, du d'Albufera sur les campagnes en Espagne, depuis 1808 jusqu'en 1814*, Paris, A. Bossange, Bossange père, F. Didot 1828.

Dumoulin, —. Surveyor and draughtsman, 'Capitaine d'Artillerie'. Credited with plate 14 in the Atlas volume of *Mémoires du Maréchal Suchet, du d'Albufera sur les campagnes en Espagne, depuis 1808 jusqu'en 1814*, Paris, A. Bossange, Bossange père, F. Didot 1828.

Dumoulin. See **Vincendon-Dumoulin**.

Du Moy, *Lieutenant-General —, comte* de. Atlas pour les campagnes d'Allemagne, 1762; Mémoires sur la direction genérale depuis la mer jusqu'au Haut Rhin, 1771.

Dun, R.G. [& Co.]. Publisher of New York. *The strategic trade centres of Asia*, 1916.

Dun, S. See **Dunn**, Samuel.

Dunaway, W.A. *Map of LaSalle County, Illinois*, Ottawa, Maierhofer and Briel, 1895.

Dunbar, —. Chart of Portland, Dorset, 1790.

Dunbar, *Colonel* David. Sketch map of part of New Hampshire, 1730.

Dunbar, Samuel. Map of the Town of Taunton, 1836.

Dunbar, William (1749-1810). United States Surveyor-General. District of Natchez, Lousiana, 1798; Washita River, Louisiana, 1804.

Dunbibin[e], Daniel. Pilot of North Carolina. *The Chart of the Coast of America From Cape Hateras to Cape Roman...*, *c.*1761, published later in John Norman's *The American Pilot*, 1791, 1792, 1798.

Duncan, —. See **Ogles, Duncan & Co.**

Duncan, —. Tramways of Birmingham & South Staffordshire, 1855; Tramways of London, *c.*1855.

Duncan, *Lieutenant Colonel* A.W. *A Map of the Island of Grenada...*, 1882 (from the surveys of Gavin Smith of 1801).

Duncan, *Captain* Charles, *RN*. Sketches of the coast of British Columbia, 1786-1789; charts west coast of America, 1788, published Alexander Dalrymple 1789; *Sketch of the Entrance of the Strait of Juan de Fuca*, 1788 A. Dalrymple 1790.

Duncan, D. See **Duncan**, W. & D.

Duncan, James. Publisher of Paternoster Row, London. Scotch Itinerary, 1805; London publisher of John Lothian's *County Atlas of Scotland*, 1827-1829 (sold as individual sheets from 1827); *A New Atlas of England and Wales*, 1833 and later (re-using maps by W. Ebden, engraved by Hoare & Reeves).

Duncan, Robert. *Survey of Nargen Is.*, 1809, published by the Admiralty, 1810.

Duncan, W. & D. Engravers for W.H. Macfarlane's *Greenock and Its Environs*, 1842.

Duncan, William. 'Principal Draughtsman Quartermaster-General's Department'. Dublin to Wexford, 1806; *County Dublin* (8 sheets), 1821; *Ireland Divided into Provinces and Counties...*, 1822; *Ireland*, 1823; Dublin Castle, 1847; *Ireland Shewing the Several Counties, Large Rivers, Canals &c....*, Dublin 1848.

Duncker, Johann Heinrich. Printer for P. Clüver's *Germania Antiqua...*, Wolfenbüttel, 1663; P. Clüver's *Introductio in universum geographiam*, Braunschweig 1672; Orbis Terrarum, 1677.

Dundas, A. Triangulation Province of Wellington [New Zealand], 1872.

Dundas, *Lieutenant* Colin M., *RN*. Easter Is., 1868 MS, published by the Admiralty 1869.

Dundas, *Lieutenant* [later *General Sir*] David (1735-1820). Quartermaster General, worked with William Roy on the survey of Scotland, 1747-1755; also produced manuscript maps of parts of Southern England, including Portland to Southampton, 1757.

Dundas, *Commander* Frederick G., *RN*. Commanded an expedition to Kenya, mapped for the Royal Geographical Society by E.G. Ravenstein, 1892; River Juba, 1893 MS; Delta River Niger, 1895.

Dundas, William (*fl.*1748-1757). Surveyor, worked in England and Scotland. Portsea Lake and Bridge, 1756 MS (with Daniel Slack).

Dundee, —. Engraved for John Seller, *Chart of the South Sea*, 1675.

Duner, N. *Karta öfver Spetsbergen*, 1861-1864 (with A.E. Nordenskiöld).

Dunett, —. Lithographer for Charles Gould's *Map of a portion of Western Tasmania....*, 1860.

Dunham, A. Middlesex County [New Jersey], 1781 MS (with J. Rue).

Dunham, F.H. Atlas Plainfield, New Jersey, 1894.

Dunham, G.T. County Stephenson, Illinois, 1859.

Dunham, J.R. Atlas Hamilton County, Nebraska, 1888.

Dunker, —. *Karte...von Stettin*, *c.*1897.

Dunker, father and son.
• **Dunker**, Balthasar Anton (1746-1807). Born in Saal (Sweden, now Germany). Painter, engraver and poet in Bern. *Plan perspectif d'une grande partie des cantons de Lucerne, d'Uri, de Schweitz...*, 1777.
•• **Dunker**, Philipp Heinrich (*c.*1780-1836). Son of Balthasar. Painter and engraver, born Bern, worked in Nuremberg.

Dunker, W. *Specialkarte der Grafschaft Schaumburg*, 1867.

Dunlap, Lauren. Territory of Dakota, 1885.

Dunlap, William. Printer and publisher of Philadelphia. *An exact plan of the harbour of Louisbourg*, 1758 later published in *Father Abraham's* [Abraham Weatherwise] *Almanack for 1759*, Boston 1758; *Father Abraham's Almanac. For the year of our Lord 1761*, Philadelphia, Dunlap 1760.

Dunlevie, G.G. 'Surveyor & Draftsman, Crown Land Department, Toronto'. *Plan of Lower Canada*, 1853; *Map of the Counties of Gaspé and Bonaventure...* [Quebec], Toronto 1857.

Dunlop, Robert G. Republics of Central America, 1847.

Dunn, E.J. Geological maps: Beechworth, Victoria, 1871; Cape Colony, 1874; *Geological Sketch map of Cape Colony...*, 1875; South Africa, 1876, 1887.

Dunn [Dun], Samuel (*d.*1794). London mathematician and publisher, teacher of mathematics from 1751, 'Boards young gentlemen and teacheth Penmanship, Merchants Accounts, Navigation, Fortification, Astronomy &c.', Chelsea, London. *A Map of the World*, 1744; Directory for the East Indies, 1767; General Atlas, Sayer & Jefferys 1768; British Isles, 1774; Scotland, 1774; *A New Atlas of the Mundane System of Geography...*, 1774, 1778, 1788, 1800; *American Military Pocket Atlas*, 1776; drew maps for Henry Gregory's *A New Directory for the East Indies*, 1780;

World (2 sheets), 1780-1781; *America North and South and the West Indies*, 1786; *Dunn's East India Directory*, 1787; England and Wales, 1788; *A General Map of the World, or Terraqueous Globe*, 1794.

Dunnavant. See **Ritchie** & Dunnavant.

Dünnhaupt, Carl. With H. Müller, bought the publishing company of Carl Flemming at Glogau from Georg Flemming in 1888 and ran it until 1898, when it became a shareholding company. Stock eventually passed to Ullstein und Westermann in 1932.

Dunnica, W.F. *City of Glasgow, Howard County, Missouri*, *c.*1860.

Dunning, A.G. Ancient Classical & Scriptural Geography, New York 1850.

Dunning. See **Prior** & Dunning.

Dunnington. See **Gittings** & Dunnington.

Dunnover, P. See **Dunoyer**, Peter.

Dunod, F.J., *SJ*. Sequanorum Provinciae Maximae, 1735.

Du Noyer, George Victor (1817-1869). Geologist of Dublin. Ireland, 1854; Geological Survey of Ireland, 1863.

Dunoyer [Dunnover], Peter. Book and mapseller 'at the sign of Erasmus's Head', London. Advertised David Mortier's stock, 1709; map & plan of Preston, 1715.

Dunscombe, Clement. Engineer, working in Cape Colony. Sewage plans for Cape Town, Wynberg 1891.

Dunstable, John. Latit. & longit. ali. urbium, *c.*1450 MS.

Dunstall, John. Prospect of the City of London, *c.*1667.

Dunthorne, James. Draughtsman. Copford, Essex, 1810 MS.

Duntze, A. Hansestadt Bremen, 1851.

Düntzfeld, J.F. See **Duentzfeld**, Johann Friedrich.

Dunwoody, *Colonel* H.H.C. Military Telegraph Lines in Cuba, Washington DC 1899.

Dupain de Montesson (1720-after 1790). Geographer of Paris.

Dupain-Triel, family. Worked with the Cassini family on the great survey of France, *c.*1760-*c.*1781, contributing to over 30 sheets as surveyors, checkers and engravers. **N.B.** In most cases the works are attributed to the family rather than individuals. **Ref.** PACHA, B. and MIRAN, L. *Cartes et plans imprimés de 1564 à 1815* (1996) *passim.*

• **Dupain-Triel**, — *père*. 'Géographe du roi et de Monsieur, Cloître Notre-Dame, Paris' (1782). Contributed to at least 6 sheets of the *Carte de France*, often noted as engraver. Named as the publisher of M. du Carla's *Expression des nivellemens, ou méthode nouvelle pour marquer rigoureusement sur les cartes terrestres & marines les hauteurs & les configurations du terrain*, Paris 1782 (with hypothetical map).

• **Dupain-Triel**, Jean-Louis (1722-1805). 'Ingénieur et géographe du Roi, rue des Noyers près St. Yves' (1780); 'Cloître notre Dame' (1789). *Navigation Intérieure du royaume de France...*, Paris, 1780; *Plan de Toulouse*, 1780; *Navigation intérieure du royaume de France*, 1781, 1789; contour map *La France considérée dans les différentes hauteurs...*, 1791 (the first hypsometric map); *Carte de la France*, 1798-1799. **Ref.** WALLIS, H. & ROBINSON, A.H. *Cartographical innovations* (Tring 1987) p.145.

• **Dupain-Triel**, —, *fils*. 'Ingenieur géographe du roi', 'Cloître Notre Dame' (1776-1777); 'Rue des Noyers pres St. Yves' (1780). Credited with work on at least 18 sheets of the Carte de France, 1765-1778; *Carte de la généralité de Bourges...*, 1780. **N.B.** Compare with Jean-Louis, above.

Duparc, L. Carte géologique du Massif du Mont Blanc, Geneva 1898.

Dupard, A.M. Plan de Lima, 1859.

Duparquet, *Père* —. *Carte...de la Côte du Loango*, 1875; Okavango River [Bechuanaland], *c.*1880; *Carte de l'Ovampo*, 1881.

Du Perac [D'Perac; du Pérace; Dupeyrac], Stefano [Etienne] (*c.*1525-1604). Engraver, artist, architect and mapseller of Paris; in

Rome, 1559-1582. Mural map of France [in the Vatican], published Jolivet 1560; Naples, 1566; *Urbis Romae sciographia...* (8 sheets), 1575; Jerusalem, [n.d.].

Duperrey, Louis Isidore (1786-1865). Commander of the French corvette *Coquille* on its scientific voyage to the Pacific. Charts for Louis-Claude Desaulces de Freycinet, 1818-1820; *Voyage autour du monde...sur la corvette...La Coquille pendant les années 1822-1825*, Paris, A. Bertrand 1827 (with accompanying *Hydrographie: Atlas*). **Ref.** HOOKER, B. *The Map Collector* 43 (1988) pp.20-21.

Duperrier, —. 'Ingenieur géographe' of Rouen. *Carte de la Province de Normandie...*, Rouen 1767.

Duperron, Jean. *Kaart van het Eyland Ceylon*, 1789 MS.

Du Petit-Thouars, *Admiral* Abel Aubert (1793-1864). French hydrographer. Circumnavigation in the *Vénus*, 1837-1839. *Plan de la Rade...des Isles de St. Pierre et Miquelon*, 1818-1819; Plan de St. Pierre, 1824; *Voyage autour du monde..., 1836-1839*, Paris 1841; *Atlas Hydrographique*, 1845; *Plan de la Baie des Iles à la Nouvelle Zélande*, Dépôt de la Marine 1845 (engraved by C.E. Collin). **Ref.** HOOKER, B. *The Map Collector* 43 (1988) pp.23-24.

Dupeyrac, E. See **Du Perac**, Stefano.

Dupin, —, *le jeune*. French engraver. Nicolas Louis de La Caille's *Carte du Cap de Bonne Espérance*, 1752.

Dupin, *Baron* Charles (*b*.1784). Mathematician, statistician and political economist, admitted to the Académie des Sciences in 1818, professor in the Conservatoire des Arts et Métiers, 1819, elected to the Chamber of Deputies 1828 and became senator in 1853. Maps showing moral and demographic statistics including *Carte figurative de l'instruction populaire de la France* (the first known choropleth map), included in *Forces productives et commerciales de la France* (2 volumes), Paris 1827; *Forces commerciale et productive de la France*, 1843. **Ref.** ROBINSON, A.H. *Early thematic mapping in the history of cartography* (1982) pp.156-158.

Du Pinet, Antoine, *sieur* de Norroy (*c*.1510-*c*.1584). *Plantz, pourtraitz et descriptions de plusieurs villes et forteresses tant de l'Europe, Asie & Afrique...*, Lyon, J. D'Ogerolles 1564 (used later by Belleforest, and Braun and Hogenberg).

Du Plat. See **Plat**.

Du Plessis, D.M. See **Martineau** du Plessis.

Du Plouich, V. See **Plouich**.

Dupon, Jean. See **Dupont**, Jean.

Duponchel, Adolphe (*b*.1821). French engineer and geographer. Géographie du departement de l'Herault, [n.d.].

Dupond, Guillaume. Chart of the coast of Africa, Alexander Dalrymple 1755.

Dupont, Edouard (*b*.1841). Belgian geologist. Lettres Congo, 1889; Cartes géologique.

Dupont, G. Engineer. *An Actual Survey and itinerary of the road from Calais to Paris...*, London 1814 (with L. Hebert).

Dupont, Jean. Chartmaker of Dieppe. Chart of the west coast of Africa, 1625.

Dupont, Paul. Chemins de fer d'Europe Centrale, 1857.

Duportail, *Major General* Louis le Beque de Presle (1743-1802). French military engineer, served in North America from 1777; returned to France on his retirement in 1783 becoming Minister of war in 1790. Manuscript military plans, contributed to design for fortifications at West Point; *Plan du Camp de Vallie Forge*, *c*.1777; Plan of the Position of Valley Forge, 1778; plan of fortifications at Paulus Hook [n.d.]. **Ref.** GUTHORN, P.J. *American maps and map makers of the Revolution* (1966) pp.16.

Du Pratz, le P. See **Le Page** du Pratz.

Dupré, E. *Atlas of the City and County of St. Louis*, 1838.

Du Pré, Galliot. Publisher of Paris, Oronce Fine's *Nova, et integra universi orbis descriptio*, 1531.

Du Pré [Dubré] Jean. Printer of Lyon.

Untitled mappamundi in *Mer des Hystoires*, Lyon 1491.

Dupré, Julius V. Atlas of Milwaukee, 1881 and later.

Dupuis, —. Engraver of 'rue Mignon St André des Arts vis-à-vis l'Imprimeur au Parlement'. *Plan de St. Jean de Luz...*, 1784. **NB.** Compare with Louis André **Dupuis** (below).

Dupuis, André. Plan de Jérusalem, Nantes 1841.

Du Puis, C. Sonora, 1794.

Dupuis, Louis André. Engraver. Switzerland, 1769; Joseph-J.-F. Ferraris's *Carte chorographique des Pays-Bas...*, 1777; *Cours du Danube*, 1785; *...la Mer Baltique*, 1785; *...la Mer Méditérranée*, 1785; *...la Mer Noire*, 1785; *Marseilles*, 1787.

Dupuy, family. Papermakers of the Auvergne, the business passed from father to son for over a century. Their paper was used by many of the foremost French mapmakers of the eighteenth century, including the Imprimerie royale, and the Cassini family for the *Carte de France*. **Ref.** MIRAN, L. in: PACHA, B. & MIRAN, L. *Cartes et plans imprimés de 1564 à 1815* (Bibliothèque Nationale de France 1996) pp.62-64 et *passim*.

• **Dupuy**, Thomas [I] (1642-1731). Acquired the papermill at La Grandrive in 1676, supplying paper to the major centres in France, including Lyons and Paris.

•• **Dupuy de la Grandrive**, Jean-Joseph (1692-1747). Worked with his father Thomas from 1731.

••• **Dupuy de la Grandrive**, Thomas [II] (1716-1780). Succeeded his father Jean-Joseph in the family business, 1747.

•••• **Dupuy de la Grandrive**, Thomas-Marie-Joseph (1746-1823). Succeeded to the family business in 1780, worked with his son from 1810-1823.

••••• **Dupuy de la Grandrive**, Pierre-François-Thomas (1775-1836). Worked with his father Thomas-Marie-Joseph, taking charge in 1823.

• **Dupuy**, Jean-François (*d.*1716). Brother of Thomas I and Claude (below). Worked with Claude at 'rue Saint-Jacques à l'enseigne des Armes de Hollande', Paris, as agents for the paper made by brother Thomas.

• **Dupuy**, Claude. Brother of Thomas and Jean-François. Worked with Jean-François in Paris as paper seller.

Dupuy, August Myionnet. *Union des deux océans, Atlantique et Pacifique*, 1855.

DuPuy, W.H. *Atlas*, New York, Phillips & Hunt 1886.

Durado, Fernão Vaz. See **Dourado**, Fernão Vaz.

'Dura Europos' shield (4th-century AD). Soldier's parchment shield found at Dura Europos, northern Mesopotamia, contains map of north shore of the Black Sea with place-names in Greek, distances in Roman miles. Now in the Bibliothèque Nationale, Paris.

Durand, —. Engraver and mapseller of Paris. Mantouë, 1734; *Horizon de Paris* [world map], Paris 1740; *Carte Topographique de la baye, Ville et Faubourg de Cartagene...*, 1741; engraver of four sheets of the Cassini *Carte de France*, 1758-1763; co-publisher with Pissot of Gilles Robert's *Atlas portatif Universel et Militaire*, 1748; engraved *Bataille de Chotusitz* [Bohemia 1741], published in *Atlas topographique et militaire*, Paris, Julien 1758.

Durand, Alfred. Diego Suarez, 1890.

Durand, Auguste. Publisher of Paris. *Atlas Géographique de la République du Pérou...*, 1861, 1865.

Durand, F. *Chemin de fer de Dunkerque à Furnes*, 1866.

Durand, G. *Plan de Niort* [France],1868.

Durand, *Major General Sir* Henry Marion, *RE* (1812-1871). Surveys used for John Walker Jr's Northern Punjab, 1846.

Durand, Jean Baptiste Léonard (1742-1812). Diplomat and traveller. Atlas du voyage au Sénégal, 1802.

Durand-Dubraye [d'Ubraye], Louis Alexandre Jean François Léopold de. Charts of New Zealand: *Plan de la Rivière Kawa-Kawa*, 1840; *Plan du Port Akaroa*, 1840;

Plan des Baies de Tokolabo et de Koko-Rarata, 1840; *Plan des Iles Chatam*, 1840, all with Joseph Marie Fournier for the Dépôt Général de la Marine.

Durant, J. Engraver of Jean-Baptiste Tavernier's *Carte des Isles du Iapon*, in *Recueil de plusieurs relations*, Paris 1679.

Duranty, —. Plan de Vavao, used by J.S.C. Dumont d'Urville, 1842-1848.

D'Urban, *Lieutenant General Sir* Benjamin (1777-1849). First governor British Guiana, 1831; governor Cape of Good Hope, 1833-1837 (town of Durban, Natal named after him). Settlements in British Guiana, 1828.

Durban, William. Channel between Sardinia, Sicily & Africa, 1810.

Durbin, John Price (1800-1876). Methodist clergyman of Kentucky. Jerusalem, Philadelphia 1850 (after Frederick Catherwood).

Dureau de la Malle, [Adrien Joseph] Charles Auguste (1777-1857). French geographer and historian. Géographie de la Mer Noire, 1807.

Durell, *Lieutenant* (later *Vice-Admiral*) Philip. Plan of Guantanimo [Cuba], 1740; *Porto Bello*, 1740; *Cumberland Harbour* [Cuba], 1741; *A Plan of the Harbour, town and castles of Carthagena...*, 1741; Louisbourg Harbour, 1745; Dunkirk, 1764 MS.

Durell, *Captain* Thomas. *Harbour & Island of Canso*, 1732; *A Chart of the Sea Coast of Nova Scotia, Acadia and Cape Breton...*, 1736; Sea coast New England, Nova Scotia, Newfoundland, *c.*1740.

Dürer [Duerer], Albrecht (1471-1528). Artist and engraver of Nuremberg. *Imagines coeli septentrionales*, 1515 (woodcut map of eastern celestial hemisphere); supplied decoration for Johann Stabius' globe, 1515 (now known only from pulls taken by Adam von Bartsch 1781, and Karl Ritter 1847); Underweysung, 1525 (contains instructions for drawing globe gores); Hemisphaerium australe, 1527.

Duret-Noel, — (*c.*1599-1650). French cosmographer and mathematician.

Durham, John, father and son.
• **Durham**, John [I] (*fl.*1796-1837). Surveyor of Dunstable, Bedfordshire, produced manuscript maps of Midland counties, including estate plans of Broughton, 1837.
•• **Durham**, John [II] (*fl.*1827-1861).
• **Durham, Son & Hurgood**. Estate surveys at Stony Stratford [Buckinghamshire], 19th-century.

Durham, Thomas. Cartographer. *The Isle of Man Described...*, 1595 (used by John Speed, 1611).

Durheim, C. Malta, *c.*1840.

Durheim, Carl (1810-1890). Lithographer and photographer of Bern, Switzerland. Made several maps of parts of Switzerland.

Durieu, Antoine. Traité de Turin, 1754 MS.

Durkee, H.R. Assisted B. Froiseth in the preparation of *New Mining Map of Utah*, New York, American photo lith. co. 1871.

Durnford, E.P. Survey St. Nicholas, Cape Verde Is., 1823; Plan Is. St. Mary Madagascar, 1827.

Durnford, *Lieutenant-Colonel* Elias. Engineer, Lieutenant Governor of West Florida. Manuscript plans of Pensacola, 1765; River Mobile, *c.*1770; River Mississippi, 1771; Rivers Iberville &c., *c.*1771; battle plan of Walmstock, 1777-1780; *Plan of the Town of Plymouth Dock*, *c.*1790.

Duroch, —. Charts for J.S.C. Dumont d'Urville, 1842-1848.

Durocher, —. French surveyor. From 1754-1761 he worked with others on sheets 42, 44, 77, 78, 111 and 143 (north, and north-east France) of the Cassini *Carte de France*.

Duroi, B. See **Du Roy**, Bernard.

Duroslan, —. Baie d'Antongil [Madagascar], 1770.

Du Roy [Duroi; Duroij], Bernard. Surveyor, produced manuscript and printed maps of Utrecht, 1670-1674.

Duruy, Victor (1811-1894). *Atlas historique de la France*, *c.*1857; *Carte comparative des départements*, 1865.

D'Urville, J.S.C.D. See **Dumont** d'Urville, Jules Sébastian César.

Dury, Andrew (*fl.*1742-1778). Publisher, printer, engraver, mapseller and surveyor at 'the Indian Queen in Duke's Court, St. Martin's Lane' (1742); at 'The Map & Printshop N°·92 under the Royal Exchange, Cornhill' (1768), London. Worked on county surveys with John Andrews. John Rocque's Dublin, 1757; *A New General & Universal Atlas...*, 1761; *Plans of the Principal Cities of Great Britain and Ireland*, 1764; *Map of the King of Sardinia's Dominions*, Genoa 1765; *Topographical map of Hartfordshire* (9 sheets), 1766 published W. Faden 1782, single sheet version, A. Dury 1777, W. Faden 1782 (with J. Andrews); *A Topographical Map of the County of Kent* (25 sheets), 1769, published R. Sayer and J. Bennett 1779, Laurie and Whittle 1794 (surveyed with J. Andrews and W. Herbert); Isle of Wight (4 sheets), London, A. Dury 1769 (surveyed with P. Bell, engraved by J. Andrews); Wiltshire (18 sheets), 1773; 65 miles round London, 1774-1777; Portefeuille... d'Italie, 1774; published John Montresor's *A Plan of the City of New-York*, 1775; Benjamin Easburn's *Plan of the City of Philadelphia*, 1776; Thomas Kitchin's Atlas (with Robert Sayer); Roads of Italy, 1777.
• **Dury**, *Mrs.* —. Widow of, and successor to Andrew, published George Taylor's Louth, 1778.

Dusault, *Sieur* —. L'Isle Dauphine, 1717.

Du Sauzet, A.G. *Geography epitomized*, 1727.

Du Sauzet, Henri. Publisher in Amsterdam. *Atlas portatif*, 1734-1738; re-issued Mercator-Cloppenburg maps of 1630; European city plans, 1739.

Dusch, Alexander von. Publisher of Karlsruhe. German edition of A. Le Sage's *Atlas historique, généalogique, chronologique et géographique*, 1825 (first published Paris 1802); Politisch-geographische Übersicht, 1831.

Dusent, C.C. See **Duysend**, Cornelis Claes.

Du Simitiere, P.E. See **Simitière**, Pierre Eugène de.

Dussieux, Louis Etienne (1815-1894). French geographer. Géographie historique de France, 1844; Cours de géographie; Atlas générale, 1848; Cartes du Canada, 1851; *Atlas générale de géographie physique*, 1854; Atlas de géographie, 1882.

Dussy, E. Engraver and draughtsman. For Gilles and Didier Robert de Vaugondy, 1754-1778; Nicolas Louis de la Caille, 1760; Louis Charles Desnos, 1767; Mexico, 1784.

Du Temple, V. *Plan de la Ville de Saint-Malo*, *c.*1865.

Du Temporal, J. See **Temporal**, Jean.

Du Temps, J. See **Templois**, Jean du.

Dutens, Louis (1730-1812). Itinéraire des routes... de l'Europe, 1768-1771.

Du Tertre, Jean Baptiste. Histoire générale des Antilles, 1667; maps of St. Christophe, Guadeloupe, Martinique, St. Croix, 1667.

Du Tertre, Jean Baptiste. Printer of J. Ogee's *Carte Géométrique de la Province de Bretagne...*, 1771 (engraved by Nion, C.D. Beauvais and J. Dezauche).

Duthie, James (*fl.*1808). Land surveyor. Kincardine, for John Thomson's *Atlas of Scotland*, 1832.

Du Tralage [du Trallage], Jean Nicolas, *sieur* de Tillemon[t] (*d.* 1699). French geographer and map collector. Revised Vincenzo Maria Coronelli's maps for Jean Baptiste Nolin, 1688-1689; *Partie occidentale [-orientale] du Canada*, 1688, 1689; *Le Globe terrestre*, *c.*1690; *Duché de Bretagne*, 1703.

Dutreuil de Rhins, Jules Léon (1846-1894). French merchant seaman, geographer and explorer in Asia and Africa. Commissioned to produce a new map of Indochina. From 1884-1890 worked with the 'Dépôt des cartes et plans de la Marine' on surveys in central and east Asia. Murdered in 1894 by Tibetan bandits. *Carte de l'Indochine orientale*, 1881; *Carte du cours de l'Ogooué*, 1884; later work published by Grenard in: *Mission scientifique*

J.S.C. Dumont D'Urville (1790-1842). Lithograph by Lemercier after A. Maurin 1833. (Taken from Map Collectors' Circle)

dans la Haute-Asie (3 volumes), 1897-1898. **Ref.** BROC, N. *Dictionnaire illustré des explorateurs Français du XIXe siècle: II Asie* (1992) pp.168-170 [in French].

Dutton, Clarence Edward (1841-1912). American soldier and geologist of Connecticut. *Topographical and Geological Atlas of the District of the High Plateaus of Utah*, New York, Julius Bien 1879; *Grand Canyon District Atlas*, New York, Julius Bien 1881 (for the United States Geological Survey); Mt. Taylor and Zuni Plateau, 1886.

Dutton, H.R. Surveyor. *Brooklyn*, 1857.

Dutton, S. Publisher of Philadelphia. *Maps of the States of Ohio, Indiana and Illinois...*, 1835.

Dutton, Walter (*fl.c.*1740-*c.*1785). Surveyor, produced manuscript maps of parts of western England. Estates in Buckinghamshire, 1769 MS.

Duval, —. *Eske-Fiord* [Iceland], 1854.

Duval, Ferdinand. Atlas municipal, Paris 1878.

Duval, Henri Louis Nicolas (1783-1854). Cartographer and printseller. *Atlas universel des sciences*, 1837-1844.

Duval, Jules (1813-1870). French geographer and publisher.

Duval, Marie Françoise. Engraver for J.A. Dezauche's Histoire sainte, 1783.

Duval, P.S. Printer and lithographer of Philadelphia, Steam Lithographic Press (1849); at 22 & 24 S. 5th St. (1861). E.D.H.E. Napier's Peninsula War, 1842; Plan of Wilmington, 1850; military map of the United States, 1861.

Duval [Du Vall; Du Val d'Abbeville], Pierre (1618-1683). Nephew and pupil of Nicolas Sanson, born Abbeville. 'Géographe ordinaire du Roy'; secretary to bishop of Aire. 'En l'Isle du palais sur le grand cours de l'eau' (1653); 'Rue Barillerie pres le Palais' (1655); 'Proche le For l'Evesque en l'isle du Palais' (1664); 'En l'Isle du Palais sur le Quay de l'Orloge proche le coin de la Rue de Harley à l'ancien Buis' (from 1665). Published a large number of maps which each appear in many different states and in more than one of his publications. He also compiled atlases in several editions, map games or 'jeu de l'oie' and town plans. He died in Paris and was succeeded by his widow and two daughters. *Cartes géographiques methodiquement divisées*, 1645 and later (included maps by others); *Le Jeu du Monde*, 1645; *Le Jeu des princes de l'Europe*, [n.d.]; *L'Evesché d'Aire* and *Le Duché d'Aiguillon*, both published in J. Janssonius's *Novus Atlas*, 1645 (later used in Blaeu's *Atlas Maior*, 1662); *Cartes géographiques*, 1653 and later; *L'Amerique*, 1656; *Cartes de géographie revües et augmentées...*, 1657 and later; *Le Monde Terrestre*, 1659 and later; *Le Jeu de France*, 1659; *La Géographie universelle*, 1660 and later; *Cartes de géographie*, 1662; *Le monde ou la géographie universelle*, 1662 and later; *La géographie françoise*, 1663 and later; *La Carte generale et les Cartes particulieres des costes de la Mer Mediterranée* (12 charts), Paris 1665; *Cartes Des Isles Britaniques...*, 1665; *Cartes des Indes Orientales*, 1665; *Diverses cartes et tables pour la géographie ancienne...*, 1665; *Canal des eaux de la Montagne Noire... pour le nouveau canal de Languedoc*, Paris 1666; *Le Globe Terrestre*, 1666; *Cartes et tables de géographie des provinces échues à la reine très chrétienne*, 1667; L'A.B.C. du monde, 1670; Provinces Unies, 1672; *Carte Universelle du Commerce*, 1674; *Carte Générale du Monde*, Paris 1676; *Cartes de Géographie...*, 1677; *Diverses Cartes et Tables*, 1677; *La Géographie Ancienne...*, 1677; *La Géographie Universelle*, 1682; *Carte Universelle du Monde*, 1684. **Ref.** BURDEN, P.D. *The mapping of North America* (1996) lists the states of several Duval maps of America; SHIRLEY, R. *Printed maps of the British Isles 1650-1750* (1988) pp.53-55; PASTOUREAU, M. *Imago Mundi* 32 (1980) p.63 [in French].

• **Duval**, Marie Desmarests [Desmaretz]. Widow of above. Together with daughters Marie-Angélique and Michèle Duval, re-issued his work including 4-sheet Amérique, 1684; *Il regno di Navarra et di Aragona con la Catalogna*, 1685; re-issued Placide's maps, 1692-1693.

•• **Duval**, *Mlle.* —. One of the daughters of Pierre and Marie (above), inherited business at father's last address, later at 'Rue St. Jacques au Dauphin'. Re-issued her father's work including *L'Amérique*, 1684 (first published 1664); Placide de Ste Hélène's Siam, 1686; Danube, 1703.

Du Verdier (1544-1600). Le Voyage de France, 1655, 1673.

Duveyrier, Henri (1842-1892). Exploration du Sahara, 1859-1861; Paris, 1864; *Carte murale de l'Afrique* (4 sheets), 1879.

Duviert, Joachim. Dutch artist and traveller. 125 drawings of French towns in the form of bird's-eye views, 1600-1614.

Du Villard, Jean. See **Villard**, Jean du.

Du Voisin, L. See **La** Popellinière.

Duvotenay, Thunot (1796-1875). Geographer and publisher. Atlas historique, 1840 (with A.H. Dufour); *Atlas des campagnes de la Révolution française*, 1846; *Atlas géographique et historique de la Suisse*, 1851; Atlas histórico, 1852; *Atlas de France et ses colonies*, 1860 (with A.-H. Dufour); United States, 1875.

Pierre Duval (1618-1683). Copperplate engraved by Langlois. (Taken from Map Collectors' Circle*)*

Duxmoulin, C.A.V. See **Vincendon-**Dumoulin.

Duyckinck, Gerardus. Publisher of New York. Francis Maerschalck's *A Plan of the City of New York*, 1755.

Duynen, Johannes van. Engineer and surveyor. Belegging van Hulst, Joan Bleau *c.*1649.

Duysend [Dusent], Cornelis Claes. Engraver and publisher of Leiden. Engraved the title to Johannes de Laet's West Indies, 1630; H. Hondius's Picardy, 1644.

Dvořák, S. See **Dworzak**, Samuel.

Dvorzacky, Ferenc Antal. Hungarian cartographer in service of count István Eszterházy in 1760s. **Ref.** FODOR, F. *A magyar térképirás* [Hungarian cartography], Budapest 1952-1954 [in Hungarian].

Dwelling, Daniel. See **Daniel**, John.

Dwight, Theodore, *Junior* (1796-1866). *Gazetteer of the United States*, Hartford 1833

Dworzak [Dvořák], Samuel (*d.*1689). Engraver of Prague. Moravia, 1677.

Dyamatier, Jehan. Publisher of Lyons. *Mer des Hystoires*, 1506 (printed by Claude Daoust).

Dybwad, Jacob. Publisher. *Veikart over Udarbeidet efter de bedste Kilder af. J. Waligorski og N. Wergeland* [Norway and Sweden] (2 sheets), c.1868.

Dyck [Dijck], Heyman van den. Dutch land surveyor. *Voorne Caert- Boeck...*, 1701.

Dyckerhoff, Jacob Friedrich. *Karte der Gegend von Mannheim*, 1814.

Dyé, Alfred H. (1874-1926). French hydrographer. Worked on surveys of the Congo-Nile, 1896-1899; hydrographic survey of the coast of Morocco, 1905-1907.

Dyer, —. See **Longmans.**

Dyer, R. Estate surveyor. Little Yeldham, 1766 MS; Layer Breton, 1767 MS.

Dyer, Thomas. Estate plans at Westmeon, 1809 MS.

Dyke, Hamilton Moore (1817-1898). Outline map Basuto country, 1848 MS.

Dymock [Dymork], Abraham (*fl.*1801-1831). Estate surveyor, working in Essex, Wiltshire and Berkshire. Manuscript surveys include Gt. Dunmow [Essex], 1801; Manor Fifield and East Overton [Wiltshire], 1811.

Dymock, Joseph (*fl.c.*1818-1848). Surveyor, produced manuscript maps of parts of Berkshire, Gloucestershire and Oxfordshire including plans of West Hanney, 1839.

Dymond, G.W. *Worlebury near Weston-super-Mare*, 1872.

Dymork, A. See **Dymock**, Abraham.

Dyonnet, Charles. Engraver of 'rue St. Jacques 220', Paris. Worked for A.H. Dufour, 1838-1863; G.H. Dufour's *Karte der Schweiz...* (21 sheets), 1842 and later; Thunot Duvotenay, 1846; Dépôt de la Marine, 1850-1861; Plan de Paris, 1850; A.R. Frémin's *Carte du Chemin de Fer de Paris à Rouen et au Havre...*, Paris, Auguste Logerot [n.d.].

Dzialowski, Zygmunta. *Carte Archéologique de la Prusse Occidentale*, 1880.